AF334448

Biomag 96
Volume II

Springer
New York
Berlin
Heidelberg
Barcelona
Hong Kong
London
Milan
Paris
Singapore
Tokyo

Cheryl J. Aine   Yoshio Okada
Gerhard Stroink   Stephen J. Swithenby
Charles C. Wood

Editors

Biomag 96

Proceedings of the Tenth International Conference on Biomagnetism

Volume II

Springer

Cheryl J. Aine
Radiology Service Albuquerque
VA Medical Center
Building 49, 114M
1501 San Pedro SE
Albuquerque, NM 87108
USA

Yoshio Okada
Department of New Mexico School
  of Medicine
Albuquerque, NM 87131
USA

Gerhard Stroink
Department of Physics
Dalhousie University
Halifax, Nova Scotia, Canada
B3H 3J5

Stephen J. Swithenby
Department of Physics and Astronomy
The Open University
Milton Keynes MK7 6AA, UK

Charles C. Wood
Biophysics Group (P-21), MS D454
Los Alamos National Laboratory
P.O. Box 1663
Los Alamos, NM 87545
USA

Library of Congress Cataloging-in-Publication Data
International conference on Biomagnetism (10th : 1999 : Sante Fe,
  N.M.)
    Biomag 96 : proceedings of the Tenth International Conference on
  Biomagnetism / editors, Cheryl J. Aine . . . [et al.].
      p.   cm.
    Includes bibliographical references and index.
    ISBN 0-387-98915-3 (hardcover : alk. paper)
    1. Biomagnetism Congresses.   I. Aine, Cheryl J.   II. Title.
  QP345.I56   1999
  571.4'7—dc21   99-42829

Printed on acid-free paper.

Production managed by Jenny Wolkowicki; manufacturing supervised by Joseph Quatela.
Camera-ready copy provided by the editors.
Printed and bound by Edwards Brothers, Inc., Ann Arbor, MI.
Printed in the United States of America.

9 8 7 6 5 4 3 2 1

ISBN 0-387-98915-3 Springer-Verlag New York Berlin Heidelberg   SPIN 10742866

Contents
Volume II

V Magnetoencephalography (MEG): Basic Research

Basic Human MEG

MEG Measures Of Brain Oscillatory Responses

Clinical MEG: Epilepsy

VII   Multimodality Comparisons And Integration

Contents
Volume I

II Models for the Biomagnetic Forward and Inverse Problems

III   Magnetocardiography (MCG)

IV. Biomagnetic Measures of the Gastro-Intestinal System, Susceptometry and Other Biomagnetic Measures

BIOMAGNETIC RESEARCH IN GASTROENTEROLOGY

Baffa, O.[1], Basile, M.[2], Bradshaw, A.[3], Forsman, M.[4], Nakaya, Y.[5] and Richards, W.O.[3,6*]

Departamento de Física e Matemática, FFCLRP-Universidade de São Paulo, 14040-901 Ribeirão Preto-SP, Brazil[1], Istituto di Tecnologie Avanzate Biomediche-Universitá "G. D'Annunzio", 66.100 Chieti, Italy[2], Living State Physics Group, Vanderbilt University, Nashville, TN, USA[3], Department of Applied Electronics, Chalmers University of Technology, S-412 96 Gothenburg, Sweden[4], Department of Nutrition, School of Medicine, The University of Tokushima, Japan[5] and Department of Surgery at the Veterans Affairs Medical Center and Department of Surgery School of Medicine, Vanderbilt University, Nashville, TN, USA[6]

Introduction:

The gastrointestinal system is responsible for providing the body with a constant supply of nutrients, water and electrolytes. The food must be digested - chemically and physically broken down to the molecular level - in order to be absorbed by the small intestine. Since both processes are rate dependent, the flow rate of the food (chime) must be compatible with the absorption rate of the small and large intestine.

Motility is an important attribute of the gastrointestinal tract and as with blood flow through the heart, it is controlled by electrical signals. In the case of the heart, electrical measurements are straightforward to perform and can yield useful clinical information. However, in the case of gastrointestinal motility, qualitative and semi-qualitative methods based mainly on the use of X-rays and radioisotopes have been employed. For quantitative evaluations the direct measurement of parameters, such as pressure, flow and force were the ones performed more routinely in clinical and basic research. Only recently have electrical measurements been investigated further to extract information about motility. Still the primary methods to investigate motility rely on X-rays, scintigraphy, and catheters inserted into the gastrointestinal tract.

The use of magnetic measurements to study the gastrointestinal tract can be divided in two categories, one concerned with magnetic fields produced intrinsically by the electrical currents of organs and the other with magnetic fields produced by magnetic materials that are ingested. In the first group, magnetic measurements are used to image electrical currents associated with the basic electric rhythms (BER) of the gastrointestinal tract, and two new words -magnetogastrogram and magnetoenterogram- were coined to describe these recordings. In the second group, the fields produced by magnetic markers (MM), or tracers (MT) can be followed by measurements at the surface of the torso. In this context, MM are localized magnetic particles, while MT are dispersed magnetic particles.

Information can be gathered regarding position, time course, quantity, state of order, among others parameters related to gastrointestinal motility. Biomagnetic techniques represent in some cases an alternative and in others a unique way to study the gastrointestinal tract. It is believed that this approach, combined with the already established methods, could contribute to new knowledge in this area. Reflecting this belief, the two last conferences of this series have seen growing interest in this topic, leading to the present chapter. In particular, during the 8th International Conference on Biomagnetism there was a special session on biomagnetic applications to gastroenterology, fully documented in the proceedings book [1] . In

---

* Authors are listed in alphabetical order, ordering of appearance has no meaning to their importance in this paper

the following sections, a survey of the work performed by some of the groups working in the area will be reviewed and updated.

Magnetic tracers to study GI motility

a-Studies Using Magnetic Markers

The initial studies using MM were methodologically simple. In 1957 a magnetometer was used to study the motility of the small bowel by having human subjects ingest a magnetic stirring bar (ALNICO V, 5 mm diameter and 12 mm long length) and detecting fluctuations of the magnetic fields produced by changes in its orientation of less than 1 degree relative to the detector [2]. A   study was carried out involving 180 subjects, subjected   to various physiological conditions, including cold pressor, hyperventilation, exercise and drugs.

A different marker was used by the biomagnetism group at the University of Chieti [3], where a magnetized steel sphere of 2 mm O.D. was inserted in a tube, yielding an apparent density $D = 1.9$ g/cm^3. The MM was ingested and was followed by magnetic measurements made with a SQUID biogradiometer. Fig. 1 shows the typical arrangement for the measurements. By fitting the isofield lines with the isofield of a magnetic dipole, it was possible to locate the MM and to overlay the location onto MRI images to determine its position in the intestine. Repeated mapping allowed the measurement of segmental transit time as shown in Fig. 2. Using this marker, mean oroanal transit time was (56 ± 5) hours, the mouth to caecum transit time was (13 ± 1.7) hours, and the total colonic transit time was (43.5 ± 5) hours [4].

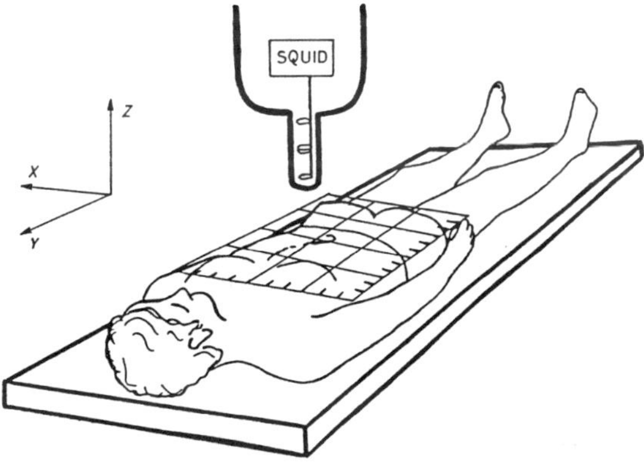

Fig.1-Typical arrangement for measurements of the biomagnetic fields produced by electrical currents of the GI tract and ingested magnetic particles. The SQUID-based biogradiometer can in some experiments be replaced by a fluxgate or active induction coils. Shielding can also be employed to improve signal-to-noise ratio.

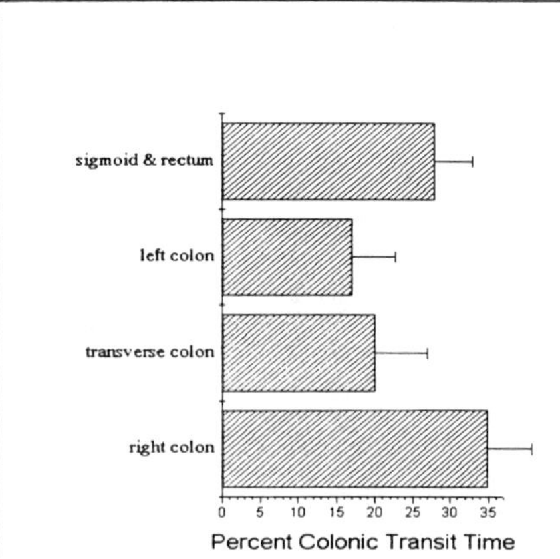

Fig.2-Proportion of time spent by the MM in different colonic regions determined by magnetic methods.

The biomagnetism group at the Chalmers University of Technology in Sweden used a spherical permanent magnet made with epoxy and microspheres (D = 1.4 g/cm^3). A fluxgate gradiometer system, originally designed for magnetopneumography (MPG) research, was employed to study GI transit time. The test meal (cheese sandwich and milk) was ingested together with MM and other radiopaque markers. The investigation involved modeling and *in vivo* measurements, where four subjects took part. They ingested the magnetic marker together with radiopaque markers. In a one-day measurement, the subject ingested a magnetic marker in the early morning. The magnetic field was then sampled outside the subject at 15-minute intervals, 3 to 15 hours after ingestion. The MPG instrument was used to measure field maps in two planes. A nonlinear least-mean-square procedure was applied to the field maps to estimate the six free parameters, comprising the location and the moment of the magnetic dipole. Radiographs validated the estimated marker positions and showed that the magnetic marker behaved similarly to the radiopaque markers in the gut as shown in Fig. 3 [5]. The modeling calculations confirmed adequate precision of the MM localization.

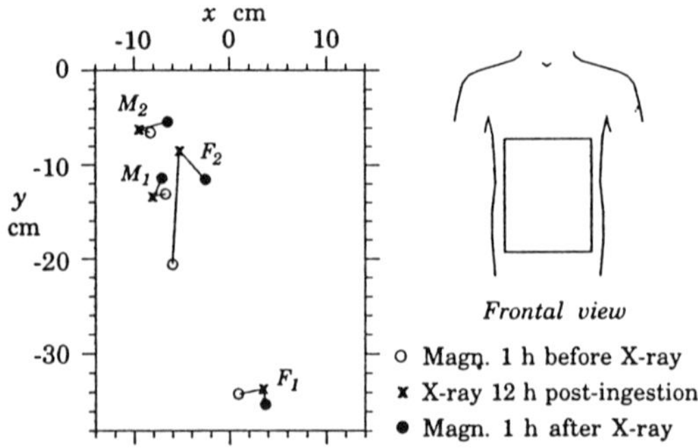

Fig.3-Location of the MM determined by X-ray and magnetic measurements. Crosses indicate the position determined by x-rays and open circles indicate the position found by fitting of the magnetic data. The radiographical determination was done one hour before and after the magnetic measurement.

MM has been also useful to study the pharmacodynamics of drug delivery. The group from PTB-Berlin prepared insoluble sucrose pellets coated with magnetite and pellets of epoxy mixed with magnetite (70 % w/w). These pills were magnetized in a 0.1 T magnetic field and they were followed by magnetic field mapping using a seven-channel system [6].

More recently the group at Tokushima University used a 16 channel high Tc SQUID magnetometer placed inside a small shielded enclosure to study stomach motility, by recording fields produced by a 1.3 mm diameter steel ball placed at the antrum. Studies were performed under different fasting conditions and are reported in more detail in these proceedings [7].

b-Studies Using Magnetic Tracers

Magnetic tracers can be ingested with a test meal, endowing the GI tract with a strong magnetic signal. Depending on the material used as a tracer either the magnetic permeability or the remanent magnetization can be measured. Thus, changes in magnetic signal due to magnetic permeability or remanent magnetization would be the correlates of specific processes under study in particular regions of the GI tract.

Frei et al [8] followed by Benmair et al [9] were the first to use ferromagnetic tracers of high magnetic permeability to study gastric emptying. They used an astatic pair of coils to detect the magnetic field produced by an excitation coil powered by 60 Hz voltage mains. Changes of magnetic reluctance of one pair of coils, due to the presence of the high permeability material produces a voltage signal. The tracer used was a magnesium ferrite, that proved to be innocuous in a previous test as an X-ray contrast agent. The test meal was made by mixing 50 grams of ferrite with 200 grams of chocolate pudding, giving a final ferrite concentration of 20 %. The subject laid in a wood bed with the detector positioned dorsally, and data was acquired as a function of time. A total of 53 normal subjects and patients were studied.

Manganese ferrite powder mixed with an yogurt was used also as a MT by the biomagnetism group at the University of São Paulo-Ribeirão Preto to study different functions and parameters of gastrointestinal tract. In a first study, gastric emptying was measured using an AC biosusceptometer (ACB) that, due to improvements in the instrumentation, was easy to operate and could detect test meals with a 3% ferrite concentration [10]. The instrument can be viewed as two air core transformers, one working as a reference and the other as a measuring transformer. To maximize the signal-to-noise ratio the detection coils are positioned as near the subject as possible, and phase sensitive detection is employed. It can be shown that the signal amplitude is very sensitive to the distance of the magnetic material from the detector ($\sim r^{-6}$) and this can be exploited to measure gastric contractions, as described below.

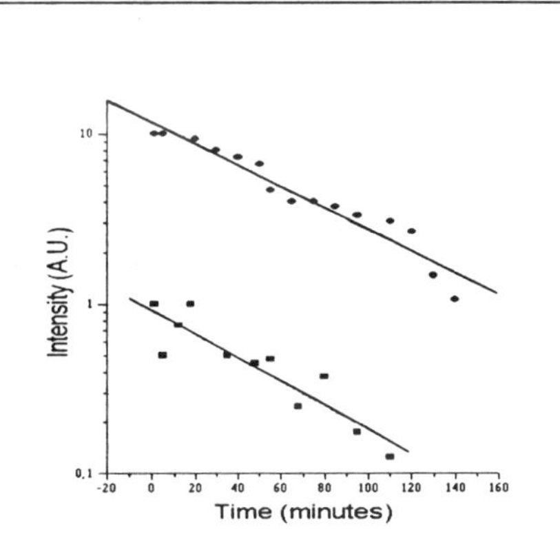

Fig.4-Gastric emptying measured by scintigaphy and the AC biosusceptometer. The test meal was labeled with technetium and ferrite.

In Fig. 4 a measurement of gastric emptying with the AC biosusceptometer is shown in comparison with a simultaneous measurement of the same function with a scintigraphic technique. The test meal was double labeled with ^{99m}Tc and ferrite.

This same instrument was used to measure the orocaecal transit time [11,12] with advantages over the method of the hydrogen expired breath, produced by a lactulose test meal. In Fig. 5 a typical curve showing the arrival of the food head at the caecum area is presented. It is important to notice that the biomagnetic method gave meaningful results in situations that the hydrogen test failed, either due to the absence of or a high basal H_2 level, thus being insensitive to interference of the colonic bacteria status and small gut infection [13].

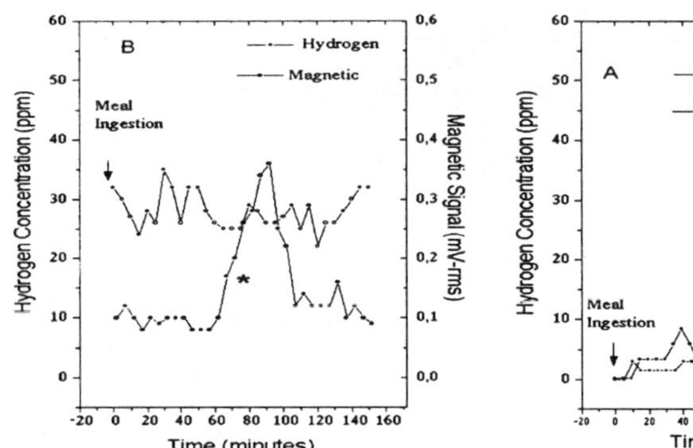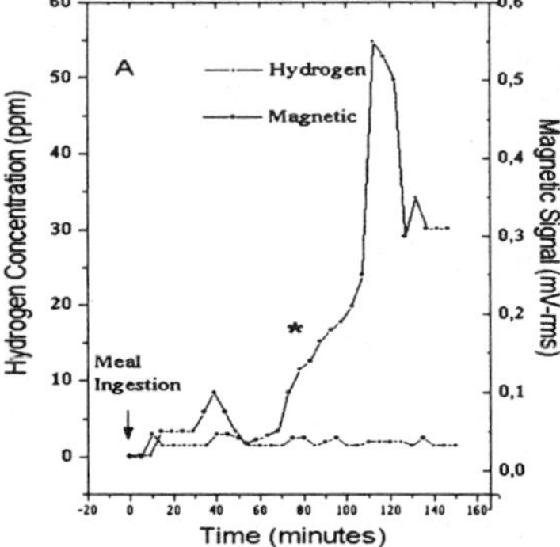

Fig.5-Orocaecal transit time determined by AC biosusceptometer compared to the hydrogen breath test in two cases where the hydrogen test fails.

Partial esophagus transit time (ETT) was studied with the same marker and a variation of the test meal. A bolus test meal, consisting of 5 ml of yogurt mixed with 5 g of ferrite powder, was followed by placing one gradiometer at the proximal esophagus (furcula) and the other on the mid distal esophagus (xiphoid process). The ETTs obtained by the biomagnetic method were compared in 6 subjects with ETTs obtained by the scintigraphy method, and were found to be in good agreement [14].

Another application of the AC biosusceptometer (ACB) was the study of the contractions of the antrum [15]. The instrument was placed near the umbilicus after the subject had ingested the magnetic test meal, and the mechanical contractions associated with the stomach slow waves (SSW) could be detected. The ACB could detect the SSW because these waves produce undulations of the stomach wall, varying the distance of the test meal to the detector and, consequently, the signal intensity. Simultaneous measurements were performed with electrogastrography (EGG) and scintigraphy, showing good agreement between the two methods. The effect of drugs known to affect the GI motility was also studied. It could be shown that the amplitude of the antral contraction were reduced immediately after the administration of Buscopan and recovered to 50% of its initial value after 40 minutes.

In a second category of experiments the test meal was magnetized and the remanent field was measured using a fluxgate magnetometer [16]. The decay of magnetization clearly shows the SSW and contains information relevant to the mixing function of the stomach, although some modeling assumptions are necessary to ascertain this relationship.

The group at Chalmers University used the same principle to measure gastric emptying. In these experiments the subjects ingested a solid test meal containing ferrimagnetic particles (γ-Fe_2O_3, 400 mg), The strength of the remanent field was measured with fluxgate magnetometers outside the stomach (fig. 6) and the amount of magnetic tracer was estimated [17].

The procedure was repeated 17 times up to two hours postprandially. The first measurement was taken about ten minutes after the beginning of ingestion. The amount of gastric content was then estimated every fifth minute during the first hour, and every tenth minute during the second hour.

In vivo measurements were carried out on 16 healthy male volunteers. The estimated retained magnetic tracer after the two-hour measurement time was (31 ± 2) % (mean ± SD) and the lag phase time was (28 ± 10) minutes. The corresponding scintigraphic curve from 16 males showed (40 ± 14) % retained isotope after two hours.

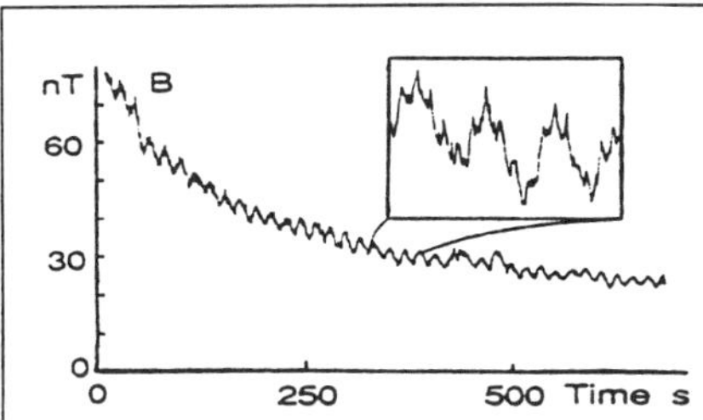

Fig.6-A typical decay curve of the remanent magnetization of a test meal in the stomach, the inset shows the characteristic 3 per minute SSW.

The magnetic detection of electrical activity of the gastrointestinal tract

Although signals from magnetic markers in the stomach were detected as early as 1957, the detection of slow wave magnetic fields produced by the electrical currents occurring in the gastrointestinal tract had to wait until very recently when high sensitivity detectors (SQUIDs) were employed to detect these fields. Measurements were initially performed in humans in an unshielded environment [18] and more recently in shielded enclosures [19] in animals and humans. The gastric electrical activity known as Basic Electric Rhythm (BER) was first detected by the group at the University of Chieti using a second order biogradiometer in an unshielded environment. The characteristic BER of 3 oscillations per minute could be observed directly in the magnetic field recording and by Fourier analysis. Fig. 7 shows their first measurements.

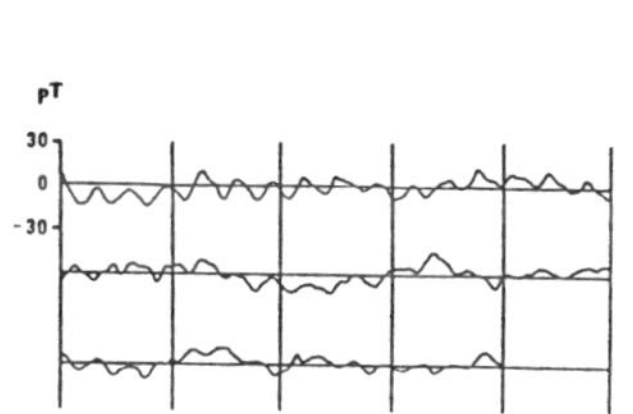
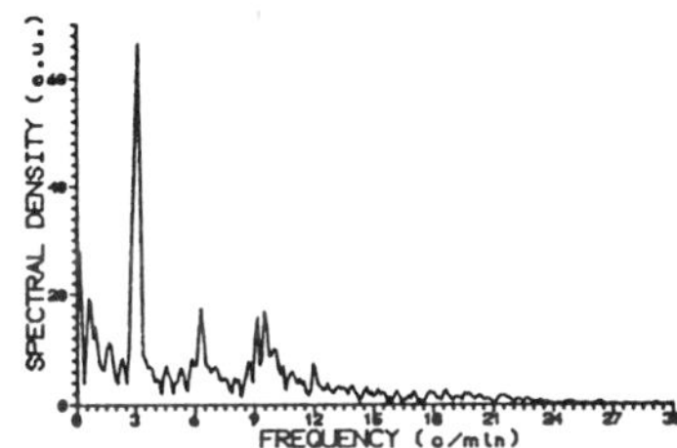

Fig.7- Magnetic signal (MGG) produced by the stomach and its Fourier transform.

The magnetogastrogram of normal volunteers has also been studied at University of São Paulo, before and after 100 ml of whole milk was ingested. Comparisons were made with the EGG, showing a strong correlation between both signals. Statistical analysis of the frequency of the MGG before and after the test meal showed a decrease in frequency ($p < 0.01$), confirming previous studies done with the EGG.

The biomagnetism group at the Vanderbilt University is also pursuing research in this subfield and measurements were performed in the small intestine, which are quite different from those recorded from the stomach. Electrically, the gastric signals are easier to detect because there is less internal fat near the stomach; however, small bowel signals are virtually impossible to see electrically without extensive adaptive filtering. Magnetically, the small bowel signals are quite clear. In initial studies, animals had their gut exposed by surgery and the magnetic signals associated with the small intestine electrical activity was measured. In a very interesting work it was shown that the magnetic measurements could diagnose the intestinal ischemia induced in rabbits by the administration of thrombin. This experiment simulated the clinically important case of mesenteric vessel thrombosis, which has no method of early detection and is associated with high mortality. More details will be given in one of the following sections.

Sources of Gastrointestinal Biomagnetic Fields: A Modeling Perspective

Two types of spontaneous gastrointestinal electrical activity may be recorded with external SQUID magnetometers. A low-frequency slow wave potential known as the Basic Electrical Rhythm (BER) mediates a higher-frequency spiking action potential that is associated with the contraction of the bowel.

The BER oscillates continuously with periods typically ranging from 3-25 cycles-per-minute (cpm) depending upon the location of the activity and the animal under investigation. In humans, the stomach BER is around 3 cpm, and the small bowel exhibits a natural gradient in its frequency. The proximal duodenum oscillates near 12 cpm and progressively more distal sites oscillate at lower frequencies. In the terminal ileum, the oscillation occurs at about 8 cpm. The normal frequency gradient is not linear; it is stepwise. The duodenal slow wave frequency of about 12 cpm persists until the mid-jejunum, where the frequency drops abruptly to another plateau. The frequency plateaus become successively shorter at more distal locations.

In diseased states, the BER changes. Experimental studies with rabbits have shown a decrease of as much as 40% to 50% in BER frequency during ischemia [20].

The small intestine consists of two layers of muscle fibers. The thick inner layer of fibers is oriented circumferentially. Contraction of these fibers is associated with the basic electrical activity and serves to mix intestinal contents. The thinner outer layer of fibers runs longitudinally along the bowel and is responsible for the longitudinal bowel contractions associated with peristalsis [21].

Electrical activity in the stomach arises at one location in the greater curve of the antrum known as the pacemaker site. Slow waves generated by the pacemaker are passively propagated via cell-to-cell coupling toward the pylorus at a rate of about 3 waves per minute. Originally, researchers believed that the small intestine also contained one pacemaker site in the upper duodenum with slow waves passively propagated down the entire length of intestine. However, the observed natural gradient in BER frequency along the intestine would suggest that certain numbers of slow waves would have to be occasionally dropped as they propagated down the bowel. Diamant and Bortoff [22] provided the first experimental evidence that multiple pacemakers are present in the small intestine. With electrodes spaced 2 cm apart down a length of intact intestine, they recorded the previously-established stepwise decrease in frequency with distance along the bowel. They subsequently cut the bowel in 2 cm segments and repeated the recording. With the bowel cut into segments, they observed a frequency decrease that was nearly linear. They also noticed that near electrodes in a transition region where the frequency changed in the intact bowel preparation, a "waxing and waning" characteristic was evident. Further investigation proved that the waxing and waning patterns resulted from the superposition of activity on either side of the electrode in the transition region. These results led them to suggest that the intestine consists of a series of electrically coupled oscillators where each oscillator has its own intrinsic "driving" frequency. The intrinsic frequencies of the oscillators decrease in a linear fashion with distance along the gut, but because of the electrical coupling, large regions of intestine may be "entrained" by the activity in a fairly localized region, thus creating a frequency plateau. When the electrical coupling is insufficient, or when the intrinsic frequency of oscillators in the chain becomes too low, the next oscillator in the series takes over as a new pacemaker oscillator. Thus, the small intestine consists of several pacemaker oscillators at different frequencies separated by regions of propagation of the electrical activity at the same frequency.

Modeling efforts for describing gastrointestinal electrical activity began in the late sixties with Nelsen and Becker's observation that the slow wave activity is consistent with coupled relaxation oscillators described by the Van der Pol differential equation [23]. By assigning intrinsic frequencies and coupling constants to a group of Van der Pol oscillators, Sarna performed a computer simulation of gastric electrical activity [24]. Smout et al. described the cutaneous electrogastrogram using a series of depolarization-repolarization dipole pairs spaced to result in an oscillatory behavior at the abdominal surface 5 cm away [25]. Recently, Familoni et al. combined the relaxation oscillator method and the dipole models to obtain a series of coupled dipoles describing the electrical activity underlying the cutaneous electrogastrogram [26].

We extended this type of model for forward and inverse calculations of the magnetic field from gastrointestinal electrical activity. A 2-dimensional "chain" model for the gastrointestinal system was developed using pacemaker and propagation oscillating dipoles. Sine waves were used to simulate the oscillatory activity since sinusoids can be described succinctly by their amplitudes, frequencies, and phases, and since the observed basic electrical activity is approximately sinusoidal. In the region representing the stomach, the pacemaker dipole was located along the greater curve of the pylorus oriented toward the pylorus in the direction of electrical propagation in the stomach. The propagation of the electrical activity was represented by four additional dipoles along a line from the antrum to the pylorus whose phases were shifted in a manner consistent with uniform propagation. For the small intestine, six dipoles act as pacemakers. The first is located in the upper duodenum with a frequency of 12 cpm, and the other five are located at progressively more aboral locations with frequencies of 11.5, 11, 10, 9, and the last pacemaker dipole in the terminal ileum is assigned a frequency of 8 cpm. Two dipoles are placed immediately following each of the pacemakers with the same frequency but with phases that shift in a manner consistent with uniform propagation of intestinal electrical activity. The arrangement of dipoles is illustrated in Fig. 8.

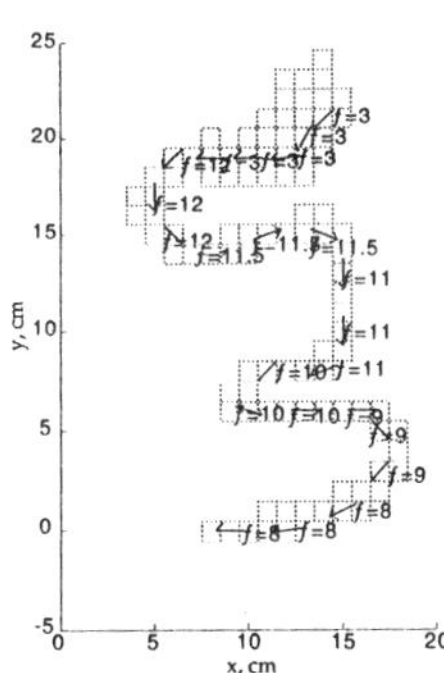

Fig. 8. Arrangement of dipoles in the "chain gut" model. Frequencies of the dipoles are indicates. Phases of same-frequency dipoles shift in a manner consistent with uniform propagation.

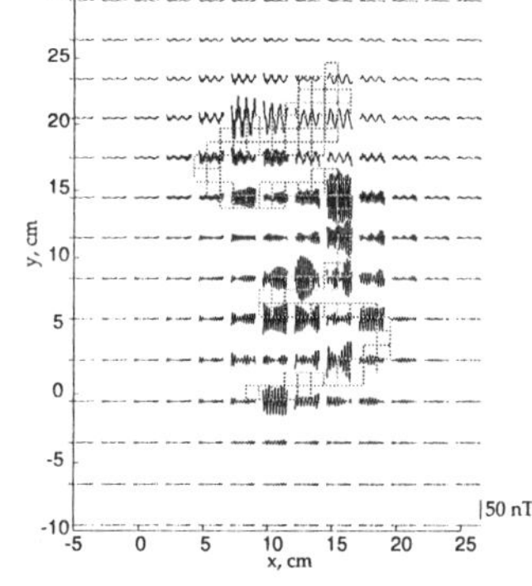

Fig. 9. Magnetic field distribution in a plane 1 cm above the source plane. Sixty seconds of data calculated for each channel are plotted. Variations in the frequency of the electrical activity are evident in the magnetic field data.

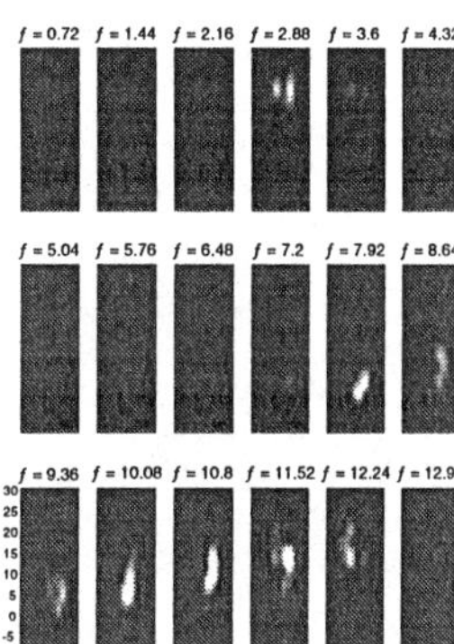

Fig.10. Spatial maps of the frequency content in the magnetic field of Fig. 2 for 18 different frequencies. Gray-scale intensity is proportional to strength of activity.

Sixty seconds of magnetic field data that would be recorded by a hypothetical 182-channel magnetometer array whose coils are arranged in the plane 1 cm above the source plane are plotted in Fig. 9. The magnetic field recorded in a single magnetometer channel contains contributions from a number of underlying sources. Above the gastric region, a 3 cpm characteristic is evident in the magnetic field corresponding to the frequency of the gastric dipoles. The frequency of the magnetic field along the intestinal tract also reflects the frequency of the underlying sources, and in several locations the magnetic field contains several frequencies from dipoles with different frequencies.

The amount of activity at a particular frequency in a specific region can be visualized by use of spectral analysis of the magnetic field data. Fig. 10 shows spatial maps of the magnitude of spectral decompositions of the magnetic field data in each channel for 18 frequencies from 0.72 cpm to 12.96 cpm. The grayscale intensity indicates the amount of activity. Gastric activity is clearly seen in the upper part of the map at f = 2.88 cpm where the stomach is located, and the small bowel activity is indicated in maps from 7.92 to 12.24 cpm near the locations of small bowel dipoles as indicated in Fig. 8. These type of maps could be useful for identifying diseased states of the bowel. For example, an ischemic episode with activity at a lower than normal frequency would result in bright spots in the spatial maps of frequency content at abnormal frequencies.

Future modeling efforts should incorporate more accurate source representations than the sine wave approximation used here. The picture presented by this 2-dimensional model was enlightening but potentially deceptive if it is compared with the actual situation. The 3-dimensional nature of the GI tract must be accounted for when attempting to compare models with actual data. In addition, models should begin to compare the external magnetic fields with external potentials to demonstrate the superiority of magnetic measurements because of the intervening layers of low-conductivity fat present in the abdomen. Clearly, these models must include accurate conductivity and geometrical considerations to precisely determine volume conductor effects.

Noninvasive measurement of gastrointestinal magnetic fields for early detection of ischemia

Previous sections have described in detail the electrical activity of the gastrointestinal tract and some of the advantages of measuring these signals noninvasively using SQUID magnetometers. From a clinical prospective, the foremost reason to be excited about the measurement of gastrointestinal magnetic fields is the high signal-to-noise ratio of magnetic field recordings. Cutaneous electrodes have many drawbacks including an extremely low signal-to-noise ratio and the inability to localize or discriminate dipole sources within the abdomen because of the layers of insulators present within the abdominal wall.

Clinical Problems

Acute mesenteric ischemia of the gastrointestinal tract is a serious condition associated with as much as a 90% mortality and significant morbidity in those that survive [27]. While the patient with myocardial ischemia is diagnosed expeditiously with the changes seen on the electrocardiogram, the patient with mesenteric ischemia has no such diagnostic tool, and thus the diagnosis can only be made at exploratory laparotomy. The ideal test for mesenteric ischemia diagnosis would be noninvasive and fast, would detect ischemia prior to irreversible changes, would have high sensitivity and specificity, would be available for multiple studies, and would be easily performed in critically ill patients.

Animal Studies

CHANGES IN BASIC ELECTRICAL RHYTHM (BER) RECORDED USING A SQUID MAGNETOMETER DURING ACUTE ISCHEMIA.
One of the first studies done at Vanderbilt University utilized a rabbit model in which the animal was anesthetized and the bowel exteriorized into a warm, plexiglas dish. Bipolar electrodes were placed on the bowel and the magnetic fields of the exteriorized small intestines were recorded using microSQUID. After stabilization for 15 minutes the mesenteric artery was thrombosed using intra arterial injection of thrombin to cause complete ischemia. The BER frequency determined by electrodes was identical to that determined by the noncontact SQUID magnetometer. After 5 minutes of injection, there was a 27% drop in frequency (p ‹ 0.05) [28].

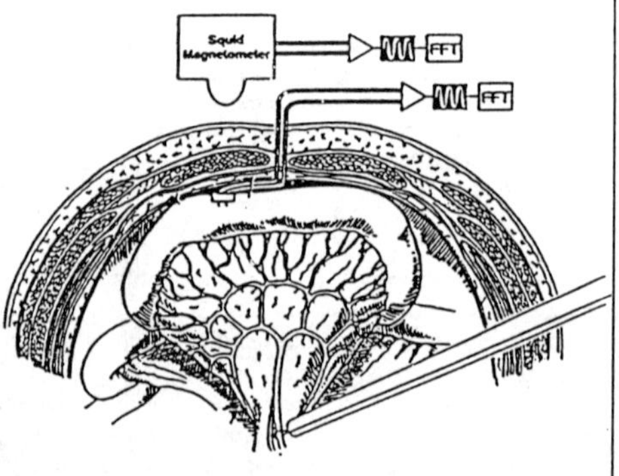

CORRELATION BETWEEN PATHOLOGY AND ELECTRICAL ACTIVITY DURING ACUTE ISCHEMIA. The next study looked at the timing of BER changes during arterial ischemia and whether or not there are irreversible pathologic changes during these early abnormalities of BER frequency. Mesenteric ischemia was induced by ligating the superior mesenteric artery in anesthetized rabbits. There was a significant decrease in BER frequency 20 minutes after the onset of ischemia [21.5 ± 1.9 cycles per minute (cpm) to 14.2 ± 1.5 cpm (p= 0.02 Student's T-test)]. Microscopic evaluation revealed no pathologic changes until 60 minutes of complete arterial ischemia. Sixty minutes after the occlusion of the SMA the pathologic changes of sloughing of the mucosal villi and

Figure 11: Diagram of set-up for noninvasive diagnosis of intestinal ischemia. Trough a midline laparotomy, a loop of ileum is isolated and the segmental mesenteric artery is dissected and encircled with a monofilament snare that is exteriorized. Bipolar electrodes are sutured onto the serosal surface of the bowel, and the bowel itself is sutured to the abdominal wall to ensure the same loop of intestine is measured by the SQUID magnetometer. After closing the abdomen, the SQUID is approached to the abdominal wall of the anesthetized animal. The eight rabbits were studied for a 15 minutes baseline period, with continuous recordings of BER from the electrodes and SQUID magnetometer. The segmental mesenteric artery was occluded for 15 minutes while continuous recordings of BER were taken by both methods.

edema were minimal. These studies show that the electrical changes occur well before the onset of pathologic changes [29].

NONINVASIVE DIAGNOSIS OF MESENTERIC ISCHEMIA IN AN ANIMAL MODEL. Another study utilized an anesthetized rabbit shown in Fig. 11 [19]. The snare around the superior mesenteric artery was used to occlude the blood supply to the intestine for 30 minutes and then loosened to allow reperfusion of the bowel. There was an immediate and dramatic drop in BER frequency during occlusion which returned to baseline levels after 15 minutes of reperfusion (Fig. 12).

The frequency determined by the noninvasive SQUID magnetometer was highly correlated (r =.91) to those obtained with invasive serosal electrodes (Fig. 13). This study demonstrated that magnetic field measurements could be obtained noninvasively in the animal model and could detect

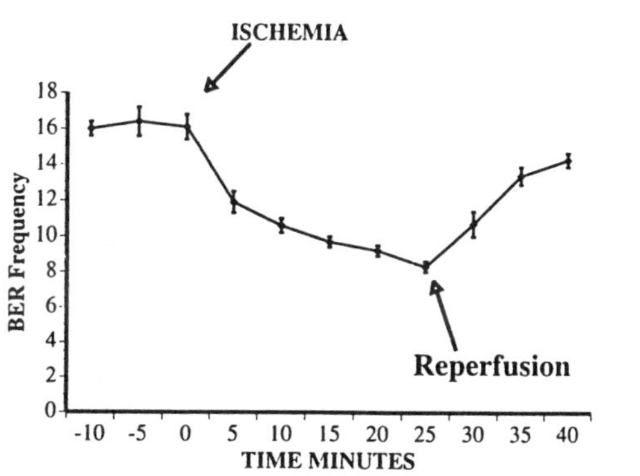

Fig.12-BER frequency determined with the SQUID magnetometer in ten animals before, during, and after snare occlusion of the main superior mesenteric artery. There was a significant decrease in frequency 5 minutes after occlusion of the superior mesenteric artery. On reperfusion, BER frequency rises to near baseline values 15 minutes after the start of reperfusion.

changes in BER frequency during periods of ischemia. Moreover these changes in electrical activity appear to reflect the viability of smooth muscle within the intestine.

MAGNETIC FIELD MEASUREMENTS DURING MESENTERIC VENOUS THROMBOSIS. While arterial thrombosis or embolism represents most of the patients with mesenteric ischemia, there is a significant subset of patients who have thrombosis of the superior mesenteric vein. We performed studies to determine whether or not changes in BER frequency could be identified noninvasively using a SQUID magnetometer in an animal model of mesenteric vein thrombosis [30]. Anesthetized rabbits were prepared in a fashion similar to that described for the arterial occlusion studies. The superior mesenteric vein was isolated, and after 15 minutes of stable recordings thrombin was injected into the superior mesenteric vein to produce complete thrombosis of the venous system. The BER dropped from 16.42 ± 0.69 cpm to 8.80 ± 0.74 cpm at 30 minutes and to 6.82 ± 0.7 after 90 minutes ($p< 0.0001$ paired T-Test). Pathology at this time shows ischemic mucosa with coagulative necrosis and sloughing of the epithelium. These studies suggest that the changes in BER frequency correlate to the viability of the smooth muscle in the intestine and that these changes occur

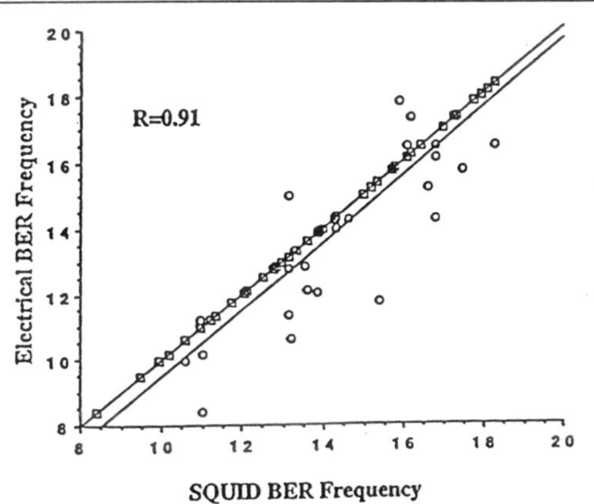

Figure 13: The frequency of the BER with the frequency determined by fast-Fourier transformation recorded by the electrode is plotted on the abscissa. The magnetic field BER frequencies obtained at the same time are plotted on the ordinate axis. There was a high level of correlation between the simultaneous recordings. (r=0.91).

before irreversible pathologic changes within the smooth muscle. Moreover, these changes could be identified noninvasively through the abdominal wall in our animal model.

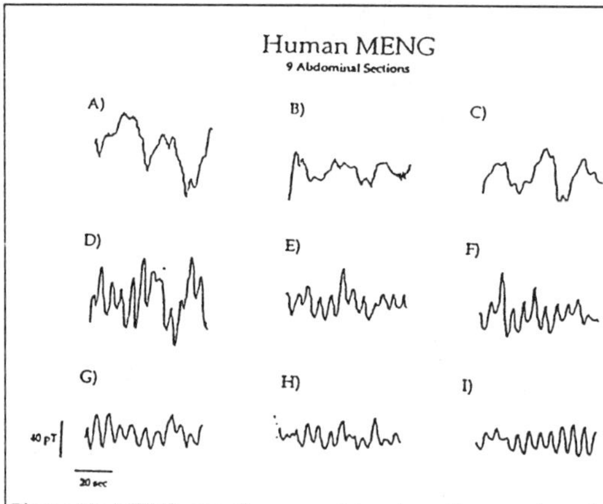

Figure 14: MENG recordings were taken from the recorder using the MAGNES™ 37 channel magnetometer from 9 locations in the same volunteer; A) Right upper quadrant, B) epigastrium, C) left upper quadrant, D) right mid quadrant, E) periumbilical, F) left mid quadrant, G) right lower quadrant, H) hypogastrium, I) left lowes quadrant. The recording were sampled at 1041 Hz and are shown after median filtering of 1040 point or 1 second wide. All recordings were obtaneid while the volunteers suspended respiration, recordings identified by A,B and C refers to gastric and D,E,F,G,H and I to small bowel in the same volunteer the small intestine.

Human Studies

A number of studies have been performed looking at the magnetic field measurements of the gastrointestinal tract in human volunteers [31]. Twenty-one human volunteers (7 male; 14 female) ages 26 - 55 (mean 33) have been studied to examine the magnetic field activity within the abdomen. Cyclical wave forms consistent with gastric and small bowel BER have been identified in every volunteer [32]. There has been great progress made in developing algorithms to exclude the magnetocardiogram contamination.

Although we were able to measure the magnetic field activity in every volunteer, in order to identify segments of ischemic bowel it will be necessary to identify the spatial distribution of electrical activity within the gastrointestinal tract. The stomach has a BER frequency of three cycles per minute that originates on the greater curvature and propagates towards the pylorus while the duodenum has a BER frequency of 12 cpm. There is a stepwise decrease in BER frequency known as the frequency gradient within the small intestine. The intestinal BER frequency gradually decreases as you move aborally in the GI tract (9 cpm in the distal ileum). A mapping study was performed in order to identify this frequency gradient within the small bowel. The volunteer was asked to lie quietly while the 37 channel magnetometer was positioned in 9 regions over the abdomen. While the volunteer suspended respirations, recordings were obtained and from each of these recordings we could identify BER activity. Recordings were subjected to autoregressive spectral analysis to identify the dominant frequencies within each of these sections. In the upper quadrants, the BER ranged between 2.5 and 2.7 cpm. In the right middle quadrant recordings, a peak of 12 cpm can be seen corresponding to the duodenal frequency. In the right lower quadrant, the BER frequency is 9.7 cpm (Fig. 14 and 15). This corresponds to the expected locations within the abdomen for gastric and small bowel electrical activity [33]. This demonstrates the potential usefulness of magnetic field recordings at multiple sites to identify the spatial variations in frequency within the gastrointestinal tract. This line of study has been extremely exciting and shows great promise for clinical applicability. There are however significant questions that remain to be answered prior to the deployment of these devices and techniques in clinical practice.

Questions and Aims

1) Do other conditions such as infection, peritonitis, or bowel obstruction affect BER frequency ?
2) Can small areas of ischemic bowel be detected within large amounts of normal activity ?
3) Can we detect the absence of signal in dead bowel if other normal signals exist around it ?
4) Since many of these patients are clinically ill and are on ventilators or in the emergency room, it will be necessary to perform measurements in an unshielded hospital environment. This is a very noisy environment and multiple technical factors will have to be overcome in order to make this a useful clinical instrument.

5) In all medical meetings today there is a great amount of talk about cost containment and the issues of cost effectiveness. Although on the surface, this appears to be a costly technology there is no competing technology capable of measuring small bowel electrical activity. Thus, new technology will have to show that it is cost effective in order for hospitals to purchase this type of equipment.

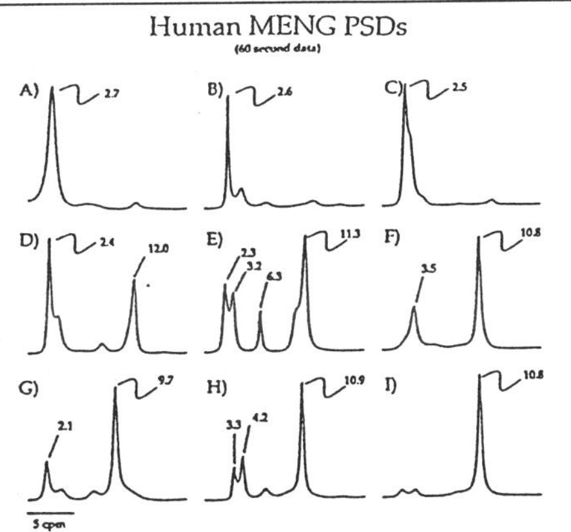

Fig. 15-Autoregressive spectral analysis was used on 60 second periods of data to determine the dominant frequency of the BER. The 9 graphs correspond to the locations described in fig. 14. Since none of the 60 second recordings were taken simultaneously, there is temporal variation between each of the 9 recordings. The Frequency of gastric BER (**A,B,C**) ranged between 2.5-2.7 cpm. In (**D**) the right upper quadrant recordings a peak at 12.0 cpm can be seen that corresponds to the predicted BER frequency of the duodenum. In the middle and lower quadrants the BER frequency was found to range from (**E**) 11.3 cpm to (**F**) 10.8 cpm to (**G**) 9.7 cpm to (**H**) 10.9 cpm to (**I**) 10.8 cpm that corresponds to the predicted frequency gradient within the small intestine.

Summary

Animal studies of acute mesenteric ischemia demonstrate that changes in BER frequency occur prior to pathologic changes and reflect the viability of the bowel. Magnetic field measurements have a high signal-to-noise ratio, are highly correlated with invasive serosal electrode recordings, and can detect ischemic bowel noninvasively. Human studies have been started and show great promise. The BER of the stomach and small intestine can be recorded noninvasively. Further work to create an optimized instrument for noninvasive magnetic field diagnosis of acute mesenteric ischemia is ongoing.

Concluding Remarks:

The future of the biomagnetic methodology in gastroenterology is linked to numerous facts. First, it is necessary to find applications where biomagnetic measurements can provide unique information, or the information provided by existing methods are poor when compared to that obtained by biomagnetic measurements. Second, the instrumentation used should be affordable to hospitals, and if possible of simple operation and low cost. Third, other aspects of the G.I. tract should be investigated.

It is strong believed that there is an enormous space for improvements in the examples above mentioned that could lead to new applications and refinements in the present research and diagnose methods in gastroenterology.

References:

[1] Hoke, M., Erné, S.N., Okada, Y.C., and Romani, G.L. Biomagnetism: Clinical Aspects, Amsterdam, Excerpta Medica, 1992.

[2] Wenger, M. A., Henderson, E.B., and Dinning, J.S. Magnetic method for recording gastric motility, science, 1957, 990-991.

[3] Di Luzio, S., Comani, S., Romani, G.L., Basile, M., Del Gratta, C., and Pizzela, V. A biomagnetic method for studying gastro-intestinal activity, Il Nuovo Cimento, 1989, 11D(12): 1853-1859.

[4] Basili, M., Neri, M., Carriero, A., Casciardi, S., Comani, S., Del Gratta, C., Donato, L. G., Di Luzio, S., Macri, M.A., Pasquarelli, A., Pizzela, V., and Romani, G. L. Measurement of segmental transit time in man, Digestive Disease Sciences, 1992, 37(10):1537-1543.

[5] Forsman, M., Hultin, L., and Abrahamsson, H. Measurements of gastrointestinal transit using fluxgate magnetometers, In: Baumgartner, C., Deecke, L., Stroink, G. and Williamson, S.J. Biomagnetism: Fundamental Research and Clinical Applications. Studies in Applied Electromagnetics and Mechanics, Amsterdam, Elsevier Science Publisher, 1995.

[5] Forsman, M., Hultin, L., and Abrahamsson, H. Measurements of gastrointestinal transit using fluxgate magnetometers, In: Baumgartner, C., Deecke, L., Stroink, G. and Williamson, S.J. Biomagnetism: Fundamental Research and Clinical Applications. Studies in Applied Electromagnetics and Mechanics, Amsterdam, Elsevier Science Publisher, 1995.

[6]-Weitschies W., Wedemeyer J., Stehr R., and Trahms L. Magnetic markers as a noninvasive tool to monitor gastrointestinal transit. IEEE Transactions on Biomedical Engineering, 1994, 41:192-195.

[7] Nomura, M., Saijyo, T., Haruta, Y., Itozaki, H., Toyoda, H., Nakaya, Y., Ito, S., and Kado, H. Biomagnetic measurement of gastric mechanical motility using a high-temperature SQUID imaging. This volume.

[8] Frei E.H., Benmair Y., Yerashalmi S., and Dreyfuss F. Measurements of the emptying of the stomach with a magnetic tracer, IEEE Transactions on Magnetics, 1970, Vol MAG-6:348-349.

[9] Benmair, Y., Dreyfuss, F., Fischel, B., Frei, E.H., and Gilat, T. Study of gastric emptying using a ferromagnetic tracer. Gastroenterology, 1977,73:1041-1045.

[10] Miranda J.R.A., Baffa O., Oliveira R.B., and Mitiko M. N. An AC biosusceptometer to study gastric emptying, Medical Physics, 1992, 19(2): 445-449.

[11] Oliveira, R.B., Miranda, J.R.A., Baffa, O., Cambrea, C.R., and Troncon, L.E.A. A biomagnetic technique for orocaecal transit time measurement, In: Hoke, M., Erné, S.N., Okada, Y., Romani, G.L. Biomagnetism: Clinical Aspects, Excerpta Medica, Amsterdam,1992.

[12] Baffa O., Oliveira R. B., Arruda Miranda J. R., and Troncon L. E. A. Analysis and development of AC biosusceptometer for orocaecal transit time measurements, Medical & Biological Engineering & Computing, 1995, 33:353-357.

[13] Oliveira, R.B., Baffa, O., Troncon, L. E. A Miranda J. R. A., and Cambrea C. R.. Evaluation of a biomagnetic technique for measurement of orocaecal transit time, European Journal of Gastroenterology and Hepatology, 1996, in press.

[14] Daghastanli, N.A., Braga, F.J.H.N., Oliveira, R.B., and Baffa, O. Esophageal transit time evaluated by means of the biomagnetic method, This volume.

[15] Miranda, J.R.A., Baffa, O., Oliveira, R.B., Braga, F.J.H., and Sousa, P.L. Detection of Stomach Contractions Using a Biosusceptometer, This volume.

[16] Baffa, O., Oliveira, R.B., Miranda, J.R.A., and Sousa, P.L. Mixing power of food in the stomach evaluated by a biomagnetic technique, In: Baumgartner, C., Deecke, L., Stroink, G. and Williamson, S.J. Biomagnetism: Fundamental Research and Clinical Applications. Studies in Applied Electromagnetics and Mechanics, Amsterdam, Elsevier Science Publisher, 1995.

[17] Forsman M. Assessment of gastric movements by monitoring magnetic field decline of magnetized trace particles, Physics in Medicine and Biology, 1994,39a: p. 62.

[18] Comani, S., Basile, M., Casciardi, S., Del Gratta, C., Di Luzio, S., Erné, S.N., Macri, M., Neri, M., Peresson, M., and Romani, G. L. Extracorporeal direct magnetic measurement of gastric activity. In: Hoke, M., Erné, S.N., Okada, Y., Romani, G.L. Biomagnetism: Clinical Aspects, Excerpta Medica, Amsterdam,1992.

[19] Saijyo, T., Nomura, M., Haruta, Y., Itozaki, Y., Toyoda, H., Nakaya, Y., Ito, S., and Kado, H. Magnetogastrograms using 64-channel SQUID system: study on clinical usefulness, This volume.

[20] Richards W.O., Garrard C.L., Allos S.H., Bradshaw L.A., Staton D.J., and Wikswo J.P., Jr. Noninvasive diagnosis of mesenteric ischemia using a SQUID magnetometer, Annals of Surgery, 1995, 221(6): 696-705.

[21] Liu L.W.C., and Huizinga, J.D. Electrical coupling of circular muscle to longitudinal muscle and interstitial cells of Cajal in canine colon, Journal of Physiology (Lond.), 1993, 470: 445-461.

[22] Diamant N.E., and Bortoff A. Nature of the intestinal slow-wave frequency gradient, American Journal of Physiology, 1969, 216(2): 301-307.

[23] Nelsen T.S., and Becker J.C. Simulation of the electrical and mechanical gradient of the small intestine, American Journal of Physiology, 1968, 214: 749-757.

[24] Sarna S.K., Daniel E.E., and Kingma Y.J. Simulation of the electrical activity of the stomach by an array of relaxation oscillators, Digestive Diseases Science, 1972, 17: 299-310.

[25] Smout A.J.P.M., Van der Schee E.J., and Grashuis J.L. What is measured in electrogastrography ? Digestive Disease Sciences, 1980, 25: 179-187.

[26] Familoni B.O., Abell T.L., and Bowes K.L. A model of gastric electrical activity in health and disease, IEEE Transactions on Biomedical Engineering, 1995, 42(7): 647-657.

[27] Williams L. Mesenteric Ischemia. In: Sawyers J, Williams L, The Acute Abdomen. Philadelphia: W.B. Saunders, 1988:331-353. Surgical Clinics of North America; vol 68.

[28] Golzarian J., Staton D.J., Wikswo J.P., Jr., Friedman R.N., Richards W.O. Diagnosing intestinal ischemia using a noncontact Superconducting Quantum Interference Device, American Journal of Surgery , 1994, 167:586-592.

[29] Garrard C.L., Halter S., Richards W.O. Correlation between pathology and electrical activity during acute intestinal ischemia, Surgery Forum , 1994, 45:368-371.

[30] Allos S., Sorrell M., Slater S., Staton D., Wikswo J.P., Richards W. SQUID magnetometer diagnosis of mesenteric vein thrombosis, Gastroenterology , 1995, 108(4):A269.

[31] Richards W.O., Staton D., Golzarian J., Friedman R.N., Wikswo J.P., Jr. Non-invasive SQUID magnetometer measurement of human gastric and small bowel electrical activity. In: Baumgartner, C., Deecke, L., Stroink, G. and Williamson, S.J. Biomagnetism: Fundamental Research and Clinical Applications. Studies in Applied Electromagnetics and Mechanics, Amsterdam, Elsevier Science Publisher, 1995.

[32] Garrard C.L., Wikswo J.P., Jr., Staton D., Golzarian J., Gallen C., Richards W.O. Noninvasive measurement of small bowel electrical activity, Gastroenterology , 1994, 106(4):A502.

[33] Richards W.O., Bradshaw L..A., Staton, D.J., Garrard, C. L., Liu, F., Buchanan, S., and Wikswo J.P., Jr. Magnetoenterography (MENG): Noninvasive measurement of bioelectric activity in human small intestine, Submitted Digestive Diseases and Sciences, 1996.

Acknowledgments:

The work presented in this chapter is the result of the endeavor of several research groups, to whom many contributors have collaborated, we thank all of them listed in the above referenced papers. We are also grateful to the partial financial support of the agencies: Department of Veterans Affairs Research Service, FAPESP, CNPq, CAPES, TWAS, the Veterans Administration, the NIH, Conductus, Inc, National Research Council of Italy and Swedish National Board of Industrial and Technical Development (NUTEK).

Magnetic Field Measurement of Rabbit Colonic Electrical Activity

Bradshaw, L.A., Allos, S.H., Wikswo, J.P., Jr. and Richards, W.O.

Vanderbilt University, Nashville, TN USA

Introduction

The smooth muscle of the gastrointestinal (GI) system produces two types of electrical activity that may be detected electrically or magnetically. The slow-wave, or basic electrical rhythm (BER) is a sinusoidal type of activity that oscillates about 3 times per minute in the human stomach and 8-12 times per minute in the small intestine. The electrical response activity is a higher-frequency spiking potential that is strongly associated with segmental contractions of the bowel that propel intestinal contents along the GI tract. Our previous studies on intestinal electrical activity have shown that magnetic measurements taken outside the abdomen correlate strongly with recordings from electrodes placed on the serosal surface of the bowel, and that external magnetic measurements are superior to cutaneous electrode measurements because of the attenuation and smearing of the potential by low-conductivity layers in the abdomen [1, 2].

Electrode measurements of colonic electrical activity have found that, in contrast to the well-defined frequency patterns of the gastric and small bowel systems, colonic activity is highly variable in both frequency and amplitude [3, 4]. Two possible explanations for this observation have been suggested. First, gap junctional coupling between cells in the colon may be weaker gap junctions may be less prevalent than in other parts of the gastrointestinal system [5]. In that case the electrical activity generated in one section of the colon would be somewhat less likely to be correlated with activity generated in another section than would activity in the intestine, for example. Another possible reason for the large variability in colonic electrical activity is the fact that the network of interstitial cells of Cajal (ICC), which acts as a pacemaker for the electrical activity, is less dense in the colon than in other parts of the GI tract [6]. Thus, electrical activity generated by the ICC network does not propagate into colonic smooth muscle in the same way that it does in the case of the intestine. Therefore, one may expect to see a variety of frequencies in colonic electrical activity.

In this study, we want to demonstrate the usefulness of a biomagnetometer system for the detection of colonic electrical activity. We expect that the frequencies of colonic activity that we record with serosal electrodes and with a magnetometer outside the subject will vary in a manner similar to the previous electrode studies. Further, we expect that frequencies detected by the magnetometer will correlate well with frequencies detected with the internal electrode.

Methods

Five male New Zealand white rabbits were anesthetized with Ketamine (30 mg/kg) and Acepromazine (5 mg/kg). The rabbits were fed a nonmagnetic diet for at least 72 hours before the experiment. A midline laparotomy was performed and a segment of the colon was exteriorized. A serosal electrode platform was sutured to the colon segment, and then the colon was replaced in the abdomen. The electrode platform consisted of eight bipolar pairs of nonmagnetic silver electrodes separated by 1 cm embedded in Silastic® silicone adhesive (Dow Corning, cat. no. 891). The laparotomy incision was closed and the animal was placed with the section of the abdomen containing the electrode platform directly under the magnetometer.

The magnetometer was the four-channel high-resolution µSQUID designed for tissue studies. The pick-up coils in µSQUID are 3 mm in diameter arranged at the corners of 4.4 mm square. The coils may be lowered to within 1.5 mm of the room-temperature sample.

Simultaneous recordings of the external magnetic fields and internal serosal potentials were obtained for 10 minutes. Since colonic electrical activity is known to be highly variable, we divided the analysis of these signals into long-term analysis and short-term or instantaneous analysis. For the long-term analysis, the frequency content of both magnetic and electric signals were analyzed using the Fast Fourier Transform (FFT) based power spectral density (PSD) of the entire 10 minute data set, and the frequency content common to both electric and magnetic signals was determined using the cross spectral density (CSD). Welch's averaged periodogram method was used for computation of both the PSD and the CSD. Because it is not clear that the electrical activity necessarily occurs at a single frequency, we applied a peak detection algorithm to determine the peaks in the PSDs and CSDs. The cross spectrum

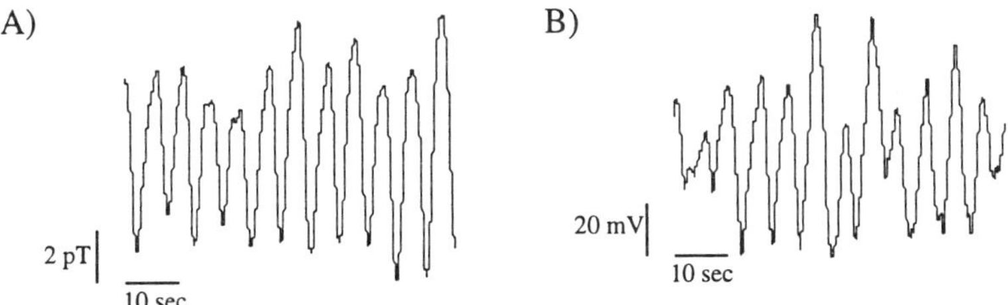

Figure 1. (A) Colonic electrical activity recorded by an external SQUID magnetometer. (B) Colonic activity recorded by serosal electrode. The two recordings were obtained simultaneously.

itself indicates how the electric and magnetic recordings are correlated at different frequencies, but to further test our hypothesis that electrodes and magnetometers record the same activity, we determined the degree of correlation between peaks in the magnetic PSD and the electric PSD, between peaks in the magnetic PSD and the CSD, and between peaks in the electric PSD and the CSD.

A high degree of variability is expected from colonic electrical activity, so we wished to further analyze the spectral content of short-duration data samples. Since the frequency resolution of the FFT is inversely proportional to the sample duration, short sample durations result in unacceptably poor frequency resolution. Thus, we applied an autoregressive (AR) method to the spectral analysis of successive 30 second time segments within the 10 minute recording period. The AR method can provide adequate frequency resolution for short-duration samples [7]. Again, the peak detection algorithm was applied to the AR PSDs, and the degree of correlation between peaks in the electric PSD and the magnetic PSD was determined. Thus, the short-term analysis will give us an idea of the instantaneous correlation of the electric and magnetic signals while the long-term analysis provides a picture of the overall correlation.

Results

Figure 1 shows 60 second tracings of typical simultaneously-recorded serosal potentials and external magnetic fields from one of the studies. A clear oscillatory rhythm is evident in both traces. Visual inspection suggests that the frequency of the oscillation is around 11.5 cpm, possibly modulated by a lower-frequency activity. The power spectral densities of the entire 10 minute electric and magnetic signals as well as the cross spectrum of the electric and magnetic signals are shown in Figure 2. The largest peaks in the power spectra are seen at 11.4 cpm in the mag-

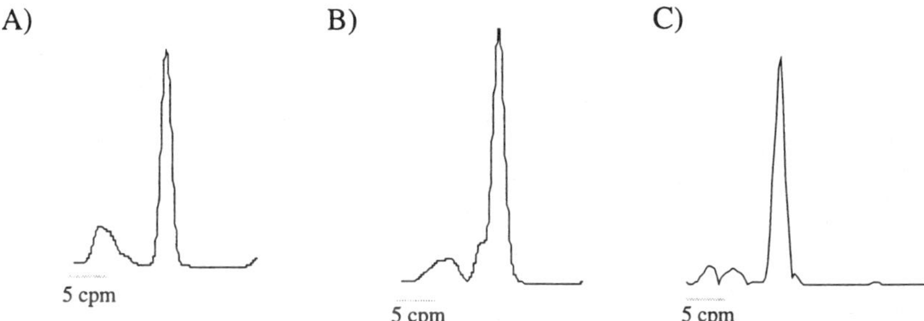

Figure 2. (A) PSD of the entire 10-minute external SQUID recording. (B) PSD of the entire 10-minute serosal electrode recording. (C) CSD of the electrode and SQUID recordings.

netic signal and 12.1 cpm in the serosal electric signal. Broad smaller peaks in the spectra appear in the 3-6 cpm range. The cross spectrum has its largest peak at 11.7 cpm, with smaller peaks at 3.0 and 5.7 cpm, indicating the these

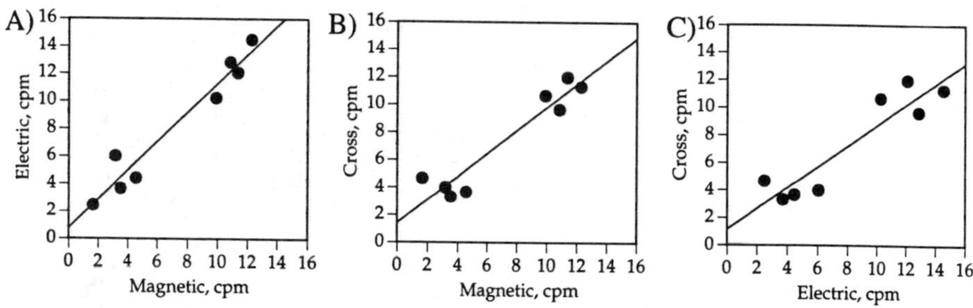

Figure 3. Peak frequencies in (A) electric vs. magnetic recordings, (B) cross spectrum vs. magnetic recordings, and (C) cross spectrum vs. electric recording. The peak frequencies determined by each method are strongly correlated (r = 0.97 for magnetic/electric; r = 0.95 for cross/magnetic; r = 0.92 for cross/electric).

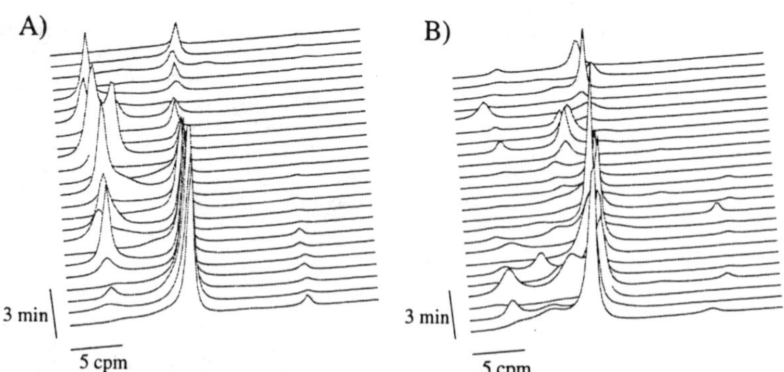

Figure 4. AR power spectral densities at successive 30 second intervals. (A) AR PSDs of the signals from the SQUID magnetometer; (B) AR PSDs of the serosal electric signal.

frequencies are common to both SQUID and electrode measurements.

Peaks such as these were seen in all experiments. During one of the studies, the magnetic signal was lost due to environmental disturbance, so that study was excluded from the correlation analysis. The other four experiments showed peaks in the 2-6 cpm range as well as the 10-14 cpm range. Peaks in the electric signal are plotted against peaks in the magnetic signal in Figure A, and peaks in the cross spectra are plotted against peaks in the magnetic and electric signals in Figure B and C respectively. The correlation of frequencies detected by each analysis method is quite strong (r = 0.97 for electric vs. magnetic, r = 0.95 for cross vs. magnetic, r = 0.92 for cross vs. electric).

Changes in the raw data during different time intervals were observed. Though activity was usually seen consistently in the 8-12 cpm range with only slight variation, lower frequency activity in the 2-6 cpm range was occasionally seen particularly in magnetic recordings. Figure 4 shows plots of the AR power spectra at successive 30 second intervals over the entire 10-minute duration of the same experiment represented in Figure 1. Although consistent peaks are seen in the 8-12 cpm range, lower frequency activity in the 2-6 cpm range is also apparent during some intervals, particularly in the magnetic recordings. From these traces, we see that the frequency peaks change with time, and that the magnetometer detects electrical activity that is not as clearly detected in the serosal electrodes.

The correlation of the short-term peaks in the electric and magnetic signals is somewhat smaller than correlation of the long-term peaks. Figure 5 shows the peaks in the short-term electric signals plotted against the peaks in the short-term magnetic signals. The instantaneous correlation coefficient (the correlation of the peaks in 30 second duration samples) is 0.82.

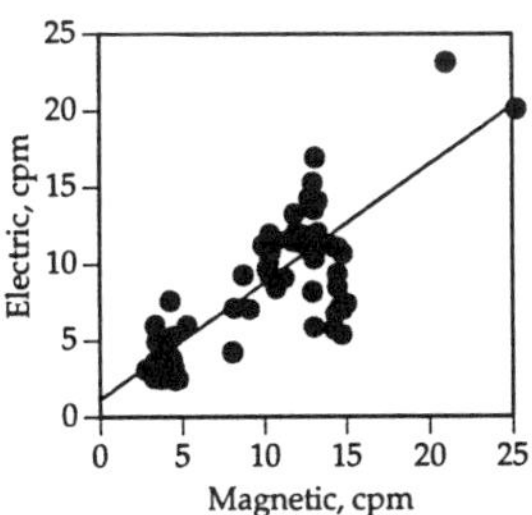

Figure 5. Instantaneous correlation of the serosal electric and magnetic signals. The peaks detected in successive 30 second intervals are shown with peaks in the electric signals plotted against peaks in the magnetic signals.

Discussion

This work represents the first report of the magnetic detection of colonic electrical activity. The electrical activity of the colon was detected with invasive serosal electrodes at a frequency of 8.3 ± 1.7 cpm, and by a noninvasive SQUID magnetometer at a frequency of 7.2 ± 1.5 cpm. Activity was typically seen in the 8-12 cpm range, and less dominant activity was observed in the 2-6 cpm range. As in previous electrode studies, the colonic electrical rhythm was seen to be highly variable in its frequency and amplitude.

SQUID recordings were very similar to serosal electrode data. The frequency of the electrical activity recorded in the SQUID was strongly correlated with that recorded in the electrodes (r = 0.97). Cross spectral peaks also correlated well with both electric and magnetic peaks (r = 0.92 and r = 0.95, respectively). Short-term correlation was somewhat smaller (r = 0.82) due to the fact that magnetometers record activity over a larger region than the serosal electrodes and so include contributions from other electrically active sources.

This work suggests that noninvasive SQUID magnetometers may be used to characterize colonic electrical activity that was previously possible only with invasive serosal electrodes.

References

[1] Bradshaw, L. A., Allos, S. H., Wikswo, J. P., Jr., and Richards, W. O. Transabdominal magnetic detection of small bowel electrical activity. In preparation.

[2] Bradshaw, L. A. *Measurement and Modeling of Gastrointestinal Electrical Activity*. Ph.D. Dissertation, Vanderbilt University, 1995.

[3] Riezzo, G., Pezzolla, F., Maselli, M. A., and Giorgio, I. Electrical activity recorded from abdominal surface before and after right hemicolectomy in man. *Digestion*, 1994, 44: 185-190.

[4] Chambers, M. M., Bowes, K. L., Kingma, Y. J., Bannister C., and Cote, K. R. In vitro electrical activity in human colon. *Gastroenterology*, 1981, 81: 502-508.

[5] Liu, L. W. C., Sperelakis, N., and Huizinga, J. D. Pacemaker activity and intercellular communication in the gastrointestinal musculature. In: Huizinga, J. D. *Pacemaker Activity and Intercellular Communication*, Boca Raton, CRC Press, 1995.

[6] Christensen, J., Rick, G. A., and Lowe, L. S. Distributions of interstitial cells of Cajal in stomach and colon of cat, dog, ferret, opossum, rat, guinea pig, and rabbit. *J. Autonomic Nervous System*, 1992, 37: 47-56.

[7] Marple, S. L. *Digital Spectral Analysis with Applications*, Englewood Cliffs, Prentice-Hall, 1987.

Acknowledgments

This work was supported in part by grants from the Veterans Administration and by an NIH SBIR grant to Conductus, Inc.

Esophageal Transit Time Evaluated by Means of the Biomagnetic and Scintigraphic Methods

Daghastanli, N.A.[1], Braga, F.J.H.N.[1,2], Oliveira, R.B.[2] and Baffa, O.[1]

Physics Department-FFCLRP,USP[1]-14040-901-Ribeirão Preto-SP, Brazil and Faculdade de Medicina de Ribeirão Preto[2], 14049-900 Ribeirão Preto-SP, Brazil

Introduction:

Esophageal motor and transit disorders are important causes of dysphagia and chest pain. Scintigraphy is a well-established method to assess esophageal transit in healthy persons as well as in patients with esophageal motility disorders [1,2]. Studies have been performed showing important aspects in health and pathologic situations, such as Chagas' disease and other pathological situations. However, scintigraphy has important limitations: the burden of ionizing radiation and the use of expensive equipment. On the other hand, the biomagnetic method has some advantages[3-6]. In such a way, pregnant women could be evaluated; additional costs with radiation protection could be eliminated and the biomagnetic method could be used in many institutions that are not well equiped as a consequence of the lack of money. Recently, a simple and easy to perform magnetic method employing an inexpensive biosusceptometer was developed for measurements of the gastric emptying and small bowel transit. We herein describe a method to measure the esophageal transit time (ETT) employing a susceptometric technique[7,8], and the initial results of its use. The aim of this study is to compare scintigraphic and biomagnetic thoracic ETT (between proximal and mid-distal esophagus).

Materials and Methods:

In vivo studies were perfomed after approval of the experimental protocols by the ethics committee of the University Hospital. We have studied 11 normal volunteers (8 males, 3 females; aged 20-31 years old). A semi-liquid meal test was used in both cases (10 ml yoghurt). For scintigraphic studies the test meal was labeled with 350 MBq of ^{99m}Tc-phytate; 100 frames (0.2 s/frame) were acquired in a Siemens-Orbiter Scintillation Camera linked to a computer. Two Regions-of-Interest (ROIs) were drawn to generate time/activity curves: one on the proximal esophagus (furcula) and the other on the mid-distal part of the organ (xiphoid   process). For biomagnetic studies the test meal was labeled with 5 g of ferrite and swallowing was monitored by 2 gradiometers, each of them placed on the very same areas where ROIs were to be drawn (external marks were used). ROI size was determined after evaluation of the point spread function (PSF) and its width at the half maximum. Scintigraphic and biomagnetic measurements were performed in the same day. The gradiometers were coupled to an AC biosusceptometer previously described [7,8]. Acquisition of the biomagnetic signal was done by a 12 bits resolution A/D board at 10 Hz frequency. Data analysis and visualization were performed in a PC microcomputer. In both experiments the subjects stood in the upright position. The signal was filtered using the MatLab software.

Results and Discussion:

Fig. 1 shows both typical biomagnetic and scintigraphic curves. The radioactive counting at the two regions shows two clearly defined positive peaks. The first maximum was due to the passage of the meal by the furcula and the second by the xiphoid. The time difference between the two maxima was used to determine ETT. The voltage produced by the biosusceptometer due to the presence of the magnetic test meal is shown in B The passage of the test meal in front of the detectors produced two negative peaks, that allowed the determination of ETT. The ETT is the elapsing time between the two peaks in these curves. Fig. 2 shows a histogram comparing ETT measurements in each volunteer. Statistical analysis shows that the ETT determined by scintigraphy (mean = 4.2; SD = 1.1) and susceptometry (mean = 4.6; SD = 1.0) are not different (paired t-Test, p > 0.05). In a general way ETT determined by the biomagnetic method is slightly larger than the one determined by scintigraphy. This could be the consequence of the a larger density of the meal when labeled with ferrite as compared to the one with liquid (^{99m}Tc-phytate).

One possibility to check the density effect of the test meal on the prolongation of ETT measured by susceptometry is to make measurements with the subject laying supine; this approach would eliminate gravity effects that are more prominent with the subject in the upright position. A technical difficulty with the present susceptometric set-up was the proximity of gradiometers: magnetic fields of one gradiometer were coupled to the other, depending on distance. The solution was to place the gradiometers as far as possible from each other at an appropriate angle.

The gradiometric system presently in use for ETT assessment was developed to evaluate stomach emptying and orocecal transit time using a magnetic test meal[9]. In these cases the meal is closer to the anterior abdominal wall and the best approach is to have magnetization and detection at this plane. However, the anatomy of the esophagus makes it to be closer to the anterior wall at the furcula but is becomes progressively posterior as it gets closer to the stomach. In this case, an alternative to the gradiometric system could be a simple transformer where the nucleus would be the esophagus. This approach is already being tested with very promising results.

Another point that is under investigation is how to turn the method observer independent. The extrema of the signal were determined by inspection and ETT calculated. A fitting procedure or a derivative of the signal can be employed to detect the extrema automatically. The second procedure has been used with good results, specially when the signal-to-noise ratio is high.

ETT determinations is an important parameter in many clinical situations. Diseases may causes several signs and symptoms due to esophageal disfunction and medical approach may be clinical or surgical. Thus, it is clear that a precise-and better, inexpensive- evaluation of the organ's function is highly desirable. In summary, these initial results suggest that ETT can be evaluated by means of the biomagnetic method and that it is possible to improve it.

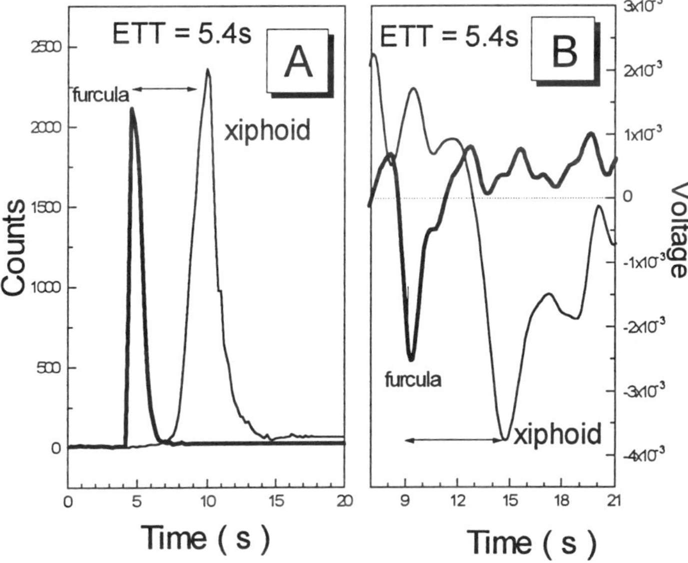

Fig. 1-Typical scintigraphic (A, volunteer 11) and biomagnetic (B, volunteer 10) recordings of the ingestion of a bolus of meal. Solid line shows the data collection at the furcula and narrow line at the xiphoid process.

volunteers	scintigraphic (s)	biomagnetic (s)
1	2.4	2.6
2	4.2	3.7
3	5.0	5.1
4	3.6	4.8
5	3.6	3.9
6	4.2	4.9
7	3.6	5.1
8	4.2	3.8
9	4.2	5.0
10	6.0	5.4
11	5.4	6.3

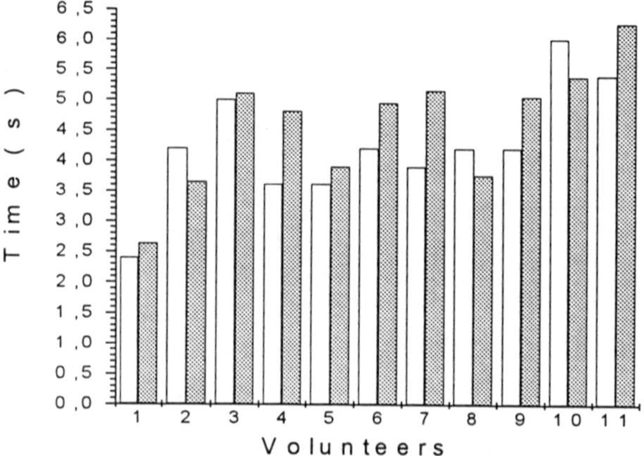

Fig. 2-Table and histogram comparing ETT measured by the biomagnetic (shaded) and scintigraphy methods.

References :

[1] Oliveira, R.B., Rezende, J., Dantas, R.O., Iazigi, N. The Spectrum of esophageal Motor Disorders in Chagas'Diseases. Am. J. Gastroenterology, 1995, 90: 1119-24.

[2] Russel C.O.H., Hill,L.D., Holmes, E.R. III, et al. Radionuclide Transit : A sensitive screening test for Esophageal dysfunction. Gastroenterology, 1981, 80:887-92.
[3] Forsman, M., Assessment of gastric movements by monitoring magnetic field decline of magnetized trace particles , Physics in Medicine and Biology, 1994, 39a: 62.

[4] Baffa, O., A brief account of biomagnetic research in gastroenterology, Physics in Medicine and Biology, 1994, 39a: 50.

[5] Forsman, M., Hultin, L. and Abrahamsson, H. Measurement of gastrointestinal transit time using fluxgate mangnetometers, In: Baumgartner C, Deecke L, Stroink G, Williamson SJ Biomagnetism: Fundamental research and clinical applications. Studies in applied electromagnetics and mechanics, Vol 7, Amsterdam, Elsevier Sci. Publ., 1995.

[6] Basile, M., Neri, M., Carriero, A., Casciardi, S., Comani, S., Del Gratta, C., Donato, L. G., Di Luzio, S., Macri, M.A., Pasquarelli, A., Pizzella V. and Romani, G.L. Measurement of segmental transit time trough the gut in man, Digestive Diseases Sciences, 1992, 37 (10):1537-1543.

[7] Miranda J.R.A., Baffa O., Oliveira R.B. and Matsuda N. M. An AC biosusceptometer to study gastric emptying, Medical Physics, 1992, 19(2): 445-449.

[8] Baffa O., Oliveira R. B., Arruda Miranda J. R. and Troncon L. E. A. Analysis and development of AC biosusceptomter for orocaecal transit time measurements, 1995, Medical & Biological Engineering & Computing, 1995, 33:353-357.

[9] Miranda, J.R.A.1,2,3, Baffa, O.1, Oliveira, R.B.3, Braga, F.J.H.N.3, and Sousa, P.L.1. Detection of Stomach Contraction Using a Biosusceptometer, 10th Int. Conf. on Biomag.: Biomag'96. Santa Fe-USA. Feb/16-21/96.

Acknowledgments:

The authors are grateful to J.G. Guasti Jr for discussions and his endeavor in development the AC biosusceptometer, Prof. N. Iazigi for his support and to E. de Paula, E. Navas, L. Aziani and M. Oliva for technical help. Partial financial support was received from Conselho Nacional de Desenvolvimento Científico e Tecnológico (CNPq), Coordenação de Aperfeiçoamento do Pessoal de Ensino Superior (CAPES), Fundação de Amparo à Pesquisa do Estado de São Paulo (FAPESP), Third World Academy of Sciences (TWAS) and FAEPA -HCFMRP-USP.

Measurements of Colonic Motility Using Ferrimagnetic Powder

Forsman, M.[1] and Abrahamsson, H.[2]

[1]Lindholmen Development, Box 8714, S-402 75 Gothenburg and Department of Applied Electronics, Chalmers University of Technology, S-412 96 Gothenburg; [2]Department of Internal Medicine, University of Gothenburg, Gothenburg, Sweden

Introduction

Biomagnetic methods for measurements of gastrointestinal motility may constitute a future clinical complement to invasive or radiological methods. The biomagnetic alternative is especially attractive for measurements in pregnant women, children and healthy persons. In previous studies, a swallowed magnetic marker was successfully localized inside the large intestine using a SQUID-magnetometer [1] and fluxgates [2]. A disadvantage, revealed when the transit of the magnetic marker was compared to that of ten radiopaque markers, was that a bulk of markers ingested simultaneously might disperse widely in the gut [2]. Hence, a single marker might poorly represent the mean progression. The aim of this study was to evaluate the usefulness of ferrimagnetic powder, mixed in a solid meal, to study gastrointestinal transit and intraluminal movements.

To locate the intraluminal tracer, the abdomen was first briefly exposed to a magnetic field, then the remanent field was measured, and the location of the particles was visually estimated from the field maps. Since the magnetization aligns the magnetic moments of the tracer particles, the movements of the particles can be assessed via the post-magnetization field strength decay. This decay is due to rotations of the particles, i.e., a consequence of the forces that mix and propel the chyme. The principle, originally proposed by Cohen [3], was recently implemented in measurements of intragastric movements [4,5]. Information about particle movements that can be assessed in this way can not be obtained from currently used clinical methods without using high radiation doses.

Methods

Ten male volunteers, aged 21 to 30 years, with no history of intestinal diseases, gave their informed consent to participate in the study. Maghemite powder (Bayferrox 8140, Bayer) was used as a tracer. This ferrimagnetic material is also known as γ iron oxide, γ-Fe_2O_3. The solubility of these particles was tested *in vitro*. The particles were added to hydrochloric acid, pH=1. No solubility could be observed in one week. During preparation of the test meal, 2 grams of the tracer were mixed with moderately heated pancake batter, which were then baked into pancakes. The mass of one portion was 130-140 g and the approximate energy value was 1340 kJ. The particles were sufficiently small so that none of the subjects could feel them while chewing.

The measurement took place 24 hours after the ingestion, and was divided into two parts. The location of the tracer particles was first sought for. During the measurement the subject rested on a pneumatically driven bench, allowing computer controlled horizontal movements [6]. The bed was initially positioned by the guidance of an optical beam, so that the horizontal origin was located at the xiphoid process. The intraluminal particles were then briefly magnetized by a homogeneous field pulse (40 mT, 2s), which was generated by two coils in a Helmholtz configuration. The vertical magnetic flux density, B_z, was then measured with first-order fluxgate gradiometers (baselength=140 mm, noise amplitude=45 pT/$\sqrt{Hz}$). B_z was measured in a grid in two planes, in six-twelve transversal scans, 40 mm apart. Each scan contained 40 points, with an inter-sampling distance of 15 mm. The anterior sensor was adjusted as close to the skin as possible. The distance between the proximal coil and the skin was about 35 mm. The posterior sensor was fixed with the proximal coil 62 mm beneath the surface of the bench.

After inspection of the anterior field map (an example is shown in Fig. 1, left) the bed was positioned to measure B_z at the point of maximum field strength. The abdomen was then magnetized a second time, after which a baseline was sampled as the subject was moved 0.5 m away from the sensors in the beginning and at the end of a 12-minute sampling period. The magnetic signals were digitized at a frequency of 3.3 Hz. To reject disturbances induced by respiration, a digital low pass filter with a cut off frequency of 0.2 Hz and no delay was applied to the data. The respiration was registered in order to facilitate a comparison with the magnetic signals, although these had been filtered. The respiration was registered by using a small air-filled cushion, placed under a rubber band, near the umbilicus. The cushion was coupled with a tube to a pressure sensor, the signal of which was amplified, lowpass filtered (1 Hz), digitized at 3.3 Hz and digitally highpass filtered (0.08 Hz and no delay). The subjects were asked to keep quiet and to avoid deep breaths during the measurement.

An *in vitro* experiment was carried out to control the remanent field stability. The test meal was placed on the bench in a plastic pot. A magnetic measurement was then performed as described above.

The obtained field maps were inspected visually. The decay of the sampled field strength was plotted as function of time. For each subject, mono- and bi-exponential models were fitted to the anterior and posterior field amplitude decays. The nonlinear regression was carried out by using a least mean square procedure. The rms residual was calculated as a percentage of the rms field and used to compare the results of the two models. Normalized mean curves were calculated for anterior and posterior respectively and modeled as described above.

Results

The field maps of 5 out of the 10 subjects showed two or more peaks, on both planes. One subject's field maps are shown in Fig. 1. The height, sharpness, and location of the major anterior peak show that a large part of the tracer is located in the transverse colon (tc). The maximums are 192 and 133 nT anterior and posterior, respectively. There is also an accumulation of particles in the recto-sigmoid region (r-s). The ratio between the anterior and posterior peak values for the second peak is 73 to 64, i.e. close to unity.

Table 1. Locations of the major part of the test meal, 24 hours post-ingestion, in the ten subjects.

Colonic region	# Classifications
ascending colon	1
transverse colon	5.5
descending colon	1
recto-sigmoid	2.5

The major part of the test meal in each person was classified into the four colonic regions (Table 1). In subject (S1) the contents were distributed equally in tc and r-s, therefore S1 contributed a "classification half" to each of the two regions in Table 1.

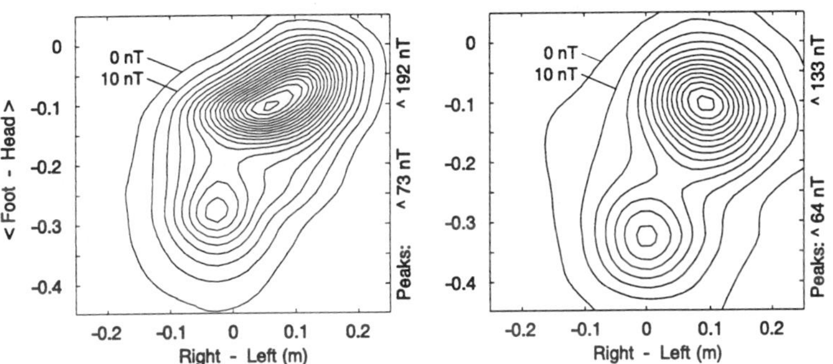

Fig. 1. Anterior (left) and posterior (right) field maps outside the colon with magnetized ferrimagnetic powder, 24 hours after ingestion. Origin was located at the xiphoid process.

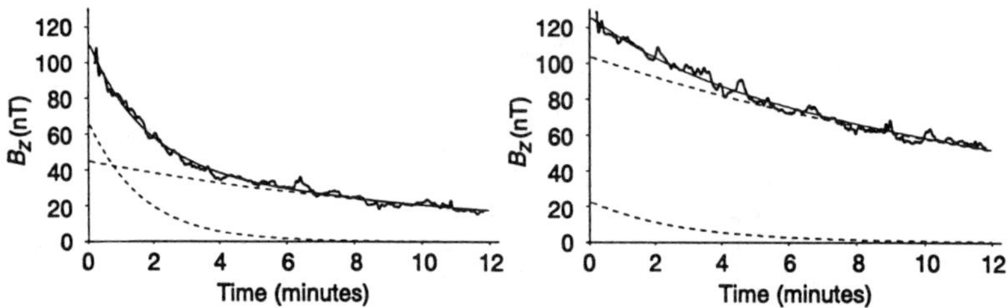

Fig. 2. Anterior field curves after magnetization from Subject S1. The left diagram illustrates the decay (thick line) outside the transverse colon, 120 mm caudally and 20 mm to the left of the xiphoid process. The thin line shows the result of an applied bi-exponential model, where the dashed lines are the two exponential components. The right diagram illustrates the decay (thick line) outside the recto-sigmoid region, 300 mm caudally and 30 mm to the right of the xiphoid process.

The *in vitro* experiment showed that the remanent field of the test meal was very stable. The remanent field of the magnetized meal decreased 0.5-1% during the 12-minute period *in vitro*. The lower and upper signals were similar, with B_z at approximately 380 nT. In six repeated measurements, the peak-to-peak noise amplitude never reached above 2 nT.

Two field decay measurements were performed in subject S1, for whom approximately equal accumulations were found in tc and in r-s. The decay rate above tc was twice that above r-s, see Fig. 2. The curves in Fig. 2 also represent the two different types that were seen among the subjects. The residual rms of the applied bi-exponential function was 5.1% of the signal rms (Fig 2. left). In Fig. 2, right, the residual/signal rms was 3.4%, which was almost the same as for the residual of a single-exponential model (3.9%).

The mean residual rms to signal rms ratios from the decay curves were 5.6% anterior and 4.1% posterior. All raw curves correlated with the measured respiration, i.e. higher/lower amplitudes anterior/posterior during inhalation. For a few subjects some respiration disturbances could be seen also after the decay curves were digitally filtered. The mean decay over twelve minutes from the ten subjects (n = 11, since two decay curves were measured for S1), was well approximated by a bi-exponential function (Fig. 3). The residuals were 1.7% anterior and 1.0% posterior. The anterior decay was faster for all subjects.

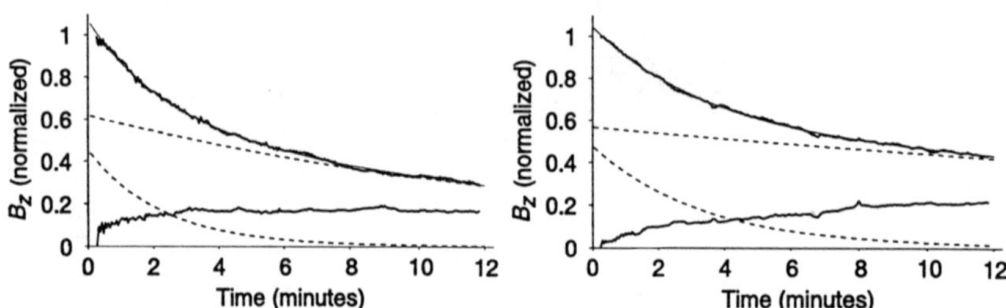

Fig. 3. Mean decay curves anterior (left) and posterior (right). Eleven curves were included in the two mean curves, the SD is shown as the lower thick line the diagrams. The SD is zero at start, since all curves were normalized prior to the averaging. The bi-exponential model is illustrated as in Fig. 2.

Discussion

The location of the test meal was easy to estimate from the field maps. Comparison of the anterior and posterior B_z facilitated estimation of the depth of the source. We did not use any inverse solution. However, it would be possible to calculate a compensated (for the source not being central between the measurement planes) source value at each (x,y) point in the maps. An integration of such values should hint whether some particles have been lost in the feces or not, which is hard to tell directly from the maps. The test meal main accumulations (Table 1) are in line with what could be expected from 24-hour post-ingestion radiological measurements on healthy males [7].

With powder as a tracer, we have overcome the problem that a single marker may not be representative for the mean progression in the colon after a certain time. Fig. 1. clearly shows that it is possible to localize a dispersed test meal. In this study the progression was represented by an "infinite" number of markers. Another difficulty remains: to relate the results from biomagnetic measurements to transit in specific structures. However, in the clinical situation, the most important question is if a transit delay involves the proximal colon, as in "slow transit costipation." This would be easy to answer with the present method.

The ratio of signal to instrumental and external noise was high in these measurements. This was confirmed by the *in vitro* measurements and by the low mean rms value of the individual regressions of the decay curves. The flux density is proportional to the inverse cube of the distance to the source-particles. Therefore the respiration disturbance was a significant source of error in the raw signal, especially for subjects who mixed normal breaths with deep ones. Most of this error was rejected by filtering.

The two mean curves (Fig. 3) were well described by the four parameters of the bi-exponential model. Also the individual curves approximated a bi-exponential decay, although examples where a single exponential fit was sufficient were seen as well (e.g. Fig. 2, right). A theory for the bi-exponential model is that the particles separate into two compartments: a semifluid compartment (chyme), where the particles are rotated quickly, and a more solid compartment where the particles rotate slowly. The factor parameters of the fitted bi-exponential function give a ratio between the amounts of particles in the two compartments.

We conclude that the present noninvasive method is valid in measuring colonic transit. The possibility to identify intraluminal movements has a potential to be useful for evaluation of motility in constipated patients.

References

[1] Basile, M., Neri, M., Carriero, A., Casciardi, S., Comani, S., Del Gratta, C., Di Donato, L., Di Luzio, S., Macri, M.A., Pasquarelli, A., Pizzella, V., and Romani, G.L. Measurements of segmental transit through the gut in man, Dig Dis Sci, 1992, 37;10:1537-1543.
[2] Forsman M., Hultin L., and Abrahamsson, H. Measurements of gastrointestinal transit using fluxgate magnetometers, In: Baumgartner, C., Decke, L., Stroink, G., and Williamson, S.J. Biomagnetism: Fundamental Research and Clinical Applications, Amsterdam, IOS Press,1995, 739-742.
[3] Frei, E.H., Benmayr, Y., Yerashalmi, S., Dreyfuss, F. Measurements of the emptying of the stomach with a magnetic tracer. *IEEE Trans. Mag.*, 1970, *MAG*-6(2): 348-349.
[4] Baffa, O., Oliveira, R. B., Miranda, J. R. A., and Sousa, P. L. Mixing power of food in the stomach evaluated by a biomagnetic technique, In: Baumgartner, C., Decke, L., Stroink, G., and Williamson, S.J. Biomagnetism: Fundamental Research and Clinical Applications, Amsterdam, IOS Press,1995, 753-756.
[5] Forsman, M. Assessment of gastric movements by monitoring magnetic field decline of magnetized trace particles, Physics in Medicine and Biology, 1994, 39a: 62.
[6] Kalliomäki, K., Aittoniemi, K., Kalliomäki, P.-L., Moilanen, M. Measurement of lung-retained contaminants in vivo among workers exposed to metal aerosols. *Am. Ind. Hyg. Assoc. J.* 1981, 42:234-238.
[7] Abrahamsson, H., Antov, S., and Bosaeus, Gastrointestinal and colonic segmental transit time evaluated by a single abdominal x-ray in healthy subjects and constipated patients, I. Scand J Gastroenterol, 1988, 23; suppl. 152: 72-80.

Detection of Stomach Contractions Using a Biosusceptometer

Miranda, J.R.A.[1,2,3], Baffa, O.[1], Oliveira, R.B.[4], Sousa, P.L.[1] and
Braga F.J.H.[4]

[1]*Departamento de Física e Matemática-FFCLR-USP-14040-901-Ribeirão Preto-SP, Brazil;*
[2]*Instituto de Física de São Carlos-USP, 13560-970 São Carlos-SP, Brazil;* [3]*Departamento de
Física e Biofísica, Instituto de Biociencias-UNESP, 18618-000 Botucatu-SP, Brazil;* [4]*Faculdade
de Medicina de Ribeirão Preto, 14049-900 Ribeirão Preto-SP, Brazil*

Introduction:

Symptoms presumably related to gastrointestinal (GI) motor disorders are among the most frequently encountered conditions in clinical practice [1]. Dearanged gastric antral contractility (GAC) is considered to play an important role in pathogenesis of several gastrointestinal motor disorders [2]. Intraluminal manometry, the standard method for assessment of GAC is both invasive and somewhat inaccurate [3]. More recently scintigraphic techniques to measure GAC were developed [4], but radiation burden and high cost of equipment limits its wide use. Magnetic Resonance Imaging (MRI) is also being investigated to measure GAC [5] with the advantage of no radiation exposure, but requiring even more expensive equipment than scintigraphic techniques. Biomagnetic methods based on the detection of magnetic flux variation produced by an ingested test meal labeled with magnetic particles can be an attractive alternative for GAC assessment.

Materials and Methods:

In vivo studies were initiated after approval of the experimental protocols by the ethics committee of the University Hospital (Medical School Ribeirão Preto). The subjects ingested a 900 kcal mixed meal followed, within a 1 hour period after, by 100 ml of a chocolate flavored custard (80 kcal) containing 3 g of manganese ferrite ($MnFe_2O_4$). The measurements were performed within 1 hour after the ingestion of the magnetic test meal and each recording was 2 to 10 minutes long. The subject stood in the upright position during the measurements, having the magnetic detector with a slight pressure against the abdominal wall over the epigastrium. In two studies, less than 370 MBq (10 mCi) of the radioisotope ^{99m}Tc-phytate was also added to the custard and GAC was simultaneously assessed by scintigraphy and biomagnetic methods. In these experiments 238 frames (1 s/frame) were acquired posteriorly in a Siemens-Orbiter Scintillation Camera linked to a computer and the magnetic signal acquired anteriorly. A region of interest (ROI) was drawn in the antrum area to generate time activity curves. Although the gamma camera has a great deal of iron, the spatial discrimination of the gradiometric system was good enough to allow simultaneous measurements. One hundred forty tracings of GAC were obtained by the magnetic method from 27 healthy volunteers (5 women and 22 men, median age 28 years, ranging from 20 to 40 years).

The magnetic detector employed in this study is an AC biosusceptometer (ACB), based on a differential (first order gradiometer) coil arrangement of two air core transformers, having one transformer for reference, and the other working as a measuring transformer on which the magnetic tracer having a high magnetic permeability acts like an external nucleus. The coils were mounted coaxily in a PVC tube and fixed with epoxy resin with a final diameter of 3.5 cm and 20 cm long. The primaries or excitation coils are fed by a power amplifier and the imbalance voltage produced by the test meal at the secondaries or detection coils, is detected by a lock-in amplifier. A 8 pole low pass filter (HP 5489A) with a cutoff frequency set at 3 Hz was used. The signal was digitized at 10 Hz frequency yielding data files up to 6 Kb. The ACB was previously used in studies of gastric emptying (6) and orocaecal transit time measurements [7].

Results and Discussion:

Fig. 1.a shows a typical raw tracing of the GAC. The 3 cycles/minute oscillation can be clearly recognized. In some measurements (fig. 1.b) the signal appeared superimposed to a higher frequency signal due to respiration; asking the subject to stop breathing for some time would remove this high frequency oscillation. Figs. 1c and 1.d show the Fourier transform of signals shown in fig. 1.a and 1.b., respectively, the 50 mHz power amplitude is clearly from the background signals. A typical oscillatory pattern of about 50 mHz (3 cycles/minute) was conspicuous in 120 of the magnetic measurements. In 17 of the remaining tracings the 3 cycles/minute was not recognizable in the crude tracings, but it was possible to distinguish 50 mHz signal from the background after Fourier transform. In only three measurements this signal was not detected by this method, indicating a rate of failure of only 2 %.

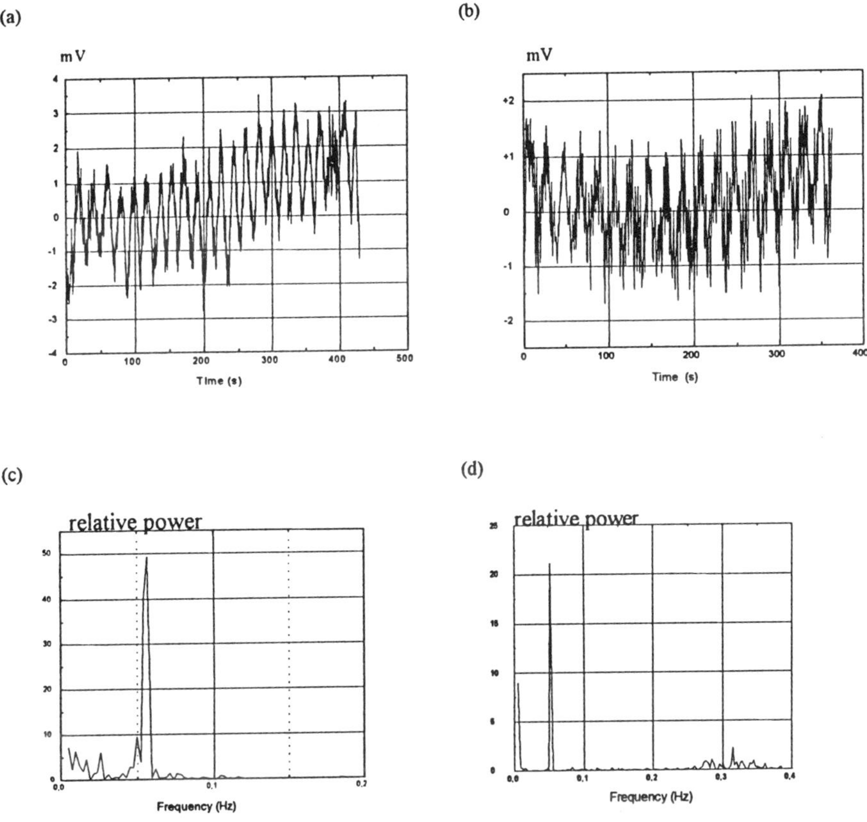

Fig. 1-Typical GAC tracing detected by the biomagnetic method, a clean signal (a) and in presence of high intensity breathing (b). The plot of the power spectrum obtained after Fourier transforms is shown in (c) and (d), respectivaly.

The question whether or not the magnetic signal was associated with the gastric contraction was further investigated by comparing GAC detected by the scintigraphic and biomagnetic techniques. Simultaneous recording of the radioactive counting of a region located at the distal antrum was made with the ACB. Fig. 2 shows one of these simultaneous recordings of the GAC. As the antrum contracts food is moved away and the radioactive counting varies proportionally to the cross section seen in the images. A cross correlation between the two tracings is given in Fig. 2.b. The coincidence between the two measurements allow us to infer that the magnetic measurements are actually reporting the mechanical activity of the stomach. The mean frequency value (mean= 53.71 mHz, SD= 3.55 mHz) measured by the biomagnetic methods is in very good agreement to the values that have been measured by EGG in a normal population of volunteers.

(a)

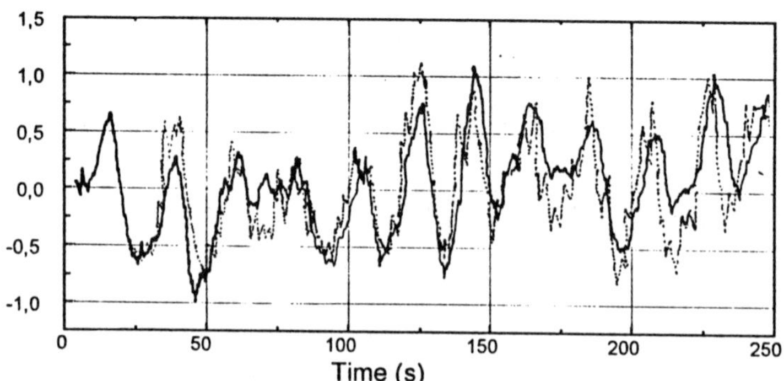

(b)

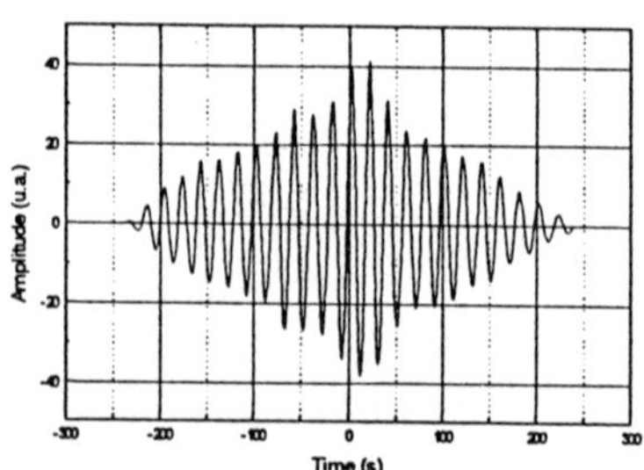

Fig. 2-Simultaneous recording of the GAC by the scintigraphic (dashed) and biomagnetic (continuous line) methods (a). The radioactive counting and ACB amplitude were normalizad to the initial amplitude for better comparison. In (b) the cross correlation function between the two signals is shown.

In the present study frequency is most easy quantity to be measured. It is possible that combining this method with another technique that could give the distance of the test food or antrum wall to the detector, the

622

voltage amplitude detected could be translated in mechanical displacement of the antrum. These results show that a biomagnetic technique may be successfully employed to record GAC. The finding that the profiles of the variations of the magnetic activity were quite similar to those recorded by a scintigraphic method simultaneously applied, supports the notion that the magnetic signal detected is tied to the mechanical counterpart gastric antrum activity.

In summary, the biomagnetic method employed here is very straighforward and could detect the GAC very easily, having none of the drawbacks present in the EGG and other measuring methods.

References:

[1] Drossaman D.A. Clinical research in the functional digestive disorders, Gastroenterology, 1987, 92:1267-1269.

[2] Malagelada J.R. and Stanghellini V. Manometric evaluation of functional upper gut symptoms Gastroenterology, 1985, 88:1223-1231.

[3] Wingate, D.L. Small bowel manometry, The American Journal of Gastroenterology, 1995, 90(4):536-539.

[4] Urbain J-L.C., Vekemans, M.C., Bouillon, R., Bex, M. and Mayeur, S.M. Characterization of gastric antral motility disturbances in diabetes using a scintigraphic technique. Journal of Nuclear Medicine, 1993, 34:576-581.

[5] Fraser R., Schwizer, W., Borovicka, J., Asal, K. and Fried, M. Gastric motility measured by MRI, Digestive Disease Sciences, 1994, 39(12):20S-23S.

[6] Miranda J.R.A., Baffa O., Oliveira R.B. and Mitiko M. N. An AC biosusceptometer to study gastric emptying, Medical Physics, 1992, 19(2): 445-449.

[7] Baffa O., Oliveira R. B., Arruda Miranda J. R. and Troncon L. E. A. Analysis and development of AC biosusceptomter for orocaecal transit time measurements, Medical & Biological Engineering & Computing, 1995, 33:353-357.

Acknowledgments:

The authors are grateful to J.G. Guasti Jr for discussions and his endeavor in development the BAC, and to E. de Paula, E. Navas, L. Aziani and M. Oliva for technical help. Partial financial support was received from Conselho Nacional de Desenvolvimento Científico e Tecnológico (CNPq), Coordenação de Aperfeiçoamento do Pessoal de Ensino Superior (CAPES), Fundação de Amparo à Pesquisa do Estado de São Paulo (FAPESP), Third World Academy of Sciences (TWAS) and FAEPA -HCFMRP-USP

Biomagnetic Measurement of Gastric Mechanical Motility Using a High-Temperature SQUID Imaging

Nomura, M.[1], Saijyo, T.[1], Haruta, Y.[2], Itozaki, H.[2], Toyoda, H.[2], Nakaya, Y.[3], Ito, S.[1] and Kado, H.[2]

[1]Second Department of Internal Medicine and [3]Department of Nutrition, The University of Tokushima, Tokushima, 770, Japan; [2]Superconducting Sensor Laboratory, Chiba, 270-13, Japan

INTRODUCTION

Biomagnetic measurement using superconducting quantum interference device (SQUID) is an effective non-invasive tool for detecting gastric function. In the present study, the weak magnetic field originating from gastric mechanical motility was detected using a newly developed high temperature (-196°C) SQUID system by liquid nitrogen.

SUBJECTS AND METHODS

Two healthy volunteers were selected for the present study. The magnetic field was recorded in the magnetically shielded room using a 16-channel high temperature SQUID system developed by Superconducting Sensor Laboratory in Japan (Figure 1). A magnetized steel ball (OD, 1.3 mm) attached to the tip of a stomach tube (12 F), was inserted from nose, positioned fluoroscopically in the antrum. The magnetic fields were recorded in fasting or fed (600-ml liquid meal) states. A magnetic sensor was placed above the epigastric region and magnetic signals produced by gastric mechanical motility were measured. Three recording bandwidths (DC-50 Hz, DC-1 Hz and DC-0.1 Hz) were used, and fast Fourier transform analysis was performed using spectral analyzer (HP3562A DYNAMIC SIGNAL ANALYZER). After recording the magnetic signal, the stomach tube with steel ball was removed.

Magnetically shielded room   Electronic equipments

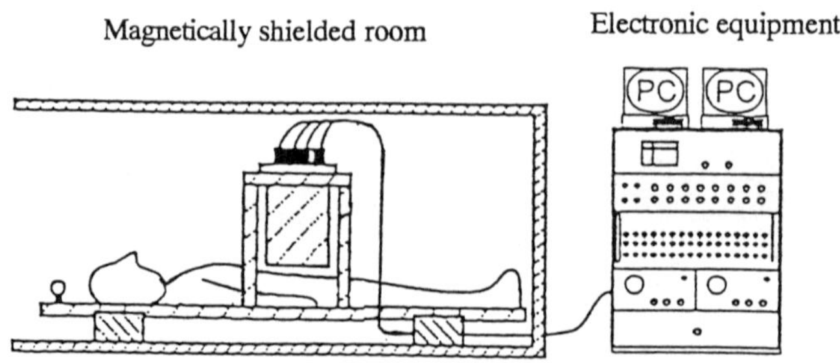

Figure 1 System diagram of 16-channel high temperature SQUID system

RESULTS

1. Back ground noise of a 16-channel high temperature SQUID system

Figure 2 shows the frequency spectra of the back ground noise of a 16-channel high- temperature SQUID system. The noise levels were 0.4 pT/Hz$^{1/2}$ and 4 pT/Hz$^{1/2}$ at 500 mHz and 50 mHz, respectively.

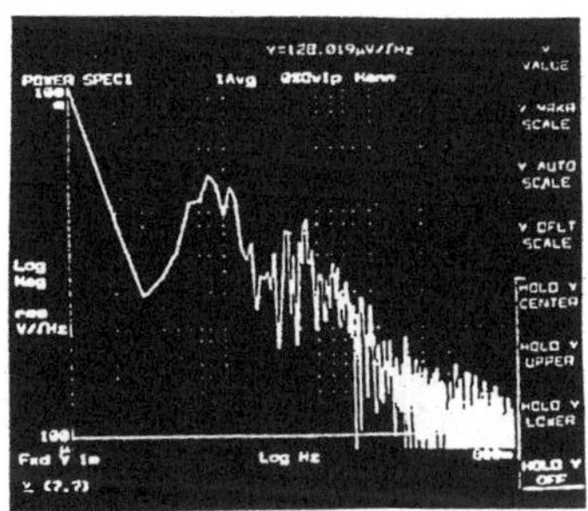

**Figure 2 Frequency spectra of the back ground noise of
16-channel SQUID system**

2. Magnetic fields originating from gastric mechanical motility

Figure 3 shows the magnetic signal from the steel ball in the stomach at fasting state. The recording bandwidth was DC-1 Hz. Magnetic signal around 13-18 cycle/min was detected. After the magnetic recording filtered with bandwidth of DC-0.1 Hz, the regular 3 cycle/min signal, which indicated gastric motility, was clearly visible. Figure 4 shows the magnetic signal at fasting and fed states using bandwidth of DC-0.1 Hz. Three cycle/min magnetic signal was visible at each state. The amplitude of the magnetic field at the fed state was stronger than that of fast state.

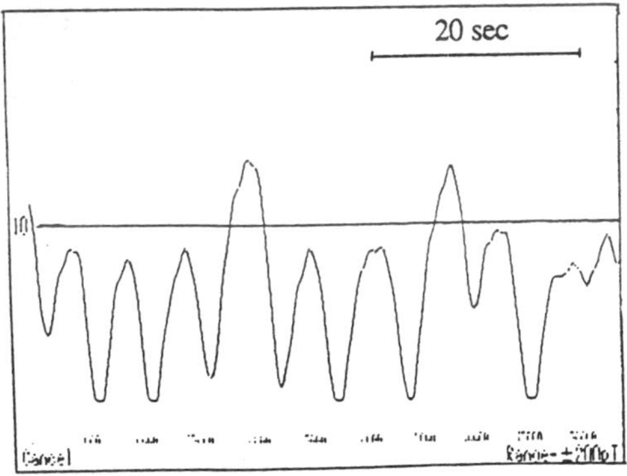

**Figure 3 Magnetic signal at fasting state
(bandwidth filter; DC-1 Hz)**

3. Fast Fourier transform analysis in fasting and fed states

Figures 5 shows the frequency spectra in fasting and fed states, respectively. The two spectral peaks were visible at 0.05 Hz and 0.15 Hz, and amplitude of magnetic field at 0.05 Hz in fed state was stronger than in fasting state. Table 1 shows the magnetic intensity and peak frequency at fasting and fed states at 0.05 Hz. The magnetic field was greater after ingestion of food than in the fasting state (p<0.01). The peak frequency of gastric motility was not statistically different between the two states (n.s.).

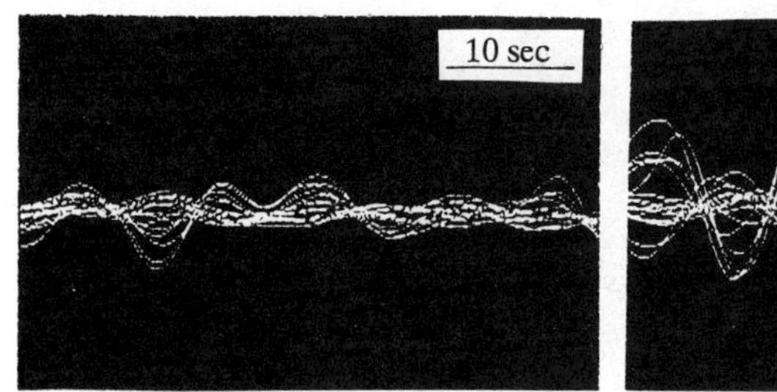
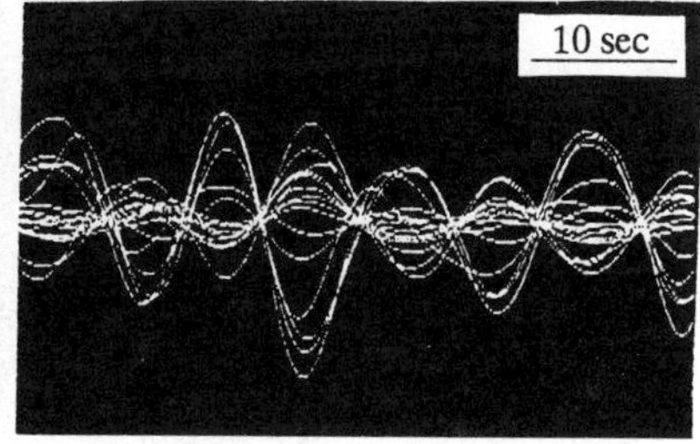

a   b

**Figure 4 Magnetic signal at fasting (a) and fed (b) states
(bandwidth filter; DC-0.1 Hz)**

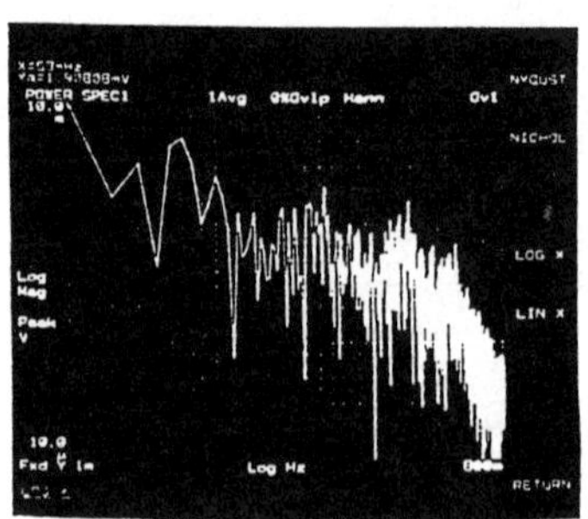
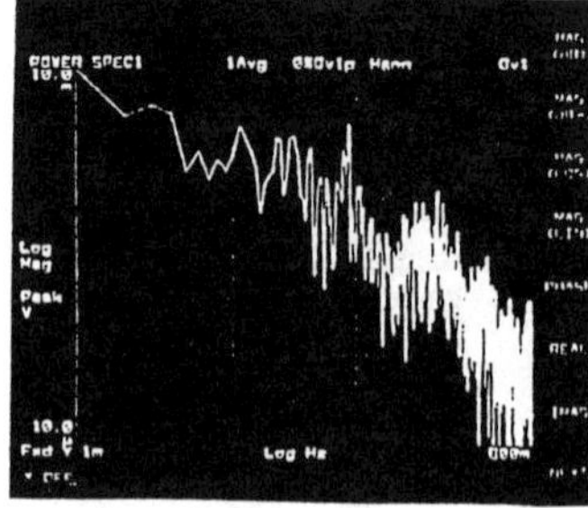

a   b

Figure 5 Frequency spectra in fasting (a) and fed states (b).

Table 1 Magnetic intensity and peak frequency at fasting and fed states

	Fasting state	Fed state
Magnetic intensity (pT)	9.0±4.0	21.2±3.8
Peak frequency (cycle/min)	3.1±0.2	3.2±0.1

(Values are mean± SEM. pT; pico-tesla)

DISCUSSION

We succeeded in recording gastric motility using a high-temperature SQUID system. This is the first multi-channel high-Tc SQUID system and its operation is much simpler than that of the conventional SQUID system that uses liquid helium.

The magnetogastrogram was recorded by Comani et al.[1] and Basile et al.[2] They detected the 3 cycle/min magnetic signals originated from the stomach without using magnetized tracer. In the present study, we recorded the magnetic signal from the magnetized tracer positioned in the stomach in order to detect the mechanical gastric movement.

An electrogastrogram is the summation of gastro-intestinal electrical activity. On the other hand, the

626

magnetogastrogram without using magnetized tracer reflect only the gastric electrical activity because the SQUID system is superior in positional resolution to the electrogastrogram. Moreover, the magnetic signals generated by a magnetized-steel ball in the stomach reflect the mechanical gastric movement.

Using fast Fourier transform analysis, two spectral peaks (3 cycle/min and 13-18 cycle/min peaks) were detected. Three cycle/min magnetic signals reflected the gastric mechanical movement and 13-18 cycle/min signals represent the respiratory movement. These results coincide well with the previous reports[1-3].

Gastric electrical activity might disagree with gastric mechanical movement in some diseases. This new method represents gastric motility more specifically, and it is expected to contribute to diagnosis and management of various mechanical gastric disorders[4,5].

REFERENCES

[1] Comani, S., Basile, M., Casciardi, S., Gratta, CD., Luizo, CD., Erne, SN., Macri, M., Neri, M., Peresson, M., Romani, GL. Extraporeal direct magnetic measurement of gastric activity. In: Hoke, M. et al. Biomagnetism: Clinical aspects. Amsterdam, London, New York, Tokyo, Elsevier Science Publishers, pp639, 1992.

[2] Basile, M. The biomagnetic approach to the study of gastrointestinal activity. In: Hoke, M. et al. Biomagnetism:Clinical aspects. Amsterdam, London, New York, Tokyo, Elsevier Science Publishers, pp613, 1992.

[3] Alvarez, W.C. The electrogastrogram and what it shows. JAMA, 1922, 78:1116.

[4] Nomura, M., Saijyo, T., Itozaki, H., Toyoda, H., Nakaya, Y., Ito, S., Kado, H. Magnetic measurement of gastric activity using a high-temperature superconducting quantum interference device (SQUID). Gastroenterol 1995, 108, 108:A658.

[5] Saijyo, T., Nomura, M., Haruta, Y., Itozaki, H., Toyoda, H., Nakaya, Y., Ito, S., Kada, H. Biomagnetic measurement of gastric electrical activity by 64-channel magnetogastrographic imaging. Gastroenterol 1995, 108:A680.

Single and Multichannel Magnetic Field Measurements of Gastrointestinal Activity in the Pre- and Post-Prandial States

Petrie, R.J.[1], van Leeuwen, P.[2], Brandts, B.[3], Turnbull G.[4], Veldhuyzen van Zanten, S.J.0.[4] and Stroink, G.[1]

[1]Dalhousie University, Dept. of Physics, Halifax, Canada; [2]EFMT, Bochum, German; [3]Ruhr University, Dept. of Physiology, Bochum, Germany; [4]Dalhousie University, Dept. of Medicine, Division of Gastroenterology, Halifax, Canada

Introduction

Current techniques for assessing gastrointestinal activity are highly invasive. These techniques such as radiology, scintigraphy and endoscopy yield results that are qualitative and non-specific. A method of evaluating GI function that is both non-invasive and quantitative is needed.

Magnetic signals corresponding to electrical depolarization of gastrointestinal smooth muscle can be measured with SQUID gradiometers [1]. The purpose of this research is to investigate GI motility by studying these magnetic signals. Motility is a term that encompasses electrical activity and innervation of smooth muscle, contraction of muscle and propulsion of GI contents [2]. This work specifically focuses on two aspects of motility: contractions resulting from action potentials and slow wave depolarization of smooth muscle.

The goal in studying contractile activity is to identify action potentials in the magnetic signal by correlating these signals with contractions measured simultaneously using a manometry catheter. Slow wave depolarization was studied by making measurements over the abdominal surface with a multichannel gradiometer system. We have previously shown that sequential, single channel measurements over the abdominal region are not sufficient to represent spatial information about gastrointestinal function. Both the pre- and post-prandial states were studied. These signals were then examined with respect to frequency content, spatial distribution and time dependence.

Methods

I. Contractile Activity: Two subjects participated in this research. Neither had a history of gastrointestinal dysfunction. Subjects were adult male (average age 42 years). Prior to recording, a manometry catheter was inserted through the nasal passages of the subject approximately 60 cm into the GI tract. Both the nostril and pharyngeal regions were anesthetized with Viscous 2% (2g/100ml) Xylocaine oral topical anesthetic and Xylocaine endotracheal spray (20mg measured dose) respectively. The catheter was designed to record intraluminal pressure at six different locations (10 cm apart) along the length of the tube. Once in place the subject was positioned under the single channel, second order SQUID gradiometer (noise=40 fT/$\sqrt{Hz}$, sensing coil diameter= 2.54cm, baseline=4cm) so that the probe was located approximately above tbe pyloric sphincter.

Measurements were made for thirty minutes in both the pre- and post- prandial states. The magnetic signal was low pass filtered at 16Hz. The pre-prandial state was assured by the subjects fasting for eight hours before recording while the fed motility state was induced by ingestion of a standard test meal: 250ml orange juice, 130ml yogurt, 1 bran muffin (7 g protein, 295 cal).

II. Slow Wave Depolarization of Smooth Muscle: Subjects participating in this study were healthy adults (three females and two males) without history of GI dysfunction. Mean age: 32 years. Subjects presented having fasted for eight hours. Both pre- and post- prandial motility states were studied.

GI slow wave depolarization was measured with a 37 channel, first order gradiometer system (Siemens Krenikon, noise=30 fT/$\sqrt{Hz}$, baseline=7cm, diameter of sensor array=19cm). The probe was positioned along the central axis of the body, half way between the sternum and line joining the iliac crests. Subjects were recorded for 60 minutes in each of the motility states.

Analysis: All multichannel data were first analyzed for frequency content using power spectral analysis. Bands exhibiting significant power were identified. For each of these bands isoharmonic maps were generated. Isoharmonic maps spatially display the average power distribution over the region corresponding to the sensor array for a 10 minute period of time. The power at the sites with maximum power was then plotted as a function of time

over the 60 minute record. The result is information about the spectral, spatial and temporal dependence of magnetic
GI signals.

Results and Discussion

1. Contractile Activity: Gastric and duodenal slow waves or basic electrical rhythm (BER) could be
identified in the magnetic record in all subjects (Fig. 1) both visually and through spectral analysis. It is clear that
the gastric signal (2.8 ± 0.1 cpm) dominated during the first 10 minutes after ingestion. The duodenal signal (12.2 ± 0.1 cpm) does not become of significant amplitude until 16 minutes after ingestion.

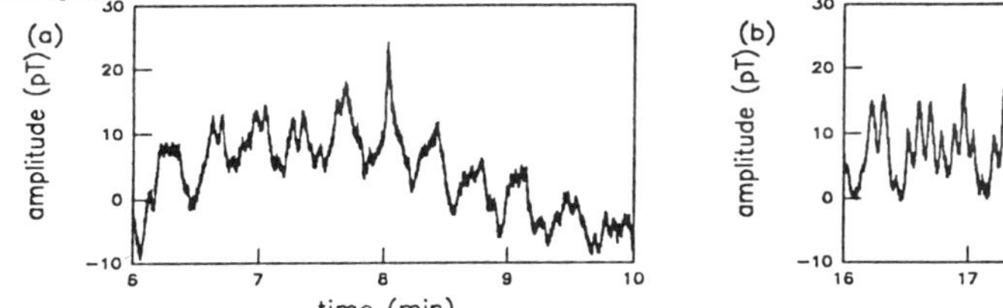

Figure 1: Example of basic electrical rhythm (post- prandial state) of the (a) stomach and (b) duodenum. The
frequency of the gastric and duodenal slow waves is 2.8 ± 0.1 cpm and 12.2 ± 0.1 cpm respectively.

Correlation of the magnetic signals with the internal GI pressure data (Fig. 2) did not identify action
potentials. Some difficulties have been encountered with this protocol. It is now evident that a radiological scan of
the catheter location is necessary to guarantee its correct positioning prior to measurement. Also, it is clear that
cardiac artifacts in the magnetic signal may partially obscure action potentials in the smooth muscle.

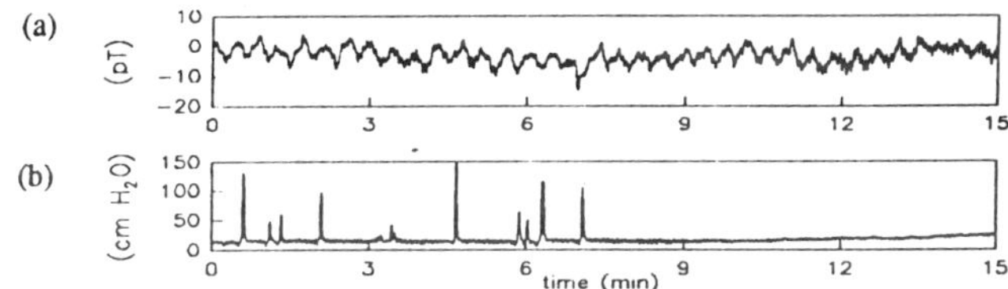

Figure 2: (a) Magnetic and (b) one (of six) internal pressure recordings (post-prandial) used in correlation study
for identifying electrical action potentials that result in contraction of intestinal smooth muscle.

Thus, although action potentials have been observed magnetically in vivo in rabbit small intestine [3], we
did not observe spiking activity in the magnetic measurement of human GI smooth muscle.

II. *Slow Wave Depolarization of Smooth Muscle:* From power spectral estimation several different
frequency bands were identified for further investigation. There was some minor variance in these bands between
individuals. The bands of relevance to this paper are: 45 - 55 mHz, (corresponding to the 3 cpm frequency of gastric
smooth muscle depolarization), 190 - 205mHz (12 cpm frequency of duodenum), 140 - 160mHz (9 cpm frequency of
distal jejunum) and 90 - 105mHz (6 cpm frequency of distal ileum). The integrated power in each band was calculated
for every channel in the array. The spatial distribution plots of signal power in these bands over the abdominal
region are shown in Fig 3.

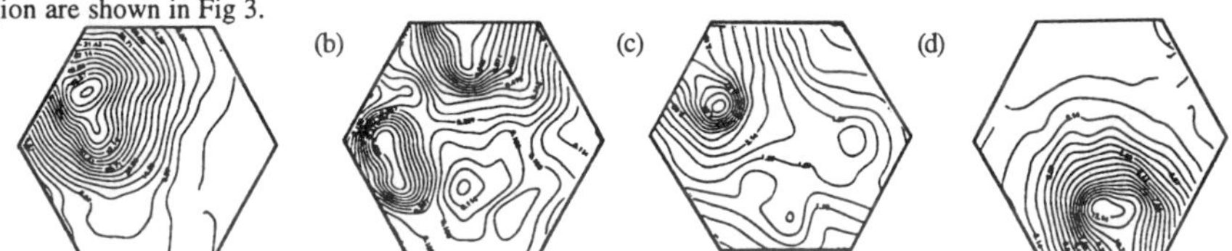

Figure 3: Typical spatial distribution of post - prandial, magnetic, GI activity in the (a) 45 - 55 mHz (b) 90 -
105 mHz (c) 140 - 160 mHz and (d) 190 - 205mHz bands in healthy subjects.

629

It is clear that the maxima of different frequencies are at different abdominal locations. Maximum power indicates strong GI activity underneath the probe at this location. This is what is expected from GI physiology which states that different segments of the gastrointestinal tract function at different frequencies. For brevity, only the post prandial - state is shown. However, individuals exhibit maps of similar morphology in both the pre-prandial states but the magnitude of the signals is less. These results suggest that magnetic field measurements can spatially distinguish gastric and small bowel activity.

It is important to investigate how this activity behaves as a function of time. Iso◊◊harmonic maps were generated over 10 minute intervals during the 60 minute recording period. No significant changes in morphology were observed. Magnitude of power relative to the maximum power at the site of maximum activity was observed to be time dependent (Fig. 4)

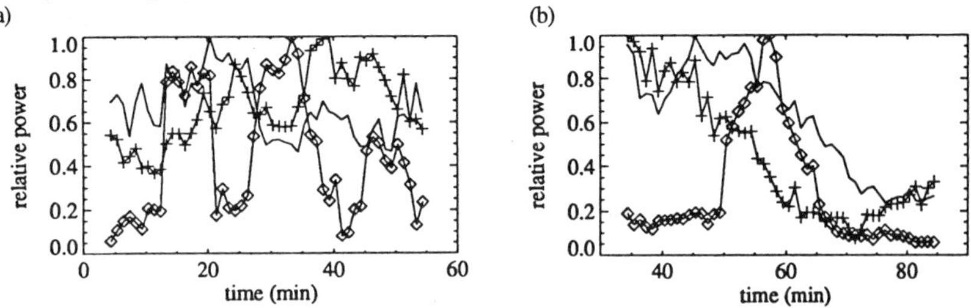

Figure 4: Pre- (a) and post-prandial (b) plots of relative power at sites of GI activity over the 60 minute recording period. Shown is 45 - 55 mHz (------), 190 - 205mHz (◊◊◊◊◊◊) and 140 - 150 mHz (++++++).

During fasting-the signals exhibit no well established pattern. These same frequency bands show that maximum power occurs at different times after ingestion for different segments of the GI tract. The gastric signal reaches maximum power 45 minutes after ingestion. The duodenal signal peaks 57 minutes after ingestion. From this we see that magnetic signals can be followed over a period of time that is typical of GI function.

Conclusion:

It has been shown that magnetic field measurements are capable of recording gastrointestinal activity generated by electric depolarization of smooth muscle. Gastric and small bowel slow wave activity can be distinguished through spectral analysis. Although slow waves have been recorded using this technique, simultaneous magnetic and internal pressure measurements have not yielded observation of action potentials.

Changes in gastrointestinal activity can be measured and quantified in terms of spatial and temporal distribution of power in the various GI frequency bands. Gastric and small bowel activity although not demonstrating a clear pattern in the pre-prandial state, exhibit peak activity at distinct times after ingestion.

We conclude that a magnetic multichannel system can be used to repetitively and non-invasively measure and analyze magnetic gastrointestinal signals.

References:

[1] Richards,W.O., Staton, D., Golzarian, J., Friedman, R.N. and Wikswo, J.P. Jr. Non-invasive SQUID magnetometer measurement of human gastric and small bowel electrical activity, In: Baumgartner, C., Deecke, L., Stroink, G. and Williamson, S.J. Biomagnetism: Fundamental Research and Clinical Applications, Amsterdam, I.O.S. Press, 1995.

[2] McCallum, R.W. and Champion, M.C., Eds., Gastrointestinal Motility Disorders: Diagnosis and Treatment. Baltimore, Williams and Wilkins, 1990.

[3] Staton, D., Golzarian, J., Wikswo, J.P. Jr., Friedman, R.N. and Richards, W.O. Measurement of small bowel activity in vivo using a high resolution SQUID magnetometer, In: Baumgartner, C., Deecke, L., Stroink G. and Williamson, S.J. Biomagnetism: Fundamental Research and Clinical Applications, Amsterdam, I.O.S. Press, 1995.

Acknowledgments:

This research has been funded by the Natural Sciences and Engineering Research Council and the Camp Hill Medical Centre Research Fund.

Magnetogastrograms Using 64-Channel SQUID System: Study on Clinical Usefulness

Saijyo, T.[1], Nomura, M.[1], Haruta, Y.[2], Itozaki, H.[2], Toyoda, H.[2], Nakaya, Y.[3], Ito, S.[1] and Kado, H.[2]

[1]Second Department of Internal Medicine and [3]Department of Nutrition, The University of Tokushima, Tokushima, 770, Japan; [2]Superconducting Sensor Laboratory, Chiba, 270-13, Japan

INTRODUCTION

Magnetogastrograms (MGGs) using a superconducting quantum interference device (SQUID) gradiometer provides excellent temporal and spatial resolution, and are effective non-invasive tool for detecting functional disorder. It is reported that slow motility waves originate high on the greater curvature of the stomach, an area that acts as a pacemaker, which coordinate contractions of the gastric smooth muscle [1, 2]. The gastric electrical activity produces a weak magnetic field. The purpose of the present study was to assess the magnetic field due to gastric electrical activity and to localize the gastric pacemaker site using a 64-channel SQUID system.

SUBJECTS AND METHODS

Two healthy volunteers were selected for the study. The MGG was recorded at 64 points over the epigastric region with a first-derivative dc-SQUID gradiometer at Superconducting Sensor Laboratory in Chiba, and the magnetic field tangentially to the abdominal wall was measured. The MGGs were recorded in a magnetically shielded room in three states (fasting, 60 min after oral medication with cisapride, and 10 min after ingestion of a 500-kcal meal). The recording bandwidths were 50 mHz - 30 Hz. Fast Fourier transform analysis was performed at each state and the peak frequency spectra and magnetic fields were compared among states using spectral analyzer (HP356A Dynamic Signal Analyzer). Magnetic isofield maps were constructed and the locations of gastric electrical activities were computed by the least squares method. These localized activities were projected onto the magnetic resonance image providing anatomical localization in the stomach.

RESULTS

1) MGGs at 64 recording positions in fasting and after fed state

Figures 1 show the MGGs in fasting (panel a) and fed (panel b) states, respectively. The magnetocardiograms were visible on the low frequency wave at each recording position. After the magnetic recording was filtered out with bandwidth at 50 mHz - 100 mHz, the regular 3-cycle/min signal was clearly visible (Figures 2 a, b). The magnitude of the magnetic field after fed state was stronger than that of fasting state.

2) Fast Fourier transform analysis in three states

Figures 3 shows an example of the frequency spectra in three states; fasting (panel a), 60 min after oral medication with cisapride (panel b), and 10 min after ingestion of a 500-kcal meal (panel c), respectively. Two frequency components were visible at each state. Peak frequency were at 52 mHz, 57 mHz and 50mHz, and the magnitude of magnetic field were 274 fT, 385 fT and 760fT at each state, respectively. This component corresponds to gastric electrical activity. Another component around 150 mHz represent the breathing frequency.

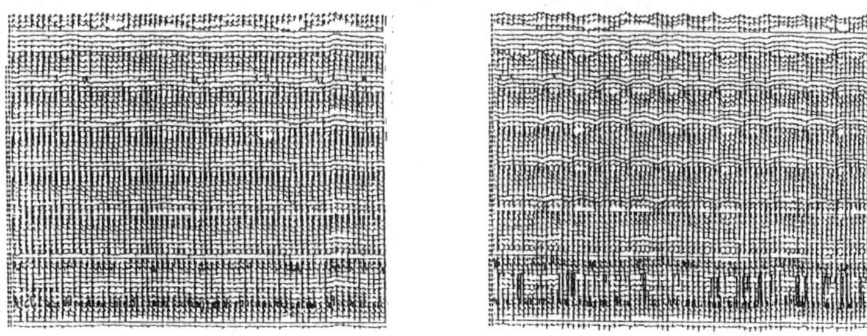

a   b

Figure 1 MCGs in fasting (a) and fed (b) states

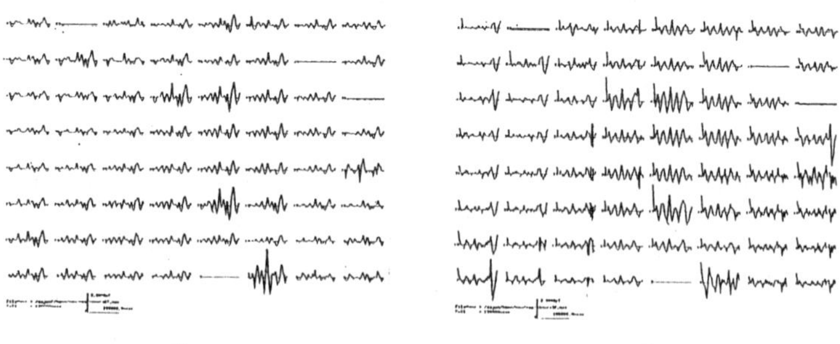

a   b

Figure 2 Filtered magnetogastrographic signals at fasting (a) and fed states

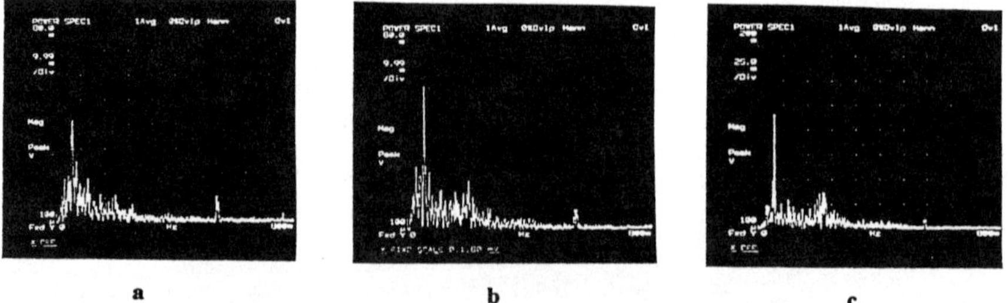

a   b   c

Figure 3 Fast Fourier transform analysis at each stage

(a): fasting state, (b):after medicating cisapride, (c):after taking meal. The 3 cycle/min signal indicated gastric motility, and around 15 cycle/min signal, which reflected respiratory movement, were detected

Figure 4 shows the magnitude of magnetic field (panel a) and peak frequency spectra (panel b) among three states at around 50 mHz. The magnetic field increased after ingestion of cisapride (2.7 ± 0.1 vs. 4.4 ± 0.4

pico tesla; p<0.05). Moreover, the magnetic field after the meal was 1.25-fold greater than that after administration of cisapride (4.4 ±0.4 vs. 5.5±2.5 pico tesla; p<0.05). The frequency spectra of gastric electrical activity were 3.0-3.4 cycles/min but there were no statistically significant differences among the three states.

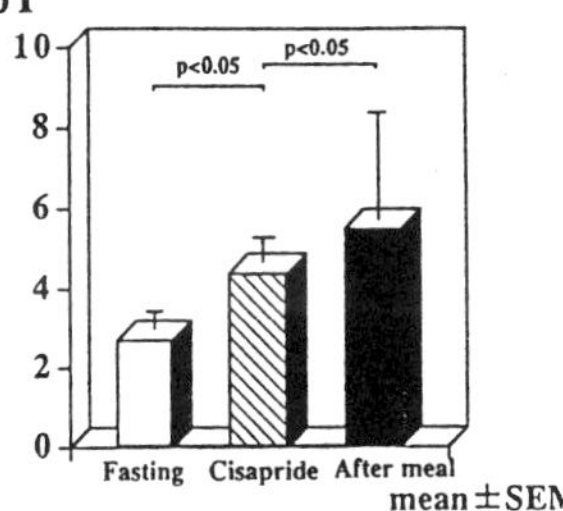

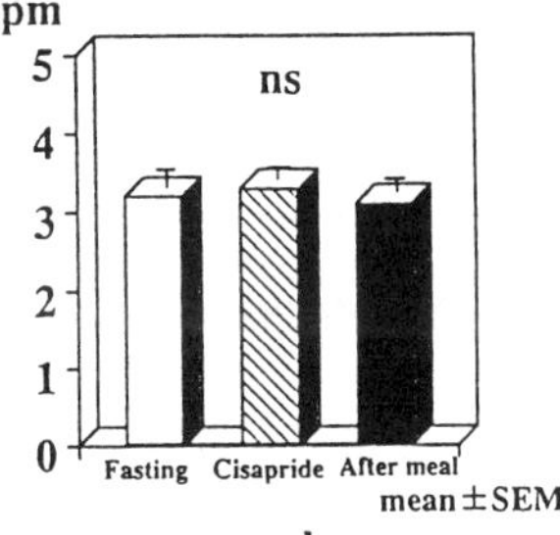

Figure 4 Magnitude of magnetic field (a) and peak frequency spectrum (b) among three states

3) *Localized activities onto the MRI*

Figure 5 shows the magnetic isofield maps after meal. The isofield maps were shown every 20msec. In Figure 6, the localized activities were projected onto the MRI in order to provide anatomical localization by the least squares method. The initial electrical activities were on the stomach.

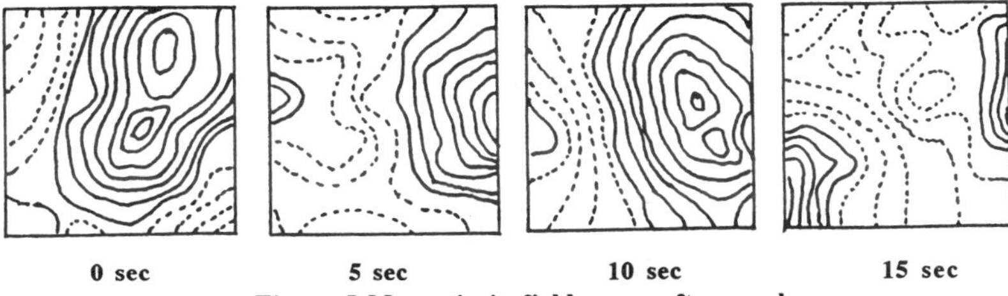

Figure 5 Magnetic isofield maps after meal

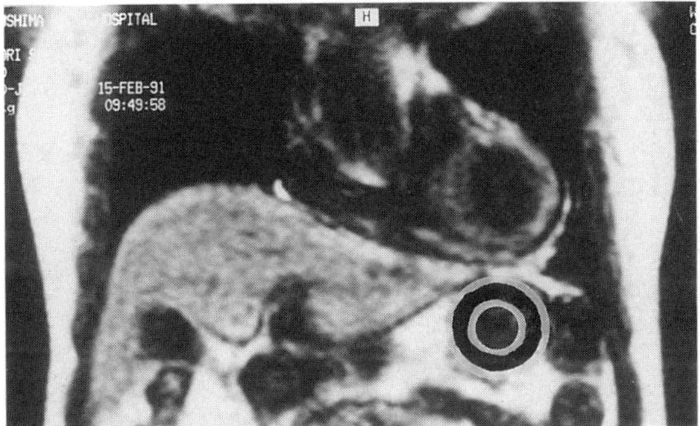

Figure 6 Magnetic resonance image of abdomen
(superimposed gastric electrical activities)

DISCUSSION

We succeeded in recording gastric motility using 64-channel SQUID gradiometer. The isofield map also showed gastric motility with 3 cycle/min in agreement with previous studies [3,4]. After ingestion of meals the amplitude of magnetic field was increased but activation sequences and frequency of slow wave motility was similar with those of control. An electrogastrogram indicates the summation of gastro-intestinal electrical activity but the magnetic field generated in the stomach reflects only gastric electrical activity [5, 6]. Multichannel recordings also showed the location of pacemaker site and slow wave motility sequentially. In this study we only measured in healthy volunteer. The present study shows that the MGGs is useful for detecting gastric electrical activity in various clinical conditions noninvasively and it is expected to contribute to the detection of gastric electrical activity propagation and disorder.

REFERENCES

[1] Alvarez, W.C. New methods of studying gastric peristalsis. Am J Physiol, 1922, 201:287.

[2] Brown, B.H., Smallwood R.H., Duthie H.L., et al. Intestinal smooth muscle electrical potentials recorded from surface electrodes. Med Biol Eng, 1975, 13:97.

[3] Basile, M. The biomagnetic approach to the study of gastrointestinal activity. Biomagnetism:Clinical aspects. Amsterdam, London, New York, Tokyo, Elsevier Science Publishers, pp613, 1992.

[4] Comani, S., Basile, M., Casciardi, S., Gratta, CD., Luizo, CD., Erne, SN., Macri, M., Neri, M., Peresson, M., Romani, GL. Extraporeal direct magnetic measurement of gastric activity. In: Hoke, M. et al. Biomagnetism: Clinical aspects. Amsterdam, London, New York, Tokyo, Elsevier Science Publishers, pp639, 1992.

[5] Nomura, M., Saijyo, T., Itozaki, H., Toyoda, H., Nakaya, Y., Ito, S., Kado, H. Magnetic measurement of gastric activity using a high-temperature superconducting quantum interference device (SQUID). Gastroenterol 1995, 108, 108:A658.

[6] Saijyo, T., Nomura, M., Haruta, Y., Itozaki, H., Toyoda, H., Nakaya, Y., Ito, S., Kada, H. Biomagnetic measurement of gastric electrical activity by 64-channel magnetogastrographic imaging. Gastroenterol 1995, 108:A680.

Noninvasive Measurement of the Vector Magnetic Field from Human Gastrointestinal Sources

Staton, D.J.[1], Allos, S.H.[3], Henry, V.K.[3], Bradshaw, L.A.[1], Ladipo, J.K.[3], Richards, W.O.[2,3] and Wikswo, J.P., Jr.[1]

[1]Dept. of Physics and Astronomy, Vanderbilt University; [2]Depts. of Surgery, Dept. of Veterans Affairs Medical Ctr. and [3]Vanderbilt University Medical Ctr., Nashville, TN USA

INTRODUCTION

A vector SQUID gradiometer (Conductus Instruments, Inc., San Diego, CA) was used to measure magnetic fields associated with bioelectric sources in the stomach and small intestine of normal, healthy, human volunteers. These measurements are an extension to previous work [1,2,3] in which a single component of the magnetic field was measured. Magnetic recordings associated with gastrointestinal current sources, known as the Magneto-ENteroGram (MENG), contain frequency information about a specific type of activity known as the Basic Electrical Rhythm (BER). In normal subjects, the BER frequency of the stomach is approximately 3 cycles per minute (cpm), and varies in the small intestine from 12 cpm in the duodenum to 8 cpm in the terminal ileum. Certain disease conditions can change BER frequency [4], and it is therefore of interest to localize sources of particular frequencies. One of the advantages of using a vector instrument to measure these magnetic fields is that we are sensitive to current sources of all geometric orientations in the abdomen. One of the challenges of using a single channel vector instrument is to decouple the contributions from multiple sources in the magnetic recordings. Since the multiple current sources in the abdomen are of varying frequency, are spatially distributed, and are generally of different orientations, it follows that at a given location, the vector magnetic fields from each current source should point in different directions. The purpose of this preliminary study is to measure the vector MENG in order to examine the possibility that using all components of the magnetic field could be more suitable than using only a single component for localizing current sources in the gastrointestinal tract of humans.

METHODS

Ten normal, healthy volunteers (6 male, 4 female) were subjected to an overnight fast. Each volunteer was dressed in nonmagnetic clothing and placed in a supine position in a magnetically shielded room. After several deep breaths, respirations were suspended during recording to eliminate motion artifact. Recordings were taken at 10 positions over the abdominal surface using the following geometry: the gradiometer was positioned with the normal channel oriented along the axis normal to the frontal plane of the volunteer; for this channel and the two remaining tangential channels, the gradient was measured along the normal direction with a 10 cm long baseline. Data were collected using a DC to 5 Hz bandwidth and 250 Hz sampling frequency. The cardiac contributions in the vector magnetic recordings were digitally removed using a running boxcar median filter of width one second (250 data points). The frequency content of the magnetic recordings was analyzed by calculating the approximated power spectra, using a maximum entropy method, also referred to as autoregressive (AR) filtering.

To test the hypothesis that distributed sources in the abdomen produce magnetic fields of varying frequency and direction, we examined projections of the vector magnetic recordings in many directions. This was done by taking various linear combinations of the three simultaneous recordings from the vector gradiometers. Let the recorded magnetic field be represented by $\mathbf{B} = B_x\mathbf{i} + B_y\mathbf{j} + B_z\mathbf{k}$, and the arbitrary direction of interest be denoted by the unit vector $\hat{\mathbf{v}} = \sin\theta\cos\varphi\,\mathbf{i} + \sin\theta\sin\varphi\,\mathbf{j} + \cos\theta\,\mathbf{k}$, where θ is the polar angle and φ is the azimuth angle describing the orientation of $\hat{\mathbf{v}}$ in spherical coordinates. The "virtual" magnetometer recording in the direction $\hat{\mathbf{v}}$, denoted by B_v, is thus the dot product $B_v = \mathbf{B} \cdot \hat{\mathbf{v}}$. We calculated the projection B_v in 32 directions (corresponding to the vertices of a rhombic triacontahedron) which represented a uniform discretization of the possible combination of angles in spherical coordinates. We then visually examined the projected magnetic component to search for the minimal and maximal components for each gastric and intestinal frequency.

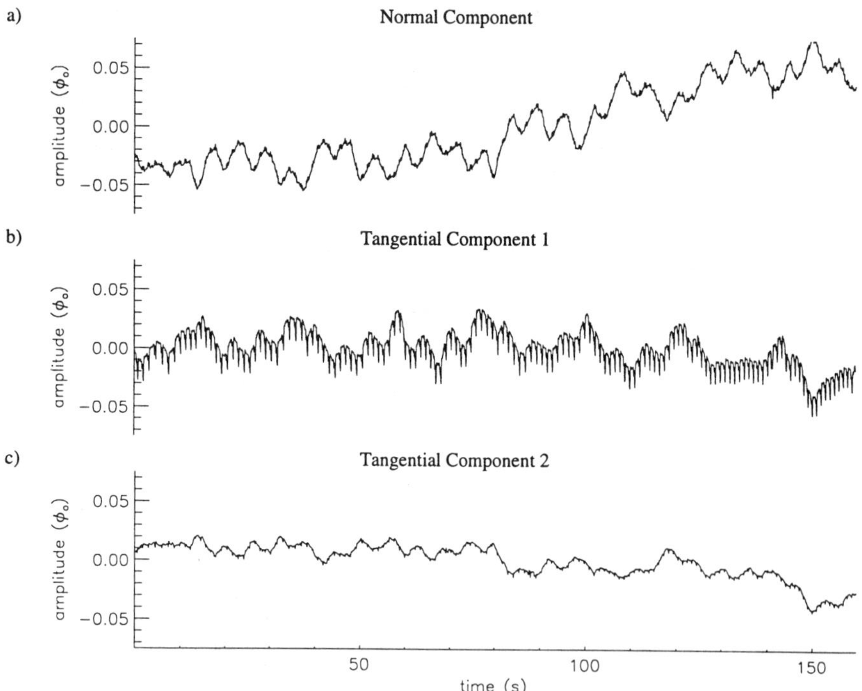

Figure 1. Simultaneously recorded magnetic fields of combined human a,b,c) gastric, intestinal , and b) cardiac electrical activity from three channels of a vector SQUID gradiometer (10 cm baseline). These raw traces were collected from the lower abdominal region (CLQ) using a 250 Hz sampling frequency and DC-5 Hz recording bandwidth. The orthogonal components of these magnetic recordings are oriented a) normally and b,c) tangentially to the frontal plane of the human volunteer.

RESULTS

Magnetic fields associated with gastrointestinal activity were recorded in all ten volunteers. A representative, unprocessed vector magnetic recording is shown in Figure 1. The three components contain contributions from multiple current sources located in the heart, stomach, and small bowel, and also environmental magnetic noise. The measurement was made in the lower abdominal region of a human volunteer. These recordings are typical in that at most locations over the abdomens of all the volunteers, the magnetometer was sensitive to multiple current sources of different BER frequency. The power spectra of the representative vector traces are shown in Figure 2. These spectra show contributions from two primary current sources of frequency 2.7 cycles per minute (cpm) and 9.9 cpm, corresponding to stomach and small bowel, respectively, as well as a weak component at 5 to 6 cpm that may be associated with the colon.

Shown in Figure 3 are projections of the magnetic field that were chosen to optimize the signals from the stomach, the small bowel, and noise. In this particular case, the unit vector projections corresponding to sources in the stomach and intestine appear to be approximately orthogonal The angles are as follows: stomach $\theta = 37.37°$, $\varphi = 108.0°$; intestine $\theta = 37.37°$, $\varphi = 36.0°$; noise $\theta = 100.8°$, $\varphi = 72.0°$. The digitized angles are determined by the rhombic triacontahedron that was used, so even greater separation of the signals may be possible using more optimized unit projections. Note that this procedure is suited for systems such as the gut, where fixed-orientation current sources are turned on and off periodically, with sources at different locations having different frequencies, so

636

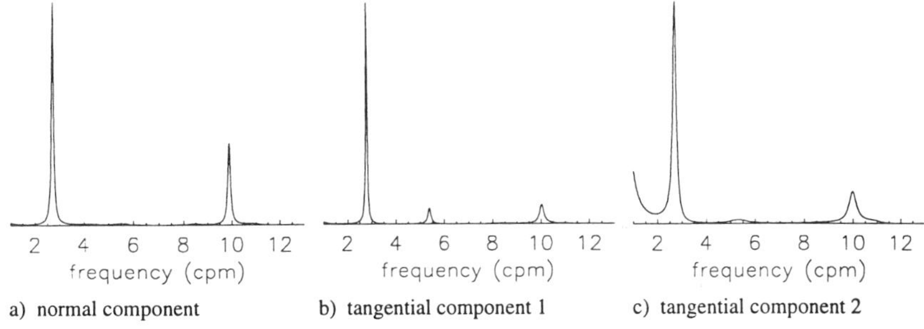

Figure 2. Normalized power spectra estimates (maximum entropy method) of the measured magnetic fields shown in Fig. 1. Frequency is given in cycles per minute (cpm). These spectra show contributions from two primary current sources of frequency 2.7 cycles per minute (cpm) and 9.9 cpm, corresponding to stomach and small bowel, respectively, as well as a weak component at 5 to 6 cpm that may be associated with the colon.

that a single direction is more applicable for a particular measurement location to separate the signals produced by different sources. In other situations, such as the magnetocardiogram, the orientation of the fields changes over a limited range during the cardiac cycle; it may be possible, however, to determine the field component for which the fetal R-wave was optimized over the maternal R-wave.

DISCUSSION

Although little is known about current sources in the gastrointestinal tract, it does not seem likely that the magnetic field vector from a given gastrointestinal source is statically directed in particular direction and varies only in field strength. It is more likely that the field vector corresponding to a particular source varies in direction and magnitude with time. We are presently investigating methods for modeling the recorded magnetic fields to determine the temporal variation in direction and magnitude of the fields corresponding to each source.

In most other biomagnetic measurements, the current sources are highly localized; for example, MEG measures the response of specific regions of the cerebral cortex responding to a particular stimulus, and the MCG measures electrical activity contained within the heart. In these situations, the challenge is to place enough magnetometer sensors in the immediate vicinity of the source. While there are contributions to the field from the unrelated activity of adjacent regions of the brain, the problem is not usually the separation of two competing signal sources. In the gastrointestinal measurements, the sources are distributed over the entire abdomen, and one challenge is to separate the signals that overlap in both space and time. Our power spectral analyses and the vector measurements can be combined into a powerful method to separate the contributions; Figure 3 shows that it is possible to do this without the need for extensive mapping of the magnetic fields. It would be potentially advantageous, in terms of both cost and convenience, if a small number of vector SQUID magnetometers could be used to record the MENG, rather than using a larger array of SQUIDs that measured only one component of the magnetic field. Our preliminary studies demonstrate that the vector MENG can be recorded with high quality, and that the components can contain differing information. The proposed research is directed toward exploring the potential advantages of the vector approach to provide the requisite clinical data at low cost.

REFERENCES

[1] Richards WO, Staton DJ, Golzarian J, Friedman RN, Wikswo JP, Jr.. (Ed) Deecke L, Baumgartner C., Stroink G., and Williamson S.J. Biomagnetism: Fundamental Research and Clinical Applications - Proceedings of the 9th International conference on Biomagnetism.

[2] Garrard CL, Wikswo JP, Jr., Staton DJ, Golzarian J, Gallen C, Richards WO. Gastroenterology, Vol. 106, No. 4, Part 2, A502 (1994).

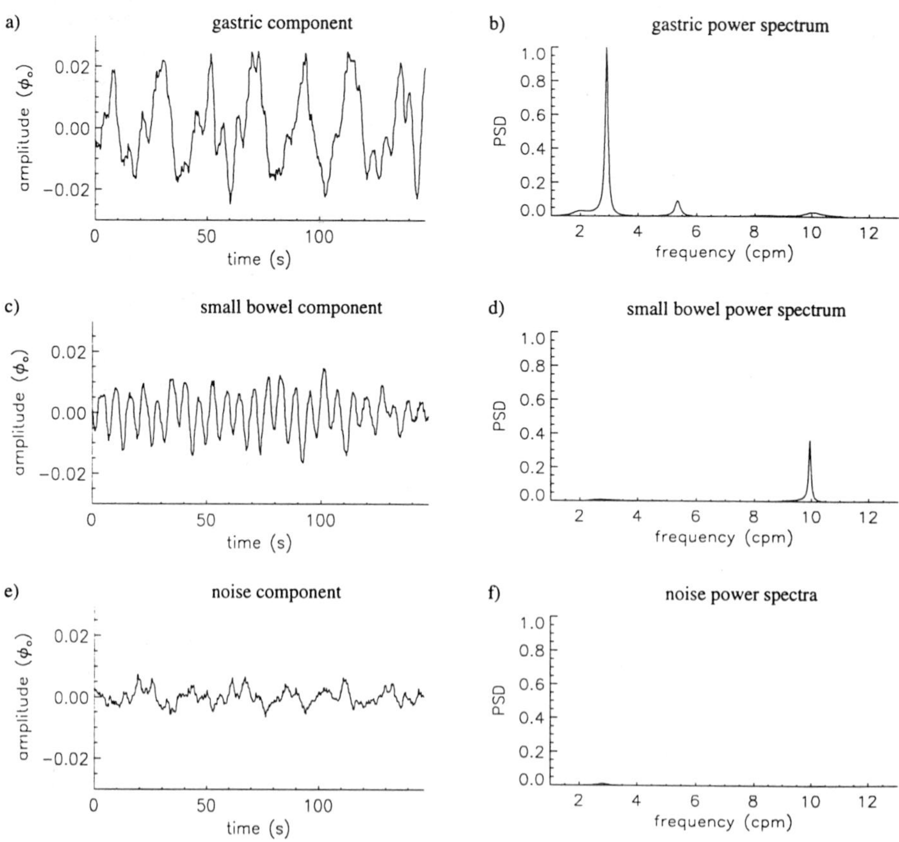

Figure 3. Components of the vector magnetic field and power spectra corresponding to a,b) gastric and c,d) small bowel electrical activity, and e,f) noise in the measurement. Power spectra densities are normalized relative to the gastric power spectra.

[3] Comani S, Basile M, Casciardi S, Del Gratta C, Di Luzio S, Erne SN, Macri M, Neri M, Peresson M, Romani GL. (Ed) Biomagnetism: Clinical Aspects, pp. 639-642, Amsterdam, Elsevier Sci. Publ., 1992.

[4] Richards WO, Garrard LC, Allos SH, Bradshaw LA, Staton DJ, Wikswo JP, Jr.. Annals of Surgery, Vol. 221, No. 6 (1995).

Human Alveolar Macrophage Activation in Healthy Subjects and in COPD-Patients Evaluated by Magnetopneumography

Barth, W.[1], Möller, W.[2], Meier-Sydow, J.[3], Pohlit, W.[1], Stahlhofen, W.[2]

[1]*Institut für Biophysik, J. W.-Goethe Universität, Paul-Ehrlich-Str. 20, D-60596 Frankfurt,*
[2]*GSF - Institut für Inhalationsbiologie, Ingolstädter Landstr. 1, D-85764 Oberschleissheim and*
[3]*Abteilung für Pneumologie, Medizinische Klinik II, D-60590 Frankfurt, Germany*

Introduction

Magnetopneumography (MPG) has been used to study long-term particle clearance of the human lungs in healthy and disease [1, 2]. Ferrimagnetic iron-oxide test particles are inhaled and used as a tracer. The particles have a sufficient stability in order to monitor alveolar clearance over a year without any burden to the subjects. Within hours after particle deposition they are phagocytized by alveolar macrophages (AM). This moves the particles to the intracellular environment, now being under the cellular motile activity. After magnetization the particles form small remanent magnets, all being oriented parallel to the external field. This allows to detect a small macroscopic lung field (LF). The LF decays within ≈20 minutes. Because of the remanent magnets the decay must be due to a stochastic rotation of the particles, finally leading to zero LF. It has been shown that this decay, called *relaxation,* reflects intracellular phagosome motion and thereby is a monitor for the motile activity of the AM-cells. In this study intracellular phagosome motion was investigated in healthy non-smokers (NS) and smokers (S) and in patients with chronic obstructive pulmonary disease (COPD), answering the question whether smoking might affect cellular motility. Further investigations have shown that relaxation can be influenced by inert particle exposure [3]. This study shows that an AM-activation was induced by the deposition of the magnetic test-particles, that depends on cigarette smoking. MPG provides a method to detect AM-activation *in vivo* without medical treatment, like a bronchoalveolar lavage (BAL).

Materials and methods

Preparation of magnetic tracer particles, inhalation and detection in the human lungs
Monodisperse ferrimagnetic iron oxide particles (Fe_3O_4) were used in this study as a tracer. The particles were produced by nebulizing a colloidal non-magnetic iron oxide solution (Fe_2O_3) with a spinning top aerosol generator (STAG) [4]. The particles were reduced into the magnetic form of iron oxide (Fe_3O_4) in a furnace at 800 °C. About 40 breaths under controlled breathing conditions (1 l tidal volume, 250 cm^3/s air flow) were used for the deposition of about 0.5 mg of particles. The particles are spherical with 1.3 μm geometric (2.9 μm aerodynamic) diameter and monodisperse (geometric standard deviation $\sigma_g < 1.1$). Under this standardized conditions the particles are mainly deposited in the alveolar region of the human lungs [2]. The MPG-system for detection of the particles in the human lungs consists of a 40 cm diameter magnet. A capacitor battery (1 mF, 1000 V) is discharged into the magnet, producing a 100 mT, 30 ms magnetic field pulse. This field is sufficient to magnetize the particles and to align the remanent magnetic dipoles parallel to the magnetizing field. All micro-magnets together produce a macroscopic LF (≈100 pT), being detected by a sensitive superconducting SQUID-device. The whole system is located within a magnetically shielded room, where a lower detection limit of 20 μg of Fe_3O_4-particles within the whole lung is achieved.

Relaxation

Relaxation is a monitor for intracellular phagosome motion. Relaxation describes a stochastic particle twisting process and was model as rotational Brownian movement [5], where thermal energy kT was replaced by a cellular energy E_Z. When using spherical particles, the decay should follow an exponential function according to:

$$\frac{B(t)}{B_0} = e^{-t/\tau_r}, \qquad \tau_r = \frac{\kappa V \eta}{2 E_Z}, \tag{1}$$

where V is the volume of the particles, η is the surrounding viscosity and κ is the rotational shape factor (κ = 6 for spheres). *In vivo* and *in vitro* investigations of relaxation with spherical particles yield deviations from a single exponential behaviour [6]. An initial fast decaying phase (over the first 30 s) was followed by a slower exponential phase lasting up to 20 min. Therefore the decay was fitted by a double exponential function:

$$\frac{B(t)}{B_0} = (1 - A_{r\_s})\exp(-\frac{t}{\tau_f}) + A_{r\_s}\exp(-\frac{t}{\tau_s}), \tag{2}$$

where τ_f and τ_s are the time-constants for the fast and for the slow relaxation phase, respectively, A_{r_s} describes the fraction of relaxation following the slow phase. Further studies have shown that both, the fast and the slow decaying fraction depend on the particle size [6]. For comparison with other studies the time constant for the initial phase, τ_a, was analyzed.

Measurement protocol

9 healthy non-smokers (NS, age: 53 ± 5 years) and 8 healthy smokers (S, 53 ± 9 years) were involved into the study. At the beginning of the study a detailed questionnaire was performed, allowing to exclude subjects with chronic diseases or persons who are under a permanent drug administration. All subjects gave their informed consent according to a protocol approved by the Ethical Committee of the University Hospital in Frankfurt. In detailed lung function tests hyperreactivity of the subjects was excluded. Cigarette smoke consumption was classified in pack-years according to the number of pack's of cigarettes smoked per day times the years of cigarette smoking. NS never smoked during their live span, S are active smokers with a mean smoke-consumption of 45 ± 23 pack-years (PY). Additionally 19 COPD-patients (age: 60 ± 8 years) were investigated, 2 were NS, 3 were S (32 ± 18 PY) and 14 were ex-smokers (ES, 36 ± 26 PY). MPG-measurements (including relaxation) were performed directly (30 min) after inhalation, as well as 1 day, 1 week, 1 month, 4 month and 9 month after inhalation.

Results and discussion

The particles are phagocytized within hours after deposition by AM-cells. Our studies have shown that non-phagocytized (free) particles do not contribute to relaxation nor to particle rotation [7]. The non-rotatable (fixed) fraction of particles was interpreted being non-phagocytized (free) and thereby subtracted from the relaxation curves. Relaxation in humans shows an influence on the time after particle deposition, as demonstrated in Fig. 1a. Directly after inhalation the decay is fastest while it slows down with increasing time after particle deposition. Earlier investigations have shown, that an inert particle deposition can induce an AM-activation with a faster relaxation [3]. From this we follow that the initially faster relaxation is a measure of a <u>non-specific AM-activation</u> due to the deposition of the iron-oxide test-particles. The corresponding change of the relaxation time-constants is shown in fig 1b, where the fast as well as the slow time-constant increase with time after particle deposition and thereby demonstrate the deceleration of the decay. From one week after particle deposition till the end of the study (9 month) relaxation was stable. Therefore the measurements obtained from the period between 1 week and 9 month were used to form mean values for the fast (τ_f), the initial (τ_a) and the slow (τ_s) decaying constant and for the fraction of relaxation showing the slow decay (A_{r_s}). The initially faster decay was described by a AM-activation factor, calculated from the ratio of the time-constant in the stable phase compared to the time-constant measured directly after particle exposure (τ_a/τ_{a_1}). An AM-activation detected by a faster decay of relaxation is reflected by a value > 1.

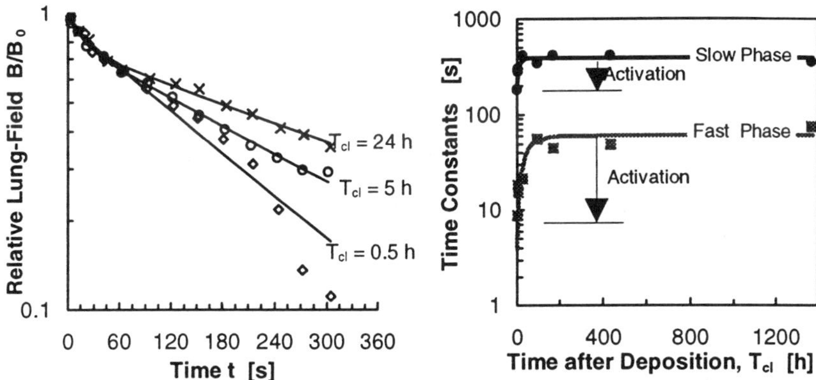

Figure 1: a) *In vivo* relaxation of the ingested magnetic microparticles within macrophage cells depending on the time after particle deposition, T_{cl}. b) Change of the fast and the slow relaxation time constant with time after particle deposition due to an AM-activation

In Table 1 the results for the mean values of relaxation in the stable phase (1 week - 9 month) are shown for healthy subjects and for patients with COPD. Relaxation shows no differences between NS and S, although there seems to be a tendency for a slower decay in smoking subjects. This results are surprising because we expected a strong influence of cigarette smoke on AM-motile activity and thereby on the relaxation process. Two in vitro studies on AM-relaxation in smokers and non-smokers are available. In both studies macrophages were harvested by broncho-alveolar lavage (BAL) from the human lungs and then incubated together with magnetic micro particles. After 24 h relaxation was measured in a cytomagnetometric system. In the study by Valberg et al. [8] relaxation was faster in smokers compared to non-smokers. In this study the cigarette consumption is not clearly reported, but the subjects seem to be younger persons. The faster relaxation was interpreted as a result of the permanent cigarette smoke exposure, causing a permanent AM-activation. In the second study by Gehr et al. [9] relaxation showed no differences between NS and S. The subjects in this study had a mean smoke consumption of 8 PY. In both studies no investigations on non-rotatable (free) particles were performed. These corrections significantly influence the results on relaxation. Our data suggest that even with a very high cigarette smoke consumption of 45 PY cell organelle motions are not influenced. The organism seem to respond to the permanent cigarette smoke exposure with an increased number of AM cells in the lungs. But it is known that several biochemical and immunological factors are raised in S compared to NS. This activation's seem to have no correlation to cytoskeletal activities. Beside the missing corrections on non-rotatable particles the differences found in one of the *in vitro*-studies additionally might be influenced by the BAL-procedure and by incubation. In COPD relaxation is faster only by tendency (τ_a: p=0.06, τ_s: p=0.27).

Cytoskeletal AM-activation measured by relaxation in healthy non-smokers and smokers and in patients with COPD is shown in Table 1. In healthy non-smokers AM-activation is very strong, directly after particle deposition the decay is accelerated twofold compared to the stable phase. A comparable AM-activation is not present in smokers. We must conclude that the missing activation in S either shows an adaptation of the AM-cells to the permanent cigarette smoke exposure or reflects a damage in cellular defense reaction due to cigarette exposure. In COPD-patients AM-activation is also present, but less than strong than in NS. Because most of the COPD-patients are ex-smokers this might reflect to a beginning back-formation of the smokers lung to non-smoking conditions.

Table 1: Parameters ($\pm$ SD) of relaxation and cellular energy together with AM-activation in healthy non-smokers (NS) and smokers (S) and in COPD-patients

Subjects	A_{r_s}	initial τ_a [s]	slow τ_s [s]	activation
NS (n=9)	0.77 ± 0.03	65.2 ± 17.2	250 ± 54	2.1 ± 1.2
S (n=8)	0.74 ± 0.05	65.9 ± 15.7	272 ± 65	1.0 ± 0.4
p (NS-S)	0.17	0.93	0.44	0.05
COPD(NS) (n=2)	0.79 ± 0.03	56.9 ± 22.9	253 ± 41	1.6 ± 0.9
COPD(ES) (n=14)	0.74 ± 0.07	55.3 ± 15.1	239 ± 66	1.7 ± 0.9

Our data show that intracellular phagosome motion is not influenced by permanent cigarette smoke consumption *in vivo*. From this one can understand that patients with COPD, a disease that mostly results form cigarette smoking over decades of years, show only little differences in phagosome motion compared to healthy subjects. COPD primarily occurs in the airways of the lungs. But the MPG-method is sensitive to the alveolar region of the lungs. In this case the results are in agreement with expectations. The investigations show that MPG is not only a tool to study long-term particle clearance in the human lungs in healthy and disease, but also a tool to monitor motile functions of AM cells *in vivo* in the lungs. AM-motile activity is an important factor in the potency of the cellular defense reaction of the human lungs against foreign substances. Reduced AM-motility might enhance the possibility of lung infections and thereby might be a co-factor for certain lung diseases.

Acknowledgement: This work was supported by the Deutsche Forschungsgemeinschaft (Po 108/7-1) and by the CEC (FI4P-CT95-0026).

References

[1] Cohen, D., S.F. Arai, J.D. Brain (1979), Smoking Impairs Long-Term Dust Clearance from the Lung, *Science*, Vol. 294, 514-517.

[2] Stahlhofen W, Möller W (1993) Behaviour of magnetic micro-particles in the human lung, *Radiation and Environmental Biophysics*, Vol 32, 221-238

[3] Stahlhofen W, Möller W (1994) In vivo and in vitro studies of the cellular defense system of the human lung, *Toxicology Letters*, Vol. 72, 127-136.

[4] Möller W, Stahlhofen W, Roth C (1990) Improved spinning top aerosol generator for the production of high concentrated ferrimagnetic aerosols, *Journal of Aerosol Science*, Suppl. 1, Vol. 21, S435-S438

[5] Nemoto, I. (1982), A Model for Magnetization and Relaxation of Ferrimagnetic Particles in the Lung, *IEEE Transactions on Biomedical Engineering*, Vol. BME-29, 745-752.

[6] Möller W, Guzijan V, Pohlit W, Stahlhofen W, Wenisch T, Wiegand J (1992) Cytomagnetometry with ferrimagnetic micro-particles - influence of particle size and dispersity, *Journal of Aerosol Science*, Vol. 23, Suppl. 1, S519-S522

[7] Barth W, Möller W, Pohlit W, Stahlhofen W, Wiegand, W (1994) Magnetopneumographic estimation of particle phagocytosis in the human lungs, Journal of Aerosol Science, Vol. 25, Suppl. 1, S491-492

[8] Valberg. P.A., W.A. Jensen, R.M. Rose (1990), Cell Organelle Motions in Bronchoalveolar Lavage Macrophages from Smokers and Nonsmokers, *American Review of Respiratory Disease*, Vol. 141, 1272-1279.

[9] Gehr, P., M. Klauser, V. Im Hof (1989), Intracellular Motility of Pulmonary Macrophages from Smokers and Nonsmokers, *Advances in Biomagnetism*, Editors: S.J. Williamson, M. Hoke, G. Stroink, M. Kotani, Plenum Press New York and London, 473 - 476.

Magnetic Markers Linked to Monoclonal Antibodies Allow *in vitro* Evaluation of Cell Population Size

Del Gratta, C.[1,2], Della Penna, S.[2], Battista, P.[3], Di Donato, L.[1], Vitullo, P.[3], Romani, G.L.[1,2], and Di Luzio, S.[1,2]

[1]*Institute of Medical Physics;* [2]*Institute of Advanced Biomedical Technologies;* [3]*Institute of Human Pathology and Social Medicine, Gabriele D'Annunzio University, Chieti, Italy*

INTRODUCTION

Biomagnetic instrumentation has been used for the detection of magnetic particles in the human body. Recently, macroscopic markers were used for specific physiological measurements, like monitoring bowel movements. At a molecular level, liver biosusceptometry allows the measurement of iron concentration in the liver of patients affected by diseases such as hemocromatosis and thalassemia. The motivation of the work to be described hereafter is the assessment of a method utilising biomagnetic instrumentation and microscopic markers, smaller than cellular size, to be directed at specific cell populations. These markers may be linked to target cells via specific antibodies.

Monoclonal antibodies (Mabs) directed against human lymphocytes markers and tumors associated antigens (TAAs) have proven to be extremely useful in defining either subpopulations of lymphocytes or differences between human malignant and normal cell populations. While different diagnostic methods, utilising Mabs, are available in vitro, the use of Mabs for the immunodiagnosis *in vivo* requires their conjugation with a radioactive marker such as [125]I. The use of magnetic markers linked to Mabs would result in a relative non-invasiveness for patient studies, and a high sensitivity due to the high field sensitivity of biomagnetic instrumentation.

The assessment of our method requires an interdisciplinary research effort, with contributions from the biological as well as from the physical side; the former concerning cell targeting and bio-compatibility, and the latter concerning the measurement of very weak magnetic fields and their analysis, like for instance the solution of the magnetostatic inverse problem.

Our research programme on this topic ranges from feasibility studies [1] to the final goal of in-vivo experimentation. We report here on an in-vitro evaluation of the specificity of the method [2].

MATERIALS AND METHODS

The biomagnetic sensor is equiped with a rf-SQUID coupled to a second-order gradiometer with 1.5 cm pick-up coil and 7 cm baseline. The peak-to-peak noise in the recording bandwidth of DC-5Hz is 2 pT. The SQUID output is analog filtered with 48 db/oct slope and then digitally converted for digital off-line analysis and data storage at a sampling rate of 100Hz.

Application of an external magnetising field was necessary for signal enhancement. We used a pair of Helmoltz coils placed so that their axis coincided with the gradiometer axis. The radius of the coils was about 60 cm allowing a comparatively large region of homogeneous field for the measurements. The coils were driven by a stable DC power supply unit and it was possible to apply a field of up to 4.5×10^{-4} T in the region of measurement. The fluctuations of the field due to thermal noise in the copper wires raises the overall noise of the experimental set up, resulting in a peak-to-peak noise of 4 pT.

The markers consisted of small beads 4.5 μm in diameter with a ferrite polycristalline core surrounded by a polystyrene shell and are commercially available as a product for cell separation from Dynal Norway AS. The volume magnetic susceptibility was measured by means of our instrumentation and was found to be 1.0 SI units. The collective behaviour of these beads is referred to as superparamagnetic since they feature a high susceptibility without the characteristics of ferromagnetic materials; in particular these beads do not show any hysteresis or saturation, at least in the low field range of our experiments. Moreover, their field scales linearly with concentration, as it was experimentally checked [1].

The field generated by the markers linked to the cells in a preparation was measured by scanning with the sensor an area of about 14 cm × 22 cm over the preparation. The preparation was moved under the sensor on a wooden table monitored by a fiberoptic system allowing the determination of the position along a scan with a precision of 3 mm.

Details of the immunoreactions are described elsewhere [2]. We used cultures of normal human lymphocytes and colon carcinoma cells. Sample preparations were obtained by cytospin and consisted of two circular spots 7.2 mm in diameter per slide. Preparations were exposed to beads linked to appropriate markers or antibodies and afterwards copiously washed to remove unbound beads. Negative control preparations were also used to test the specificity of the reactions.

The model for the forward field calculation, consisted of two thin, uniformly magnetised disks. We considered the field contribution of an infinitesimal element of the disk, assuming that is consisted of a single layer of beads. The total field of a circular disk was obtained by approximating the integral by means of a seven point integration technique. The effect of finite coil dimensions was taken into account by means of the same approximation. This yielded theoretical maps of the mean magnetic field at the gradiometer location in the measurement plane. Before comparing the theoretical maps to the experimental ones, net field maps were obtained by subtracting from each experimental map a map of an analogous preparation, non exposed to the beads. This was done in order to eliminate the diamagnetic contribution of the glass and of the biological substance. The net maps were finally fitted, by means of the least squares method, with the theoretical maps, where the free parameter was the susceptibility of the distribution. The negative control maps were analysed in the same way.

Two further steps must be taken to obtain the number of cells from the estimated mean susceptibility. The comparison of the latter with the susceptibility of a bead yields the number of beads. The number of cells may then be obtained if the mean number of beads attached to a cells is known. We called this figure the targeting ratio (TR). In our case it had to be estimated independently and this was done by visual inspection of the preparations.

RESULTS AND DISCUSSION

In the following two ways of counting the cells in a preparation are compared: 1) the mean number of cells per unit area - obtained by visual inspection - times the area of the preparation; and 2) the total number of beads divided by the TR. The first figure is indicated in the text whereas the second is indicated in the last column of the tables.

<u>Preparations containing lymphocytes:</u>
The measurements were performed in an applied field of 4.63×10^{-4} T. The mean TR value was estimated to be 9.2. The mean number of cells for the three preparations was 68,000, which gives, considering the usual percentatges of T- and B-lymphocytes of 75% and 15%-19% respectively, 51,000 and 10,200-12,900 cells respectively.

Table 1: T- lymphocyte count

N.	Susceptibility	Beads	Target cells
1	0.152 ± 0.008	$585,000 \pm 88,000$	$64,000 \pm 10,000$
2	0.154 ± 0.008	$595,000 \pm 89,000$	$65,000 \pm 10,000$
3	0.117 ± 0.006	$450,000 \pm 67,000$	$49,000 \pm 7,000$

Table 2: B- lymphocyte count

N.	Susceptibility	Beads	Target cells
1	0.050 ± 0.002	$191,000 \pm 29,000$	$21,000 \pm 3,000$
2	0.047 ± 0.002	$181,000 \pm 27,000$	$20,000 \pm 3,000$
3	0.048 ± 0.002	$185,000 \pm 28,000$	$20,000 \pm 3,000$

<u>Preparations containing tumor cells:</u>
The measurements were performed in an applied field of 4.66×10^{-4} T. The mean number of cells for the three preparations was 61,000. The TR value was estimated to be 2.8.

Table 3: colon carcinoma cell count

N.	Susceptibility	Beads	Target cells
1	0.058 ± 0.003	$180,000 \pm 27,000$	$64,000 \pm 8,000$
2	0.041 ± 0.002	$130,000 \pm 20,000$	$46,000 \pm 6,000$
3	0.058 ± 0.003	$180,000 \pm 27,000$	$64,000 \pm 8,000$

In addition to these, a single preparation containing a greater number of the same cells was measured in the same conditions. The number of cells was 195,000, the TR value was estimated to be about 2.7.

Table 4: colon carcinoma cell count

N.	Susceptibility	Beads	Target cells
1	0.19 ± 0.01	$610,000 \pm 90,000$	$225,000 \pm 35,000$

The most important conclusion that can be drawn from this series of experiments is that the specificity requirement was met, both from a qualitative and from a quantitative point of view. From the qualitative point of view, the negative control preparations showed no magnetic signal above the noise level. From a quantitative point of view, when performed in triplicate, results were repeatable; the mismatch between calculated and estimated values of population was low in the preparations of tumor cells, being between 5% and 15% , but was higher in the case of lymphocyte preparations, although the population ratio value was adequate. In this later case, we believe that the direct counting procedure was not accurate. Overall, the population values obtained are acceptable, considering the large errors that are intrinsic to cell counting procedures. Finally, the peroxidase method, conducted in parallel proved to be less specific than the use of Mabs.

The direct counting of the cells was perhaps the weakest part of our procedure; a more precise counting method could be useful only for evaluating the precision of the biomagnetic method. With a more accurate counting the reported errors may be found to be lower. Also, the determination of the targeting ratio may seem awkward; however, it is likely that TR may be fixed once and for all in various experimental situations: this is suggested by the estimation of the TR regarding the tumor cell preparation. Moreover, the knowledge of TR is not mandatory when only qualitative results are needed. The objective of extending this method to in vivo applications, requires improved performances of the instrumentation as well as of the data analysis system. In this case, the marker distribution may be located at larger distances from the sensor and also have an irregular shape. Regarding the former problem, enhanced sensitivity may be obtained with a more sensitive magnetic sensor and a larger magnetising field. Regarding the latter problem, data analysis procedures must be used that do not require a precalculated field pattern. Finally a parallel reasearch should be carried on to assess the biological issues of in vivo marker administration and cell targeting.

The experiments described in this paper were performed with a rather simple and out of date instrumentation, being equipped with a rfSQUID coupled to a wire wound gradiometer. However, the use of a rfSQUID proved to be successful due to its noise properties in the low part of the spectrum. Present day technique instruments equipped with dcSQUIDs featuring a higher 1/f noise level would require the application of a slowly varying magnetising field. The use of multichannel instrumentation could speed up the measurements. Applied magnetising fields of larger intensity and appropriate spatial distribution may improve overall and local sensitivity of the instrumentation. From the point of view of data analysis, more elaborate susceptibility distributions may be retrieved from the data by solving the magnetostatic inverse problem, leading to the imaging of the target population.

ACKNOWLEDGEMENTS

Work partially supported by the National Research Council of Italy (CNR), under the Progetto Finalizzato "A.C.R.O". P.B. is partially supported by MURST-40% funding.

REFERENCES

[1] Macrì, M.-A., Del Gratta, C., Di Donato, L., Di Luzio, S., Romani, G.L., Della Penna, S., and Pasquarelli, A., 1994, Biomagnetic measurements utilising a superparamagnetic marker: a feasibility study, Nuovo Cimento, D16:425-432.
[2] Del Gratta, C., Della Penna, S., Battista, P., Di Donato, L., Vitullo, P., Romani, G.L., and Di Luzio, S., 1995, Detection and counting of specific cell populations by means of magnetic markers linked to monoclonal antibodies, Phys. Med. Biol., 40:671-681.

Iron Quantification by Non-Invasive Biomagnetic Organ Susceptometry

Engelhardt, R., Fischer, R., Nielsen, P. and Gabbe, E.E.

Abteilung Medizinische Biochemie, UKE, University Hamburg, Germany

Introduction:

In the last years (1989 - 1996) SQUID biosusceptometry has been used to assess iron stores in more than 600 patients suspected for iron overload. In the following we report on applications to 3 major clinical problems in iron overload diseases. From more than 1600 biomagnetic measurements in humans, we determined the iron concentration in liver and spleen tissue with respect to hereditary hemochromatosis (I), transfusional hemosiderosis (II) and screening for hereditary hemochromatosis in prospective blood donors (III).

The iron concentration in different organs is an important parameter in the diagnosis and staging of human iron overload diseases such as genetic hemochromatosis (GHC) or transfusional hemosiderosis.

In hemochromatosis, the inborn regulation defect of intestinal iron absorption leads to a progressive iron overload, mainly of the liver. The biochemical parameter such as serum iron (> 200 µg/dl), transferrin saturation (> 60 %), serum ferritin (> 300 µg/l) are typically increased in homozygous patients. Undiagnosed and therefore untreated patients have a high risk to develop severe organ damages such as liver cirrhosis, diabetes and cardiomyopathy, especially when higher liver iron concentrations above 2 mg Fe/g_{liver} are present. The standard therapy is the iron depletion by weekly phlebotomies of 500 ml blood, until normal liver iron concentration (0.1 - 0.5 mg Fe/g_{liver}) is achieved.

Patients with ß-thalassemia major or other transfusion dependent chronic anemias need regular blood transfusions. This therapy also results in a severe accumulation of iron in the body. Although the distribution of iron in non-parenchymal and parenchymal cells of the liver, spleen and other tissue is quite remarkably different from genetic hemochromatosis, the degree of iron-loading in this secondary hemosiderosis is also a factor which is correlated with the expectation of life. A severe siderosis can result in a cardiomyopathy, diabetes or liver cirrhosis. In order to prevent organ damages and prolong life, the daily use of subcutaneous infusions of the iron chelating compound *deferoxamine* (DFO) is recommended in patients under regular blood transfusions. Preliminary data indicate that the oral iron chelator *deferiprone* (DFP, former names L1, CP20) may have similar effectiveness in promoting iron excretion in iron-loaded patients.

Currently, the classical liver biopsy with histologic determination of Prussian blue stainable iron in liver slices and the chemical determination of iron in the liver biopsy specimens is still the standard method in the diagnosis of a clinical relevant iron overload. During the past decade, several methods for non-invasive liver iron quantification (CT, MRI, nuclear resonance scattering, SQUID biomagnetic susceptometry) have been studied mostly in smaller groups of patients. Among this, only MRI and biomagnetic susceptometry seemed to have the principle capability to measure liver iron concentration sensitive enough to be of clinical use.

In the present study, the benefit of non-invasive iron quantification by biomagnetic organ susceptometry (BOS) under clinical conditions was studied in a large group of patients suspected for iron overload diseases.

Methods:

Biomagnetic organ susceptometry was performed by lowering the patient in the known localized magnetic dc-field (B_{max} = 20 mT, gradiometer 1storder) of the Hamburg Biosusceptometer (BTi) with water as the reference medium. The fluxintegral contributions for organs and torso were calculated in advance for certain distances and radii (3 - 25 cm) as ellipsoids or hemispheres (liver and spleen) and as hemispheres or cylinders (torso tissue). The calculated fluxintegrals were fitted over distances and radii and the coefficients were tabulated and stored in the computer for the analyzing of measurements.

In BOS, the magnetic flux change (rf-SQUID voltage output) detected by two 2ndorder gradiometer pickup coils was fitted by tissue (ribs, muscle, fat; near field) and liver or spleen (far field) flux integral contributions simultaneously in a 2-layer model [2]. From bedside sonographic imaging, the actual distance from the respective organ to the torso surface and the organ volumes were determined. The total body iron stores were estimated by multiplying the biomagnetically measured liver and spleen iron concentrations (the major storage organs), with the respective organ volumes obtained by sonography.

Results and discussion:
I. Hereditary hemochromatosis

In hereditary hemochromatosis it is most important to prevent a marked iron overload in the affected individuals before the onset of irreversible clinical symptoms. The gene frequency for hemochromatosis (HFE-gen) in caucasoids is about 10 % and between 0.2 - 0.6 % of the general population are affected in a homozygous (clinical relevant) form. One approach for a screening study for hemochromatosis is to measure liver iron concentration in subjects with pre-known increased parameters of iron metabolism (serum iron, serum ferritin). Due to absolute non-invasive character of the BOS-method, the indication for such a magnetic biopsy can be made very widely. Another way is to check family members of subjects with known homozygous hemochromatosis.

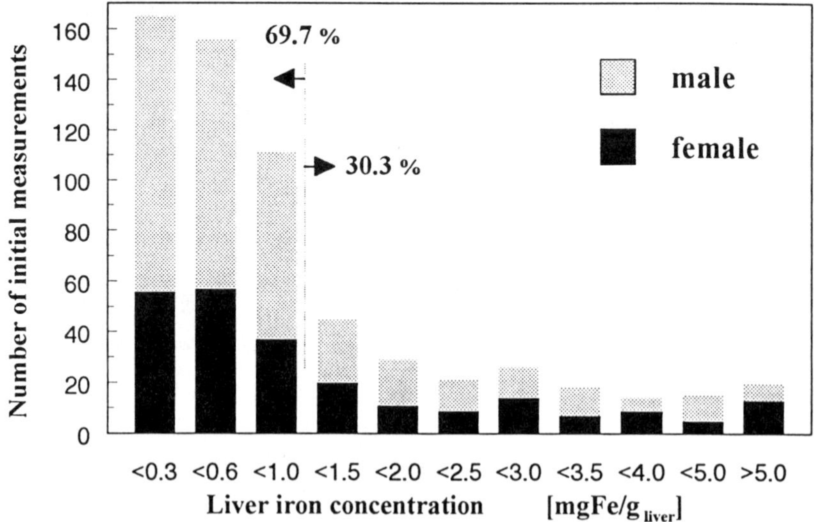

Fig. 1 Frequency distribution of liver iron concentrations measured by SQUID biosusceptometry in patients suspected for iron overload (n = 621).

Under the mentioned approach, more than 600 patients, suspected for hemochromatotic iron overload, were investigated with the biomagnetic iron quantification (Fig. 1). Pre-treated homozygotes or patients with secondary siderosis were excluded from this study. Out of these patients, 86 were diagnosed as homozygotes for hereditary hemochromatosis. As a consequence of this diagnosis, these patients underwent an iron depletion therapy by weekly phlebotomies, until the iron concentration in the liver reaches normal values (0.1 - 0.5 mg Fe/g$_{liver}$). The effect of phlebotomies (roughly 0.05 mgFe/g$_{liver}$ for a 500 ml venesection) was monitored during the therapy by BOS [3]. After the successful depletion of the excessive iron stores, a maintenance therapy with regular phlebotomies must be performed life-long. For this phase of long-term treatment of hemochromatosis patients, the BOS-method is a valuable tool to avoid reverted iron accumulation.

Following family members of 86 identified homozygous carriers, 73 relatives were found (parents, children) which are obligate heterozygote due to the autosomal recessive mode of inheritance in this disease. The results from BOS demonstrated the significant lower amount of liver siderosis in this group.

In the remaining group, in which no relevant liver siderosis was present, a homozygous idiopathic hemochromatosis was excluded although, one or two blood parameters of iron metabolism were occasionally or permanently increased (serum-iron, serum-ferritin, transferrin-iron saturation). In one part of this group, secondary factors were found (acute liver cell damage, oestrogen medication, vitamin-B12-deficiency etc) which are known to influence parameters of iron metabolism. In the absence of secondary factors, a subgroup of 151 subjects was classified as suspected heterozygous carriers.

Subjects, in which a conclusive diagnosis (homozygous/heterozygous for hemochromatosis) was not possible at the moment, should be followed within the next years in order to detect or exclude changes in the liver iron concentration.

II. Transfusional hemosiderosis

Monitoring of iron chelation therapy is the main purpose of SQUID biosusceptometry in patients with posttransfusional siderosis and/or iron loading anemias (ß-thalassemia major, - intermedia, sickle cell anemia) were monitored by BOS (n = 39, age: 4-33 y). Liver and spleen iron concentrations ranged from 0.4 to 13 $mgFe/g_{liver}$ and 0.3 to 3.7 $mgFe/g_{spleen}$ respectively. The correlation of serum ferritin with the liver iron concentration was very poor in patients under treatment with parenteral DFO-injections (n = 33) and/or oral DFP (19) as shown in Fig. 2. The organ volumes were measured by ultra sound scanning. The total body iron stores were estimated by assuming that 90 % of the iron stores are present in liver and spleen.

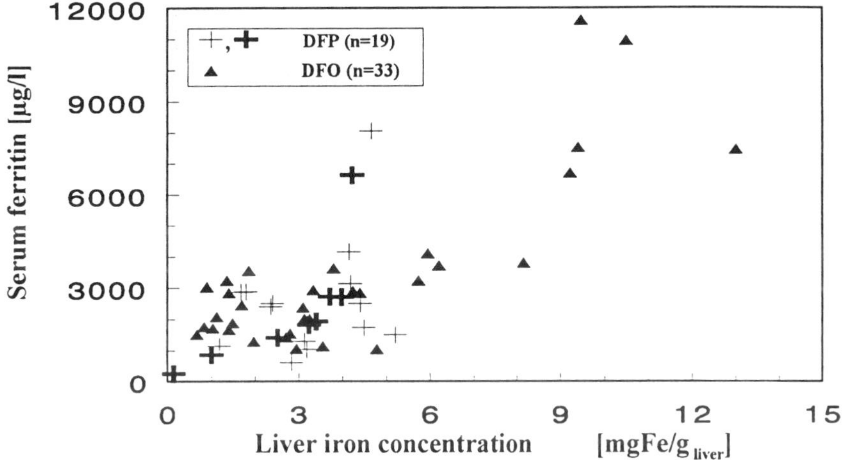

Fig. 2 Serum ferritin versus liver iron concentration measured by BOS in patients with secondary iron overload. Crosses represent patients under treatment with DFP (thin crosses for 0.2 - 2 years, bold crosses for 2- 5.5 years.

Repeated measurements of liver and spleen iron concentrations as well as sonographical determination of liver and spleen volume were available in six patients under DFP treatment for 3 - 15 month. In this group, detailed information was obtained on the whole body iron store (5 - 22 g) and the iron excretion rates (10 - 43 mg/d) for each patient. As indicated by decreasing liver iron concentrations, 4 of 6 subjects showed a negative iron balance with iron excretion rates of 70 - 84 mg/d, which could be assessed by a 2-compartment model, taken into account the blood transfusion rate (influx of iron) and the actual iron stores [4].

The non-invasive measurement of liver (spleen) iron offers some advantages in the therapy control of patients under chelation therapy. From BOS and sonographic imaging, a precise and reliable information on the iron balance is available and an individual iron chelation therapy can be assessed. This is of special value in the clinical evaluation of new oral iron chelators. For the individual patient, the reliable information on his actual iron overload may help to increase the acceptance of chelation therapy and thus may stabilize the compliance.

III. Screening in prospective blood donors

In a study on the prevalence of hereditary hemochromatosis in North-Germany, 2812 prospective blood donors (mean age: 26y, female = 1409, male = 1403) from our local blood transfusion department were screened. The screening was performed in a 3-step filter for elevated serum iron (> 180 µg/dl), serum ferritin (> 100 µg/l, female; > 200 µg/l, male), transferrin saturation (> 50 %) and erythrocyte ferritin (> 50 ag/RBC).

At the 3rd filter level, 63 subjects suspected for iron overload were measured by BOS. The 3rd filter was passed by 20 subjects with iron concentration > 0.5 mgFe/g$_{liver}$.

The final diagnosis was made by taking into account the increased liver iron determined by BOS and the finding of an increased intestinal iron absorption (> 50 %) from a 10 μmol ^{59}Fe(II)ascorbate test dose. As a result of this study, 7 cases of homozygous hemochromatosis (without clinical symptoms) were found. This confirms the current frequency estimation (0.2 - 0.6 % in Western countries) also for North-Germany. The benefit from the screening by BOS is the improved health care in the presymptomatic state of hemochromatosis. Moreover, these persons suffering sofar only from a disregulation of iron absorption, may be engaged as "super blood donors".

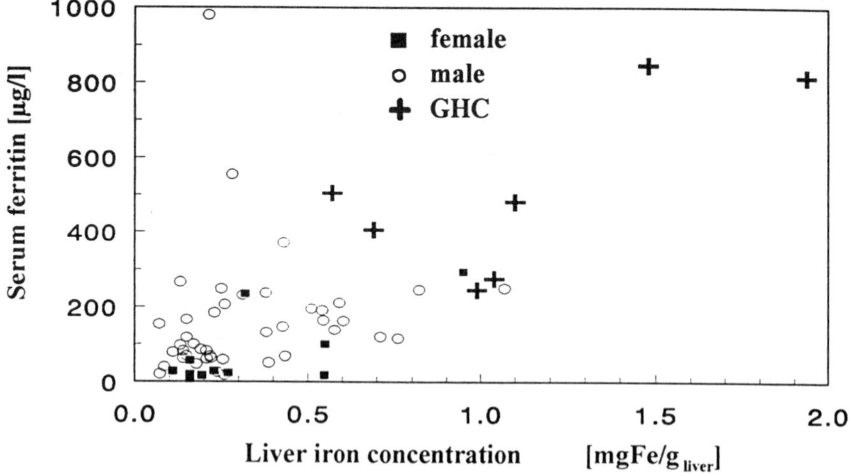

Fig. 3 Serum ferritin versus liver iron concentration in the third-step-filter. The crosses indicate subjects diagnosed as homozygous hemochromatosis.

In conclusion, the Hamburg Biosusceptometer is a dedicated system for non-invasive liver and spleen iron quantification over the whole range of possible iron stores (deficient to overload). For the diagnosis of genetic hemochromatosis, BOS substitutes the conventional liver needle biopsy completely. The method is precise, practicable and convenient for patients from 3 to 80 years and gives the clinicians relevant information for diagnoses, therapy and prognosis in iron overload diseases.

The results of our studies should have a strong impact on the design of the next generation biosusceptometer. It should be optimized for a better spatial resolution and for measurements in the low iron concentration range, because about 70 % of our patients had liver iron concentrations below 1 mgFe/g$_{liver}$ (Fig. 1).

References:

[1] Powell, L.W. In: Brock. J.H., Halliday, J.W., Pippard, M.J. and Powell, L.W. Iron metabolism in health and disease, London, Saunders,1994.

[2] Fischer, R., Eich, E, Engelhardt, R., Heinrich, H.C., Kessler, M. and Nielsen, P. The calibration problem in liver iron susceptometry . In: Williamson, S.J. et al. (ed), Advances in Biomagnetism, Plenum Press, New York, pp 501-504, 1989.

[3] Fischer, R., Engelhardt, R., Nielsen, P., Gabbe, E.E., Heinrich, H.C., Schmiegel, W.H. and Wurbs, D. Liver iron quantification in the diagnosis and therapy control of iron overload patients. In: Hoke, M. et al. (ed), Biomagnetism: Clinical Aspects, Elsevier, Amsterdam, pp 585-588. 1992.

[4] Nielsen, P., Fischer, R., Engelhardt, R., Tondüry, P., Gabbe, E.E. and Janka, E.E. Liver iron stores in patients with secondary haemosiderosis under iron chelation therapy with deferoxamine or deferiprone. Brit.J. Haematol.1995, 91:827-833.

The Use of Biomagnetic Liver Susceptometry in the Ferrara-Hamburg-Turin Study on Thalassemia

Fischer, R.[1], Piga, A.[2], Tricta, F.[2], Nielsen, P.[1], Engelhardt, R.[1], Garofalo, F.[2], Di Palma, A.[3] and Vullo, C.[3]

[1]Medizinische Biochemie, Universität Hamburg, Germany; [2]Centro Microcitemie, Università di Torino, Italy; [3]Divisione Pediatrica, Arcispedale S. Anna, Ferrara, Italy

Introduction

Thalassemia is a genetic disease which requires lifelong blood transfusions resulting in progressive iron overload. This disease is rare in Northern Europe, but frequent in Mediterranian and Asian countries (e.g. 6000 patients in Italy). Current treatment for iron overload is by subcutaneous administration of the iron chelator deferoxamine (DFO) or in clinical trials, by the oral iron chelator deferiprone (DFP). In some patients where histocompatible donors are available also bone marrow transplantation (BMT) can be used. Most of these transfusion-dependent thalassemic patients have been chelated for more than 14 years at an age of 20. In these patients it has become particularly important to determine as precisely as possible the level of iron stores so as to take into account:

- the individual adjustment of chelating therapy;
- the prevention of the toxic effects of chelation attributed to DFO, mainly in young patients;
- the assessment of compliance with iron chelation therapy, as this is strictly correlated to survival.

Patients and Methods

In the present work we report on measurements of the iron stores by SQUID biomagnetic liver susceptometry (BLS) in 175 thalassemic patients from Northern Italy (Turin and Ferrara) (age: mean 18 y, range 4 - 38 y). Biomagnetic spleen susceptometry (BSS) was also possible in 50 % of the 73 nonsplectomized patients with enlarged organs. The main diagnoses were ß-thalassemia major (n=150), combined sickle cell anemia / ß-thalassemia (n=4), and ex-thalassemia after bone marrow transplantation (BMT) 1 to 7 years ago (n=18).

Measurements took place at 6 instants from December 94 to Oct 95 at the Hamburg Biosusceptometer. The system (Ferritometer: BTi, San Diego) with 2nd order gradiometers (2 channels) and rf-SQUIDS in the nonuniform field of 1st order field coil gradiometers (2 channels) has been described elsewhere [1]. Sonography (positioning, liver/spleen contour and volume estimation) in one patient and biomagnetometry in another were accomplished simultaneously at different places with 2 patients/hour. The patient's position was transfered from one place to the other by laser-cross alignment.

Full data analysis was done by forward modeling of the magnetic flux change from spherical organ/torso geometries as function of detector coil distance. Magnetic volume susceptibilities of liver/spleen (i.e. liver/spleen iron concentration) and body (torso) tissue were fitted to SQUID voltages as function of precalculated fluxintegrals (spherical torso-liver model) [2]. For online analysis the magnetic body susceptibility (χ_{tis}) was derived in 1st order approximation from the patient's body mass index (BMI = height/weight^2) as: χ_{tis} = (-9.130 + 0.805/ [1+226/(BMI-13.3)^2])$\cdot 10^{-6}$ [SI] (unpublished data). The liver iron concentration was calculated from a linear fit to SQUID voltages as function of fluxintegrals from a hemispherical liver after subtraction of the magnetic tissue contribution from a cylindrical torso (range: 13 to 31 kg/m^2, corresponding with χ_{tis} = (-9.13 to -8.66)$\cdot 10^{-6}$ SI).

Clinical evaluation of biomagnetometry liver and spleen iron concentration was made by serum ferritin (SF), total transfused iron input, total infused DFO, mean annual compliance, total chelated iron extrapolated from periodical measurement of urinary iron excretion, and by liver biopsy with histological determination of the hepatic iron score (grading: 1 - 20). Mean annual compliance was calculated as mean annual number of daily DFO-infusions from the total number of DFO-infusions(~ 1.5g DFO/infusion) during the years of chelation therapy done by each patient. As each of these 7 parameters were not known in all patients at the same time a group of patients with ß-thalassemia major (n=36) was selected where a linear cross-correlation could be done.

Results and Discussion

Liver iron concentrations ranged from 170 (normal: 100-400) to 9600 $\mu g/g_{liver}$ (severe iron overload) with a median value of 1390 $\mu g/g_{liver}$ and an interquartile range (50%) from 910 to 1990 $\mu g/g_{liver}$. Spleen iron concentrations ranged from 280 to 2350 $\mu g/g_{liver}$ with a meadian value of 1320 $\mu g/g_{liver}$ and an interquartile range from 990 to 1530 $\mu g/g_{liver}$. The ratio of spleen/liver iron concentration was found to be 1.21 ± 0.76. Approximatively, one could apply liver iron concentration values to those spleens which were not measured. At least 75 % of our patients have iron stores well below the levels associated with complications typical in thalassemics (e. g. myocardiopathy, liver fibrosis and cirrhosis, diabetes mellitus, hypothyroidism). Online liver results differed by 5 ± 11 % from more precise results obtained by spherical torso-liver modeling. This was different in BSS where geometry mapping for small-sized spleen ellipsoids had to be done first by sonography in order to apply individual fluxintegrals.

Serum ferritin is the routine biochemical parameter used in the monitoring of individual therapy in thalassemic patients. Although this parameter is strongly influenced by inflammation and tissue damage processes, it is still used to estimate individual body iron stores and to monitor iron mobilization. The limited use of serum ferritin determinations for this purpose was already shown in biomagnetic measurements by Brittenham et al [3] and Nielsen et al. [4] and is demonstrated on a large scale in Fig. 1. Standard logarithmic transformation has been applied to the skewed distributions of serum ferritin and liver iron concentrations. The poor correlation coefficient of r=0.52 indicates that only 27% of serum ferritin values are within the ± 1 s.d. range of the respective regression line. Even in the interquartile range of liver siderosis (910 - 1990 $\mu g/g_{liver}$), serum ferritin values ranged from 81 to 5329 $\mu g/l$ (normal: 35 - 235).

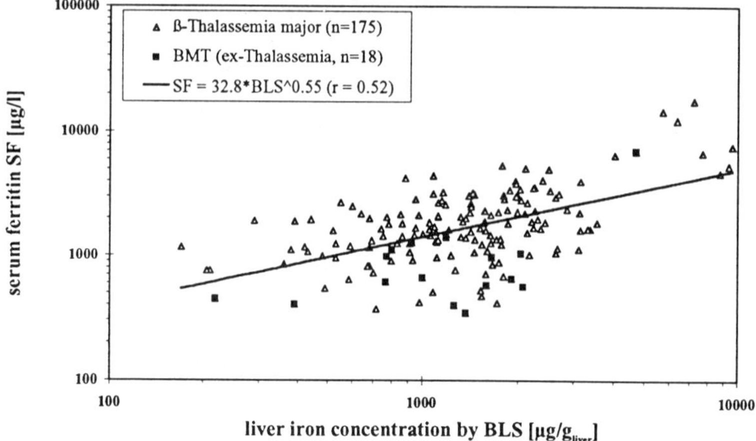

Fig. 1   Correlation between serum ferritin and liver iron concentration in patients with ß-thalassemia major (*triangles*) and in ex-thalassemic patients after bone marrow transplantation (BMT: *squares*). Liver iron was measured by SQUID biomagnetic susceptometry (BLS).

Patients after bone marrow transplantation have still elevated iron stores between 218 and 4700 $\mu g/g_{liver}$, whereas the corresponding serum ferritin values are relatively low (see Fig. 1). The reduction of iron stores after BMT needs further investigation. Biomagnetometry may be a very useful tool to study this problem in relation to the other influencing parameters as blood transfusions, DFO infusions, and time after BMT.

In 30 % of the patients (n=50) also liver biopsies were executed. Qualitative estimation of the degree of liver iron siderosis was performed by standard staining techniques in the corresponding histological sections. Hepatic iron scores from 3 to 18 were registered. The interquartile range was between 8 and 13 with a median value of 11. For the linear correlation with liver iron concentrations measured by BLS a correlation coefficient of r = 0.65

was achieved. The physico-chemical measurement of iron concentration in biopsy specimens was lacking from several problems: (i) the size of several biopsies was below 3-5 mg$_{w.w.}$, (ii) a variety of patients are affected by hepatitis with fibrotic and cirrhotic infiltration of liver tissue, (iii) the portion of the biopsy specimen assigned for reliable chemical analysis was not inspected by a pathologist with respect to cirrhosis, substantial scarring, tissue from the capsule, or iron-free foci [5]. Only a poor correlation was achieved with LIC-BLS, which was even worse than with hepatic iron scoring. However, the discrepancy between in-vitro physico-chemical quantitation of biopsies and in-vivo measurement of liver iron stores by SQUID biosusceptometry in thalassemic patients needs further exploration. As the calibration of the Hamburg Biosusceptometer is based on physical procedures only (air-water susceptibility difference of an object with known geometry), and as this is in contrast to findings in genetic hemochromatosis [6], the iron quantitation in the liver biopsy as the accepted reference standard may be question in this kind of patients. Moreover, a recent investigation in autopsy livers from two patients with ß-thalassemia has shown a striking variability of hepatic iron between samples [7].

In the selected group of 36 patients with ß-thalassemia major a linear cross-correlation was performed with all parameters typically used in therapy monitoring of these patients (see Tab. 1). After several years of therapy well chelated patients are in a steady-state with the total amount of transfused blood/iron (tot. trsf. Fe), the total amount of infused deferoxamine (total DFO), and the total amount of chelated iron (tot. chel. Fe). This is indicated by their high correlation coefficients between each two of the respective parameters in Tab. 1. The correlation of liver iron concentration measured by SQUID biosusceptometry (LIC-BLS) with all other parameters was surprising:

- after years of chelation therapy, the total amount of transfused blood (iron) does not determine the degree of iron overload any longer (r = 0.26);
- on the other side, the chelation therapy (i. e. total amount of DFO infusions) does influence the liver iron stores to a certain degree (r = -0.64);
- urinary iron excretion correlates only poorly with liver iron stores (r = -0.24), which can be explained by fecal iron as main excretion pathway in patients kept at high hemoglobin values [8];
- the relative close correlation coefficient with serum ferritin (r = 0.76) in contrast to the correlation test with all thalassemic patients (r = 0.52!) is explained by the skewed distribution of serum ferritin, where only few patients with high ferritin values determine the slope of the corresponding regression line;
- correlation with grading of liver iron overload by hepatic scoring (r = 0.68) was similar to the correlation test in the larger group of biopsied patients (r = 0.65!);
- surprisingly, the closest correlation was found for the mean daily number of DFO infusions during the years of individual chelation therapy, i. e. compliance. The relation for compliance as function of liver iron concentration was found as COMP = 1 - 0.2·LIC [mg Fe/g$_{liver}$]. Values of compliance better than 0.9 were found only for patients with LIC-BLS < 1000 µg Fe/g$_{liver}$.

Tab. 1 Linear cross-correlation between clinical parameters used for therapy monitoring in selected patients with ß-thalassemia major (n=36) and iron concentrations measured by biomagnetic liver susceptometry [LIC-BLS](r* significant at p < 0.05).

Parameter	tot. trsf. Fe	total DFO	compliance	tot. chel. Fe	ferritin	hep. score	LIC-BLS
tot. trsf. Fe	1	0.89*	0.14	0.90*	0.18	0.12	-0.26
total DFO	0.89*	1	0.55*	0.82*	0.51*	0.37*	-0.64*
compliance	0.14	0.55*	1	0.12	0.76*	0.66*	-0.94*
tot. chel. Fe	0.90*	0.82*	0.12	1	0.16	0.17	-0.24
ferritin	0.18	0.51*	0.76*	0.16	1	0.50*	0.76*
hep. score	0.12	0.37*	0.66*	0.17	0.50*	1	0.68*
LIC-BLS	0.26	0.64*	0.94*	0.24	0.76*	0.68*	1

The survival of patients is strongly correlated to compliance [9, 10], e. g. a mean annual compliance of at least 0.68 DFO infusions per day (~ 1.0 g DFO/d) is necessary to achieve a survival of 95 % beyond an age of 20 years. This corresponds with a liver iron concentration of less than 2000 $\mu g/g_{liver}$. As the accurate determination of compliance is difficult to achieve, annual measurement of liver iron concentration by SQUID biosusceptometry may become a simple and useful method for this purpose. Moreover, BLS may also be used as a reliable method for the cross-check of compliance and thus, may improve the patient's compliance. The closest correlation between compliance and biomagnetometry could be explained by the high specificity and sensitivity of the SQUID biomagnetometry method.

In conclusion, the study has shown that SQUID biomagnetic liver susceptometry can be applied routinely to all kinds of thalassemic patients, even in measurement sessions where a high daily throughput of patients is required. In contrast to standard current methods, more reliable information can be obtained from BLS. Annual assessment of liver iron by BLS could become the method of choice for evaluation of the iron burden in transfusion dependent thalassemia patients and for the assessment of chelating therapy. This study may help to enhance the clinical acceptance of this technology and may help to develop a next instrument generation at one of the thalassemia centers in Italy.

Acknowledgements

We thank our co-workers R. Kongi, I. Zimmermann, C. Albert, S. Hoppe, B. Probst, K. Reller, B. Dresow, E.E. Gabbe (Hamburg), M.R. Gamberini, R. Govoni, M. Sprocati (Ferrara), and U. Bertola, R. Pimazzoni, L. Sacchetti (Torino), who were involved in the measurements and in taking care of the patients with enthusiam. Moreover, we thank our patients from Ferrara and Torino for accepting the challenge to go for a SQUID measurement to Amburgo. We would also like to thank the patient's associations for thalassemia: Associazione del Bambino Talassemico and Red Cross of Torino, who supported this investigation.

References

[1] Paulson, D.N., Fagaly, R.L., Toussaint, R.M., and Fischer, R. Biomagnetic susceptometer with SQUID instrumentation, IEEE Transactions on Magnetics, 1990, MAG-27: 3249-3252.

[2] Fischer, R., Eich, E., Engelhardt, R., Heinrich, H.C., Kessler, M., and Nielsen, P. The calibration problem in liver iron susceptometry, In: Williamson, S.J., et al. Advances in Biomagnetism, New York, Plenum Press, 1990.

[3] Brittenham, G.M, Cohen, A.R., McLaren, C.E., Martin, M.B., Griffith, P.M., Nienhuis, A.W., Young, N.S., Allen, C.J., Farrell, D.E., and Harris, J.W. Hepatic iron stores and plasma ferritin concentration in patients with sickle cell anemia and thalassemia major. American Journal of Hematology, 1993, 42: 81-85.

[4] Nielsen, P., Fischer, R., Engelhardt, R., Tondüry, P. Gabbe, E.E., and Janka, G.E. Liver iron stores in patients with secondary haemochromatosis under iron chelation therapy with deferoxamine or deferiprone, British Journal of Haematology, 1995, 91: 827-833.

[5] Ludwig, J., Batts, K.P., Moyer, T.P., Baldus, W.P., Fairbanks, V.F. Liver biopsy diagnosis of homozygous hemochromatosis: a diagnostic algorithm. Mayo Clinic Proceedings 1993, 68: 263-267

[6] Fischer, R., Engelhardt, R., Nielsen, P., Gabbe, E.E., Heinrich, H.C., Schmiegel, W.H., and Wurbs, D. Liver iron quantification in the diagnosis and therapy control of iron overload patients, In: Hoke, M., et al. Advances in Biomagnetism '91, Amsterdam, Elsevier, 1992.

[7] Ambu, R., Crisponi, G., Sciot, R., VanEyken, P., Parodo, G., Iannelli, S., Marongiu, F., Silvagni, R., Nurchi, V., Costa, V., Faa, G., and Desmet, V.J. Uneven hepatic iron and phosphorous distribution in beta-thalassemia, Journal of Hepatology, 1995, 23: 544-549.

[8] Di Palma, A., Moratelli, A., Tolomelli, P., Giuberti, P., Tenan, R., Fagioli, F., Grandi, L., Toffoli, C., Atti, G., and Vullo, C. Pattern of iron excretion in relation to haemoglobin level and iron load in 8 Haematological patients following the administration of subcutaneous deferioxamine. European Journal of Haematology, 1994, 53: 197-200.

[9] Brittenham, G.M, Griffith, P.M., Nienhuis, A.W., McLaren, Young, N.S., Tucker, E.E., Allen, C.J., Farrell, D.E., and Harris, J.W. Efficacy of deferoxamine in preventing complications of iron overload in patients with thalassemia major. New England Journal of Medicine, 1994, 331: 567-573.

[10] Gabutti, V. and Piga, A. Results of long-term iron chelating therapy. In: Hershko, C. Chelators in Medicine, Basel, Karger, 1996 (in press).

Detection of Biodegradation of Ferrofluid by Relaxation Measurements

Kötitz, R.[1], Trahms, L.[1], Pfefferer, D.[2], Kresse, M.[2], Semmler, W.[2] and Weitschies, W.[2]

[1]*Physikalisch-Technische Bundesanstalt, Berlin, Germany;* [2]*Institut für Diagnostikforschung GmbH, Berlin, Germany*

Introduction

Ferrofluids are colloidal solutions of ferro- or ferrimagnetic particles of typically 10 nm size, coated with a surfactant and dispersed in a carrier liquid. Due to their size these particles are single domain and magnetized to saturation. Ferrofluids are superparamagnetic i.e. gain a very high magnetization in the presence of a magnetizing field. Thus the high sensitivity of SQUID sensors allows the detection of low concentrations of these particles. Fig.1 gives an overview of the components and some relevant parameters of ferrofluids.

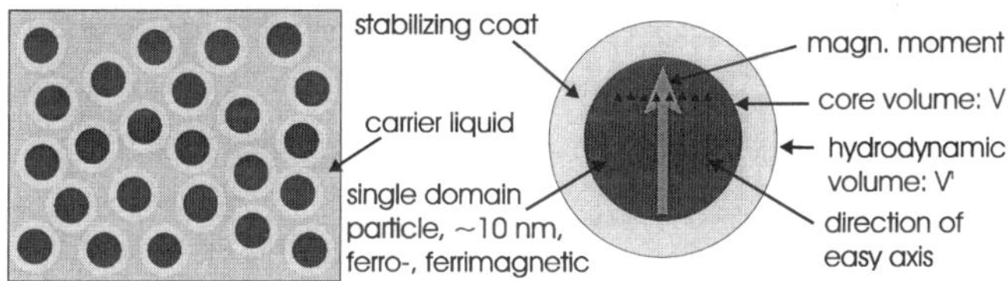

Fig.1 Some basics about ferrofluids

Ferrofluids were developed in the 1960's mainly for industrial applications and are now widely used in a variety of technical applications. During the past years their application in biomedicine has become increasingly important. When bound with specific substances they are commonly used as carriers for drug targeting and for magnetic cell separation. Furthermore ferrofluids are established as contrast agents in MRI investigations.

A new application is their use as magnetic markers for biomagnetic measurements [1,2]. After the injection of dilute ferrofluids into the body it is useful to determine the distribution of magnetic particles in the body tissue. This is usually done in analogy to susceptibility maps where the magnetization of the sample is measured in the presence of a magnetizing field. However biomagnetic susceptibility measurements are often complicated by the presence of diamagnetic tissue or paramagnetic contamination which especially at low ferrofluid concentrations mask the superparamagnetism of the injected particles [2]. Moreover when using SQUID sensors usually the signal hast to be modulated either by sample or sensor movement or variation of the magnetizing field, in order to generate a time dependent signal.

Here we present **MagnetoRelaXometry** (MRX) as a novel measurement technique which is based on the relaxation processes in ferrofluids. The principles of MRX are shown in Table 1.

Table 1: Principles of MRX

0. incorporation of fine magnetic particles into the object under study
1. magnetization of the object for a time of t_{mag} in a field of H_{mag}
2. switching off the magnetizing field, H=0
3. measurement of the relaxation of the sample magnetization, M=M(t)

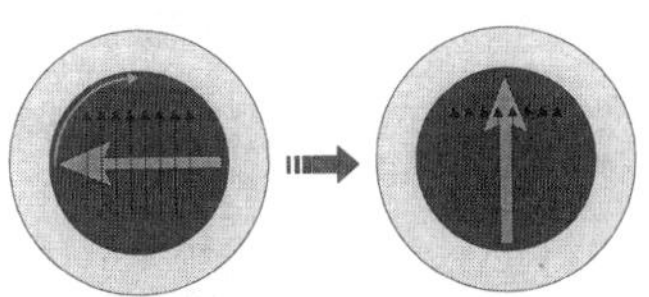

Fig. 2 Principle of Néel relaxation

Here we focus on the Néel relaxation mechanism which is characterized by the rotation of the internal magnetization vector inside the particle [3]. This is schematically shown in Fig. 2. The Néel relaxation process is characterized by a time constant of $\tau_N = \tau_0 \exp(KV/k_BT)$, where τ_0 is usually taken as 10^{-9} s, V is the core volume of the magnetic particle and K the anisotropy constant of the bulk material, k_B Boltzmann's constant and T temperature.

Due to the distribution of particle sizes, the Néel relaxation of magnetic systems after magnetization in a weak field for a finite time obeys a universal time dependence[5]:

$$M \sim \log(1 + t_{mag}/t) \qquad (1)$$

So we face a measurement signal which is determined only by the magnetization parameters. Thus signal processing methods like triggering, averaging and curve fitting can easily be adapted to this universal time dependence to increase the signal to noise ratio.

Since the measurement signal has an inherent time dependence SQUID detection is possible without modulation of the sample signal. In contrast to susceptibility investigations MRX measurements are not affected by para- or diamagnetic components within the body [2] and therefore yield a signal specific only for the applied marker.

In a simple phantom experiment it was shown that the spatial distribution of ferrofluid can be reconstructed by measuring its Néel-relaxation [1]. In this paper we report the use of MRX for studying the accumulation and degradation of ferrofluids in biological tissue.

Methods

We used 2 water based ferrofluids consisting of magnetite cores of about 10 nm with different coatings (dextran and silan) resulting in hydrodynamic sizes of 40 - 60 nm. The measurements were performed with the PTB 37 channel dc SQUID Magnetometer [4] in the Berlin Magnetically Shielded Room (BMSR). The application of a magnetizing field inside the shielded chamber causes problems since (i) the shielded room must not be magnetized permanently and (ii) the ferromagnetic material of the chamber walls generates relaxation signals that interfere with the sample signal. Therefore a specially designed twin coil was used which consists of a magnetizing and a compensation coil connected in series opposition. The magnetic fields of the individual coils add at the location of the sample and cancel at a distance. The design is depicted in Fig. 3.

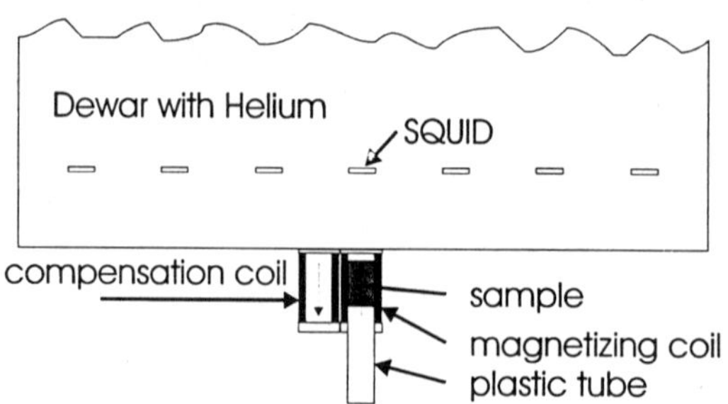

Fig. 3 Setup for magnetizing the samples

The current feeding the coil is software controlled. All samples were magnetized for 10 s in a field of 5 mT. Relaxation was measured between t_0 = 200 ms and t_e = 90 s after switching off the field. An electronic gradiometer cancels artifacts, that are still generated by the relaxation of the chamber wall.

Results

In order to estimate the sensitivity of the MRX technique different dosages of dextran coated magnetite (100, 50, 10 and 2 µmol Fe/kg body weight, and for control, 0 µmol Fe/kg body weight) were injected into rats (Han-Wistar SPF, 100 g, male). Samples of liver and spleen were harvested 4 h post injection (p.i.) and investigated by MRX. Fig. 4 shows the relaxation signal of a liver sample observed for the lowest applied dosage of 2 µmol Fe/kg. The amplitude of the relaxation signal is about 3 pT / 515 mg. Note that this is the result of a single shot, filtered by a low pass of 0.5 Hz.

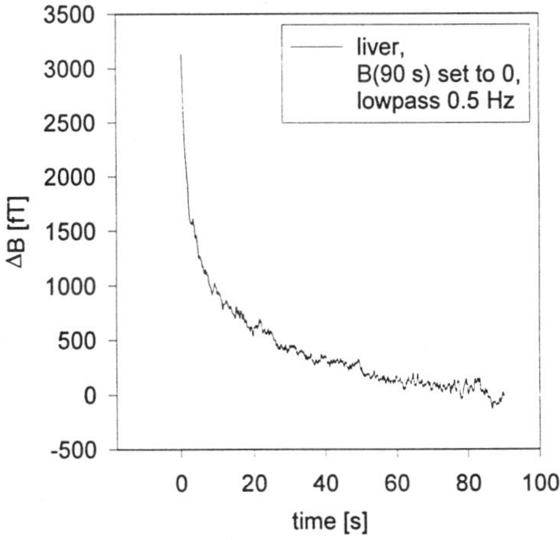

Fig. 4 Relaxation of liver sample after application of 2 μmol Fe/kg ferrofluid

There was no signal detectable for reference samples of untreated animals. This confirms the specificity of the relaxation signal which is independent of the natural iron background of the organs under investigation. For comparison the relaxation amplitude ΔB was reduced to the mass of the samples. As expected the time dependence of the relaxation was similar for all samples containing ferrofluid, according to eq. (1).

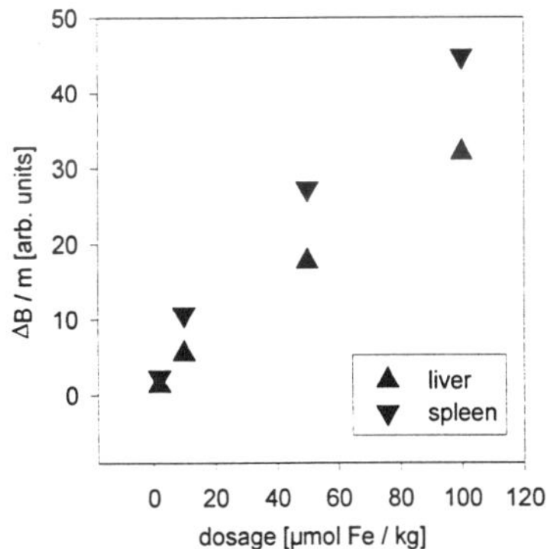

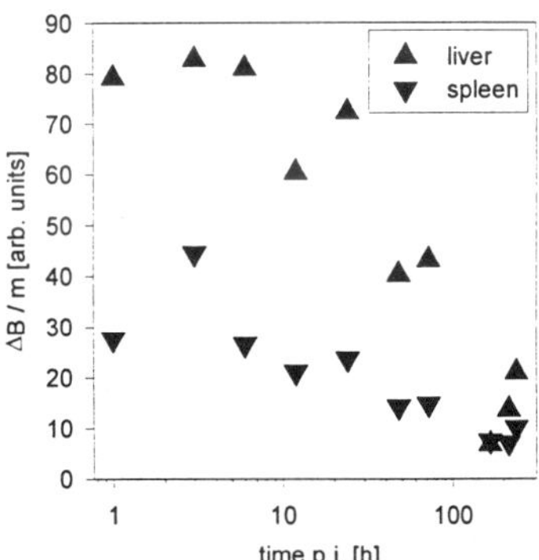

Fig.5 Dependence of reduced relaxation amplitude on applied dosage ferrofluid

Fig. 6 Dependence of reduced relaxation amplitude on time after application of 20 μmol Fe / kg ferrofluid

Fig. 5 shows that the observed reduced relaxation amplitudes $\Delta B = B(t_0) - B(t_e)$ in both organs were proportional to the applied dosages. Since it can be assumed that the relaxation amplitude is proportional to the amount of particles in the sample it is a measure for the accumulation of ferrofluid in the particular organ.

In order to study the biodegradation of the ferrofluid particles, samples of rat liver and spleen from 1 h p.i. to 10 d p.i. were investigated by MRX. The injected ferrofluid consisted of a suspension of silan coated magnetite. The given dosage was 20 μmol Fe / kg body weight. The result is depicted in Fig. 6.

In both organs the maximum relaxation signal was found 3 h after injection. For later time points the relaxation signals decrease. This indicates a biodegradation with a time constant of about 3 days. It is similar for both organs.

However the relaxation amplitude and hence the amount of particles accumulated in the different organs varies for particles with different coatings. Silan coated particles accumulate to a higher extend in the liver than in the spleen whereas the spleen accumulates more dextran coated particles per gram of organ than the liver. This is possibly due to the different surface properties of the coating.

Discussion

The MRX method presented here is a novel tool for the highly sensitive determination of the accumulation, organ distribution and biodegradation of injected ferrofluids in biological tissue. Due to the specificity of the relaxation signal there is no interference with diamagnetic tissue susceptibilty or paramagnetic components. Consequently MRX is not influenced by the high natural iron background of the organs under investigation. The use of radioactive markers is avoided.

The application of the MRX technique for an *in vivo* study will be a further challenge.

References

[1] Kötitz, R., Trahms, L., Koch, H., and Weitschies, W. In: Baumgartner, C., Deecke, L., Stroink, G., and Williamson, S.J. Biomagnetism: Fundamental Research and Clinical Applications, Amsterdam, Elsevier Science, 1995, 785-788.

[2] Thomas, I.M., and Friedman, R.N. In: Baumgartner, C., Deecke, L., Stroink, G., and Williamson, S.J. Biomagnetism: Fundamental Research and Clinical Applications, Amsterdam, Elsevier Science, 1995, 809-813.

[3] Kötitz, R., Fannin, P.C., and Trahms, L., J. Mag. Magn. Mat., 1995, 149:42-46.

[4] Koch, H., Cantor, R., Drung, D., Erné, S.N., Matthies, K.P., Peters, M., Ryhänen, T., Scheer H.J., and Hahlbohm, H.D. IEEE Trans. Magn., 1991, MAG-27: 2793.

[5] Berkov, D.V., and Kötitz, R. Journal of Physics C, accepted for publication.

Distribution and Biotransformation Dynamics of Magnetic Contrast Agents in Organism on Evidence Derived from SQUID-Magnetometry

Mikhailik, O.M.[1], Pankratov, Yu.V.[1], Dudchenko, A.K.[1], Sosnitzky, V.N.[2], Budnik, N.N.[2], Minov, Yu.D.[2], Sutkovoj, P.I.[2] and Bakai, E.A.[1]

[1]*Research SONAR Center for Biotechnical Systems, Ukrainian Acad. Sci., Kiev, Ukraine;*
[2]*Institute of Cybernetics, Ukrainian Acad. Sci., Kiev, Ukraine*

Introduction

The development of magnetic contrast agents is the important and promising field of ferromagnetics use in medicine [1]. In order to outline contrast effect it is possible to apply different techniques, such as MRI and SQUID-magnetometry techniques. SQUID-magnetometry could permit accurate noninvasive localization of contrast agents in biological tissues (Magnetic Source Imaging) [2]. Storage of magnetic particles in tissue changes it magnetic susceptibility which becomes evident when measuring magnetic signal. In MRI the contrast effect is based on the effect of proton relaxation time variation by their interaction with magnetic field generated by particles. Thus the important advantage of the high sensitive magnetometry method as compared to that of nuclear magnetic resonance is the direct coupling of measured magnetic field with particle distribution in space. The lack of qualitative and the absence of quantitative data about the aspects associated with the finely dispersed ferromagnetics injection into organism hampers their use as contrast agents for magnetic tomography. The aim of the present study is the investigation of distribution and biotransformation dynamics of finely dispersed iron and magnetite as a base for magnetic contrast agents in organism of experimental animals by supersensitive magnetometry when coupled with low temperature ESR spectroscopy.

Metods

Finely dispersed iron (FDI) or magnetite suspensions in physiological solution were injected into the eye sine of mice at a dose of 200 mg/kg per animal body weight. The mean diameter of particles with different surface structure was about 80 mm. The surface of particles was stabilised via plasmocondevsation of γ aminopropiltriethoxysilane (sample I), graft polimerization of 2-hydroxyethyl methacrylate, acrylamide and N,N'-methylen *bis*-acrylamide (1:1:0.1) (Specimen II) and divinylbenzene (Specimen III). The magnetic field maps have been registered 15 mm over the mice. The measurements were made with single channel SQUID-magnetometer under the following parameters: sensitivity - 30 fT/Hz$^{1/2}$, frequency band 0.1-100 Hz. The ESR spectra have been obtained on a EPA-10 mini spectrometer (St. Petersburg Instruments, Russia) at temperatures varying from 100 to 200 K and a super high frequency of 9.30 GHz. Metallic iron concentration has been quantifed from the signals of magnetic particles aggregates (g-factors 5-6 and 2-3 and line width 2000-5000 Oe)[3], ferritin concentration - from the signals with g-factors 4.0-4.5 and 2.0-2.5 and line width 300-800 Oe. Low molecular iron concentration has been determined from iron nitrosyl complexes spectrum (g-factor 2.03) after tissue treatment with nitrite. Ferromagnetic signals have been calibrated against ferromagnetic suspensions while paramagnetic ones - with a CuEDTA standard solution and ferritin ESR signals - against Mossbauer data. The Mossbauer spectra have been registered on a conventional constant acceleration NP255 spectrometer with a 150 mCi ^{57}Co source in a chromium matrix.

Results and discussion

Experiments with animals proved the method of high-sensitive magnetometry to be able to register the picture of magnetic field formed by finely dispersed iron distributed in animal organisms (Fig.1A). Registered outside of the body magnetic field is determined mainly not only by concentrations of exogenic magnetic particles but also by their aggregative state. According to our model of ferromagnetic resonance in the aggregates of small

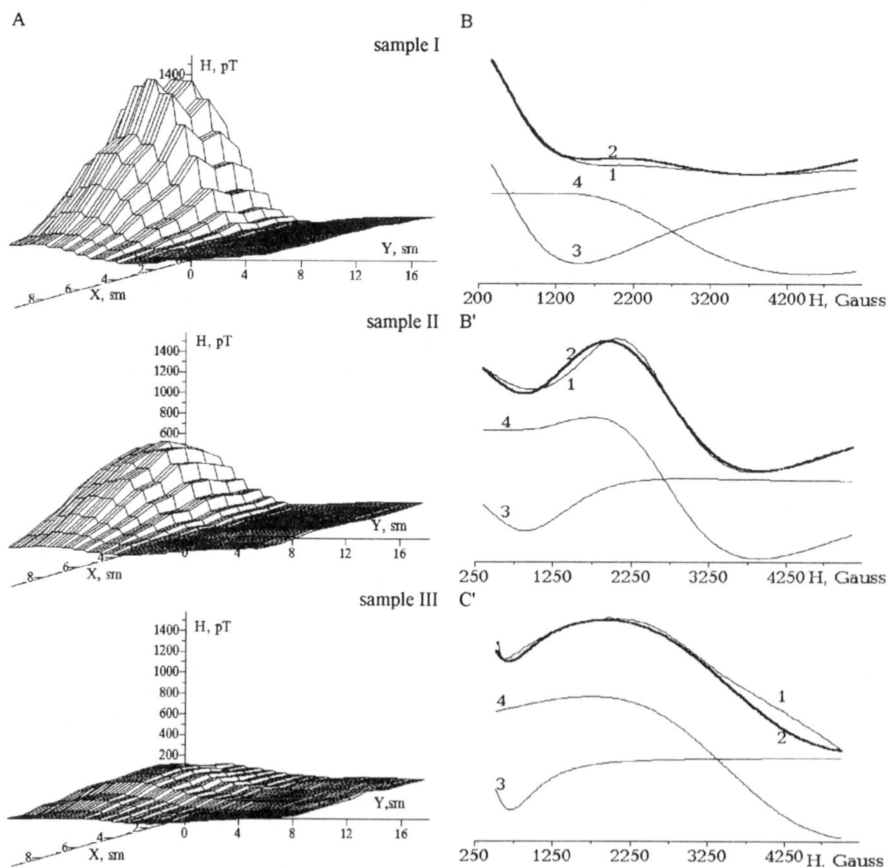

Fig. 1. Magnetic field maps (A) 15 mm over the mice bodies 5 min after intravenous injection of FDI and ESR spectra (B) of suspensions of FDI I,II,III (1 - experimental ESR data; 2 - the sum of curves corresponding to chain (3), spherical (4) aggregates, "flat" layers (5) of metallic iron particles, frequency 9.30 GHz, microwave power 5 mW, modulation amplitude 10 G, T=150K).

magnetic particles the ESR signals corresponding to finely dispersed magnetic particles have been   represented (with an error of 2 - 10%) as the sum of the spectra for particle aggregates of three types: chain, spherical and "flat" layers of particles sorbed on the surface of tissues [3] (Fig.1B).

The type of aggregates is responsible for a g-factor value   whereas their width is dependent on the thickness of a non magnetic layer on the particle surface. The field created by spherical aggregates is much less then the field generated by the chain ones because of random orientation of magnetic moments of particles in spherical aggregates. That is why the same dose injected FDI with greater relative amount chain aggregates generates the higher magnetic field outside the body (Fig. 1).

Using our model of ferromagnetic resonance in the aggregates of small magnetic particles one can obtain quantitative data on content of metallic phase in each form of aggregates. In our experiments, concentration of iron particles in animal organs was 0.1-0.3 mg/ml while magnetic field strength was in a range of 200-1500 pT without external magnetizing field. In experiments on MRI contrasting [4] accumulation of magnetite

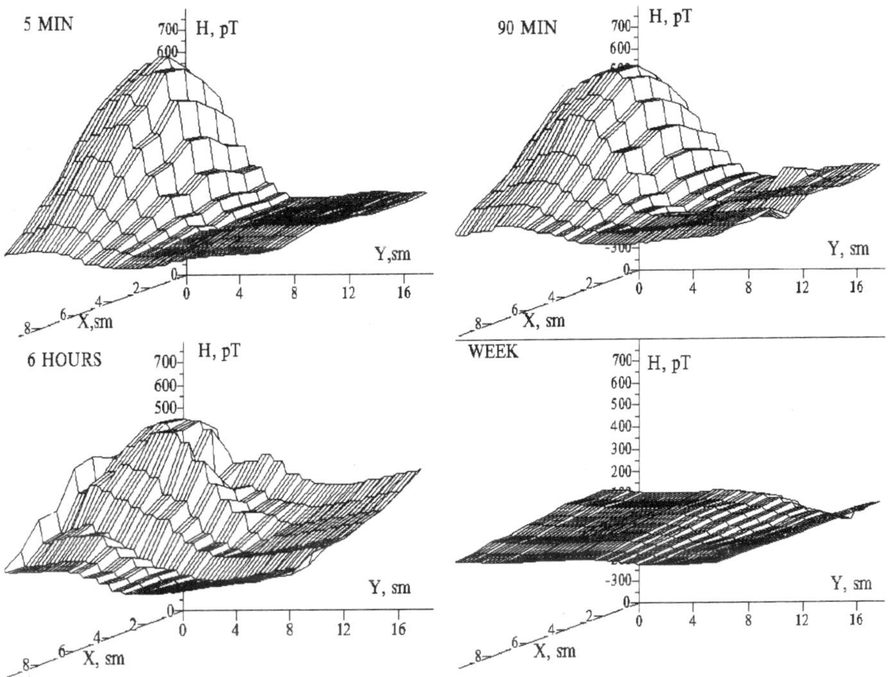

Fig. 2. Magnetic field maps 15 mm over the mice bodies in different times after intravenous injection of FDI suspension in a dose of 200 mg per kg body weight, sample II.

microspheres with concentration 0.13 mg/ml was shown to reduce the T2 relaxation time of liver tissue by 30%. One could believe that SQUID magnetometry could be a real competitor to NMR tomography in detecting of small amounts of ferromagnetic contrast agents.

Our experiments with animals proved the method of high-sensitive magnetometry to be able to register the picture of magnetic field formed by finely dispersed iron and magnetite distributed in animal organisms. It allows also to trace the dynamics of their redistribution even in organism of small animals (mice) (Fig.1). After a week magnetic field over the body decreases almost to the background picture thus supporting the biodegradation of ferromagnetic particles in the organism (Fig.2). ESR results also have verified the rather fast biodegradation of finely dispersed metal iron in organism. In all tissues under investigations a far greater exogenic iron fraction has been detected in ferritin and only a small one (not more than 10%) in 'free' iron pool and a still lesser fraction (not more than 1%) in transferrin. The iron content in ferritin increased approximately 2-10 times as compared to that of check test whereas the concentration of low molecular iron complexes only 1.5 - 3.5 times. Hence any detrimental effects that may be exerted by FDI on a living body due to intensification of free radical oxidation processes are limited by the time of biotransformation.

We have studied also the dependence of the ESR spectra of ferritin in tissues on orientation in a magnetic field after sample freezing in rather weak magnetic field (250 Gauss) (Fig. 3). The variations in the position and shape of lines in ferritin spectra upon rotation of samples about a magnetic field suggest an anisotropic distribution of magnetic moments in the sample volume. Consequently one could believe that ferritin molecules with high iron content can form ordered aggregate structures displaying strong magnetic anisotropy and therefore create rather strong local magnetic field. The local magnetic field of ferritin aggregates have been estimated from experimental ESR spectra using approaches of our model for ferromagnetic resonance in magnetic particles aggregates and was shown to exceed 1000 Oe. The matter is that the shift of low-field component (L) with respect to high-field component (H) may be represented as shift due to the local magnetic field while maximum position and line width of high-field component should be accounted for magnetic anisotropy. Ferritin molecules display magnetic anisotropy, they can be in magnetically ordered or superparamagnetic state depending on temperature and degree of saturation with iron [6]. So, one could suggest that magnetic field generated by ferritin, as in the case of any magnetic particles, is determined mainly not only by concentrations but also by aggregative state and magnetic ordering. These circumstances should be taken into account on iron quantification by tissue susceptometry. Evidently, it the aggregation of ferritin molecules that is the cause of nonlinear dependence of magnetic biopsy data versus liver iron concentration observed in patients with heavy liver iron overload [7].

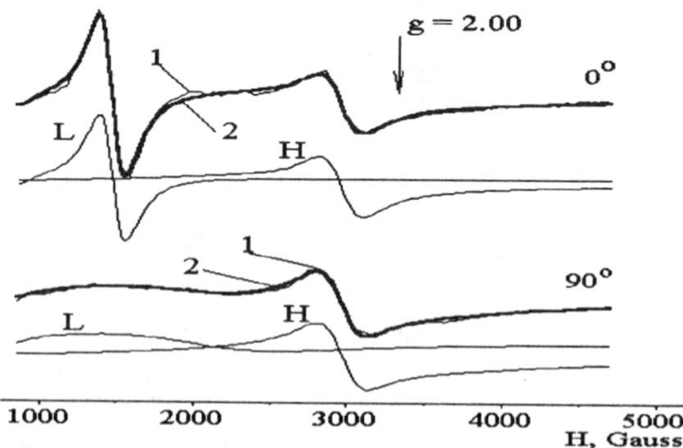

Fig.3. ESR spectra performed on mice liver sample rapidly frozen in a magnetic field of 250 Oe depending on the angle between magnetic field vectors upon sample freezing and spectrum registration moments: (1), thin lines - experimental data; (2), thick lines - the sum of curves corresponding to low-field (L) and high-field (H) components; frequency 9.30 GHz, microwave power 50 mW, T=150 K, modulation amplitude 10 G.

Acknowledgments

This work was supported by the Ukrainian Academy of Sciences as well as by Ukrainian State Committee for Sciences and Technology.

References

[1] Weissleder, R. . Elizondo, G., Wittenberg, J. et al. Ultrasmall superparamagnetic iron oxide: characterization of a new class of contrast agents for MR imagine, Radiology, 1990, 175: 489-93.

[2] Thomas, I.M., Ma, Y.P., Tan, S. et al. Spatial resolution and sensitivity of magnetic susceptibility imaging, IEEE Trans. App. Superconductivity, 1993, 3: 1937-1940.

[3] Mikhailik, O.M., Pankratov, Yu.V. and Bakai, E.A. Biotransformation of intravenously injected finely dispersed iron powders, J. Magn. Magn. Mater., 1993, 122: 379-382.

[4] Majumdar, S., Zoghbi, S.S., Gore, J.C. Pharmacokinetics of superparamagnetic iron-oxide MR contrast agents in the rat, Invest.Radiol.,1990, 25: 771-7.

[5] Smith, A.G., Carthew, P., Francis, J.E. et al. Characterization and accumulation of ferritin in hepatocyte nuclei of mice with iron overload, Hepatology, 1990, 12: 1399-1405.

[6] Kaufman K.S, Papaefthymiou G.C., Frankel R.B., Rosenthal A. Nature of iron deposits on the cardiac walls in β-thalassemia by mossbauer spectroscopy, Bioch.Biophys.Acta, 1980, 629:522-529.

[7] Fisher, R., Engelhardt, R., Nielsen, P. et al. Liver iron quantification in the diagnosis and therapy control of iron overload patients, In: Hoke, M. et al. Biomagnetism: clinical aspects, Amsterdam, Elsvier Science Publishers, 1992.

Human Alveolar Long-Term Clearance in Healthy and Disease Evaluated by Magnetopneumography

Möller, W.[1], Barth, W.[2], Meier-Sydow, J.[3], Pohlit, W.[2], Siekmeier, R.[3] and Stahlhofen, W.[1]

[1]GSF-Institut für Inhalationsbiologie, D-85758 Oberschleissheim; [2]J.-W.-Goethe Universität, Institut für Biophysik, D-60596 Frankfurt, Germany; [3]Medizinische Klinik II, Abt. für Pneumologie, Universität Frankfurt, D-60590 Frankfurt, Germany

Introduction

Magnetopneumography (MPG) was used to investigate alveolar long-term clearance of the human lungs in healthy and disease. Further investigations have shown that this method can be used to record alveolar clearance of the human lungs non-invasively over a period of more than one year [1, 2]. The method of magnetopneumography prevents ethical objections when doing the investigations on humans. Another method for the investigation of alveolar clearance is the radioactive labeling technique. But radionuclides with long half-lives ($\approx$0.5 year) are necessary for alveolar clearance studies. Therefore the radioactive labeling technique is only applicable in animal studies. Further studies have shown that alveolar clearance might be impaired by cigarette smoking. Therefore the data were analyzed with respect to smoking conditions. Additionally patients with chronic obstructive lung disease (COPD), sarcoidosis (SARK) and idiopathic lung fibrosis (FIBR) were investigated. This study gives new information on disturbed alveolar defense mechanisms, that might be a co-factor in the formation of the disease.

Methods

0.5 mg of spherical monodisperse ferrimagnetic iron-oxide particles (Fe_3O_4, 2.9 µm aerodynamic, 1.35 µm geometric diameter, $\sigma_g < 1.1$) were deposited in the lungs by voluntary inhalation. The particles are produced by a Spinning Top Aerosol Generator (STAG) [3]. Particles with 3 µm aerodynamic diameter are deposited mainly in the alveolar region of the lungs. Therefore no or only little fast clearance (within one day) was expected. After deposition the particles were magnetized in an external 100 mT, 30 ms field pulse in homogeneous field technique [2]. All particles together produce a weak remanent magnetic field (rmf), being detected by a SQUID-device. The whole MPG-system is located within a magnetically shielded room, allowing a lower detection limit of 20 µg magnetite in the whole lungs. Before inhalation the natural ferromagnetic contamination of the lung was recorded for each subject and subtracted from all further measurements. For statistical analysis the groups were compared with a ttest, using SAS-software package.

Subjects

A group of younger (age 20 - 39 years) and a group of older (age 40 - 65 years) healthy subjects were involved in the study, both being subdivided into never-smokers (NS) and asymptomatic smokers (S). The mean age and the cigarette smoke consumption are reported in Table 1. Cigarette smoke consumption was classified in pack-years (PY) according to the number of pack's of cigarettes smoked per day times the years of cigarette smoking. At the beginning of the study lung function tests and a detailed questionnaire were performed, allowing to exclude subjects with chronic diseases or persons who are under a permanent drug administration. All subjects gave their informed consent according to a protocol approved by the Ethical Committee of the University Hospital in Frankfurt. Among COPD-patients more than 50 % are ex-smokers (ES). MPG-measurements were performed directly (30 min) as well as 2 days, 1 week, 1 month, 5 months and 9 months after particle inhalation.

Table 1: Age and cigarette smoke consumption (in pack-years, PY) of the subjects involved into the MPG-study on alveolar long-term clearance

Group	N	Age [Yrs]	+/-SD	PY	+/-SD
NS (20-39)	10	29	3	0.0	0.0
S (20-39)	9	31	4	11.7	10.9
NS (40-65)	9	54	5	0.0	0.0
S (40-65)	9	53	9	44.9	22.6
COPD (NS)	1	53		0.0	0.0
COPD (S)	3	61	11	32.0	18
COPD (ES)	6	61	8	42.0	25.6
FIBR (NS)	5	49	13	0.0	0.0
FIBR (S)	7	49	17	14.7	6.4
SARK (NS)	10	46	13	0.0	0.0
SARK (S)	5	43	8	18.3	12.9

Results and discussion

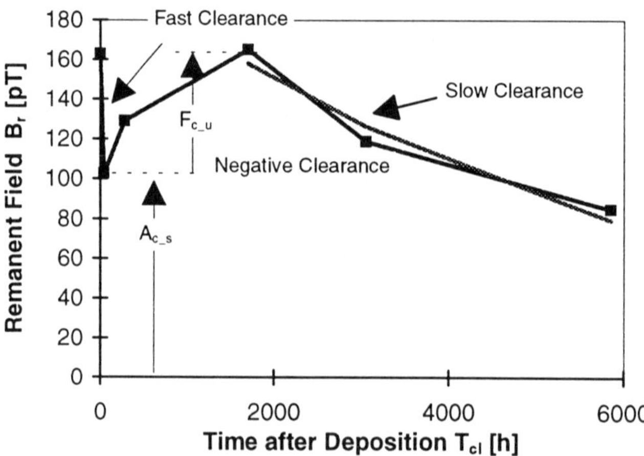

Figure 1: Typical course of the rmf of a healthy subject after deposition of 1.3 µm magnetic tracer particles.

A typical course of the rmf after particle deposition in the human lungs is shown in Fig. 1. We can clearly distinguish **three** phases:
i) a *fast clearance* phase occurring within the first day,
ii) a phase with an increase of the signal, called *'negative' clearance*, happening between day 2 and the first two months, and
iii) the *slow clearance* phase, beginning after 1-2 months.
The following parameters were estimated from this curve:
1) the *fraction of slow cleared material*, called A_{c_s},
2) the *strength of 'negative' clearance*, F_{c_u}, describes the ratio between the highest rmf after 1-2 month and the rmf after 1 day (fast clearance completed), and
3) the *half-time of slow clearance*, T_{c_s}, using all rmf's from 1 month and later.

This data are shown for healthy subjects (NS and S) and for patients with sarcoidosis, fibrosis and with chronic obstructive pulmonary disease in Tab. 2.

Table 2: Parameters of clearance measurements in healthy subjects, subdivided into non-smokers (NS) and smokers (S) and in patients.

Group	N	$A_{c\_s} \pm SD$	$F_{c\_u} \pm SD$	$T_{c\_s} \pm SD$ [days]
NS (20-39)	10	0.940±0.164	1.173±0.179	124.5±66.4
S (20-39)	9	0.864±0.143	1.139±0.175	208.4±82.4
NS (40-65)	11	0.865±0.199	1.584±0.743	161.9±119.5
S (40-65)	9	0.804±0.185	1.165±0.174	459.2±334.0
SARK (NS&S)	15	0.863±0.178	1.308±0.230	330.3±219.5
FIBR (NS&S)	12	0.677±0.334	1.278±0.455	602.3±435.5
COPD (NS&S)	15	0.494±0.200	1.800±0.615	259.7±70.7 (N=10)

Fast Clearance

Despite to the assumption, that all particles would penetrate to the alveolar region, several subjects showed a small fraction of particles being cleared within the first two days. Because the design of the study was concentrated to the slow clearance phase the time course of fast clearance could not be analyzed. The mean amount of slow cleared particles (A_{c_s}) is shown in Tab. 2. In younger healthy subjects more than 90 % of the particles remain in the lungs after 2 days and follow the slow clearance mechanism. There is a tendency to higher fraction of fast cleared particles ($1-A_{c\_s}$) in smokers and in older subjects (40-65). Fast cleared particles result from a deposition in the ciliated (bronchial) regions of the lung. An increasing amount of fast cleared particles demonstrates an increased fraction of bronchial deposition in this subjects. In patients with COPD 50 % of the particles were cleared within the first two days, none of the patients was without a fast clearance phase. This demonstrates a high bronchial deposition due to <u>bronchial obstructions</u>, that are known from an increased airway-resistance. The increased fast clearance phase in FIBR is also a measure for bronchial obstructions. The results in healthy smokers as well as in older subjects give indications for beginning airway obstructions.

Negative Clearance

The phenomenon of *'negative'* clearance, that describes an increase of the rmf of the lungs, is typical for MPG-studies using a magnetic tracer. In earlier studies [4] it was estimated that particle translocation might be the reason for this process. An analysis with clustered steel-spheres shows that the formation of aggregates results in an amplification of the rmf due to a mutual magnetization. A doublet has double the rmf compared to two separated spheres. From this analysis we follow that particle aggregation additionally can induce a negative clearance effect. Particle aggregates can be formed when macrophages die an release particles. Other particle-loaden macrophages can then phagocytize the particles. The results for negative clearance are reported in Tab. 2. There is no difference between NS and S. The effect is strongest in COPD and seem to correlate with bronchial obstructions and the high bronchial deposition.

Slow Clearance

Half-times for alveolar *long-term clearance* in younger and older subjects are shown in Tab. 2 and Fig. 2. In younger as well as in older smokers long-term clearance is significantly impaired (p<0.05). I two subjects of the group S(40-65) no clearance could be detected within the period of investigation. In this case half-time was set to 1000 days. Alveolar clearance capacity seems to decay with increasing PY. A correlation between clearance half-time T_{c_s} and pack year PY yields:

without NS $\quad T_{c\_s} = 191\pm71$ days + 5.0±2.1 (days/pack-year) * PY, p=0.026

with NS $\quad\quad T_{c\_s} = 159\pm31$ days + 5.7±1.3 (days/pack-year) * PY, p<0.01.

An analysis without NS yields an overestimation for PY = 0, additionally including NS gives a higher significance. Finally we can conclude that cigarette smoke consumption reduces alveolar clearance by: ≈ **6 days/pack-year**.

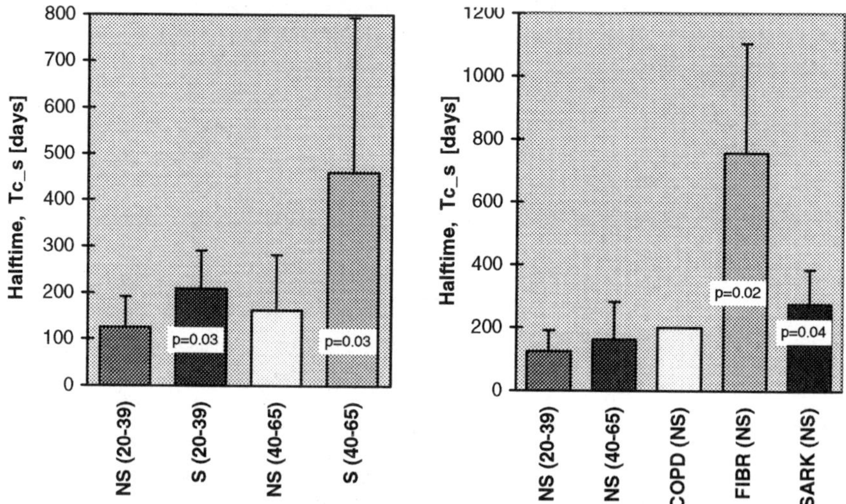

Figure 2: Half-times for alveolar clearance in healthy NS and S in different age groups (left) and in non-smoking patients (right).

Alveolar Clearance is highly impaired in FIBR-patients. Because of the significant influence of cigarette smoking in healthy subjects, only non-smoking patients were compared to older healthy subjects NS (40-65). Half of the FIBR(NS) showed no clearance within the period of observation. In FIBR alveolar clearance is strongly disturbed and thereby might be a co-factor for inducing this disease. SARK also show an influence on alveolar clearance, but the effect is less pronounced compared to healthy S. Despite the fact that COPD have high PY, alveolar clearance is faster compared to healthy S. This gives a hint that part of the influence of cigarette smoking on alveolar clearance might be reversible, after subjects stop smoking. This results show that MPG provides a sensitive and non-invasive method to study early influences on alveolar clearance and to monitor a disturbed lung defense in disease.

Acknowledgement: This study was supported by the DFG (Po 108/7-1) and by the CEC (FI4P-CT95-0026).

References

[1] Cohen, D., S.F. Arai, J.D. Brain (1979), Smoking Impairs Long-Term Dust Clearance from the Lung, *Science*, Vol. 294, 514-517.

[2] Stahlhofen W, Möller W (1993) Behaviour of magnetic micro-particles in the human lung, *Radiation and Environmental Biophysics*, Vol 32, 221-238

[3] Möller W, Stahlhofen W, Roth C (1990) Improved spinning top aerosol generator for the production of high concentrated ferrimagnetic aerosols, *Journal of Aerosol Science*, Suppl. 1, Vol. 21, S435-S438

[4] Freedman, A.P., S.E. Robinson, M.R. Street (1988), Magnetopneumographic Study of Human Alveolar Clearance in Health and Disease, *Annals of Occupational Hygiene*, Vol. 32, Supplement I, 809-820 (Inhaled Particles VI).

A Model of Phagosome Motion within Cells Incorporating Viscoelasticity Based on Cytomagnetometric Measurements

Nemoto, I.

Tokyo Denki University, Hatoyama, Saitama, Japan

Introduction

Cytomagnetometry was proposed more than ten years ago [1][2] but has not yet acquired the citizenship in biophysics. The main reason seems to be that the method is macroscopic and crude while there are advanced microscopic biophysical techniques available to handle the molecules themselves in the cells. Indeed, these new techniques have made it possible to measure even the force produced by one actin molecule. However, there are still quite limited possibilities for measuring the activities within the cells *in vivo*. We believe that cytomagnetometry using magnetic particles provides such possibilities; the movement of the intracellular structures can be detected non-invasively by measuring the magnetic field from the cells, and furthermore, magnetic particles within the cells can be manipulated by an external magnetic field easily.

Fig. 1 shows the basic idea of cytomagnetometry. Magnetic particles are introduced into the cells such as pulmonary macrophages by phagocytosis. The paticles then exist within the organelles called phagosomes. (They change names as the cellular process proceeds, but we simply call them 'phagosomes' for simplicity.) A strong field magnetizes the particles and the phagosomes act like small magnets. A weak field is measured from the cells which decays with a typical half time of 1-2 minutes. This magnetic relaxation is due to the randomization of the direction of the magnetic moments of phagosomes caused by their mechanical motions. Therefore, the relaxation reflects the mechanism of the intracellular motion and also the rheological properties of the cytoplasms. Manipulation of the phagosomes from outside can be done by applying a magnetic field. Experiments may be devised to estimate the values of the parameters involved in the intracellular motion (especially of the organelles), such as the energy (E_r), the viscosity and the elasticity. Phagosome motion is important not only in its own sake but also in its relationship to the cellular motion in general including muscle contraction[3].

Viscoelasticity and relaxation

Cytomagnetometry has shown that the movement of the phagosomes are impeded not only by the apparent viscosity but also by elasticity[2][4]. These studies suggested that the relaxation curve might be influenced by the elasticity as well. Indeed, if the particles are aligned by a strong field for a certain period of time (instead of by pulse

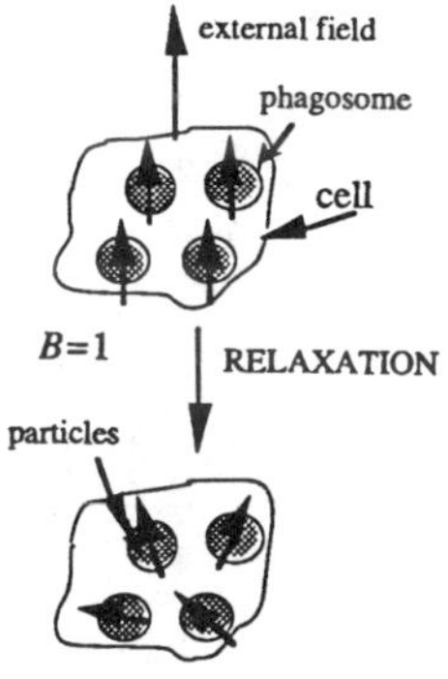

Fig. 1 Basic idea of cytomagnetometry (measurement of relaxation).

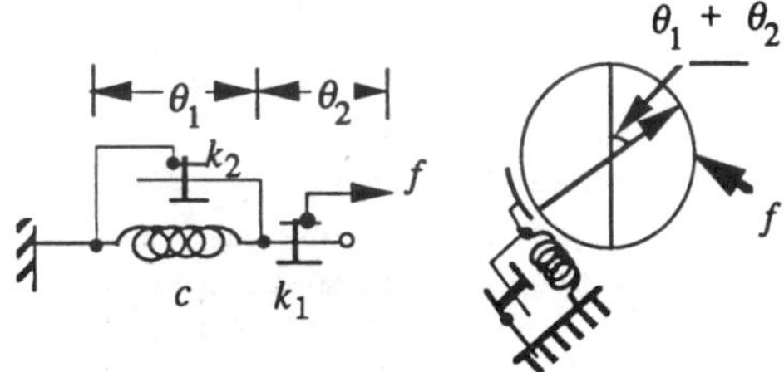

Fig. 2 The viscoelasticity model against the rotary motion of the phagosome.

field), the relaxation rate, especially in its beginning, may be considerably increased by the elastic recoil. Therefore, in relaxation measurements, pulse magnetization has been mostly used. However, we also have to consider possible effects of elasticity on the movement of the phagosomes produced by the (seemingly) random forces responsible for the relaxation. For this purpose, we have the 1-dimensional model of visco-elastic body pulled by an external force, be it the still undefinite random force responsible for the relaxation or the magnetic torque produced by an external field (Fig. 2). The equation of motion for this model is written

$$\frac{d\theta_1}{dt} = -\frac{c}{k_2}(\theta_1 - \overline{\theta_1}) + \frac{f}{k_2}, \quad \frac{d\theta_2}{dt} = \frac{f}{k_1} \tag{1}$$

where θ_1 and θ_2 are defined as shown in Fig. 1 such that $\theta = \theta_1 + \theta_2$ is the the direction of the magnetic moment of the phagosome measured from that of the measurement. $\overline{\theta_1}$ is the value of θ_1 for which the elastic body is in equilibrium. For the relaxation process, we take f to be the formal differentiation of the Wiener process $W(t)$ of which the mean is 0 and the variance is αt where α is some constant to which we will give a specific form later. Then the solution of the stochastic differential equation is formally written as

$$\theta_3(t) = \overline{\theta_1}\left(1 - e^{-\frac{c}{k_2}t}\right) + \int_0^t e^{-\frac{c}{k_2}(t-s)} dW(s), \quad \theta_2(t) = \int_0^t \frac{1}{k_1} dW(s) \tag{2}$$

where the integrals are the Wiener integrals. Note that the initial condition $\theta_1(0) = \theta_2(0) = 0$ is used. The mean and the variance of the solution of this equation can be calculated to yield

$$E[\theta] = 0, \quad V[\theta(t)] = E[\overline{\theta_1}^2](1 - e^{-\beta t})^2 + \alpha\left\{\frac{1 - e^{-2\beta t}}{2\beta k_2^2} + \frac{2(1 - e^{-\beta t})}{\beta k_1 k_2} + \frac{t}{k_1^2}\right\} \tag{3}$$

where $\beta = c/k_2$. Since θ is disbributed normally with mean 0 and variance $V[\theta(t)]$, the cell field is given by

$$B(t) = E[\cos\theta(t)] = \frac{2}{\sqrt{2\pi V[\theta(t)]}} \int_0^\infty \cos\xi \exp\left\{-\frac{\xi^2}{2V[\theta(t)]}\right\} d\xi = \exp\{-V[\theta(t)]/2\} \tag{4}$$

which shows that when $t \to \infty$, the relaxation follows an exponential decay but that for finite t, the relaxation curve is certainly influenced by elasticity.

We next want to relate α to E_r which we previously defined to be the energy (corresponding to kT in case of thermal Brownina motion) responsible for the relaxation[1]. In our model, the mean square distance during the infinitesimal time Δt was given by $E[\Theta^2] = 4\Delta t E_r/v$ where v is the viscosity factor. Here we have

$$\lim_{\Delta t \to 0} \frac{V[\theta(t)]}{\Delta t} = \alpha\kappa^{-2}, \quad \kappa = \left(\frac{1}{k_1} + \frac{1}{k_2}\right)^{-1} \tag{5}$$

Comparing these two, we get $\alpha = 4E_r\kappa^2/v$. We assume that v can be replaced by κ and thus $\alpha = 4E_r\kappa$.

Pulse magnetization and complete alignment

We employ two methods of magnetizing the particles; pulse magnetization and complete alignment. If the particles are homogeneous, there would be no change in the magnetization of the whole ensemble of the phagosomes. However, if there is visoelasticity, the relaxation curve would be different because the elastic body is in general, not in equilibrium at any time, that is to say $\overline{\theta_1}$ can be disbritubed. The variance of $\overline{\theta_1}$ should be equal to the variance of the solution of the equation $d\xi = -(c/k_2)\xi\,dt + (1/k_1)dW_t$ at the limit $t \to \infty$, which is

$$\lim_{t \to \infty} V[\xi(t)] = \lim_{t \to \infty}\left[V[\xi(0)]e^{-2\beta t} + \frac{\alpha(1 - e^{-2\beta t})}{2\beta k_2^2}\right] = \frac{\alpha}{2\beta k_2^2} \tag{6}$$

With the complete alignment scheme, we can put $E[\overline{\theta_1}^2] = 0$ because after a sufficient period of alignment, the elastic body will have nearly reached its equilibrium state. In summary we have two relaxation curves corresponding to two kinds of variance; $V[\theta^P(t)]$ for pulse magnetization and $V[\theta^a(t)]$ for complete aignment:

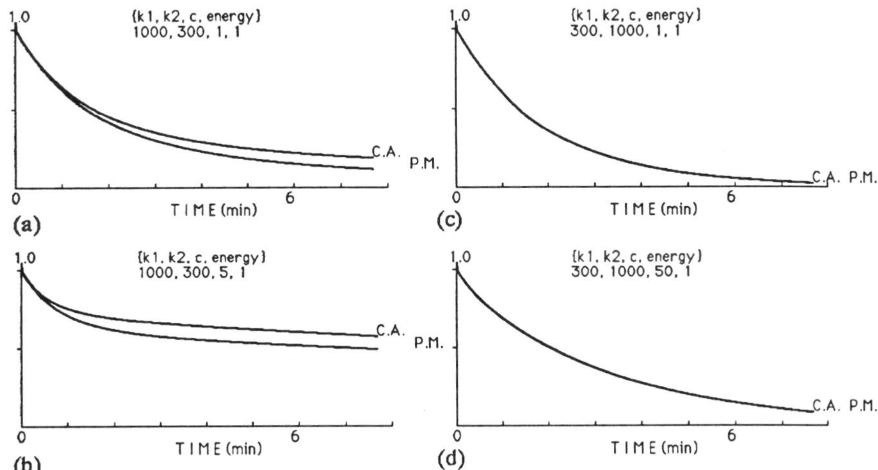

Fig. 3. Relaxation curves calculated from eq.(7). C.A. and P.M. stand for complete-alignment and pulse-magnetization, respectively. The parameter values are normalized by E_r.

$$V[\theta^p(t)] = 4E_r\kappa\left\{\frac{(1-e^{-\beta t})}{\beta k_2^2} + \frac{2(1-e^{-\beta t})}{\beta k_1 k_2} + \frac{t}{k_1^2}\right\} \quad V[\theta^a(t)] = 4E_r\kappa\left\{\frac{(1-e^{-2\beta t})}{2\beta k_2^2} + \frac{2(1-e^{-\beta t})}{\beta k_1 k_2} + \frac{t}{k_1^2}\right\}. \tag{7}$$

Fig. 3 shows a few examples of relaxation curves corresponding to the variances calculated by eq. (7). It is seen that when $k_1 > k_2$, the effect of elasticity and the difference between the two magneitzing schemes are apparent.

Viscoelastic recoil

Assume that we apply a medium field H (i.e., not sufficient to alter the magnetization of each particle but sufficient to overcome the randomizing force) perpendicularly to the primary-magnetization direction right after the primary magneti-zation for T_2 seconds and remove the field. Then the equations describing theese processes are:

$$\text{for } 0 \le t \le T_2, \quad \frac{d\theta_1}{dt} = -\frac{c}{k_2}\theta_1 + \frac{1}{k_2}\mu H\cos(\theta_1+\theta_2), \quad \frac{d\theta_2}{dt} = \frac{1}{k_1}\mu H\cos(\theta_1+\theta_2), \quad \text{and}$$

$$\text{for } t \ge T_2, \quad \frac{d\theta_3}{dt} = -\frac{c}{k_1}\theta_3 + \frac{1}{k_2}\frac{dW}{dt}, \quad \frac{d\theta_2}{dt} = \frac{1}{k_1}\frac{dW}{dt}. \tag{8}$$

Here we assume for simpicity that $E[\overline{\theta_1}^2] = 0$. μ is the magnetic moment of the phagosome. The first part is a nonlinear differential equation and can only be solved numerically. The second part is the stochastic differential equation already mentioned above. The initial condition for the firstp part is given by $\theta_1(0) = \theta_2(0) = 0$ and that of the second part is given by the solution of the first part at T_2. The numerical solution for the first part would be inconvenient for parameter estimation by fitting the measurement data. Therefore, we approximate the first part by

$$\frac{d\theta_1}{dt} = -\frac{c}{k_2}\theta_1 + \frac{1}{k_2}\mu H\left(1 - 0.5\cdot\theta_1^2(1+k_2/k_1)^2\right), \quad \frac{d\theta_2}{dt} = \frac{1}{k_1}\mu H\left(1 - 0.5\cdot\theta_2^2(1+k_1/k_2)^2\right) \tag{9}$$

The two variables are now separate, which makes the solution much easier. Its rigorous solution is given by

$$\theta_i(t) = -\frac{r_{1i}r_{2i}(\exp(r_{1i}t) - \exp(r_{2i}t))}{a_i(r_{2i}\exp(r_{1i}t) - r_{1i}\exp(r_{2i}t))}, \qquad i = 1, 2 \tag{10}$$

where

$$r_{1i} = \frac{b_i + \sqrt{b_i^2 - 4a_i d_i}}{2}, \quad r_{2i} = \frac{b_i - \sqrt{b_i^2 - 4a_i d_i}}{2}, \quad a_1 = -\frac{\mu H}{2k_2}\left(1 + \frac{k_2}{k_1}\right)^2, a_{2=} = -\frac{\mu H}{2k_1}\left(1 + \frac{k_1}{k_2}\right)^2$$

$$b_1 = -\lambda, \quad b_2 = 0, \quad d_1 = \mu H / k_2, \quad d_2 = \mu H / k_1 .$$

Fig. 4 shows the whole time courses described by the above differential equation (the solid curves) and the one using eq. (9) for the vertical magnetization process (the black circles). The parameter values are normalized by E_r taken to be the unit value. These curves (especially the form of the curves in (d)) are similar to those experienced in cytomagnetometric measurements. The parameter values shown are normalized with respect to E_r which is taken to be 1. A typical value $\sim 5\times 10^{-18} J$ for E_r [1] and $k_i = 100, c = 10$ would yield the viscocity coefficient of $\sim 10^2$ Ns/m2, and the rigidity moduous of ~ 1.5 N/m2.

Concluding Remark

A one-dimensional model of cell field behavior incorporating viscoelasticity was proposed. Preliminary analysis of the experimental data with this model suggests the involvement of microtubules and intermediate filaments in producing the elasticity, but we need further verification for the estimation. The proposed 1-dimensional model should be extended to the spherical model formulated in [1], but it is not trivial to incorporate viscoelasticity in the spherical model. The model of interacation between filaments and phagosome membranes proposed in [3] should also be modified to take into account the viscoelasticity.

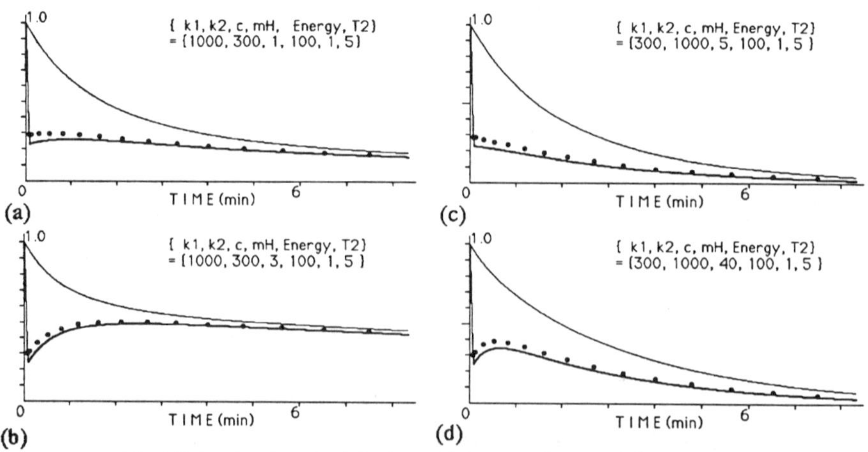

Fig. 4. Relaxation curves and elastic-recoil curves prediected by the model. See Text.

References

[1] Nemoto, I., A model of magnetization and relaxation of ferrimagnetic particles in the lung, IEEE Trans. Biomed. BME-29: 745-752 (1982).

[2] Valberg, P.A. and Butler, J.P., Magnetic particle motions within living cells:Biophys. J. 52:537-561(1987).

[3] Nemoto, I., A model of phagosome motion within cells based on cytomagnetometric measurements, Math Biosci., 118:97-117(1993).

[4] Wiegand, J., Moeller, W., Nemoto, I. and Stahlhofen,W., Magnetic microparticles can measure intracellular viscoelasticity in vivo, IN:Baumgartner, C., Deecke, L., Stroink, G. and Williamson, S.J. Biomagnetism, Amsterdam, Elsevier, 1995.

Magnetic Susceptibility Tomography with Nonuniform Field

Parente Ribeiro, E.[1,2], Wikswo, Jr., J.P.[2], Costa Ribeiro, P.[1] and Szczupak, J.[1]

[1]PUC-Rio, Rio de Janeiro, Brazil; [2]Vanderbilt University, Nashville, TN, USA

Abstract

Imaging systems that generate slice views of the body are called tomographic scanners. Several modalities of scanners that generate such cross sectional images have been developed for biomedical applications, such as X-ray tomography, PET and MRI. Magnetic susceptibility tomography (MST) may be a new tool for visualizing the interior of the body. Previous works have shown that it is possible to reconstruct a $4 \times 4 \times 4$ susceptibility distribution from the perturbation produced under a uniform field. In this paper the solution to the inverse problem is addressed in the framework of reconstruction from projections. A nonuniform field is considered together with uniform field.

Introduction

The number of tomographic scanners available in almost every hospital illustrates the importance of such tools for physicians. From the more popular x-ray computerized tomography (CT), magnetic resonance imaging (MRI), positron emission tomography (PET) to the emerging electrical impedance tomography (EIT), each modality employs a different method to provide 2D slice views of the body [1]. All of them share the same underlying mathematical problem of reconstructing the object data from external measurements made at different 'angles', also referred to as reconstruction from projections [2,3].

Magnetic susceptibility measurements on humans have been made for some time [4,5,6]. It has found useful clinical application, particularly in liver iron overload assessment [7] and it has been used in gastroenterology research to track paramagnetic tracers. Imaging of planar samples has already been demonstrated [8] but the possibility of 3D reconstruction without the assumption of a particular model was only recently established [9,10]. A $4 \times 4 \times 4$ voxel model representing a 3D susceptibility distribution could be successfully reconstructed using a uniform magnetizing field, in the presence of a small amount of noise. The problem is very ill-posed and an adequate geometry and applied field must be identified to enable convergence at higher resolutions and higher noise levels.

Methods

When an object is exposed to a magnetic field of strength H, it develops a magnetization M. The magnetic susceptibility χ (a dimensionless quantity in S.I.) is M/H. The magnetic flux density B is given by $B=\mu_o(H+M)$, where μ_o is the permeability of free space. It can also be expressed in term of the susceptibility: $B=\mu_o(1+\chi)H$. The permeability is B/H and the quantity $(1+\chi)$ is the relative permeability (μ_r). If the orientation of the magnetization is opposite to the applied field, the object has negative susceptibility and is said to be diamagnetic. Water, which is the major constituent of the body, has $\chi=-9 \times 10^{-6}$. If the magnetization has the same direction as the applied field and it is small ($0<\chi<10^{-3}$) the object is paramagnetic. Ferromagnetic materials have high permeability, in the range $\sim 10^3 < \mu_r <10^5$ and they may exhibit histeresis. For diamagnetic and paramagnetic materials (which includes biological tissue) the Born approximation holds, which means that the magnetization of any one region is too weak to affect an adjacent region.

The magnetic flux density $B(r)$ at point r in space due to an object with susceptibility $\chi(r')$ and volume v' is given by the dipole field equation

$$B(r) = \frac{\mu_o}{4\pi} \int_{v'} \frac{3\chi(r')H_a(r') \cdot (r-r')}{|r-r'|^5}(r-r') - \frac{\chi(r')H_a(r')}{|r-r'|^3} dv'^3 \ . \tag{1}$$

This can be rewritten as

$$B(r) = \int_{v'} G(r,r',H_a)\chi(r')dv' \ , \tag{2}$$

where

$$G(r,r',H_a) = \frac{\mu_o}{4\pi} \left\{ \frac{3H_a(r') \cdot (r-r')}{|r-r'|^5}(r-r') - \frac{H_a(r')}{|r-r'|^3} \right\} \ . \tag{3}$$

The measured magnetic perturbation is thus the projection of the susceptibility distribution on the Green's function. One can collect several projections by varying r and H_a, i.e., taking measurements in several locations and with different applied fields, and attempt to estimate the original susceptibility distribution. The first question which arises is whether this mapping is one-to-one or, in other words, whether the inverse solution is unique. A necessary and sufficient condition for that to happen is that the null space of the mapping N be equal to $\{0\}$, i.e., there must exist no magnetically silent source configurations.

If only a uniform applied field is considered, there are only three linear independent options for H_a because G is linear with respect to H_a. Furthermore, any concentric sphere with uniformly distributed susceptibility has the same multipole expansion coefficients (being all equal to zero but the dipole term). Hence, sources of different size will produce the same dipolar field pattern, creating an infinite set of magnetically silent configurations. This will not be the case if a nonuniform field is also applied.

To tackle the problem numerically, the source object is discretized into n elements of volume v_i', where $i = 1,2,\ldots,n$ and the measurements are made at m different points r_j, where $j = 1,2,\ldots,m$. Equation 2 becomes

$$B(r_j) = \sum_{i=1}^{m} G(r_j, r_i', H_a) \chi(r_i'). \tag{4}$$

In matrix notation, a $(n \times 1)$ column matrix $\mathbf{X}$ denotes the susceptibility distribution. A $(m \times n)$ matrix $\mathbf{G}$ represents the Green's function and the magnetic flux density is described by a $(m \times 1)$ column matrix $\mathbf{B}$. Only the z component is considered. Equation (4) is then $\mathbf{B=GX}$.

If $m=n$, the number of measurements equals the number of unknowns and the system of equations will be exactly determined. However due to the nature of the Green's functions, it will be very ill-conditioned. An exact solution could be found, but if noise is present in the measurement, the solution will have large errors. To improve matrix conditioning, more measurements are incorporated so that the system becomes over-determined ($m>n$) and a least squares solution can be attempted.

In the forward problem, the magnetic field matrix $\mathbf{B}$ is calculated using a sample-generated susceptibility distribution $\mathbf{X}$, and then $\mathbf{B}$ is truncated to a fixed number of digits. This is equivalent to adding noise N: $\mathbf{B=GX+N}$.

The generalized inverse $\mathbf{G}^{\#}$ is evaluated using singular value decomposition (SVD) and the inverse problem reduces to $\mathbf{X}^e = \mathbf{G}^{\#}\mathbf{B}$. The error is the difference between the estimated and actual distributions. The relative root mean square error $rrmse$, obtained by dividing the rms error by the mean value of the original distribution, is used to evaluate the reconstruction quality.

In Fig. 1a, the $rrmse$ of a $4 \times 4 \times 4$ random susceptibility cube reconstructed from magnetic data with different truncation levels (3, 4 and 5 digits) is plotted together with the condition number of the matrix $\mathbf{G}$, which is the ratio between the largest and smallest singular values. This is a number between unity (well conditioned) and infinity (singular). A large number indicates an ill-conditioned matrix. High errors are encountered when the problem becomes ill-posed. The curves correlate well and they should be even more similar if each error plot is averaged over an ensemble of different source distributions.

When resolution is increased, the matrix becomes more ill-conditioned. Even with a large number of measurements and a uniform field applied along the other two axes, the condition number does not seem to go below a certain level ($\approx 5 \times 10^6$), as can be seen on Fig. 1b.

To improve the conditioning of the problem, we analyzed the use of a nonuniform field. The field due to a single coil encircling the sample was considered. Contrary to the uniform field, such a field is not rejected by the gradiometric detecting device. This rejection is important since the susceptibility values are very small, making the field perturbation several orders of magnitude smaller than the applied field. Were such an instrument actually built, a cancellation scheme would have to be utilized to subtract the field coupled directly to the SQUID.

Results

Several field configurations were analyzed for $8 \times 8 \times 8$ susceptibility voxels by varying the position and direction of the sample and the field. Measurements were considered to be taken at 0.6 units above and below the sample. Two scenarios were investigated: In the first, the external uniform and nonuniform field was applied only on the same direction as the detection device. In the second, the field was applied in all three directions. They are summarized on Fig. 3. A reduction of the condition number to 5×10^5 was achieved.

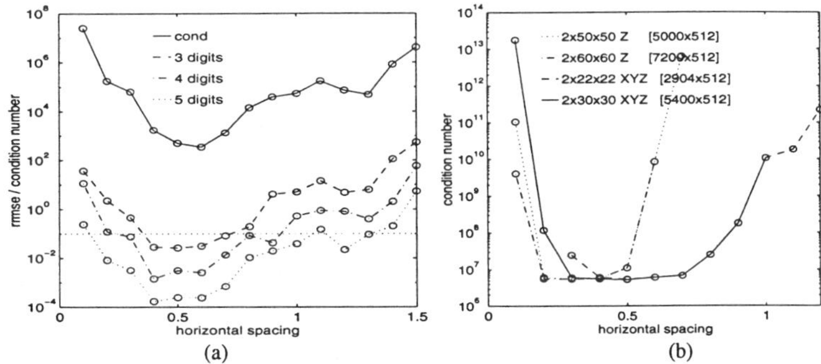

(a)   (b)

Fig. 1 - (a) The matrix G describing the forward problem of a sample of $4 \times 4 \times 4$ susceptibility voxels exposed to uniform field (unitary spacing). An array of 8×8 measurement locations at 1 unit above and below the sample is considered. The calculated magnetic field is truncated with different numbers of significant digits (3,4,5) prior to inversion. The condition number is also plotted as a function of measurement point spacing. The horizontal line establishes a threshold of acceptable noise (subjectively) to be 10%. (b) The condition number for uniform field reconstruction of a $8 \times 8 \times 8$ susceptibility cube. Measurements are made on plane 0.6 units above and below. The legend shows the number of measurements, the direction of applied field and size of Green's matrix (inside brackets).

The condition number could be further reduced to 9.5×10^4 by also taking some measurements on the sides of the sample. This configuration was used to simulate the forward and the inverse problem for different digits of precision in the measured field. A susceptibility cube with values selected to be either 0 or 1 was used. Figure 4 shows the original planes and the reconstructed solutions for 5, 4 and 3 digits of precision.

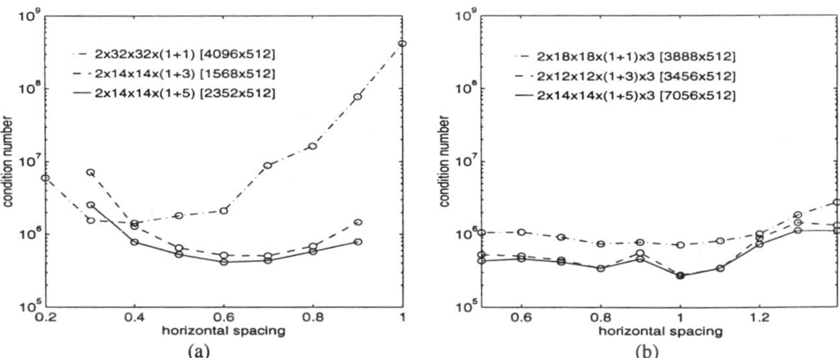

(a)   (b)

Fig. 3 - The condition number for an $8 \times 8 \times 8$ voxel susceptibility distribution, with different geometries. A uniform field and the field due to a single coil with diameter larger than the sample are considered. (a) Fields are applied only on one direction. (b) Fields are applied on all three directions. Each plot shows three traces corresponding to: coil placed at the center, coil placed at three axial positions (-2,0,2) and coil placed at five axial positions (-2,-1,0,1,2). Inside the brackets is the matrix size for each measurement.

673

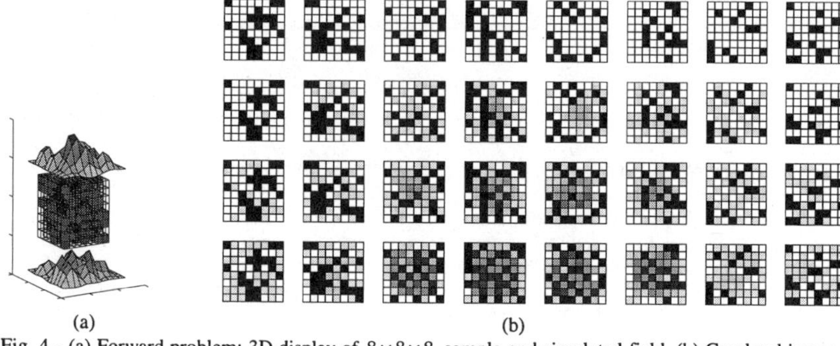

(a)   (b)

Fig. 4 - (a) Forward problem: 3D display of $8 \times 8 \times 8$ sample and simulated field. (b) Graylevel image of each sample plane. The first row is the original distribution. The following rows are reconstruction's from data with 5, 4 and 3 digits of precision, respectively. The *rrmse* for each reconstruction is 8%, 38% and 257%.

Discussion

We have established a mathematical framework to treat the three dimensional magnetic susceptibility inverse problem as a reconstruction from projections. Condition number helped explain why under certain configurations the inverse solution exhibited large error.

An improvement of about one order of magnitude on the condition number was verified by also employing together with the uniform field the nonuniform field generated by a single coil enclosing the sample. That additional information was confirmed to be linearly independent from the information obtained with the uniform field only. Reconstruction of the $8 \times 8 \times 8$ susceptibility distribution from the measured magnetic perturbation in the presence of noise was demonstrated to be possible from magnetic data with small noise (5 digits). Novel nonuniform configurations may provide further increase in resolution.

This three dimensional imaging technique could greatly improve the accuracy of already established applications of susceptometry such as quantitative liver iron measurements [7], and may also be able to image different tissues types, launching new applications with appealing imaging and non-invasive characteristics.

References:

[1] Webb, S. The Physics of Medical Imaging, Bristol, Institute of Physics Publishing, 1988.

[2] Natterer, F. The Mathematics of Computerized Tomography, John Wiley,1986.

[3] Herman, G.T. Image Reconstruction from Projections, Academic Press, 1980.

[4] Wikswo, J.P., Jr., Opfer, J.E., Fairbank, W.M. Noninvasive magnetic detection of cardiac mechanical activity: Experiments, Medical Physics, 1980, 7:307-313.

[5] Rassi, D., Hoare, A.,Samadian, V., Melville, D. SQUID measurements and computational modeling of a simple thorax phantom. In: Atsumi, K., Kotani, M., Ueno, S., Katila, T., Williamson, S.J. Biomagnetism'87. Tokio, Tokio Denki University Press, 1988.

[6] Di Luci, S., Obletter, G., Comani, S., Del Gratta, C., Romani, G.L. Magnetic mapping of dc fields related to tissue susceptibility in the human body. In: Williamson, S.J., Hoke, M., Stroink, G., Kotani, M. Advances in Biomagnetism. New York, Plenum Press, 55-508, 1989.

[7] Fischer, R., Heinrich, H.C. Biosusceptometry: Current status of clinical diagnostics and biomagnetic research, In: Hoke, M., Erne S.N., Okada, Y.C., Romani, G.L. Biomagnetism: Clinical Aspects, 1992.

[8] Roth, B.J., Sepulveda, N.G., Wikswo, J.P., Jr. Using a magnetometer to image a two-dimensional current distribution, J. Appl. Phys. 65, 361-372, 1989.

[9] Wikswo, J.P., Jr., Sepulveda, N.G., Thomas, I.M. Three-dimensional biomagnetic imaging with magnetic susceptibility tomography, In: Baumgartner, C., Deecke, L., Stroink, G., Williamson, S.J. Biomagnetism: Fundamental research and clinical applications. Amesterdam, IOS Press, 780-784, 1995.

[10] Sepulveda, N.G., Thomas, I.M., Wikswo, J.P., Jr. Magnetic susceptibility tomography for three-dimensional imaging of diamagnetic and paramagnetic objects, IEEE Transactions on Magnetics, 1994, 30:5062-5069.

MCG Studies in Rats Using Ferrofluids

Saligram, U.[1,2], Moran, J.E.[1] and Tepley, N.[1,2]

[1]Henry Ford Hospital, Detroit, MI, USA; [2]Oakland University, Rochester, MI, USA

INTRODUCTION

Ferrofluids are colloidal suspensions of single-domain ferrite particles in a liquid carrier. The liquid carrier used in biological applications is water. The particles are coated with a proprietary stabilizing dispersing agent which prevents particle agglomeration.

Ferrofluids have a broad range of biological applications. They have been used *in vivo* in animal studies to target chemotherapeutic agents to specific anatomical sites [1]. They have also been used in blood flow studies as a magnetic tracer [2,3]. In a previous study [2] with Wistar rats an amplification in Magnetocardiogram (MCG) was observed after introducing ferrofluid into the blood circulation. However this apparent amplification was observed in only one of the four rats injected with ferrofluid and this led to the speculation that this particular animal might have had damaged cardiac tissue, and the ferrofluid from the blood had leaked into the tissue. Based on this assumption additional studies were carried out. In these studies the various factors that could be responsible for this apparent increase in amplitude were examine. The ferrofluid used in this study EMG 507, is manufactured by FERROFLUIDICS CORPORATION, Nashau, NH. The liquid carrier is water; average particle size is 100°A; saturation magnetization is 100 Gauss; viscosity is ~2CP @ 27°C; particle concentration by volume is 1.8%; initial susceptibility is 0.38; density is 1.15 gm/ml; ionic nature of the surfactant is anionic and the pH is in the range 6-8. The manufacturer claims that the ferrofluids are non-toxic.

METHODS

Male wistar rats with mean weight of 250 gms were used. A seven channel BTi model 607 magnetometer was used to gather data (0.1 - 50 Hz). The sampling rate was 250 Hz Each rat was anesthetized using sodium pentobarbital (40 mg/kg). The magnetometer was positioned over the rat such that the channel at the center of the hexagonal array was directly above the heart. This magnetometer geometry was kept the same in all the studies. Thirty seconds of baseline MCG data were collected.

The right femoral artery was cannulated. The rat was then placed on a table inside the shielded room. 0.25 cc of ferrofluid diluted with saline (1 part ferrofluid in 20 parts saline) was introduced into the blood stream through the cannulated artery. Thirty seconds of data were then collected immediately after the introduction of ferrofluid. Ten second epochs of data were collected at intervals of 5 minutes. This study was a repeat of the previous study [2], referred to earlier.

A second study was then carried out, in which a fine needle was used to puncture the cardiac tissue. Thirty seconds of data were collected immediately after the puncture. Ten seconds of data were collected 5 minutes after the puncture, in order to determine if the act of puncturing displaced the heart from its normal position in the thoracic cavity, thereby causing changes in the amplitude of the MCG. 0.25 cc of ferrofluid diluted with saline (1 part ferrofluid in 20 parts saline) was then injected directly into the heart. The site of injection was similar in all the studies. Thirty seconds of data were collected immediately after the injection. Five minutes after the injection 10 second epochs of data were collected at 5 minute intervals. This five minute time interval was used to minimize the effect of the possible displacement of the heart due to the injection on the MCG signal amplitude. (If the heart is displaced due to the injection, its position with respect to each of the 7 magnetometer channels changes. Some of the channels would then show an increase in the amplitude of the signal that they record, and this increase would be due to the heart shifting closer to these channels and not due to the ferrofluid. Even though the heart may be displaced due to the injection, it is expected to return to its original position within a few seconds. The five minute interval after the injection allows eneough time for this to happen, thereby minimising the effect on the amplitude of the signal due to the shifting of the heart.) One hour's worth of data were collected in each study. The effect of an external magnetic field was studied using a circular magnet with its poles along the axis. The magnet (pole strength 1100 gauss-cm^2) was placed under the rat's heart at a vertical distance of 24 inches (field due to the magnet at this point was 7.8 mgauss), with its north pole facing up. Thirty seconds of data were collected at intervals of two minutes. This was repeated with the south pole facing up.

A third study using saline instead of ferrofluid was carried out. In this study 0.25 cc of saline was injected into the heart after the initial puncture. The protocol used to collect data in this study was the same as that used for ferrofluid study.

An average of 12 epochs of data were collected in each of the above studies.
For the time line of each of the studies refer to Fig . 1.

DATA ANALYSIS

The MCG signals were digitized using a HP 694A multiprogrammer. The primary comparison was the analysis of difference in signal amplitude before and after introduction of ferrofluid and saline. The baseline signal amplitude was compared to the following :
1) Signal amplitude after ferrofluid was introduced into the animals bloodstream through the cannulated femoral artery.
2) Signal amplitude after cardiac tissue puncture.
3) Signal amplitude after ferrofluid was injected into the heart.
4) Signal amplitude after saline was injected into the heart.

The mean amplitude of the MCG signal was calculated for each of the seven channels for each of the cases listed above. The mean amplitude of the baseline signal was then subtracted from the mean amplitude in each of the above cases. Standard deviation was also calculated for each of the channels for each run and for all the runs in each of the cases listed above. A paired t-test was done for each channel to determine if the amplitude difference from the baseline was significant. The testing level for each of these tests was set at .007. In the case where a magnet was used to create an external field, the data were filtered using a high pass filter of 1 Hz to remove the artifact due to breathing. This artifact was introduced into the signal by the presence of ferrofluid in the thoracic cavity.

RESULTS

No discernible change in the amplitude in case 1, i.e; ferrofluid introduced into the bloodstream through the cannulated artery without cardiac puncture was observed. Four rats were used in this study [Fig . 2].

No discernible change in the amplitude in case 2, i.e; after cardiac puncture was observed. Four rats were used in this study [Fig . 3].

A statistically significant ($p <= 0.007$) increase in the amplitude of the signal was found in case 3, i.e; after ferrofluid was injected into the heart. Seventeen rats were used in this study. Of the seventeen rats thirteen showed a significant increase in amplitude from the base line amplitude [Fig . 4]. Four rats did not show a significant increase in amplitude. One of the rats died due to an overdose of sodium pentobarbital.

No discernible increase in amplitude was observed in case 4, i.e; after saline was injected into the heart. Five rats were used in this study [Fig. 5].

DISCUSSION

The increase in the amplitude of the MCG signal only when ferrofluid is injected into the heart and not when it is introduced into the blood stream, justifies our assumption that the increase in amplitude is caused by ferrofluid leaking into the damaged cardiac tissue. The major difficulty with the procedure is the inability to induce uniform tissue damage in size, severity and location in every animal. This would make it difficult to study the effect of ferrofluid on tissue at the same location in every trial. This is an important consideration for this study because the site of tissue damage may have an effect on the magnitude of the increase in the signal amplitude.

This property of ferrofluids i.e; increasing the amplitude of electrophysiological signals, could probably be used to study weak signals from other electrophysiologically active tissues which would normally go undetected. In the case of cardiac tissue, the increase in amplitude is observed only after the tissue is damaged; this may or may not be the case for other tissue types. If ferrofluids leak into damaged tissue only, and if the tissue is electrophysiologically active, the disruption in the normal tissue function due

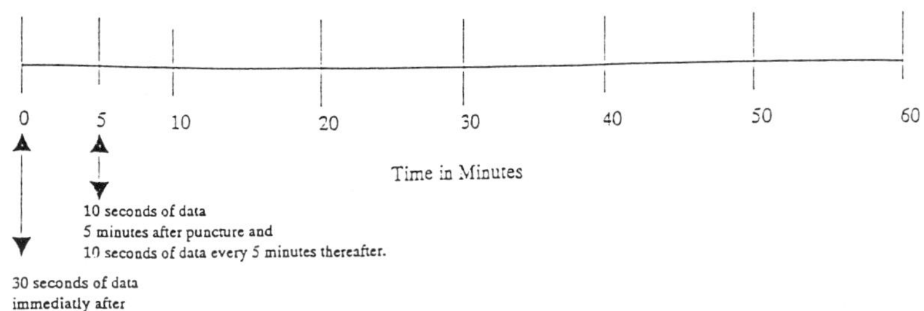

Fig. 1 Time line : 30 seconds of data were collected immediately following puncture of cardiac tissue or injection of ferrofluid or saline. 10 seconds of data were collected 5 minutes later and 10 seconds of data were collected every 5 minutes thereafter. Total time of each study was 60 minutes.

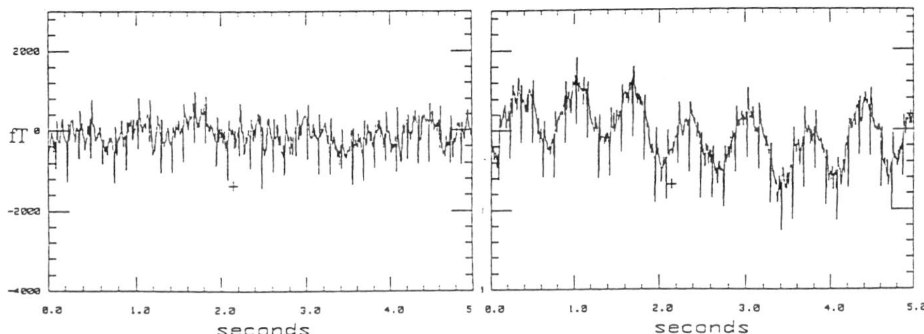

Fig. 2 MCG of rat after cardiac tissue was punctured.

Fig. 3 MCG of rat after ferrofluid was introduced into the bloodstream through cannulated femoral artery.

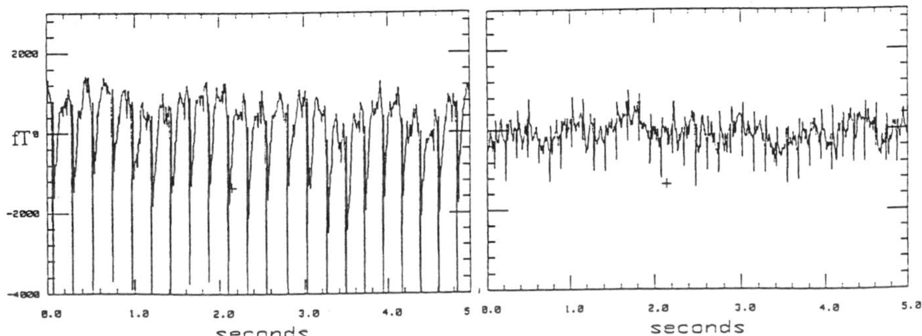

Fig. 4 MCG of rat after ferrofluid was injected directly into the heart.

Fig. 5 MCG of rat after saline was injected directly into the heart.

to the damage could possibly be detected in the presence of ferrofluids. In addition repair and restoration of normal tissue function could be monitored. Using the spatial information of the relaxation processes of the magnetite particles, registered by the neuromagnetometer, a spatial image of the ferrofluid distribution in biological tissue can be determined. This could then be used to determine areas of damage. On the other hand if the amplitude increase due to the presence of ferrofluids is not restricted to electrophysiological signals from damaged tissue, then it could prove useful in studying the electrophysiological functions of normal tissue in greater detail. Ferrofluids could possibly be reconcentrated using an external magnetic field and used repeatedly. It could also be concentrated in specific regions to study that particular tissue.

The size of the magnetite particles and the charge of the surfactant coating may contribute to the actual mechanism responsible for the amplitude increase described in this study. Since the actual mechanism is not thoroughly understood at this time, it is not possible to draw any conclusions regarding the various factors that may be contributing to the increase in signal amplitude. The effect of the above two properties of ferrofluids may in turn be different for different tissue types.

REFERENCES:

[1] Senyei, A., Widder, K., Czeslinsky, G. Magnetic Guidance of Drug Carrying Microspheres, Journal of Applied Physics, 1978 49: 3578-3583.
[2] Chopp, M., Chen, Q., Moran, J.E., Tepley, N. Biomagnetic Measurements Utilizing Ferrofluids, Proceedings   of the 8th International Conference on Biomagnetism.
[3] NewBower, Ronald.S. Magnetic Fluids in the Blood, IEEE Transactions on Magnetics, 1973, MAG-9, 3: 447-450.

This research supported by NIH/NINDS, grant no. 1R01 NS30914.

Magnetic Field and Electric Potential of Excited Plant Cell Chara Corallina: Calculation and Comparison with Experiment

Šlibar[1], M., Trontelj[1], Z., Jazbinšek[1], V., Thiel[2], G., Müller[3], W.

[1] Institute of Mathematics, Physics and Mechanics, University of Ljubljana, SI–1111 Ljubljana, Slovenia; [2] Pflanzenphysiologisches Institut, Universität Göttingen, D–37073 Göttingen, Germany; [3] Physikalisch–Technische Bundesanstalt, Institut Berlin, D–10587 Berlin, Germany

Introduction

It has been shown that the propagation of a single action potential in the internodal cell of green algae Chara corallina can be also detected magnetically [1, 2]. From this experiment some parameters, like the velocity of the depolarization wave along the excited cell, the length of the depolarized area and the peak value of the intracellular current were determined.

The relatively simple geometry and structure of the internodal cell of Chara corallina offer also a possibility to calculate the magnetic field distribution in the cellular vicinity if the transmembrane potential is known from measurements. In the inverse calculation the transmembrane potential can be obtained from the measured magnetic field. The results of these calculations will be shown.

Similarities in the electrophysiological behavior between these large single plant cells and large animal nerve axons are of considerable help here. Both, the cylindrical symmetry and quasi–stationary equations for electric and magnetic quantities can be applied in the Chara corralina case. So we can start with results obtained by Clark and Plonsey [3] and by Woosley, Roth and Wikswo [4]. They have formulated relations between the action potential, cellular currents and magnetic field for the single nerve axon.

Calculations

The potential calculation is based on the Laplace equation

$$\Delta \Phi = 0.$$

with the boundary conditions: The transmembrane potential Φ_m is equal to the difference of intracellular Φ_i and extracellular potential Φ_e

$$\Phi_m(z) = \Phi_i(a, z) - \Phi_e(a, z),$$

and the normal component of current density at the membrane surface is continuos

$$\mathbf{n} \cdot \mathbf{J}_i(a, z) = \mathbf{n} \cdot \mathbf{J}_e(a, z),$$

where $\mathbf{n}$ is the normal to the mebrane surface, $\mathbf{J}_i$ and $\mathbf{J}_e$ are intracellular and extracellular curent densities, respectively. The cylindrical symmetry and the meaning of all quantities are evident from Fig. 1.

Solution of Laplace equation for Φ_i and Φ_e was obtained by Clark and Plonsey [3]

$$\Phi_i(\rho, z) = \frac{1}{2\pi} \int_{-\infty}^{\infty} \frac{I_0(|k|\rho)}{\beta(|k|a)I_0(|k|a)} \phi_m(k)e^{-ikz}\, dk,$$

$$\Phi_e(\rho, z) = \frac{1}{2\pi} \int_{-\infty}^{\infty} \frac{K_0(|k|\rho)}{\alpha(|k|a)K_0(|k|a)} \phi_m(k)e^{-ikz}\, dk,$$

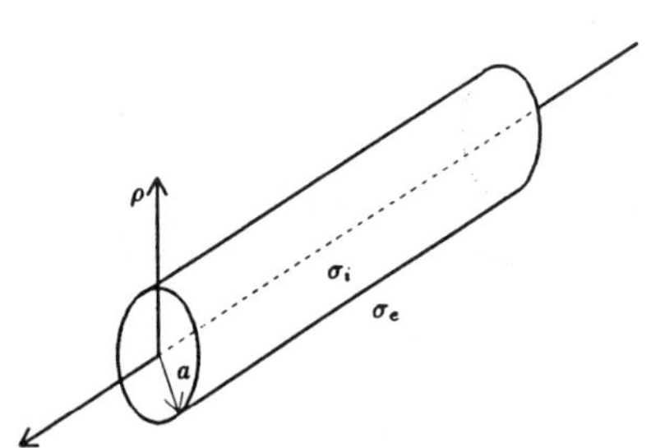
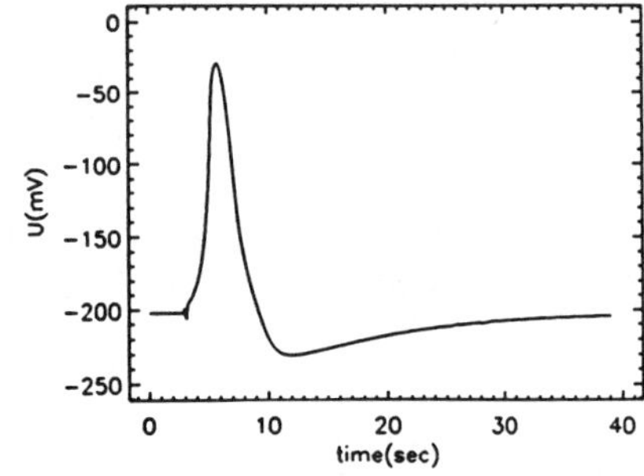

Figure 1: The cell model. $\sigma_i = 1.2\,\Omega^{-1}m^{-1}$, $\sigma_e = 0.025\,\Omega^{-1}m^{-1}$ and $a = 1\,mm$.

Figure 2: The whole sequence of measured action potential.

with

$$\alpha(|k|a) = -(1 + \gamma(|k|a)), \quad \beta(|k|a) = 1 + \frac{1}{\gamma(|k|a)}, \quad \gamma(|k|a)) = \frac{\sigma_e K_1(|k|a)I_0(|k|a)}{\sigma_i K_0(|k|a)I_1(|k|a)},$$

where I_0, I_1, K_0 and K_1 are modified Bessel functions, and $\phi_m(k)$, the spatial Fourier transform of the transmembrane potential

$$\phi_m(k) = \int_{-\infty}^{\infty} \Phi_m(z)e^{+ikz}\,dz.$$

Further Fourier transformations give, as pointed out in [4]

$$\phi_i(\rho, k) = \frac{I_0(|k|\rho)}{\beta(|k|a)I_0(|k|a)}\phi_m(k),$$

and

$$\phi_e(\rho, k) = \frac{K_0(|k|\rho)}{\alpha(|k|a)K_0(|k|a)}\phi_m(k).$$

The magnetic field $\mathbf{B}_i$ due to the intracellular current density is obtained by the application of the law of Biot-Savart (or the Amper's law) and can be formulated, as it is shown in [4]

$$\mathbf{B}_i(\rho, z) = \int_{-\infty}^{\infty} \mathbf{G}(\rho, a, z - z')J_i^z(a, z')\,dz',$$

where $\mathbf{G}$ is the proper Green's function and J_i^z the intracellular current density component along the cell. More useful is the expression in the k space, since it is directly related to $\phi_m(k)$

$$\mathcal{B}_i(\rho, k) = i\mu_0 a\sigma_i k \frac{I_1(|k|a)K_1(|k|\rho)}{\beta(|k|a)}\phi_m(k).$$

Similarly, $\mathbf{B}_e(\rho, z)$ and $\mathcal{B}_e(\rho, k)$ due to the extracellular current density are obtained. The sum of the both $\mathcal{B}_i(\rho, k)$ and $\mathcal{B}_e(\rho, k)$ gives the $\mathcal{B}(\rho, k)$ which is connected via the Fourier transformation to the $\mathbf{B}(\rho, z)$. The whole transformation path in both directions from $\Phi_m(z)$ to $\mathbf{B}(\rho, z)$ is shown schematically on Fig. 3. The results of calculated and measured action potentials and magnetic field are shown on Figs. 4, 5, 6, 7.

680

$$\Phi_m(z) \underset{F^{-1}}{\overset{F}{\rightleftarrows}} \phi(k) \rightleftarrows \mathcal{B}(\rho,k) \underset{F}{\overset{F^{-1}}{\rightleftarrows}} \mathrm{B}(\rho,z)$$

Figure 3: Calculation steps from the transmembrane potential to the magnetic field. F denotes the Fourrier transformation, F^{-1} the inverse Fourier transformation.

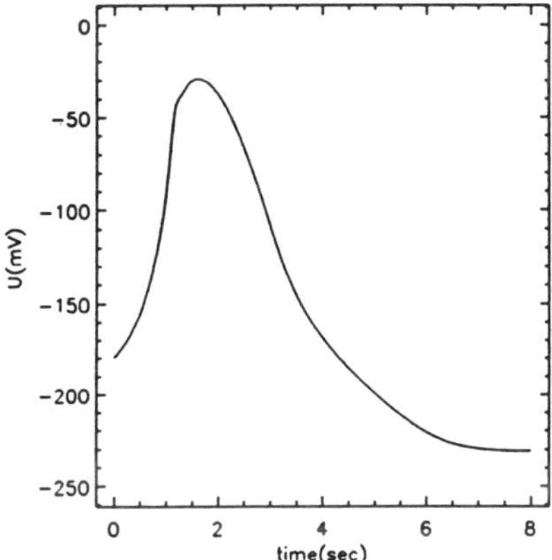

Figure 4: The first 8 seconds of measured action potential in the Chara corralina internodal cell.

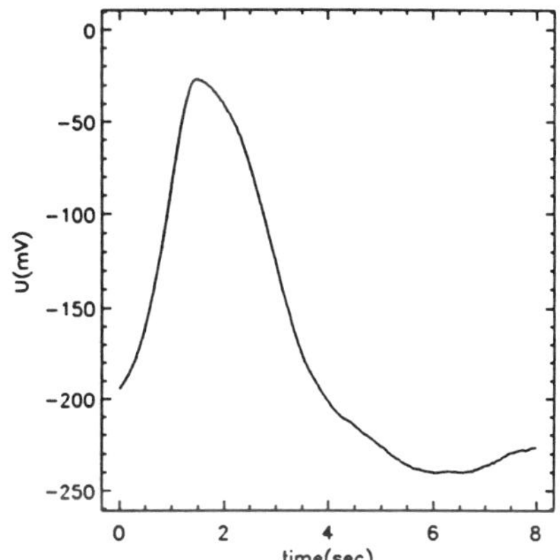

Figure 5: Action potential in the Chara corralina internodal cell calculated from the measured magnetic field shown in Fig. 6.

Figure 6: Measured magnetic field $6\,cm$ away from the Chara corallina internodal cell.

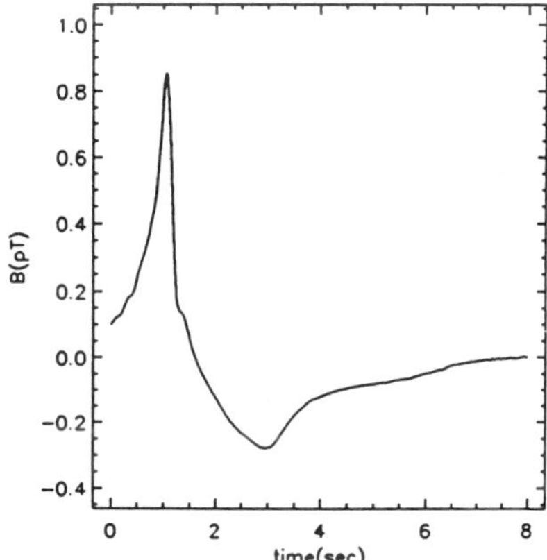

Figure 7: Magnetic field $6\,cm$ away from the Chara corallina internodal cell calculated from the measured action potential shown in Fig. 4.

Discussion and conclusions

Comparing the measured and calculated magnetic field and action potential for the internodal cell of green algae Chara corralina we notice a relatively good agreement between the calculated and measured signals, however, there are differences in the shape of signals.

We have to remember that mainly the intracellular current determines the measured magnetic signal in our experiment [1, 2]. The cylindrical symmetry, though valid within narrow area arround the Chara corralina internodal cell when considering the extracellular current density contribution, is correctly applied in our case.

It has to be mentioned that the internodal cell of Chara corralina consists actually of two membranes: The outer plasmalemma and the inner tonoplast which separates the cytoplasm from the inner vacuole. We believe our approximation where only the plasmalemma takes complete responsibility for the ionic exchange between extracellular and intracellular space is to big extent correct. That agrees with experimental results of plant electrophysiologists [5]. However, for the more precise calculation both membranes have to be considered separately and not only integrally as we did in our case. It is to expect that this way the improved agreement between experimental and calculated signal shape will be achieved.

We can conclude that expressions for electric potentials and magnetic fields of single large nerve axon, described in [3] and [4], can be applied also for the case of single internodal cell of green algae Chara corallina at least for the contributions of intracellular current density.

References

[1] Trontelj, Z., Zorec, R., Jazbinšek, V., and Erné, S.N. Magnetic detection of a single action potential in Chara corralina internodal cell, Biophysical Journal, 1994, 66:1694-1696.

[2] Jazbinšek, V., Trontelj, Z., Zorec, R., Erné, S.N., Lužnik, J., Pirnat, J., and Jagličić, Z. Magnetically detected intracellular current in single cell of the green algae Chara corralina, In: Baumgartner, C., Deecke, L., Stroink, G., and Williamson, S.J. Biomagnetism: Fundamental Research and Clinical Applications, Amsterdam, Elsevier, 1995.

[3] Clark, J., and Plonsey, R. A mathematical evaluation of the core conductor model, Biophysical Journal, 1966, 6: 95–112.

[4] Woosley, J.K., Roth, B.J., and Wikswo jr., J.P. The magnetic field of a single axon: A volume conductor model, Mathematical Biosciences, 1985, 76:1–36.

[5] Beilby, M.J., Electrophysiology of Giant Algal Cells, Methods in Enzymology, 1989, 174:403–442.

System for Magnetic Susceptibility Investigations of Human Blood and Liver

Sosnitsky, V.N., Budnik, N.N., Minov, Yu.D., Sutkovoj, P.I. and Vojtovich, I.D.

Biomagnetic Laboratory, Institute of Cybernetics, Kiev, Ukraine

Introduction

Recent years there is growing evidence of iron overloading involvement in the pathology of atherosclerosis, cancer, hepatitis, diabetes, diseases of joints. Reactions of free radical oxidation are involved in the pathology of most, if not all, human diseases. Free radical oxidation seems to be involved also in disease development in "iron overloaded" patients with high levels of the so-called "free iron" in tissues (the catalysis of peroxidation). This idea is supported, in particular, by severe cardiovascular disorders observed when excess ascorbic acid is given to iron-overloaded patients. This as well as our observations suggest the problem of exogenic iron role in steel and cast iron welders' pathology development.

We have created computer-aided systems in which these and some other problems are solved. One of the systems is used for recording of magnetic fields of the cardiovascular system, the other -- for investigation of magnetic susceptibility of various organs and blood. Signal processing and representation are performed by specially designed algorithms and programs.

Susceptibility Biomagnetic System

The biomagnetic system used to register the signals coming from blood and human body organs (induced by a test magnetic field) has the same basic components as the MCG system [1,2]. A movable patient's bed permits to move a patient in vertical and horizontal directions. The test magnetic field is created by the system of square Helmholtz coils. The case of each coil is made from the aluminum details when side equals 2 m. The distance between coils is 108.9 cm. The parameters of one coil are as follows: the numbers of turns, resistance and inductance are equal to 72, 14 Ω and 53 mH, respectively. The coils provide the alternating (14 Hz) test magnetic field of about 78 μT in the system center. The coils are driven by an alternating low-frequency signal from generator by the power amplifier. The equality of the currents in the coils is caused by them serial connection. It provides heterogeneity of magnetic field at least than 10^{-5} within 70×70×120 mm volume (in the center of the system). The patient's bed and the cryostat are situated in the middle of the coils' system so that the bottom coil of gradientometer was allocated exactly in the center of the system of the Helmholtz coils. A rubber bag is fastened to

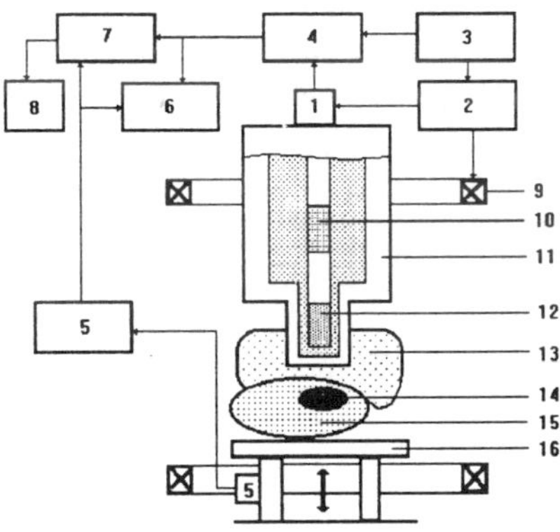

Fig.1. Cryoelectronic system for magnetic susceptibility measurement of human blood and liver: 1 - SQUID electronic; 2 - power amplifier; 3 - generator; 4 - lock-in amplifier; 5 - position transducer; 6-2-channel chart recorder; 7 - ADC; 8 - PC; 9 - Helmholtz's coils; 10 - SQUID-detector; 11 - radiotransparent helium cryostat; 12 - gradientometer; 13 - water bag; 14 - liver; 15 - patient's body; 16 - pneumatic driven bed

the cryostat tail and this is one for a parasitic paramagnetic air signal to be removed. The upper part of the bag is a Plexiglas disk having $\varnothing$45 cm (with the circular region for the tail, where the thickness was reduced) which rigidly fastened to the cryostat. The bag is filled with the distilled water. Under the movement of a patient upwards the volume of the bag is decreased and the water is poured into expanding reservoir by the connective pipe and lets a patient be taken down at the speed of 0.5 cm/s. The movement of the bed is caused by the pneumatic driver. The bed distance is registered by a position transducer. Despite of the undertaking efforts to receive a high degree of balance of antenna and homogeneity of the magnetic field , the signal created by the current in the coil fill a full dynamic range of SQUID-magnetometer at the value of this current beginning from 0.2 A. To eliminate this effect the compensation of the signal created by the magnetic field is used. Here a part of the signal from the output of the power amplifier is fed in the feedback circuit of the magnetometer. The output signal of the magnetometer is fed by lock-in amplifier to chart recorder or by ADC on the PC. The design of the system is presented in Fig.1.

Magnetic biopsy of liver.

Method of measurement. A test magnetic field with the frequency of 14 Hz is used under the magnetic biopsy. Before measuring the amplitude and phase of a signal which enters from an output of a power amplifier in a circuit of feedback of the magnetometer, are fitted so that the value of a "parasitic" signal for the frequency of 14 Hz (and its harmonic) is minimized from the output of the magnetometer. A measuring procedure is as follows. A patient idles (on a bed) at an angle of $45°$, where a liver is in an upper position. The bed is lifted in an upper position, then the bag fits tightly the body of the patient. Thus, a "parasitic" signal because of air is removed. Under measuring the bed is lowered to a distance of 5 cm for a time of 10 s. The output signal of the magneto-meter, which is proportional to the magnetic susceptibility of the liver, is input of a lock-in amplifier. A signal, which is proportionate to a distance between the liver and a bottom coil of the gradientometer, is registered from an output of a transducer of bed displacement. These signals are input of a 2-channel chart recorder. Further, the mentioned signals are sent conveyed to PC through ADC. The measurements are performed a few times to improve the SNR. Strictly speaking, the signal from the output of the lock-in amplifier is proportionate to the difference of magnetic susceptibilities of the liver and distilled water. The desired concentration is found by the expression $C=K[dV(C)-dV(0)]$, where K, $dV(C)$ and $dV(0)$ are the calibrating coefficient, the change of the signal from the liver depending on the distance, and the change of signal for the case of iron absence, respectively.

The value of the change of the signal depending on the distance is determined as a difference of the signal between two points X_1 and X_2, where the dependence of the signal on the distance is about linear $dV=V(X_1)-V(X_2)$. Choice of these points is complied with by trials to obtain the above mentioned most linear dependence. For this case, points have corresponded to the distances of 2 cm and 5 cm between the bottom coil of the gradientometer and surface of patient body.

The "parasitic" signals, which are present in the process of the measurements, can be divided into two classes. The signal from the Helmholtz coils and the vortex currents, generated by test magnetic field, fall into first class. These signals are minimized by mechanical balancing of the gradientometer and, finally, by fitting the compensation of both amplitude and phase in the feedback circuit of the magnetometer. The above mentioned signals do not depend on the distance to the liver and, therefore, are removed by data processing since they give the same contribution in the output signal for both a neighboring and outlying position of the liver.

The signals from the bed and body of patient fall into second class of the "parasitic" signals. These signals depend on the distance to the liver and, therefore, give a contribution in the difference dV. However, the signal from bed is taken into account by the calibration. This is also true for almost all parts of body surrounding the liver. The magnetic susceptibility of these parts equals approximately the susceptibility of distilled water. Under calibrating the signals from both lungs and blood circulating in the liver are not taken into account. The signal from lungs (its susceptibility differs from the susceptibility of water because of the presence of air in lungs) is very small that is argued by the corresponding calculations [3]. The liver contains 15-18 ml of blood for 100 g of the tissue of the liver, that give the equivalent concentration of iron in 78-94 μg/g. If it is necessary, this value can be subtracted from the measured signal.

The calibration has been carried out by glass (250 ml) and rubber (500 ml) capacities, which have been filled up by a solution of $FeCl_3 6H_2O$ with equivalent concentration of iron of 0.1; 0.5; 1 and 5 mg/ml. Comparing the received result with a theoretical dependence of the output signal of the magnetometer on the magnetic susceptibility, we obtain a calibration characteristic for arbitrary concentration of iron in the liver. Here it was enough to measure two values of the concentration, since both the theoretical calculations [3] and experimental results show a good linear relation between the output signal and the concentration of iron. It can be concluded analyzing the shape of this characteristic that the concentration of iron of 0.1 mg/g and higher can be measured by the described device. The calibration signal is accumulated for several measurements, then this signal is averaged and the value dV is calculated. The calibration coefficient is computed by $K = [C_1 - C_2]/ [dV(C_1) - dV(C_2)]$, where C_1-C_2 and $dV(C_{1,2})$ mean the difference of the concentrations for two measurements and the corresponding difference of the output signal, respectively. Such calibration by a "chemical" standard should be implemented after each engaging of the device. However, it is inconvenient for clinic conditions. An "electric" calibration by a coil, which is fed by rectangular impulses of a current of 14 Hz, is more convenient. The amplitude of the impulses in the coil is calibrated so that the signal from the output of the magnetometer equals the signal from the capacity with the known concentration of iron.

The main systematic error changes of dependence on the concrete patient and relates to an uncertainty of a thickness of the skin-liver layer. According to the published results, the nominal value of the thickness is equal to 20 mm, a range is within 15 and 25 mm. This range leads to an uncertainty of a value of iron concentration of 2.5 mg/g; this value is completely improper. To decrease this error the ultrasonic technique should be used, which has a precision of determination of the mentioned thickness of 1mm, that allows to diminish the error by about 10.

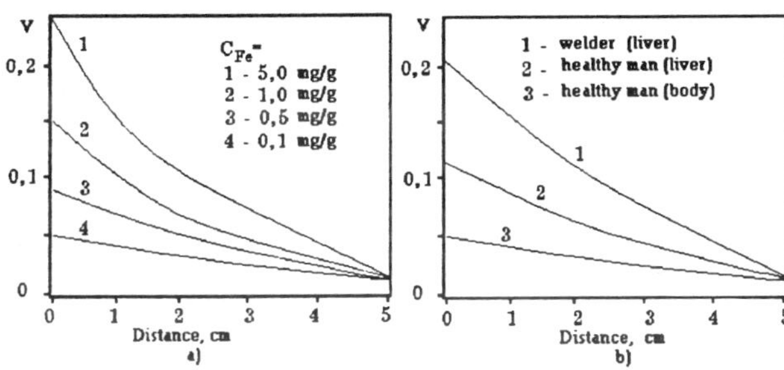

Fig. 2. The dependence of the output signal of biosusceptometer on a distance to the set of the rubber capacities with different calibration values of iron concentration (a), the liver of a healthy person and welder (b).

It is seen from Fig.2a that the iron concentration higher than 0.1 mg/g liver can be measured by the described measuring system. This means that a normal concentration (0.1-1 mg/g) and an overloading (more than 1 mg/g) can be recorded. Fig.2b presents the results of the measurements of the magnetic susceptibility of the liver of both a healthy man and a welder suffering from occupational diseases. A concentration of 0.8 mg/g liver has been derived for the healthy man. It corresponds to the normal value. The iron concentration for a group of healthy subjects varies within 0.5-0.8 mg/g. A high iron concentration about 3 mg/g has been registered for the welder. The distance to object is measured from its upper position. The signal from the body (curve 3) has been received for the position of the patient when the liver is situated in the lower position.

Magnetoplethysmography (MPG)

MPG means the measurement of the blood volume changing for a cardiocycle from human organs by a signal from blood in external test magnetic field. This signal can be elicited due to a difference between the magnetic susceptibilities of blood $\chi_b = 5 \cdot 10^{-6}$ and body (χ_{water}). The change of the blood volume in heart for cardiocycle leads to the change of magnetic field in the bottom coil of the gradientometer. The essence of the method of MPG developed in [4] is based on the application of a DC test magnetic field with a value of 135 A/m (170 μT). We use the AC field with a frequency of 114 Hz, then the noise of the magnetometer and various external noises are considerably decreased. Furthermore, a "parasitic" MCG signal has no influence for this frequency.

Along with obtaining the MPG the system is used for the magnetic biopsy of the liver, where technical changes are introduced as follows. The LFF with a cut-off frequency of 30 Hz is included after the lock-in detector. A capacitance of 22 µF is included serially to the Helmholtz coils for compensating its inductance. This makes it possible to increase a current in the coils up to 1,1 A. Here the ECG is used as an input signal for the reference channel instead of the signal of a bed displacement. Before measuring, the amplitude and phase of a signal of 114 Hz (which is sent from the power amplifier output to the NF circuit of the magnetometer) are fitted in an appropriate manner.

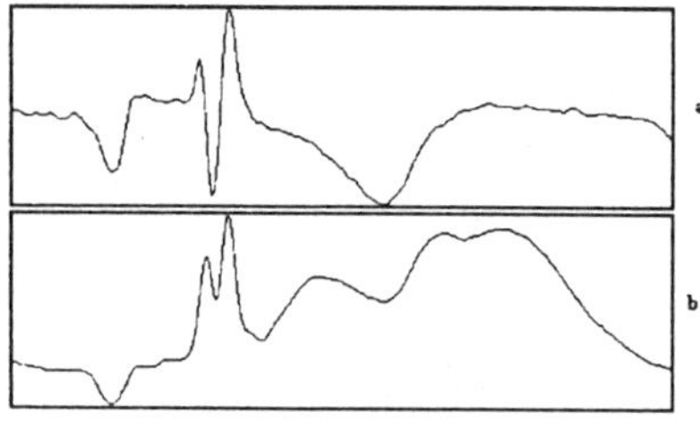

<table>
<tr><td>

Fig.3. The magnetoplethysmogram of heart of a healthy subject (a) and a welder with iron overloading (b)

</td><td>

Fig.4. The magnetoplethysmogram of liver of a welder (ECG is presented for reference).

</td></tr>
</table>

The MPG of healthy man and welder having an augmented content of iron in blood presented in Fig. 3. This iron overloading is registered by biochemical analysis too. The iron content level in blood is so high among some welders that it became possible for us to have registered liver MPG for the first time (Fig.4).

Conclusion

Computer-aided systems for recording, processing and representation of own or induced magnetic signals emitted by various human organs are created. Relaxation-oscillation (RO) SQUID-magnetometers are the basis of these systems [1]. When ROs are used in a SQUID-magnetometer, it becomes possible to refuse from the modulation-demodulation scheme applied in the conventional analog mode. The result is that we can pass from the microvolt signal registration to the millivolt pulse signals. There appears the ability to simplify the electronics, to enhance SR and noise-immunity, to widen the dynamic range, to speed up the searching process. This circumstance provides the reliable operation of the device in the presence of the high industrial magnetic noise level without the special means for magnetic vacuum creation (magnetic shielding room, active Helmholtz coil compensation).

Over the last several years a group of scientists from Institute for Occupational Health of the Ukrainian Academy of Medical Sciences, Biomagnetic Laboratory of Institute of Cybernetics and Laboratory for Finely Dispersed Ferromagnetics of the Research SONAR Center for Biotechnical Systems of the National Academy of Sciences of the Ukraine is engaged in research of studying the impairments in welders' health due to exposure to welding aerosols. The methods that were use: standard biochemical, immunological methods as well as biomagnetometry.

References

[1] Budnik, N.N., Minov, Yu.D., Sosnitsky, V.N., Sutkovoj, P.I., Vojtovich, I.D. Pulse-relaxation oscillation SQUID-magnetometer, Proc. XIII IMEKO World Congress, Torino(Italy), 1994, III, 2383-2387.
[2] Budnik, N.N., Gapelyuk, A.V., Minov, Yu.D., Primin, M.A., Sosnitsky, V.N., Vojtovich, I.D. Human biomagnetic signal measurement for diagnosing some diseases, Ibid., 2509-2514.
[3] Romanovich, S., Sosnitsky, V., Vojtovich, I. Investigation of biomagnetic field using two-dimensional model of secondary sources, Proc. 9th Int. Conf. Biomagnetism, Vienna (Austria), 1993, 455-456.
[4] Maniewski, R., Katila, T., Poutanen, T. et al., Magnetic measurement of cardiac mechanical activity, IEEET, 1988, BME-35, 662-670.

Registration of Iron Overload by Means of SQUID-Magnetometry When Coupled with Low-Temperature ESR Spectroscopy

Sosnitsky, V.N.[1], Mikhailik, O.M.[2], Lubyanova, I.P.[3], Gapelyuk, A.V.[1], Budnik, N.N.[1], Minov, Yu.D.[1], Sutkovoj, P.I.[1], Pankratov Yu.V.[2] and Bakai, E.A.[2]

[1]Institute of Cybernetics, Ukrainian Acad. Sci., Kiev, Ukraine; [2]Research SONAR Center for Biotechnical Systems, Ukrainian Acad. Sci., Kiev, Ukraine; [3]Institute for Occupational Health, Ukrainian Acad. Med. Sci., Kiev, Ukraine

Introduction

At present new noninvasive biomagnetic methods of tissue iron quantification for primary and secondary hemochromatosis diagnosis are gaining fast development mainly as liver susceptometry [1]. However, often iron accumulation in other organs (spleen, heart, pancreas etc.) is also of importance. Studies have been performed on ferromagnetic contaminants in the lung and estimation of lung clearance, it was determined that in a year after contamination of the lungs by magnetite ranged from 10% to 50% elimination of ferromagnetic materials occurred [2,3]. Whether these particles enter the blood flow, whether they accumulate in other organs have not yet been clear. Our examinations of more than 2000 of iron welders have revealed the possibility of iron overloading syndrome development during iron inhalation intake from welding aerosol. In hemosiderotic patients accumulation of iron in cardiac muscle results in impairment of cardiac activity, enlargement of myocardium mass of the left ventricle [4], this was found to be also true for welders with pneumoconiosis [5]. The purpose of the present work was to elucidate the peculiarities of iron state in patients with iron body overload resulting from inhalation exposure to iron containing welding aerosols by means of magnetocardiography (MCG), magnetic-susceptibility plethysmography (MSPG) and low temperature electron spin resonance spectroscopy (ESR) methods.

Methods

Men-welders with occupational diseases of bronchopulmonary system were examined. The diagnosis of iron overload was confirmed primarily by biochemical parameters (serum iron and serum iron binding capacity). In the process of MCG studies a component of the magnetic field perpendicular to the chest surface in 36 points of right-angle measuring grid with a step of 4 cm was registered. The measurements were made with single channel SQUID-magnetometer under the following parameters: sensitivity - 30 $fT/Hz^{1/2}$, frequency band 0.1-100 Hz. Recording duration for each point amounted to 30 sec, discretization frequency - 500 Hz. MSPG were registered over the end of left ventricular in the alternating magnetic field of about 0.1 mT with a frequency of 114 Hz to decrease the noise of magnetometer as well as external noises. The ESR spectra of blood and serum samples were registered at microwave radiation frequency ranged from 9.28 to 9.30 GHz, microwave power - 5-50 mW, magnetic field - 500-5500 Oe, temperature 100-150K. The levels of serum transferrin, iron in the form of serum ferritin as well as the degree of serum transferrin and ferritin saturation with iron were calculated on evidence derived from low temperature ESR data with calibration of ESR ferritin spectrum according to gamma-resonance spectroscopy data.

Results and discussion

Among welders with pneumoconiosis patients with iron serum concentration above 37 μmol/l were registered frequently than in patients with chronic bronchitis. Total serum iron binding capacity in welders amounts to 91±5 μM. The major differences in morphology of the MCG curves in the group of welders relative to the control group were observed at ST-T interval corresponding to the period of mechanical systole at the absence of heart electric activity (Fig.1), MCG curves showed the appreciable deviation from magnetic isoline (10-30 pT) in comparison with the control one (1-2 pT). This deviation was chosen as an index A_d (Fig.1b). The average integral coefficient for all 36 points $<A_d>$ obtained when analyzing magnetograms of 10 men from the control group which were never exposed to iron inhalation and had no diseases of cardiovascular system did not exceed 0,5 relative units. This index in 18 examined welders was higher than 0,5 and amounted to 4.1 relative units.

Evidently, the welders MCG signal is superposition of the signal due to electric cardiac activity and magnetic susceptibility signal directly related to cardiac volume changes and movements only with the Earth magnetic field as magnetizing one. In the magnetic field of the Earth a magnetic susceptibility signal of about 1 pT at most can be expected [6]. The changes of the amplitude approximately by ± 2 pT when changing the direction of the magnetized external field strength 135 A/m were also observed [7] and explained as occurring due to moving of blood with paramagnetic species as a result of changes in heart volume. When the heart is exposed to magnetic field 520 A/m the maximum amplitude of the observed signal is about 8 pT [8]. In our case the signal amplitude up to 30 pT was observed only at the Earth magnetic field.

The considerable increase of $<A_d>$ and A_{dmax} was noticed in welders with pneumoconiosis and increased liver echogenicity. It is likely to suppose that the registered magnetic signal is associated with deposition of magnetic iron species in the lungs or liver. However, according to Maniewski's estimations [6] magnetic susceptibility plethysmography signal contributes 85 to 90% of the total magnetic susceptibility signal measured above the heart and the signals caused by displacement of the chest surface and by the motion of the chest wall-lung boundary contribute 15 to 20 and 5 to 10%, respectively. Even considerable increase of concentrations of paramagnetic blood constituents as transferrin and low-molecular iron complexes cannot elucidate observed ten-fold increase of the magnetic component amplitude. Four cases of patients with high level of serum iron and transferrin in blood but with normal signal amplitude at TP interval support the idea of the other source of magnetic component which is registered in a number of subjects. At the same time 1/3 of welders with high A_d showed absence of changes in the lung roentgenogram typical for pneumoconiosis in welders. In 1/3 of welders of this group echogenic liver properties were not changed, a high magnetic component was registered in two welders with normal echogenic liver properties. Thus, there is a reason to suppose that intensive signal in the area of quadrants C3-C5, D3-D5, E3-E5 according to the measuring grid can be associated with high concentration of ferritin in the blood and/or ferromagnetic inclusions entering the blood flow and/or cardiac muscle from the lungs of the examined welders.

The examination of blood and serum samples of patients with high MCG deviation from the magnetic isoline at T-P interval (Fig.3) and their comparison with the spectra of ferritin (Fig. 3f) has shown that in addition to the components which can be referred to ferritin [9] new ones are registered. Similar spectra were registered in tissues of experimental animals after biodegradation of intravenously administered finely dispersed metallic iron. It is important to note that the intensity of new registered components in the whole blood significantly exceeds that in serum samples. Thus, the sources of these signals may partially deposit with blood elements when receiving the serum and the index content of iron in the blood serum does not reflect the true content of non-heme iron in the whole blood. The changes in the position and width of the absorption lines under rotation of the blood samples frozen in the residual field of ESR-spectrometer (250 Oe) (Fig. 3a-3e) testify sufficiently high magnetic properties of the absorbing species. The clusters and small particles of oxidized iron with a few magnetic subgrids and their aggregates can be considered as these particles [10]. One could attribute these lines to ferritin molecules aggregates displaying strong magnetic anisotropy and/or to iron magnetic impurities that penetrate from the lungs into the blood flow.

MPCG curves for examined patients with iron overload are of rather complex character. It was found that parameters iron in the form of serum ferritin and degree of serum ferritin saturation with iron are in interrelation with MPCG curves. High (up to 30 pT) amplitudes of plethysmograms (Fig. 2c) suggest the idea about the presence of species with high magnetic properties in the blood of some patients and impulsive deviations toward positive magnetic fields in a phase of heart contraction seems to be an evidence of magnetic species accumulation in a heart muscle (Fig. 2b). It is noteworthy that for patients with such kind of deviations in MPCG increased organ echogenicity was registered and the total iron binding serum ferritin ability was low. The highest amplitude of MPCG curve (Fig.2b) was observed in a patient with maximum iron concentration in serum ferritin (5600 μg Fe/l) and highest degree of serum saturation with iron (47%). It seems that the low ability to apoferritin synthesis leads to accumulation of iron in organs whereas high capacity for ferritin production to some extent prevents from deposition of iron in tissues. As in the case of liver magnetic biopsy [6] common parameter as transferrin level as well as the degree of its saturation with iron have no diagnostic value for estimation of nonheme iron in organs whereas amplitudes of MPCG curves and the parameters A_d correlates with observed frequency of pneumoconiosis and chance of increased liver echogenicity (Fig.4).

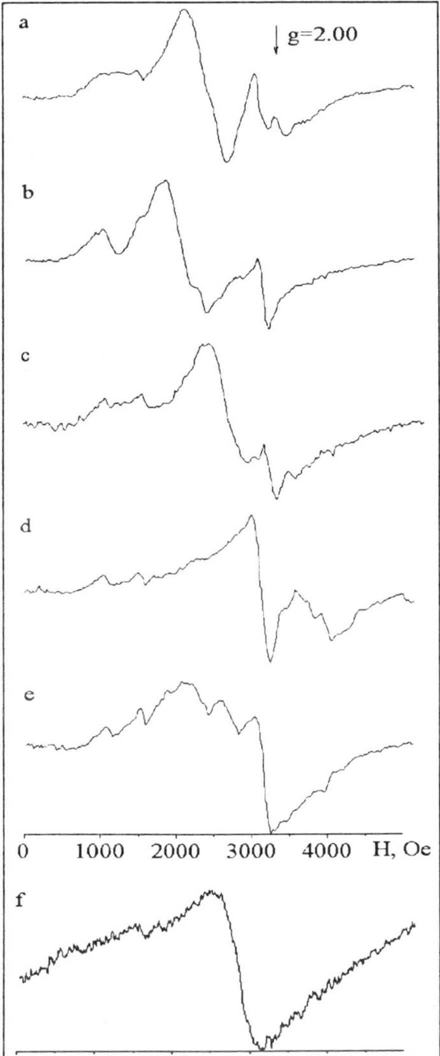

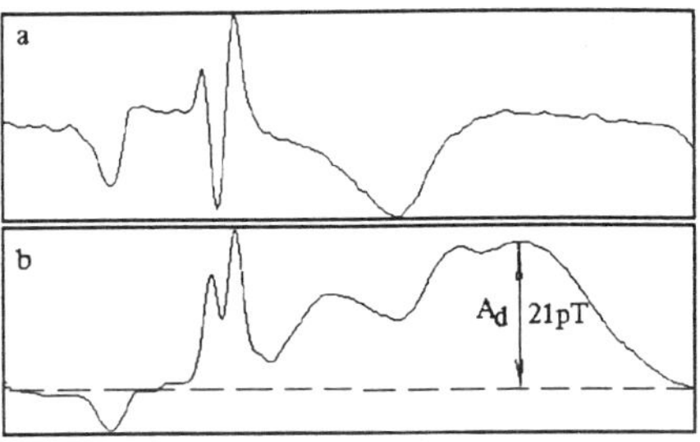

Fig.1. Magnetocardiograms of a healthy patient (a) and of a welder with iron overload (b) in C4 points of the spatial grid.

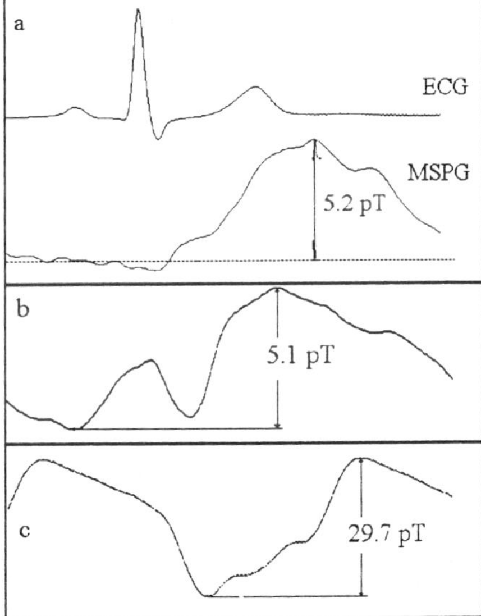

Fig. 2. Electrocardiogram (ECG) and magnetic-susceptibility plethysmograms (MSPG) for a healthy subject (a) and for patients with iron overload as a result of iron inhalation entry, iron in serum ferritin 220 and 5600 μg/l, respectively (b and c).

Fig. 3. The ESR spectra performed on blood sample of a welder with maximum deviation (A_d=31pT) from isoline in MCG (a) and after sample fast freezing in a magnetic field of 250 Oe at the angles between the magnetic field vectors during sample freezing and spectrum registration 0° (b), 30° (c), 60° (d), 90° (e), 180° (a) and ESR spectrum of highly saturated ferritin (f) performed on serum sample of a welder with heavy iron overload (5600μg/l iron in ferritin); ν=9.3 GHz, microwave power 50 mW, HF-modulation amplitude 10 G, T=124K.

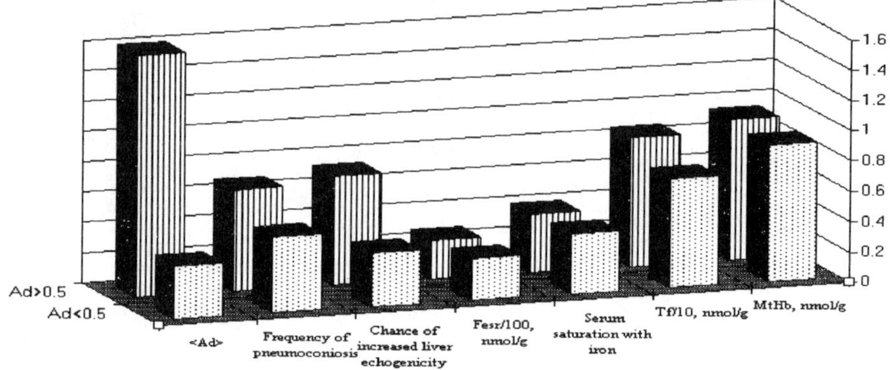

Fig. 4. Average deviation from isoline Ad (rel. un.) at ST-T interval of magnetocardiogrames and some characteristics of iron metabolism for two groups of patients.

Conclusion can be made that common calibration in biomagnetometry for quantitative iron determination using paramagnetic iron compounds and magnetic susceptibility determined for ferritin with low degree of saturation with iron is not appropriate for cases of significant iron overload in the presence of magnetic species. The index - content of iron in the blood serum - does not reflect the true content of non-heme iron in the whole blood under inhalation iron entry, the parameters iron in the form of serum ferritin and degree of serum ferritin saturation with iron are in interrelation with MSPG curves. To elucidate the nature of the revealed magnetic signals both in magnetograms and in the ESR spectra is of importance for solving calibration problem in the case of body iron overload. Not only the enhancement of 'free' iron concentration but also the local magnetic field of iron magnetic species could be considered as an important factor biological action (in particular, as promoter of lipid peroxidation) and one of the main mechanisms of pathology development. In this context the development of methods ferromagnetic species detection in tissues seems rather important. MCG and MSPG when coupled with ESR spectroscopy has much potential for yielding information about kinds and content of iron pools in organism under the iron overload.

References

[1] Fisher, R. and Heinrich. H.C. Biosusceptometry - current status of clinical diagnostics and biomagnetic research, In: Hoke, M. et al. Biomagnetism: clinical aspects, Elsvier Science Publishers, 1992.

[2] Cohen, D. Ferromagnetic Contamination in the Lungs and Other Organs of Human Body, Science, 1973, 180: 745-748.

[3] Kalliomaki, P.L., Aittoniemi, K. and Kalliomaki, K. In: Williamson, S.J. et al. (eds) Biomagnetism: An Interdisciplinary Approach, New York, Plenum Press, pp. 559-568.

[4] Yeliseev, O. M. Heart lesion in hemochromatosis, Ter. Arkhiv., 1988, No 8: 141-147 (in Russian).

[5] Krasnyuk, E. P., Lubyanova, I. P., Koshkina et al. The state of hemodynamics of chronic bronchitis and pneumoconiosis in welders, Vrachebnoye delo, 1988, No 3: 113-115 (in Russian).

[6] Maniewski, R. Magnetic studies on mechnical activity of the heart, Clin. Rev. Biomed. Eng., 1991, 19: 203-229.

[7] Katila, T., Maniewski, R., Tuomisto, T. et al. Magnetic measurement of cardiac volume changes, IEEET, 1982, BME-29: 18-25.

[8] Manievski R., Katila, T., Poutanen, T., Magnetic measurements of cardiac mechanical activity, IEEE Trans. Biomed. Eng., 1988, 35: 662.

[9] Weir I D, Peters T J, Gibson J F. Electron Spin Resonance studies of spleen ferritin and hemosiderin. Biochim. Biophys. Acta 1985; N 828: 298-305.

[10] Vonsovsky, S.V. Magnetism, Moskow, Nauka, 1971 (in Russian).

V. Magnetoencephalography (MEG): Basic Research

MEG Studies of Cognition

Salmelin, R.

*Low Temperature Laboratory, Helsinki University of Technology,
Espoo, Finland*

Introduction

In this paper, the vast potential of MEG in revealing cortical processes of cognition will be discussed through experiments on picture naming, visual word recognition in normal and dyslexic subjects, and visual and motor imagery.

Methods

The measurements were done with a Neuromag-122™ whole-head magnetometer [1]. In this device, the 122 sensors are arranged in a helmet-shaped holder. In each block, there are two figure-of-eight planar pickup coils which are most sensitive to currents in the two orthogonal directions along the plane of the helmet. These gradiometers detect maximum signal just above a current source, where the field gradient is largest.

In addition to the on-line averaged evoked responses, also the on-going cortical activity was recorded and stored on a magneto-optical disk for later off-line analysis. Cortical rhythms often show systematic event-related changes. The oscillations are, however, not exactly time-locked to the stimulation or task and, thus, will not show in ordinary averaged responses. The modulation of cortical rhythms was quantified using the Temporal Spectral Evolution (TSE) method [2], where the signal is first filtered through a passband suggested by the frequency spectrum, typically around 10 and 20 Hz, and then its absolute value is taken. When this signal is averaged with respect to stimulus/task onset, one gets the average amplitude level of the narrow-band oscillations as a function of time. The cortical rhythms are modulated over periods of a few seconds whereas the time-locked evoked responses typically have a time scale of less than half a second.

Picture Naming

In this study the subjects were shown pictures of everyday objects. The measurement started with a passive viewing condition. In the overt naming task, the subjects were told to name the objects aloud as soon as possible. During the covert naming task, they were asked to name the pictures mentally, not preparing for vocalization [3].

Figure 1 shows the cortical dynamics of picture naming in a single subject. The occipital visual cortex was activated first. During the next 200 ms, bilateral signals occured over the temporo-parieto-occipital junctions. The early frontal source presumably indicates motor preparation for mouth movements. About 500 ms after picture onset, signals were observed in the inferior parts of the frontal lobes; the left-hemisphere region likely represents Broca's area. A site close to the vertex, slightly frontal to the foot area, then became active, apparently reflecting involvement of the supplementary motor area in coordination of vocalization. The early visual response and the signal in the right temporo-parieto-occipital junction were the same for all tasks. All other sources showed naming-specific responses, most strongly those in the left hemisphere. The traditional view maintains that the left hemisphere dominates in language tasks. In this experiment, however, signal processing during picture naming clearly advanced symmetrically in both hemispheres, from the visual to language- and eventually to vocalization-related areas.

Figure 2 illustrates the results of the TSE analysis of the 10-Hz activity during passive viewing and covert and overt picture naming, in the posterior, midline, and frontal areas. In fact, our original assumption was that very little would be seen in the phase-locked responses because of the large variation in latencies one might expect in this kind of task and that one would rather have to rely on the TSE analysis. During passive viewing, rhythmic activity was reduced only in the posterior region, that is, only the visual areas were strongly involved in this task. This effect was sharpened by covert naming. Overt naming led to a prominent suppression in the back of the head, lasting for about 3 s. It was followed by suppression in the frontal sites and, finally, along the central sulcus. The gradient isocontour maps of the overt naming condition illustrate how, immediately after stimulus onset, the parieto-occipital rhythms are strongly diminished. At about 500 ms, during the maximum suppression of the posterior rhythms, 10-Hz activity in the hand areas is briefly enhanced. About 900 ms after stimulus onset, suppression spreads to the inferior frontal cortex in both hemispheres. The signal processing thus seems to advance from visual to language and eventually to motor areas, in general agreement with the time-locked responses.

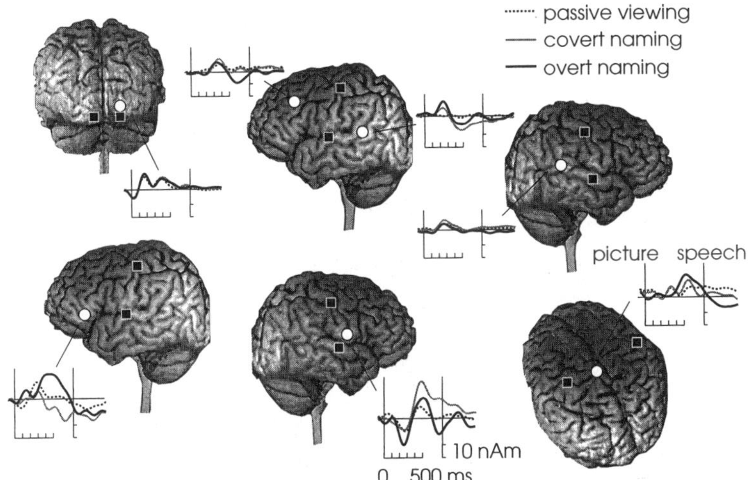

Fig. 1. Activation sequence in one subject during picture naming. The black squares indicate the visual, auditory, and hand somatosensory cortices for reference. The orange dots show the active cortical sites identified during the overt naming condition. The curves depict the temporal behaviour of the source areas over a period of about 1 s. The picture was shown at the first vertical line and vocalization started at the other vertical line.

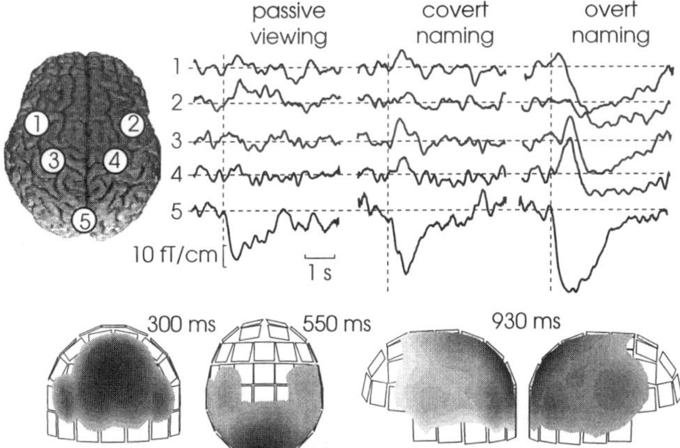

Fig. 2. TSE curves of 10-Hz oscillations during passive viewing and covert and overt picture naming in the numbered brain areas, plotted over a period of 5 s, starting 1 s before picture onset. The base levels are shown by the thin horizontal lines. Levels below and above zero indicate suppression and enhancement of rhythmic activity, respectively. The gradient isocontours below show the spatial extent of the maximum suppressions/enhancements at selected time instants during overt picture naming. Adapted from [3].

Visual Word Recognition in Normal and Dyslexic Subjects

Six subjects with developmental dyslexia and 8 control subjects participated in the experiment [4]. The dyslexic and control groups were matched for age and level of education and all subjects were also tested behaviourally. These dyslexics have eventually learned to read fluently but are still quite slow compared to normals.

The subjects were shown 7-8 letter Finnish words. Four categories alternated randomly. There were concrete words, abstract words, pronounceable pseudowords with no meaning, and meaningless random letter strings with the same amount of consonants and vowels as in the other classes. To keep the subjects alert, they were instructed to say aloud the rarely occurring word 'giraffe' which was not included in the analysis.

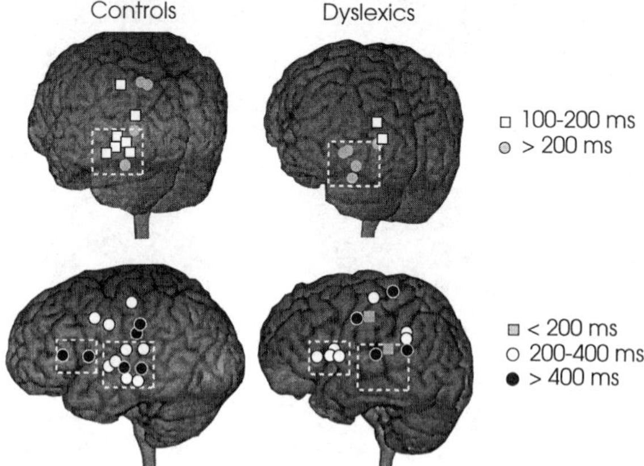

Fig. 3. Cortical areas activated by visually presented words in control *(left)* and dyslexic subjects *(right)*. Only the left posterior *(above)* and lateral views, summarizing areas with clear differences between the two groups (dashed rectangles), are displayed. The symbols indicate the different time windows.

The dominant cortical areas and the latencies of maximum activation were determined for all subjects. Figure 3 summarizes the active cortical areas for all control subjects on the left and for all dyslexics on the right. Activation within 200 ms after stimulus onset occurred predominantly in the posterior parts of the brain. After 200 ms, activity spread to the temporal and frontal areas. Controls and dyslexics showed clearly different early activation in the left inferior temporo-occipital cortex. In controls, strong, transient signals were recorded in this area within a very narrow time window about 180 ms after word presentation. In the dyslexics, however, the activation peaked only at about 500 ms, if at all. The inferior temporal sulcus or fusiform gyrus has recently been reported to show letter-string specific responses, in addition to the previously found face-specific signals [6, 7]. Thus, it appears that the coherent response to letter strings is missing or delayed in the language-dominant left hemisphere of the dyslexics.

After 200 ms, temporal and frontal cortices became involved. In controls, the left temporal lobe showed strong activation between 200 and 400 ms but, in dyslexics, this area was practically silent. This result supports the view that phonological processing in the left temporal lobe is impaired in dyslexics. Interestingly, left inferior frontal cortex, approximately agreeing with Broca's area, showed activation within 400 ms after word presentation in dyslexics but not in controls. These signals raise the possibility that the dyslexics, in order to compensate for the deficits in the automatic word recognition, in fact try to 'guess' the words, thus adopting rather a top-down than a bottom-up approach.

Our results suggest that, in addition to the phonological impairment, the visual word form perception is impaired in dyslexics, which could be the immediate reason for the manifest difficulties with written words. The obvious question is whether the visual and phonological problems are independent or whether the phonological deficit could in fact lead to the observed visual impairment. We first learn to listen to words and only much later make the connection between the symbolic written words and the original phonological code. Impaired phonological processing could thus impede the formation of a word recognition unit which would automatically set letters apart from other objects and facilitate fast reading.

Visual Imagery

There is an on-going debate of whether the same brain areas are involved both in actual sensory perception or motor performance and in imagery. Simple visual stimulation results in a brief suppression of the posterior cortical rhythms, also illustrated in Fig. 1. In fact, the posterior 10-Hz activity does not originate only in the calcarine fissure. Figure 4 shows how another source cluster is found along the midline, in the parieto-occipital sulcus, clearly distinct from the primary visual cortex. Rhythmic activity in both of these areas is suppressed by visual stimulation.

But what happens when, instead of an actual visual stimulus, one imagines a picture in the 'mind's eye'? In this study, the subjects had their eyes closed and they were given a name of a letter through earphones and asked to visualize its upper-case form [8]. After 3 s, they were asked a question about that letter, for example, whether the letter was symmetrical with respect to the vertical axis. To answer yes or no, they lifted the left or right index finger. Figure 4 shows the averaged modulation of the parieto-occipital 10-Hz rhythm as a function of time in two subjects. In Subject 1, rhythmic activity remains suppressed throughout the imagery task, whereas Subject 2 shows two distinct suppressions, one after the letter is given through the earphones and an even stronger one when he has to decide about its visual properties. The gray curves show the result of a control task which required auditory instead of visual imagery and, obviously, did not reduce the parieto-occipital 10-Hz rhythm. The same cortical areas are thus affected in a very similar way both by actual visual stimulation and by visual imagery.

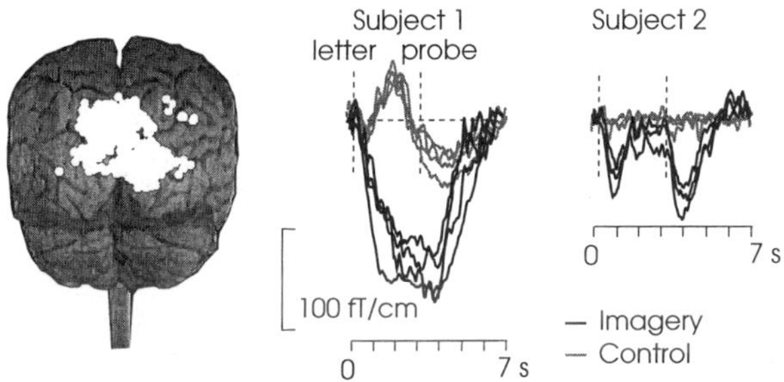

Fig. 4. Sources of posterior 10-Hz rhythm, with two distinct clusters. The curves show suppression of the 10-Hz rhythm during visual imagery and during a control task requiring auditory imagery in two subjects, with replications. Adapted from [8].

Motor Imagery

In this study, the right median nerve was stimulated at the wrist to induce a sharp thumb twitch. Just like voluntary movement, electric median nerve stimulation is also followed by bursts of 20-Hz activity [2, 9]. The 20-Hz oscillations originate predominantly in the motor cortex [2, 10]. When the stimulated right hand was at rest, there was a typical burst of 20-Hz activity in the left central sulcus, with the TSE curve peaking at around 400 ms after stimulus onset. When the subject continuously explored an object with the stimulated hand, the 20-Hz burst was completely suppressed. When he stretched his fingers, keeping them tense, the rhythms were also somewhat reduced. But, most interestingly, just imagining the exploration task resulted in a prominent suppression of the post-stimulus 20-Hz activity, without any appreciable signal in the simultaneously recorded electromyogram. Thus, the primary motor cortex is involved in motor imagery [11].

Another interesting feature is the phase-locked evoked response. The sharp 20-ms response is followed by a 30-ms deflection with opposite polarity and a strong 60-ms component. The 30-ms deflection is suppressed and even reversed by simultaneous exploring movements. But, neither stretching nor imagined exploration affect the early components in any appreciable way. Thus, we have an interesting situation where the somato*sensory* evoked response is not changed by motor imagery whereas the simultaneously recorded spontaneous rhythmic activity of the *motor* cortex is strongly affected.

Conclusion

Clearly, whole-head neuromagnetometers have made MEG studies of cognition truly feasible. Both phase-locked evoked responses and modulation of cortical rhythms, which is also less sensitive to slight variation between repetitions, are extremely useful in these studies. MEG provides the accurate timing which is quintessential for understanding the functional role of distinct cortical areas in cognition, even if the localization accuracy certainly suffers from the inverse problem. And finally, analysis of individual data sets, instead of group averages, allows one to fully appreciate – and also suffer from – differences between individual activation patterns which derive, at least partly, from distinct strategies in cognitive processing.

References

[1] Ahonen, A.I., Hämäläinen, M.S., Kajola, M.J., Knuutila, J.E.T., Laine, P.P., Lounasmaa, O.V., Parkkonen, L.T., Simola, J.T., and Tesche, C.D. 122-channel SQUID instrument for investigating the magnetic signals from the human brain, Physica Scripta, 1993, T49: 198-205.

[2] Salmelin, R., and Hari, R. Spatiotemporal characteristics of sensorimotor MEG rhythms related to thumb movement, Neuroscience, 1994, 60: 537-550.

[3] Salmelin, R., Hari, R., Lounasmaa, O., and Sams, M. Dynamics of brain activation during picture naming, Nature, 1994, 368: 463-465.

[4] Salmelin, R., Service, E., Kiesilä, P., and Uutela, K. Cortical dynamics of word recognition in normal and dyslexic subjects: a neuromagnetic study, Society for Neuroscience Abstracts, 1995, 21: 1763.

[6] Nobre, A.C., Allison, T., and McCarthy, G. Word recognition in the human inferior temporal lobe, Nature, 1994, 372: 260-263.

[7] Allison, T., McCarthy, G., Nobre, A., Puce, A., and Belger, A. Human extrastriate visual cortex and the perception of faces, words, numbers, and colors, Cerebral Cortex, 1994, 5: 544-554.

[8] Salenius, S., Kajola, M., Thompson, W.L., Kosslyn, S., and Hari, R. Reactivity of magnetic parieto-occipital alpha rhythm during visual imagery, Electroencephalography and clinical Neurophysiology, 1995, 95: 453-462.

[9] Salenius, S., Schnitzler, A., Salmelin, R., Jousmäki, V., and Hari, R. Reactivity of human cortical sensorimotor rhythms associated with median nerve stimulation, Society for Neuroscience Abstracts, 1995, 21: 115.

[10] Salmelin, R., Hämäläinen, M., Kajola, M., and Hari, R. Functional segregation of movement-related rhythmic activity in the human brain, NeuroImage, 1995, 2: 237-243.

[11] Schnitzler, A., Salenius, S., Salmelin, R., Jousmäki, V., and Hari, R. Involvement of primary sensorimotor cortex in motor imagery: a neuromagnetic study, Society for Neuroscience Abstracts, 1995, 21: 518.

The Magnetic Field Components Evoked by an Optical Illusion

Adachi, Y.[1], Kohda, M.[2], Morita, T.[2] and Takebayashi, M.[1]

[1]Electrotechnical Laboratory, Amagasaki, Hyogo, Japan; [2]Dept. of Engineering, Osada University, Suita, Osaka, Japan

Introduction

We often misperceive some physical phenomena whose spatio-temporal properties may be beyond the ability of our given sensory system. When two sensory stimuli are presented in a short stimulus onset asynchrony (SOA), the perception of the preceding stimulus is influenced by the following stimulus. It is well known as the backward masking that can evoke misperception[1]. This is observed in visual, auditory and tactile perception. Backward masking is thought of as the phenomenon that the responses to the following stimulus influence those of the preceding stimulus inversely in the signal processing system.

When two visual stimulus dots located at separate sites out of the central part of visual field are flashed serially in a short time, we often perceive the position of the preceding stimulus to be closer to the position of the following stimulus. This misperception is one kind of backward masking called optical saltation. The origin of this optical illusion should be in the information process of the central nervous system (CNS) because the preceding dot can be perceived on the blind spot on which we can usually perceive nothing under some spatial and temporal conditions.

There are two well-known facts that the optical illusion does not occur with a longer SOA or a larger spatial distance between two stimuli, and that the responses of specific adjacent retinal cells activate specific adjacent cells in both the laterale geniculate body and primary visual cortex[2]. These facts suggest that the neural responses caused by a single stimulus need some spatio-temporal extension for normal perception in the tissue of the brain, and that the illusion can occur when the extensions of each response overlap partly and interact with each other.

In this paper, we examined responses related to this optical illusion in the human brain by three-dimensional analysis of neuromagnetic signals that has high temporal resolution in attempts to discover the neurological origin of backward masking.

Methods

Two healthy men aged 23 and 25 years were examined as subjects. The subject sat in a magnetically shielded room without illumination. The flashes and fixation point were presented into the shielded room through optical fibers from red light-emitting diodes (λ=660 nm). The diameter of the stimulus dot was 1.2 mm and the duration of each visual stimulus was 10 msec.

Figure 1(a) shows the arrangement of the subject and the stimuli. Two stimulus dots L1 (upper dot) and L2 (lower dot) were arranged on a vertical line horizontally apart from the fixation point which was at the visual angle of 15 degrees. The visual angle between L1 and L2 was 5 degrees. All stimulus dots were 40 cm away from subject's eye. The subject was instructed to watch the fixation point at the center of his visual field with his dominant eye during the measuring so as not to allow eye movement. His non-dominant eye was covered with an eye mask. The stimuli were given on the opposite side of the stimulated eye.

The temporal arrangement of stimuli is illustrated in Fig. 1(b). One sequence consisted of five flashes. SOA between flashes c and d was 60 msec and an illusion was induced in this period. Flash b was the control stimulus so that the subject could make certain of the true location of L1 before the illusion occurred. Flash a was the sign of the biginning of the sequence. SOAs of a-b and b-c were 1000 msec. Flash e was given 1600 msec after the illusional stimulus (flash c and d), in order to inform the subject of the end of the sequence. After flash e, subjects were allowed to blink and relax until the next sequence. The interval between sequences was 2 sec. The illusional stimulus (flash c and d) were delivered every 5.6 sec in this experiment. The SOA between flashes c and d) and the spatial arrangement of L1 and L2 were decided according to reference[1], and it was confirmed in pre-psychophysical experiments that the subject was able to perceive the illusion under these conditions.

The neuromagnetic signals elicited from the brain were measured with a 122-channel whole-cortex magnetometer that consists of planar first-order SQUID gradiometers[3]. The head of the subject was positioned in the

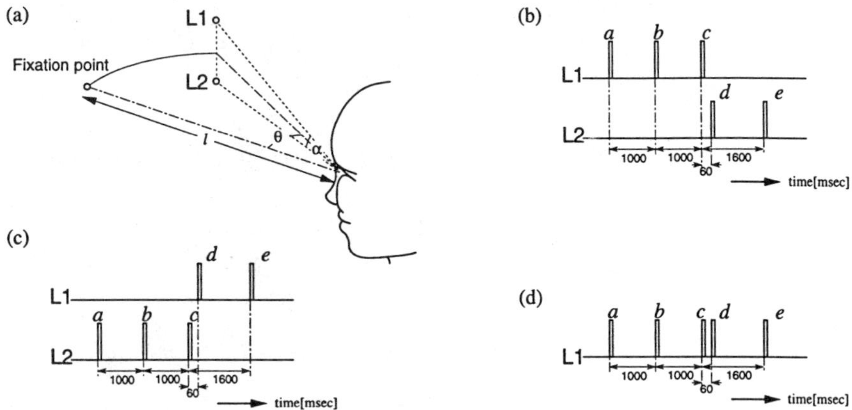

Fig. 1 (a) The spatial arrangement of the light stimuli. In this paper θ=15°, α=5° and l=40 cm. (b)(c)(d) Schematic illustrations of the temporal arrangements of stimulus sequences.

helmet-shape magnetometer array and supported with the head fixer, made of non-magnetic material, to prevent movement. The detected signals from the magnetometer were filtered with frequency limits of 0.03-100 Hz and digitized with sampling rate of 400 Hz and stored on a workstation for following off-line analysis. In the off-line analysis the signals were filtered with digital band-pass filter of 0.3-30 Hz. The responses to each stimulus were averaged in order to improve the signal-noise ratio.

In order to measure the response to the single L2 stimulus, we conducted a similar experiment as shown in Fig. 1(c). The subject could still perceive the illusion with this order. In addition we measured with the stimulus sequence that consisted of only L1 as shown in Fig. 1(d).

The electro-oculogram (EOG) was recorded to monitor ocular and muscle artifacts during the stimulating process. Any epochs which included such artifacts were to be rejected.

Results

Figure 2 shows the grand-average of subject 1's responses to flash b (single L1 stimulus; thin line) and flash c (illusional stimulus; bold line). The averaged responses were analyzed for 1000 msec, between 100 msec prior and 900 msec after the each stimulus. The responses to flash c and to flash d overlapped and merged, because SOA between them was 60 msec. It is obvious that the primary and/or secondary visual cortices in the contralateral occipital lobe were activated (in latency 100-250 msec) because the planar first-order gradients straight above the activated location of the cortex delivered the largest amplitude of the signal[3].

The MEG signals from the channel eliciting the largest peak amplitude (channel A) are shown in Fig. 2(b). The response to the single L2 stimulus at channel A is also shown in Fig. 2(b) (dashed line). The responses to the double L1 stimulus (Fig. 1(d)) at channel A are shown in Fig. 2(c). The illusional stimulus elicited the slow components with latency ca. 370 msec above the right temporal lobe (channel B) as shown in Fig. 2(d). The equivalent current dipoles (ECDs) for the three major peaks of the responses seen in Fig. 2(b), corresponding to the single L1 stimulus and the illusional (L1->L2) stimulus, were estimated. The locations of ECDs are shown in Fig. 3.

The contralateral occipital lobe components and the slow components above the right temporal lobe were obtained from subject 2 as well as from subject 1. However, the occipital components from subject 2 were not as clear as those from subject 1.

Discussion

The presence of the slow components whose latencies were 250 msec or more was reported in several odd ball paradigm experiments. These components are called P300 and P3. The origin of the P3 magnetic fields is under discussion. The deep source of the P3 was suggested to be around the hippocampus[4] or the temporal lobe[5]. It is plausible that the right temporal slow components evoked by the illusional stimulus might be associated with P3 components according to their locations and latencies.

It was difficult for the subject to perceive the 50 msec interval between flash c and flash d clearly in the double L1 sequence. More often the subject perceived the series of two L1 stimuli as mostly continuous. In Fig. 2(c), the observed responses to the double L1 stimulus were identical to ones to the preceding L1 only. In the case of the L1->L2 stimulus the subject could easily perceive the difference in location of the two flashes and judge which flash was preceding. Nevertheless the component for the following L2 stimulus was missing in the MEG signal corresponding to the L1->L2 stimulus as well as the double L1 stimulus. The incoming signal in the visual cortex induced by the two flashes was comparable to that caused by the single preceding flash. It is supposed from these results that the subject perceives two flashes by signal processing in the mid-brain, including the laterale geniculate body and superior colliculus, before cortical activation. If the two flashes were perceived by the visual cortex or higher cortical structure, the two components corresponding to the preceding L1 stimulus (flash c) and the following L2 stimulus (flash d)

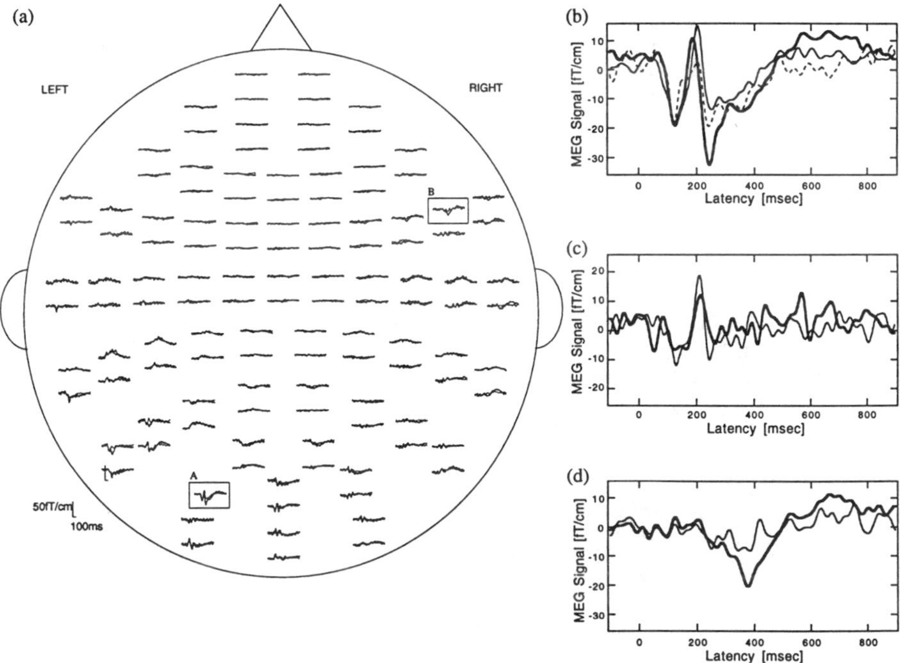

Fig. 2 (a) The MEG signal of subject 1 caused by the single L1 stimulus (thin line) and by the illusional stimulus (bold line). The stimuli were given on the right visual field of the left eye. Upper channels and lower channels at each measuring point are the latitudinal derivatives and the longitudinal derivatives respectively. (b) The thin line, bold line and dashed line correspond to the single L1 stimulus, the illusional stimulus and the single L2 stimulus respectively. (c) The response caused by the L1->L1 stimulus (bold line) at channel A. The thin line represents the response to the single L1 stimulus in the double L1 sequence. (d) The slow component at channel B.

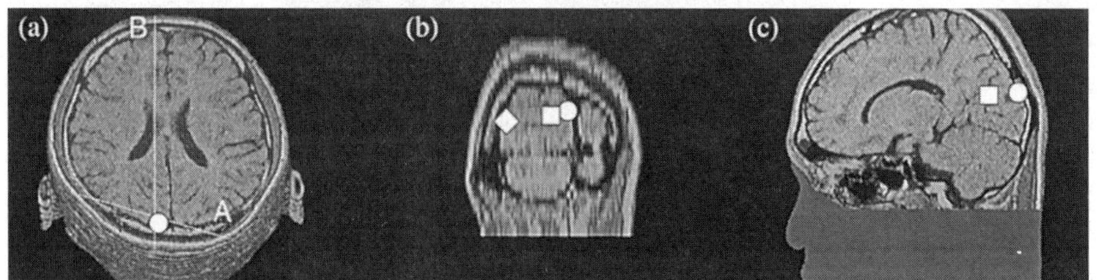

Fig. 3 The location of the ECDs for the three peaks of the response to the illusional stimulus (circle, 120 msec; square, 180 msec; diamond, 240 msec). We can see the shifts in location of the activation from the primary visual cortex to the secondary visual cortex in this period. (b) and (c) show the cross section cutting through plane A and B in (a) respectively. The locations of the sources of the response to the single L1 didn't differ from that of the L1->L2 significantly. The MR images were acquired with Yokogawa 0.5 T MRI system, MR Vectra at Fukai Hospital.

should be observed in the visual cortex. Consequently, it is probable that the optical illusion caused by the interaction of the two responses also occurs in the mid-brain.

The latencies of the three third major peaks of the signal from channel A, as caused by the single L1 and L2 stimuli and the illusional stimulus (Fig. 2 (b)), were nearly equivalent, all being approx. 250 msec. Furthermore, the amplitude of the peak caused by the illusional stimulus was equal to the linear sum of the amplitudes of the peaks caused by the single L1 and L2 stimulus. Accordingly it is suggested that the information processing of both the preceding and the following stimuli might be carried out at the same time in the secondary visual cortex, and that this could be associated with evoking the optical illusion. However, we need to confirm the reproducibility of these results not only by increasing the number of subjects, but also by measuring the changes in response under conditions with various SOAs and spatial distances between the two stimuli.

We are grateful to Dr. Akiba and Dr. Sato for advice and encouragement, Ms. Nakamura for fruitful discussion and Dr. Kakui for his continuous encouragement.

References:

[1] Geldard, F.A., Sherrick, C.E. Space, Time and touch, Scientific American, 1986, 255: 84-89

[2] Nieuwenhuys, R., Voogd, J., and Hujizen, Chr. The Human Central Nervous System-A Synopsis and Atlas, Springer-Verlag, Berlin, Heidelberg, New York, 1989

[3] Hämäläinen, M., Hari, R., Ilmoniemi R.J. et al Magnetoencephalography-theory, instrumentation, and applications to noninvasive studies of the working human brain, Rev. Mod. Phys., 1993, 65: 413-498

[4] Rogers, R.L., Papanicolaou, A.C., Baumann, S.B. et al Late magnetic fields and positive evoked potentials following infrequent and unpredictable omissions of visual stimuli, Electroenceph. Clin. Neurophysiol., 1992, 83: 146-152

[5] Okada, Y.C., Kaufman, L. and Williamson, S.J. The hippocampal formation as a source of the slow endogeneous potentials, Electroenceph. Clin.Neurophysiol., 1983, 55: 417-426

Spatiotemporal Imaging of Human Cortical Areas Sensitive to Visual Motion

Ahlfors, S.P.[1], Aronen, H.J.[2], Belliveau, J.W.[3], Dale, A.M.[4], Huotilainen, M.[5], Ilmoniemi, R.J.[6], Korvenoja, A.[2], Liu, A.K.[3], Simpson, G.V.[1], Tootell, R.B.H.[3] and Virtanen, J.[5]

[1]Albert Einstein College of Medicine, Bronx, NY; [2]Radiology, Helsinki University Central Hospital, Finland; [3]Massachusetts General Hospital, NMR Center, Charlestown, MA; [4]University of Oslo, Norway; [5]Cognitive Brain Research Unit, University of Helsinki, Finland; [6]BioMag Laboratory, Helsinki University Central Hospital, Finland

Introduction

Movement in the visual field is an important part of perceptual information. Previous imaging studies have identified a number of cortical areas sensitive to visual motion [1,2], including what is considered to be the human homologue of area MT/V5 [3,4]. Our study sought to reproduce and expand upon a previous functional magnetic resonance imaging (fMRI) investigation [4] and to examine the timing of activity in the network of regions participating in the processing of visual motion information with magnetoencephalography (MEG).

We measured brain activity in the same subjects using both MEG and fMRI, thereby combining the good spatial resolution of fMRI and the millisecond-scale temporal information obtained with MEG [5,6]. Special considerations, however, are necessary, when combining these brain imaging methods, because at present, the exact relationship between the hemodynamic (as measured with fMRI) and electrophysiological (MEG) responses is not known.

Methods

Four subjects (all male, aged 25-45 years) were recorded both with whole-head 122-channel MEG (Neuromag Ltd, dc-SQUIDs, planar first-order gradiometers, analog filtering 0.03-100 Hz, sampling at 397 Hz) and with functional MRI (GE 1.5 T with Advanced NMR echo planar, asymmetric spin echo sequence, head coil, bite bar, 64 sets of 25 slices were collected in each run, TR = 7 s, slice thickness 5 mm). The computer-generated visual stimuli consisted of expanding and contracting concentric low-contrast rings. About 400 epochs of MEG/EEG responses were averaged, triggered from the reversal of the direction of motion occuring every 3 s. The speed of motion was 2.4 deg/s. The analysis period was –200 to 800 ms, the averaged responses were lowpass filtered at 0-40 Hz using FFT, and the zero level was adjusted using the prestimulus baseline –100 to 0 ms. The fMRI measure was the Fourier component of the fMRI signals in response to the stimulus pattern being alternatingly moving and stationary in cycles of either 30 or 40 s.

Sources of the MEG responses were modeled with multiple equivalent current dipoles. As a first attempt, the locations of the dipoles were determined by a non-linear least-squares fit to the MEG data only. A single dipole was fitted at one instant of time (or a range of latencies around the peak signal), and the contribution from this source was projected out using the signal-space projection technique [7]. Then, another single dipole was fitted to the part of the data that could not be explained by the first one. These two dipoles were used as the initial guess for a two-dipole fit performed to the original data. The orientations of these MEG-derived dipoles were assumed not to change in time. This process signal-space projection followed by a multi-dipole fit was repeated, adding dipoles until a satisfactory explanation was obtained for the measured responses. To determine the locations and orientations, usually only two or three dipoles were fitted at a time to part of the data (in space and time, for example, to sensors over one hemisphere), and only in the end, the optimal amplitudes for all the dipoles were obtained using the whole set of data.

The multiple dipole model becomes prohibitively difficult to solve when the complexity of the source configuration increases, partly because the non-linear least-squares minimization algorithm tends to stop in a local, sub-optimal minimum, partly because a dipole may not be an adequate model for the source. Therefore, we examined solutions in which the locations of the dipoles were constrained to the loci of the fMRI data. The orientation of these dipoles was allowed to vary and was optimized, together with the dipole moment, for each time instant separately using the linear pseudo-inverse method. Some sets of fMRI locations had to be represented by a single dipole, because nearby dipoles often resulted in an unstable solution with a quadrupole-like combination of dipoles with large opposing amplitudes, the field patterns of which mostly cancel each other.

Locations of active regions were identified and visualized within each subject's high-resolution anatomical MRI, co-registered with MEG using the nasion and preauricular points as fiducial points.

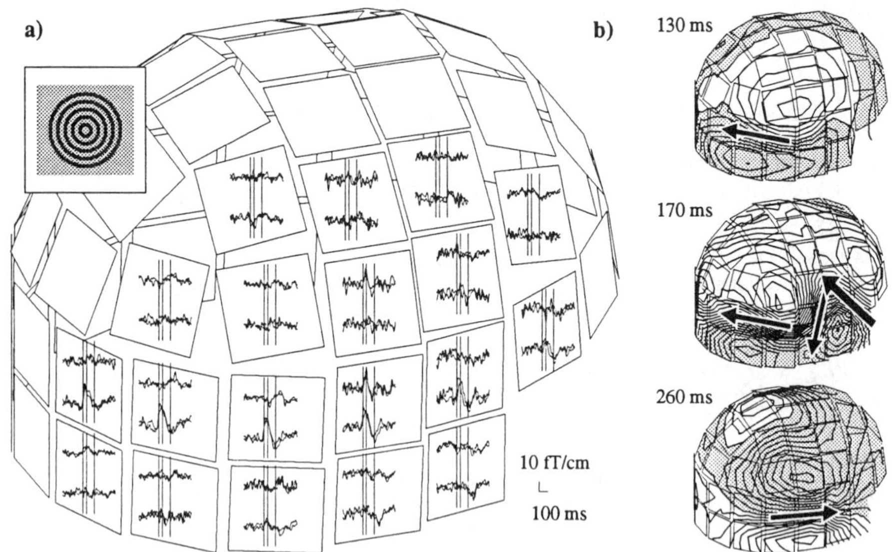

Fig. 1 a) Visual evoked magnetic fields as a function of time over the right hemisphere, in response to the reversal of the direction of the motion of concentric rings (inset at top left). The superimposed curves show responses to two types of reversal (expanding to contracting, and vice versa). The vertical markers indicate latencies 130, 170, and 260 ms. The rectangles depict the locations of sensor pairs. The upper and lower waveforms correspond to $\partial B_z/\partial x$ and $\partial B_z/\partial y$, where B_z is the normal component of the magnetic field; x, y, and z refer to the local coordinate system of the sensor units. b) Isocontour maps of the estimated B_z at three latencies. The arrows indicate the location and orientation of equivalent current dipoles determined by a least-squares fit to the MEG signals. Subject SA.

Results

Averaged MEG responses are shown in Fig. 1a. The largest deflections peak at 170 and 260 ms after the reversal of the direction of motion. The spatial distribution of the measured magnetic field at three different latencies is depicted in Fig. 1b. At 130 ms the field map is characteristic of a field that would be generated by a single source, but at 170 ms the field pattern is more complex, indicating multiple simultaneously active brain regions. A multi-dipole analysis of the case shown Fig. 1 suggests at least three sources at 170 ms (two of them located anteriorly to the one that is active at 130 ms) and one dipole at 260 ms which is located close to the middle dipole at 170 ms, but with an orthogonal orientation.

The fMRI data showed in all subjects a pattern of activity consisting of three regions: one in the lateral surface of the occipital lobe, another in the occipito-temporal junction, and a third one in the vicinity of the superior temporal sulcus near the posterior termination of the Sylvian fissure. In some subjects, fMRI activity was also found in inferior occipital lobe and parietal lobe.

Fig. 2 depicts the amplitudes of source activity when the locations of dipoles were constrained to fMRI loci. This combined MEG-fMRI analysis suggests that there are temporally overlapping activities in several cortical regions. Dipole #1 has the most uncertainty with respect to the cortical region it represents, since it is probably accounting for activity from multiple sources near the pole and inferior surface of the occipital lobe. Attempts to model these regions with multiple dipoles resulted in large interactions due to their proximity, and thus they are represented by a single source in Fig. 2. The first areas to be activated are those represented by dipole #1, and dipoles #4,5 which may correspond to the human MT/V5 regions bilaterally. These regions onset at about 140 ms, followed by activation of the posterior dipoles #2,3 and by the anterior dipoles #8,9 (not identified with fMRI, see below). The temporal data illustrate that, in addition to staggered onset times, all of these areas have transient responses that last for hundreds of milliseconds but do not continue for the entire duration of the motion stimulation. However, the fMRI-derived anterior motion area represented by dipoles #6,7 has a later onset (after 250 ms) which

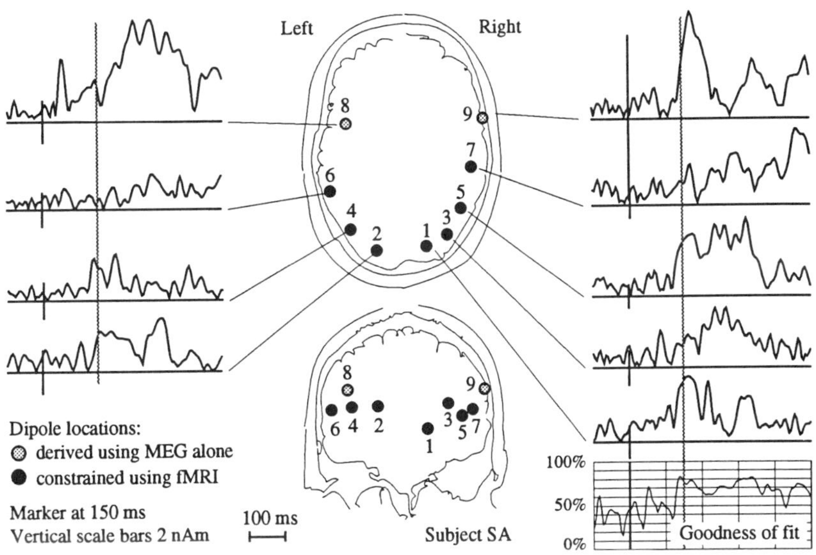

Fig. 2 Source activity for the MEG response to the reversal of motion. Locations of equivalent current dipoles derived from the MEG data and regions of hemodynamic activity as measured with fMRI are shown within the outline of the head. The amplitude waveforms of the dipole moments were scaled individually by a factor proportional to the inverse square of the distance between the dipole and the array of sensors, thereby approximately normalizing the noise level.

occurs after the other areas have reached maximal or near maximal amplitude, and then sustains its activity throughout the 800 ms analysis period.

While the fMRI based sources explained the majority of the MEG data, they did not adequately account for the most anterior regions of MEG activity. MEG derived sources in the region at or near the precentral sulcus were needed to explain this remaining MEG data (indicated by the gray dots #8 and #9 in Fig. 2). These have prominent peaks at 180 ms (right, #9) and at around 350 ms (left, #8; this dipole also has a sharp peak very early, at 40 ms, but that was not seen in other subjects).

Discussion

The combined MEG and fMRI measurements indicated a complex spatial and temporal pattern of activity, involving a number of cortical regions related to processing of visual motion information. The fMRI data consistently showed activity in three areas in the posterior lateral surface of each hemisphere. Previous PET and fMRI studies have given evidence that two of these regions might correspond to the human homologues of the monkey areas V3A and MT/V5 [3,4]; the third one is close to the regions which in previous PET studies have been reported to be activated by vestibular [8,9] and visual motion [1,2] stimuli. In addition, MEG indicated activity in a region near the precentral sulcus, which may correspond to the frontal eye field. No activity was detected there with fMRI, and further studies will be employed to examine this region. The fMRI-constrained dipole analysis of the MEG responses suggested that there is temporally overlapping transient activity in MT/V5 and regions posterior to it, as well as the in the region of the frontal eye field. These regions have early, yet slightly differing onset times and are not sustained throughout the recording epoch. Whereas the fMRI-derived anterior motion area showed later onset and sustained activity.

The exact relationship between hemodynamics and electrophysiology, as measured with fMRI and MEG, respectively, is not known at present. There are several possible causes for the differences in the locations of fMRI activity and MEG sources. Besides technical matters, like measurement noise and coregistration error, and changes in the state of the subject during the separate recording sessions, there are some fundamental problems in comparing

703

the two measurements. First, fMRI is sensitive to activity all over the brain, whereas MEG is most sensitive to activity in sulci near the surface of the head and relatively insensitive to radially oriented, deep, or mutually cancelling sources. Second, because the time scales of the two physiological phenomena are different (milliseconds for MEG, seconds for fMRI), it is difficult to ensure that sensory stimuli and the task of the subject contribute equally in the two measurement modalities. In our study, MEG reflects the responses to transient reversal of motion and continuous movement over the period leading to the next reversal. The baseline referent for MEG was the 100 ms preceding a reversal and therefore includes responses to sustained motion. The fMRI also measures responses to transient reversals of motion and sustained motion. However, the referent for these motion responses was the same pattern in a stationary state. This may partly explain the difference between the locations of fMRI activity and the MEG sources in our experiments.

Our data indicate that an assumption of a one-to-one mapping of fMRI loci and MEG sources does not necessarily obtain. Thus a further exploration of the source configuration is warranted, and may otherwise give misleading results. For example, if a true source is missing from the fMRI-derived dipoles, the activity corresponding to the missing source may be assigned to nearby dipoles. On the other hand, if too many nearby sources are included in the MEG analysis, the inverse solution may have large, highly correlated source waveforms, the field patterns of which mostly cancel out. In the latter case, a smaller number of non-interfering dipoles often gives a more acceptable solution, with physiologically reasonable primary current amplitudes.

Our results demonstrate the usefulness and importance of the combined MEG-fMRI approach. First, fMRI can suggest a structure in the underlying source configuration for the MEG which may be difficult to deduce from the MEG data alone. Second, a change in the orientation of a source dipole as a function of time can separate contributions from different neural populations within a single region in the fMRI. Third, a source may be detected with MEG but not with fMRI, or vice versa.

References

[1] Dupont, P., Orban, G.A., De Bruyn, B., Verbruggen, A., and Mortelmans, L. Many areas in the human brain respond to visual motion, Journal of Neurophysiology, 1994, 72: 1420-1424

[2] Cheng, K., Fujita, H., Kanno, I., Miura, S., and Tanaka, K. Human cortical regions activated by wide-field visual motion: an $H_2^{15}O$ PET study, Journal of Neurophysiology, 1995, 74: 413-427

[3] Zeki, S., Watson, J.D.G., Lueck, C.J., Friston, K.J., Kennard, C., and Frackowiak, R.S.J. A direct demonstration of functional specialization in human visual cortex, Journal of Neuroscience, 1991, 11: 641-649.

[4] Tootell, R.B.H., Reppas, J.B., Kwong, K.K., Malach, R., Born, R.T., Brady, T.J., Rosen, B.R., and Belliveau, J.W. Functional analysis of human MT and related visual cortical areas using magnetic resonance imaging, Journal of Neuroscience, 1995, 15: 3215-3230.

[5] Simpson, G.V., Pflieger, M.E., Foxe, J.J., Ahlfors, S.P., Vaughan, Jr., H.G., Hrabe, J., Ilmoniemi, R.J., and Lantos, G. Dynamic neuroimaging of brain function, Journal of Clinical Neurophysiology, 1995, 12: 432-449

[6] George, J.S., Aine, C.J., Mosher, J.C., Schmidt, D.M., Ranken, D.M., Schlitt, H.A., Wood, C.C., Lewine, J.D., Sanders, J.A., and Belliveau, J.W. Mapping function in the human brain with magnetoencephalography, anatomical magnetic resonance imaging, and functional magnetic resonance imaging, Journal of Clinical Neurophysiology, 1995, 12: 406-431

[7] Tesche, C.D., Uusitalo, M.A., Ilmoniemi, R.J., Huotilainen, M., Kajola, M., and Salonen, O. Signal-space projections of MEG data characterize both distributed and well-localized neuronal sources, Electroencephalography and Clinical Neurophysiology, 1995, 95:189-200.

[8] Friberg, L., Olsen, T.S., Roland, P.E., Paulson, O.B., and Lassen, N.A. Focal increase of blood flow in the cerebral cortex of man during vestibular stimulation, Brain, 1985, 108:609-623.

[9] Bottini, G., Sterzi, R., Paulesu, E., Vallar, G., Cappa, S.F., Erminio, F., Passingham, R.E., Frith, C.D., and Frackowiak, R.S.J. Identification of the central vestibular projections in man: a positron emission tomography activation study, Experimental Brain Research, 1994, 99:164-169.

Acknowledgements

We thank Drs. R. Näätanen, B. Rosen, M. Sereno, and C.-G. Standertskjöld-Nordenstam. This study was supported by grants from the United States Public Health Service: NS27900, MH50054, EY07980, by grants form the Human Frontier Science Program, and by the Academy of Finland, and conducted during the tenure of an established investigatorship from the American Heart Association to JWB.

Task-Specific Magnetic Fields from the Left Human Prefrontal Cortex

Basile, L.F.H.[1], Simos, P.G.[1], Tarkka, I.M.[2] and Papanicolaou, A.C.[1]

[1]Deparment of Neurosurgery, University of Texas, Houston, USA; [2]Baylor College of Medicine, Houston, USA

Introduction

Recent studies have provided evidence for task-specific activity in distinct areas of the frontal cortex [1,2,3]. In a previous study, we used magnetoencephalography (MEG) to test the hypothesis of functional specialization of the frontal cortex [1] by asking subjects to perform two visual recognition tasks. A spatial recognition task was associated with sources of magnetic fields in the right frontal lobe, which were localized more anteriorly and inferiorly, than the sources of activity obtained during a visual pattern recognition task. Both tasks involved presentation of common visual stimuli and required common motor responses. These results were analogous to the findings from a multi-unit recording study in the monkey [3]. That study showed concentrations of neurons in dorsolateral and inferior convexity prefrontal areas, that responded preferentially during the performance of a visuo-spatial and a visual-pattern delayed response task, respectively. Another experiment from our laboratory (unpublished data) provided evidence for different sources of activity in the right frontal lobe, when a familiar-person recognition task was compared to a visual pattern recognition task. Sources in more medial areas of the right prefrontal cortex were obtained when subjects performed the familiar-person recognition task. A study by Petrides et al. [2], using positron emission tomography in humans, also reported regional prefrontal cortical activity corresponding to the performance of two different tasks.

The purpose of the present experiment was to extend our results on task-specific frontal cortical magnetic activity with the use of a verbal task. In this case, we expected to obtain consistent sources of magnetic fields from the left frontal cortex in right handed subjects. The expected hemispheric asymmetry in activity would be consistent with the view that certain portions of the left frontal cortex play an essential role in the performance of even simple verbal tasks. To enhance the possibility of observing this asymmetry in activity, we eliminated the requirement of an overt motor response, which in our previous experiments might have resulted in the bilateral activity seen in most subjects. In fact, results from a pilot study in which a button press response was added to the present task, confirmed our suspicion regarding bilateral frontal activity.

Methods

Six normal subjects, three males and three females, ranging in age between 24 and 54 years, volunteered to participate in two or three two-hour experimental sessions. All subjects had normal vision and no history of neurological disorders. They were required to sign consent forms and they were paid for their participation. Stimuli were presented on the screen of a computer that also sent trigger pulses to start the MEG data acquisition. The stimuli consisted of 216 three-to-ten letter words which were drawn white on a black background, and subtended a maximum of approximately 3 ° of visual angle. All words were concrete nouns, that belonged to nine well defined semantic categories (e.g., tools, fruits, vehicles, etc.). The exposure duration of each stimulus was 370 ms. Each trial consisted of the successive presentation of two words. The inter-stimulus interval was 2000 ms, whereas the inter-trial interval varied randomly between 2000 and 5000 ms. Each experimental run consisted of a series of 120 trials. Across the different runs, the percentage of word pairs matching semantically ranged from 45 to 55%. The order of trials containing matches or mismatches was randomized within each run. Subjects were instructed to fixate their gaze on a small white square of light, that was continuously present in the center of the screen during the interstimulus interval. They were asked to decide whether the first word in a trial matched the second one, i.e., whether the two words belonged to the same semantic category. Finally, they were required to count silently the number of matches and report their count after each experimental run. Specific instructions were given to the subjects to avoid blinking, swallowing, or moving their tongue, face or mandibles until shortly after the end of each trial.

Magnetic evoked fields were recorded with a 7-channel Neuromagnetometer (Biomagnetic Technologies, Inc., Model 607), using second-order gradiometers with a 4 cm baseline and 1.8 cm coil diameter. Recordings were done in a magnetically shielded room. Each average was computed from a minimum of 60 epochs, from the total of 120 recorded during each run. The MEG analog filtering was set at 0 to 50 Hz and the digitization rate was 200 Hz. A Nicolet Compact Four system and Ag/AgCl electrodes were used for acquisition of the electro-oculogram (EOG). The EOG was recorded for the purpose of subsequent exclusion of MEG trials containing eye movement or blink artifacts. Before averaging, trials on which the EOG signal exceeded a peak-to-peak amplitude of 70 μV were rejected. The magnetometer itself was also a source of some artifacts. Those artifacts were easily recognizable, and they typically consisted of transient high amplitude noise bursts usually confined to a single MEG channel. Visual inspection of all MEG epochs ensured that trials contaminated with this type of artifact were not used in subsequent analysis. When more than 60 epochs for a given condition and channel contained artifacts (leaving less than 60 epochs to be averaged), that channel was eliminated from further analysis. The position of the magnetometer relative to the subject's head was automatically recorded with a BTi probe position indicator, consisting of a radio-frequency transmitting device attached to the dewar and three sensors placed on the head. Error tolerance for the location of sensors was less than 3 mm and the accuracy 0.1 mm. Each subject's fiduciary points, the nasion and preauricular points, served as basis for the MEG Cartesian coordinate system. The Y axis was determined by the two preauricular points, the X axis by the nasion and the mid distance between the two preauricular points, and the Z axis was normal to the X-Y plane, passing through the X-Y intersection (the center of the head). Five to six placements of the magnetometer on each side of the head covered the frontal, centro-parietal and temporal areas. Following artifact elimination and averaging, the MEG data were baseline adjusted relative to the mean field amplitude during the 500 millisecond period preceding the onset of the first word. The averages were then digitally filtered (bandpass 0 to 5 Hz, Butterworth 4 pole filter, 24dB/octave roll-off) and used to create isofield contour maps.   The isofield contour maps were constructed at 5 ms intervals, from 100 to 2700 ms after the onset of the first word. The intracranial sources of the observed fields were modeled as equivalent current dipoles (ECDs) using a finite difference version of the Levenberg-Marquardt algorithm incorporated into the BTi software. The dipole fitting method was applied at every 5 ms, from 100 to 2700 ms of the trial duration. The resulting dipole parameters (spatial coordinates, orientation and strength) were computed with reference to the best-fitting sphere, the curvature of which was defined by the surface of each subject's scalp at the region of the head containing the field extrema. The criterion we used for accepting solutions as satisfactory, was a correlation of at least .85 between the measured fields and the predicted fields, at each time point. The predicted fields were calculated as solutions of the forward problem, from a set of hypothetical dipole parameters. The hypothetical parameters were set by the BTi software as initial values in the iterative dipole-fitting process. Magnetic resonance images (MRIs) from five subjects were used to identify the brain locations of the calculated dipolar sources. The procedure for transforming the MEG coordinates into MRI coordinates involved the use of lipid markers in the nasion and preauricular points during the MRI session. The MRI coordinate system was then aligned to the MEG coordinate system using a computer algorithm.

Results

Clear dipole-like isofield maps were observed in all six subjects on the left fronto-temporal region of the scalp. On the right side, only the dipole-like maps corresponding to the event-related fields were observed. The sources of these fields were restricted to temporal lobe regions and will not be further discussed. The middle row in Figure 1 shows representative isofield contour maps from two subjects during the course of the experimental trial. For illustration purposes these maps were selected at the time of highest peak-to-peak (i.e., outward and inward flux) field amplitude. A common angle of view relative to the subjects' coordinate systems (65 degrees anteriorly in the X-Y plane, starting from the left preauricular point, and 70 degrees relative to the Z axis) was used in the figure. The shaded areas represent magnetic flux entering the head, and white areas represent flux exiting the head. The similarity in the maps between the subjects is remarkable, despite the fact that we observed a large degree of variability in the temporal pattern of their corresponding average fields (see waveform morphology on the top of the Figure). As it was also observed in our previous experiment [1], the spatial distribution of extrema for a given subject was approximately the same throughout the trial duration.

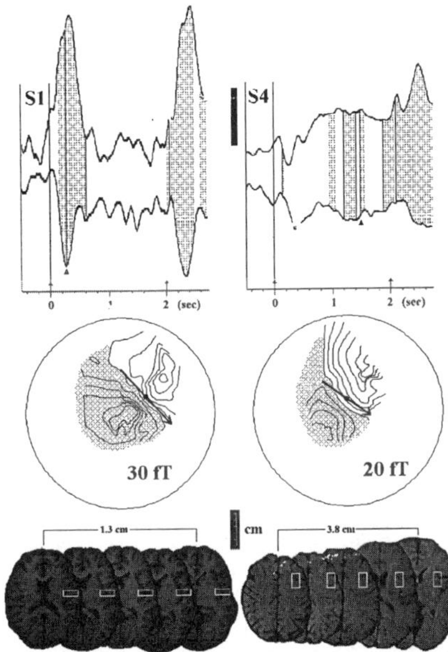

Figure 1 - Magnetic field extrema, isofield contour maps and range of dipole positions for subjects S1 and S4, who represent the extreme examples of phasic and tonic types of frontal activity.

The upper part of the Figure shows the MEG waveforms recorded at the positions of the magnetic field extrema, for the two representative subjects. One recording contains the highest amplitude average field exiting the head (upward deflection) whereas the other contains the highest amplitude average field entering the head (downward deflection). Subjects S1, S2 and S3 presented a more phasic pattern of activity, as opposed to subjects S4, S5 and S6, who showed more sustained activity throughout the trial. Since we had never observed those extreme cases of phasic activity in the previous studies, we repeated the recordings for two subjects, S1 and S3. The results were practically identical across the two replications, leading to the conclusion that the observed temporal pattern of field strength is a characteristic of each individual. Intracranial sources of the measured fields, in the form of satisfactory dipole solutions located in the frontal lobes, were obtained only for the left hemisphere in all six subjects. This was the case for several time points during the trial. The number of solutions and their distribution in time varied across subjects. We observed time windows with good solutions from as early as the onset of the first word (indicated by arrowheads in the Figure), to as late as after the offset of the second word. The durations of these time windows ranged from a few tens to many hundreds of milliseconds. The shaded areas in the waveforms of the Figure mark the time periods during which dipolar maps and corresponding satisfactory dipole solutions were found. In all subjects, there were relatively small time-dependent changes in dipole location. The bottom part of the figure shows the cubic regions containing the entire range of dipole locations for the two subjects. These regions were defined by the ranges of dipole coordinates in the x, y and z dimensions of our coordinate system, across all time windows with satisfactory solutions for a given subject. The actual positions of dipoles obtained during the task were points near or underneath the cortex corresponding to Brodmann's areas 6 and 44, in the left posterior frontal lobe (in the right side of MRI scans in the figure).

Discussion

The purpose of the present study was to use elicited magnetic fields as a means of detecting task-specific regional activity from the frontal cortex. This attempt was successful in that all six subjects displayed dipole-like sources restricted to posterior regions of the left frontal lobe (in the vicinity of Brodmann's areas 6 and 44), during a verbal task. We interpret the dipoles as unevenly weighed centers of tangential current distribution, therefore as indirect measures of activity in extended cortical patches. This interpretation is based on two assumptions. The first comes from studies of local blood flow changes in humans. They show that the metabolic changes occur in areas much larger than those corresponding to a few cortical columns [4]. This was indeed the case in experiments using tasks that engage the frontal cortex [2,4,5]. The second assumption is based on the fact that, given that there is activity in an extended cortical surface, a dipole representing it is expected to lie in any sub-region of this surface. That is, the dipole can lie in its center, borders, or underneath the cortical surface, depending on the particular geometry of the cortical currents in question [6]. Specifically, it is only the portion of cortex showing an asymmetry in current distribution in the total active surface, that will result in a dipole-like field pattern on the scalp [7].

An unexpected result in this study was the variability across subjects in the temporal pattern of elicited magnetic activity, and it is possible that this pattern reflects an underlying variability in the distribution of more phasically or tonically active cells ('delay cells') across subjects, and thus their contribution to the externally recorded fields. Alternatively, the variability that we noted may reflect a fundamental difference across subjects in the timing of the engagement of their cortical areas into larger functional circuits. This last possibility could, in principle, be related to different personal strategies in the performance of the task. To our knowledge, a combination of those explanations cannot be ruled out in the present.

MEG has proven to be useful as a non-invasive method for functionally mapping the human frontal cortex. In the present case, posterior left frontal cortex activity, specific to the performance of a verbal task was obtained in all subjects

References

[1] Basile, L.F.H.; Rogers, R.L.; Bourbon, W.T. and Papanicolaou, A.C.. Slow magnetic fields from human frontal cortex. Electroenceph. Clin. Neurophysiol., 1994, 90:157-165.
[2] Petrides, M.; Alivisatos, B.; Evans, A.C. and Meyer, E. Dissociation of human mid-dorsolateral from posterior dorsolateral frontal cortex in memory processing. Proc. Natl. Acad. Sci. USA, 1993a, 90:873-877.
[3] Wilson, F.A.W.; Scalaidhe, S.P.O. and Goldman-Rakic, P.S. Dissociations of object and spatial processing domains in primate prefrontal cortex. Science, 1993, 260:1995-1958.
[4] Roland, P.E. and Friberg, L. Localization of cortical areas activated by thinking. J. Neurophysiol., 1985, 53:1219-1243.
[5] Petrides, M.; Alivisatos, B.; Meyer, E. and Evans, A.C. Functional activation of the human frontal cortex during the performance of verbal working memory tasks. Proc. Natl. Acad. Sci. USA, 1993b, 90:878-882.
[6] Kaufman, L.; Schwartz, B.; Salustri, C. and Williamson, S.J. Modulation of spontaneous brain activity during mental imagery. Journal of Cognitive Neuroscience, 1990, 2:124-132.
[7] Kaufman, L.; Kaufman, J.H. and Wang, J. On cortical folds and neuromagnetic fields. Electroenceph. clin. Neurophysiol., 1991, 79:211-226.

Searching for the Oddball Response in the EEG and MEG

Berg, P.[1], Kakigi, R.[2], Koyama, S.[2], Dobel, C.[1], Zobel, E.[1], and Scherg, M.[3]

[1]University of Konstanz, Konstanz, Germany; [2]National Institute for Physiological Sciences, Okazaki, Japan; [3]University of Heidelberg, Heidelberg, Germany

Introduction

We start out from the observation that the P300 component is often visible in single trial EEG data. When the data are appropriately filtered (we used 1-8 Hz band-pass), the signal becomes clearer and the auditory N100/P200 is often also apparent. Using a new spatio-temporal pattern search method, integrated into the FOCUS program [1], an objective test was available to see if such patterns could be detected in the continuously recorded raw data on the basis of the average waveforms. By applying the search to both EEG and MEG data from the same subject, a comparison between the two measurement procedures could be made. In order to test reliability of the EEG oddball response over sessions, averages from one session were used to detect responses in other sessions.

Spatio-temporal pattern search method

Instead of searching for abstract characteristics of a pattern (e.g. a generic spike-wave, an eyeblink) on a single channel [2], pattern search is based on a template taken from a multiple channel data set. For example, it can be used to detect further occurrences of spikes in a data set after one occurrence is chosen as a template. The method is illustrated in Fig. 1 using a template pattern consisting of an average ERP. The template is characterized in terms of the first 5 PCA components (topographies) and their waveforms. Pattern search is performed by multiplying the topography matrix with a test window in the data, generating 5 test waveforms. For a pattern match these waveforms should be similar to the template waveforms. A pattern match is detected when the correlation between the 5 concatenated test and template waveforms exceeds a threshold. The test window is then adjusted in time to find the maximum correlation, determining the final match.

Methods

One male 49-year-old subject performed oddball tasks in multiple sessions for both EEG and MEG over a period of 3 months.

Pure tones were presented binaurally through headphones (EEG) or a loudspeaker (MEG). Tones were 50 ms in duration, 80 dB SPL, with 5 ms rise and fall times. Tone pitches were 800 Hz and 1200 Hz. Stimuli were presented in blocks of 200, with an ISI of 1.5 s. Each block consisted of a pseudorandom sequence containing 20% of one stimulus ('rare') and 80% of the other ('frequent'). In separate blocks the subject was required to respond with either a left or a right hand button press, to either the rare or the frequent tone.

In 8 sessions on separate days with four measurements per session, 64-channel EEG was recorded continuously during the oddball tasks. EEG electrodes were distributed as evenly as possible over the head, including infra-orbital, F9, F10, mastoid, and cerebellar placements, and referred to Cz. EEG data were converted to average reference. Similarly, 74-channel MEG was recorded during 3 sessions. Dual 37-channel MEG sensor arrays (BTI Magnes) were centered over T3 and T4. Data for EEG and MEG were each divided into 32 measurements containing responses to 200 stimuli. Averages over the time range 200 ms before to 1000 ms after the stimulus were computed for each stimulus type in each measurement, using an amplitude threshold for artifact rejection. Separately for each measurement, using the average as a template, a pattern search was made on the 1-8 Hz band-pass filtered data using a 50% correlation threshold. The proportions of correct detections, 'good' false alarms (where the response to the wrong stimulus was detected), and 'bad' false alarms (where the detected pattern was not synchronized to either stimulus type) were recorded (cf. Table 1).

Characterization of the template pattern

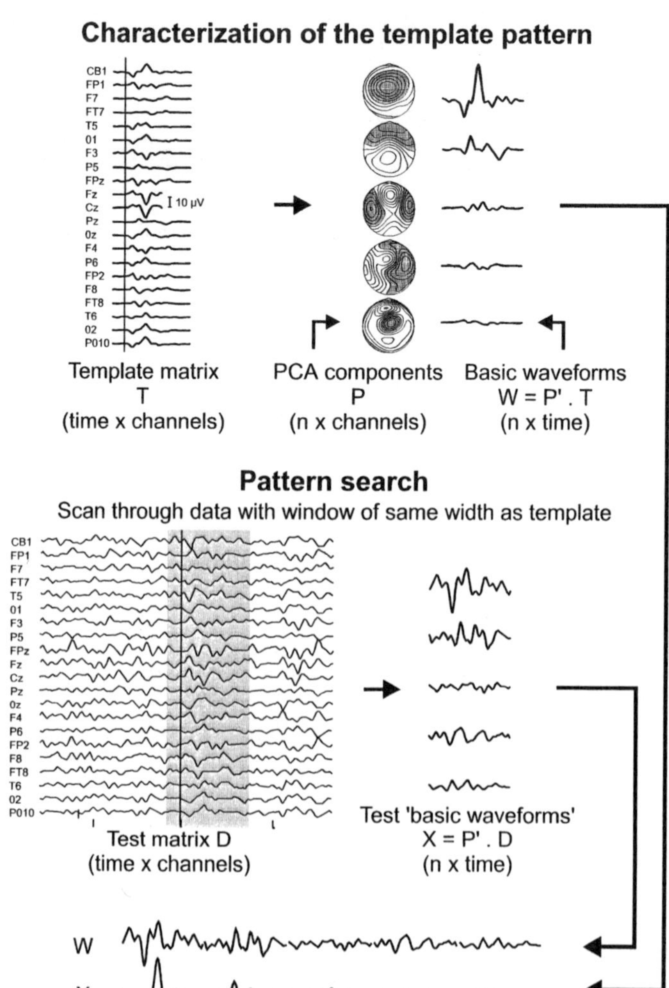

Fig. 1. Steps to the pattern search method. Top: The template matrix (in this case an EEG average to the rare stimulus, width 1200 ms) is characterized in terms of the first 5 PCA components (**P**), represented by the maps, and the basic waveforms (**W**) associated with each component. Bottom: In a scan through the data with a moving window, test waveforms (**X**) are generated. When the correlation between the concatenated waveforms exceeds a threshold, we have a pattern match. In the raw data illustrated, such a match is shown. Triggers are shown by vertical lines beneath the curves (4 triggers: frequent tone, button press, rare tone, rare tone). The oddball response is visible in the raw data, particularly at the Cz electrode, in the shaded region and in the response to the following trigger.

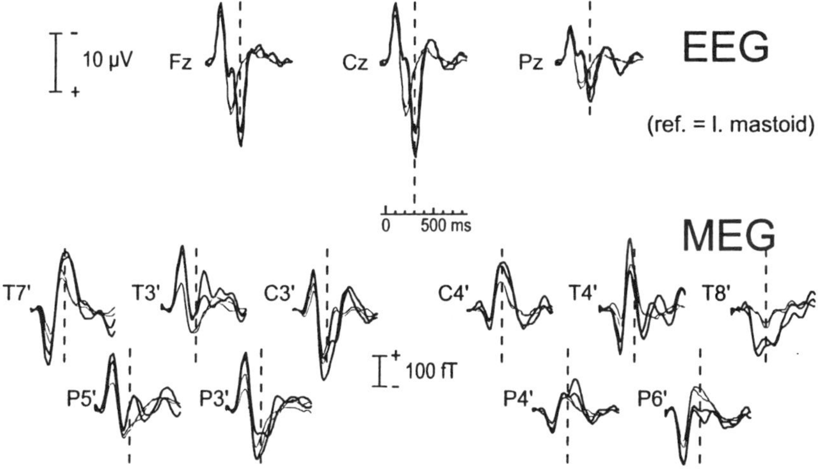

Fig. 2. Selected averages for the EEG (top) and MEG (bottom) to the stimulus that did not require a button press. Vertical dotted lines mark the latency of the peak of the EEG P300. Separate averages over odd (left hand) and even (right hand button press) days are overplotted for responses to rare (thick lines) and frequent (thin lines) tones. MEG labels show the 10-10 equivalent sensor locations.

For the EEG, averages from day 1 were used as templates for pattern search on days 3, 5, and 7. Similarly, averages from day 2 were used as templates for search on days 4, 6, and 8. Twenty-four independent pattern searches were made.

Results

Fig. 2 shows selected means for rare and frequent stimuli in the EEG and the MEG. In the EEG this subject had a large amplitude P300 (>20 µV between Cz and left mastoid) at latency 310 ms with a peak between Cz and Fz. In the MEG, peaks occur both before and after this latency, indicating a complex generator structure.

Pattern recognition results are shown in Table 1. Values are given in terms of percentage of the number of to-be-detected responses. Hit rates for both EEG and MEG were the same for the oddball response. False alarm rates were generally low, but significantly lower for the MEG. For responses to frequent stimuli, hit rates were higher for the MEG. MEG pattern search obtained **no** bad false alarms, i.e. if a match was obtained, it was always synchronized to the occurrence of a stimulus.

Table 1. Mean percentage correct detections and false alarms for pattern detection of EEG and MEG responses to rare and frequent stimuli. The third row shows the results of one-tailed t-test comparisons (df=31) between EEG and MEG.

	rare stimuli			frequent stimuli		
	correct	good FA	bad FA	correct	good FA	bad FA
EEG	71.0	10.5	2.5	22.0	1.7	0.7
MEG	72.0	7.2	0.0	32.0	1.7	0.0
t-test	N.S.	$p < .05$	$p<.001$	$p<.001$	N.S.	$p<.001$

Pattern search using a template from one day to detect oddball responses on data from another day resulted in a 61% hit rate (0.14 good false alarms, 0.04 bad false alarms).

Discussion

The fact that templates recorded on one day could be used to detect 61% of the oddball responses on data recorded on other days indicates that to a large extent the same brain processes were elicited in each session. Therefore, differences in results for EEG and MEG are probably due to the differences in signals picked up by the type of measurement rather than to differences in signals generated in the subject's brain.

In the MEG recordings, the pattern search detected no 'bad' false alarms and more responses to frequent stimuli than in the EEG. Detection of responses to frequent stimuli is probably better than the EEG because the main generators are tangential sources which are detected optimally by the MEG sensors. In other EEG data with visual stimuli (Berg et al., in preparation), in which large radial sources are probably activated in the secondary visual areas, the pattern detection rate, at 34%, was similar to the MEG detection rate for auditory stimuli.

It is remarkable that for both EEG and MEG, the pattern search could detect more than 70% of the oddball responses, indicating both a high efficiency of the search method, and a high replicability of brain responses to the rare stimuli. In other subjects with smaller P300 responses we have observed lower detection rates (mean 24%). In the same subjects' responses to novel auditory stimuli, which produced large amplitude responses, the hit rates were much higher (mean 70%). The presence of artifact signals such as alpha waves and eye movements were observed to reduce the hit rate. These results indicate that the success of pattern search depends on signal strength and signal-to-noise ratio.

Although the MEG did not display a 'P300' at the latency obtained in the EEG, several consistent peaks could be observed up to 600 ms after the stimulus, suggesting a degree of complexity of the oddball response often not recognized in the EEG literature.

References

[1] Scherg, M., and Ebersole, J.S. Neurophysiol. Clin., 1994, 24: 51-60.

[2] Gotman, J., and Wang, L. Electroenceph. clin. Neurophysiol., 1992, 83: 12-18.

Acknowledgements

We would like to acknowledge support for the first author from the Ministry of Education, Science, and Culture of Japan to carry out the MEG study.

Magnetoencephalographic Measurements of a Developing Infant: A Pilot Study

Bowyer, S.M.[1,2], Moran, J.[1] and Tepley, N.[1,2]

[1]Henry Ford Hospital, Detroit, MI, USA; [2]Oakland University, Rochester, MI, USA

Introduction

MEG time-series recording of spontaneous activity is being used in our laboratory to study the adult brain in association with the mechanisms for stroke [7] and Migraine [9, 1]. In the present study MEG is used to study the development of the Central Nervous System (CNS) in a developing infant. In the past electroencephalography (EEG) has been used for similar studies. EEG has recorded the organization of the frequencies associated with CNS development [2, 4, 6]. MEG studies are much more easily conducted on a sleeping infant. One feature of the data obtained, sleep spindles are bursts of activity in the 11-15 Hz range lasting from tenths of seconds to approximately four seconds [5]. Sleep spindles of early childhood first occur during the third month of life and are an indication of normal sleep patterns being established. This particular activity in the 11-15 Hz tends toward an adult 14 Hz rhythm between 18 and 24 months of age, when a more mature (shorter spindle train length with a less spike-like negative component) sleep spindle activity is developing [3]. Early childhood sleep spindles have been seen on EEG and are considered to be a good indication of a normal developing Central Nervous System [6, 8]. Sleep spindles are observed in the frontal or central cortex during stage 2, and stage 3 sleep [5]. There tends to be an asynchrony in the appearance of sleep spindles between hemispheres in infants under one year of age [6, 8]. The absence or decrease of early childhood sleep spindle activity indicates an abnormality of the CNS such as cerebral dysfunction [3, 6, 8]. A second feature of the data obtained is the low frequency activity in the 4-10 Hz range. Low frequencies tend to decrease over time in the maturing CNS [5]. Magnetoencephalographic measurements were performed on a developing infant during this longitudinal study.

Method

The subject was a healthy male, full-term, infant (birth weight: 6 lbs. 4 ozs.). There were no known abnormalities or defects of the Central Nervous System and no family history of neurological disease. No medication was administered at any time during the study. Informed consent was obtained from the parent.

The subject was studied once a week from 18 weeks of age to 65 weeks of age. During the first half of the study, from 18 to 43 weeks of age, mother's milk was used to nurse the infant to sleep. Formula was used thereafter to nurse the infant to sleep (44 to 65 weeks). After he fell asleep the infant was placed under a 7 channel Superconducting Neuromagnetometer (Bti model 607), inside of a magnetically shielded room. The probe was positioned over the left hemisphere of the head first then over the right hemisphere. In reference to the International 10-20 system of electrode placement, the central coil in the 7 channel array was centrally located over the frontal-central region in the T4 location on the right side and in the T3 location on the left side of the head. (Fig. 1)   Six minutes of data were collected on one side of the infants head. He was then rotated 180° and repositioned under the probe while he slept. Three more minutes of data were collected on this side for a total of six minutes of data. Data was stored on optic disk for analysis later.

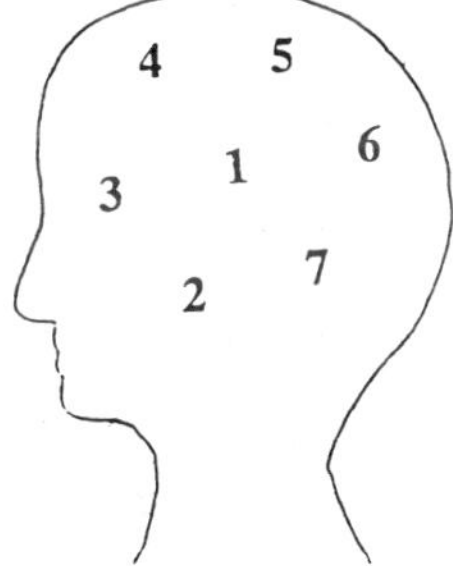

Fig. 1 Magnetometer channel locations

Results

Only the data for the channel over the frontocentral region of the infant's brain was used. This was where the sleep spindles were most prominent. Fig. 2 and 3 display the real time data showing typical sleep spindle bursts (underlined) as their length decreased over time.

Real Time Plot for Week 19

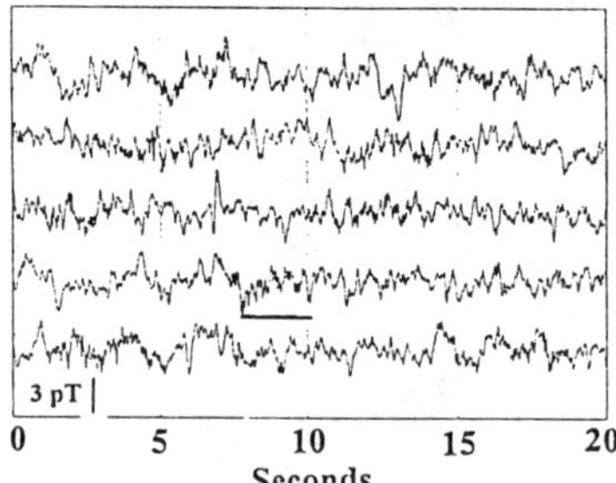

Fig. 2 Raw data collected from the right hemisphere during week 19. Each horizontal line represents 20 seconds of continuous data. The above plot represents 80 seconds of data. The sleep spindle burst is underlined.

Real Time Plot for Week 57

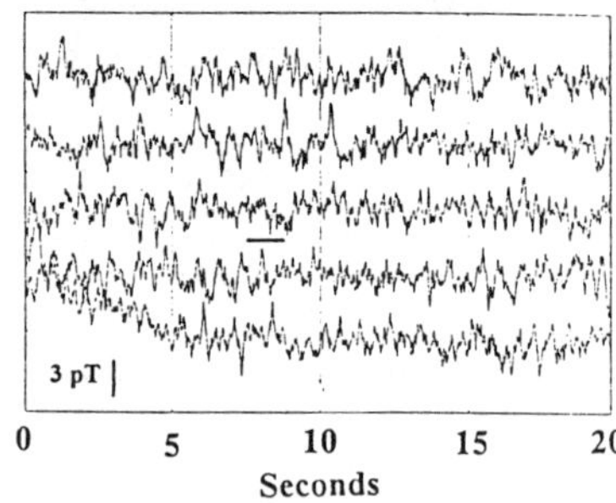

Fig. 3 Raw data collected from the right hemisphere during week 57. Each horizontal line represents 20 seconds of continuous data. The above plot represents 80 seconds of data. The sleep spindle burst is underlined.

Fig. 4 and 5 are FFTs performed on week 19 and 57 showing the peak frequency increased from 13.23 Hz in week 19 to 14.04 Hz in week 57.

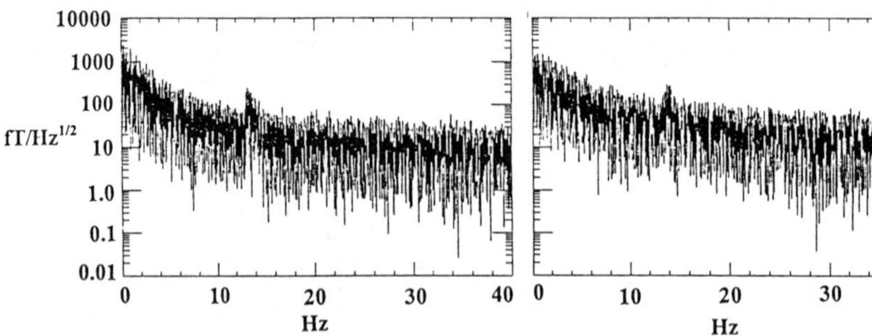

Fig. 4 FFT of data for week 19

Fig. 5 FFT of data for week 57

The six minutes of data were broken up into 18 epochs of 20 seconds, each was spectrally analyzed by a Fast Fourier Transform (FFT). Analysis was performed on the frequency range from 3.5 Hz to 15.5 Hz. The data was converted into a percentage of the daily spectrum. An overall time averaged FFT was obtained. The average of the sequential FFTs was used to quantify the spectrum for each date. To quantify the changes in the study we calculated the percent changes in the relative spectral composition of the daily spectrum from this average spectra. The Frequencies were classified into two ranges (4-10

Hz) and 11-15 Hz) and these bands were grouped together and graphed over time. Linear regressions were plotted over the data spectra. A linear increase in the percent change over time was found for the frequency range 11-15 Hz and a decrease for the frequency range 4-10 Hz. These findings were significant in each hemisphere and for both frequency ranges: sleep spindle band (11-15 Hz), and low frequency (4-10 Hz). Table I displays the correlations and significant p-values.

TABLE I

		correlation	p-value
Sleep spindle 11-15 Hz:	Right hemisphere	0.504	0.002
	Left hemisphere	0.523	0.003
Low Frequency 4-10 Hz:	Right Hemisphere	0.578	0.0003
	Left Hemisphere	0.589	0.0005

Fig. 6 through 9 display graphically the percent change over time of the daily averaged FFT from the overall time averaged FFT.

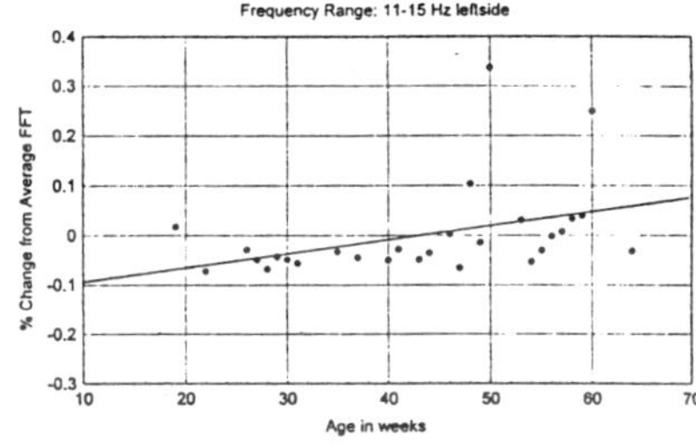

Fig. 6 Percent change in Sleep
spindle activity of the left hemisphere.

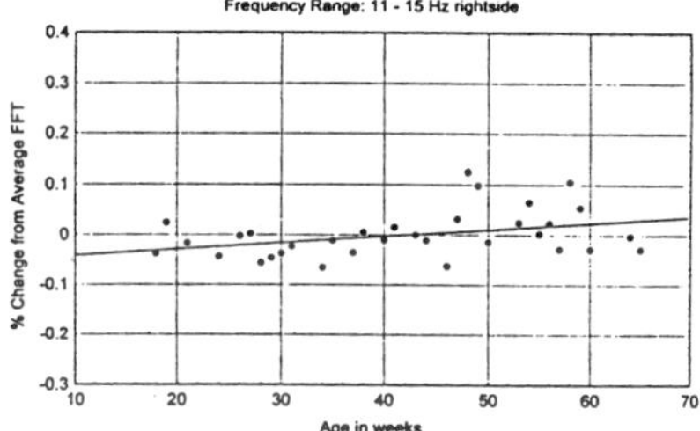

Fig. 7 Percent change in Sleep
spindle activity of the right hemisphere.

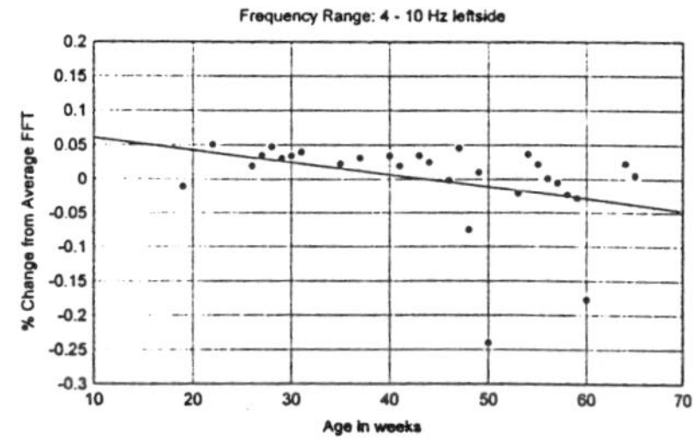

Fig. 8 Percent change in Low Frequency
activity of the left hemisphere.

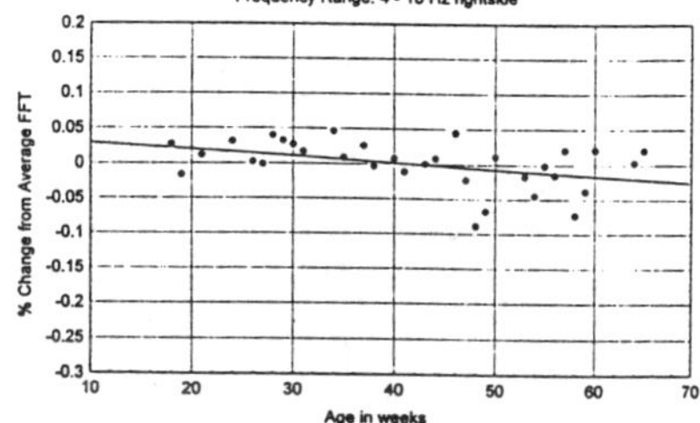

Fig. 9 Percent change in Low Frequency
activity of the right hemisphere.

Discussion

Magnetoencephalography can be used to determine the significance of the CNS developments over time in a developing infant. This procedure has a number of advantages over

715

Electroencephalography. MEG is less time consuming; it takes approximately 20 minutes to place a full array of electrodes on an infant, while it takes less then 2 minutes to position the probe over the area of interest on the infants head. Electrodes are not required to collect data for MEG making it easier for the technician to prepare the infant for the study. Infants can peacefully fall asleep as opposed to being forcibly restrained in papooses used to facilitate the electrode placement. In papooses they tend to scream and cry placing undue stress on them prior to the studies.

The results showed a difference in the increase and decrease of the linear regressions from hemisphere to hemisphere. These results can be explained by observing that the infant was always studied on the left side first. During this time he was most likely entering into a state of arousal conducive to sleep spindle activity. (i.e. light sleep, stage 2) During the second half of the study he fell into a deeper sleep in which sleep spindle activity decreases or ceases. Since we cannot collect simultaneous data on both sides of the head we are unable to make a conclusion regarding the asyncronisity of sleep spindle appearance.

The data in this study was based on 6 minutes of light, daytime sleep and can not truly be compared to the over-night studies completed in EEG. In addition the stages of sleep were not scored. One of the purposes of this study was to see if the results from prior EEG studies could be replicated. We have shown that similar results can be obtained using MEG. These developmental changes need to be further studied to determine their significance over a larger sample. To the best of our knowledge this is the first study of this kind performed on MEG. This can be a clinically useful technique for obtaining information about the developing Central Nervous System.

Acknowledgment

Thank you to Michael for being a good subject, and to Uma Saligram and Yeaton Clifton for their help in collecting the data. This research supported by NIH/NINDS Grant No 1R01 NS30914

References

[1] Barkley, G., Tepley, N., Nagel-Leiby, S., Moran, J., Simkins, R., Welch, K.M.A. Magnetoencephalograhic Studies of Migraine. Headache Journal, 1990; 30 (7), 428-434.

[2] Ellingson, R. Development of Sleep Spindle Bursts During the First Year of Life. Sleep, 1982 5 (1): 39-46

[3] Jankel,W.R., Niedermeyer, E. Sleep Spindles. Journal of Clinical Neurophysiology, 1985, 2 (1): 1-35.

[4] Louis, J., Zhang, J.X., Reuol, M., Debilly, G., Challamel, M.J. Ontogenesis of nocturnal organization of sleep spindles: a longitudinal study during the first 6 months of life. Electroencephalography and Clinical Neurophysiology, 1992, 83: 289-296.

[5] Niedermeyer, E. Lopes Da Silva, F. Electroencephalography, Basic principles, clinical , applications and related fields, Maryland, Williams and Wilkins 1993.

[6] Shibagaki, M. Kiyono, s., Kazuyoshi, W. Spindle Evolution in Normal and Mentally Retarded Children: A Review. Sleep, 1982, 5 (1): 47-57.

[7] Tepley, N. Spreading Cortical Depression and Related DC Phenomena, In: Hoke, M., Ernd, S. N., Okada, Y. C., Romani, G. L. Biomagnetism: Clinical Aspects, Amsterdam, Excerpta Medica 1992, 329-335.

[8] Tyner, F. S., Knott, J.R., Mayer, Jr., W. B. Fundamentals of EEG Technology. Volume 1: Basic Concepts and Methods. New York, Raven Press p 242-245.

[9] Welch, K.M.A., Barkley, G., Tepley, N., Ramadan, N., Central Neurogenic Mechanisms of Migraine. Neurology, 1993; 43 (suppl 3): S21-S25.

The Steady State Visual Evoked Response and Estimates of Phase Velocity

Burkitt, G.R., Silberstein, R.B. and Wood, A.W.

Swinburne Centre for Applied Neurosciences, Swinburne University of Technology, Melbourne, Australia

Introduction

Several models of brain function predict the existence of wave like phenomena where electrical activity travels from one cortical region to another [1-3]. A model which can explain the rhythmogenesis of the MEG and EEG should approximate neural mass dynamics at a spatial scale consistent with that of scalp recordings. In this respect, Nunez's global resonance model is of particular relevance since it allows direct comparison between scalp recorded electrical activity and theory [3]. This model predicts waves mediated by the rapidly conducting cortico-cortical fibres with speeds between 6 to 9 m/sec. Hence, a specific test of the model can be made by measuring the velocity of MEG and EEG signals and comparing estimated values with the above range. A direct approach is to use the Steady State Visual Evoked Response (SSVER) and measure the spatial change in phase across the scalp. In this paper the SSVER phase is measured at regular intervals on the scalp and the resulting phase topography yields an estimate of phase velocity at the stimulus frequency. In addition, estimates of wavenumber are obtained from the same measurement.

Method

Concurrent MEG and EEG signals were recorded from midline scalp sites while 21 subjects (mean age = 26.2 yrs, SD = 7.1 yrs) with normal or corrected vision viewed a sinusoidally alternating checkerboard. Subjects lay face down on a cushioned wooden bench while the stimulus was observed through a 45° mirror positioned directly beneath. All MEG recordings (unshielded) were performed with a BTi biomagnetometer (model 601) consisting of a single channel DC SQUID and second-order gradiometer with a 19.8 mm diameter pick-up coil and 50.4 mm baseline. The distance from the bottom of the pick-up coil to the bottom outside surface of the dewar was 12 mm. Accurate alignment of the dewar tail over each recording position was assisted by grid markings on a customised electrocap which also served to locate the EEG electrodes. MEG sites were marked at 1.5 cm intervals anterior from the inion while 10 EEG electrodes were placed at 3 cm intervals. To reduce volume conduction distortion of the EEG signal, the midline EEG electrodes were arranged as a bipolar chain. Checkerboard stimuli were presented for approximately two minutes at each MEG scalp site. Between trials subjects were permitted to move and if necessary the individual was rested for a short time. Over the duration of a session, 8 to 9 MEG scalp sites were recorded from midline.

Maximum and minimum luminances of the checkerboard pattern were 8 and 0.7 cd/m^2 respectively, thus the spatial contrast for the checkerboard was 84%. The viewing distance was 1.08 metres yielding a screen size of 10.9° (horizontal) by 8.25° (vertical) and the check size was set to 11'.

It has been suggested that alpha rhythm frequencies contribute to global resonance and travelling waves [3], thus the checkerboard alternation rate was matched to the peak alpha frequency of each subject. Peak alpha frequencies were determined at the beginning of the recording session from 5 minutes of spontaneous EEG (eyes closed) recorded from a bipolar electrode at Oz.

The SQUID output was amplified to give a sensitivity of 3.65 pT/Volt. EEG signals were amplified by 50 000. All signals were band-pass filtered (MEG: 1 to 40 Hz, EEG: 0.1 to 40 Hz) and digitised to 12 bit accuracy (±5 Volts) at a rate of 160 Hz. MEG signals underwent additional filtering to reduce noise at 50 Hz using Twin-T notch filters centred at 50, 100 and 150 Hz and an inverse Chebyshev filter with a stop-band notch at 50 Hz. Checkerboard patterns were generated on a NEC 3FG multisync monitor with a vertical frame rate of 83 Hz. The alternation rate of the checkerboard was controlled with a sinusoidal frequency generator consisting of a phase-locked loop, ROM sine wave lookup table and an 8 bit digital-to-analog converter.

SSVER phase was determined at the alternation frequency of the stimulus by evaluating the cosine and sine Fourier coefficients during each stimulus cycle. High frequency resolution and noise resistance was then obtained by collapsing all the coefficients calculated over the recording period into a single average.

Results

Typical examples of the Steady State Visual Evoked Field (SSVEF) and Steady State Visual Evoked Potential (SSVEP) obtained from conventional averaging methods are illustrated in Fig. 1. MEG sites extended only half the distance of the midline due to the positioning limitations of the probe and the prolonged time required to record from separate sites.

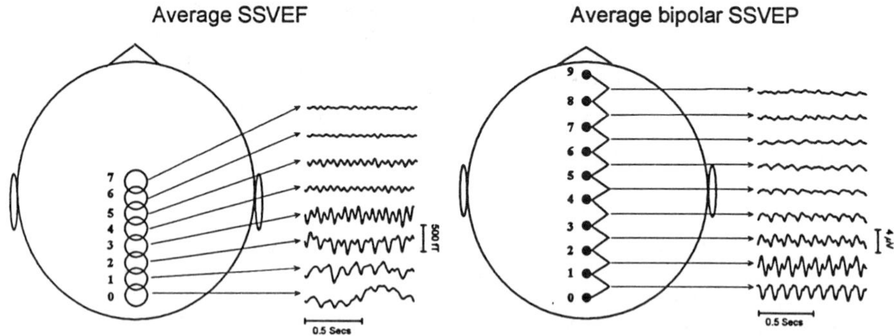

Fig. 1. The average SSVEF and SSVEP recorded from midline sites. In these illustrations, the average SSVEF (left) was obtained from 80 epochs (subject AW) while the average bipolar SSVEP (right) was obtained from 200 epochs (subject GH). The SSVEF demonstrates greater variability compared to the SSVEP since a smaller number of stimulus cycles were recorded during magnetic measurements.

In all subjects studied, the spatial variation of the SSVER was examined by plotting the phase at each midline recording site. Two distinctly different spatio-temporal features were observed. These comprised a smooth variation of phase with distance and a sharp phase discontinuity of approximately π radians along the midline. When present, the discontinuity occurred at parietal or central sites. Discontinuous variations may be due either tangential dipoles or standing wave nodes. As this paper addresses the issue of phase velocity estimation, it will be restricted to a consideration of smooth variations of phase with distance.

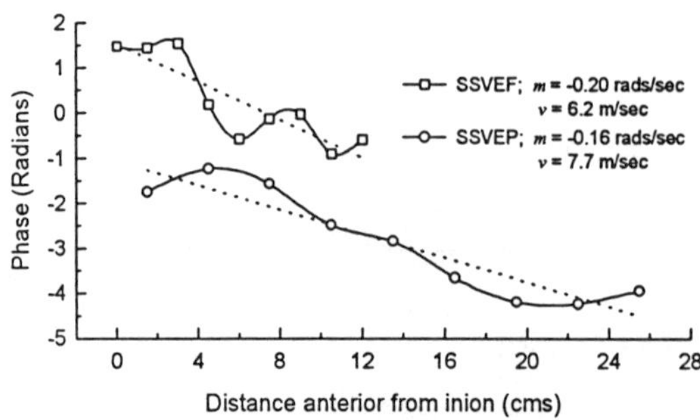

Fig. 2. Linear regression analysis (dotted line) of the phase topography (solid line) from one subject. MEG and EEG phase measurement sites along the midline are shown by open squares and circles respectively.

The spatial gradient of smooth phase variations was estimated from linear regression methods and only those gradients with a correlation coefficient exceeding an arbitrary level of 0.9 were included in the subsequent analysis. It was then possible to estimate phase velocity by treating the changes in phase as a time delay elapsed over a given scalp distance. Phase velocities were calculated from the best fit gradient of the spatial phase variation. Fig. 2. illustrates the linear regression technique with associated phase gradient and velocity estimates.

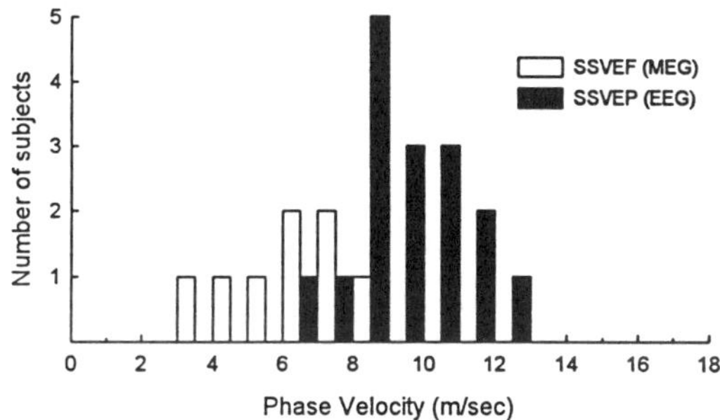

Fig. 3. Histogram of phase velocity estimates obtained from the SSVEF (open bars) and bipolar SSVEP (solid bars). Phase velocities from these two methods overlap within a range of 6 to 8 m/sec but are statistically different (p = 2.4×10⁻⁶, unpaired t-test). Note that all values were corrected for fissures with a constant of 2.2.

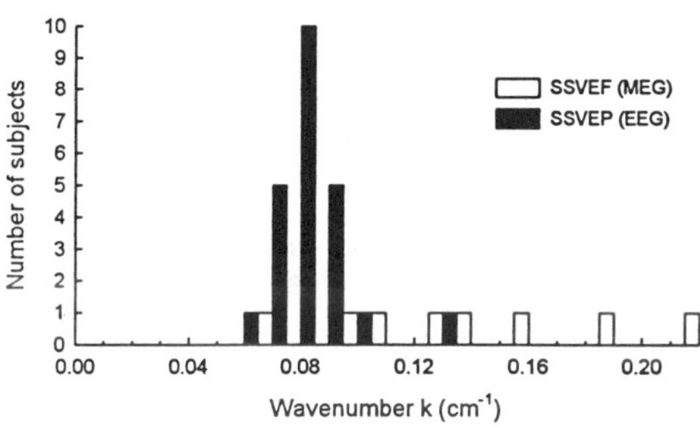

Fig. 4. Histogram of wavenumber estimates obtained from the SSVEF (open bars) and bipolar SSVEP (solid bars). Bipolar SSVEP wavenumbers are clustered between 0.06 and 0.10 cm^{-1}. No distinct wavenumber peak can be observed from SSVEF phase variations. Note that all values were corrected for fissures with a constant of 2.2

The phase velocity is given by;

$$v = \frac{2\pi f c}{m}$$

where v is the phase velocity (m/sec), f is the stimulus frequency (Hz), c is a folding ratio (2.2) to correct for the distances hidden within cortical fissures and m is the phase gradient (rads/m).

A histogram of all estimated phase velocity values is illustrated in Fig. 3. The SSVEF histogram (open bars) is characterised by a smaller number of subjects since fewer measurements met acceptable signal to noise criteria. Also note that SSVEF values show a tendency towards lower values compared to the SSVEP. The bipolar SSVEP (solid bars) generated the most sharply peaked distribution with all values bound between 6 and 13 m/sec. The mean SSVEF phase velocity was 5.1 ± 1.7 (SD) m/sec while mean SSVEP phase velocity was 8.7 ± 1.4 (SD) m/sec.

Gradients measured as radians per cm (cm^{-1}) can be treated as estimates of wavenumber (k). The distribution of wavenumber obtained from the SSVEF and bipolar SSVEP is illustrated in Fig. 4. The bipolar SSVEP is characterised by a sharp peak at 0.08 cm^{-1} while the SSVEF exhibits a broader distribution.

Discussion

The SSVER phase topography show features that are consistent with travelling waves. For instance, most estimates of phase velocity range from 5 to 10 m/sec, a range very similar to those estimated for cortico-cortical fibres (6 to 9 m/sec) from axon diameter studies of the corpus callosum [4, 5] and the frontal lobe [6, 7]. Previous data providing estimates of scalp velocity suggest wide variation ranging from as low as 3 m/sec to as high as 20 m/sec [8, 9]. This contrasts with the present finding of a narrow velocity range. The difference may be attributable to volume conduction distortion of the scalp topography since past estimates of velocity were calculated from monopolar EEG recordings. The monopolar EEG method is prone to volume conducted spread and reference electrode effects and this can reduce the apparent phase differences between electrode sites thereby resulting in larger estimates of phase velocity. In the current study, these difficulties were reduced by use of MEG and bipolar

EEG recording methods. In this respect, MEG and bipolar EEG methods appear well-suited to accurate measurement of the spatial changes in SSVER phase and estimates of phase velocity.

More speculatively, the mean wavenumber (0.075 ± 0.014 cm^{-1}, corrected for fissures) derived from Fig 4. is within the range anticipated (0.1 to 0.07 cm^{-1}) for global standing waves [10]. It is interesting to note that the corresponding wavelength (83 ± 15 cm) is of a similar magnitude to the sagittal circumference of the neocortex (~100 cm, corrected for fissures). Hence, travelling waves evoked by visual stimuli may extend over large distances. Such a finding is consistent with a global wave model of the EEG and MEG where periodic boundary conditions of the closed surface of the neocortex determine the wavenumber [3].

While the phase velocity and wavenumber estimates from SSVER phase topography are close to expected ranges, their accuracy is compromised by limited knowledge of the neocortical surface distance. All distances across the scalp were corrected for cortical folding by estimating the true brain surface distance divided by the exposed brain surface distance. Perimeter measurements of brain sections suggest that a folding ratio of 2.2 is appropriate [10] although it is probable considerable inter-subject variation exists. Furthermore, it is known that the ratio depends on the direction across the neocortex in which the perimeter is measured. Folding ratios that take into account such variation suggest a range of 2.0 to 2.5 [10]. At this stage, it is not clear whether a folding ratio of 2.2 is suitable for the midline surface of the neocortex but this potential inaccuracy can be avoided by measuring the sagittal neocortical perimeter with MRI methods.

In general, it was found that the SSVEF phase velocity estimates were slightly lower than SSVEP. The reasons for this finding are uncertain. One possibility is that the MEG is more sensitive to spatio-temporal noise arising from environmental magnetic fields. The difference may also be due to the volume conduction processes to which MEG is less sensitive. Further work is required to differentiate between these two possibilities.

In summary, the present findings indicate that features of the SSVER topography are compatible with a global model of EEG and MEG. It is not suggested that travelling wave mechanisms are the sole process which give rise to SSVER phase variations but merely that travelling waves may contribute to the observed data.

References

[1] Wilson, H.R., and Cowan, J.D. A mathematical theory of the functional dynamics of cortical and thalamic nervous tissue, Kybernetik, 1973, 13: 55-80.

[2] van Rotterdam, A., Lopes da Silva, F.H., van den Ende, J., Viergever, M.A., and Hermans, A.J. A model of the spatial-temporal characteristics of the alpha rhythm, Bull. Math. Biology., 1982, 44: 283-305.

[3] Nunez, P.L. Neocortical dynamics and human EEG rhythms, New York, Oxford University Press, 1995.

[4] Tomasch, J. Size distribution and number of fibres in the human corpus callosum, Anat. Rec., 1954, 119: 119-135.

[5] Aboitiz, F., Scheibel, A.B., Fisher, R.S., and Zaidel E. Fibre composition of the human corpus callosum, Brain Res., 1992, 598: 143-53.

[6] Bishop, G.H., and Smith, J.M. The size of nerve fibres supplying cerebral cortex, Exp. Neurol., 1964, 9: 483-501.

[7] Swadlow, H.A., Rosene, D.C., and Waxman, S.G. Characteristics of interhemispheric impulse conduction between prelunate gyri of the Rhesus monkey, Exp. Brain Res., 1978, 33: 445-467.

[8] Thatcher, R.W. Krause, P.J., and Hrybyk, M. Cortico-cortical associations and EEG coherence: A two-compartmental model, Electroenceph. Clin. Neurophysiol., 1986, 64: 123-143.

[9] Hughes, J.R., Kuruvilla, A., and Kino, J.J. Topographic analysis of visual evoked potentials from flash and pattern reversal stimuli: Evidence for "Travelling Waves", Brain Topography, 1992, 4: 215-228.

[10] Nunez, P.L. Electric Fields of the Brain, New York, Oxford University Press, 1981.

Nonlinear Spatio-temporal Interactions and Neural Connections in Human Vision Using Transient And M-Sequence Stimuli

Chen, H.-W., Aine, C.J., Flynn, E.R., and Wood, C.C.

Biophysics Group, MS D-454, Los Alamos National Laboratory
Los Alamos, NM 87545, U.S.A.

INTRODUCTION

There are two classes of connections among neurons in cat and monkey cortices [1]: (a) the traditional receptive field connections (feedforward connections), and (b) the dynamic association of neurons in different cortical areas which form functionally coherent assemblies (reciprocal connections). The feedforward connections have been well studied using traditional single-unit recording techniques. As pointed out by Singer [1], there are two strategies for characterizing the dynamic association of neurons in different cortical areas (reciprocal connections): 1) examining changes in discharge rate, and 2) examining temporal synchrony (the stimulus evoked coherent oscillations are in the gamma range 30-110 Hz). The first was evidenced by spatio-temporal interactions beyond the classical receptive fields (RFs) for V1 cortical neurons. The second was revealed recently using multiple-unit recording techniques and cross-correlation analysis, which have shown that cortical neurons can synchronize their discharges very precisely (e.g., [2]). The reciprocal connections can be categorized as tangential intra-areal, interhemispheric, and/or feedback (between cortex and LGN) connections, and may play an important role in integrating and binding widely distributed and parallel representations in different cortical areas.

Reciprocal connections, in essence, are the dynamic wiring (connections) of the neural network circuitry [1]. Given the high complexity of the neural circuitry in human brain, it is quite a challenge to study the dynamic wiring of highly parallel and widely distributed neural networks. The measurements of stimulus evoked coherent oscillations provide indirect evidence of dynamic wiring. In this study, in addition to the coherent oscillation measurements, we provide two more techniques for testing possible dynamic wiring: 1) measurements of spatio-temporal interactions beyond the classical receptive fields, and 2) neural structural testing using nonlinear systems analysis.

Spatio-temporal interaction and its relation to reciprocal connections. Animal studies have revealed the existence of facilitatory and/or inhibitory integration fields (IF) beyond the classical receptive fields (RFs) for retinal ganglion cells and striate cortical neurons. The extent of the integration fields is about 2-5 times the classical receptive field size [3]. Polat and Sagi [4] have recently found, in human vision, suppression and facilitation lateral interactions between spatial frequency (SF) channels revealed by lateral masking psychophysical experiments conducted at the fovea. The mechanism mediating these interactions at different eccentricities is, however, still an open question. One hypothesis for the interactions is based on the existence of multiple spatial frequency (SF) channels in the human visual system. Each channel may involve many neurons, and the receptive field of an SF channel may be defined as overlapping receptive fields of all the involved neurons (i.e., the RF of an SF channel is larger than that of a single neuron). Therefore, the spatial lateral interaction beyond the classical RFs of single neurons may be explained by the interactions between neighboring SF channels (feedforward connections) having spatially overlapping receptive fields. Another hypothesis is that the multiple SF channels are parallel and distributed in different cortical areas, and they are connected and bind via dynamic wiring. In this case, the distance of spatial lateral interactions would be large. Indeed, Ts'o and coworkers [5] have found excitatory horizontal intracortical connections from cat primary visual cortex, and suggested that these connections might contribute to properties beyond classical RF analysis. Therefore, measurements of spatio-temporal interactions may provide information on both traditional feedforward and reciprocal connections.

Nonlinear Systems analysis and Neural Circuitry testing. Several approaches to studying unknown biological systems consider the system initially as a "black box." In one approach, knowledge of the input-output (I/O) behaviors of the system are acquired by identifying the system's I/O transfer function. For example, the Laplace transfer function of a linear system is directly related to the I/O relationship of the system described by a linear differential equation. Unlike analyses of simple linear systems, analyses for nonlinear systems present a greater challenge since higher-order transfer functions are required for full characterization of a nonlinear system's I/O behaviors. One method used to characterize a nonlinear system (which is generally time-invariant or stationary, causal, and continuous), expresses the I/O relationship in terms of Volterra or Wiener functional expansions [6]. Such expansions are fully specified by a set of Volterra (or Wiener) kernels which can be estimated from the I/O measurements of the system under study. The system kernels provide a characterization of the system I/O relationship, and thus allow one to predict the response of a physical or physiological system to an arbitrary stimulus. Furthermore, the measured lower-order (the 1st- and 2nd-order) system kernels can provide information about the internal structure of the practical system under study [7]-[9], which suggests a number of potential applications in the areas of neural system modeling and identification. For example, we are interested in characterizing the I/O relationship of populations of neurons in cortical areas of human brain, obtained from noninvasive electromagnetic measurements. Neuronal responses can be modeled as circuitry, and nonlinear analysis may ultimately allow one to characterize the structure of neural system circuitry, using system kernels.

We have recently developed techniques using sparse-stimulation and short m-sequence stimuli [10], and have successfully measured the 1st- and 2nd-order kernels from neuromagnetic responses. The advantages for using this method are: 1) easier and fast kernel calculation; 2) short data acquisition times and less measurement errors; 3) short m-sequences with different lags (delays) are orthogonal to one another which permit their use in studying multiple input systems (i.e., the simultaneous presentation of m-sequences). Furthermore, the results from MEG experiments [10] indicate that the S/N ratio of responses evoked by m-sequences can be 2-3 times higher than responses evoked by transient target stimuli when stimulus presentation time is equated.

In this study, we have measured spatio-temporal interactions in human vision using transient and m-sequence stimuli. The estimated cross-correlations (system kernels) obtained using m-sequence stimuli were examined in hope of identifying possible neural connections (structures) and mechanisms underlying both feedforward and reciprocal connections. Measurements of coherent oscillations using transient stimuli will also be briefly discussed.

METHODS

Evoked neuromagnetic fields were recorded with a BTi 7-channel SQUID-coupled gradiometer system in a magnetically-shielded chamber. Four subjects with normal vision (HWC, CA, WEN, and JE) participated in this study. In the transient studies, we presented two pattern onset target stimuli (circular sinusoidal gratings, 266 ms duration and ~1 Hz rate of presentation) either simultaneously or sequentially with different SFs, to a specific eccentricity. These paired stimuli were presented at different eccentricities (0.8°, 1.6°, 3°, and 6°) in the visual field. The spatial separation between them was varied in order to quantify the nonlinear summation area. The nonlinear summation index was calculated as the peak response to the simultaneous presentation of stimuli divided by the sum of the peak responses to the two separate stimulus presentations. In the m-sequence studies, the paired m-sequence stimuli were presented at the same locations as the two target stimuli used for the transient study. Both m-sequence and transient target stimuli were presented for each placement of the dewar, for comparison. Three different m-sequence stimuli were used: 1) *Counter-phase stimuli*. Paired target patterns were counter-phase modulated by two orthogonal short binary m-sequences between two states (in-phase and off-phase). Subjects rested for ~5-10s after each of 6 binary m-sequence trials. 2) *In-phase stimuli*. Paired target stimuli were modulated between two different contrast levels (in-phase) of the same patterns by two orthogonal binary short m-sequences. The key feature of this stimulus is that it can be decomposed as a steady grating pattern component and a counter-phase modulated grating pattern component. 3) *Unpatterned stimuli*. Paired unpatterned stimuli were modulated above and below the background mean luminance by two orthogonal short binary m-sequences. MEG responses were digitized at 200 Hz.

The nonlinear dynamics and interactions between the spatial locations were revealed by the cross-correlations between the response and the input m-sequences.

RESULTS

The dashed curves in Fig. 1 are evoked neuromagnetic responses to the simultaneous presentation of the two stimuli, whereas the solid curves represent the sum of the responses to the individual presentations of the two stimuli. The dashed and solid curves should be approximately equal in peak amplitude if the spatial interactions between the stimuli were linear. On the other hand, a (large) difference between the two curves suggests a (strong) nonlinear interaction. Figs. 1a and 1c show responses of 2 sensors for stimuli 0.93° apart for subject WEN. Responses in Figs. 1b and 1d are responses to the stimulation with greater separation (1.6°). The results indicate stronger nonlinear interactions for stimuli which are closer together. Figs. 1e and 1f compare responses to paired stimuli (99% contrast) at different eccentricities for subject CA: in Fig. 1e the stimulus eccentricity and distance were 0.8° and 0.61°, respectively; whereas the stimulus eccentricity and distance were 3° and 2.3° for the responses in Fig. 1f, respectively. Figs. 1g and 1h are responses to the paired stimuli (4.5 c/d, 3° eccentricity) at two different

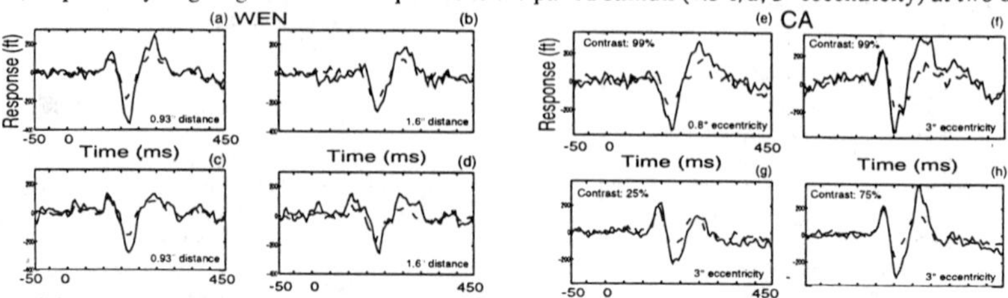

Figure 1. Spatio-temporal interactions. 1a-d: Subject WEN, the stimuli were 5.9 c/d, 99% contrast, and 1.6° eccentricity. 1e-h: Subject CA, the distance between the paired sitimuli is 0.61° in (e), and 2.3° in (f), (g), and (h).

contrast levels 25% and 75%, respectively for CA.

Fig. 2 shows measured 2nd-order cross-correlations for subjects WEN and HWC using counter-phase and in-phase stimuli. The contours of these cross-correlations provide information on neural circuitry structures. The kind of contour picture in Fig. 2a suggests simple feedforward (cascade) connections, whereas the more complex pictures in Figs. 2b and 2c suggest involvement of more complex neural circuitry. Fig. 3 shows the power spectra (using short-time FFT, duration: 115 ms, step: 20 ms) of waveforms (subject CA) stimulated by transient stimuli (50% contrast) at 3 different SFs (2.0, 4.5, and 8.5 c/d). The equipment and system noise (60 Hz) was suppressed using a threshold technique and low-pass filtering. Stronger 40 Hz coherent oscillations under stimulus conditions of SF=4.5 c/d (Fig. 3c) is evidenced, compared with other SF conditions (Figs. 3a and 3d).

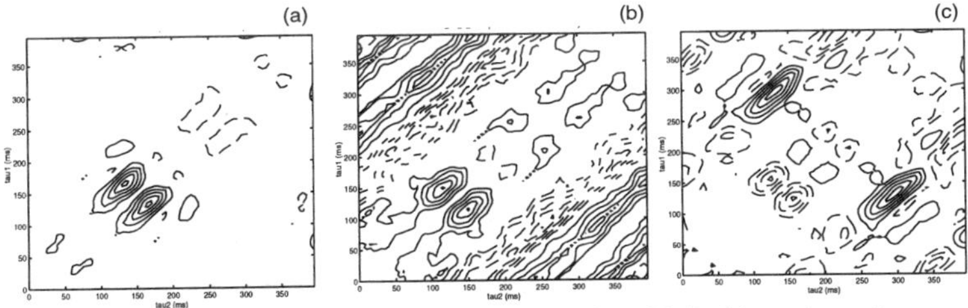

Figure 2. Measured 2nd-order cross-correlations (system kernels). Positive and negative contour values are shown as solid- and dashed-lines, respectively. (a): Subject HWC, counter-phase stimuli, 1.6° eccentricity, 25% contrast, 4.5 c/d. (b): Subject WEN, counter-phase stimuli, 1.6° eccentricity, 25% contrast, 2.0 c/d. (c): Subject HWC, in-phase stimuli, 3° eccentricity, contrast change from 20% to 80%, 4.5 c/d.

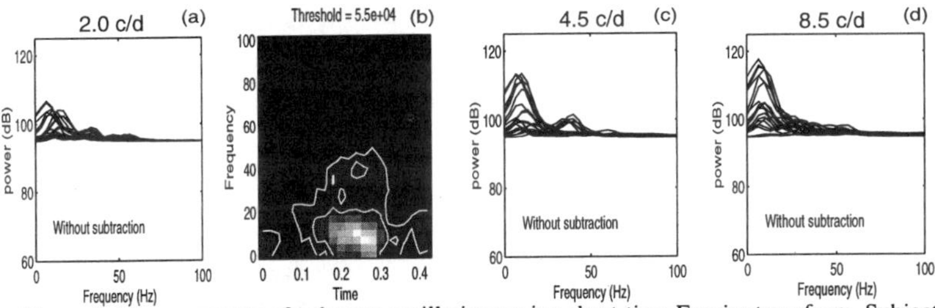

Figure 3. Measurements of coherent oscillations using short-time Fourier transform. Subject CA, 1.6° eccentricity, 50% contrast. (a): SF=2.0 c/d. (b) and (c): SF=4.5 c/d. (d): SF=8.5 c/d.

For a more quantitative analysis of nonlinear spatial summation, we introduce a summation Index (SI) by dividing the peak amplitude (e.g., the dashed curve in Fig. 1a) of the response to paired stimuli by the peak amplitude (e.g., the solid curve in Fig. 1a) of the response to the sum of the two individual stimulation conditions. An index value SI=1 indicates a linear summation, whereas SI < 1 indicates a nonlinear suppressive interaction. For example, an SI < 0.5 means that at least one of the peak amplitudes for single stimulus presentations was larger than that of the simultaneous stimulation. SIs were calculated for the strong second negative peak of the waveforms in Figs. 1a-d for subjects WEN and CA (not shown in Fig. 1), and are shown in Table 1.

Table 1. Summation Index (SI) for two stimulus distances and their ratio

Subject	SI for Distance 0.93°	SI for Distance 1.6°	Ratio
CA (Sensor #1)	0.57	0.76	0.75
CA (Sensor #6)	0.44	0.58	0.76
WEN (Sensor #2)	0.52	0.71	0.73
WEN (Sensor #7)	0.55	0.72	0.76

DISCUSSION AND CONCLUSIONS

The results in Fig. 1 indicate that the response waveforms evoked by transient stimuli showed strong nonlinear summation for shorter separations between the two stimuli. Both subjects (in Table 1) showed strong nonlinear summations for shorter distances between the two stimuli. The interactions were weaker when the distance increased (as indicated by larger SI values). As discussed above, the distance of spatial interactions may be larger for reciprocal connections than for feedforward connections (SF channels). Therefore, it is possible that the interactions resulted from both feedforward connections and reciprocal connections when the two stimuli were closer. As the distance increases, the involvement of feedforward connections decreases, and thus the spatial interactions may be dominated by reciprocal connections. It was also found in Table 1 that the SI ratio between different separations was quite consistent between subjects. Figs. 1e and 1f indicate that the foveal responses have smaller spatial interaction areas than peripheral responses. The SI values for the 1st negative peak and the 2nd positive peak in Fig. 1g (25% contrast, 3° eccentricity) are 0.48 and 0.77, and are 0.50 and 0.46 in Fig. 1h (75% contrast, 3° eccentricity), respectively. It appears that the interaction for the 2nd positive peak is more sensitive to the contrast change. It is less likely that stimulus contrast would change the RF size of an SF channel, therefore, the interaction evidenced at the higher contrast level in the 2nd positive peak may primarily result from the reciprocal connection (Singer [1] and Eckhorn et al., [2] have shown that coherence oscillations were sensitive to visual stimulus parameters such as position, orientation, movement direction and velocity, etc. Here, we show that the reciprocal connection may also be sensitive to the stimulus contrast). The SI values for the 1st positive peak in Figs. 1f, 1g, and 1h are 1.06, 0.84, and 0.92 respectively, which indicates that the spatial summation for this peak is more linear.

From the m-sequence studies, the 1st- and 2nd-order cross-correlations were measured using unpatterned, counter-phase, and in-phase stimulation conditions. In general, strong 2nd- (nonlinear component) but weak 1st-order (linear component) cross-correlations for the unpatterned and counter-phase stimulus conditions resulted, whereas a strong 1st-order cross-correlation was apparent when in-phase stimuli were used. The estimated cross-correlations have been further used for testing possible neural feedforward and reciprocal connections (structures) and mechanisms underlying spatial nonlinear interactions. The contours of Fig. 2a is typical for the 2nd-order unpatterned kernels and some counter-phase kernels under specific stimulus conditions. This kind of kernel picture suggests simple feedforward (cascade) connections, as corroborated by computer simulations [10]. On the other hand, the more complex pictures of the 2nd-order counter-phase and in-phase kernels (under certain stimulus conditions) in Figs. 2b and 2c suggest involvement of more complex neural circuitry (parallel, and/or feedback, etc.). Ongoing efforts have been directed to find appropriate neural circuitry models (e.g., the recurrent thalamic-cortical-thalamic feedback connection and/or cortico-cortical reciprocal connection, etc.) which can generate these kernel pictures using computer simulations.

Fig. 3(b) and (c) indicate possible 40 Hz coherent oscillations under stimulus conditions of SF=4.5 c/d and 50% contrast. The 40 Hz component is notably weaker under other SFs (Figs. 3a and 3d) and contrast conditions (not shown), which suggests that the coherent oscillations (and thus reciprocal connections) may be sensitive to stimulus parameters such as SFs and contrast levels. These results agree with our results from studies of spatio-temporal interactions, as discussed above.

In summary, the three methods (measurements of spatio-temporal interactions, system kernels, and coherent oscillations) used in this study may provide reliable measurements for understanding complex neural circuitry structures with reciprocal connections.

REFERENCES

[1] Singer, W. Development and plasticity of cortical processing architectures. *Science*, 1995, 270:758-764.

[2] Eckhorn, R., Bauer, R., Jordan, W., Brosch, M., Kruse, W., Munk, M., and Reitboeck, H. Coherent oscillations: a mechanism of feature linking in the visual cortex? *Biol. Cybern.*, 1988, 60:121-130.

[3] Li, C.Y. and Li, W. Extensive integration field beyond the classical receptive field of cat's striate cortical neurons--classification and tuning properties. *Vision Res.*, 1994, 34:2337-2355.

[4] Polat, U. and Sagi, D. Lateral interactions between spatial channels: Suppression and facilitation revealed by lateral masking experiments. *Vision Res.*, 1993, 33:993-999.

[5] Ts'o, D., Gilbert, CD, and Wiesel, T. Relationships between horizontal interactions and functional architecture in cat striate cortex as revealed by cross-correlation analysis. *Neurosci.*, 1986, 6:1160-1170.

[6] Marmarelis, P.Z. and Marmarelis, V.Z. Analysis of physiological system: the white noise approach. New York: Plenum Press, 1978.

[7] Chen, H.-W., Ishii, N., and Suzumura, N. Structural classification of non-linear systems by input and output measurements. *Int J Systems Sci.*, 1986, 17:741-774.

[8] Chen, H.-W., Jacobson, L.D., and Gaska, J.P. Structural classification of multi-input nonlinear systems. *Biological Cybernetics*, 1990, 63:341-357.

[9] Chen, H.-W. Modeling and identification of parallel nonlinear systems: structural classification and parameter estimation methods. *Proceedings of The IEEE*, 1995, 83:39-66.

[10] Chen, H.-W., Aine, C.J., Best, E., Ranken, D., Harrison, R.R., Flynn, E.R., and Wood, C.C. Nonlinear analysis of biological systems using short m-sequences and sparse-stimulation techniques. *Annals of Biomedical Engineering*, 1996, 24:4 (In press).

ACKNOWLEDGMENT

This work was supported in part by a grant from NEI (EY08610) and by the United States Department of Energy (DOE) Contract W-7405-ENG-36.

Spatial Frequency tuning functions and Contrast Sensitivity at Different Eccentricities in the Visual Field

Chen, H.-W., Aine, C.J., Flynn, E.R., and Wood, C.C.

Biophysics Group, D454, Los Alamos National Laboratory
Los Alamos, NM 87545, U.S.A.

INTRODUCTION

Threshold and suprathreshold vision. The human luminance spatial frequency contrast sensitivity function (CSF) has been well studied using psychophysical measurements by detecting spatial frequency (SF) grating patterns at *threshold* [1], [2]. *Threshold* CSFs at different eccentricities have proven to be quite useful in both basic and clinical vision research. However, near threshold, the CSF is measured at a linear area of the saturating contrast-response curve. In contrast, most of our everyday vision may be at suprathreshold levels, and thus may function most of the time at the nonlinear area of the contrast-response curve. Furthermore, since the CSF is measured near threshold, it is quite possible that only the most sensitive retinal or LGN cells may contribute to the measured CSF. Since the M (magnocellular) cells in retina and LGN are about 10 times more sensitive than the P (parvocellular) cells at low to medium spatial frequencies [3], [4], the measured threshold CSF may reflect more M cell contributions. There have been recent attempts to measure suprathreshold contrast responses using contrast matching techniques [5], [6], [7]. These results show that the contrast response functions at suprathreshold levels are flatter than CSF curves measured at threshold. However, whether the results from contrast matching reflect the spatial frequency tuning functions at suprathreshold for human vision is still unclear since contrast matching is quite different from threshold measurements [4].

Contrast-response (CR) functions. In monkey studies, the contrast-response (CR) functions at different contrast levels have been measured for M and P cells in the retina and LGN as well as cortical neurons in V1 and MT [3], [8]. A Michaelis-Menten-like equation can be used to describe the contrast-response function well:

$$R(c) = R_{max} c^n / (c^n + c^n_{50}) + M; \qquad (1)$$

where R is the amplitude of the response, c is contrast, R_{max} the maximum response, c_{50} the contrast at which the response reaches half of its maximum value, n the steepness (gain) of the curve, and M the dc level (spontaneous discharge). Equation (1) provides a quantitative way to describe the contrast-response relationship. For example, the parameter c_{50} can be used as an index for contrast sensitivity. The lower the c_{50} value, the higher contrast sensitivity the CR function will be. For the M and P cells in retina, the average c_{50} values were 0.13 and 1.74 respectively [3], and for cortical neurons on average, the c_{50} value was 0.33 for V1 and 0.07 for MT [8].

In this study, in order to better characterize the CSF at normal contrast levels, we measured the SF tuning functions as well as the CR functions at different *suprathreshold* contrast levels and different eccentricities of the visual field using noninvasive MEG techniques. Traditionally, peak analysis has been used for the averaged transient evoked waveform [9]. However, sometimes, it is difficult to define a single peak when the signal-to-noise ratio of the waveforms is low (the noise will cause multiple peaks), or when the waveforms are complex due to overlapping components. In this study, in addition to peak analysis, we have also developed more reliable "averaged power" analysis methods using FFT: the averaged power can be calculated from the entire waveform (or portion of).

Although equation (1) can generally fit the measured data well, the fitting errors will be large if the CR functions are super-saturated (i.e., the response decreases when contrast increases, which may be caused by inhibition between mechanisms at high contrast levels). In order to reduce the fitting errors and to obtain a quantitative measure of the inhibition at high contrast, we introduce a fifth exponential parameter, in addition to the 4 parameters in equation (1):

$$R(c) = e^{-ac} R_{max} c^n / (c^n + c^n_{50}) + M; \qquad (2)$$

where, a is the exponential parameter. Equation (2) reduces to equation (1) when $a=0$, and the larger the a value, the stronger the super-saturation.

METHODS

Experiments. Evoked magnetic fields were recorded with a BTi 7-channel SQUID-coupled gradiometer system in a magnetically-shielded chamber. Four subjects with normal vision participated in this study. Transient target stimuli (circular sinusoid, 266 ms duration and ~1 Hz rate of presentation) were presented one at a time in a sequential fashion to 3 different eccentricities (1.6°, 3°, and 6°). Three SFs (2.0, 4.5 and 8.5 c/d) were presented at

1.6° eccentricity, three SFs (1.5, 4.5, 8.5 c/d) were presented at 3° eccentricity, and three SFs (1.0, 4.5, 8.5 c/d) were presented at 6° eccentricity (The low SFs differed slightly across eccentricities in order to have at least 1.5 cycles of grating for each target). For each specific SF and eccentricity, there were 5 different contrast levels (10%, 25%, 50%, 75%, and 99%). Each of the 5 contrast conditions in a block of trials contained 20 stimulus presentations, making a total of 100 presentations for each block of trials (less than 2 min for a block of trials). Subjects rested for about 10s after 5 blocks of trials which produced a total of 100 responses for each contrast level. The presentations were randomly interleaved for the 5 contrast conditions. The responses to the 3 different SFs at 3 different eccentricities were recorded from the same dewar location. Cortical magnification factor was also considered. The sizes of the stimuli at 1.6°, 3°, and 6° eccentricities were calculated to be 0.76°, 1°, and 1.5° respectively.

Data analysis. In peak analysis, the SF tuning curves and the c_{50} values were obtained from the peak amplitude of the 1st negative peak (with latency ~160-190 ms) of the response waveforms for different stimulus conditions. In averaged power analysis, the SF tuning curves and the c_{50} values were obtained from the averaged power of the whole waveform (0-500 ms) or from a fraction of the waveform (90 ms duration around the 1st negative peak (with latency ~180 ms). The averaged power was calculated by first taking the Fourier transform of the waveform, and then averaging the spectral power of different temporal frequencies.

For each eccentricity, there are 5 SF tuning curves at 5 different contrast levels, and each curve has 3 data points at 3 different SFs (i.e., 3x5=15 data points). To more quantitatively describe potential contributions of the M and P systems to the responses, we propose a two-stream model whose parameters were estimated by fitting the model to the data using a least-square-error program (Matlab, Signal processing toolbox). The model is expressed as:

$$R(sf, c) = M(sf)CR_m(c) + P(sf)CR_p(c) \qquad (3)$$

where $R(sf, c)$ is the response at 3 different SFs (sf) and 5 different contrast levels (c); $M(sf)$ and $P(sf)$ are SF tuning functions for the M- and P-like systems, respectively; and $CR_m(c)$ and $CR_p(c)$ are the contrast-response (weighting) functions for the M- and P-like systems, respectively. There are 3 unknown parameters (at 3 SFs) for each $M(sf)$ and $P(sf)$, and 4 unknown parameters (at 4 contrast levels) for each $CR_m(c)$ and $CR_p(c)$. The values at contrast level *0.1* can be normalized as a unit for $CR_m(c)$ and $CR_p(c)$ (i.e., $CR_m(0.1) = Cr_p(0.1) = 1.0$). The total unknowns in equation (3) is 14, and the independent data values in the data set is 15. Therefore, we can uniquely estimate the 14 unknowns using a least-square-error fitting program. Equation (2), then, was used to estimate the 5 parameters *(R_{max}, c_{50}, n, M, and a)* from the obtained $CR_m(c)$ and $Cr_p(c)$ for the M- and P-like systems.

RESULTS

Both peak analysis and averaged power analysis were used to analyze the data measured from 4 subjects in order to construct SF tuning curves at 5 different contrast levels CR (contrast-response) functions for each specific SF. The CR functions can then be fitted by equation (2), resulting in estimations of the 5 parameters (R_{max}, c_{50}, n, M, and a). The results (SF tuning curves) from peak analysis for HWC were plotted in Figs. 1a-c, and those from averaged power analysis were plotted in Figs. 1d-f. The SF tuning curves in Figs. 1a-c are similar to those in Figs. 1d-f, which suggests that both peak analysis and averaged power analysis methods provided similar results. The CR functions and estimated 5 parameters at 6° eccentricity are shown in Figs. 1g-i.

The estimated SF tuning curves from averaged power analysis were further fitted by a two-stream (M & P) model (equation (3)) for all the 4 subjects at 3° eccentricity. The 1st-row in Fig. 2 shows the fitting accuracy: the minimum-square-error (MSE) is 9.3% for HWC (Fig. 2a), 5.6% for CA (Fig. 2d), 11.7% for WEN (Fig. 2g), and 6.9% for JE (Fig. 2j). The estimated SF tuning functions for the M and P systems ($M(sf)$ and $P(sf)$) were shown in the 2nd-row of Fig. 2. The estimated $CR_m(c)$ and $CR_p(c)$ (the contrast-response (weighting) functions for the M and P systems) were shown in the 3rd-row of Fig. 2 as the 'o' and '+' symbols, respectively. Equation (2), was then used to fit and estimate the 5 parameters from the obtained $CR_m(c)$ and $Cr_p(c)$. The fitting curves are shown in the 3rd-row of Fig. 2 as the dashed curves. The estimated parameters are listed in Tables 1 and 2. The fitting results in the 2nd- and 3rd-rows of Fig. 2 show similar features for all the 4 subjects.

Table 1. Parameters estimated from $CR_m(c)$.

Subjects	R_{max}	c_{50}	n	M	a	MSE
HWC	135	0.21	6.0	0.0	2.29	3.6%
CA	63	0.19	5.5	0.0	1.29	3.7%
WEN	2	0.11	10.0	1.0	2.74	8.6%
JE	983	0.33	7.0	2.0	10.24	9.1%

Table 2. Parameters estimated from $CR_p(c)$

Subjects	R_{max}	c_{50}	n	M	a	MSE
HWC	480	0.81	3.2	1.0	1.81	0.4%
CA	44	0.39	5.5	0.0	0.15	6.1%
WEN	8	0.47	10.0	1.0	1.48	4.4%
JE	664	0.66	7.0	0.0	3.93	1.5%

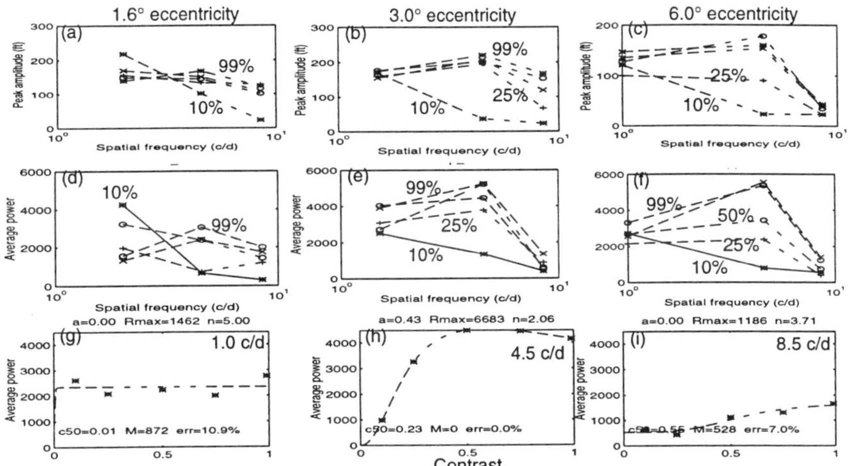

Figure 1. Spatial frequency tuning functions and contrast-response functions (Subject: HWC).
1a-c: Peak analysis (the 1st negative peak, latency ~160-190 ms); (a) 1.6 degree eccentricity;
(b) 3 degree eccentricity; (c) 6 degree eccentricity. Figs. 1d-f: Averaged power analysis
(waveform duration: 0-500 ms). 1g-i: Contrast-response functions fitted by equation (2);
6 degree eccentricity.

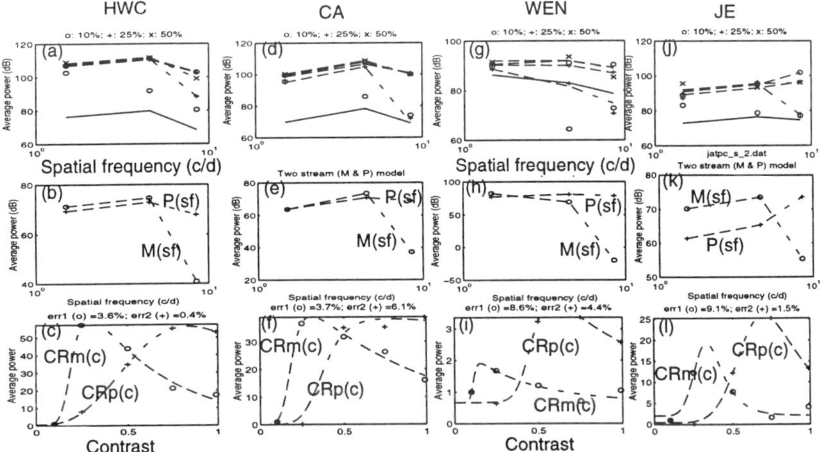

Figure 2. Data from averaged power anaysis (waveform duration: 90 ms; latency: ~180 ms) and fitted
by a two-stream (M & P) model (equation (3)). Four subjects: HWC, CA, WEN, and JE. 1st column:
HWC; 2nd column: CA; 3rd column: WEN; 4th column: JE. 1st row: Data were plotted as symbols;
the curves were predictions from the model (equation (3)). 2nd row: Estimated M(sf) and P(sf) for
the 4 subjects. 3rd row: estimated CRm(c) and CRp(c) for the 4 subjects.

DISCUSSION AND CONCLUSIONS

The SF tuning curves (obtained from both peak analysis and the averaged power analysis in Fig. 1) for high contrast levels at 1.6° eccentricity, were very flat with peak SFs > 8.5 c/d. In comparison, for stimuli at 3° and 6° eccentricities, the curves show a broad bandpass property with a peak SF about 3-5 c/d. For the lowest contrast level (10%), the curve shows a lower SF peak (< 1.5 c/d). This result is similar to results from other laboratories where CSF was determined using psychophysical measurements at threshold for a similar background mean luminance. Therefore, these results suggest that human eyes, under suprathreshold conditions, may have higher spatial resolution (higher SF gain) than what the CSF functions determined at threshold would predict. Furthermore, when comparing foveal with parafoveal conditions, our results also show that more foveal stimulation results in wider SF bandwidth.

These results are consistent with the two stream (M and P) theory: at low contrast levels, the majority of responding cells may be M-type (which prefer low contrast and low SFs), resulting in lower peaks in the SF tuning curves. With an increase in contrast level, more P-type cells (which prefer high contrast levels and high SFs) may be activated, resulting in curves with peaks shifted in the direction of higher SFs. As predicted by the two-stream model (the 2nd-row of Fig. 2), the M system shows a narrow-band low-pass SF tuning property, whereas the P system shows a broad-band band-pass SF tuning property, which is consistent for all the 4 subjects.

The measured CR functions (the 3rd-row of Fig. 1) also show that, at low SF the c_{50} value is low, indicating a high contrast sensitivity and earlier saturation of the CR function. As shown in Tables 1 and 2 (also in the 3rd-row of Fig. 2), the c_{50} value of $CR_m(c)$ (the M system) is lower (i.e., higher contrast sensitivity) than that of $CR_p(c)$ (the P system). These results are consistent with magno-parvocellular distinctions derived from single-unit monkey studies. The results in Tables 1 and 2 (and the 3rd-row of Fig. 2) indicate that the techniques developed in this study may provide useful information about the relative strengths of the M and P systems at different eccentricities. Therefore, based on these results, it appears that the c_{50} value and the estimated $CR_m(c)$ and $Cr_p(c)$ functions may provide information of the relative involvement of magno-like and parvo-like systems in humans. It is worth noting that (as shown in the 3rd-row of Fig. 2 and Tables 1 and 2) both M and P systems show super-saturation at high contrast levels as evidenced by the exponential a values in Tables, which may be caused by inhibition between mechanisms at high contrast levels. The super-saturation is generally stronger (higher a values) for the M-like system.

Single-unit studies in nonhuman primates indicate that different cortical areas may be differentially innervated by M or P streams. Based on our previous retinotopy studies using source localization strategies and our current results (i.e., SF tuning curves and CR functions), noninvasive MEG techniques may be able to shed more light on functions of neural groups in different cortical areas related to the two stream (M vs. P) theory.

REFERENCES

[1] Rovamo, J., Virsu, V. and Nasanen, R. Cortical magnification factor predicts the photopic contrast sensitivity of peripheral vision. *Nature,* 1978, 271:54-56.

[2] Kelly, D.H. Retinal inhomogeneity. I. Spatiotemporal contrast sensitivity. *J. Opt. Soc. Am.* 1984, A1:107-113.

[3] Kaplan, E. and Shapley, R.M. The primate retina contains two types of ganglion cells, with high and low contrast sensitivity. *Proc. Natl. Acad. Sci.,* 1986, 83:2755-2757.

[4] De Valois, R. and De Valois, K. Spatial Vision. Oxford Univ. Press, 1988.

[5] Georgeson, M.A., and Sullivan, G.D. Contrast constancy: deblurring in human vision by spatial frequency channels. *J. Physiol.,* 1975, 252:627-656.

[6] Bowker, D.O. Suprathreshold spatiotemporal response characteristics of the human visual system. *J. Opt. Soc. Am.* 1983, 73:436-440.

[7] Davis, E.T. Modeling shifts in perceived spatial frequency between the fovea and the periphery. *J. Opt. Soc. Am.* 1990, A7:286-296.

[8] Sclar, G., Maunsell, J.H.R. and Lennie, P. Coding of image contrast in central visual pathways of the macaque monkey. *Vision Res.,* 1990, 30:1-10.

[9] Regan, D. Human Brain Electrophysiology--Evoked Potential and Evoked Magnetic Fields in Science and Medicine. New York: Elsevier Science Publishing Co., 1989.

ACKNOWLEDGMENT

This work was supported in part by a grant from NEI (EY08610) and by the United States Department of Energy (DOE) Contract W-7405-ENG-36.

Neuromagnetic Fields Evoked by Sine Tones, Sine Tone Complexes, Vowel Formants, and Two-Formant Vowels

Diesch, E., Luce, T.A., and Eyferth, K.

Department of Psychology, Technical University of Berlin, Berlin, Germany

Introduction

In an earlier study [1], stimulus related differences were found in source location both among N1m sources and sustained field (SF) sources elicited by synthetic five-formant vowels which did not seem to correspond in any straightforward manner to the tonotopic functional organization of the auditory cortex established with sine tones [2]. This apparent discrepancy may be due to spectral complexity differences as well as the dimensionality difference between multivariate vowel sets and univariate sine tones. More generally, the question arises as to how the response of the auditory cortex to spectrally complex acoustic stimuli like vowels may be related to the responses to their spectral components. Spectral components of vowels may be considered at the level of harmonic structure and at the level of formant structure. In order to further investigate this issue, the magnetic field response to sine tones and vowels radically reduced in complexity and dimensionality was studied in two experiments.

Methods

In the first experiment, 14 subjects were presented with a 600 Hz and a 2100 Hz sine tone and the first (F1) and second formant (F2) of a synthetic two-formant instance of the vowel /æ/. F1 and F2 formant frequency values were 600 Hz and 2100 Hz. Additionally, the subjects received the two-tone complex and the two-formant vowel generated by additive superposition of the respective components. In the second experiment, a 200 Hz (F1a), a 400 Hz (F1b), a 800 Hz (F1c), a 2600 Hz vowel formant (F2), and three two-formant vowels generated by superimposing F1a, F1b, and F1c with F2 were presented. The two-formant vowels were perceived as /i/, /e/, and /æ/ and the single formants F1a and F1b as /u/ and /o/. 15 subjects were investigated. In both experiments, stimulus duration was 600 msec. Component stimuli were equated in subjective loudness, formants and vowels in fundamental frequency. Stimuli were delivered to the right ear in a random sequence with an ISI of 2 sec +/- 100 msec. There were 150 replications of each stimulus. In the first experiment, subjects were asked to count rare light flashes, in the second experiment, to count an additional rare vowel.

The magnetic field was recorded over the left hemisphere with a 37-channel magnetometer [3] installed in a magnetically shielded room. The bandwidth of the recording was 0.16-64 Hz. After artefact rejection, 1200 msec epochs including a 200 msec prestimulus baseline interval were selectively averaged and digitally lowpass filtered at 20 Hz for N1m analysis and 5 Hz for SF analysis. The model of a single spatio-temporal equivalent current dipole in a spherical volume conductor was used for source analysis. Source parameters were computed in a head-centered coordinate system. Spatio-temporal dipoles were fitted to a 20 msec interval around the N1m rms peak and to the SF in the 400-600 msec latency interval. Source solution were accepted only if polarity reversal was evident in the field pattern and the goodness of fit was 0.90 or better. In experiment 2, satisfactory single-dipole source solutions for the SF could be found only for a minority of subjects and stimulus conditions. Multi-dipole source solutions proved instable and variable across subjects. Therefore, in order to separate source contributions to the SF and to isolate the predominant contribution, SF waveforms in experiment 2 were submitted to principal components analysis (PCA) and the PCA waveform component accounted for by the first PCA factor was entered into the dipole source analysis.

The source parameters were submitted to repeated measures ANOVAs. Two types of analysis were performed. The first type of analysis sougt to determine the frequency dependence of source parameters. Thus, the component and composite stimulus sources, respectively, were compared among each other. The second type of analysis investigated whether the source fitted to the field elicited by a composite stimulus was equal to the model source fitted to the linear sum of the fields elicited by the stimulus components. With some kind of nonlinear addition, a systematic deviation of the source activated by composite stimuli from the the source predicted by the linear addition model should be observed. Thus, the composite stimulus sources were compared with the expected composite stimulus.

Results

A representative example of the evoked magnetic field waveforms and field distributions recorded in experiments 1 and 2 is presented in Fig. 1. The field average shows a pronounced N1m and a distinct SF component. The field distribution of both the N1m and the SF can be accounted for by a single equivalent current dipole. In experiment 1, the

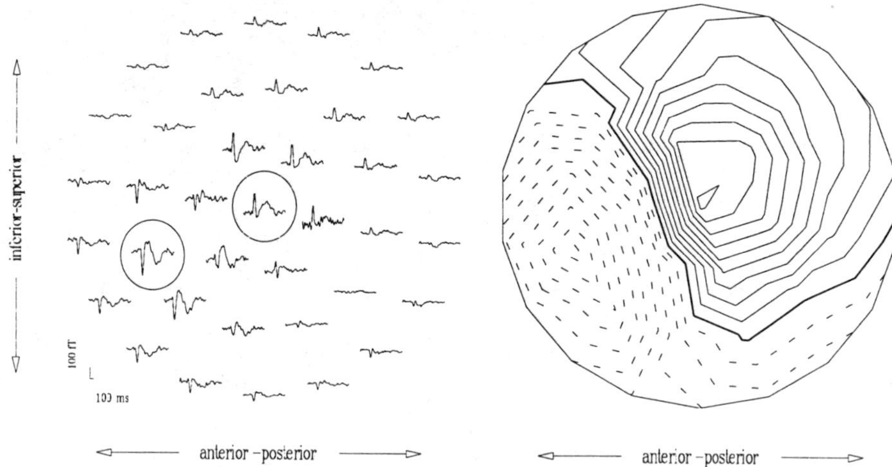

Fig. 1. Left: The 37 averaged evoked magnetic field waveforms recorded in response to the 600/ 2100 Hz compound tone in one of the subjects of experiment 1. The two waveforms with across-channels maximum and minimum N1m amplitude, respectively, are marked by circles. Right: The field distribution at the maximum N1m field power latency.

composite stimulus N1m sources were shorter in latency ($F[1,11] = 58.78$, $P < 0.001$), weaker in dipole moment amplitude ($F[1.11] = 55.72$, $P < 0.001$), and more lateral in source location ($F[1,11] = 10.50$, $P < 0.01$) than the expected composite stimulus N1m sources. 2100 Hz N1m sources were more medial in location ($F[1,11] = 5.67$, $P < 0.05$) than 600 Hz N1m sources. Tone stimulus N1m sources were more posterior ($F[1,11] = 5.99$, $P < 0.05$) and more medial in location ($F[1,11] = 9.79$, $P < 0.01$) and shorter in latency ($F[] = 13.66$, $P < 0.01$) than formant stimulus N1m sources. The composite stimulus SF sources were weaker in dipole moment amplitude ($F[1,10] = 18.50$, $P < 0.001$) than the expected composite stimulus SF sources. The tone stimulus SF dipole moments were weaker than the formant stimulus SF sources ($F[1,10] = 16.59$, $P < 0.001$). The SF source location pattern mirrored the N1m location pattern, but the SF source location differences fell short of statistical significance. Fig. 2 displays the N1m and SF source topographies.

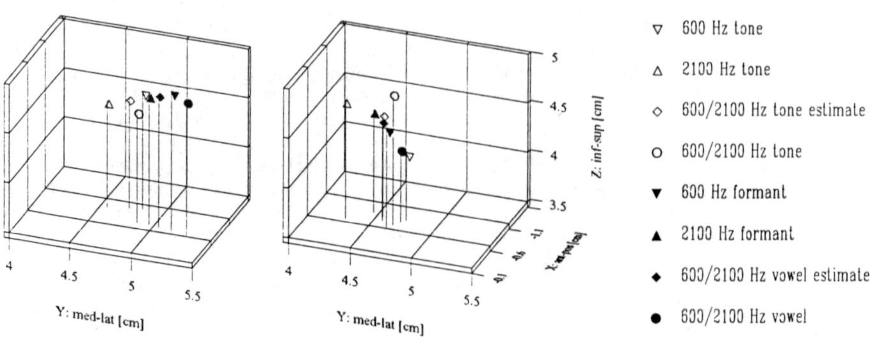

Fig. 2. Across-subjects average N1m (left) and SF (right) dipole source locations in experiment 1.

730

As in experiment 1, in experiment 2 the dipole moment of composite stimulus N1m sources was smaller than the dipole moment of expected composite stimulus N1m sources (F[1,7] = 20.68, P < 0.01). Apart from this finding, results were rather different for experiment 2. Composite stimulus N1m source location was somewhat more lateral than expected composite stimulus N1m source location, but the difference failed to attain statistical significance. The location of one-formant N1m sources was a function of formant frequency (F[3,21] = 9.67, P < 0.002) and the location of two-formant vowel N1m sources showed a similar first formant frequency dependence (F[2,22] = 4.83, P < 0.05). N1m sources associated with lower first-formant frequency were located more posterior than N1m sources associated with higher frequencies. Two-formant vowels were shorter in latency than one-formant vowels (F[1,11] = 61.15, P < 0.001). The latency of one-formant vowels was strongly related to formant frequency (F[3,21] = 19.20, P < 0.001). Analysis of the first PCA waveform component of the N1m analysis interval yielded comparable results. In the analysis of the first PCA waveform component of the SF the composite stimulus SF sources were weaker in dipole moment amplitude (F[1,10] = 18.50, P < 0.001) than the expected composite stimulus SF sources. As N1m source locations, SF source locations were related to first-formant frequency (F[2,14] = 3.97, P < 0.05). However, the SF frequency dependence was almost a mirror image of the N1m frequency dependence. Fig. 3 displays the average N1m and SF source locations. Fig. 4 shows individual subjects data for N1m source location on the anterior-posterior axis.

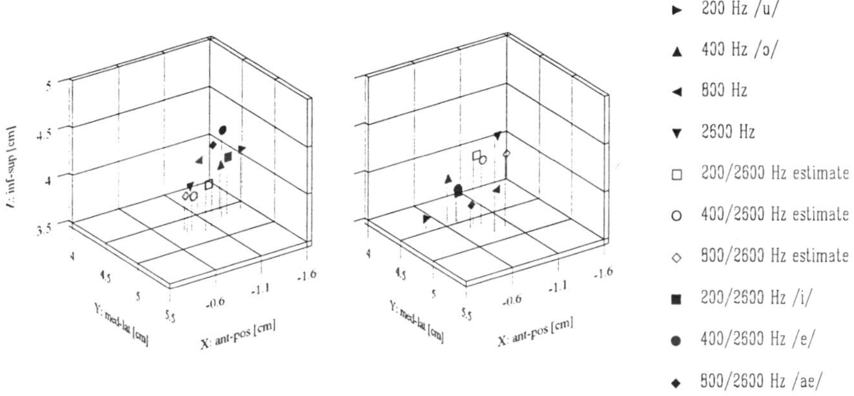

Fig. 3. Across-subjects average N1m (left) and SF (right) dipole source locations in experiment 2.

Comparison of the source location vs. frequency results suggests that different cortical fields may have contributed towards the generation of the evoked response in experiments 1 and 2. For ten subjects who participated in both experiments, this hypothesis could be tested by comparing centroid source locations. The centroid source locations were slightly more posterior and medial in experiment 2 than in experiment 1 both for N1m and SF. However, as none of the comparisons attained statistical significance, a gross anatomical hypothesis is rendered implausible.

Discussion

The latency of compound stimulus N1m sources was shorter than the latency of component stimulus N1m sources. The dipole moment of compound stimulus N1m sources was smaller than the sum of the dipole moments of stimulus component N1m sources. Both findings strongly suggest that the cortical response to two-tone stimuli and two-formant vowels cannot be accounted for in terms of the linear addition of the responses to sinusoidal components and single formants, respectively.

With regard to the tone activated sources, the finding in the first experiment that the 2100 Hz sources were more medial than the 600 Hz sources is consistent with the established tonotopic map of sound frequency [2]. Furthermore, that formant and vowel sources were more lateral than tone and compound-tone sources is compatible with the hypothesis of a cortical pitch map [3]. However, unless we allow for a timbre effect, the more medial location of the 2100 Hz formant source, compared with the 600 Hz formant source, seems incompatible with the pitch-map hypothesis.

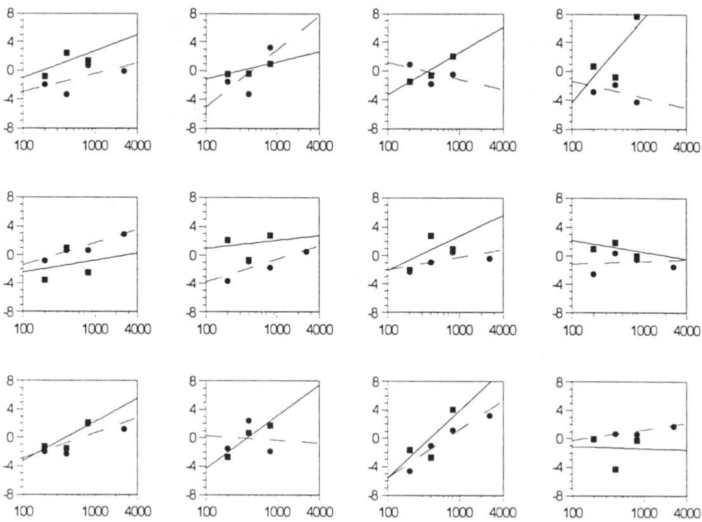

Fig. 4. Anterior-posterior N1m source location coordinate values as function of first-formant frequency. Circles: one-formant stimuli. Squares: two-formant vowels. Frequency is scaled in log Hz, source location in mm (ipsative values).

It seems likely that the tonotopic maps of formant frequency observed in the second experiment are spurious, i.e. not really maps of formant frequency. If they were, then, given their orientation, one would expect these maps to occupy an area of cortical space separate from the area occupied by the sources in the first experiment. However, no difference in source location was found between the first and the second experiment. A hint towards an account may be taken from the finding that, in cat auditory cortex, units are more narrowly tuned in the center than the periphery of the isofrequency strips oriented perpendicular to the frequency axis [5]. Vowel recognition depends on the extraction of the spectral envelope. One step of a variety of methods to do this is to use broadly tuned filters. Whereas frequency difference limens are correlated with frequency, the spacing of the harmonics of the fundamental frequency of a vowel is constant across the frequency range. Thus, the source topography in the second experiment may reflect the degree of frequency resolution diminution that is required to abstract from the harmonic structure of a formant peak.

References

[1] Diesch, E., Eulitz, C., Ross, B., Hampson, S. The neurotopography of vowels as mirrored by evoked magnetic field measurements. Brain and Lang. (in press)
[2] Pantev, C., Hoke, M., Lehnertz, K., Lütkenhöner, B., Anogianakis, G., Wittkowski, W. Tonotopic organization of the human auditory cortex revealed by transient auditory evoked magnetic fields. Electroencephal.Clin. Neurophysiol., 1988, 69: 160-170.
[3] Drung, D., Zimmermann, R., Cantor, R., Erné, S.N., Koch, H., Matthies, K.P., Peters, M., Scheer, H.J., Stollfuß, D. A 37-channel DC SQUID magnetometer system. Clin. Phys. Physiol. Meas., 1991, 12, suppl. B: 21-29.
[4] Pantev, C., Hoke, M., Lütkenhöner, B., Lehnertz, K. Tonotopic organization of the auditory cortex: Pitch versus frequency representation. Science, 1989, 246: 486-488.
[5] Heil, P., Rajan, R., Irvine, D.R.F. Sensitivity of neurons in cat primary auditory cortex to tones and frequency-modulated stimuli. II: Organization of response properties along the 'isofrequency' dimension. Hear. Res., 1992, 63: 135-156.

Acknowledgements

The research was supported by the *Deutsche Forschungsgemeinschaft*. The support provided by the staff of the Biomagnetism laboratory of the *Physikalisch-Technische Bundesanstalt* is gratefully acknowledged.

Source Location and Orientation, Dipole Moment, and Latency of Magnetic Mismatch Fields May Mirror the Perceptual Salience of Phonetic Contrasts

Diesch, E., Luce, T.A., and Eyferth, K.

Department of Psychology, Technical University of Berlin, Berlin, Germany

Introduction

Evoked magnetic field measurements have provided evidence for topographic differences between vowels and consonants with regard to the N1m component of the magnetically evoked response [1]. However, it has not yet been shown, whether the magnetic mismatch field (MMF) is sensitive to the distinction between vowels and consonants. The mismatch response reflects the detection of a difference between current stimulus input and a trace in sensory memory [2]. Mismatch responses are elicited by rare deviant stimuli presented in the context of frequent standard stimuli. For the electric counterpart of the MMF, the mismatch negativity, it has been shown that the amplitude of the mismatch response is correlated with the discriminability of the eliciting stimulus contrast [3,4]. It is relatively easy to discriminate perceptually among vowels even when they are presented in noise [5]. Under comparable conditions of ambient noise, it may be difficult to discriminate among plosive stop consonants differing in place of articulation [6]. Thus, it is to be expected that MMF sources activated by vowel and consonant contrasts, respectively, differ in dipole moment amplitude.

Methods

In two experiments, the MMF elicited by vowel and stop plosive consonant place-of-articulation contrasts was studied. In experiment 1, vowels occurred in isolation. In experiment 2, they were embedded within a syllabic context. In experiment 1, the vowels [a], [i], and [u] and the consonant-vowel syllables [ba], [da], and [ga] were presented. In vowel blocks, [a] was the standard (80%) and [i] and [u] were deviants (10% each). In consonant blocks, [da] was the standard and [ba] and [ga] were deviants (10% each). The sequence of blocks was counterbalanced and the within-block stimulus sequence was randomized. Half of the subjects received natural, the other half synthetic speech stimuli. Stimulus duration was 114 msec for the natural and 100 msec for the synthetic stimuli. In experiment 2, subjects were presented with one random stimulus sequence including [da] as standard (80%) and [di], [du], [ba], and [ga] as deviants (5% each). All stimuli were synthesized. Stimulus duration was 150 msec. The stimuli were equated for fundamental frequency and subjective loudness and delivered at an individually adjusted listening level trough a magnetically neutral transmission system to the subject's right ear. A constant offset-to-onset ISI of 400 msec was used. In both experiments, random intermittent light flashes were presented which the subjects were instructed to count. 13 subjects participated in the first and 12 subjects in the second experiment.

The magnetic field was recorded over the left hemisphere with a 37-channel magnetometer [7] inside a magnetically shielded room. The recording was continuous with a passband of 0.16-64 Hz. The ECG and EOG were monitored through separate electrical leads. The position and orientation of the dewar relative to the subject's head was obtained by measuring the magnetic field produced by five small induction coil triplets attached to the scalp. The data was segmented offline into 500 msec epochs (including a prestimulus baseline interval of 100 msec) and, after artefact rejection, selectively averaged. Only responses to standards immediately preceding a deviant were included in the standard averages. In order to improve the signal-to-noise ratio and to facilitate source analysis, the averages were digitally bandpass filtered with cut-offs at 0.5 Hz and 10 Hz. MMF waveforms were obtained by subtracting the standard from the deviant averages. Single spatio-temporal equivalent dipole sources were fitted for the 20 msec interval centered on the MMF rms peaks. On the basis of the induction coil measurements, an orthogonal right-handed headframe was defined with positive x projecting from the inion through the nasion and positive z projecting through the 10-20 location Cz. Source solutions were accepted if and only if the MMF field pattern showed polarity reversal, the distance of the source to the midsagittal xz-plane was larger than 25 mm, the distance to the horizontal xy-plane was larger than 10 mm, and the goodness of fit attained was 0.90 or better. The source parameters were submitted to a repeated measures ANOVA. The Greenhouse-Geisser epsilon correction was applied where appropriate.

Results

In both experiments, pronounced MMF deflections were registered in the vowel deviant and difference waveforms. Generally, the MMF deflections occurring in the consonant deviant and difference waveforms were distinctly smaller in amplitude. Figs. 1 and 2 present an example.

733

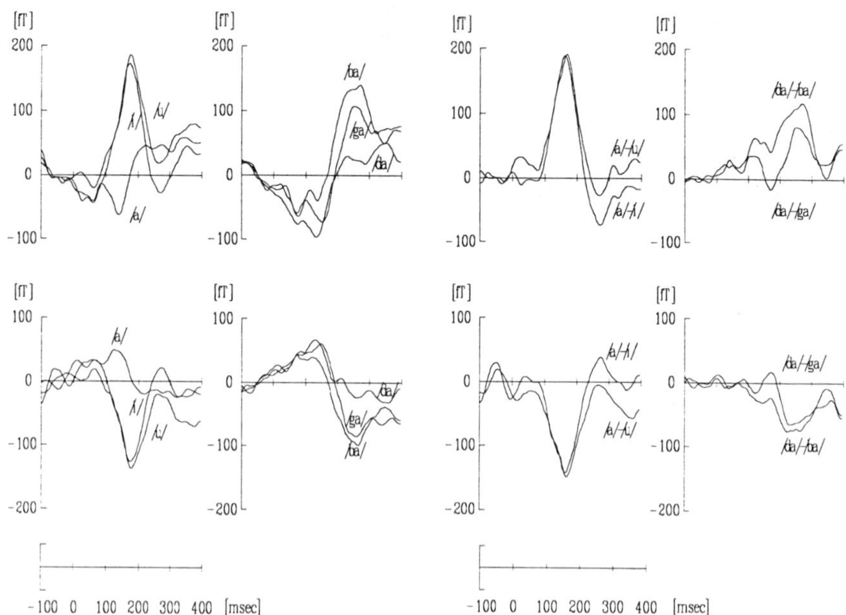

Fig. 1. Left part: Maximum and minimum averaged evoked magnetic field waveforms recorded in response to the standard (vowel condition: [a], consonant condition: [da]) and deviant stimuli (vowel condition: [i], [u], consonant condition: [ba], [ga]) in one subject in experiment 1. Right part: The difference waveforms computed by subtraction of the respective standard waveform from the deviant waveforms. Flux entering the skull is positive, flux leaving the skull negative.

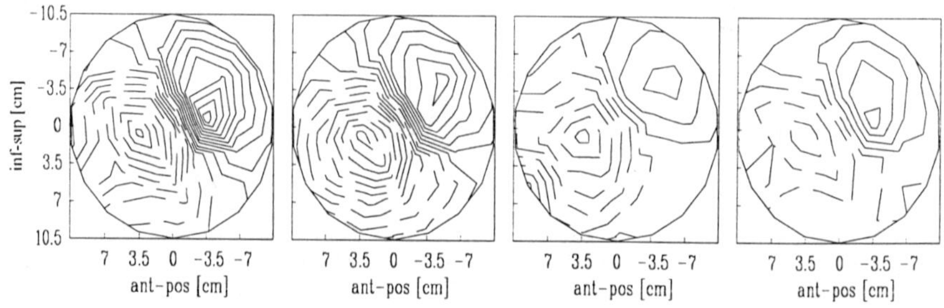

Fig. 2. The field distribution of the MMF difference waveforms presented in Fig. 1 at their respective rms peak latency. The spacing of the isocontour lines is 20 fT. Ant-pos: anterior-posterior, med-lat: medial-lateral, inf-sup: inferior-superior. From left to right: [a]-[i], [a]-[u], [da]-[ba], [da]-[ga]. Solid lines mark magnetic flux leaving the head, dashed lines magnetic flux entering the head.

734

Data of three subjects in the first experiment and one subject in the second experiment were discarded, because they failed to meet the acceptance criteria defined above. The remaining subjects exhibited mismatch field distributions which could be accounted for by single equivalent current dipoles.

In experiment 1, no difference was found between natural and synthetic speech stimuli. Therefore, the data from the two stimulus conditions were pooled. The vowel contrasts were found to elicit MMFs with shorter latency (F[1,9] = 19.67, P < 0.002) and stronger dipole moment (F[1,9]=16.94, P < 0.003) than the place-of-articulation consonant contrasts. MMF average peak latencies were 150 msec for [a]-[i], 157 msec for [a]-[u], 198 msec for [da]-[ga], and 189 msec for [da]-[ga]. The corresponding average dipole moment amplitudes were 26, 20, 13, and 9 nAm. Significant spatial separation was obtained between the source locations of vowel and the source locations of consonant contrast MMFs (F[1,9] = 8.43, P < 0.02). The consonant contrast MMF sources were located more anteriorly than the vowel contrast MMF sources. Post-hoc comparisons using Tamhane's T² showed that the effect was due mainly to the 10 mm distance, on the average, between the [da]-[ga] MMF sources and the vowel MMF sourcess along the anterior-posterior axis.

The results of experiment 2 were similar with respect to latency and dipole moment. Vowel contrast MMFs occurred with shorter latency (F[1,10] = 12.9, P < 0.01) and had stronger dipole moment amplitudes (F[1,10] = 14.1, P < 0.01) than consonant contrast MMFs. On the average, the MMF peaked at 142 msec for da]-[di] with a dipole moment amplitude of 16 nAm, at 175 msec for [da]-[du] with 15 nAm, at 232 msec for [da]-[ba] with 7 nAm, and at 200 msec for [da]-[ga] with 8 nAm. Vowel contrast MMF sources were located more laterally than consonant contrast MMF sources (F[1,10] = 5.56, P < 0.05). The average location difference between vowel and consonant sources was 9 mm. The MMF sources activated by the [da]-[ga] contrast were more anterior than the MMF source activated by the [da]-[ba] contrast (F[2,20] = 5.12, P < 0.05). On the average, this location difference amounted to 5 mm. Vowel and consonant contrast MMFs differed in their inclination towards the anterior-posterior axis of the head frame (F[1,10] = 10.4, P < 0.01). Again, post-hoc comparisons with Tamhane's T² showed that the effect was mainly due to the difference between [da]-[ga] and the vowel contrasts.

The spatial configuration of the average MMF sources was rather similar across the two experiments (Fig. 3). Vowel sources were spatially separated from consonant sources, but were not spatially separated from one another. The consonant sources were spatially separate, with the [da]-[ga] source in a more anterior location than the [da]-[ba] source.

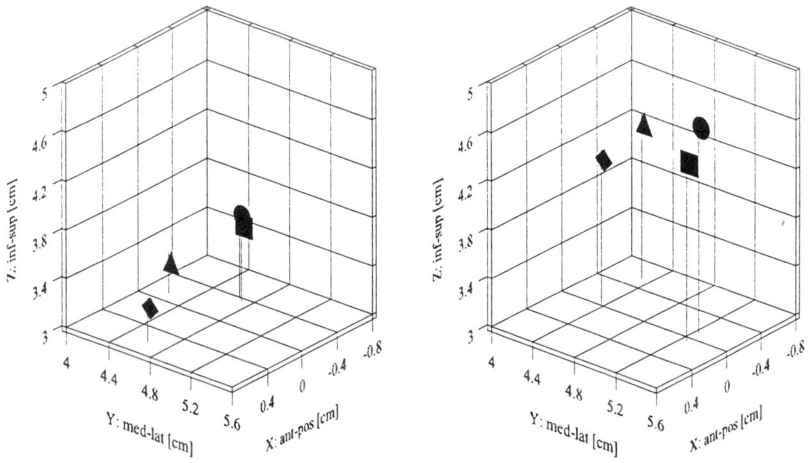

Fig.3. Across-subjects mean mismatch field source locations in head frame coordinates. Left: Experiment 1: circle: [a]-[i], square: [a]-[u], triangle: [da]-[ba], diamond: [da]-[ga]. Right: Experiment 2: circle: [da]-[di], square: [da]-[du], triangle: [da]-[ba], diamond: [da]-[ga]. Ant-pos: anterior-posterior, med-lat: medial-lateral, inf-sup: inferior-superior.

However, there were some differences. Vowel and consonant sources were spatially separated along the anterior-posterior axis in experiment 1, but along the medial-lateral axis in experiment 2. Furthermore, compared with the first experiment, the vowel sources seemed slightly more anterior and the consonant sources somewhat more posterior in the second experiment. These differences may be due either to stimulus factors, to subject factors, or both. Another difference between the two studies concerns the dipole moment amplitudes which were smaller for vowel and consonant contrasts in the second experiment. Most likely, this differences is due to the stimulus protocols used. Whereas vowel and consonant contrasts were presented in separate blocks in the first experiment, they were presented together in one random sequence in the second.

Discussion

Vowel contrasts and plosive stop consonant place-of-articulation contrasts elicited magnetic mismatch fields. Their field distributions were bipolar in structure and could be accounted for by single equivalent current dipoles. In both experiments, the consonant contrasts elicited MMFs with later latencies and smaller dipole moments than did the vowel contrasts. Vowel MMF sources were spatially seperated from consonant MMF sources. There was spatial separation between the consonant MMF sources, but not between the vowel sources. The pattern of these findings corresponds to the difficulty-of-discrimination pattern for vowels [5] and consonants [6]. The rate of confusion of [ba] and [ga] with [da] is much higher than the rate of confusion of [i] and [u] with [a]. Confusion of [da] with [ga] is more likely to occur than confusion of [da] with [ba]. It seems plausible to relate the latency and dipole amplitude findings to these perceptual data: The brain may take longer to generate a signal indicative of a weak stimulus contrast and the signal it generates may be smaller in amplitude. The source location findings may require another account. It has been shown that the topography of the electric mismatch negativity depends on the kind of stimulus contrast (frequency, intensity, duration) it is elicited by [8]. Accordingly, vowel and consonant MMF sources may be spatially separated because vowel and consonant contrasts differ in quality. Different consonant MMF sources may be spatially separated, because they are activated by different feature contrasts. However, this hypothesis leaves the lack of spatial separation between the vowel MMF sources without explanation. Thus, even the spatial separation effects may have to be accounted for in terms of the ease-of-discrimination hypothesis. A larger number of cells and synaptic connections may be involved in the perceptual extraction of the more subtle stimulus contrasts. As computation in neural nets involves activation and inhibition, this would not necessarily result in larger amplitude evoked magnetic fields. However, unless an increase of activity spreads in concentric circles across the cortical surface, there would be a shift in the location of an equivalent current dipole.

References

[1] Kuriki, S., Murase, M. Neuromagnetic study of the auditory responses in right and left hemispheres of the human brain evoked by pure tones and speech sounds, Exp. Brain Res., 1989, 77:127-134.

[2] Näätänen, R. Attention and brain function. Hillsdale, New Jersey, LEA, 1992.

[3] Sams, M., Paavilainen, P., Alho, K., Näätänen, R. Auditory frequency discrimination and event-related potentials. Electroencephal. Clin. Neurophysiol. 1985, 62: 437-448.

[4] Näätänen, R., Schröger, E., Karakas, S., Tervaniemi, M., Paavilainen, P. Development of a memory trace for a complex sound in the human brain. Neuroreport, 1993, 4: 503-506.

[5] Pickett, J.M. Perception of vowels heard in noises of various spectra. J. Acoust. Soc. Am., 1957, 29: 613-620.

[6] Miller, G.A., Nicely, P. An analysis of perceptual confusions among some English consonants. J. Acoust. Soc. Am., 1955, 27, 338-352.

[7] Drung, D., Zimmermann, R., Cantor, R., Erné, S.N., Koch, H., Matthies, K.P., Peters, M., Scheer, H.J., Stollfuß, D. A 37-channel DC SQUID magnetometer system. Clin. Phys. Physiol. Meas., 1991, 12, suppl. B: 21-29.

[8] Giard, M.H., Lavikainen, J., Reinikainen, K., Perrin, F., Bertrand, O., Pernier, J., Näätänen, R. Separate representation of stimulus frequency, intensity, and duration in auditory sensory memory: An event-related potential and dipole-source analysis.

Acknowledgements

The research was supported by the *Deutsche Forschungsgemeinschaft*. The support given by the staff of the Biomagnetism laboratory of the *Physikalisch-Technische Bundesanstalt* is gratefully acknowledged.

Auditory Evoked Magnetic Fields After Stimulation with Test-Tones and Words - Studies on Experimentees with Good Hearing and with Long-Term Hearing Impediments

Emmerich, E.[2], Meyer, R.[1], Nowak, H.[3], Huonker, R.[3], Gießler, F.[3], Meißner, W.[1] and Beleites, E.[1]

[1]Clinical Department of Audiology; [2]Institute of Physiology I and [3]Biomagnetic Center, Friedrich-Schiller-University Jena, Germany

Introduction:

The aim of the present study was to evaluate results of the distorted central auditory process. Localization of human cortical responses to acoustical stimulation with tones and speech signals began with magnetoencephalographic (MEG) studies on primary auditory cortex performed by R. HARI and her working group [1,2]. The MEG studies in connection with MRT-picture have demonstrated that the sources of the magnetic fields (peak-latencies 100 ms) are in the primary auditory cortex .

Our questions are partly based on clinical studies of U.S. IVARSSON et al. and N. KRAUS et al. about speech recognition after noise exposure [3,4]. The clear results of pure tone audiometry differ from those of speech audiometry in patients with noise induced hearing loss.

Method:

1. Clinical audiological tests on patients and on controls in the deparment of Audiology

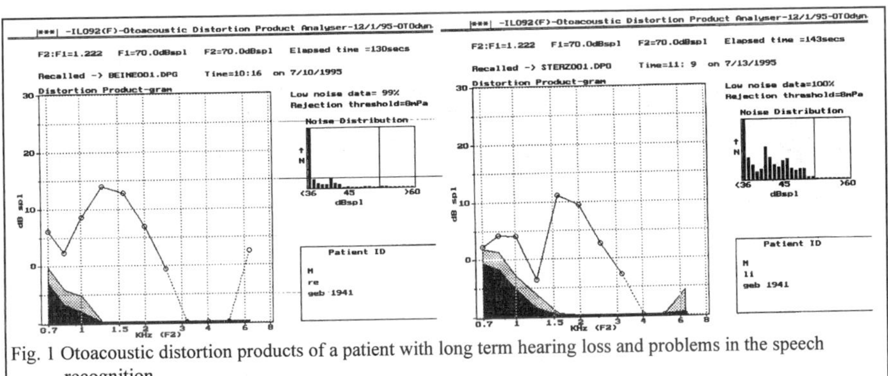

Fig. 1 Otoacoustic distortion products of a patient with long term hearing loss and problems in the speech recognition

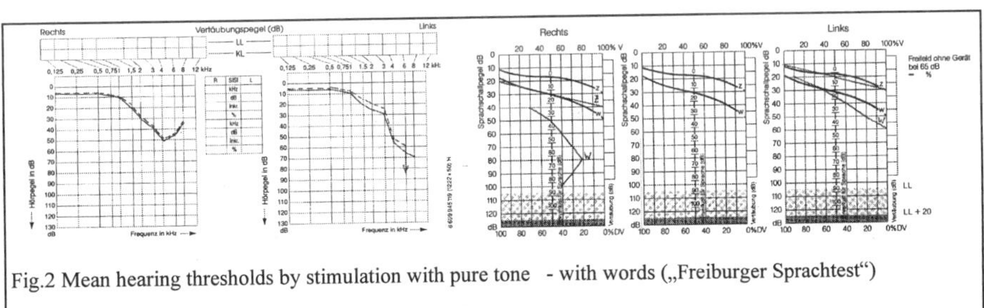

Fig.2 Mean hearing thresholds by stimulation with pure tone   - with words („Freiburger Sprachtest")

The auditory evoked magnetic fields (AEF) and auditory evoked potentials (AEP) of 12 young healthy subjects with good hearing (average age 24, 10 male, 2 female) and patients with long term hearing loss (average age 45, 3 male) were derived using test-tones with *varying intensity and frequency* (60, 80, 100 dB SPL; 0,5, 1 and 2 kHz) with the help of the 50-channel Philips system [5]. Through the application of combinations of test-tone - test-word (sensical and non-sensical words) central auditory processes and word recognition were studied. The strength and position of a current dipole is reconstructed by the magnetic field distribution and in this way the source localization in the MRT- picture can be carried out. The highly time-sensitive localization results of the MEG combined with the MRT give us pictures for the functionally determined magnetic field source localization. Subsequently the AEF of hearing-impaired test subjects was derived using the test-tones and test-tone - test-word combinations. Using the AEF dipole localization and current density distribution were calculated and applied to the MRT-pictures. The stimuli were applied to subjects contralaterally to the 31-channel derivation of the AEF through a tested "Tube and Funnel System". The results of test from „Tube and Funnel Systems" are demonstrated in Fig.3a and b.

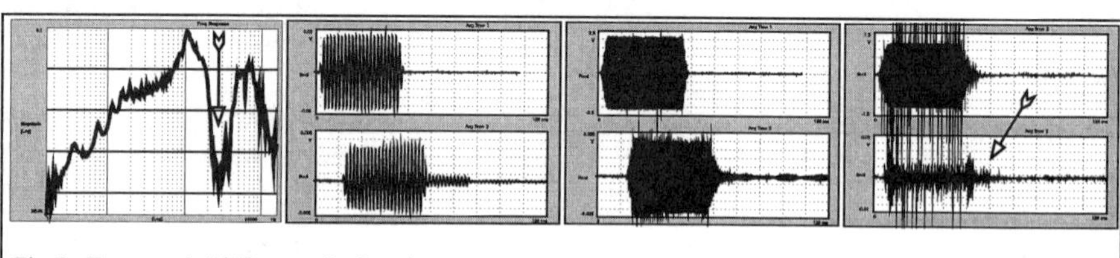

Fig. 3a Test tone 1-5 kHz: transfer function,
- upper trace: sound pressure at loudspeaker output outside MSR,
- lower trace: sound pressure at the ear inside MSR

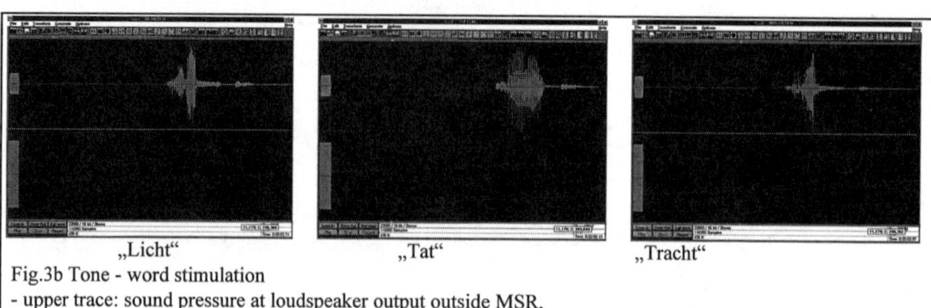

Fig.3b Tone - word stimulation
- upper trace: sound pressure at loudspeaker output outside MSR,
- lower trace: sound pressure at the ear inside MSR

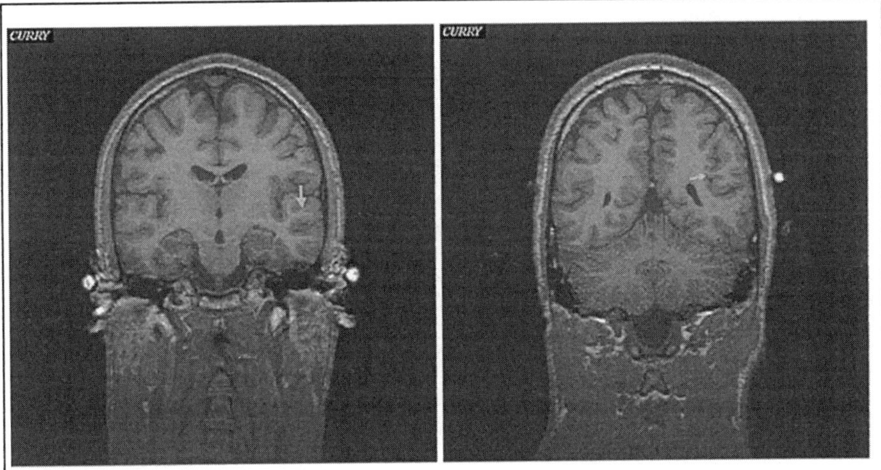

Fig.4 Source localisation pure tone stimulation   and onset word - controls

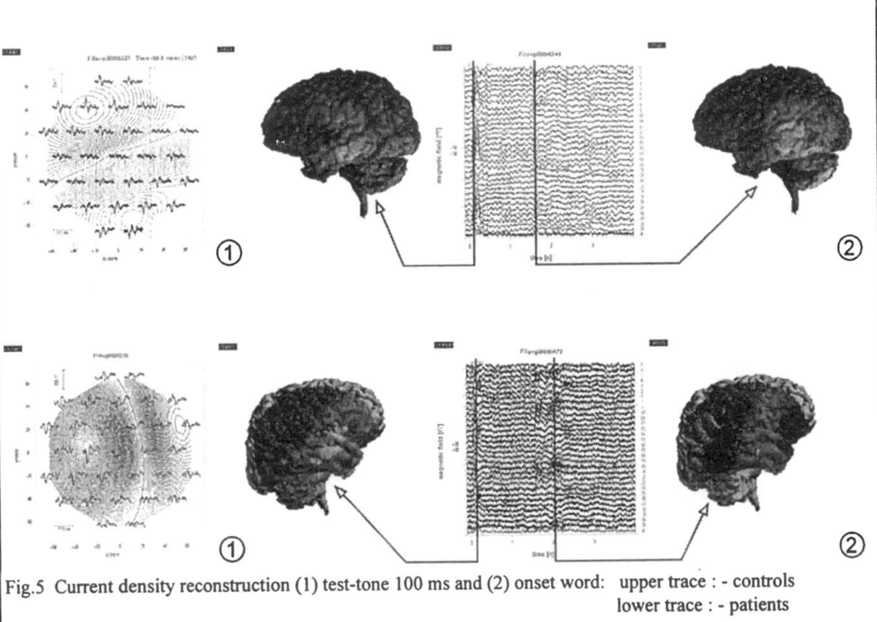

Fig.5 Current density reconstruction (1) test-tone 100 ms and (2) onset word: upper trace : - controls
lower trace : - patients

Scale: gray - maximum curent density

Results and discussion :

It is apparent that the amplitudes of AEF components $\leq$ 100 ms depend on stimulus intensity. It is also clear that the dipoles depiction of the components near 100 ms are reproducible with a large degree of stability, and that late components > 200 ms of these AEFs are displaced in the localization. Latent shifts of the AEF components could be found, as documented in the literature i.e. by CH. PANTEV et al. [5]. The results of current density reconstruction show that for more complex stimuli this localization of sources is a possible method.
We found different maximum neuronal activity by stimulation with pure tone and words. The results are demonstrated in Fig. 4 for controls.Changes in the source localization and current density distribution of patients are shown in Fig 5. Here we can see first results for patients. The clinical checkup of that patient shows in accordance with the literature H.G. DIEROFF, [7], that there is a difference in pure tone and speech audiometry. This phenomenon is generally known as late-onset auditory deprivation and discussed as process of inactivation in the brainstem and in the auditory cortex (G.M. GERKEN, [8] and R.J. SALVI, [9]) .
Our results after word stimulation on patiens suggested that it is impossible to find a focus of neuronal activity in our patient in contrast to good hearing control persons. Identification of cortical areas after the different auditory stimulation may allow a specification of the role that these regions play in perception of pure tone stimuli and word stimuli.

References :

[1] HARI, R. "The Neuromagnetic Method in the Study of the Human Auditory Cortex" in F.Grandori, M.Hoke and G.L.Romani (Eds.), Auditory evoked magnetic Fields and Potentials Advances in Audiology, Vol.6, Karger, Basel, 222-282, 1990
[2] McEVOY, L.; MÄKELÄ, J.P.; HÄMÄLÄINEN, M.; HARI, R. "Effect of interaural time differences on middle-latency and late auditory evoked magnetic fields" Hear. Res. 78: 249-257, 1994
[3] KRAUS, N.; McGEE, T.; SHARA, A.; CARRELL,L.; NICOL,T. "Mismatch Negativity Event-Related Potential Elicited by Speech Stimuli" Ear and Hearing 13: 158-164, 1992
[4]IVARSSON, U. and ARLINGER, ST. „ Speech recognition in noise before and after work-days noise exposure „ Scand. Audiol. 23: 159- 163, 1995
[5] DÖSSEL,O.; DAVID, B.; FUCHS, M.; KRÜGER, J.; LÜDEKE, K.M´.; WISCHMANN, H.A. „ A 31 - channel SQUID systems for biomagnetic imaging." Appl. Superconductivity, vol.1, 10-12: 1813-
[6] PANTEV,C.; EULITZ,C.; ELBERT,T.; HOKE,M. " The auditory evoked sustained field: origin and frequency dependence" Elektroencephalogr. Clin. Neurophysiol. 90: 82-90, 1994
[7] DIEROFF, H.G, Lärmschwerhörigkeit, 3.Auflage Gustav Fischer, Jena -Stuttgart, 1994
[8] GERKEN, G.M., Alteration of central auditory processing of brief stimuli: A review and a neural model. J. Acoust. Soc. Am. 93, 2038-2049, 1993
[9] SALVI, R.J.; SAUNDERS, S.S.; GRATTON, M.A.; AREHOLE, S.; POWERS, N.: „Enhanced evoked response amplitudes in the inferior colliculus of the chinchilla following acoustic trauma." Hear. Res.50, 245-258, 1990

Acknowledgment :

This work was partly supported by BMBF (FKZ 01 ZZ 9104)

Analysis of Movement-Related Brain Activities
Elicited by External Instructions

Endo, H., Takeda, T., Kizuka, T., Masuda, T., and Kumagai, T.
National Institute of Bioscience and Human-Technology, Tsukuba, Japan

Introduction

MEG recording has made it possible to identify activities in the sensorimotor cortex. A few studies have been done about a self-paced voluntary movement [1][2][3], an event-related activity with sensory guidance [4] and the functional organization [5][6]. However, it is still not clear how the temporal feature of each activity in the primary motor cortex is affected by the motivation that starts movement. Because sources of activity in the primary motor and the primary sensory cortex are located on the opposite wall in the central sulcus and are active during movement, the estimation of their activities has been difficult and no one has succeeded to clarify the temporal features of each. We have been studying movement-related brain activities and found that if the movement is triggered by an external Go-stimulus, a steeper change in magnetic fields is observed between the Go-stimulus and the EMG onset [7]. We named this activity as the motor field (MF) refered to the peak of the readiness field at the EMG onset [1]. In this study, we focused on the MF and identified the temporal aspects of the motor activities elicited by different conditions of the movement onset using LED visual stimuli.

Methods

Experimental procedure

The experiments were carried out on 4 right-handed subjects. The subjects were asked to move a right index finger briskly in the flexional-abductional direction. The subjects were asked to react to the Go-stimulus as quickly as possible. Fifty movements were performed in each experimental condition. Movement-related brain activities were investigated under the following 4 conditions of the movement onset:

(1) Self-Paced Movement (SPM): The subjects start their finger movement at free intervals without any stimuli.

(2) Visually Guided Movement with an LED Go-stimulus (VGM1): The subjects start after sensing an LED Go-stimulus presented at randomized intervals around 8 s.

(3) VGM1 with an LED warning stimulus (VGM2): The subjects start after sensing the LED Go-stimulus which is preceded by an LED warning stimulus.

(4) Go/NoGo task with an LED warning stimulus (VGM3): A Go or a NoGo-stimulus is presented after the warning stimulus (probability of the Go or NoGo-stimulus is 0.5).

The magnetic fields were recorded using a 64-channel whole-cortex SQUID system (CTF Systems Inc., Canada). Data were acquired at a rate of 250 samples/s. The EMG was recorded from the first dorsal interosseous muscle. A 3-dimensional head coordinate system was defined for each subject with 3 positioning coils put on the nasion and both tragions. A head shape was digitized on the each head coordinate system.

Data analysis

Averaging was done after off-line elimination of artefact-contaminated trials. The onset of the rectified EMG burst was used as the trigger to synchronize the averaging. The data were also averaged with respect to the Go-LED onset. The first 500 ms prior to any elicited activities was used as the baseline for the averaged data. The data were filtered with a zero phase shift, 20 Hz low-pass filter.

Equivalent current dipoles were used for activity estimation with the head modeled as a spherical volume conductor. The sphere model was adjusted to fit the shape of parietal and occipital areas of each digitized head shape. The strength of dipole's current moment was determined by fitting with the least square error method (LSEM), and the nonlinear parameters such as the dipole position and orientation, were searched by the Nelder-Meade simplex method (NMSM). The dipole was assumed to be fixed at the most probable position during the analysis period.

Results

Movement with self pace

Fig. 1 shows the results of the SPM. Fig. 1 (a) is the superimposed waveforms of 64 channels' data and Fig. 1 (b) is iso-contour maps of the MF at the EMG onset and the movement evoked field (MEF) at the peak latency of the

741

MEF. Fig. 1 (c) and (d) are the positions and the strength of three dominant activities: two motor activities and one sensory activity. The activity of contralateral MF (solid line) lasted after the movement onset, but on the other hand the activity of ipsilateral MF (dotted line) dropped off after the movement onset. Because an increasing motor activity prior to the EMG onset was not observed, the MF activity estimated under the assumption of one concentrated location could not represent the activity of neurons that activate the descending motor pathway. The activity of the MEF in the contralateral sensory cortex started after a few tens ms from the EMG onset. In contrast to the strength of magnetic fields at the peak latencies, the current moment of the MEF activity was estimated larger than that of the MF.

Movement with Go-stimulus

Fig. 2 shows the superimposed waveforms of the VGM1. A steeper change of magnetic fields instead of a slow shift of the readiness field was observed just before the movement onset (Fig. 2 (a)) The average triggered by Go-LED made it clear that this change started about 100 ms after the onset of the Go-stimulus (Fig. 2 (b)). This change was spatiotemporally distinguishable from the onset of the VEF. The steeper activity of the MF was observed clearly in three among four subjects.

Fig. 3 shows the relation between the reaction time and the onset latency of this change in three Go-stimulus conditions: VGM1, VGM2 and VGM3. The onset latency was determined from the waveforms when the magnetic fields started increasing. Soild lines indicate the reaction time and dotted lines indicate the onset latency of the MF. Different lines show the experiments were done on different days. In spite of variations of the reaction time, the onset latency does not change around 100 ms.

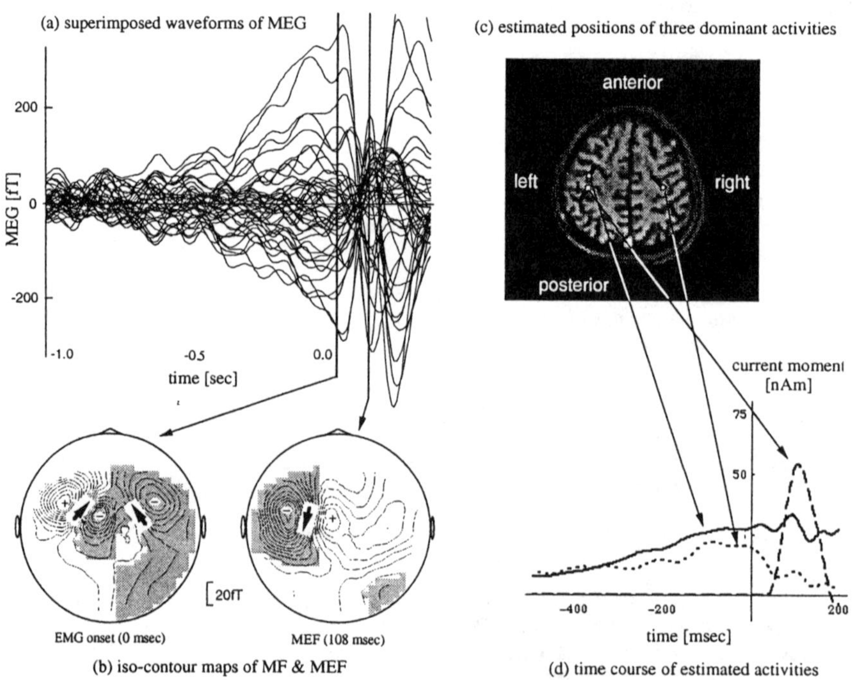

Fig. 1 Movement-related brain activites of a self-paced right finger movement. (a) Superimposed waveforms of 64 channels' data. (b) Iso-contour maps of the MF at the EMG onset and MEF at the peak latency of the MEF. (c) Estimated dipoles of three dominant activities. (d) The time course of estimated current moment by the LSEM.

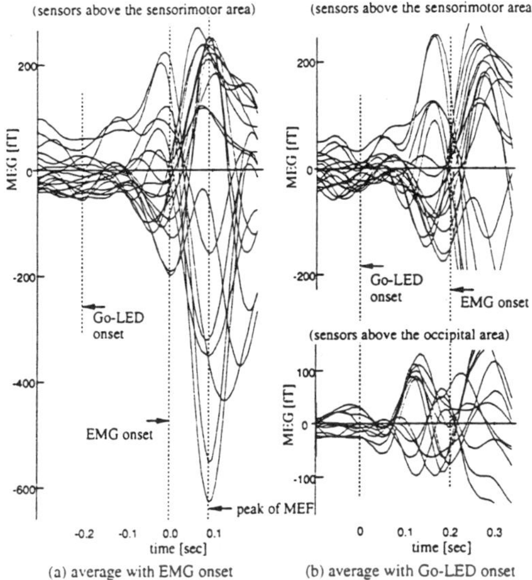

Fig. 2 MEG of the VGM1. (a) Data were averaged with respect to the movement. (b) Data were averaged with respect to the Go-stimulus. The MF shows a clear steeper change synchronize with the Go-stimulus.

In order to make each activity clear, the estimation of the time course of each activity was done in the VGM1 condition where there was no contamination of the CNV activity. Fig. 4 shows the time course of the activities before movement onset. The results of two subjects were superimposed. In the estimation, we assumed that there were 4 dipoles: 2 dipoles for the VEF and 2 for the MF. As for the sources of the VEF, the dipole positions were determined from the 1st large peak of the VEF by the warning LED stimulus. As for the motor activities, the dipole positions estimated in the SPM were used for the positions of the MF. The strength of the current moment was calculated with 4 fixed dipoles by the LSEM. The VEF started around 90 ms and the MF started around 120 ms. The MF reached its peak around 160~170 ms and dropped off. The time course of the motor activities was different from the activities estimated in the SPM. The onset and peak latency of the MF did not change according to the reaction time as observed from the waveforms and we could not get the consistent results of the relation between the activities of the MF and the reaction time.

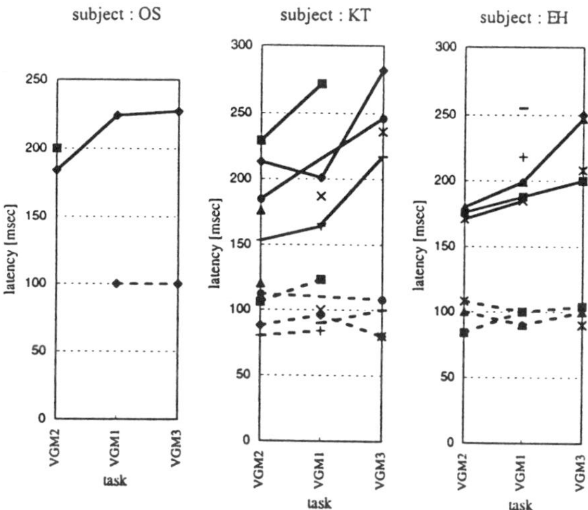

Fig. 3 The relation between the reaction time and the onset latency of the MF. The results of three subjects are plotted. Sold lines indicate the reaction time and dotted lines show the onset latency. Different lines show that the experiments were done on different days.

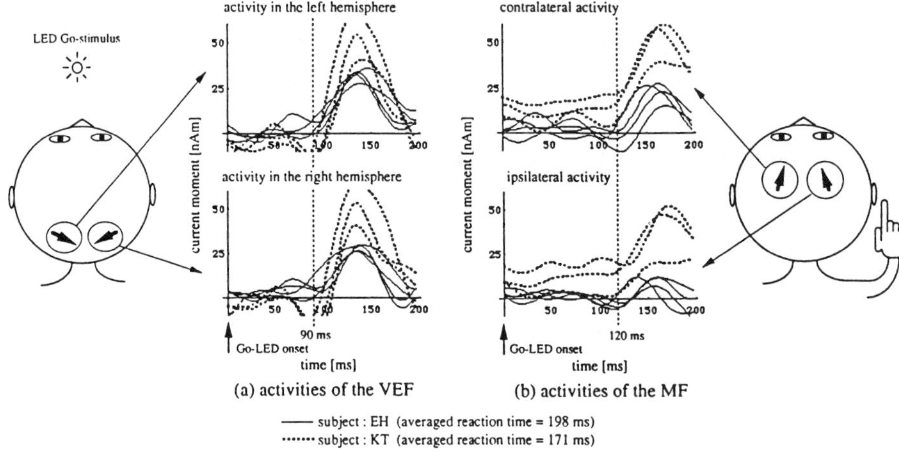

Fig. 4 The time course of estimated activities before the movement in the VGM1. (a) Activities of the VEF. (b) Activities of the MF. Upper trace: Activities in the left hemisphere. Lower trace: Activities in the right hemisphere. The activities estimated from several experiments of two subjects are plotted.

Discussion

In order to clarify how the motor activities change with the conditions of the movement onset, movement-related magnetic fields were examined in 4 kinds of movement onset conditions: one self-paced condition and three Go-stimulus conditions. In the conditions of the Go-stimulus, the steeper change of magnetic field was observed at 100 ms after the Go-stimulus and 120 ms from the results of the activity estimation. The change of the MF elicited by the Go-stimulus did not change with the onset conditions. Also, its latency did not coincide with the delay of the reaction time. As seen in the SPM, because the estimated MF activities do not synchronize with the EMG, they may not clearly represent the activity of neurons that activate the descending motor pathway. However, it is considered that there is a closely related pathway from the first stage of the stimulus processing system to the movement preparation.

References

[1] Kristeva, R., Cheyne, D., and Deecke, L. Neuromagnetic fields accompanying unilateral and bilateral voluntary movements: topography and analysis of cortical sources, Electroenceph. clin. Neurophysiol., 1991, 81: 284-298.

[2] Chiarenza, G.A., Hari, R.K., Karhu, J.J., and Tessore, S. Brain activity associated with skilled finger movements: multichannel magnetic recordings, Brain Topog., 1991, 3(4): 433-439.

[3] Nagamine, T., Toro, C., Balish, M., Deuschl, G., Wang, B., Sato, S., Shibasaki, H., and Hallett, M. Cortical magnetic and electric fields associated with voluntary finger movements, Brain Topog., 1994, 6(3): 175-183.

[4] Sasaki, K., Gemba, B., Nambu, A., and Matsuzaki, R. No-Go activity in the frontal association cortex of human subjects, Neurosci. Res., 1993, 18(3): 249-252.

[5] Cheyne, D., Kristeva, R., and Deecke, L. Homuncular organization of human motor cortex as indicated by neuromagnetic recording, Neurosci-Lett., 1991, 122: 17-20.

[6] Kristeva, R., Walter, H., Lutkenhoner, B., Hampson, S., Ross, B., Knorr, U., Steinmetz, H., and Cheyne, D. A neuromagnetic study of the functional organization of the sensorimotor cortex, Eur. J. Neurosci., 1994, 6(4): 632-639.

[7] Endo, H., Takeda, T., Kizuka, T., Kikuchi, Y., Masuda, T., and Kumagai, T. Estimation of movement-related brain activities with visual stimuli, Electroenceph. clin. Neurophysiol. Suppl., (in printing).

Visual Evoked Magnetic Fields: Bilateral Dipole Pattern for Full-Field Stimuli

Fujita, S.[1], Nakasato, N.[2,3], Seki, K.[2], Kanno, A.[3], Kawamura, T.[2], Ohtomo, S.[2], Fujiwara, S.[2,3], Tamura, I.[1] and Yoshimoto, T.[2]

[1]R&D Center, Osaka Gas Co. Ltd., Osaka, Japan; [2]Department of Neurosurgery, Tohoku University School of Medicine, Sendai, Japan; [3]MEG Laboratory, Kohnan Hospital, Sendai, Japan

Introduction

Pattern reversal stimulus is most frequently used for clinical application of visual evoked potentials (VEPs). Half-field stimulus has been employed to separate the left and right hemispheric responses. In the partial-field stimulus, however, subjects have to fix their eyes on a given point during the entire procedure; if the visual fixation is not strict, bilateral occipital responses may interfere with each other. In a clinical applications, therefore, the partial-field stimulus may not be suitable for inexperienced patients.

Recently, whole head magnetoencephalography (MEG) systems, which allow simultaneous recordings over the entire head, have been developed and used for clinical applications. In our recent studies of auditory evoked magnetic fields, a whole head MEG system revealed a clear dipole pattern of N100m responses, indicating two sources on bilateral superior temporal planes [1]. In the present study, we applied whole head MEG recordings to visual evoked fields (VEFs), in order to separate left and right hemispheric sources for full-field stimuli.

Materials and Methods

Four healthy males participated in this study. Green-black checkerboard pattern reversal stimuli were presented by a non-magnetic stimulator, which consisted of optical fiber bundles and LEDs on the full-field monoculary. The stimulus pattern was reversed once per second and individual checks, and full-field size, subtended visual angles of 1.4 and 16.8 degrees respectively at the subject's eye. The mean luminance level of the green checks was about 24 cd/m^2, with a contrast greater than 98% relative to the black checks. VEFs were measured using a whole head MEG system (CTF-Osaka Gas) linked to MRI. For source modeling purposes, three dimensional MRI head shape data were used to determine the best fit sphere for each subject. In consideration of the effect of the volume current, the equivalent current dipole was estimated using the Grynszpan-Geselowitz equation and superimposed on MRI slices for comparison with the anatomy directly.

Results

The major peaks appeared around 100msec after the pattern reversal in both hemispheres (Fig.1). The P100m peaks were identified visually as the maximum value. In all subjects, isofield maps of the P100m component indicated a two-dipole pattern over the bilateral occipital area. All the P100m dipoles estimated using a two-dipole model were localized at the lateral end of the calcarine fissures on each subject's MRI, although the individual map patterns were variable in shape (Fig.2).

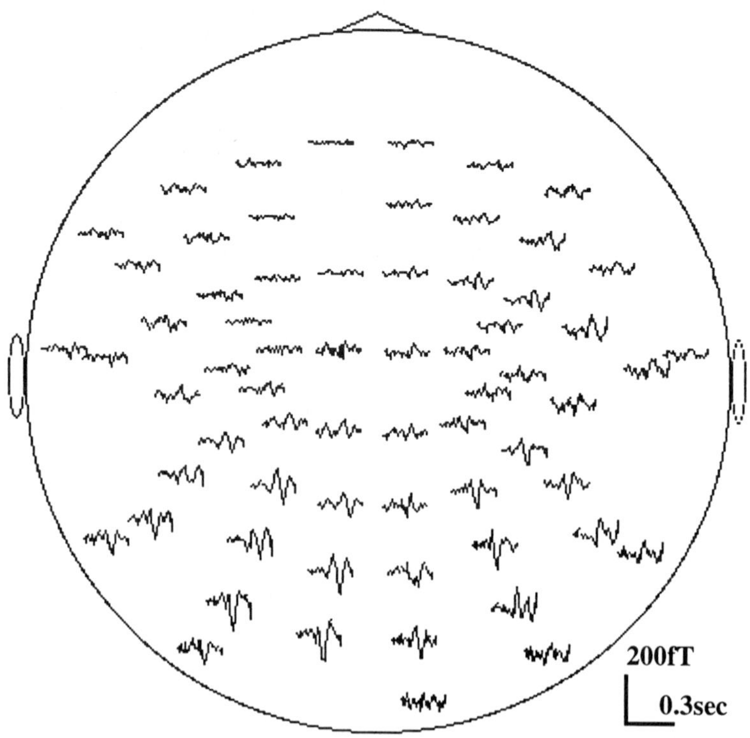

Fig.1 Waveforms of visual evoked field for the full-field stimulus. Major peaks appeared around 100msec after the pattern reversal. The positive and negative peaks were observed in both hemispheres.

Discussion

The anatomy of the occipital lobe is individually variable. Various map patterns of the P100m responses may be explained by the anatomical variations of the occipital lobes, such as depth and/or orientation of the calcarine fissures. In spite of the variation of the occipital anatomy, the estimated dipoles were always localized at the lateral end of the calcarine fissures, corresponding to our previous half-field studies [2]. For the stimulus pattern described under Methods, we believe the origin of P100m exists within the limited area of the striate cortex, because an extended origin should indicate the variation in the dipole position.

Our present study demonstrated the clinical utility of the full-field stimulus and a two-dipole model in visual evoked responses. Previous VEP studies required a partial-field stimulus; subjects have to fix their eyes on a given point. If the fixation is not strict, not only the contralateral but also the ipsilateral hemisphere generate P100m responses. In such cases, it would be necessary to use a two-dipole model instead of a single dipole model for accurate source localization. On the other hand, a two-dipole model requires a wide area measurement, because a two-dipole pattern spreads over not only occipital but also temporal and/or parietal areas. Due to the combination of the whole head measurement and a two-dipole model, bilateral P100m sources for full-field stimulus were separated

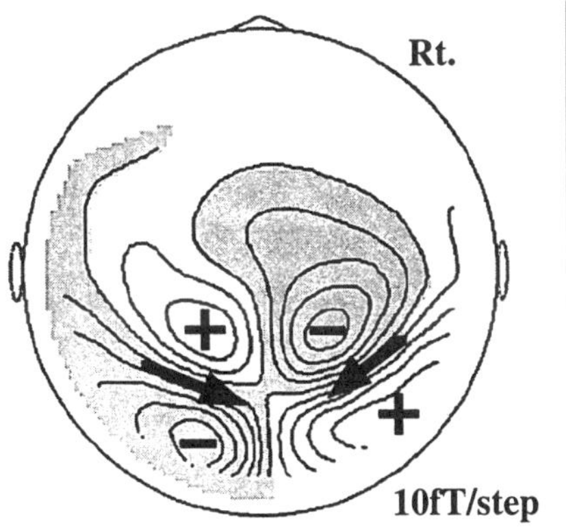
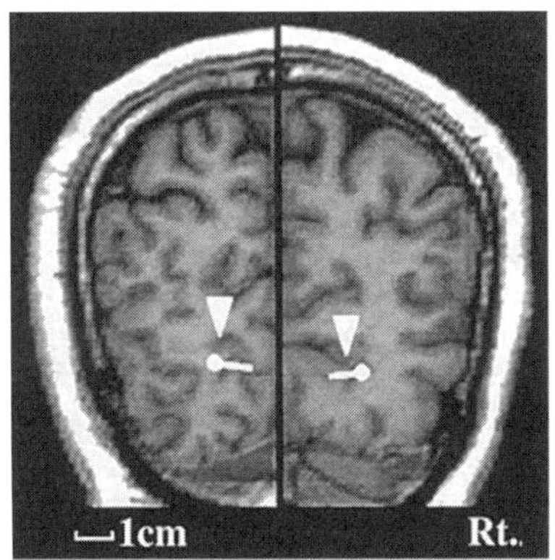

Subject 1

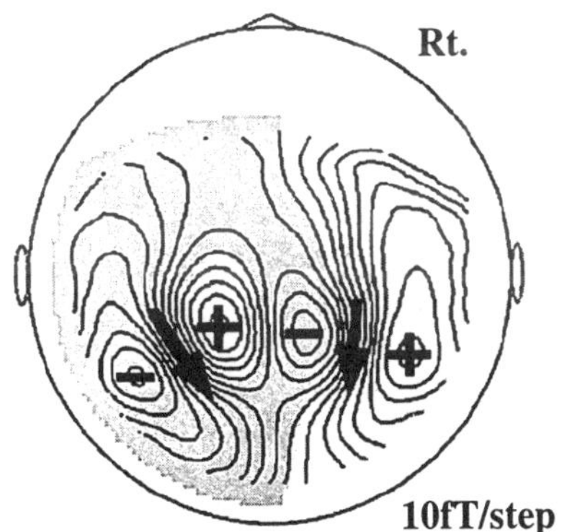
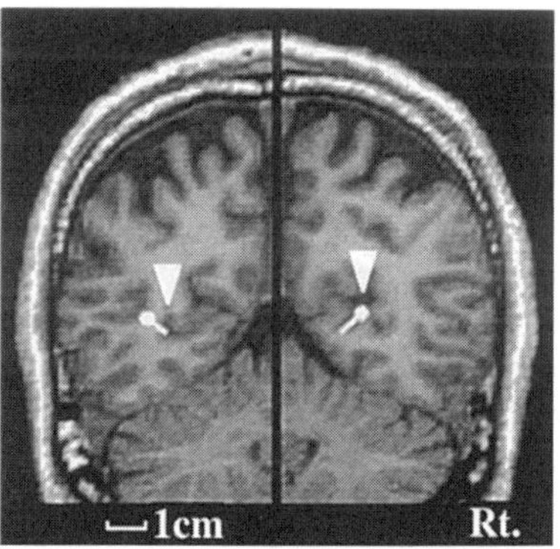

Subject 2

Fig.2 Visual evoked field for the full-field stimulus in two cases. The P100m isofield maps indicate a two-dipole patterns over the occipital area. The MRI coronal images indicate the P100m dipoles localized at the lateral end of the bilateral calcarine fissures(arrow heads). Circles and bars indicate the dipole positions and orientations respectively.

successfully. Our results accelerate further clinical application of VEFs to a wider range of patients [3,4], because the full-field VEF requires no strict fixation of the eyes.

References

[1] Nakasato, N., et al. Functional localization of bilateral auditory cortices using an MRI-linked whole head magnetoencephalography(MEG) system. Electroencephalography and Clinical Neurophysiology, 1995, 94: 183-190.

[2] Seki, K., et al. P100m of pattern reversal visual evoked magnetic field: source localization using an MRI-linked helmet-shaped magnetoencephalography system. Electroencephalography and Clinical Neurophysiology (Society proceedings), 1995, 95: 86-94.

[3] Nakasato, N., et al. Clinical application of Visual evoked fields using an MRI-linked whole head MEG system. Frontiers of Medical and Biological Engineering, 1995, in press

[4] Seki, K., et al. Visual Evoked Fields for Pattern Reversal Stimuli in Patients with Occipital Lobe Lesions. To be presented in this conference.

Evoked Magnetic Responses from Human V1: Evidence for a Chromatic-Sensitive Motion Mechanism

Fylan, F., Singh, K.D., Holliday, I.H., Anderson, S.J. and Harding, G.F.A.

Department of Vision Sciences, Aston University, Birmingham, B4 7ET, U.K.

Introduction

Within the human visual system various hierarchical processing models indicate a clear dichotomy of colour and motion processing [1] with predominantly luminance information being input to cortical motion centres via the magnocellular pathway. This separation of information is consistent with reports of poor responses from cortical motion centres to isoluminant chromatic stimuli. Recent evidence suggests that chromatic information may be used in motion analysis [2]. This study will explore the temporal response characteristics of chromatic-sensitive parvocellular units in human visual cortex by examining the variation in magnitude of the evoked magnetic response to isoluminant chromatic stimulation. Co-registration of MEG and MRI data will be examined to determine the cortical origins of the response.

Method

Five adult humans volunteered for the experiment (three females, two males, age range 27-39). The visual evoked magnetic response was recorded to isoluminant chromatic (red/green) gratings using the Aston 19-channel SQUID magnetometer [3] which was housed in a magnetically shielded room. A control stimulus comprised achromatic gratings of 80% contrast and was matched for the luminance of the chromatic gratings, at 12cd/m². The stimuli were generated using a Cambridge Research Systems VSG2/2 grating generator and displayed on an Eizo Flexscan T560i. The gratings were presented horizontally using an abrupt-onset temporal envelope of 1s duration. The interstimulus interval was 500ms, during which the gratings were replaced with a homogeneous background of the same mean hue and luminance. The spatial frequency was 2cpd and the temporal frequency was varied between 0Hz and 40Hz. The stimuli were viewed binocularly through a system of front-silvered mirrors, and subtended 4 degrees vertically by 6 degrees horizontally. To restrict input to a discrete region of cortex and to avoid activation of multiple regions of cortex due to dual representation of the horizontal and vertical meridians, the stimuli were positioned in the right lower quadrant of the visual field and displaced 0.5 degrees from the principal meridians.

The subjects were seated and stabilised using a bite bar and forehead rest and the dewar was centred above International 10-20 position Oz. Responses were sampled at 1KHz for 512ms and 100 trials were averaged for each stimulus condition. The resultant signals were bandpass filtered at 2-30Hz. For each experimental session the position of the subject's bite-bar, and therefore head, was digitised using a Polhemus Isotrak three-dimensional digitising system and converted to a standard co-ordinate system. Four bite-bar reference markers (oil filled drill holes) appeared as small, white spots on the MRI allowing accurate co-registration of MEG and MRI data. The MRIs were obtained using a 1.5T Siemens magnetic resonance scanner with 1mm x 1mm x 1.5mm voxel size. Localisation of the responses was performed using single and multiple current dipole models and Monte-Carlo analysis [4] provided 95% confidence regions which were co-registered with the subject's MRI.

Results

The response to isoluminant chromatic gratings was dominated by a single major component. For each of the average evoked responses, the global field power provided a measure of the total power in the recorded signal. The maximum global field power was plotted as a function of temporal frequency. This tuning function obtained was bimodal, peaking at 0Hz and 4Hz, shown in Fig. 1.

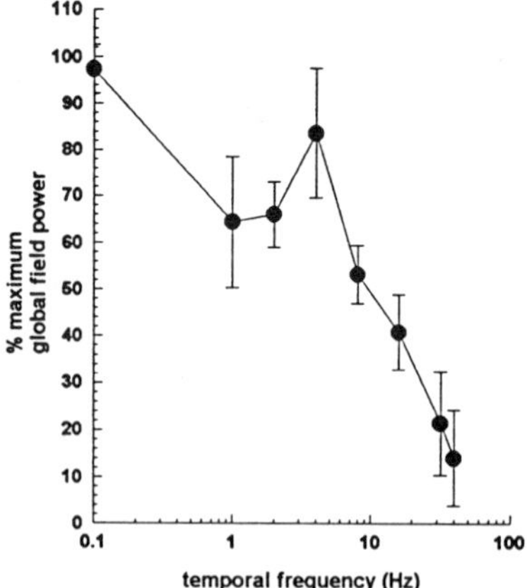

Fig. 1 Temporal frequency tuning characteristics of the chromatic evoked magnetic response (data points represent the mean of 5 subjectsand error bars show 2 standard deviations).

The latency of the major component increased gradually with temporal frequency up to approximately 16Hz and then declined rapidly to a level below that obtained for stationary patterns. To exclude the possibility that the chromatic responses arose from luminance artefacts, the experiment was repeated with an achromatic control stimulus for subject JD (Fig. 2). The control stimulus yielded low field strength signals and the latency of the response was independent of temporal frequency and at least 10ms earlier than the earliest chromatic response.

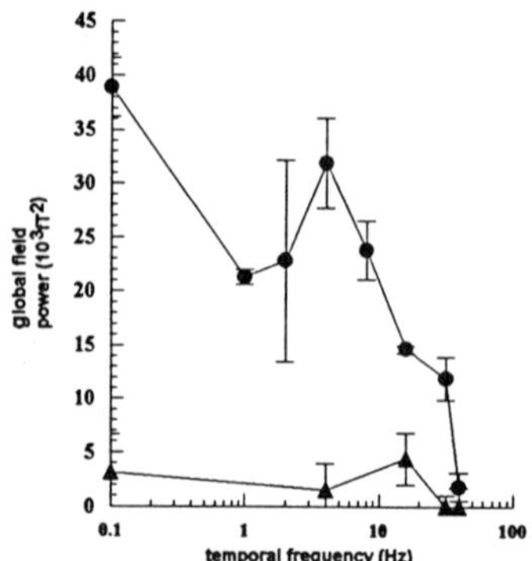

Fig. 2 Temporal frequency tuning characteristics of the response to chromatic (circles) and luminance (triangles) gratings (subject JD). Error bars show the anti-average noise data.

Each average evoked response was well modelled by a single equivalent current dipole with an orientation consistent with the anatomy of striate cortex [5]. To determine the cortical location of the underlying response generators the data was co-registered with MR images, shown in Fig. 3. This reveals the dipole confidence ellipsoids for each temporal frequency to lie along the upper surface of the calcarine fissure, consistent with a V1 origin.

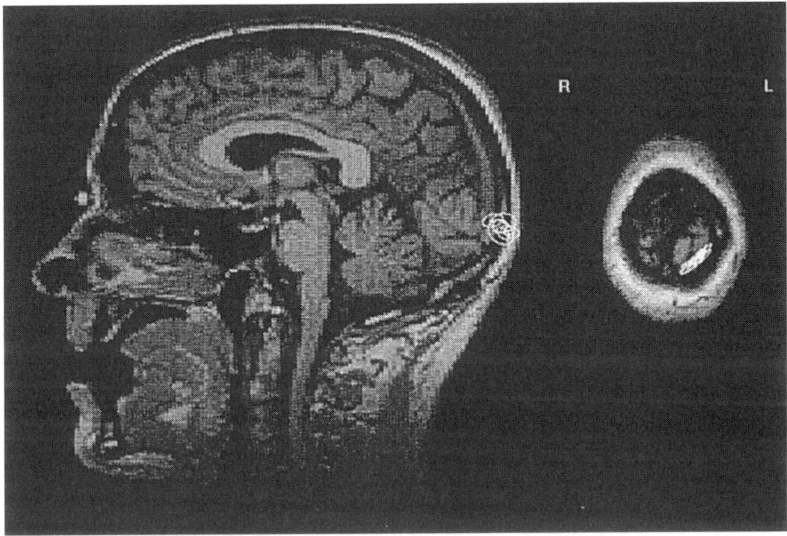

Fig. 3 95% confidence ellipsoids co-registered with sagittal and coronal MR images for the response to 2cpd red/green gratings of temporal frequency 0, 1, 2,4,8 and 16Hz (subject IH)

Discussion

The evoked magnetic response was recorded to 2cpd isoluminant chromatic gratings drifting at between 0Hz and 40Hz. The response was dominated by a major component whose latency and field power varied with the drift temporal frequency of the stimulus. The latency of the response increased with increasing temporal frequency up to 16Hz above which the latency decreased sharply. The temporal frequency tuning function was bimodal, peaking at 0Hz and at 4Hz. This consistent peak at 4Hz is surprising as psychophysical techniques do not indicate an increase in sensitivity to chromatic gratings at this temporal frequency. However, such an increase in amplitude of the VEP has also been noted previously for blue/yellow gratings [6]. The results may be indicative of two temporal frequency selective chromatic mechanisms, one sensitive to stationary targets and the other to moving targets.

To confirm the response did not arise from luminance artefacts in the stimulus the experiment was repeated with achromatic gratings. The response evoked by this luminance-modulated stimulus was of low field strength over the entire range of temporal frequencies tested. The latency of the luminance response was invariant with temporal frequency and was at least 10ms earlier than the earliest chromatic response. We therefore conclude that the responses obtained with chromatic stimuli cannot be attributed to luminance artefacts and represent the activity of chromatic-sensitive units.

Each average evoked response was well modelled by a single equivalent current dipole which was co-registered with MR images. This revealed the dipole confidence ellipsoids to lie within area V1 for each temporal frequency. This stability in the location of the response generator provides evidence to suggest that units in V1 encode the motion of chromatic targets over a wide range of temporal frequencies.

References

[1] Livingstone, M.S. & Hubel, D.H. (1988) Segregation of form, color, movement and depth: anatomy, physiology and perception. *Science* 240: 740-750.

[2] Cavanagh, P., Favreau, O.E. (1985) Color and luminance share a common motion pathway. *Vision Res.* 25: 1595-1601; Cavanagh, P., Anstis, S. (1991) The contribution of color to motion in normal and color-deficient observers. *Vision Res.* 31: 2109-2148.

[3] Matlashov, A.N., Slobodchikov, V., Bakharev, A, Zhuravlev, Y & Bodarenko, N. (1993) Biomagnetic multichannel system built with 19 cryogenic probes. In Deecke, L. et al. (eds) *Recent Advances in Biomagnetism: Proc. 9th International Conference on Biomagnetism* pp 284-285.

[4] Singh, K.D. & Harding, G.F.A. (1996) Monte-Carlo analysis and confidence region ellipsoids for equivalent current dipole solutions. This volume.

[5] Fylan F., Holliday I.E., Singh K.D., Anderson S.J. & Harding G.F.A. (1995) Cortical origins of visual evoked magnetic responses differ in humans for chromatic and luminance stimuli. *Human Brain Mapping (Suppl.)* 1:51.

[6] Rabin, J., Switkes, E., Crognale, M., Scheck, M.E. & Adams, A.J. (1994) Visual evoked potentials in three-dimensional color space: correlates of spatio-chromatic processing. *Vision Res.* 34: 2657-2671.

Acknowledgements

This research was supported by the Dr Hadwen Trust for humane research.

Evoked Magnetic Responses to Isoluminant Chromatic Stimuli from Human V1: Spatial Frequency Characteristics

Fylan, F., Singh, K.D., Holliday, I.H., Anderson, S.J. and Harding, G.F.A.

Department of Vision Sciences, Aston University, Birmingham, B4 7ET, U.K.

Introduction

In the human visual system, segregation of information on a functional basis commences as early as the retinal ganglion cells. Both chromatic and luminance information is conveyed within the parvocellular pathway to the primary visual area V1. Descriptions of the chromatic evoked response tend to concentrate on the amplitude and latency of the principle response peak rather than to characterise its distribution or to determine its cortical origins. A more intensive investigation of both the cortical origins of the response and the spatial tuning properties of these chromatic sensitive cells was therefore considered appropriate. This study will examine the magnitude of the evoked magnetic response to isoluminant chromatic gratings over a range of spatial frequencies. Co-registration of MEG and MRI data will be examined to determine the cortical origin of the response.

Method

Six adult humans volunteered for the experiment (four females, two males, age range 26-39). The visual evoked magnetic response was recorded to isoluminant chromatic (red/green) gratings using the Aston 19-channel SQUID magnetometer [1] which was housed in a magnetically shielded room. The stimuli were generated using a Cambridge Research Systems VSG2/2 grating generator and displayed on an Eizo Flexscan T560i. The mean luminance of the display was 12cd/m² and the contrast of the red and green component gratings was 100%. The gratings were presented horizontally using an abrupt-onset temporal envelope of 1s duration. The interstimulus interval was 500ms, during which the gratings were replaced with a homogeneous background of the same mean hue and luminance. The spatial frequency was varied between 0.25 and 8cpd and the temporal frequency was 0Hz. The stimuli were viewed binocularly through a system of front-silvered mirrors, and subtended 4 degrees vertically by 6 degrees horizontally. To restrict input to a discrete region of cortex and to avoid activation of multiple regions of cortex due to dual representation of the horizontal and vertical meridians, the stimuli were positioned in the right lower quadrant of the visual field and displaced 0.5 degrees from the principal meridians.

The subjects were seated and stabilised using a bite bar and forehead rest and the dewar was centred above International 10-20 position Oz. Responses were sampled at 1KHz for 512ms and 100 trials were averaged for each stimulus condition. The resultant signals were bandpass filtered at 2-30Hz. For each experimental session the position of the subject's bite-bar, and therefore head, was digitised using a Polhemus Isotrak three-dimensional digitising system and converted to a standard co-ordinate system. Four bite-bar reference markers (oil filled drill holes) appeared as small, white spots on the MRI allowing accurate co-registration of MEG and MRI data. The MRIs were obtained using a 1.5T Siemens magnetic resonance scanner with 1mm x 1mm x 1.5mm voxel size. Localisation of the responses was performed using single and multiple current dipole models and Monte-Carlo analysis [2] provided 95% confidence regions which were co-registered with the subject's MRI.

Results

The response to isoluminant chromatic gratings was dominated by a single major component whose latency increased with increasing spatial frequency, from 87ms at 0.25cpd (subject FF) to 163ms at 6cpd (subject KS). This resulted in an increase in the latency of the response of over 60ms when the spatial frequency was incresed from 0.25-6cpd, shown in Fig. 1. For each of the average evoked responses, the global field power provided a measure of the total power in the recorded signal. The maximum global field power was plotted as a function of spatial frequency, yielding a bandpass tuning curve, peaking at 1-2cpd and falling to the level of the noise at around 6cpd (Fig. 2).

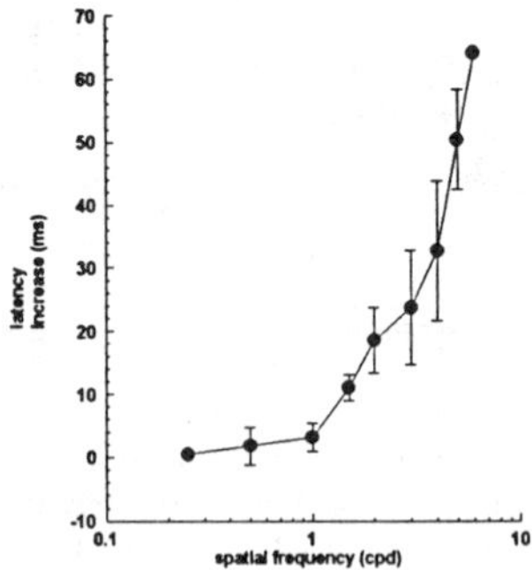

Fig. 1 Increase in latency of the major response component with increasing spatial frequency (data points are the mean of 6 subjects and error bars show 2 standard deviations).

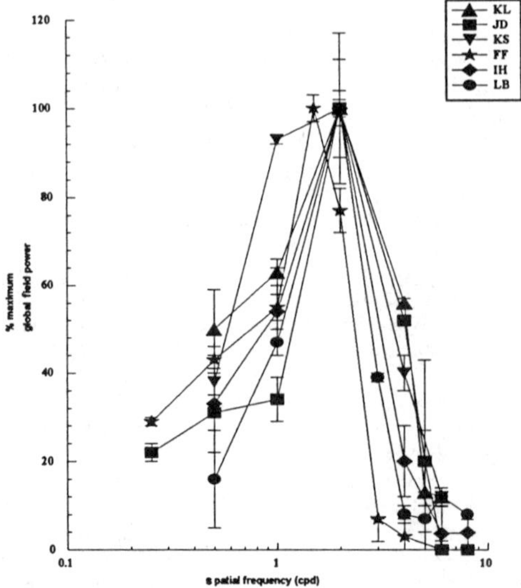

Fig. 2 Spatial frequency tuning characteristics of the chromatic evoked magnetic response. Data are shown for 6 subjects and error bars indicate the anti-average noise data.

Each average evoked response was well modelled by a single equivalent current dipole and the orientation of the dipole was consistent with the anatomy of striate cortex [3]. To determine the cortical location of the underlying response generators the data was co-registered with MR images, shown in Fig. 3. This reveals the dipole confidence ellipsoids to lie along the upper surface of the calcarine fissure, consistent with a V1 origin. The cortical location of the dipole generator was found to be invariant with spatial frequency, in agreement with the spatial frequency properties of area V1 [4].

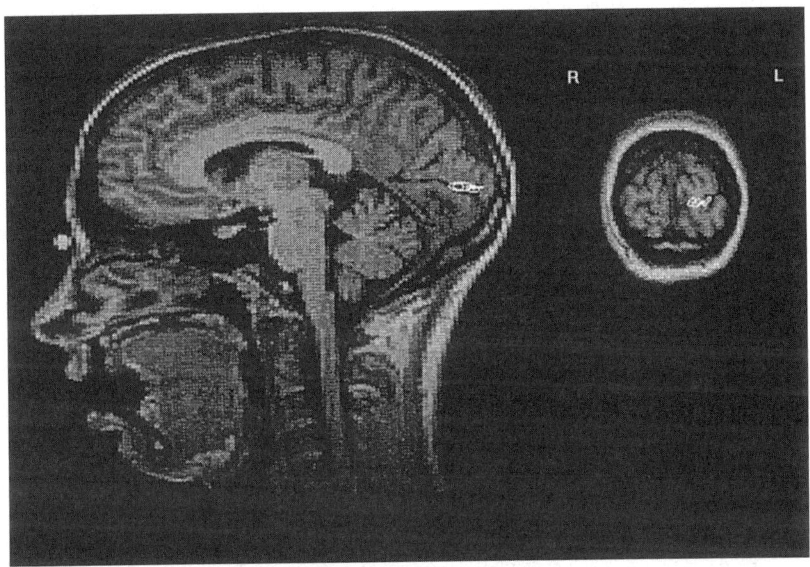

Fig. 3 95% confidence ellipsoids co-registered with sagittal and coronal MR images for dipoles to red/green gratings of spatial frequency 0.5, 1.0, 1.5 and 2cpd (subject FF).

Discussion

The evoked magnetic response was recorded to isoluminant chromatic gratings of 0.25-8cpd. The response was dominated by a single major component with a latency that increased with increasing spatial frequency, consistent with previous evoked potential studies [5]. The spatial frequency tuning response recorded here was bandpass, in contrast to the lowpass function determined psychophysically [6]. However, our experimental setup required a long viewing distance, precluding the use of more than one cycle of sinusoid for the lowest spatial frequency displayed. The sensitivity to gratings varies with the number of visible cycles, causing a reduction in field power at the lower spatial frequencies. This may also explain the reduction in visual evoked potential amplitudes at low spatial frequencies with chromatic stimuli [5].

The difference in our results from psychophysical observations can be explained by considering the neuronal activity necessary for detection by both techniques. To perceive a stimulus, relatively few neurones must signal the presence of the stimulus [7] but to record an evoked magnetic response that is sufficiently large to be distinguishable from the background noise many neurones must be simultaneously active. Hence although a reduction in field power at low spatial frequencies was caused by the low number of cycles in the display, it remains possible that the chromatic contrast sensitivity function is formed by an envelope of channels tuned to different spatial frequencies with more neurones tuned to 1-2 cycles/degree than to lower and higher spatial frequencies.

A further point of difference is that psychophysical measurements have indicated a high frequency cutoff of 11-12cpd (5) while the evoked magnetic response recorded here falls to the level of the noise at around 6cpd, a value comparable with previous evoked potential results (6). The difference between the evoked magnetic response data and psychophysical threshold measurements can again be explained by the requirement that many neurones must simultaneously respond before an evoked response can be recorded. As the sensitivity of the chromatic system decreases, fewer units will respond and while there may be sufficient numbers to allow detection of a stimulus, their may not be sufficient numbers to produce a clear evoked magnetic response. Hence MEG may provide a more accurate measure of the magnitude of cells tuned to a particular spatial frequency than detection thresholds traditionally used in psychophysics.

References

[1] Matlashov, A.N., Slobodchikov, V., Bakharev, A, Zhuravlev, Y & Bodarenko, N. (1993) Biomagnetic multichannel system built with 19 cryogenic probes. In Deecke, L. et al. (eds) *Recent Advances in Biomagnetism: Proc. 9th International Conference on Biomagnetism* pp 284-285.
[2] Singh, K.D. & Harding, G.F.A. (1996) Monte-Carlo analysis and confidence region ellipsoids for equivalent current dipole solutions. This volume.
[3] Fylan F., Holliday I.E., Singh K.D., Anderson S.J. & Harding G.F.A. (1995) Cortical origins of visual evoked magnetic responses differ in humans for chromatic and luminance stimuli. *Human Brain Mapping (Suppl.)* 1:51.
[4] Tootell, R.B.H., Silverman, R.S., Hamilton, S.L., Switkes, E. & DeValois, R. L. (1988e) Functional anatomy of macaque striate cortex. V. Spatial Frequency. *J. Neurosci.* 8: 1610-1624.
[5] Rabin, J., Switkes, E., Crognale, M., Scheck, M.E. & Adams, A.J. (1994) Visual evoked potentials in three-dimensional color space: correlates of spatio-chromatic processing. *Vision Res.* 34: 2657-2671.
[6] Mullen, K.T. (1985) The contrast sensitivity of human colour vision to red-green and blue-yellow chromatic gratings. *J. Physiol.* 359: 381-400.
[7] Newsome, W.T., Britten, K.H. & Movshon, J.A. (1989) Neuronal correlates of a perceptual decision. *Nature* 341: 52-54.

Acknowledgements

The research was supported by the Dr Hadwen Trust for humane research.

EEG and MEG Activity Accompanying Spontaneous Reversals of the Necker Cube

Gaetz, M., Weinberg, H., Jantzen, K.J., Rzempoluck, E. and Millbank, A.

Brain Behaviour Laboratory, School of Kinesiology, Simon Fraser University, Burnaby, Canada

Introduction

Synchronous oscillations between assemblies of cortical neurons have been proposed as a functional mechanism for linking or binding sensory information [1, 2]. Synchrony between groups of cells has been reported to occur at inter-electrode distances greater than one centimetre [3, 4]. Edelman's theory of neuronal group selection (TNGS) proposes that complex synchronous or *reentrant* activity serves as the basis for processing complex information [5]. Reentry is a process described as temporally continuous parallel interactions between distributed maps along ordered anatomical pathways. Reentry is not feedback, but parallel signalling in the time domain between spatially distributed maps, more similar to a process of dynamically changing correlations between distributed systems. Based on the TNGS three experimental questions were posed: 1) Can distributed cortical patterns present during perceptual reversals of a Necker cube be classified differently using a neural network ? 2) Does correlated activity increase during perception of Necker cube reversals? 3) Can distributed systems operating synchronously be identified ?

Methods

The subjects in the EEG phase were two males and two females (right handed) between 18 - 34 years of age (mean = 26.6 years). The EEG amplifiers were Nihon-Kohden model EEG-4217 interfaced with software designed to record single trials. The single trial epochs occurred between -1000 to 0 milliseconds, encapsulating a one second period prior to the trigger, and were digitised at 1024 points per second. Data collection for single trials was initiated by an electrical potential generated from the nasion electrode referenced to Cz. The EEG was filtered on-line 0.03 Hz to 70 Hz with a 60 Hz notch filter.

Five "naive" right handed adult subjects, three males and two females, participated in the second (MEG) phase of these experiments. Subjects were between 18 - 32 years of age, with a mean age of 26.4 years. A whole head CTF Systems MEG was used to record magnetic fields. The sensor array consisted of 64 first order gradiometers uniformly distributed over both hemispheres. The baseline of each 1st order gradiometer was 5 cm and each coil was 2 cm in diameter. Third gradient response was computed in software using a reference system of coils and squids. The single trials were one second epochs prior to the subject's response. The sampling rate was 250 points per second with low and high filter settings at DC and 100 Hz respectively. A 60 Hz filter and artifact rejection were used. Magnetic shielding was not necessary when third order spatial gradients were used.

Visual stimuli consisted of a square and a Necker cube, each presented on 8 1/2 x 11" sheets of white paper. Two indices of Necker cube reversal were defined: The perception of a *"shift-up"*, when the upper left-most corner of the cube face was in the upper left-most quadrant of the figure, or when perception of shift was in the opposite direction called a *"shift-down"*. The Necker cube and square were presented in random order. During perception of the control stimulus, the subject was instructed to close their eyes briefly, at approximately 3-5 second intervals, and to focus on the subjective centre of the control stimulus. While viewing the Necker cube, subjects were instructed to close their eyes for approximately 0.5 - 1.0 second immediately following perception of a reversal in the appropriate direction. Subjects were instructed to ignore reversals that occurred incompletely or in rapid succession. Consequently, the single trials included a subset of reversals that were of a subjective "high quality". Subjects were instructed to allow spontaneous perception of reversals without attempting to control them by modifying gaze. Data were collected until 20-30 EEG and over 30 MEG single trials were available for analysis. For experimental single trials, measures were taken to ensure eye closures signalled perceptual reversals (video monitoring and rapid blinking if an eye closure was not intended to mark a reversal).

EEG single trials were processed prior to neural network analysis. A subset of electrodes (T3, C3, C4, T4, T5, P3, P4, T6, O1 and O2) were selected for analysis. Correlation matrices were calculated for each single trial, resulting in a 10 x 10 matrix. Redundant information was removed from each matrix and the remaining data was transformed into a single vector (1 x 45). Vectors were subsequently organised into testing and training matrices. A subset of the MEG sensors was selected to closely approximate data analysed in the EEG phase (Fig. 1b). Correlation matrices were calculated for the subset resulting in a 32 x 32 matrix for each single trial. Redundant information was removed from each matrix and the remaining matrix values were transformed into vectors (1 x 496). The procedure for creating an MEG training matrix was similar to that described for EEG, with the exception that longer vectors were used.

Neural Network Analysis: The purpose of the neural network analysis was to discriminate EEG and MEG recorded one second preceding the Necker cube reversal from that recorded in the control condition. The network analysis used was a generalised regression neural network (GRNN) developed by Specht [35, 38]. Input to the GRNN consisted of individual exemplar row vectors from the training matrix. The GRNN estimated a number representing the probability that a single trial vector belonged to the experimental group from that training matrix. The probabilities were then grouped into control or experimental columns within subject for statistical analyses (i.e., for each subject, two statistical comparisons of probabilities were generated, *shift-up* vs. control and *shift-down* vs. control). One tailed equal variance t-tests were used to statistically test differences between GRNN classifications.

Correlation Subset Analysis: An *a priori* model was developed (see Fig. 1a and b) based on Edelman's description of reentrant pathways in visual cortex. The model started with the same electrode/sensor positions used for the GRNN analysis. Subsets of correlations chosen for analysis, (12 of 45 for EEG and 95 of 496 for MEG) were intended to spatially represent reentrant connections described by Edelman [8]. Averages were calculated across single trials, within subject and experimental condition. A sum of each average was then calculated, which resulted in a single value representing the magnitude of correlations within condition and experiment. The summed averages were analysed using the stepwise Bonferroni procedure for multiple comparisons.

Figure 1. Lines represent correlations between electrode/sensors that were included for EEG (a) and MEG (b).

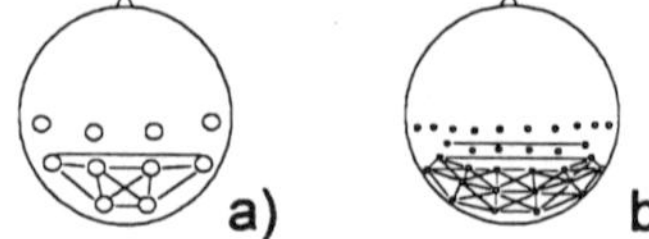

Overlapping Time Window Analysis (MEG): To determine whether distributed systems that function during perceived reversals could be detected, correlation matrices were calculated for each single trial using overlapping time intervals: 1-200, 100-300, 200-400, 300-500, 400-600, 500-700, 600-800, 700-900, 800-1000 (msec). The time window with the highest absolute sum of correlations was selected for each single trial and added to a matrix within subject and condition. Average correlation vectors were then calculated for each matrix. Average vectors from the control condition were subtracted from both experimental average vectors. Only correlations greater than -0.01 were analysed. This analysis was intended to be a conservative estimate of distributed systems with synchronous properties that produce reversals.

Results

Neural Network Analysis: EEG classifications were significant (p.< 0.05) for each of the four subjects. The *shift-up* vs. control and *shift-down* vs. control were classified significantly by the GRNN for two of four subjects. For the remaining two subjects, only one of the two comparisons classified significantly (*shift-up* vs. control for one, *shift-down* vs. control for the other). MEG classifications were significant (p.< 0.05) for both comparisons in all five subjects. The t-test p values did not indicate that *shift-up* could be classified as different than the control figure more easily than *shift-down* or vice versa for EEG or MEG.

Correlation Subset Analysis: For EEG, the summed averages were analysed using the stepwise Bonferroni t-test for multiple comparisons. The p value for the *shift-up* vs. control was p = 0.019, and *shift-down* vs. control was p = 0.029. The *shift-up* vs. control comparison was removed from the next iteration, and the remaining p value was significant at p. < 0.05.

For MEG, the summed averages were analysed using t-tests with the stepwise Bonferroni for multiple comparisons. The p value for the *shift-up* vs. control was p = 0.039. The p value for the *shift-down* vs. control was p = 0.017. The *shift-down* vs. control comparison was removed from the next iteration, and the remaining p value was significant at p.< 0.05. The correlation subsets were chosen *a priori*, therefore no further corrections were used.

Overlapping Time Window Analysis (MEG): Differences in correlation between Necker cube reversal and control average vectors are shown in Figure 2. Only correlation differences greater than .10 are shown. Differences in correlation between sensors are relatively consistent within subject and condition, but differ between individuals.

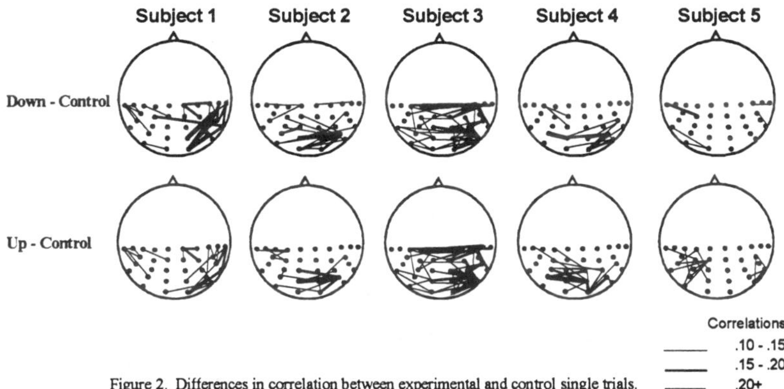

Figure 2. Differences in correlation between experimental and control single trials.

Discussion

Implicit in the TNGS is the concept that distributed groups of neurons operating within a system will display complex patterns of correlated activity. The GRNN was effective in detecting differences in correlation between experimental and control vectors, however, EEG and MEG did classify differently. The EEG classifications were less powerful than MEG; only six of eight experimental conditions were classified differently than control conditions. MEG data were classified significantly 10 out of 10 times. The magnitude of classifications was also considerably greater for the MEG.

There are *at least* three possible reasons for the difference in classifications of EEG and MEG. First, the MEG provides better spatial resolution of cortical sources. This may have increased the spatial sampling density for MEG and may have contributed to an increased signal to noise. Furthermore, the signals recorded with MEG are not degraded by bone and scalp and therefore could be expected to be differentiated to a greater extent than signals recorded at the same locations with EEG. Second, more MEG single trials were recorded. This is important because with enough information, the GRNN "is guaranteed to converge to a Bayesian classifier (the usual definition of optimality)" [7] (p.35). Finally, assuming a spherical model, the MEG is primarily sensitive to tangentially oriented sources, whereas the EEG may result from a summation of radial and tangential sources [9].

Gilbert and Wiesel [10] discuss tangentially oriented collaterals of extraordinary richness that extend over widespread areas of cortex, forming a number of distinct and repeating clusters. These horizontal cortical connections may extend beyond cortical columns and, in fact, may extend beyond hypercolumns [11]. Intercommunication of cortical clusters is an important attribute of information processing [12]. Tangential interconnections between distributed neuronal groups are presumed to be important components of cortico-cortical networks active during visual information processing, the function of which is probably non-linear. Therefore the evidence suggests that non-linear relationships between neuronal groups have an anatomical basis in the tangential network of collaterals and that MEG is more sensitive to these tangential sources.

According to the TNGS, reentry between neural groups is similar to a process of correlation, and therefore, one would expect correlations between proximal and distal recording sites to increase during increased information processing. Patterns of correlations recorded during Necker cube reversals were significantly higher than for the control stimulus. Using a moving time window analysis of MEG single trials, differences in correlation between Necker cube reversal and control average vectors were observed (Fig. 2). The correlations between sensors were relatively consistent within subject and condition, but differed between individuals. The results are considered to be consistent with reentrant properties in a distributed system that processes visual information.

GRNN classifications coupled with correlation matrix transformations were successful in demonstrating that patterns of CNS activity present during perception of Necker cube reversals were different from those preceding the perception of a two dimensional non-reversing figure. Increased correlations preceding Necker cube reversals were interpreted to be consistent with reentrant processes between distributed neuronal groups. The results provide evidence in agreement with Edelman's concept of reentry as a basis for information processing.

References

[1] Eckhorn, R., Bauer, R., Jordan, W., Brosch, M., Kruse, W., Munk, et al. Coherent oscillations: A mechanism of feature linking in the cat visual cortex?[1] Biological Cybernetics, 1988, 60:121-130.

[2] Gray, C. and Singer, W. Stimulus specific neuronal oscillations in orientation columns of cat visual cortex [2], Proceeding of the National Academy of Sciences USA, 1989, 86:1698-1702.

[3] Engel, A., Konig, P., Gray, C. and Singer, W. Stimulus-dependent neuronal oscillations in cat visual cortex: Inter-columnar interaction as determined by cross-correlation analysis [3], European Journal of Neuroscience, 1990, 2:588-606.

[4] Desmedt, J. and Tomberg, C. Transient phase-locking of 40 Hz electrical oscillations in prefrontal and parietal human cortex reflects the process of conscious somatic perception [4], Neuroscience Letters, 1994, 168:126-129.

[5] Edelman, G. Neural Darwinism: Selection and reentrant signalling in higher brain function [5], Neuron, 1993, 10:115-125.

[6] Specht, D. F. A general regression neural network [6], IEEE Transactions on Neural Networks, 1991, 568-576.

[7] Wasserman, P. Advanced Methods in Neural Computing, New York: Van Nostrand Reinhold, 1993.

[8] Edelman, G. The Remembered Present, New York, Basic Books, 1989.

[9] Cohen, D. and Cuffin, N. Demonstration of useful differences between magnetoencephalogram and electroencephalogram [9], Electroencephalography and Clinical Neurophysiology, 1983, 56:38-51.

[10] Gilbert, C. and Wiesel, T. Clustered intrinsic connections in cat visual cortex [10], Journal of Neuroscience, 1983, 3:1116-1133.

[11] Gilbert, C. Horizontal integration in the neocortex [11], Trends in Neuroscience, 1985, 8:160-165.

[12] Zeki, S., and Shipp, S. The functional logic of cortical connections [12], Nature, 1988, 335: 311-317.

MEG and EEG Responses to a Pitch Naming and Identification Task: Slow DC Shift Predicts Performance.

Gordon, R., Weinberg, H., Cheyne, D., and Jantzen, K.

Brain Behaviour Laboratory, Simon Fraser University, Burnaby, Canada

Introduction.

Damasio [1] has proposed that long-term memory storage relies on neural populations located in cortical regions responsible for perceptual processing. Memory occurs when primary and first-order sensory association areas are reactivated through conscious effort. Evidence in support of this position has recently been obtained in the visual modality [2]. In the auditory domain, there exits a population of individuals who, through extensive practice, have developed a unique long-term memory skill. Musicians who possess the skill of absolute pitch can name notes of specific frequency according to Western musical notation without reference. If Damasio's theory is correct, we would expect that individuals with absolute pitch would utilize neural generators normally reserved for sensory encoding for long-term memory storage of pitch information. The supratemporal cortex seems a likely cortical region for such processing. Also, in order to provide linguistic labels to specific frequency information, it is suspected that other neural generators may be involved. Recent MRI evidence suggests absolute pitch musicians have an enlarged left planum temporale [3]. Thus, both supratemporal cortex and left side planum temporale are hypothesized to play a role in absolute pitch processing.

Mechanisms of auditory memory storage have been indexed through the use of Mismatch Negativity, a negative potential occurring some 200 msec after stimulus onset. MMN and its magnetic counterpart, MMNm, are most easily elicited in oddball paradigms where an infrequent rare tone is presented in a series of standards. Averaged responses to the rare tone produce an enhanced negativity thought to represent neuronal populations responding to a change in stimulus characteristics [4]. The contents of the MMN storage are known to decay with time. An interstimulus interval of no more than 7-10 seconds is necessary to elicit the response [5]. If MMN generators play a role in memory storage of sensory information, it may in fact be these generators that are reactivated when discrete information such as pitch is recalled.

The current study examined MEG and EEG responses to a pitch identification task from subjects claiming to possess perfect pitch and from those who did not. It was hypothesized that neural generators responsible for absolute pitch ability would be localized to cortical areas normally reserved for processing perceptual information, i.e., MMN. As well, additional activation of left side parietal regions should also occur as an indication of linguistic processing.

Methods.

Collection parameters: Neuromagnetic activity was recorded using a 64-channel whole-cortex magnetoencephalographic system (CTF Systems Inc) without shielding. Electrical responses were recorded using a 21 channel Nihon Khoden system coupled with a 486 PC for digitization and collection of averages. Waveforms were sampled at a rate of 250 Hz for MEG data, 128 Hz for EEG. Signals were bandpassed DC to 50 Hz for MEG; .2 to 70 Hz for EEG. Linked ears were used as reference for EEG.

Subjects: Twelve subjects responded to an advertisement placed at local music schools requesting volunteers who possessed perfect pitch. Six of these subjects participated in the MEG sessions and 6 in the EEG. Subjects were given a pitch processing test consisting of a random presentation of tones spanning a chromatic two octave range. Results from this test were used to classify subjects either as good or moderate performers. Seven control subjects who did not have any formal musical training participated in both MEG and EEG conditions.

Procedure: For MEG recording, subjects were seated upright under the dewar and were instructed to view a computer monitor placed 5 meters away. Note names from Western musical notation were flashed onto the computer screen with a duration of 300 msec. Subjects were instructed to imagine the pitch of this note. Two seconds later, a binaural tone was presented via plastic earphones (to avoid magnetic artifact) that was either the same frequency of the visually presented note name (Match condition, 55% probability) or 1 semitone above or below (Mismatch condition, 45% probability). Eight tones and corresponding note names from an ascending chromatic scale (starting at Ab - 237 Hz) were used and presented in random order. Tones were sine waves of 100 msec duration with a 10 msec ramp. Once the tone was presented, a random silent interval of 1-2 seconds occurred after which the computer monitor prompted the subjects for an answer regarding the relationship between the visual prompt and tone (i.e., match or mismatch). Subjects responded verbally. The intertrial interval was variable from 7-8 seconds. 130 trials were collected. Recording epoch was 1.1 seconds triggered by the onset of the tone. Recordings were baselined 100 msec before tone onset. All stimulus parameters were identical for the EEG condition.

Results

Visual inspection revealed no differences between match and mismatch trials. Thus, data were combined and averaged across all trials. Figure 1 depicts superimposition of averaged MEG data for three subjects representative of the different skill levels. Figure 2 shows the corresponding topographic distributions. The most salient feature of these waveforms is a sustained potential occurring for the good and moderate task performers.

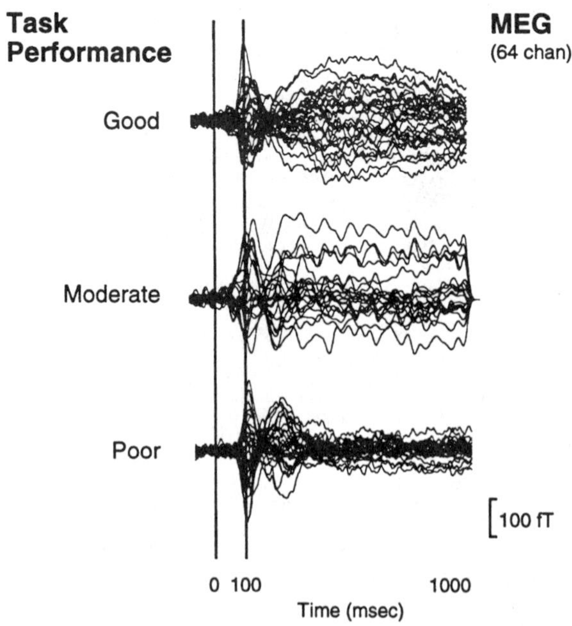

Figure 1. Superimposition of selected MEG channels. *Upper:* Averaged responses to tones for subject showing 100% accuracy on task. *Middle* Subject with 80% correct. *Lower:* Subject with chance performance. Averages contain approximately 100 artifact free trials.

This sustained potential begins to develop approximately 200 msec after the tone and, for good performers, has a broad flat peak occurring between 400-600 msec. In contrast, individuals with poor pitch processing skills produced no organized brain activity after 200 msec. For good performers, topographic distributions for the sustained potential resemble closely the N1m response although slightly more anterior (Figure 2). For moderate performers, the sustained potential resembles the N1m distribution but is missing the frontal fields seen in lateral and anterior sensors. Also noticed is that for both poor and moderate performers, a field pattern emerges at 200 msec that is not seen in good performers (see Figure 2). Given this pattern is peaking near 200 msec it is suspected to be part of the MMNm complex. EEG results (not shown) also reveal distributions consistent with MMN at 200 msec, but only for moderate and poor task performance. The topography of the sustained potential seen with EEG is characterized by a broad positivity at parietal and occipital electrodes and negativity at frontal locations. The negativity is more pronounced in the moderate performers than good performers. Noted in the MEG recordings (400 msec, Figure 2) are ingoing fields detected by left side parietal sensors for good and moderate task performance. Although there appears other asymmetries associated with sensor magnitude, these effects were not consistent between individuals nor related to task performance.

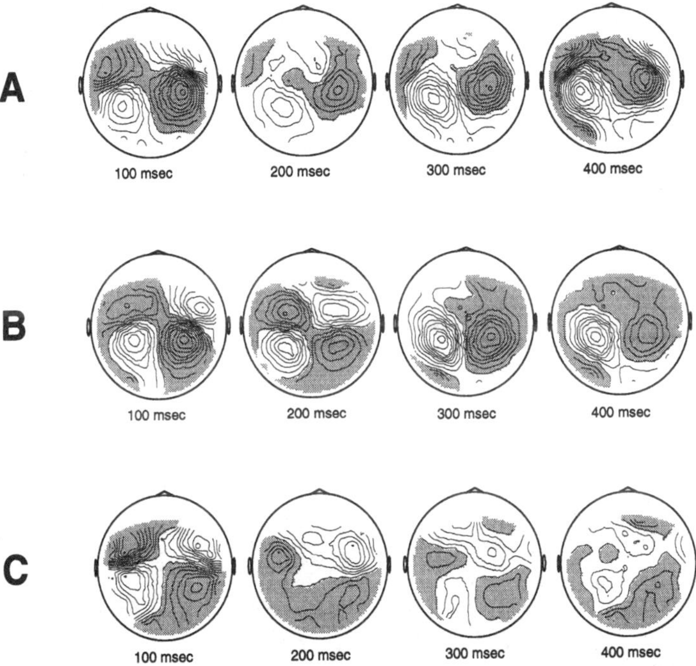

Figure 2. Isocontour plots for magnetic fields from averaged responses to tones. Time indicated is after tone onset. Gray indicates ingoing field, contours are in steps of 32 fT. **A**: Responses from subject showing good performance (100% correct) on pitch identification task. **B**: Responses from subject with moderate performance (80% correct). **C**: Responses from subject with poor task performance (chance).

Discussion

Its seems reasonable to suggest results are in agreement with Damasio's views on memory processes. Though analyses are only descriptive, topographic mapping indicates that individuals who possess tonal memory abilities produce a response that appears similar to the initial N1m topography. Although similar, it is also clear that the distribution is not identical to the N1m. Given the link of MMN generators to auditory memory [5], it is possible that it is access to these neuronal populations as opposed to N1 generators that gives rise to the absolute pitch phenomenon. What is unclear from the current study is if the sustained potential seen in good task performers is in fact similar to MMNm field generators. Dipole fits are not stable as both MEG and EEG data indicate multiple source activation. Nevertheless, the sustained potential appears to suggest an anterior movement of current sources which would be in agreement with at least one component of the MMNm [6]. Thus, one element of absolute pitch processing requires activation of supratemporal cortex.

A notable difference between good and moderate performance is the field distribution at 200 msec after stimulus onset. For moderate performance, a distinct dipolar field pattern emerges. This may in fact be a true MMNm component. Sams [5] reported MMNm for ISIs up to 12 seconds. For the current study, time between tones was variable from 11-12 seconds. It is conceivable that moderate performers utilized a strategy that relied on remembering the previous tone. Moderate performers, therefore, would be thought to possess the skill of relative pitch and not absolute pitch. Musicians possessing relative pitch skills are able to name notes but only when a known reference is provided. As the tones were presented randomly, only in rare cases did the same frequency occur twice in a row. With each tone being different from the previous one, we would expect each tone to produce a small MMN if the individual was able to retain in memory the frequency of tone heard prior. This strategy was perhaps also adopted by the poor performers. For absolute pitch subjects, however, this strategy is not necessary; with an long-term memory store of exact pitch information there is no need to remember previous tones for comparison purposes. In this manner, each tone is heard as a single entity and no MMNm is generated. Once the tone is presented, however, the absolute pitch subject is able to retroactively access memory stores of that particular frequency. This memory storage appears to be located in supratemporal cortex and likely involves similar generators initially involved in sensory encoding.

As a final note, left side, posterior sensors are active for both good and moderate performers beginning 300-400 msec after stimulus onset. Although tentative, this may be taken to reflect activity from the planum temporale region which has been implicated in absolute pitch processing [3]. Taken together, the current results support the retroactivation theory as individuals possessing absolute pitch produce a sustained endogenous potential that topographically closely resembles magnetic field patterns associated with the transient sensory response. As well, additional sources from left side parietal lobe and possibly frontal cortex appear to be involved in pitch processing ability.

References

[1] Damasio, A.R., Time-locked multiregional retroactivation: A system level proposal for the neural substrates of recall and recognition. Cognition, 1989, 33: 25-62.
[2] Rosler, F. Heil, M. and Hennighausen, E. Distinct cortical activation patterns during long-term memory retrieval of verbal, spatial and colour information. Journal of Cognitive Neuroscience, 7(1): 51-65.
[3] Schlaug, G., Jancke, L., Yanxiong, H., and Steinmetz, H. In vivo evidence of structural brain asymmetry in musicians, Science, 1995, 267: 699-700.
[4] Naatanen, R. Attention and Brain Function, New Jersey, Erlbaum, 1992.
[5] Sams, M., Hari, R. and Knuutila, J. The human auditory sensory memory trace persists for about 10 sec: neuromagnetic evidence. Journal of Cognitive Neuroscience, 5(3), 363-370.
[6] Tiitinen, H. Alho, K., Huotilainen, M., Ilmoniemi, R. J., Simola, J., and Naatanen, R. Tonotopic auditory cortex and the magnetoencephalographic (MEG) equivalent of the mismatch negativity, Psychophysiology, 30, 1993.

Magnetic Detection of Activity of Presumed Inhibitory Interneurons in the Human Somatosensory Cortex

Hashimoto, I.[1], Mashiko, T.[2] and Imada, T.[2]

[1]Department of Psychophysiology, Tokyo Institute of Psychiatry, Tokyo, Japan; [2]Information Science Research Laboratory, NTT Basic Research Laboratories, Atsugi-shi, Japan

Introduction

Previous studies of somatosensory evoked potentials (SEPs) and fields (SEFs) have shown that high-frequency signals in the range of 300-900 Hz concur with the primary response (N20 and N20m) of the somatosensory cortex [1,2]. However, the physiological mechanisms of the high-frequency oscillations remained undetermined. We addressed the issue by analyzing magnetic fields during wakefulness and sleep over the left hemisphere to right median nerve stimulation. Results obtained from this study have been described in detail elsewhere [3].

Methods

Eleven subjects (Ss) participated in the experiments; 10 Ss were studied during wakefulness and 7 Ss were tested during natural sleep. Brief electrical stimuli with a 0.2 msec duration were delivered to the right median nerve at the wrist. The stimulus intensity was about 3 times sensory threshold and elicited a mild twitch of the abductor pollicis brevis muscle. The stimuli were delivered at regular intervals with the repetition rate of 4 Hz. Magnetic recordings (bandpass 0.1-2,000 Hz) were sequentially taken over the left hemisphere with two 7-channel SQUID gradiometer systems (BTi Model 607).

An epoch of 50 msec duration (a 10 msec pre- and 40 msec post-stimulus) was digitized at a 5 kHz/channel sampling rate and 5,000 responses were averaged off-line. For isolation of the high-frequency oscillations from the underlying N20m, the wide-band (0.1-2,000 Hz) recorded responses were digitally high-pass (> 300 Hz) and low-pass (< 300 Hz) filtered. In order to define the same 3-dimensional head coordinate system for MEG and MRI measurements, the same 3 fiducial points, marked with vitamin pills, were used such that the estimated magnetic source locations could be projected on to the appropriate points on the MRI slices of the individual subjects.

Results

SEFs during wakefulness

The magnetic signals from the somatosensory cortex consisted of high amplitude N20m and P27m within 40 msec after median nerve stimulation. Superimposed upon the N20m, several small inflections could be recognized in the unfiltered original SEF records (Fig. 1A). FFT analysis of the wide-band records revealed two to four frequency peaks of signal energy. The main signal energy was distributed broadly between 20 and 300 Hz with a peak around 40-60 Hz, and a weaker signal energy with one to three peaks was found above 300 Hz. Among the weaker peaks above 300 Hz, the higher frequency peak from 580 Hz to 780 Hz carried the principal signal energy. After high-pass filtering of the original records, these small inflections buried in the predominant N20m were isolated and could be discriminated from the background noise (Fig. 1B).

High-frequency oscillations start approximately at or later than the onset of the N20m and end in the middle of the second slope. Similar to N20m, each peak of the high-frequency oscillations reversed polarity medio-laterally at exactly the same location. The field patterns at the peaks of the high-frequency oscillations and the low-pass filtered SEFs at the corresponding period showed a close similarity (Fig. 2). Estimated location of the dipole for each peak of the high-frequency oscillations agreed with activation of the 3b area of somatosensory cortex and was very close to that for N20m. The mean absolute dipole strength was 1.2nA.m for the high-frequency peaks and 17.0 nA.m for the low-pass filtered N20m at the corresponding points in time.

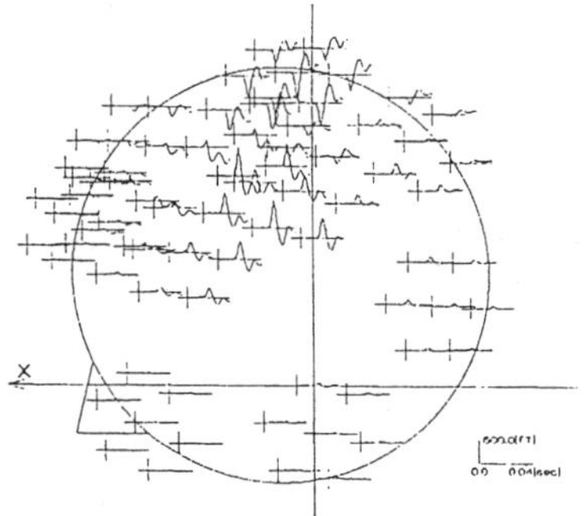

Fig. 1A: Wide-band recorded SEFs.   Fig. 1B: High-pass filtered SEFs.

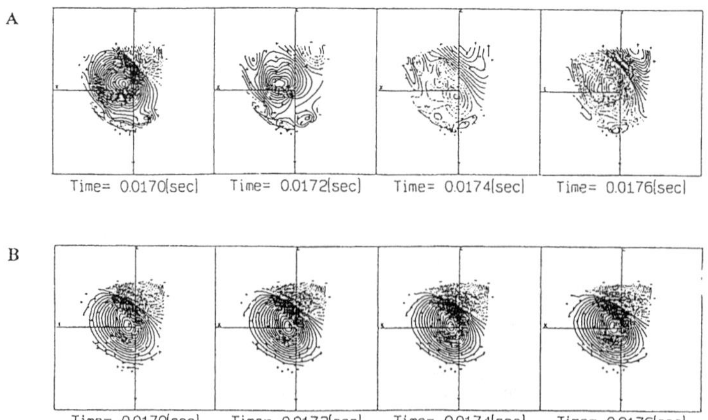

Fig. 2,   A: Isofield maps of the high-frequency oscillations over the left hemisphere.   The maps are shown at 0.2 msec intervals.   Continuous lines indicate magnetic flux out of the head and dashed lines flux into the head.   The isofield lines are separated by 2.5 fT.   B: Isofield maps of the low-pass filtered SEFs during the same period as in A.   The isofield lines are separated by 20 fT.

Although the N20m showed a longer mean latency during sleep, this difference (mean, 0.6 msec) did not reach statistical significance. However, the mean N20m peak amplitude was significantly larger during sleep (p < 0.005). While strong high-frequency oscillations were present during waking (Fig. 3A), they were reduced in amplitude (Fig. 3B).

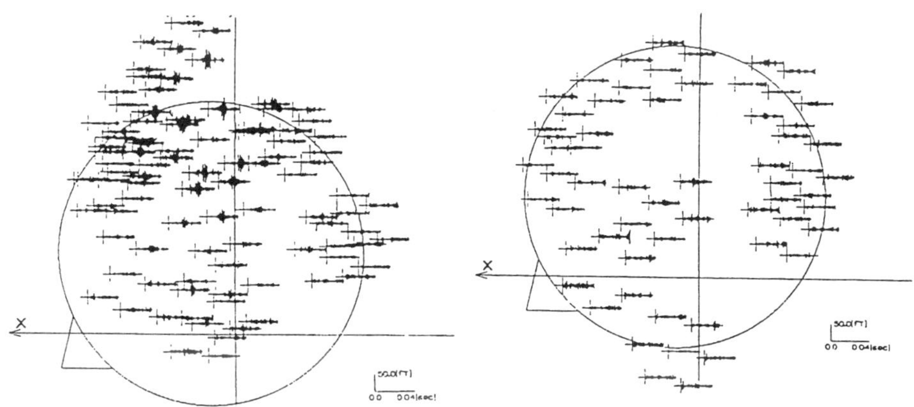

Fig. 3A: High-frequency oscillations during waking.

Fig. 3B: High-frequency oscillations during sleep. Note remarkable amplitude reduction.

Discussion

The physiological features of the high-frequency somatic signals can be summarized as follows. First, the equivalent dipoles for the high-frequency peaks and the underlying N20m are temporally overlapped and spatially colocalized within the 3b area. Second, the high-frequency oscillations present during wakefulness attenuate or disappear during sleep. Third, the underlying N20m in contrast increases its amplitude during sleep. Thus, there exists a reciprocal relation between the high-frequency oscillations and the N20m during a wake-sleep cycle. The reciprocal relation between the high-frequency oscillations and the N20m representing ensemble EPSPs of the pyramidal neurons in 3b area led us to assume that some inhibitory mechanisms are involved in the generation of high-frequency oscillations.

Morphological studies demonstrate that GABAergic inhibitory interneurons receive direct thalamic axon synapses and have their own axon terminations focused on the cell somata, proximal dendrites and initial segments of the pyramidal cells. The findings suggest that GABA mediated inhibition plays a powerful role in the shaping of the response profiles of pyramidal neurons. Furthermore, the patterns of connectivity of many GABAergic interneurons are bidirectional in the vertical direction between deep and superficial layers as well as in the horizontal direction [4]. This parallel arrangement of the interneurons provides the anatomical basis for spatial summation of activity of individual neurons. Thus, synchronous monosynaptic activation of a population of GABAergic interneurons, by summating individual intracellular currents spatiotemporally, may produce magnetic fields strong enough to be detectable outside the head.

Electrophysiological characteristics of cortical interneurons are high-frequency spike bursts (200-1,000 Hz) in response to synaptic volleys. In addition, the extracellular spike waveform originating from the interneurons is short in duration and about one-half that of the "regular spikes" produced by the pyramidal neurons, reflecting mainly different cell morphology [5].

The activity of cortical inhibitory interneurons studied in animal experiments bears striking similarities to the high-frequency oscillations in the somatosensory cortex in humans as described above. Furthermore, GABAergic interneurons are activated by acetylcholine and serotonin upon arousal, which causes enhancement of inhibition of down-stream glutamatergic pyramidal cell activity [6]. This GABA mediated feed-forward inhibition or disinbition of the pyramidal neurons may underlie the reciprocal relation of the amplitude between the high-frequency oscillations and the N20m during a wake-sleep cycle. Thus, an amplitude reduction of the high-frequency oscillations is associated with the increase in the N20m during sleep and vice versa during wakefulness. From these convergent lines of evidence as discussed above, it may be concluded that the high-frequency oscillations represent the activity of the GABAergic inhibitory interneurons in layer 4 of area 3b of the primary somatosensory cortex.

Acknowledgments

This research was supported by the Ministry of Education, Science and Culture of Japan grant 05404087.

Reference

[1]   Yamada, T., Kameyama, S., Fuchigami, Y., Nakazumi, Y., Dickins, Q. S., and Kimura, J. Changes of short latency somatosensory evoked potential in sleep, Electroenceph. clin. Neurophysiol., 1988, 70:126-136.

[2]   Curio, G., Mackert, B. M., Burghoff, M., Koetitz, R., Abraham-Fuchs, K., and Härer, W. Localization of evoked neuromagnetic 600 Hz activity in the cerebral somatosensory system, Electroenceph. clin. Neurophysiol., 1994, 91:483-487.

[3]   Hashimoto, I., Mashiko, T., and Imada, T. Somatic evoked high-frequency magnetic oscillations reflect activity of inhibitory interneurons in the human somatosensory cortex, Electroenceph. clin. Neurophysiol. (in press)

[4]   Jones, E. G. Varieties and distribution of non-pyramidal cells in the somatic sensory cortex of the squirrel monkey, J. Comp. Neurol., 1975, 160:205-268.

[5]   McCormick, D. A., Connors, B. W., Lighthall, J. W., and Prince, D. A. Comparative electrophysiology of pyramidal and sparsely spiny stellate neurons of the neocortex, J. Neurphysiol., 1985, 54:782-806.

[6]   Steriade, M., Jones, E. G., and Llinás, R. R. Thalamic oscillations and signaling. New York, John Wiley & Sons, 1990.

Effect of Color on Visual Evoked Magnetic Fields with Pattern-Reversal Stimulation

Hatanaka, K.[1], Seki, K.[2], Nakasato, N.[2], Kanno, A.[3], Ohtomo, S.[2], Fujiwara, S.[2] and Yoshimoto, T.[2]

[1]KRI, 17 Chudoji Minami-machi, Shimogyo-ku, Kyoto, Japan; [2]Department of Neurosurgery, Tohoku University School of Medicine, 1-1 Seiryo-cho, Aoba-ku, Sendai, Japan; [3]MEG Laboratory, Kohnan Hospital, 4-20-1 Nagamachiminami, Taihaku-ku, Sendai, Japan

Introduction

Visual image is supposed to be processed by parallel pathways that analyze different properties of vision such as form, depth, motion, and color. Positron emission tomography (PET) studies of functional specializations of the human visual system have demonstrated that the fusiform gyrus is playing a role in color processing[1,2]. Poor time resolution of PET, however, gave no information about the time course of the activity in the brain. Although visual evoked potentials (VEPs) studies with dipole source analysis [3,4] supplemented PET results, no direct comparison with the anatomy of the visual cortex of subjects was made. This is because spatial resolution of source localization in VEPs is low[4]. We have been investigating visual evoked magnetic fields (VEFs), which in principle have much better spatial resolution than VEPs, and found that the most prominent response (P100) evoked with pattern-reversal stimulation was located in the primary visual cortex[5].

The most important factor in visual evoked response is the type of stimulus. Earlier investigations for the effect of color on visual response [6,7] used simple colored disc shaped pattern. An abstract display of colored squares and rectangles (a ' Mondrian') was occasionally employed [1-4] because this type of stimulus is believed to excite only color component. Pattern-reversal stimulation, which consists of reversing white and black checkerboard pattern, is widely used for physiological studies as well as clinical application. Pattern-reversal stimulation modulates luminance, contrast, form, and motion properties of the vision. If we use colored checkerboard pattern, color component is also modulated. Since color of the pattern-reversal stimulation has effects on response[8,9], investigations of the difference in VEFs give much insight in the color processing mechanism. Moreover, with excellent spatial and temporal resolution of VEFs, we can naturally expect to trace the time course of the information processing in human visual cortex.

Methods

We recorded VEFs from five healthy subjects with a 64-channel whole-head SQUID magnetometer (CTF Systems Inc., Canada). Stimulus consisted of green-black or red-black checkerboard pattern-reversal stimulation. The brightness of each color was adjusted to 22 Cd/m^2 and the contrast was better than 99 %. Distance between subject's eye and the stimulator was about 1 m. Individual check size was 1° and the total stimulating field subtended rectangular visual field of 5° by 10°. Right-half visual field stimulation to right eye was carried out, and the fixation point was 1° lateral to the center of the left edge of the pattern. Stimulation rate was 2 Hz with pattern reversal period of 500 msec. An analysis time of 450 msec (150 msec pre- and 300 msec post-pattern-reversal) was used and 200 trials passed through 100 Hz low pass filter and 50 and 100 Hz notch filters were digitized at 625 Hz and averaged. For each color, two sessions were carried out to confirm reproducibility of VEFs.

Source locations of the most common components of N75, P100, and N145 were analyzed by single current dipole model. The effect of volume current was taken into account using Grynszpan and Geselowitz spherical head shape approximation. Position and radius of ' best sphere' (*i.e.* sphere which fit into subject's head in the least square sense) were determined by magnetic resonance imaging (MRI) of the subject. Conversion of VEFs coordinates system to MRI was carried out by reference to three fiducial points (two preauricular points and nasion), each marked by small coils for VEFs and oil-containing capsules for MRI. Positions and orientations of the estimated equivalent current dipoles (ECDs) were directly superposed on the MRI of the subjects.

Results

All five subjects showed well-defined P100 and smaller N75 and N145 responses for green-black pattern-reversal stimulation. As for red-black stimulation, responses differed between subjects, but generally N145 waves were most markedly observed and P100 and N75 waves were also noticed. Results for source analysis of a subject are shown in Table 1. Source locations of N145 for green-black stimulation and N75 component for red-black stimulation were not clearly defined. Note that for green-black stimulation, P100 wave has the largest magnitude and lowest fitting error while for red-black case N145 becomes major component.

Averaged parameters of N75, P100 and N145 waves are summarized in Table 2. Estimated locations and orientations of ECDs on the MRI of the subject are shown in Fig. 1. The P100 dipoles were localized in the bottom of the calcarine fissure contralateral to the stimulation. Distance between the locations of P100 dipole for green-black stimulation and red-black one is 0.58 cm. The position of N145 for red-black stimulation were also located in much the same place for P100 (distance is 0.46 cm). The N75 dipoles for green-black stimulation were located about 0.9 cm anterior and 0.5 cm lateral to P100 dipoles (distance is 1.11 cm). The localization error of the VEFs system is supposed to be about 0.3 cm. Therefore the locations of P100 are seemed to be unrelated to colors, and P100 and N145 waves originate from the same place.

Table 1. Anatomical locations of the ECDs evoked with color-black pattern-reversal stimulation of the right-half visual field (x, y, z: anterior, left lateral, and upward coordinates of the VEFs coordinate system; φ: azimuthal angle of dipole, 0 for +x direction; θ: declination angle of dipole, 0 for +z direction; Q: magnitude of current dipole)

Session	Color	Wave	Latency (msec)	x (cm)	y (cm)	z (cm)	φ (deg)	θ (deg)	Q (nAm)	Error (%)
1	Green	N 75	75.2	-3.57	2.21	5.16	62.1	67.2	11.5	22.3
		P100	107.2	-4.36	1.55	4.71	254.1	114.9	25.3	3.6
2	Green	N 75	75.2	-3.10	1.87	4.88	64.5	62.3	15.3	11.8
		P100	105.6	-4.19	1.46	4.78	254.2	113.7	26.5	4.4
3	Red	P100	94.4	-4.38	0.91	4.71	260.7	108.5	12.0	15.4
		N145	153.6	-4.68	1.22	4.67	78.2	69.2	19.9	7.7
4	Red	P100	102.4	-4.47	1.02	4.46	261.2	119.3	12.6	12.0
		N145	155.2	-4.59	1.27	4.58	77.9	70.3	21.5	5.0

Table 2. Averaged parameters of ECDs for the same subject shown in Table 1

Wave	Color	Latency (msec)	x (cm)	y (cm)	z (cm)	φ (deg)	θ (deg)	Q (nAm)	Error (%)
N 75	Green	75.2	-3.34	2.04	5.02	63.3	64.8	13.4	17.1
P100	Green	106.4	-4.28	1.51	4.75	254.2	114.3	25.9	4.0
P100	Red	98.4	-4.43	0.97	4.59	261.0	113.9	12.3	13.7
N145	Red	154.4	-4.64	1.25	4.63	78.1	69.8	20.7	6.4

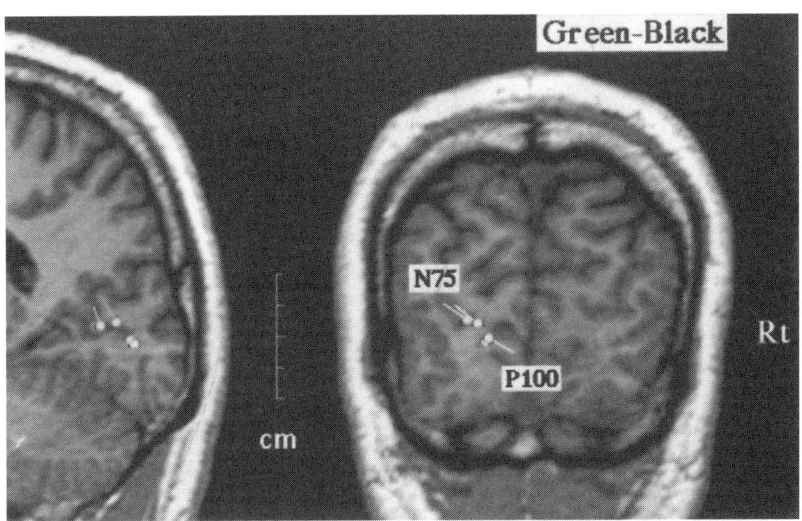

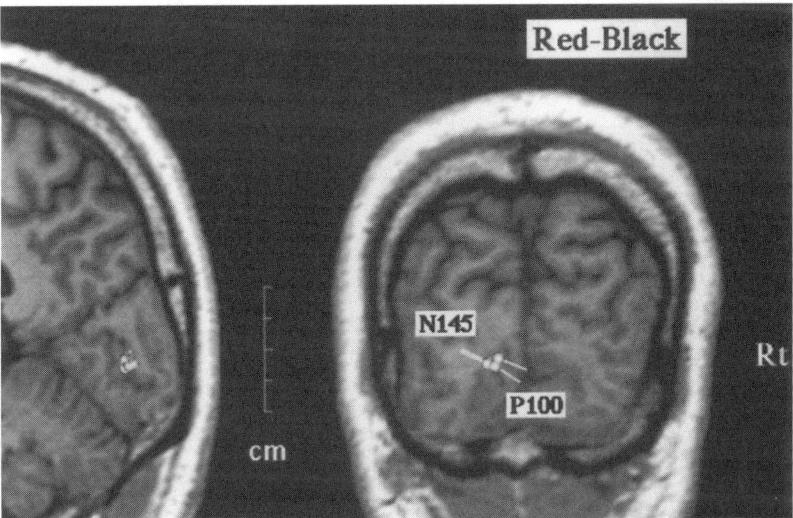

Fig. 1.
ECDs evoked by right-half visual field pattern-reversal stimulation to the right eye. Circles and bars indicate positions and orientations of the dipoles. Dipoles of successive two sessions are shown for each waves. Positions of the ECDs are along the bottom of the calcarine fissure. Note the inversions of the dipoles in time sequence of N75, P100 and N145.

Discussions

Although the precise locations of N75 and P100 ECDs might be different, three major responses of N75, P100, and N145 stayed along the bottom of the calcarine fissure (V1). They changed only directions, and during these time sequences, processing such as projection to the higher visual cortex such as V4 (color) and V5 (movement) or back-projection to V1 must occur.

We can express color-black pattern-reversal stimulation schematically as shown in Fig. 2. Although luminance is adjusted to the same intensity, chromaticity is apparently much stronger in red-black stimulation. Hence, red-black pattern-reversal stimulation has much stronger effect on color component; or we can assume red-black stimulation mainly as color (and brightness) stimulation. For green-black stimulation, the effect of color is small and the response is similar to the conventional white-black, or brightness, stimulation. Stronger response of N145 for red-black stimulation, therefore, suggests N145 is related to the color component. However, the locations of N145 dipoles are almost same as P100 component and localized in the bottom of the calcarine fissure (V1), and not in the fusiform gyrus (V4). One hypothesis we could propose is that N145 component is the result of back projection from V4 to V1. Buchner et al.[4] pointed out that activity peaked at 135 msec could be located in fusiform gyrus. Although visual stimulation is quite different, their findings may support our hypothesis since their method(VEPs) is sensitive to dipoles in gyrus (V4, 135 msec) whereas ours (VEFs) is sensitive to sources in sulcus (V1, 145 msec).

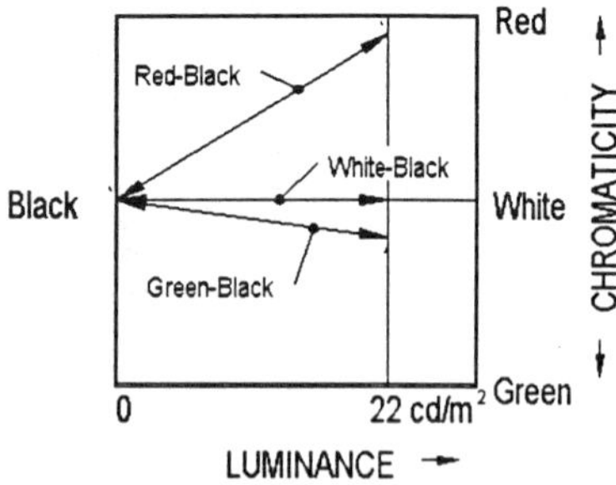

Fig. 2 Schematic representation of color-black pattern-reversal stimulation.

References

[1] Lueck, C. J., Zeki, S., Friston, K. J., Deiber, M. -P., Cope, P., Cunningham, V. J., Lammertsma, A. A., Kennard, C., and Frackowiak, R. S. J. The colour centre in the cerebral cortex of man, *Nature*, 1989, 340: 386-389.

[2] Zeki, S., Watson, J.D.G., Lueck, C.J., and Friston, K.J. A direct demonstration of functional specialization in human visual cortex, *J. Neurosci.*, 1991, 11: 641-649.

[3] Plendl, H., Paulus, W., Roberts, I. G., Bötzel, K., Towell, A., Pitman, J. R., Scherg, M., and Halliday, A. M. The time course and location of cerebral evoked activity associated with the processing of colour simuli in man, *Neuroscience Letters*, 1993, 150: 9-12.

[4] Buchner, H., Weyen, U., Frackowiak, R.S.J., Romaya, J., and Zeki. S. The timing of visual evoked potential activity in human area V4, *Proc. R. Soc. Lond.* B, 1994, 257: 99-104.

[5] Seki, K., Hatanaka, K., Nakasato, N., Otsuki, T., and Yoshimoto, T. The origin of P100 component in visual evoked response, *J. Funct. Neurophysiol.* 1990, 3:270.

[6] Paulus, W. M., Hömberg, V., Cunnigham, K., Halliday, A. M. and Rohde, N. Colour and brightness components of foveal evoked potentials in man, *Electroencephalogr. Clin. Neurophysiol.*, 1984, 58: 107-119.

[7] Krauskopf, J., Klemic, G., Lounasmaa, O. V., Travis, D., Kaufman, L. and Williamson, S. J. Neuromagnetic measurements of visual responses to chromaticity and luminance, In: Williamson, S. J. *et al. Advances in Biomagnetism*, 209-212, Plenum Press, New York, 1990.

[8] Taghavy, A., and Kügler, C. F. A. Colour-black pattern reversal visual evoked potentials (colour-black-PVEPs): neurobiological aspects and clinical applicability of a new method, *Intern. J. Neuroscience*, 1988, 43: 225-236.

[9] Hatanaka, K., Seki, K., Nakasato, N., and Yoshimoto, T. Visual evoked magnetic fields with red-black and green-black pattern-reversal stimulation, In: Hoke, M. *et al., Biomagnetism: Clinical aspects*, 197-201, 1992.

Acknowledgments

This work has been supported in part by Grants-in-Aid for Scientific Research No. 03404042 and No. 064050 from the Ministry of Education, Science and Culture of Japan.

Evoked Magnetic Fields at Semantic Judgment of Simple Japanese Sentences

Hirata, Y.[1], Sakai, T.[1], and Kuriki, S.[1,2]
[1]Research Institute for Electronic Science, Hokkaido University, Sapporo, Japan
[2]National Institute for Physiological Sciences, Okazaki, Japan

Introduction

Most studies of auditory evoked magnetic fields of the human brain have been conducted by recording responses reflecting early processes to sound such as N1m or MMF. From lesion and PET studies, it has been hypothesized that spoken words first activate the primary auditory cortex, then the information is processed at Wernicke's area, and lastly a frontal cortex processes semantic association [1][2]. Magnetoencephalography is a suitable method for tracing such activities spatio-temporally. In this study, we used some Japanese sentences, in which each word was formed from syllables consisting of the combination of a consonant-vowel or a vowel. We measured the magnetic fields from the temporal area of subjects by using the word stimuli with an equalized duration.

Experiment I

Method   Four Japanese subjects (three right-handed, one left-handed) participated in the first experiment. The combination of a noun followed by a verb was known as one of the most simplest Japanese sentences. Stimuli used in this experiment consisted of a Japanese noun (400 ms duration) followed by a 30 ms blank, a Japanese verb (600 ms), a 1.57 s blank, and a sound cue (500 Hz, 40 ms) with a repetition rate of 5.6-6.1 s

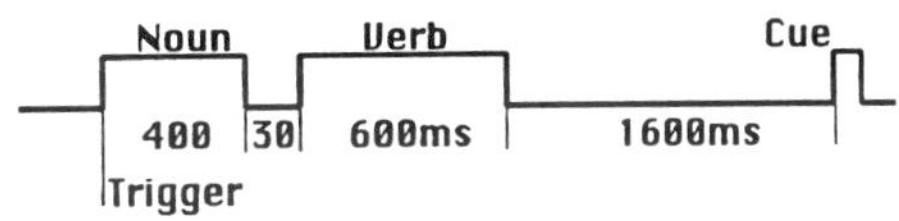

Fig. 1 . Stimulus sequence of experiment I.

(Fig. 1). In order to adjust the duration of the stimuli, two-syllable nouns and three-syllable verbs were selected. Spoken words digitized at a rate of 10 kHz were presented to both ears of the subjects through plastic ear tubes at sound intensity of 45 dB SPL. The subjects were instructed to listen to a simple Japanese sentence which consisted of a noun and verb and to judge quickly whether the sentence was congruous or not. After listening to the cue of a beep, they were required to push a left button of a mouse when the sentence was congruous, or a right button when the sentence was incongruous. One hundred responses to both congruous and incongruous stimuli were recorded in each session. This experiment was done for each subject several times by using different sentences of similar kinds. Magnetic fields evoked from the subjects' brain were recorded in a magnetically shielded room with a dual 37-channel SQUID magnetometer (BTi Inc.). which covered the right and left temporal areas and thus two hemispheres' activities were simultaneously recorded. The signals were digitized at 297.6 Hz. The period of 2.9 s from the onset of the noun including a 0.4 s pre-stimulus baseline was analyzed. The responses were averaged and digitally filtered with a bandpass of 1-40 Hz. In order to measure the reaction time of the same semantic task, an experiment using the same stimuli, except for a tone cue, was also carried out.

Results   Figure 2 shows superimposed 37 magnetic fields which were recorded from the right and left hemispheres of two subjects. The large signal at 0 ms latency is an artifact of electrical noise from a D/A converter. Although this kind of noise was also observed at other latencies of 400 ms, and 1.0 s, it was clear that the noise did not affect the baseline of the recorded fields. Two out of the four subjects showed a large amplitude of about 350 fTpp of the N1m elicited by the first syllable of the noun. However, no obvious N1m which was elicited by the syllables of the verb was observed in all the subjects. Comparing to the power (squared amplitude of the fields) of both hemispheres the left power was greater than that of the right one in several sessions. This observation is compatible with the PET study which reports the dominancy of the left hemisphere activation in lexico-semantic processes [3].

The averaged reaction time was 1170 ms from the onset of the noun to the button press time of the subjects. The simple reaction time to the movement of the index finger muscles by a magnetic transcranial stimulation of the motor cortex is about 120 ms [4]. Therefore, the semantic judgment is supposed to be completed at the latencies of around 1050 ms. In this respect, a slow wave of the fields was observed in several sessions in all subjects, where its

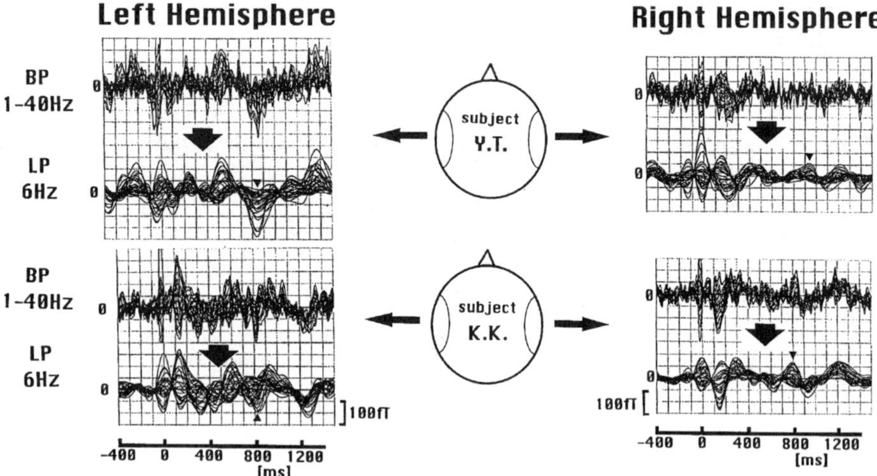

Fig. 2. Superimposed 37 waveforms (1–40 Hz) recorded from both hemispheres of two subjects. Low-pass filtered data below 6 Hz were drawn to examine the existence of the slow wave at latencies of 0.8-1 s. The wave observed at 0 ms was an electrical artifact, and it did not affect the baseline.

latency varied between 800 ms and 1.0 s (Fig. 2). The slow wave was positive in the right hemispheres and negative in the left hemispheres and was synchronized in both hemispheres. A digital bandpass filter from 1 to 6 Hz was applied to all data to confirm the existence of the wave. The wave was observed in 14 sessions of the grand total 25 sessions among all subjects. This slow wave might be related to the semantic judgment.

Experiment II

The combinations of a noun and a three-syllable verb used in the experiment I might be not appropriate to keep the subjects' judgment with equal latency. Because there was a possibility that the subjects could judge the meaning of the whole sentences when listening to the first or second syllable of the three-syllable verbs.

Method   Three right-handed subjects participated in this experiment. The selected sentences, which were also a combination of a noun and a verb, could not be judged before hearing the second syllable of the verb. The nouns had two or three or four syllables. We made 184 pairs of congruous and incongruous sentences, which differed only in their second syllables of their verbs. The subjects were required to judge quickly whether the each verb was coherent to the noun when listening to the sentence. Each stimulus sentence consisted of a noun, followed by a 1.1-1.6s blank, a two-syllable verb (650 ms), a 1.5s blank, a tone cue (500 Hz, 40 ms), and a 1.2s blank (Fig. 3). All the nouns and the verbs were synthesized by a computer software. The original duration of the synthesized verbs was 400 ms. Two syllables of the verbs were separated from each other, and the second syllable was presented 400 ms after the onset of the first syllable. The trigger point was provided at the onset of the second syllable of the verbs. The recording duration was 3.6 s, including 800 ms for the pre-trigger period. The subjects were

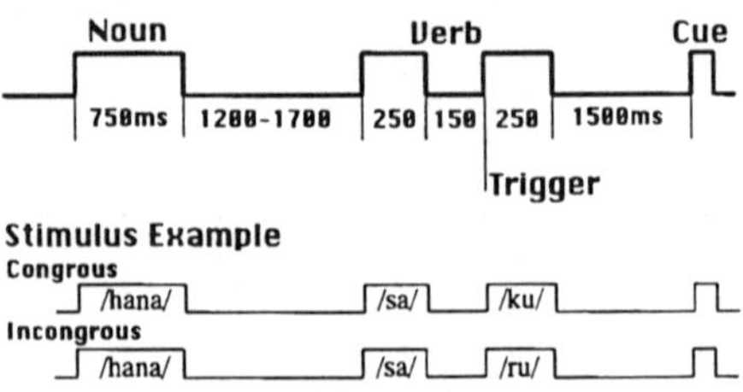

Fig. 3. Stimuli sequence of experiment II. The subjects could not be judged before hearing the second syllable of the verb.

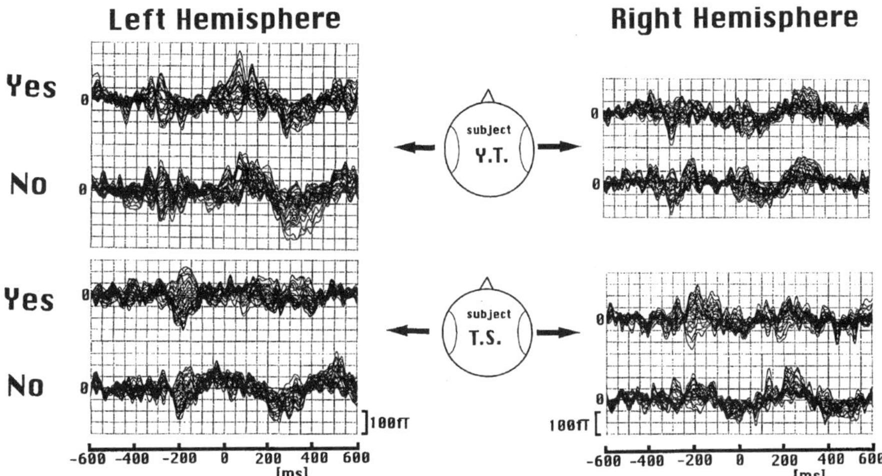

Fig. 4. Superimposed 37 waveforms (1-40 Hz) recorded from two hemispheres of the two subjects, where the responses were selectively averaged by the subjects' judgment: semantically congruous (Yes) and incongruous (No). Three evoked magnetic fields by the first syllable of the verbs were observed, and they were identified as P1m, N1m, and P2m based on the latencies from the onset of the verb. Occurrence of those which were evoked by the second syllable of the verbs was not clear. All the eight recorded responses of subject Y.T. and one of No (incongruous) magnetic fields of the subject T.S. showed slow waves at latencies between 250 and 300 ms.

instructed to follow the same procedures to Experiment I. They were also required not to press the button when they were not convinced whether the sentence was congruous or not.

Results   Each subject participated in the experiment twice. Figure 4 shows the selectively averaged magnetic fields from the left and right hemispheres by the subject's judgment of congruous (Yes) and incongruous (No). Three evoked field responses to the first syllable of the verbs were observed , and they were identified as P1m, N1m, and P2m based on their latencies from the onset of the verb. However, the occurrence of those which were evoked by the second syllable of the verbs was not clear. All the eight recorded responses of the subject Y.T. showed slow waves at latencies between 250 and 300 ms. On the other hand, the subject F.T.'s magnetic fields did not show obvious slow waves. The subject T.S. showed the slow wave in one out of two incongruous responses. The slow wave was larger in the semantically incongruous responses than congruous ones in some sessions.

By high-pass filtering above 3 Hz and readjusting their baseline with 400 ms duration before the onset of the verbs, activities corresponding to the second syllables of the verbs were also clearly observed. Hence, the observed waveform of the magnetic fields could be interpreted as the superimposition of the responses of the P1m, the N1m, and the P2m from the auditory cortex and other event related fields of the slow wave. To isolate the slow wave, the fields data were low-pass filtered below 3 Hz (Fig. 5). The resulting isofield map at latency of around 300 ms showed the magnetic flux out of the skull in the right hemisphere, and magnetic flux into the skull in the left hemisphere. The maximum flux-in and -out points were located at the anterior temporal region of the subject.

DISCUSSION

In the Wernicke-Geschwind model for language, the Wernicke's area is located in the Broadmann's area 22 [5], but we could not find any obvious activity in the posterior part of the measured region. The activity might take place in the gyrus or may not be synchronized with the stimuli. The slow wave observed in the experiment I and II seem to be the same activity because of the same semantic task. From the latency (400-600 ms) from the onset of the verb and the task in the experimental I, the slow wave might be N400, which appears in semantically incongruous stimuli. On the other hand, from the latency (250-300 ms) from the second syllable of the verbs in the experiment II,

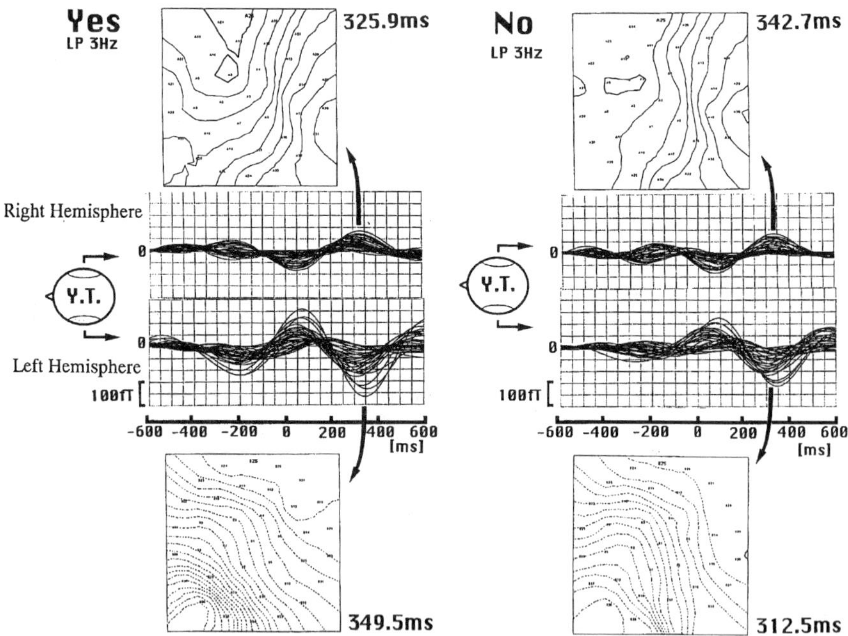

Fig. 5. The slow wave low-pass filtered (3 Hz) from the recorded magnetic fields of the subject Y.T.. The resulting isofield map at latency of around 300 ms showed the magnetic flux out of the skull in the right hemisphere, and the magnetic flux into the skull in the left hemisphere. The maximum flux-in and -out points were located at the anterior temporal region of the subject.

the slow wave might be P300, which reflects the perception and usually is observed in oddball tasks. The iso-field maps, however, indicate that the source of the slow wave is deep within the skull, and it may be located near in the hippocampus or the cingulate gyrus, whose function is related to memory.

CONCLUSION

A slow wave was observed in the two semantic experiments. It may be related with higher auditory process from its latency. However, the slow wave was observed in about half of the sessions of the experiments. Further measurements and the dipole source estimation are required to clarify the characteristic of the slow wave.

References:

[1] Geshwind, N., "Specializations of the human brain," Scientific American, 1979, **241**: 158-168.

[2] Petersen, S.E., Fox, P.T., Posner, M.I. Mintun, M., and Raichle, M.E., "Positron emission tomographic studies of the cortical anatomy of single-word processing," Nature, 1988, **331**: 585-589.

[3] Démonet, J.F., Chollet, F., Ramsay, S., Cardebat, D., Nespoulous, J.L., et al., "The anatomy of phonological and semantic processing in normal subjects," Brain, 1992, **115**: 1753-1768.

[4] Pascual-leone, A., Brasil-neto, J.O., Valls-solé, J., Cohen L.G., and Hallett M., "Simple reaction time to focal transcranial magnetic stimulation. comparison with reaction time to acoustic, visual and somatosensory," Brain, 1992, **115**:109-122.

[5] Mayeux, R., and Kandel, E.R, "Disorders of language: the aphasias," In: Kandel, ER., Schwartz, J.H., and Jessell, T.M. Principles of neural science, New York, Elsevier,1991.

Neuromagnetic Correlates of Endogenous Auditory Sensations

Hoke, E.S., Hoke, M. and Ross, B.

Institute of Experimental Audiology, University of Münster, Münster, Germany

Introduction

Endogenous auditory sensations occur without *simultaneous* reception of an external acoustic stimulus. They can occur under physiological conditions, but can also be associated with pathological processes. A commonly known sensation of the second kind is tinnitus, or ringing in the ear, a frequent symptom which is often associated with several hearing disorders. A sensation of the first kind is the auditory afterimage. Contrary to the vision system in which after-images have been known for a long time, auditory stimulation was, in general, not assumed to lead to clear-cut after-effects. An exception is the exposure to acoustic overstimulation which has been known to lead to a temporary threshold shift which is sometimes accompanied by an auditory sensation, tinnitus.

It was in 1964 when Zwicker, for the first time, described a clear-cut and reproducible auditory after-effect that followed a specific acoustic stimulation [1]. This after-effect, which he termed "negative afterimage in hearing" and which was later coined "Zwicker tone" in honour of its discoverer [2], is an endogenous auditory sensation which occurs after the offset of a band-suppressed noise that was presented for some seconds. The sensation of this monaural phenomenon is most similar to that of a pure tone with a frequency corresponding to the centre frequency of the gap and with an equivalent level of 10-15 dB above the auditory threshold. The sensation decays gradually, and it may last as long as 10 s, depending on how long the evoking noise was presented.

While the physiological basis of the after-image in vision has been uncovered some decades ago, the search for a neurophysiological correlate of the auditory afterimage has not been successful so far.

This Zwicker tone study [3] was undertaken in the search for a human model of tinnitus, because there are many similarities between the two auditory sensations:

- Beats cannot be generated by interference with an external tone [4].
- Both sensations do not follow the masking laws which exist for real sound signals (e.g., contralateral masking).
- No binaural fusion of the Zwicker tones occurs if different exciting signals are presented to the two ears.
- The loudness of both sensations is only slightly above the subjective threshold.

These many similarities led us to look for a neurophysiological correlate of the Zwicker tone, hoping that this might be helpful in our search for neurophysiological correlates of tinnitus. Studies of the neuromagnetic activity of the auditory cortex were expected to be the most promising ones because of the power they have demonstrated in localizing neurophysiological correlates of other purely subjective auditory phenomena [5].

Methods

Four female and seven male subjects, all right-handed and at the age of 21 to 30 years (25 ± 3 years, median 25 years), participated in this study. All subjects had a normal audiological status and were able to perceive the Zwicker tone.

Three different stimulation paradigms were employed successively in separarte experimental blocks of 500 s duration each. In each paradigm, noise pulses were monaurally presented with a sensation level of 45 dB. The duration of the bursts was 3 – 5 s (randomly distributed), and pauses of 1 s were interposed between successive bursts. The series of three blocks was repeated four times for each subject.

- The first paradigm was the *notched noise condition (NN)* in which bursts of notched noise were presented. The cut-off frequencies were 3.2 and 5.2 kHz, and the minimum of the spectral level was at 4 kHz. The pauses of 1 s were silent. In this condition, a Zwicker tone was supposed to be heard.
- The second paradigm was the *white noise condition (WN)* in which bursts of white noise were presented. The pauses of 1 s were silent. In this condition, no Zwicker tone was supposed to be heard.
- The last paradigm was the *white noise with interposed tone condition (WNT)* in which also bursts of white noise were presented. During the pauses, however, a pure tone was presented with an intensity of 10 dB sensation level and a frequency of 4 kHz. In this condition, an auditory sensation similar to that of a Zwicker tone was supposed to be heard.

The MEG recordings were performed in a magnetically shielded room (Vacuumschmelze) with a 37-channel neuromagnetometer (MAGNES®, Biomagnetic Technologies). The sensor array was placed over the left temporal plane. During the experimental blocks of 500 s, magnetic data were continuously recorded with a bandwidth from DC to 100 Hz and a sampling frequency of 294 Hz.

For each stimulus condition, epochs of 2 s, beginning 500 ms before the offset of the noise pulses, were averaged after artifact rejection. The averaged data were digitally low-pass filtered at 20 Hz (second-order zero phase-shift Butterworth characteristics). In order to derive correlates of the Zwicker tone, difference waveforms of the averaged and filtered MEG signals were computed i) between the NN and the WN conditions, and, for comparison, ii) between the WNT and the WN conditions.

The next steps consisted in calculating rms waveforms for the original and for the difference waveforms, and generating isofield contour plots for the original and for the difference waveforms.

Finally, source analyses based on a single moving equivalent current dipole (ECD) model in a spherical volume conductor were applied for each sampling point of the averaged original and of the difference waveforms. Only dipole localizations with a goodness of fit > 0.95 were considered.

For the difference waveforms, also mean locations were calculated for that epoch during the pause in which a clear-cut dipolar field distribution was recognizable, the goodness of fit exceeded 0.95, and the dipole coordinates remained sufficiently stable.

Results

Fig. 1 shows in the upper row the grand means of the original averaged rms waveforms for the three conditions, and in the bottom row the grand means of the difference waveforms. It is obvious that the three grand means of the original waveforms look very similar.

The N_1m deflection is clearly expressed approximately 120 ms after the offset and, weaker, after the onset of the noise burst, with almost similar amplitude in the different conditions. During the pause, the niveau of the activity is elevated, with a greater elevation in the NN and the WNT conditions as compared to the WN condition. This activity during the pause is clearly expressed in the difference waveforms, in which identical contributions like the transient components have obviously been almost totally removed by the subtraction operation.

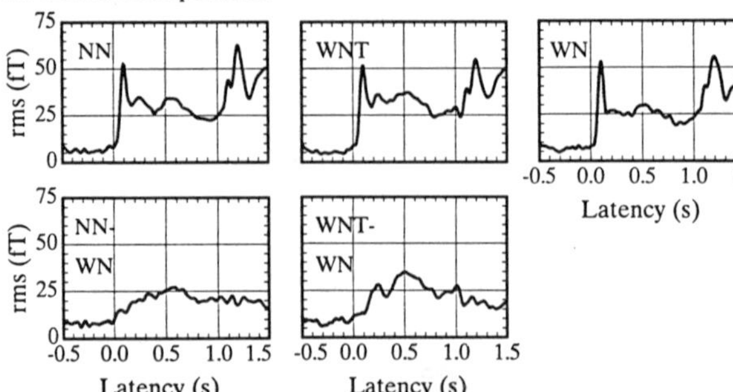

Fig. 1 Grand means of the original averaged rms waveforms for the three conditions, NN, WNT and WN, (upper row), and the grand means of the difference waveforms NN-WN and WNT-WN (bottom row). (After [3])

Fig. 2 shows the averaged magnetic signals for two channels recorded over the field extrema of the N_1m_off, again in the top row for the waveforms of the original signals and, in the bottom row, for the difference waveforms. In all of the three stimulation conditions, a clear-cut polarity reversal for the exogenous components N_1m_off and N_1m_on is evident.

In the difference waveforms, identical contributions like the exogenous components N_1m_off and N_1m_on have obviously been almost totally removed by the subtraction operation. However, there exists a post-stimulus activity which develops during the pause between pulses and which exhibits a clear-cut polarity reversal. This gradually increasing post-stimulus activity is basically similar in the two difference waveforms; it gradually increases until it assumes a flat maximum around 500-600 ms and then decreases.

Eight out of eleven subjects showed an obvious dipolar field distribution in the difference waveforms for a certain time during the pause. In the remaining three subjects, we have not been able to find a distinct dipolar field distribution.

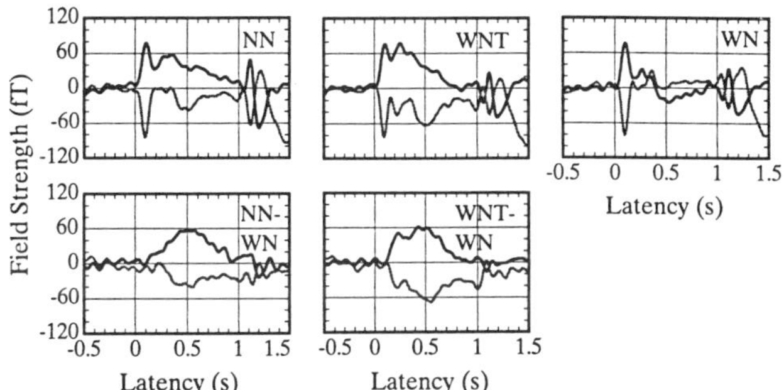

Fig. 2 Averaged magnetic signals for two channels recorded over the field extrema of the N_1m_off, again in the top row for the waveforms of the original signals, NN, WNT and WN, and, in the bottom row, for the difference waveforms, NN-WN and WNT-WN. (After [3])

Fig. 3 represents, for three subjects (p3, S14, S15), a comparison of the magnetic field distribution patterns of both the original and the difference waveforms at selected instants of time.

The first column shows the field pattern 100 ms prior to the offset of the burst. No regular pattern can be observed. The second column shows the field pattern around the maximum of the N_1m after the offset of the noise burst. On account of the similarity of this response component in the waveforms of the three stimulation conditions, averages were calculated for this component. The evoked field shows a distinct dipolar distribution.

The next two columns show the field patterns for the difference waveforms NN-WN (third column) and WNT-WN (fourth column), 500 ms after the offset of the noise. A distinct dipolar field distribution can be observed in both cases. The fifth column, finally, shows the field patterns for the white noise waveforms, again 500 ms after noise offset. In some cases a dipolar pattern of different location and different orientation is slightly expressed, whereas in other cases only a random pattern exists.

It becomes obvious that the distribution patterns of the post-stimulus activity resemble those of the transient responses; but the dipole orientation of the post-stimulus activity appears to be slightly more inclined than that of the transient responses.

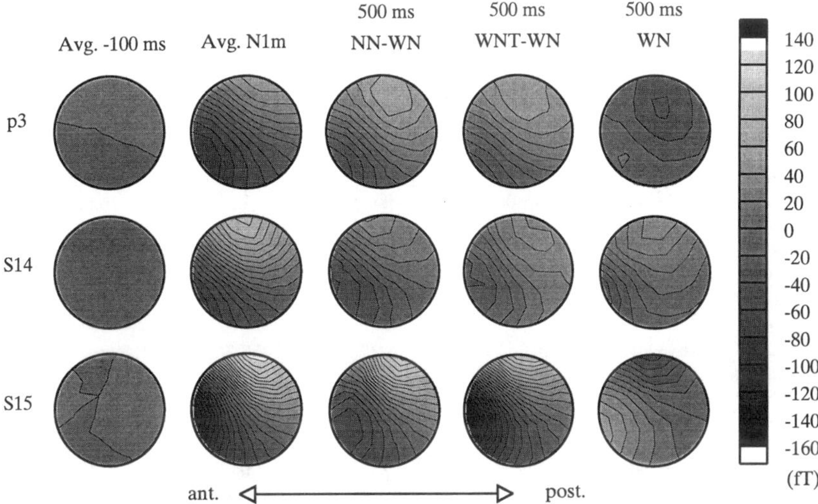

Fig. 3 Comparison of field distribution patterns of the original and the difference waveforms. (After [3])

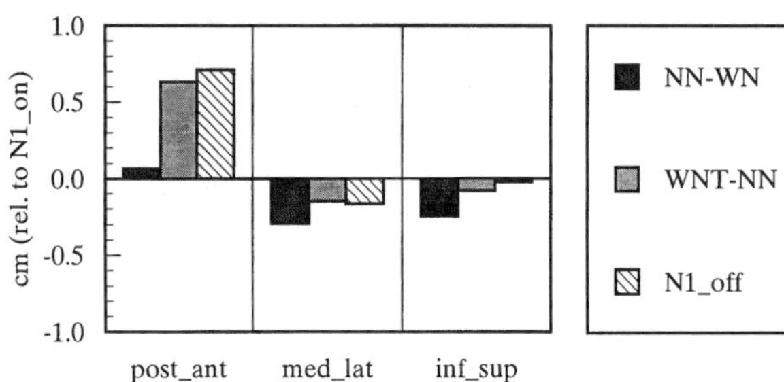

Fig. 4 Comparison of mean source locations of the N_1m_off, of the poststimulus activity during the pause in the NN-WN and in the WNT-WN conditions. (After [3])

Fig. 4 shows comparisons of mean source locations of the N_1m_off, and of the poststimulus activity during the pause in the NN-WN and WNT-WN conditions.

Source locations were normalized relative to that of the N_1m_on, and the analysis was restricted to those eight out of eleven subjects in which the prerequisites as mentioned above were fulfilled. It is obvious that the source locations of the different activities do not differ significantly, especially in the medial-lateral and inferior-superior direction.

Discussion

A neuromagnetic study in normal-hearing subjects was performed in order to look for a neurophysiological correlate of the Zwicker tone in the auditory cortex. With a stimulation paradigm especially designed for this study we have been able to derive post-stimulus activity which appears to be related to the Zwicker tone and which originates in the supratemporal auditory cortex. It is a longer lasting neuromagnetic activity which is distinctly weaker than that of the onset and offset responses to the tone burst, and which shows a clear-cut dipolar field distribution.

It appears that the assumed Zwicker tone-related activity has certain similarities with that of the sustained response. The hypothesis is put forward that during the sensation of the Zwicker tone a process takes place in the auditory cortex which is similar to that underlying the sustained response, and which gives rise to the Zwicker tone sensation. However, while the sustained response is due to neural activity generated by a *simultaneously* presented external acoustic stimulus, the assumed sustained response is due to a temporary absolute or relative reduction of neural activity originating from those regions in which the Zwicker tone exciting stimulus caused an adaptation. The relatively augmented neural activity which originates from circumscribed regions outside the spectrum of the Zwicker tone exciter and which lasts for some time might then be interpreted by higher neural levels as elicited by an external acoustic stimulus. This may cause the generation of a sustained response which, in turn, leads to the sensation of the Zwicker tone.

We have failed so far to confirm for the Zwicker tone-related sustained response a source location which is, in all subjects, systematically different to those of the transient responses to the onset and offset of the evoking noise bursts. We believe, however, that this failure is due to the weak signal amplitude of the supposed Zwicker tone-related signal.

Certain psychoacoustic similarities between the Zwicker tone and tonal tinnitus rise the question whether a similar effect might exist also in tonal tinnitus. A preliminary study with a comparable experimental design has shown that neural activity is obviously recordable which might be related to tinnitus.

References

[1] Zwicker, E. "Negative afterimage" in hearing, J. Acoust. Soc. Am., 1964, 36: 2413-2415

[2] Lummis, R.C., and Guttman, N. J. Acoust. Soc. Am., 1972, 51: 1930-1944

[3] Hoke, E.S., Hoke, M., and Ross, B. Neurophysiological correlate of the auditory after-image ("Zwicker Tone"), Audiology & Neuro-Otology (submitted)

[4] Krump, G. Dissertation, München, 1993

[5] Proefrock, E.S., and Hoke, M. In: Baumgartner, C., Deecke, L., Stroink, G., and Williamson, S.J. Biomagnetism: Fundamental Research and Clinical Applications, Amsterdam, Elsevier, 1995, 234

Acknowledgements

This work has been supported by grants from the Deutsche Forschungsgemeinschaft (Ho 847/6) and is part of a Ph.D. thesis.

Spatial and Temporal Properties of the Slow Components of the Contingent Magnetic Variation in a Warned Choice Reaction Time Task

Hultin, L.[1], Högstedt, P.[1], Pizzella, V.[2], Romani, G.L.[3], Rossini, P.[4,5] and Tecchio, F.[2]

[1]*Department of Applied Electronics, Chalmers University of Technology, Gothenburg, Sweden;* [2]*Istituto di Elettronica Stato Solido - C.N.R., Roma;* [3]*Istituto di Fisica Medica, Università G.D'Annunzio, Chieti, Italy;* [4]*Dipartimento di Neurologia, Ospedale "Fatabenefratelli", Isola Tiberina, Roma;* [5]*I.R.C.C.S. "S. Lucia," Via Ardeatina, Roma, Italy*

Introduction

Electrically recorded contingent negative variation, eCNV, has been reported as a two-component wave [1]. The early component, the O-wave (orientation-wave), is related to the processing of the warning stimulus (S1) and is sensitive to information that S1 provides with respect to the imperative stimuli (S2). The late component, the E-wave (expectation-wave), is considered to be a manifestation of at least two major processes: motor preparation and stimulus anticipation. The electrical O-wave has a frontocentral distribution, with right hemisphere preponderance, while the E-wave is located precentrally, with a preponderance contralateral to the side of the movement. The CNV is a cognitive process where the neurological responses are constantly changing during an experimental session, due to alterations in the state of alertness of the subject, the process of learning and the reorganization of mental strategy to perform the task. During an initial learning period the amplitudes of the eCNV are decreasing and the potential distribution changes. Recordings of Contingent Magnetic Variation (CMV), the neuromagnetic counterpart of the eCNV, have been reported by several authors [e.g. 2, 3]. The scope of this study was to analyze the presence, distribution, and origin of slow components of CMV in a warned choice reaction time task after the initial learning period has been performed.

Methods

Neuromagnetic signals from 12 healthy subjects were recorded using the 28-channel MEG instrument installed at IESS, Rome [4]. The paradigm used was a modified version of the classical CNV paradigm, first proposed by Walter et al. [5], and has previously been reported by Hultin et al. [2, 3]. A low-intensity optical pulse of short duration was used as warning signal. A 70-ms tone of 400 or 1400 Hz, presented to the left ear, was used as S2 (imperative stimulus). S2 followed S1 by 1500 ms and the probability for the high-pitch tone was 0.66. The subject was asked to press a nonmagnetic dummy push button with the right index finger immediately on hearing the high-pitch tone, but not to push when the low-pitch tone was heard. The EEG was recorded through Ag-AgCl electrodes at Cz, Oz, T3, and T4. Linked earlobes were used as reference, left mastoid process as ground. Electro-oculogram, EOG, was recorded from medial supra-orbital and lateral infra-orbital electrodes at the left eye. EMG signals, from the first dorsal interosseous muscle on the right hand, were rectified and then recorded. All signals were bandpass-filtered (0.048 - 64Hz) and digitized with a sampling rate of 250 Hz. Six different probe positions were obtained for each subject and seventy epochs were recorded for each position. Epochs were rejected if they contained eye artifacts, magnetic or electric artifacts, EMG activity before S2, or if the reaction time exceeded mean reaction time plus 2 SD. The training session that preceded the recordings consisted of 45 epochs.

An ellipsoidal model of the head was used for field-maps and source localization [3]. Coordinates refer to a Cartesian system corresponding to the axes of the ellipsoidal model. The y-axis is parallel, and one centimeter superior, to a line between the inion and the nasion. The positive z-axis passes through the skull at the vertex, and the positive x-axis passes through the skull at the midpoint on a line from the inion to the nasion, approximately 2 - 3 cm superior of the right pre-auricular point. Sources were modeled as two separate equivalent current dipoles. Start values for the inverse solution were given manually under visual guidance of simulated forward solution. The strength and localization of the two current dipoles were determined as the best fit to the observed data using an iterative least-mean-squares fitting Marquardt algorithm. The extent to which the fitted dipoles provided an adequate model was expressed as the degree to which the calculated field, Bc, could account for the observed data, Bo, and is referred to here as the goodness of fit, $g = 100 \left(1 - \sum (B_o - B_c)^2 / \sum B_o^2\right)$.

Results

Large interindividual differences are seen in the morphology of the CMV, as well as in the eCNV. According to the classification proposed by Tecce [6] the morphology of the CNV can be differentiated on the basis of the rise time of the ascending limb. Furthermore, large differences are seen in the early CNV complex that follows immediately on the warning stimuli. The large interindividual differences, the complex neurological process, and the limited number of subjects in this study thus reduce the possibilities of reaching general conclusions in terms of dipole localizations. The spatial and temporal morphology of the CMV obtained from the recorded subjects reveal, nevertheless, several common features. Based on the spatial and temporal properties of the recorded neuromagnetic signals, four slow components could be separated from the CMV (Fig. 1). The first three components are here considered to be subcomponents of the O-wave, and are thus labeled O1, O2 and O3. The late component is considered to be related to the expectation wave and is thus labeled E.

The first slow component, O1, followed immediately on the visual evoked response. The component peaked about 300 ms after S1 and lasted approximately 100 ms. The electric potentials showed a slight right-sided lateralization and the amplitudes of the occipital potentials were in all cases stronger than, or equal to, the amplitude at vertex (Fig. 1). The spatial morphology of magnetic flux and the obtained source localization suggested occipital activity as well as bilateral frontocentral activity (Fig 2, 3). Visual inspection and dipole localization indicated further that the activity in the sensorimotor cortex is, in most subjects, active as early as 300 ms after the warning stimuli. The second early subcomponent, O2, peaked at 460-520 ms after S1 and had a duration of up to 150 ms. During this time period a two equivalent current dipole model failed to achieve proper source localization, indicating a complex neural activity (Fig. 3). Visual inspection of the obtained field-maps indicates activity in the posterior part of the right parietal lobe, as well as bilateral frontocentral activity. The third early component, O3, peaked at 620-720 ms after S1 and was present up to 400 ms before S2. The corresponding electrical component was not clearly distinguishable in all subjects. The electrical potentials were strongest at vertex and a left-sided lateralization was seen in most subjects. Source localization suggests, in several subjects, activity in the sensorimotor areas.

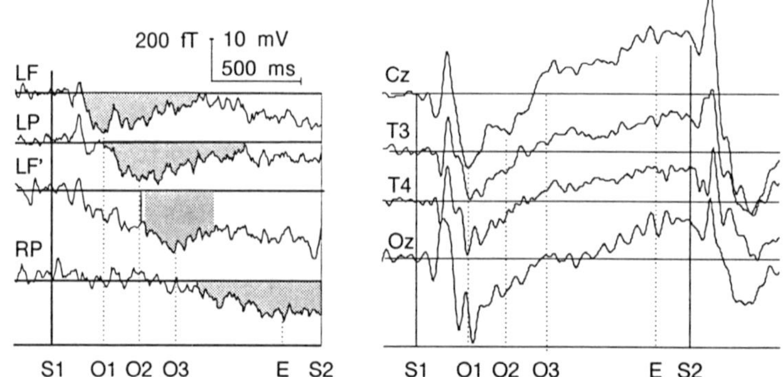

Fig. 1: One example of CMV (left) and eCNV (right), recorded in one subject (FT). The eCNV is the grand average of the electric signals recorded in six different probe positions. Negativity at the active electrode, and outgoing magnetic flux, are seen as an upward deflection. The magnetic signals are recorded over the left frontal (LF and LF'), left parietal (LP), and right parietal (RP) hemispheres. Based on the spatial and temporal properties of the recorded neuromagnetic signals four slow components (O1, O2 ,O3 , E) could be separated in the CMV.

The late component, related to the expectation wave and thus labeled E, started 500 - 700 ms before S2 and ended abruptly at the delivery of the imperative stimuli. Field-maps as well as the obtained dipole localization showed a close relation between O3 and the late component. The obtained results suggested that the dominant sources of the late component are symmetrically located in the left and right premotor areas [3].

Discussion

CNV reflects a complex cognitive process where the magnetic fields span large parts of the scalp. Thus CMV has to be recorded over the entire head simultaneously or at multiple probe positions. It is known that the amplitude of the eCNV varies during a sequence of recordings [7]. The ambiguity resulting from measuring the non-stationary CNV process at multiple probe positions must be carefully taken into consideration in the experimental setup, the analysis of the data and the interpretation of the results. The morphological structures of field-maps are only partly degraded due to this variability. The inverse solution, on the other hand, might be affected by a significant error. For future studies, recently developed MEG helmets with the ability to record magnetic fields from the entire scalp simultaneously will partly solve this problem.

As a rule of thumb, the time constant should be at least three times the interstimulus interval (1.5 seconds) to avoid distortion of the data [8]. Due to instrumental setup and noise in the very low frequency band, the measurements were performed using a time constant of 3.3 seconds. It cannot, therefore, be excluded that the time constant we used has influenced the obtained data.

The ability to separate the late component of the CNV from the early component is highly dependent on the interval between the warning and the imperative stimuli. At an inter-stimulus interval of 1.5 seconds, which has been used in this study, there might be a significant contribution from O3 at the time of the E-wave. Although the morphology of the CMV suggests that the magnetic components are more separated that their electric counterparts, and that an ISI of 1.5 seconds would be satisfactory in most cases, a longer ISI would be preferable.

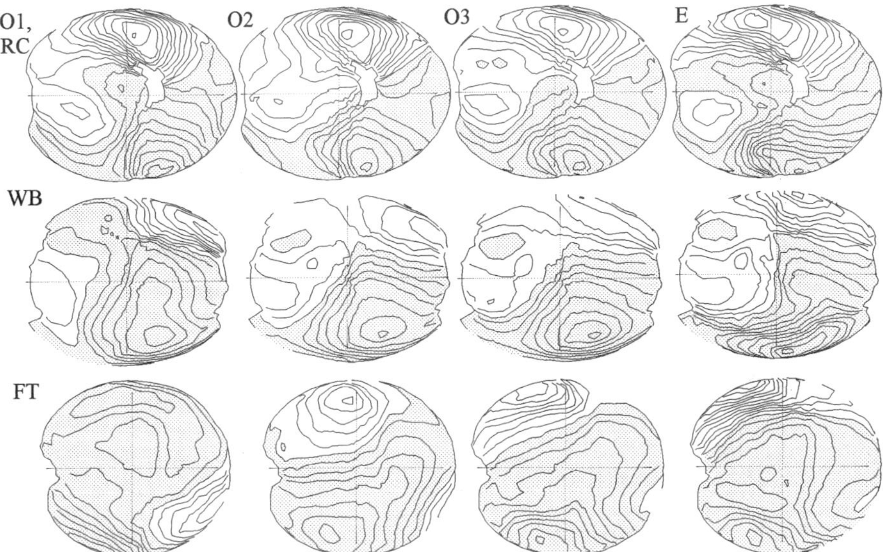

Fig. 2: Field-maps, based on the CMV, from three subjects (RC, WB and FT) at four time periods (O1, O2, O3 and E) corresponding to the slow components of the CMV. The field-maps are seen from the top with the nose directed to the left; shaded areas represent ingoing magnetic flux.

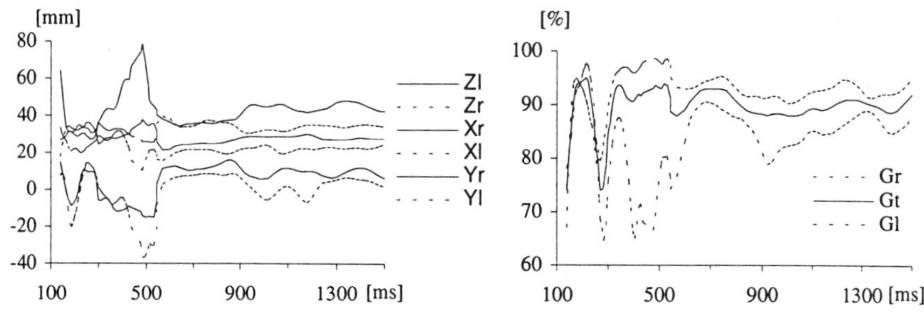

Fig. 3 Localization for least-mean-squares fitted dipoles for one subject, RC, as a function of time (left). The contralateral side and the ipsilateral side were fitted simultaneously. Position coordinates are referred to a Cartesian system corresponding to the ellipsoidal model used. Goodness of fit, G, refers to the degree to which the calculated field could account for the observed data, and is calculated for sensor positions on the contralateral side, Gl, and ipsilateral side, Gr, as well as for all sensor positions, Gt (right). The slow components of RC peaked at 320 ms (O1), 480 ms (O2), 680ms (O3) and 1400 ms (E).

References

[1] Rohrbaugh, J.W., Syndulko, K., and Lindsley, D.B. Brain wave components of the contingent negative variation in humans, Science, 1976, 191: 1055-1057.

[2] Hultin, L., Rossini, P., Romani, G.L., Högstedt, P., Stanzione, P., Tecchio, F., and Pizzella, V. A method for investigation of magnetic fields related to CNV in normal subjects and Parkinson patients, In: Deecke, L., Baumgartner, C., Stroink, G., Williamson, S.J. Biomagnetism: Fundamental research and clinical applications, Elsevier, Amsterdam. 1995.

[3] Hultin, L., P. Rossini, G. L. Romani, P. Högstedt, F. Tecchio and V. Pizzella. Neuromagnetic localization of the late component of the contingent negative variation, Electroencephalogr Clin Neurophysiol, Accepted for publication, 1996.

[4] Torrioli, G., Casciardi, S., Del Gratta, C., Foglietti, V., Gallagher, W.J., Ketchen, M.B., Kleinsasser, A.W., Pasquarelli, A., Pizzella, V., Romani, G.L., Sandstrom, R.L. (1992) Twenty-eight-channel hybrid neuromagnetometer. In: Hoke, M., Erné, S.N., Okada, Y.C., Romani, G.L (Eds.), Biomagnetism: Clinical Aspects. Elsevier, Amsterdam.

[5] Walter, G., Cooper, R., Aldrige, V. J., and McCallum, W. Contingent negative variation: an electric sign of sensorimotor association and expectancy in the human brain, Nature, 1964, 203: 380-384.

[6] Tecce, J.J. Contingent Negative variation and individual differences: a new approach in brain research, Arch Gen Psychiatry, 1971, 24: 1-16.

[7] Peters, J.F., Billinger, T.W., Knott, J.R. Event related potentials of brain (CNV and P300) in a paired associate learning paradigm, Psychophysiology, 1977, 14: 579-585.

[8] Cooper, R., Osselton, and J.W., Shaw, J.C. EEG Technology, London, Butterworth & Co., 1980.

Acknowledgments

This work was supported in part by the Swedish National Board of Industrial and Technical Development (NUTEK) under grant No. 623-90-0183, and by the National Research Council of Italy (CNR) under the Progetto Finalizzato "Superconductive and Cryogenic Technologies." The authors want to thank Dr. Alberto Pasquarelli for adapting the MEG system at CNR to the experimental setup. We also want to express our gratitude to Professor Roland Kadefors for valuable comments on the manuscript.

Dependence of SEFs on the Interstimulus Interval: A Model for SEF Generation Mechanism at the Primary Sensorimotor Cortex

Huttunen, J.[1], Wikström, H.[1], Korvenoja, A.[2], Aronen, H.[2] and Ilmoniemi, R.J.[1]

[1]BioMag Laboratory, Medical Engineering Centre; [2]Department of Radiology, Helsinki University Central Hospital, Helsinki, Finland

Introduction

Somatosensory evoked fields (SEFs) following median nerve stimulation can be recorded from the contralateral primary sensorimotor cortex (SMI), from the second somatosensory cortex (SII) bilaterally and from contralateral posterior parietal cortex (PPC) [1,2]. At SMI, the response can be divided into distinct N20m, P35m and P60m deflections, while SII responses are biphasic with peaks at 80–100 ms and about 150 ms. The latency of the PPC response is usually 80–100 ms.

It is generally agreed upon that the first cortical response, the N20m, represents summated excitatory postsynaptic potentials (EPSPs) in area 3b pyramidal neurones as these are excited at the basal dendrites by thalamocortical axons. However, the neurophysiological mechanisms underlying the later deflections are completely unknown. A better understanding of these mechanisms would obviously improve meaningful interpretations of SEF changes under various physiological or pathophysiological conditions.

The present study reports dependence of the SEFs on the interstimulus interval (ISI) in a population of healthy subjects. Based on remarkable similarities between the ISI dependence of SEFs and the ISI dependence of intracellularly recorded evoked potentials in sensorimotor cortical neurones, as reported in the literature, we present a tentative model for the generation mechanism of the SEFs from SMI.

Methods

A detailed description of the methods will be presented elsewhere [3].

SEFs were recorded in 9 healthy volunteers (20–48 years, 4 women) with a 122-channel planar gradiometer array (Neuromag Ltd.).

The right median nerve was stimulated at the wrist with 0.2-ms constant-current pulses above motor threshold. 100—200 responses were averaged on-line at least twice with each of the ISIs of 0.15, 0.3, 1, 3 and 5 s, which were presented in random order. The subjects watched a movie during the recording with no task related to stimulation.

The signals were band-pass filtered between 0.03 Hz and 310 Hz, digitized at 940 Hz and averaged on-line. Automatic rejection procedures were used to exclude blink and eye-movement artefacts.

A single equivalent current dipole (ECD) was used to model local cortical activity at the peaks of the identified deflections from SMI, SII and PPC. The spherical head model was obtained individually from each subject's magnetic resonance images. The head coordinate system was defined with a digitizer; the head and magnetometer coordinates were aligned with an automatic process defining the head position within the magnetometer dewar. ECDs explaining more than 80% of the field variance in the subgroup of channels (30—40) used for ECD fitting were accepted for analysis.

Results

With long ISIs, the response from the left (contralateral) SMI consisted of clear N20m, P35m and P60m peaks with an intervening 'negative'-going N45m, which usually did not cross the baseline at ISIs over 1 s. The field patterns at the peaks of these deflections were dipolar, the corresponding ECDs pointing in the anterior direction for N20m and in the posterior direction for P35m and P60m. The ECD locations agreed well with the hand area of SMI. The ECDs of the individual peaks were located close to each other (Table 1), P35m, however, being slightly more medially located than N20m in most subjects.

A PPC response could be identified in 7 of the 9 subjects; it is characterized by its more posterior location compared with the earlier peaks from the SMI region. A contralateral SII response was observed in all subjects and an ipsilateral in 7. The SII responses were small in 4 subjects, the corresponding dipole moments being below 10 nAm.

Table 1. Mean peak latencies and ECD coordinates for the different SEF deflections (standard deviations in parentheses). The ISI was 5 s with the exception of N45m, for which the ISI was 0.15 s. The x-axis passed from the left to the right preauricular point; the positive y-axis through the nasion; positive z was thus upwards. PPC = posterior parietal cortex, SIIc and SIIi = contra- and ipsilateral secondary somatosensory cortices, respectively.

Deflection	Latency/ms	x/mm	y/mm	z/mm
N20m	21 (1.6)	-37 (2.6)	12 (7.1)	94 (7.0)
P35m	35 (3.7)	-34 (6.2)	13 (7.3)	99 (9.1)
N45m	46 (8.2)	-37 (6.0)	16 (9.8)	95 (8.9)
P60m	69 (18.3)	-35 (5.4)	14 (7.4)	100 (8.6)
PPC	88 (19.0)	-35 (6.8)	1 (9.1)	106 (10.1)
SIIc	104 (9.5)	-29 (35.4)	19 (13.0)	68 (5.4)
SIIi	110 (7.2)	38 (11.6)	21 (20.0)	67 (6.8)

With shortening ISI, N20m remained stable from 5 to 0.3 s, but was slightly attenuated at 0.15 s (Fig.1, Table 2). P35m and P60m were largest at the 5-s ISI and showed steady declines toward shorter ISIs, being nearly undetectable at the ISI of 0.15 s. We should like to emphasize that both were diminished between 5-s and 1-s ISIs, *i.e.*, at relatively long ISIs. With the redution of P35m and P60m, the N45m showed a gradual increase with shortening ISI. At 0.15 s the N45m dominated the SMI responses.

Both the PPC and the SII responses could only be seen at ISIs of 1 s or longer. On the average, these were largest at the 5-s ISI, declining rapidly toward shorter ISIs. The P35m, P60m, PPC and SII deflections were not clearly saturated at the 5-s ISI. The ECD locations did not depend on ISI.

Table 2. The average ECD strengths in nAm (standard deviations) of the SEF deflections at different ISIs. The F statistics were obtained from an analysis of variance for repeated measures.

Defletion	5 s	3 s	1 s	3 s	0.15 s	F(4,28)	p
N20m	15 (1)	15 (2)	18 (2)	13 (2)	9 (2)	5.8	0.002
P35m	28 (2)	27 (2)	19 (2)	9 (2)	6 (2)	48.0	0.000
N45m	10 (5)	12 (5)	10 (4)	19 (5)	19 (3)	2.7	0.05
P60m	33 (1)	28 (2)	21 (2)	9 (3)	9 (3)	16.8	0.000
PPC	17 (4)	9 (3)	4 (2)	-	-	8.2	0.000
SIIc	20 (6)	12 (4)	7 (4)	6 (3)	1 (1)	5.2	0.003

Discussion

Intracortical microelectrode studies, the anatomy of thalamocortical connections, previous SEP and SEF studies all indicate that the N20m deflection represents summated excitatory postsynaptic potentials (EPSPs) from layer IV or IIIb pyramidal neurones of Brodmann's area 3b [cf. 3]. No explicit models have been presented, however, to account for the physiological mechanisms underlying any of the later deflections from SMI, or of the PPC and SII responses.

Evoked intracellular inhibitory postsynaptic potentials (IPSPs) are known to be very sensitive to repetitive stimulation with short ISIs [*e.g.*, 4–6]. This decline of IPSPs occurs already with ISIs in the range of seconds [5]; stimulation with ISIs shorter than 0.2 s can completely abolish phasic IPSPs [4]. This ISI behaviour of IPSPs is strikingly similar to the ISI behaviour of P35m and P60m. Furthermore, the latencies of P35m and P60m are consistent with latencies of fast and slow IPSPs, respectively, in intracellular recordings. The ECD orientations of P35m and P60m are consistent with current flow from cortical surface toward cortical depth at area 3b. Such direction of current flow can be caused by active current sources near the cell somata of pyramidal neurones within the middle layers of area 3b, i.e., by soma-near inhibitory currents.

In intracellular recordings from the neurones at SMI, shortening the ISI of afferent stimulation to 0.1—0.2 s results in a strengthening of a secondary EPSP, which follows the primary EPSP by about 20 ms [*e.g.*, 6]. This secondary EPSP coincides with a prominent increase of the field potentials at ISI of 0.1–0.2 s, known as the 'augmenting' response. Such behaviour is similar to the increase of N45m SEF deflection with shortening of the ISI. The nearly identical ECD location of N45m with that of N20m and the orientation of the N45m ECD are consistent with EPSPs arising in much the same neuronal population that is initially excited during N20m.

In summary, the ISI behaviour of the SEFs from SMI suggests the following tentative model: N20m reflects primary EPSP, P35m fast IPSPs, N45m secondary EPSPs and P60m slow IPSPs in the pyramidal neurones of Brodmann's area 3b. While the ISI behaviour alone obviously does not provide direct evidence to support the model, pharmacological interventions, e.g., with agonists and antagonists of γ-amino-butyric acid, could be used to test the hypothesis in humans or in an animal model.

On the basis of the ISI behaviour, similar hypotheses appear more difficult for the long-latency PPC and SII responses.

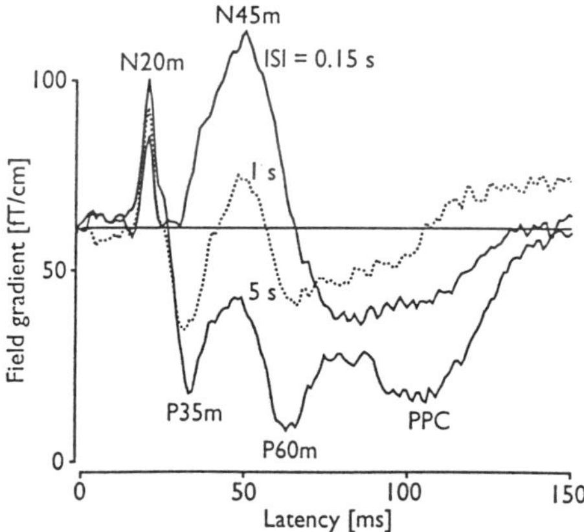

Figure 1. The longitudinal field gradients of SEFs from one channel overlying the left SMI at different ISIs. Note the marked reductions of P35m and P60m already between 5-s and 1-s ISI, and the enhancement of N45m toward the short ISIs.

References:

[1] Hari, R., Reinikainen, K., Kaukoranta, E., Hämäläinen, M., Ilmoniemi, R., Penttinen, A., Salminen, J., and Teszner, D. Somatosensory evoked cerebral magnetic fields from SI and SII in man. Electroenceph. clin. Neurophysiol., 1984, 57: 254—263.

[2] Forss, N., Hari, R., Salmelin, R., Ahonen, A., Hämäläinen, M., Kajola, M., Knuutila, M., and Simola, J. Activation of the human posterior parietal cortex by median nerve stimulation. Exp. Brain Res., 1994, 99: 309-315.

[3] Wikström, H., Huttunen , J., Korvenoja, A., Virtanen, J., Salonen, O., and Ilmoniemi, R.J. Effects of interstimulus interval on somatosensory evoked magnetic fields: a hypothesis concerning SEF generation at the primary sensorimotor cortex.. Electroenceph clin Neurophysiol., 1996, submitted for publication.

[4] Nacimiento, A.C., Lux, H.D., and Creutzfeldt, O.D. Postsynaptische Potentiale von Nervenzellen des motorischen Cortex nach elektrischer Reizung spezisfischer und unspezifischer Thalamuskerne. Pfluegers Archiv, 1964, 281: 152—169.

[5] Deisz, R.A., and Prince, D.A. Frequency-dependent depression of inhibition in guinea-pig neocortex *in vitro* by $GABA_B$ receptor feed-back on GABA release. J. Physiol., 1989, 412: 513—541.

[6] Klee, M. R. Different effects on the membrane potential of motor cortex units after thalamic and reticular stimulation, In: Purpura, D.P., and Yahr, M.D. The Thalamus, New York, Columbia University 1966, pp. 287—322.

Somatosensory Evoked Fields from SMI and SII During 'Interfering' Finger Movements and Tactile Stimulation

Huttunen, J.[1], Wikström, H.[1], Korvenoja, A.[2], Aronen, H.[2] and Ilmoniemi, R.J.[1]

[1]*BioMag Laboratory, Medical Engineering Centre;* [2]*Department of Radiology, Helsinki University Central Hospital, Helsinki, Finland*

Introdution

Ascending sensory impulses play an important role in providing feedback information necessary for the programming of movements. Therefore, it could be expected that sensory input to the cortex is facilitated from a moving limb, as demonstrated in monkeys [1].

During movements, however, the conscious perception of cutaneous stimuli is dampened. Furthermore, the early cortical somatosensory evoked potentials (SEPs) are reduced during various types of active or passive finger movements [e.g., 2]. The effect of passive movement indicates that the observed changes of SEPs could be due to occlusion or inhibition by reafferent impulses associated with the movements rather than due to the ongoing motor program. This view is supported by the fact that continuous cutaneous stimulation similarly suppresses the SEPs as does movement [3]. Hence, SEPs have shown no evidence for an enhancement of sensory input to the primary sensorimotor cortex (SMI), which would specifically correspond to enhanced input during active movements.

The somatosensory evoked fields (SEFs) bear the advantage over SEPs that the underlying generators can be better identified. Consequently, using SEFs it has been possible to realiably locate separate generators at SMI and at the second somatosensory cortex (SII) [4], and recently also at the posterior parietal cortex (PPC) [5].

The functional significance of SII has remained relatively poorly understood. However, some observations suggest that SII can play a role in performing active skilful movements requiring sensory feedback [6]. To date, the sensory responsiveness of SII neurones has been shown to be enhanced in monkeys when a behaviourally important motor response is expected [7].

In the present study, we wanted to determine whether tactile interfering stimulation and finger movements have differential effects on the sensory responsiveness of SMI and SII. To achieve this, we recorded SEFs to electrical median nerve stimulation in four conditions. In the first condition the subjects had no task related to the stimulation (resting condition). In the second condition the subjects performed active finger movements continuously during the nerve stimulation. In the third and fourth conditions interfering continuous tactile stimulation was applied either to the same or to another part of the body which was being stimulated electrically.

Methods

Somatosensory evoked fields (SEFs) to alternating bilateral median nerve stimulation at an interstimulus interval of 1.3 s (2.6 s per each nerve) were recorded with a 122-channel helmet-shaped planar gradiometer (Neuromag Ltd.) in 6 healthy volunteers. SEFs were obtained at least twice in a resting condition with the subject watching a film and being motionless, and in the following test conditions:

(1) Finger opposition: the subjects touched with the tip of the right thumb the tips of the other digits in a specified sequence,

(2) Tactile-palm: the right palm of the subject was brushed by the experimenter with a cotton pad,

(3) Tactile-face: the right lower face region of the subject was brushed in the same way.

Conditions 1–3 will be called interference conditions; also during these the subject watched a film. Because the interference condition was always applied to the right side, we could assess SEF modulations separately when the same part of the body (right median nerve) or the homologous site on the opposite side of the body (left median nerve) was stimulated electrically to elicit SEFs.

The recording pass-band was 0.03–310 Hz. Signals were digitized at 940 Hz, and averaged on-line after passing an automatic artefact rejection algorithm. The analysis time extended from 50 ms before stimulus to 350 ms beyond it.

The single-dipole model was used in the analysis. Dipole fitting to the SII responses was done after subtracting the influence of simultaneous activity at SMI from the responses with a signal-space projection method. The head was modelled with a sphere obtained from individual magnetic resonance images. The N20m, P35m and P60m SEFs from left and right SMI, a P100m generated near SMI but possibly getting contribution from the posterior parietal cortex [5], and the bilateral SII responses after stimulation on each side were analyzed. Statistical differences were evaluated with the Wilcoxon matched-pairs signed ranks test.

Results

As described in earlier studies, the SMI responses could be divided into distinct N20m, P30m and P60m peaks. The corresponding field patterns were dipolar with equivalent current dipoles (ECDs) located at the SMI hand area close to each other. The SII responses had 2 peaks at 65–100 ms and at 120–170 ms; the 1st of these was used in the analysis. P100m showed a dipolar field pattern with ECDs lying near SMI, which tended, however, to be 1–2 cm medial and posterior to the N20m ECD, thus possibly being generated in the posterior parietal cortex [5].

1. Responses from the left SMI to right median nerve stimulation.
In this case, both the electrical stimulation used to elicit SEFs and the interferences were applied to the contralateral (right) hand. Tactile interference at the palm and finger opposition had qualitatively similar effects (Fig. 1): both induced a marked reduction of P35m (mean ECD strengths 21 nAm at rest, 5 nAm during finger opposition and 8 nAm during tactile-palm, $p<0.05$ for both conditions), a smaller decrease of N20m (mean ECD strengths 14 nAm at rest, 11 nAm during finger opposition and 7 nAm during tactile-palm, $p<0.1$ for finger opposition and $p<0.05$ for tactile-palm), and no significant change in P60m. P100m was markedly reduced during finger opposition and tactile-palm (mean ECD strengths 18 nAm at rest, 9 nAm during finger opposition and 7 nAm during tactile-palm, $p<0.05$ for both). P100m was, however, difficult to define as a separate deflection in some subjects, especially during the interference conditions; the changes of P100m nevertheless reflect altered activity around 100 ms after stimulus.

Tactile interference to the right side of the face had no significant effect at the group level, although P35m was enhanced in 2 subjects. It is of interest to note that tactile interference to the palm appeared to reduce N20m slightly more than did finger opposition, although P35m was even more reduced during finger opposition than during tactile interference.

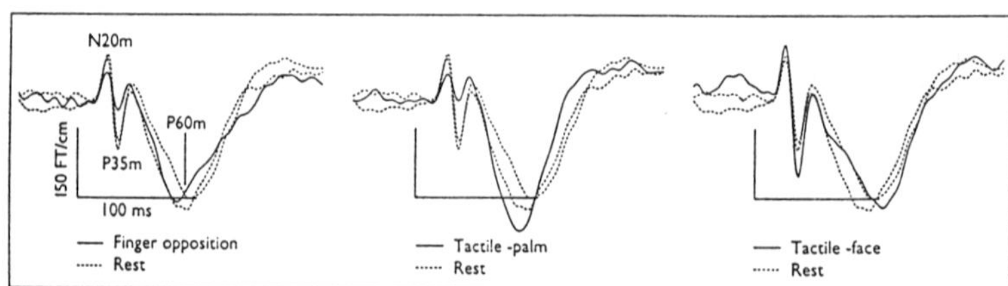

Figure 1. Longitudinal SEF gradients from the channel with maximum signals from the left SMI after right median nerve stimulation in one subject. Note reductions of N20m and especially P35m during finger opposition and tactile stimulation of the palm. In this subject, P35m was enhanced during tactile stimulation of the face.

2. Responses from the right SMI to left median nerve stimulation

In this situation, the interferences were applied to the ipsilateral hand, while SEFs were triggered by stimulation of the contralateral (left) median nerve. The SEFs were similarly affected by all 3 conditions: P35m was slightly increased (mean ECD strengths 19 nAm at rest, 27 nAm during finger opposition, 26 nAm during tactile-palm and 24 nAm during tactile-face, $p<0.05$ for all interferences), while N20m, P60m and P100m were not significantly affected.

3. Responses from SII after right median nerve stimulation

Here the interferences and electrical trigger stimuli were applied to the same hand. The SII responses from right and left SII regions (ipsi- and contralateral to median nerve stimulation) behaved in the same way in all interference conditions. During finger opposition, the SII responses were enhanced whereas both tactile interference of the palm and face markedly reduced them (Fig. 2; mean ECD strengths for the 80–100 ms deflection at the contralateral SII were 11 nAm at rest, 23 nAm during finger opposition, 7 nAm during tactile-palm and 5 nAm during tactile-face, $p<0.05$ for all interferences).

4. Responses from SII after left median nerve stimulation

When the left median nerve was stimulated, the interferences and median nerve stimulation were on the opposite sides. In this situation, tactile interference to the right palm and face again suppressed the SII responses bilaterally (mean ECD strengths at the contralateral SII were 15 nAm at rest, 1 nAm during tactile-palm and 2 nAm during tactile-face, $p<0.05$ for both interferences). However, right-sided finger opposition did not enhance the SII responses, but rather tended to reduce them (mean ECD strength 8 nAm, $p>0.1$). Responses from both left and right SII behaved again in the same way in all conditions.

In summary, at contralateral SMI, tactile interference applied to the palm and finger opposition movements markedly reduced P35m and P90m, caused a smaller reduction in N20m, and had no significant effect on P60m. Tactile interference to the right side of the face had no significant effects. At the ipsilateral SMI, tactile interference to the palm, to the face and finger opposition movements all caused a significant increase in P35m, while the other deflections were not affected. At SII, tactile interferences reduced the responses regardless of whether they were applied to the same or to the opposite side of the body compared with the electrical median nerve stimuli. However, finger opposition movements enhanced the SII responses, when performed with the same hand that was being stimulated to elicit SEFs, but not when performed with the opposite hand.

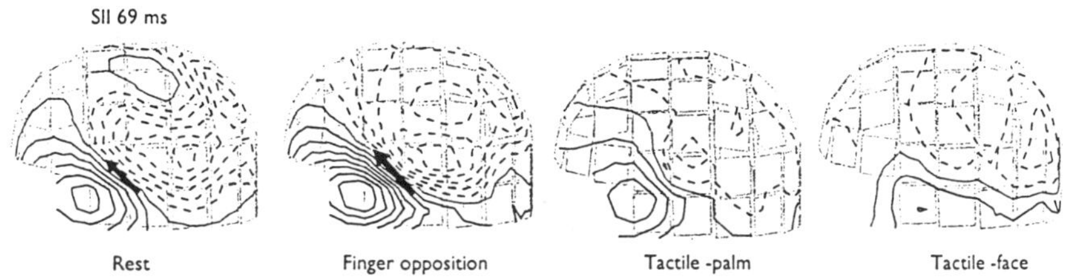

Figure 2. Isofield contour maps and surface projections of corresponding ECDs from the left SII region to right median nerve stimulation at rest, during right-hand finger opposition, tactile stimulation of the right palm and right face. Simultaneous activity at the left SMI has been subtracted. Note the increase of the response during finger opposition, but reductions during tactile interfering stimulations. In the tactile-face condition, no successful dipole fitting could be achieved in the SII region.

Discussion

Because of the similar modifications of SEFs from the contralateral SMI during finger opposition movements and tactile interference, the results do not disclose a modification of afferent input to SMI that would be specific to movement and could be interpreted as enhancement of sensory transmission from the moving limb to the SMI cortex [1]. This conclusion is in agreement with recent reports also describing similar SEF modulations at SMI during movements and during passive touching of objects [8,9]. One reason for a lack of movement-specific enhancement of SEFs at SMI could be that the responsiveness is increased only within the motor cortex, which does not necessarily participate in SEF generation.

The enhancement of P35m from the ipsilateral SMI during tactile interference was also reported recently [11]. In that study the effects in the ipsilateral hemisphere were interpreted as being due to interhemispheric interactions via callosal fibers. However, our finding that the enhancement of P35m was not caused only by tactile interference applied to the hand but also to the face, suggests that the effect could be more non-specific in nature. It is difficult to explain the effect of tactile interference to the face as being caused by transcallosal facilitation or inhibition, because the callosal fibers are known to interconnect mainly homologous parts of the body. Because all 3 interference conditions similarly enhanced the ipsilateral P35m, we suspect that this effect could result from an overall change in cortical excitability, e.g., in the average membrane potential level, possibly due to a change in the arousal of the subject.

At SII, the reductions of the responses during tactile interferences regardless of the side of median nerve stimulation and the site of the interference is consistent with the well-known property of large bilateral receptive fields of the SII neurones. During tactile interferences, the continuous afferent bombardment probably renders the neurones refractory to intervening impulses due to electrical nerve stimuli. The enhancement of the SII responses to right, but not to left, median nerve stimulation during right-sided finger movements suggests that the SII neurones are 'tuned' to respond more briskly to afferent impulses from the moving limb than from a non-moving limb. This feature indicates that SII plays an important role in integrating relevant sensory inputs to ongoing motor programs.

References:

[1] Evarts, E.V., and Fromm, C. The pyramidal neuron as summing point in a closed-loop control system in the monkey, In: Desmedt, J.E.. Cerebral Motor Control in Man: Long Loop Mechanisms, Basel, Karger, 1978, pp. 56-69.

[2] Huttunen, J., and Hömberg, V. Modification of cortical somatosensory evoked potentials during tactile exploration and simple active and passive movements, Electroenceph. clin. Neurophysiol., 1991, 81:216-223.

[3] Jones, S.J., and Power, C.N. Scalp topography of human somatosensory evoked potentials: the effect of interfering tactile stimulation applied to the hand, Electroenceph. clin. Neurophysiol., 1984, 58: 25-36.

[4] Hari, R., Reinikainen, K., Kaukoranta, E., Hämäläinen, M., Ilmoniemi, R., Penttinen, A., Salminen, J., and Teszner, D. Somatosensory evoked cerebral magnetic fields from SI and SII in man, Electroenceph. clin. Neurophysiol., 1984, 57: 254-263.

[5] Forss, N., Hari, R., Salmelin, R., Ahonen, A., Hämäläinen, M., Kajola, M., Knuutila, J., and Simola, J. Activation of human posterior parietal cortex by median nerve stimulation, Exp. Brain Res., 1994, 99: 309-315.

[6] Caselli, R.J. Ventrolateral and dorsomedial somatosensory association cortex damage produces distinct somesthetic syndromes in humans, Neurology, 1993, 43: 762-771.

[7] Poranen, A., and Hyvärinen, J. Effects of attention on multiunit responses to vibration in the somatosensory regions of the monkey's brain, Electroenceph. clin. Neurophysiol., 1982, 53:525-537.

[8] Schnitzler, A., Witte, O., Cheyne, D., Haid, G., Vrba, J., and Freund, H.-J. Modulation of somatosensory evoked magnetic fields by sensory and motor interferences, NeuroReport, 1995, 6: 1653-1658.

[10] Kakigi, R., Koyama, S., Hoshiyama, M., Watanabe, S., Shimojo, M., and Kitamura, Y. Gating of somatosensory evoked responses during active finger movements: magnetoencephalographic studies, J. Neurol. Sci., 1995, 128: 195-204.

[11] Schnitzler, A., Salmelin, R., Salenius, S., Jousmäki, V., and Hari, R. Tactile information from the human hand reaches the ipsilateral primary somatosensory cortex, Neurosci. lett., 1995, 200: 25-28.

Time Course of Neuromagnetic Signals Related to Reading Japanese Phonograms

Imada, T.[1,2], Kawakatsu, M.[3], Ikeya, A.[3], Kadota, R.[3], Miyamoto, J.[3], Nakayama, S.[3] and Kotani, M.[3]

[1]Basic Research Laboratories, Nippon Telegraph and Telephone Corporation, Atsugi, Japan; [2]Low Temperature Laboratory, Helsinki University of Technology, Espoo, Finland; [3]Dept. of Electronic Engng., Faculty of Engng., Tokyo Denki University, Tokyo, Japan

Introduction

The Japanese language has three writing systems: kanji script, or ideograms, hiragana script, or phonograms and katakana script, another type of phonograms. Modern Japanese has 71 hiragana and 71 katakana characters, which have one-to-one correspondence. A hiragana character has only one phonetic value, which represents a mora, or a syllable, in spoken Japanese. A recent positron emission tomography (PET) study [1] reported that overt reading of words written in hiragana activated bilaterally the medial and lateral occipital gyri, the posterior inferior temporal areas, Heschl's gyri, the posterior inferior frontal areas (including Broca's area), and the supplementary motor area. In addition, unilateral activation was seen in the left basal ganglia and the right thalamus. This PET study, however, did not reveal the temporal dynamics of the hiragana reading process. We thus attempted to follow the spatiotemporal activation sequence of cortical signals while subjects read hiragana.

Methods

Visually evoked cortical magnetic fields were recorded from five healthy right-handed Japanese males (22-46 years) with a whole-scalp Neuromag-122™ magnetometer [2] in a magnetically shielded room at the Helsinki University of Technology. We employed a visual attention paradigm, where the subject, looking at the fixation point in the middle of the screen, attended to the left visual field (LVF) in one block and to the right visual field (RVF) in the other block. Stimuli were presented, in random order within one block, either to the attended visual field or to the unattended visual field. To the attended visual field, two of 43 hiragana characters (Fig. 1a) were presented, and the subject was instructed to release a response key when these two hiragana had identical consonants in Japanese pronunciation, say /sa/ and /shi/, and not to release the key when they had different consonants. To the unattended visual field, two of 43 deformed hiragana characters (Fig. 1b) were presented; these could not be identified as hiragana and thus could not be read. The presentation probability was 0.4 for the hiragana with different consonants, 0.2 for the hiragana with the same consonant, and 0.4 for the deformed hiragana. Stimulus duration was 170 ms and the interstimulus interval varied randomly between 1.05 and 1.2 s.

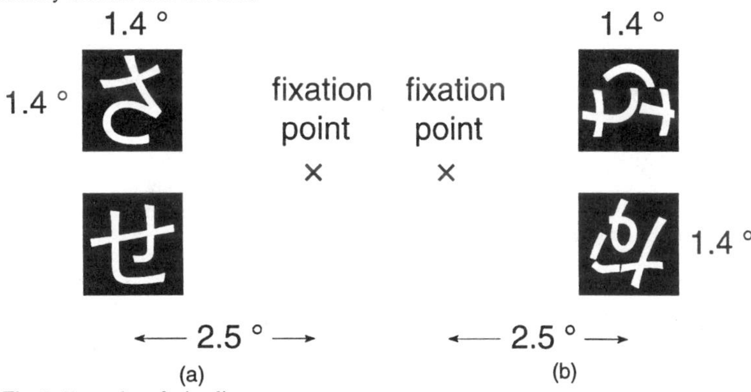

Fig. 1 Examples of stimuli.
(a) An example of the hiragana stimuli presented to LVF.
(b) An example of the deformed hiragana stimuli presented to RVF.

Responses (passband 0.03-100 Hz) were digitized at 0.5 kHz. Epochs with electro-oculograms exceeding 100 µV or with magnetic fields beyond 3000 fT/cm were rejected. More than 100 epochs were averaged separately according to the following conditions: (A) read hiragana in LVF, (B) read hiragana in RVF, (C) ignore deformed hiragana in LVF, and (D) ignore deformed hiragana in RVF. Responses to hiragana with the same consonant were not analyzed because they included motor reactions. Averaged responses were digitally lowpass-filtered at 40 Hz. The amplitude was measured with respect to a 100-ms prestimulus baseline. Equivalent current dipoles (ECDs), searched using a spherical head model, were selected according to the following four criteria [3]: *(i)* goodness-of-fit > 80%, *(ii)* 95%-confidence-volume < 268 mm^3, *(iii)* chi square probability* P_{obs} > 0.95, and *(iv)* continuous activation of the dipole for at least 10 ms. If the chi square probability is close to 1.0, the single dipole model is considered adequate.

Results

Figure 2 shows response waveforms of subject S2. In the occipital area, a strong deflection was observed between 100 and 200 ms, with larger peak amplitude in the "read" than in the "ignore" condition. In the left temporal area, from about 200 through 600 ms, the hiragana reading elicited slow activity, which was not seen in the "ignore" condition. All subjects showed the same tendency, and the maximum peak amplitude between 106 and 169 ms was significantly larger in the "read" than in the "ignore" condition for LVF stimuli (78 and 66 fT/cm; p < 0.01), but was not significant for RVF stimuli (98 and 91 fT/cm). At 80-180 ms, all subjects showed ECDs in one or more of the following locations: the contralateral occipital regions near the calcarine fissure, the bottom of the occipital lobe, the cuneus and the precuneus. From about 150 ms onwards, ECDs appeared in the ipsilateral occipital cortex. Figure 3 illustrates the time course of the activation in subject S2 in the "read" condition. At 88-138 ms, the activation moved from the area close to the right calcarine fissure to the right lateral cortex near the fusiform gyrus, where it stayed at 144-166 ms. At 156-166 ms, another activity was observed in the left lateral occipital cortex close to the left temporal cortex.

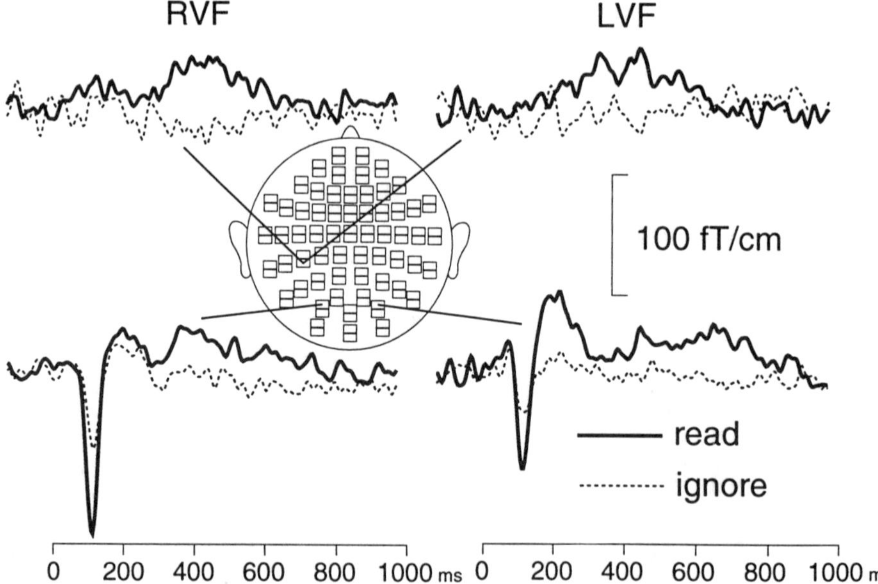

Fig. 2

Responses of subject S2 responding to attended hiragana (solid line) and ignored deformed hiragana (dashed line). The stimulus was presented to the RVF and LVF.

* $P_{obs} = P(\chi^2_{n-r} < \chi^2_{obs})$, where χ^2_{n-r} is the value of the chi square distribution with the freedom of n-r (n: the number of channels, r: the number of ECD parameters), and $\chi^2_{obs} = \Sigma^n_{i=1} (B_i - B'_i)^2 / \sigma_i^2$, where B_i is the magnetic signal measured at channel i, B'_i is the magnetic data calculated from the estimated ECD, and σ_i^2 is variance of the noise in channel i.

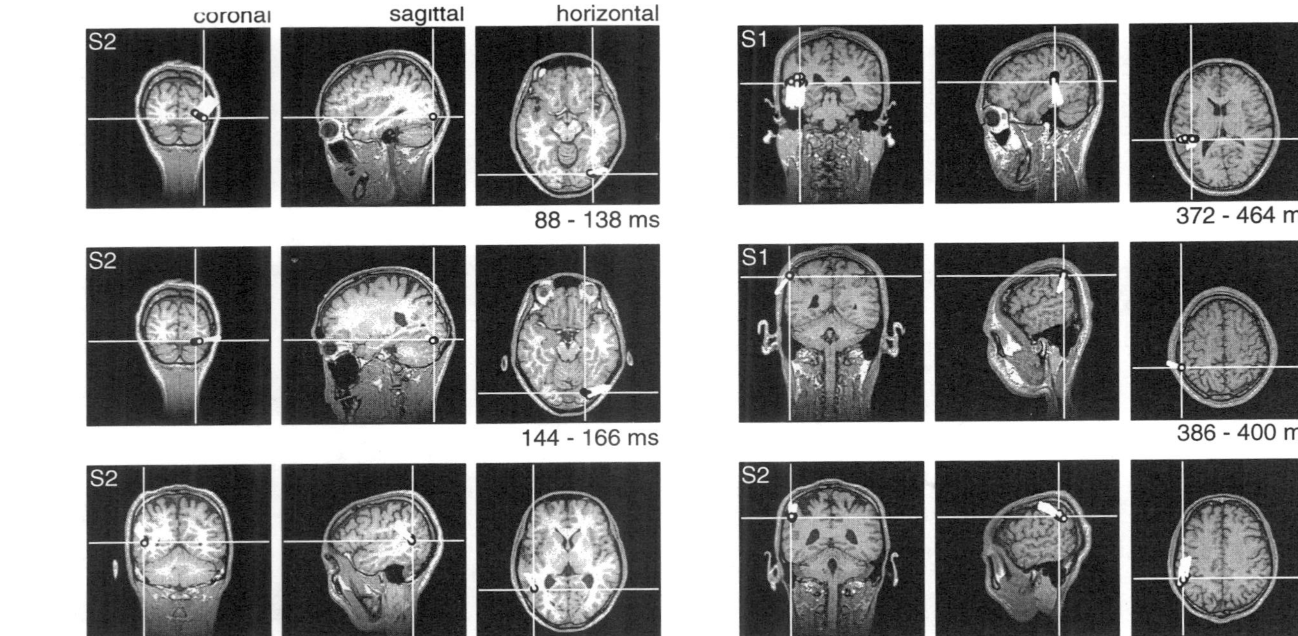

Fig. 3

ECD clusters from subject S2 until 200 ms after hiragana was presented to the LVF. The sources are projected on the subject's magnetic resonance images. The activation time period is indicated below the images. The ECD clusters are depicted by small white circles. The white lines starting at these circles indicate the direction of the ECD current .

Fig. 4

ECD clusters from subjects S1and S2 200-500 ms after the hiragana presentation. The activation time period is indicated below the images.

Magnetic resonance images were taken at the Communications Research Laboratory, Ministry of Posts and Telecommunications.

In subject S2, at 200-500 ms, the left hemisphere was most clearly activated by the hiragana characters in the LVF and RVF. In fact, for LVF stimuli, there were 16 ECD clusters in the left hemisphere while there were two ECD clusters in the right. In the same subject, the left posterior inferior frontal cortex was activated at 276-328 ms by the LVF hiragana and at 292-310 ms by the RVF hiragana. Then, the left parietal cortex near the supramarginal gyrus and the angular gyrus became active 392-462 ms after the LVF hiragana presentation and 382-398 ms after the RVF hiragana presentation.

In all subjects, the common areas activated by hiragana in the LVF and RVF were mostly in the left hemisphere; the cortices around the supramarginal and the angular gyri (S1 386-400 ms, S2 380-472 ms, S5 214-338 ms), the posterior part of the floor of the Sylvian fissure (S1 374-394 ms, S5 350-414 ms), the superior temporal sulcus (S4 438-458 ms, S5 284-342 ms), the posterior part of the inferior temporal sulcus (S4 410-480 ms), and the lateral occipital cortex (S4 126-136 ms). Figure 4 illustrates these common areas from subjects S1 and S2. Two other subjects displayed right-hemisphere activity as well: in the primary visual cortex (S4 166-214 ms and 388-398 ms) and in the cortices around the supramarginal and the angular gyri (S5 198-212 ms and 286-346 ms). The subject-by-subject comparison showed that the left parietal cortex, the left superior temporal sulcus and the posterior floor of the left Sylvian fissure were active 200-500 ms after the hiragana presentation, but were not activated by deformed hiragana in the unattended visual field. The ignore condition produced fewer ECD clusters for both LVF stimuli (read: 22.6, ignore: 3.8, p=0.03) and RVF stimuli (read: 29.4, ignore: 11.8, p=0.066). In the left hemisphere in particular, the difference was significant for both LVF stimuli (read: 15.0, ignore: 2.2, p=0.044) and RVF stimuli (read: 17.8, ignore: 7.0, p=0.04).

Discussion

The results can be summarized as follows: Reading hiragana characters activates, until about 200 ms, mainly the occipital visual cortices. After 200 ms, the activation moves to the left hemisphere, including areas related to visual and/or language processing (for example, the cortices around the supramarginal and the angular gyri, the posterior part of the floor of the Sylvian fissure, and the superior temporal sulcus). Neuropsychological studies have shown that patients with alexia and/or aphasia often show damages in these cortical regions [4,5], suggesting their involvement in reading. Thus, after preprocessing in the occipital visual cortices, hiragana reading might continue in these regions for the duration of 200-500 ms after the stimulus onset. These claims are also supported by the results in which the "ignore" condition produced far fewer ECD clusters, especially in the left hemisphere.

In the PET study of Sakurai et al. [1], however, the activity in the left angular gyrus was significantly decreased by the task of overt hiragana reading. This discrepancy might be explained by the difference between their task and ours.

The activated areas varied depending on the subject, although all areas belonged to the visual and/or language processing areas. This variability can be explained by assuming that all these areas were more or less activated in all subjects but the dipoles producing weak magnetic fields outside the head were ignored by the applied strict screening criteria.

References

[1] Sakurai, Y., Momose, T., Iwata, M., Watanabe, T., Ishikawa, T. and Kanazawa, I. Semantic process in kana word reading: activation studies with positron emission tomography. NeuroReport, 1993, 4: 327 - 330.

[2] Ahonen, A., Hämäläinen, M.S., Kajola, M.J., Knuutila, J.E.T., Laine, P.P, Lounasmaa, O.V., Parkkonen, L.T., Simola, J.T. and Tesche, C.D. 122-Channel SQUID instrument for investigating the magnetic signals from the human brain. Physica Scripta, 1993, T49: 198-205.

[3] Hämäläinen, M.S., Hari, R., Ilmoniemi, R.J., Knuutila, J., and Lounasmaa, O.V. Magnetoencephalography -- theory, instrumentation, and applications to noninvasive studies of the working human brain. Reviews of Modern Physics, 1993, 65: 413 - 497.

[4] Sasanuma, S. Kana and kanji processing in Japanese aphasics. Brain and Language, 1975, 2: 369 - 383.

[5] Iwata, M. Kanji versus Kana; Neuropsychological correlates of the Japanese writing system. Trends in Neurosciences, 1984, 7: 290 - 293.

Acknowledgement

We are very grateful to Professor Riitta Hari (Low Temperature Laboratory, Helsinki University of Technology) for her helpful discussions and valuable comments on our draft.

Electric and Magnetic Brain Responses Evoked by Deviant Tones in Melodies

Jürgens, R.[1], Lopez, L.[2,3], Diekmann, V.[1], Becker, W.[1], Ried, S.[1], Grözinger, B.[1], and Erné, S.N.[2]

[1]Sektion Neurophysiologie, Universität Ulm, 89081 Ulm, Germany; [2]Zentralinstitut für Biomedizinische Technik, Universität Ulm, 89081 Ulm, Germany; [3]Instituto Tecnologie Avanzate Biomediche, Universita "G.D'Annunzio", Chieti, Italy

Introduction

The perception of deviant tones in an otherwise stereotyped sequence of auditory stimuli creates a characteristic mismatch activity (MMA) in the auditory cortex, which typically has a latency of about 150 ms and which can be recorded with EEG or MEG (mismatch negativity, MMN or mismatch field, MMF) [1]. MMA is generally viewed as reflecting the "hits" of an ever ongoing search for novel events based on the comparison of current events with a memory trace of the preceding ones, that is, on *short term memory* [2]. Alternatively, however, one could conceive of a more generalized mechanism based on the brain's ability to compare actual with *predicted* events. To pursue this question we investigated whether detection of "wrong" tones in pieces of classical music would also evoke MMA.

Methods

20 naive subjects (Ss) took part in the experiment; eleven rated themselves as musically ungifted and/or unknowledgeable while the others were performing amateurs. All Ss were given Bentley's test of musical talents (an established procedure in German schools, comprising pitch discrimination, resolution of chords, memorizing melodies and rhythms) [3]; its score differed significantly between the gifted and ungifted groups.

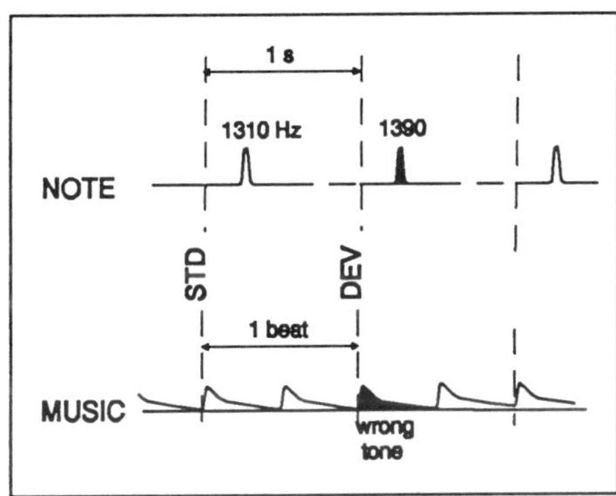

Figure 1. Principle of stimulus presentation. Traces represent sound amplitude as a function of time.

NOTE: typical odd-ball paradigm with standard (STD, 300 repetitions) and deviant (DEV, 100 randomized repetitions) single frequency tones (40 ms duration, 10 ms rise/fall time). 1 s inter-stimulus interval.

MUSIC: continuously played pieces of classical music, synthetized piano sound, interspersed with wrong tones (DEV, 1/2 tone above correct, STD tones) see text for further details.

There were three NON-MUSIC paradigms in which frequent events (75% of cases; STD) alternated randomly with rare, deviant ones (25%; DEV). Three types of stimuli were used: TONE (STD=1310 Hz, DEV= 1390 Hz, cf. Fig. 1), CHORD (three simultaneous tones; STD=1040,1310,1560 Hz, DEV=1040,1390,1560 Hz) and ARPEGGIO (three consecutive tones, frequencies as in CHORD). The MUSIC paradigms comprised a simple and well known melody (MOZART, 40 repetitions of the same piece with a total of 478 beats, of which 120 started with a wrong tone, cf. Fig. 1) and a complex melody (BACH, 15 repetitions, 508 beats, 120 DEVs). All tones were generated by a sound blaster and presented binaurally with 60 dB HL. Ss were to listen to the stimuli and to count the number of deviations.

EEG (19 channels according to 10-20 system and 6 extra electrodes around C3/C4; 64 Hz / 1.0 s; averaged ears as reference) and MEG (DORNIER System, 22 software gradiometers operated in shielded room; DC - 64 Hz) were simultaneously recorded at 500 Hz sampling rate. Experiments were split into two sessions with the MEG aimed at the right and left hemispheres, respectively. STD and DEV events were selectively averaged off-line, based on their different trigger channel attribution.

Results

In all Ss, STD and DEV stimuli elicited typical auditory N100/N100m responses in EEG and MEG; however, there was considerable idiosyncratic variability with regard to amplitude and, for N100m, also to shape. The BACH melody evoked an almost sinusoidal response due to the high tone repetition frequency; its amplitudes ranged from 0.5 to 1.5 μV and were larger for Ss with high Bentley score (ANOVA, p=0.007).

MMA could be recognized in the majority of our Ss, even for the MOZART and BACH paradigms. As an example, the responses of one S to MOZART are depicted in Figure 2. There are three N100/N100m responses corresponding to the three tones occurring during the depicted 900-ms epoch. Following the first N100/N100m wave, a clear MMA elicited by the wrong tone at the sweep onset is recognized (shaded). The wrong tone evoked also a late P300 activity (typically over electrode Pz) and a P300-related MEG response (which was lacking in a number of other Ss, however).

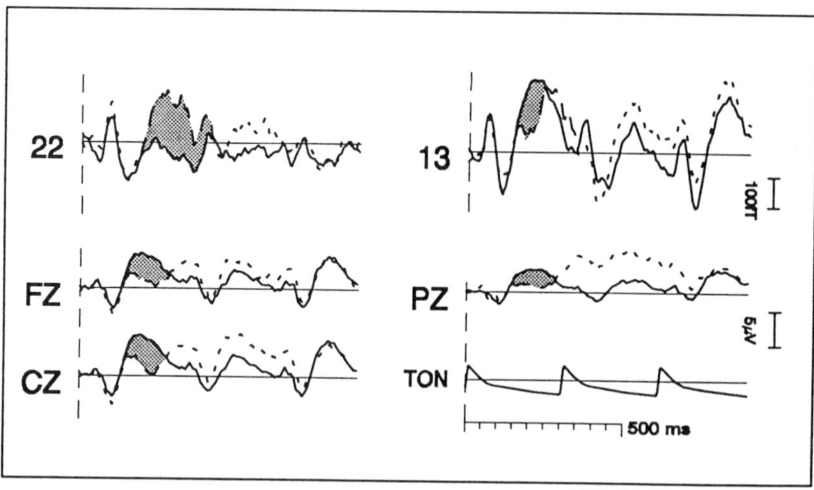

Figure 2. Averaged magnetic and electric responses to MOZART melody. One subject. EEG: positive=up. Recording position of MEG channel 13 posterior and 22 anterior to acoustic cortex. Solid lines, STD; broken lines, DEV.

MMN amplitudes exhibited a significant positive correlation with musical ability as expressed by the Bentley score (NOTE: p=0.048, CHORD: p=0.035, ARPEGGIO: p=0.0003, BACH: p=0.003, ANOVA, electrodes F3/F4).

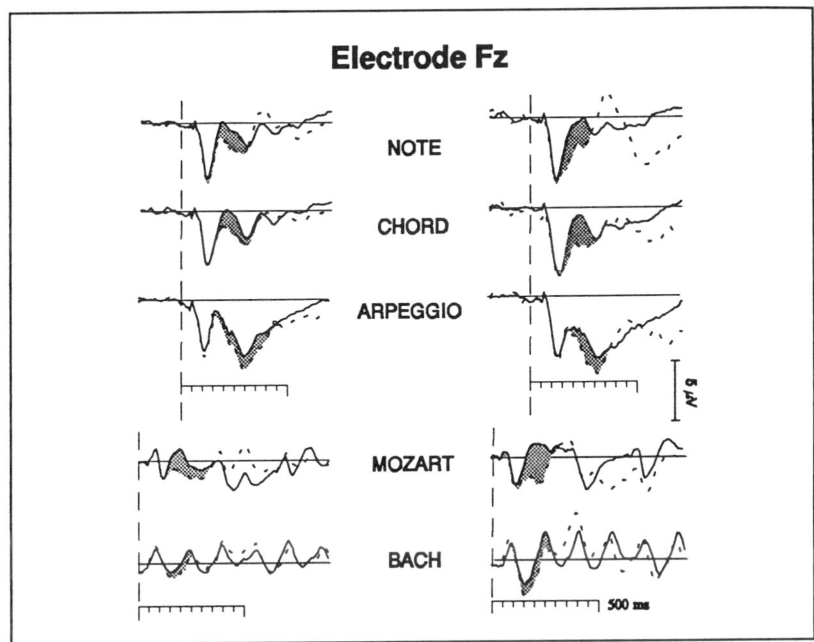

Figure 3. Grand averages of electrical responses from eight Ss with Bentley score < 37 (left column) and six Ss with a score > 40 (right column); positive=up; solid lines, STD; broken lines, DEV.

Further differences between Ss with low and high scores of musical ability were examined on the basis of grand averages from the two extreme groups of Ss. This step is illustrated in Figure 3. Whereas in the three NON-MUSIC paradigms the "low" and "high" groups showed quite similar N100 and MMN amplitudes, the musically gifted Ss produced clearly larger N100 and MMN in the MOZART and BACH conditions.

The amplitudes of the P300 potential as measured at Pz also differed between the two groups (generally larger in the gifted group as already observed by others [4], not shown). Interestingly however, these differences did not exactly mirror those of MMN; differences were largest for the perceptually difficult ARPEGGIO, whereas in MOZART, where deviants were easy to detect, the P300 amplitudes were almost equal. These results are compatible with the notion that the P300 reflects more conscious perceptual levels than MMA; conceivably, MMA corresponds to a precursor of conscious sensation [1].

Single equivalent dipole locations of N100/N100m in the NOTE paradigm calculated from a simultaneous MEG/EEG model [5] revealed a hemispheric asymmetry for musically ungifted Ss which was not observed in the gifted Ss (Figure 4). Similar differences were found also in "CHORD" and "ARPEGGIO". In "MOZART" and "BACH" no differences could be detected because of a larger scatter of localizations. Attempts to extend localization to MMA by means of the "simple" (one dipole/hemisphere, one instant) simultaneous electric/magnetic model used for N100, remained unsatisfactory, so far. MMA onset in EEG and MEG is often quite asynchronous (see Fig. 2); hence, appropriate modelling calls for spatio-temporal methods.

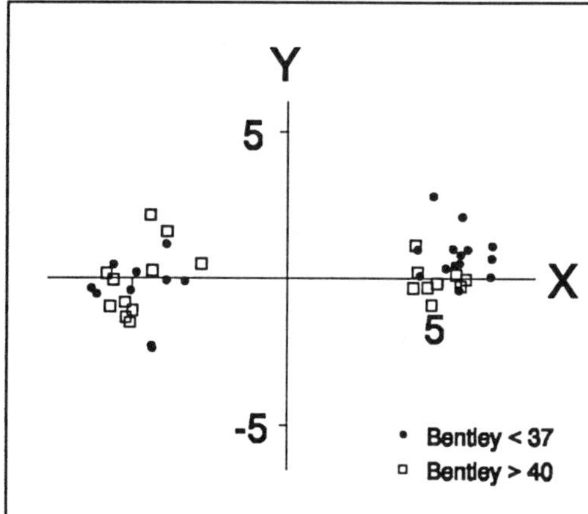

Figure 4. Equivalent current dipole (ECD) locations of N100/N100m in NOTE paradigm calculated with a simultaneous electric / magnetic model (one instant, four layer spherical model, one dipole per hemisphere; average goodness of fit, 95%). Transversal plane defined by nasion and preauricular point markers.

In the right hemisphere, ECDs of the "low musicality" group (dots) are located 6 mm more anterior (U-test, p=0.009) than in the "high" group (squares).

Discussion

The observation that wrong tones in melodies evoke MMA is difficult to explain on the basis of a comparison with a memory trace of the physical properties of the most recently perceived preceding tones. Continuously played music differs qualitatively from all NON-MUSIC paradigms including ARPEGGIO in that its "regularity" cannot be stored as a fixed reference pattern in short term memory. Therefore, we suggest that MMA does not only reflect short term memory mechanisms but also comparisons between *expected* and actual events in a more general sense. There are two conceivable processes whereby expectation could be built up in a MUSIC paradigm:

1)   By storing a complete melody in *long term memory*. Every forthcoming tone would be immediately available for comparison to actual tones; for the simple MOZART melody, such a strategy seems possible even for musically ungifted subjects.

2)   *Learned rules* would provide a set of "allowed" future tones on the basis of the recently heard tones; the selection of these rules would be determined by the perceived musical context. If this idea applies to our BACH paradigm it is interesting to note that MMA latency was still comparable to that of other paradigms, although rule application would seem to involve highly complex processes of analysis.

References

[1] Näätänen, R. and Alho, K. Mismatch negativity - a unique measure of sensory processing in audition. Intern. J. Neuroscience, 80, 317-337, 1995

[2] Sams, M., Hari, R., Rif, J. and Knuutila, J. The human auditory sensory memory trace persists about 10 sec: Neuromagnetic evidence. Journal of Cognitive Neuroscience 5 No 3, 363-370, 1993

[3] Bentley, A. A study of some aspects of musical ability amongst young children, including those unable to sing in tune. Unpublished thesis, University of Reading, U.K., library, 1963

[4] Besson, M., Faïta, F. and Requin, J. Brain waves associated with musical incongruities differ for musicians and non-musicians. Neuroscience Letters 168, 101-105, 1994

[5] Diekmann, V., Becker, W., Grözinger, B., Jürgens, R., Westphal, K.P. and Kornhuber, H.H. Localisation of focal epileptic activity with a simultaneous EEG and MEG model. In: Baumgartner et al. (Eds.) Biomagnetism: Fundamental Research and Clinical Applications, pp 23-27, Elsevier, 1995

Multidipole Estimation of the Neuromagnetic Fields Related to Auditory Discrimination

Kikuchi, Y.[1], Yoshizawa, S.[2], Kita, M.[3], Nishimura, C[4], Tanaka, M.[2],
Endo, H.[5], Kumagai, T.[5] and Takeda, T.[5]

*[1]Tokyo Medical and Dental University, 1-5-45 Yushima, Bunkyo-ku, Tokyo 113, Japan;
[2]University of Tokyo, 7-3-1 Hongo, Bunkyo-ku, Tokyo 113, Japan; [3]National Institute of
Mental Health, National Center of Neurology and Psychiatry, 1-7-3 Kohnodai, Ichikawa City
272, Japan; [4]Toho University School of Medicine, 5-21-16 Ohmori-nishi, Ohta-ku, Tokyo 143,
Japan; [5]National Institute of Bioscience and Human Technology, 1-1 Higashi, Tsukuba City,
Ibaragi 305, Japan*

Introduction

Auditory P3, which is the event-related potential (ERP) usually recorded during auditory discrimination tasks, is related to cognitive, memory and attentional process, and is also useful in clinical evaluation of the psychiatric disorders such as schizophrenia and so on. While the cognitive correlates of the P3 components, as well as their changes with respect to disease and development, have been extensively investigated, their synaptic generators remain controversial despite efforts using scalp EEG recordings in brain-lesioned subjects, animal models, and recording directly from the depth of the human brain.

MEG recordings are most useful in investigating the normal brain functions non-invasively. However, MEG measurements have been thus far made with a single- or a several-channel system ([6]), so a comprehensive magnetic field map covering a whole head had to be obtained over multiple and prolonged recording sessions. This may have been a limitation to the control of a subject's attention levels. Recently, according to the introduction of whole-head type SQUID systems, some of the problems of habituation and changing attention levels of subjects have been attenuated. We carried out MEG recordings by using a 64-channel whole head type SQUID system, and investigated the source locations of the components in the event-related field related to auditory discrimination.

Methods

Event-related fields (ERFs) were recorded from four normal-hearing adults with no neurological disorders in an auditory oddball paradigm. The 64-channel whole-head type SQUID system (CTF Inc., Canada) using first-order gradiometers with a coil diameter of 2.0 cm and a 5.0 cm baseline was used to measure external evoked magnetic fields over a whole head. The Cartesian coordinate system was defined based on three landmarks: the nasion, the left and the right preauricular points. The spatial parameters (Cartesian coordinates), as well as the strength and the orientation of the equivalent current dipoles (ECDs), were estimated from the 64 magnetic field measurements every 3.2 msec during the period between 280 msec and 450 msec post stimulus.

Pure tones of 2kHz and 1kHz were applied as a target and as a non-target respectively. The presentation of sounds was randomized with targets occurring 20% of the time and non-targets 80%. Both sounds were delivered binaurally to the subjects at about 60 dB above threshold, via plastic tube terminating in high-quality insert earphones (Cabot Inc., USA). Subjects were required to mentally count the targets as correctly as possible with their eyes closed. MEGs were bandpassed from 0 to 40Hz and digitized at 625 Hz. Furthermore, off-line digital filtering reduced the bandpass to 0.01-30 Hz, and forty responses to target rare tones were averaged for the ERFs. The entire recording procedure for each subject lasted approximately 10 minutes. Eye movement artifacts were checked by monitoring the magnetic fields from the neuromagnetic sensors placed anteriorly.

Following the experimental session, magnetic resonance images (MRIs) were taken of each subject using a 0.5 Tesla system. All images were T1 weighted to emphasize anatomy and consisted of 2.0 mm x 2.0 mm pixels. 128 images were obtained, and all images were saved to a magneto-optical disk for off-line analysis. We carried out multi-dipole (2-4 dipoles) estimation, by considering anatomical symmetries and for minimizing error, based on Grynszpan-Geselowitz's equation. The coordinates of the dipoles' gravity centers were overlaid on individual MRI to show the corresponding locations in the brain.

Results

Clear evoked magnetic wave forms to both frequent and rare tones were obtained from all the subjects. In all the subjects, the averaged fields in response to targets showed prominent with latencies longer than 300 msec as compared with those to non-targets. Observations of the isomagnetic field maps during the time generally showed that the fields directed outward from the anterior region in the left hemisphere and the posterior region in the right hemisphere, while the direction was reverse over the posterior region in the left hemisphere and the anterior region in the right hemisphere, in each subject (Fig.1).

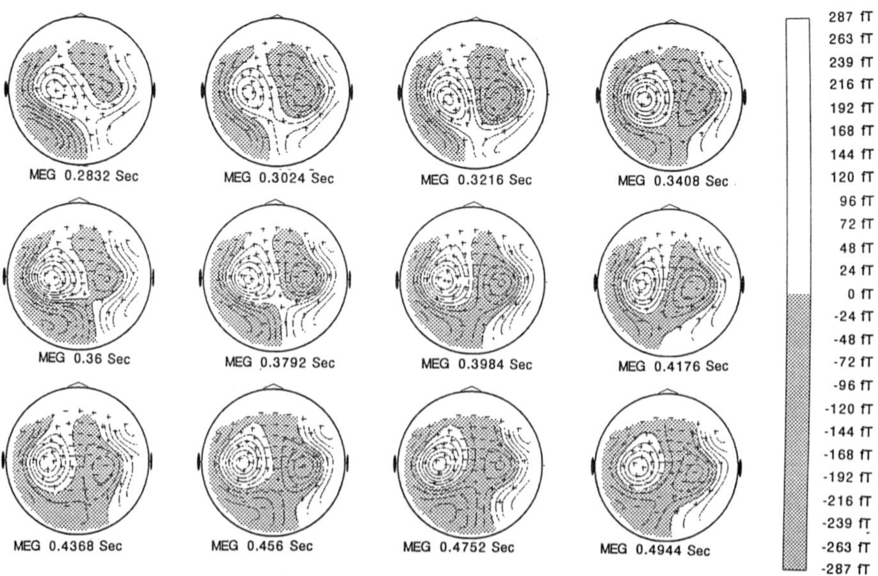

Fig. 1. Temporal change of the iso-magnetic field maps during the period between 283.2 msec and 494.4 msec. White area (+) shows the outward magnetic field, and hatched area shows the inward field.

In order to study the temporal changes of the sources responsible for the late components, dipole solutions were calculated at successive 3.2 msec intervals for the wave form between 280 msec and 450 msec post stimulus. Fig. 2 shows the averaged MRI images superimposed by the ECDs estimated in subject A, during the period between 355 msec and 432 msec post stimulus. In this case, a two dipole model was applied, and the gravity centers of ECDs were found around the hippocampus, the posterior parts of the parahippocampal gyrus, the collateral sulcus, the lingual gyrus and the fusiform gyrus. Every dipole directed upwards. In the case of subject B, ECDs were also estimated in the hippocampus, the posterior parts of the parahippocampal gyrus, the collateral sulcus, the lingual gyrus and the fusiform gyrus during the period between 280 msec and 350 msec. In addition, they were found in the superior temporal sulcus (STS) and the superior temporal plane (STP) between 350 msec and 450 msec. Their directions were identical to those of subject A. In subject C, their locations were around the STS, the STP, the collateral sulcus and the lingual gyrus between 293 msec and 304 msec, and were in the collateral sulcus, the lingual gyrus, the central sulcus, the precentral sulcus and the inferior frontal sulcus between 314 msec and 390 msec. The locations of ECDs estimated in subject D were around the central sulcus and the precentral sulcus during the period between 302 msec and 328 msec, and were around the parietal and the subparietal cortices between 338 msec and 380 msec.

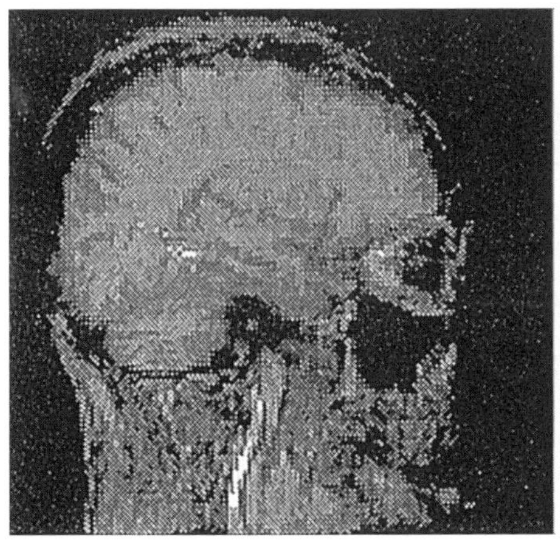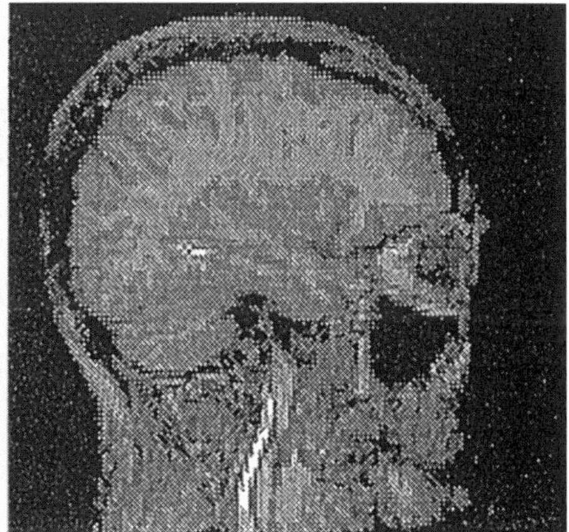

Subj.A/2kHz/355-432/L   Subj.A/2kHz/355-432/R

Fig. 2 Averaged sagittal MRIs superimposed by the gravity centers of the ECDs estimated in subject A, between 355 msec and 432 msec. White pixels represent the gravity centers. (left:left, right:right)

Discussion

The P3 is subdivided into a fronto-central P3a relatively enhanced by non-target rare or novel stimuli, and a later parietal P3b enhanced by certainty of target detection. The studies by depth electrode recordings ([1], [2] [3]) showed that the responses are classified into two patterns: The first sharp triphasic short-latency sequence resembling the scalp N2/P3a/SW pattern is observed in all recorded neocortical areas and is most often focal or large in the anterior or the posterior middle temporal gyrus, the posterior parahippocampal gyrus and the fusiform gyrus. The second pattern is a broad, usually monophasic, relatively long-latency, waveform that resembled the scalp P3b. It is localized to the hippocampus, to a part of the superior temporal sulcus, and to rhinal cortex underlying the amygdala. The N2/P3a/SW pattern is also found in supramarginal and posterior cingulate gyri, as well as in multiple frontal sites, including the anterior cingulate gyrus, gyrus rectus and dorsolateral prefrontal cortex. While the temporal order of each active area in the brain has not been so clear in the present study, the source locations obtained in normal subjects by multi-dipole estimation, were in good agreement with parts of those found in epileptic patients by the depth electrode recordings.

References

[1] Halgren, E., Baudena, P., Clarke, J.M., Heit, G., Liegeois, C., Chauvel, P., Musolino, A. Intracerebral potentials to rare and distractor auditory and visual stimuli. I. Superior temporal plane and parietal lobe. Electroenceph. clin. Neurophysiol., 94:191-220, 1995.

[2] Halgren, E., Baudena, P., Clarke, J.M., Heit, G., Marinkovic, K., Devaux, B., Vignal, J-P., Biraben, A. Intracerebral potentials to target and distractor auditory and visual stimuli. II. Medial, lateral and posterior temporal lobe. Electroenceph. clin. Neurophysiol., 94:229-250, 1995.

[3] Baudena, P., Halgren, E., Heit, G., Clarke, J.M. Intracerebral potentials to target and distractor auditory and visual stimuli. III. Frontal cortex. Electroenceph. clin. Neurophysiol., 251-264, 1995.

[4] Kikuchi, Y., Yoshizawa, S., Kita, M., Nishimura, C., Tanaka, M., Endo, H., Kumagai, T., Takeda, T. Source estimation of the auditory event-related neuromagnetic fields (ERFs) using a whole-cortex type SQUID system. "Visualization of Information Processing in the Human Brain: Recent Adavances in MEG and Functional MRI", The 10th Tokyo Institute of Psychiatry International Symposium, Tokyo, Japan, Oct. 12-13, 1995.

[5] Kikuchi, Y., Yoshizawa, S., Kita, M., Nishimura, C., Tanaka, M., Endo, H., Kumagai, T., Takeda, T. Source localization of event-related neuromagnetic fields using a whole-cortex type SQUID system. Electroenceph. clin., 97:S124, 1995.

[6] Rogers, R.L., Baumann, S.B., Papanicolau, A.C., Bourbon, T.W., Alagarsamy, S., Eisenberg, H.M. Localization of the P3 sources using magnetoencephalography and magnetic resonance imaging. Electroenceph. clin. Neurophysiol., 79:308-321, 1991.

Measurement of Gustatory-Evoked Magnetic Fields with Sharp Stimulation

Kobayakawa, T.K.[1], Endo, H.E.[1], Saito, S.S[1], Ayabe-Kanamura, S.A.[2], Kikuchi, Y.K.[3], Yamaguchi, Y.Y.[1], Ogawa, H.O.[4], and Takeda, T.T.[1]

[1]National Institute of Bioscience and Human Technology, MITI, Tsukuba, Japan;
[2]University of Tsukuba, [3]Tokyo Medical and Dental University; [4]Kumamoto University

Introduction

Non-invasive methods to measure brain function, particularly the methods of magnetoencephalography (MEG) and functional magnetic resonance imaging (fMRI), have improved rapidly and remarkably. With these methods, researchers have measured higher cortical functions of the living human brain, especially visual and auditory functions. MEG has the special advantage that it can trace the movement of cortical activity.

In the case of gustation, few experiments have addressed human gustatory-evoked potentials (GEPs) [1][2][3][4], because of difficulties involved in precise control and presentation of gustatory stimuli. In imaging studies performed using PET or fMRI, it has been reported that the insula, parietal cortex, cingulate gyrus, parahippocampal gyrus, and thalamus were activated during taste discrimination (PET)[5][6], and insula and frontal operculum were activated by simple taste stimulation (fMRI)[7]. These imaging methods, however, offer relatively low temporal resolution. MEG offers much better resolution and we have therefore employed it to trace the flow of gustatory information in the brain. As MEG requires strict stimulus control, we have developed a gustatory apparatus to present the stimulus with a short rise-time. First, we measured the gustatory-evoked magnetic fields (GEMs) and then, we sought to estimate the area of gustatory activity in the human brain using a dipole model. Next, before measuring GEMs, we presented the substance which surpress a sensitivity to a certain taste, and investigated the change of responses.

Methods

Taste Delivery System

The stimulating apparatus was composed of three sub-systems: (1) the system that supplied the tastants and deionized rinsing water, (2) the system for stimulus presentation, and (3) the system that sensed the timing of the stimulus. The supply system and the presentation system were connected by teflon tubing (Fig.1).

The deionized water and tastants were driven through the system by compressed air and separated from each other by a small amount of air injected between them. The water, tastants, and air were switched by solenoid valves controlled by a personal computer. Stimuli were presented through a hole (4 mm x 10 mm) in the wall of a teflon tube. The subject bit gently the tube, then covered the hole with his or her tongue. Liquid and air from inside the tube could then flow across the exposed position of the tongue. Since the flow of the liquid and air generated slight negative pressure, the tongue of the subject was sucked slightly into the hole and hence the liquid did not leak into the mouth.

A precise trigger was necessary for proper averaging of data. The tastant took about 27 s to go from the solenoid valve to the hole (stimulus presentation position). Arrival time would vary over too wide a range, if timing to control the solenoid valve served for triggering. To obtain a stricter trigger, an optical sensor was placed as close as possible to the subject's mouth. The tastants were colored red, whereas the air and the deionized water were uncolored. Light was generated outside the room and fed inside by a bundle of optical fibers. The light passed through the teflon tube and was modulated by the liquid or air in the vicinity of the subject's mouth. The signal was then brought out of the room via optical fibers and converted into a voltage by the photo-transistor (Fig. 2). The distance between the sensor and the hole in the tube equalled 6.5 cm. The liquid took 300 ms (sd=2.9 ms) to cover that distance. Rise-time to the 80% level of the signal equalled 16.5 ms (sd=1.49 ms). Such performance was accurate enough for strict triggering.

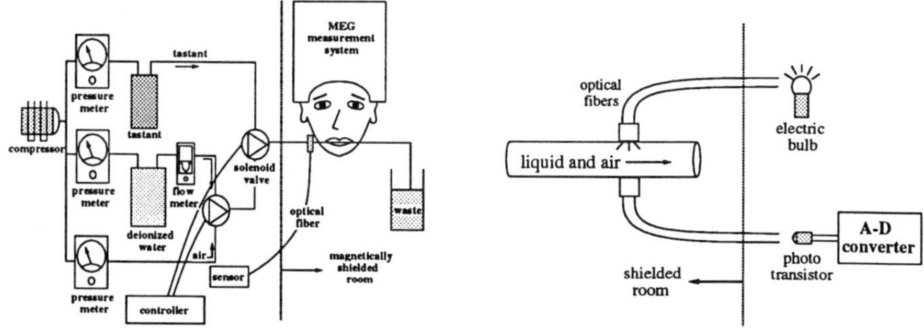

Fig. 1: The taste delivery system   Fig. 2: The optical sensor

The small amount of air that separated the tastant and water was released into the tube for as short a duration as possible to minimize the likelihood that the subject could sense a difference of pressure between the liquid and air.

Evoked Magnetic Field Measurements

For measurement of magnetic fields, we used a 64-channel whole-head SQUID system (CTF Systems Inc., Canada). This system has sensors for measuring environmental magnetic noise. Subtraction of the outer noise from the measured data leads to extraction of clearer signals from the brain.

Since magnetic substances would interfere with the measurements in the magnetically shielded room, a special wooden chair with a pneumatically operated cushion was developed [8]. In addition, the subject's jaw was supported by a wooden chin rest. The tube through which the liquids flowed was brought to the subject's tongue and held in place by a frame attached to the chair. The cushioned chair and the chin rest tended to forestall feelings of fatigue during a session.

Solutions of 1 M NaCl and of 3 mM saccharin were used as tastants. Next, we measured GEMs from saccharin after with a subject chewed gymnemic acid gum for 15 min. The tastants were presented at the central, front edge of the subject's tongue, the part known to have high sensitivity to these tastes. After a tastant was presented, deionized water was presented as a rinse. The duration of the stimulus was about 400 ms and that of the rinse was approximately 30 s. The tastant and rinse were maintained at the temperature of the tongue. Thirty trials were presented for each tastant and accompanying rinse.

Six subjects, which included 3 females, participated in the experiment. They ranged from 21 to 32 years of age. Sampling rate for the MEG signal was 250 Hz and the low-pass filter was set at 40 Hz. During recording of the MEGs, the subject watched a fixation point. After the experiment, signals from the optical sensor were used as a trigger for off-line averaging of the data. Trials contaminated by eye movements were rejected. The number of trials used for averaging was about 35.

Results

Fig. 3 shows the magnetic isocontour map of the brain surface 400 ms after presentation of saccharin. The magnetic field indicated force flow from the left-front to the left-back of the head and from the right-back to the right-front.

In this investigation, dipoles were estimated under the assumption that the shape of the brain was a simple sphere, though of course the actual shape departs from a sphere. In the future, the dipoles need to be estimated for a real brain

806

model calculated from MRI.

We estimated the dipoles on the assumption that the pattern of this map had two dipoles, which we superimposed on the MRI. In 86% experiments, the dipoles were estimated in the upper side of insula region (goodness of fit was from 80% to 96%) (Fig. 4). In five subjects, the directions of these dipoles were superior, posterior and medial.

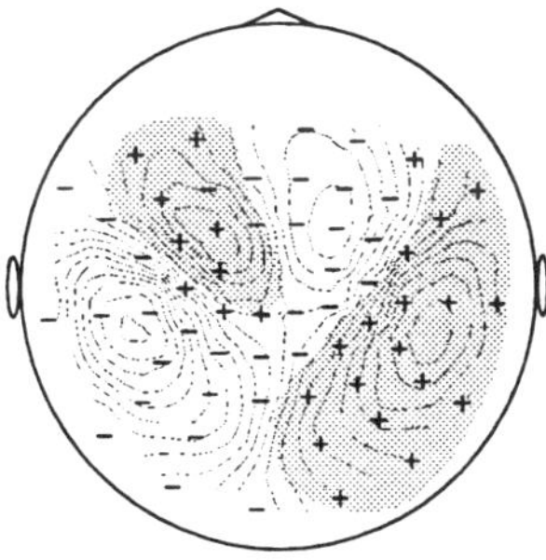

Fig. 3: Contour map of the surface

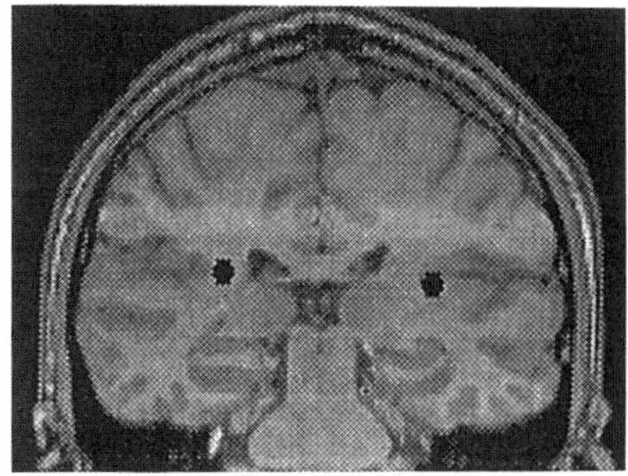

Fig. 4: Activated cortex by gustatory stimulation

In four of five subjects, the latencies at which the clear pattern appeared were much longer (from 150 ms to 200 ms) after subjects chewed gymnemic acid gum than when they did not chew the gum. Figs. 5 and 6 depict averaged GEMs of 64 channels aligned according to the stimulus onset and superimposed on the same graph in one subject. Fig. 5 shows the response of saccharin, and Fig. 6 shows that of saccharin after subject chewed gymnemic gum

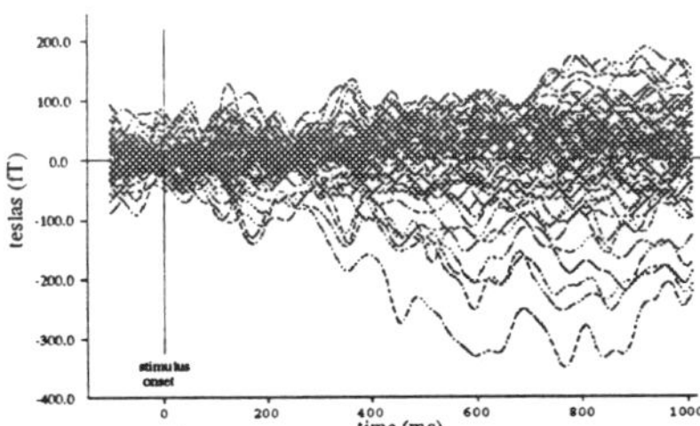

Fig. 5: response of saccharin

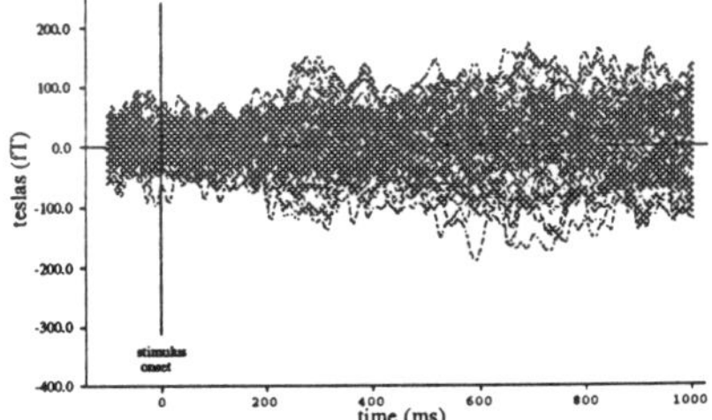

Fig. 6: response of saccharin after gymenmic acid

respectively. These figures show the amplitude of the response of saccharin after gymnemic acid presentation was much smaller than that of saccharin.

Discusstion

Previous studies using PET [5][6] reported that the insula, parietal cortex, cingulate cortex, parahippocampal gyrus, and thalamus were activated by gustatory stimulation. In those experiments, cortical activity was measured during a discrimination task. Several areas including the insula were activated by the task; some may have been related to recognition or motor activity.

In the case of the Macaque monkey, the primary gustatory cortex has been reported to be in the insula[9]. The present results from humans are consistent with that finding.

In an fMRI study[7] when a solution of NaCl was presented, both hemispheres were activated. In contrast, when a solution of sucrose was presented, just one hemisphere was predominantly activated. For NaCl, the stimulus in common in that experiment and the present one, we also found both hemispheres to be activated.

All subjects felt a sweet taste to saccharin, although some subjects also said they could feel sour, bitter or a salty taste to saccharin. After presentation of gymnemic acid, subjects reported a sour or salty taste; (of course, they did not report sweet taste). The sodium ion is included in saccharin solution and it seems this causes the sour, bitter, or salty taste. Therefore, in the experiment using gymnemic acid the response was not surpressed completely. In order to surpress the sweet taste, we would be better to use another substance (i.e. aspartame) as a sweet tastant instead of saccharin.

References

[1] Kobal, G. (1985) Gustatory evoked potentials in man. Electroenceph. clin. Neurophysiol., 62: 449-454.

[2] Plattig, K., Bekiaroglou, P. and Schell, S. (1994) Gustatory-evoked potentials in man (A long history and final results). In: K. Kurihara, N. Suzuki and H. Ogawa (Eds.), Olfaction and Taste XI, Springer-Verlag, Tokyo: 638-641.

[3] Prescott, J. (1994) Brain event-related potential to primary tastes. In: K. Kurihara, N. Suzuki and H. Ogawa (Eds.), Olfaction and Taste XI, Springer-Verlag, Tokyo: 642-645.

[4] Maetani, C., Notake, K., Takemoto, I., Hizuka, I., Hagino, J., Matsunaga, T., Yoshimura, S. and Tonoike, M. (1988) Measurement of EEG caused by gustatory stimulation. Proceedings of the 22nd. Japanese Symposium on Taste and Smell: 97-99.

[5] Fukuda, H., Yamada, K., Kinomura, S., Kawashima, R. and Cuo, S. (1991) High order gustatory projections in the human brain studies with positron emission tomography. The Salt Science Research Foundation - Annual Research Report: 337-347.

[6] Kinomura, S., Kawashima, R., Yamada, K., Ono, S., Itoh, M., Yoshioka, S., Yamaguchi, T., Matsui, H., Miyazawa, H., Itoh, H., Goto, R., Fujiwara, T., Satoh, K. and Fukuda, H. (1994) Functional anatomy of taste perception in the human brain studied with positron emission tomography. Brain Res., 659: 263-266

[7] Hirsch, J., DelaPaz, R., Relkin, N., Victor, J., Bartoshuk, L., Norgren, R. and Pritchard, T. (1994) Localization of human gustatory cortex using functional magnetic imaging. Chem. Sens., 19: 486.

[8] Takeda, T., Endo, H., Kumagai, T. and Maurizio, M (1995b) Evaluation of a whole-head type MEG system and its application. J. Jpn. Biomag. Bioelectromag. Soc., 8: 128-131.

[9] Ogawa, H. (1972) Taste response in the Macaque monkey chorda tympani. Physiol. Behav., 9: 325-331.

Source Location Outside the Auditory Cortex: Auditory Event Related Field to Tone-Duration in a Discrimination Task

Kofoed, B.[1], Bak, C.K.[1], Csepe, V.[2], Pantev, C.[3] and Saermark, K.[1]

[1]Department of Physics, Technical University, Lyngby, Denmark; [2]Department of Psychophysiology, Hungarian Academy of Sciences, Budapest, Hungary; [3]Institute of Experimental Audiology, University of Münster, Germany

Introduction

The purpose of the present experiment was to locate the sources of the "mismatch field" (MMF), when the discrimination was based on tone duration differences. In the oddball paradigm applied, standard tones of duration D1 are randomly replaced by deviant tones of duration D2<D1. It is well known, that deviants invoke a MMF approximately 200 ms after the deviation, in these experiments the termination of the the target tone. The subjects were required to attend to and silently count the number of deviants. By choosing D2 appropiately, the latency of the MMF signal can be moved away from the M200 complex, thus permitting a localization of the MMF sources. Source location of the MMF to pitch differences and to tone differences where the MMF signal appears in the M100-200 region has earlier been reported [1],[2]. In those experiments a single dipole fit has been used for the purpose of localization. In the present experiment the MUSIC-algorithm is used.

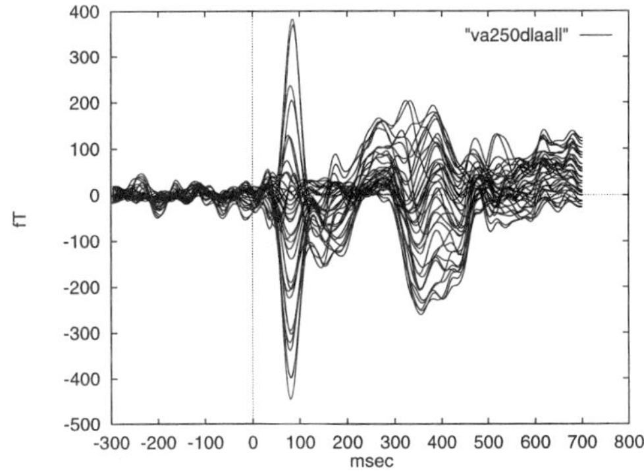

Figure 1: The figur shows the averages for the target tone (A-averages). All 37 coils are shown. The duration of the target tone is 250 ms., and the tone onset is at 0 s. The MMF-signal is seen in the interval 380-520 ms after onset of the stimulus tone. Probably more than one component is involved.

Methods

The measurements were performed by means of a BTI-37 channel SQUID system equipped with a probe position indicator. The sound was led to the right ear via a plastic tube, and the measurements were made on the left hemisphere. The duration of the frequent tone was D1=500ms, and it occurred with the probability p = 0.67. The duration of the target tone was D2=250 ms or D2=200 ms, and it occured with the probanility p = 0.33. Both stimuli were of 1kHz pitch and of intensity 70 db above the psychoacoustic threshold. The interstmulus interval [ISI] varied randomly in the range 1.25 - 2.75s or in the range 2.5 - 5.5 s.

The signals were highpass filtered, using a hardware filter with a cut-off frequency 1 Hz. The sampling rate was 298 Hz. The collected epoch was 1000 ms including 300 ms prestimulus interval. The average was performed across the epochs for the target tone (A-averages) and across the tones immediately preceeding a target epoch and across the remaining epochs. After averaging frequencies higher than 30 Hz were digitally filtered out. Inspecting the measured time-series visually (Fig.1), the MMF-signal does not look dipolar. More then one source is involved, and in time the sources are overlapping. Futhermore, the combination of highpass filtering and the size of the interstimulus interval used will affect the data in such a way, that a single dipole fit can not be applied. A method to separate more than one source, overlapping in time, is the MUsic SIgnal Classification (MUSIC) algorithm [3]. An inherent property of this method is, that source localization is not affected by filtering of the signals.

Results

Fig.1 shows the MMF signal appearing in the time interval 300 - 500ms and only in averages across the targets. The MUSIC algorithm is applied to 50 time slices around the MMF signal. Following the procedure given by Mosher J.C. et al.[3] the so called costfunction is calculated. In xy-planes, xz-planes or yz-planes contour plots for the reciprocal costfunction is made. x, y and z are the usual axes for the head. The points where the reciprocal costfunction has a maximum specify the locations for the sources.

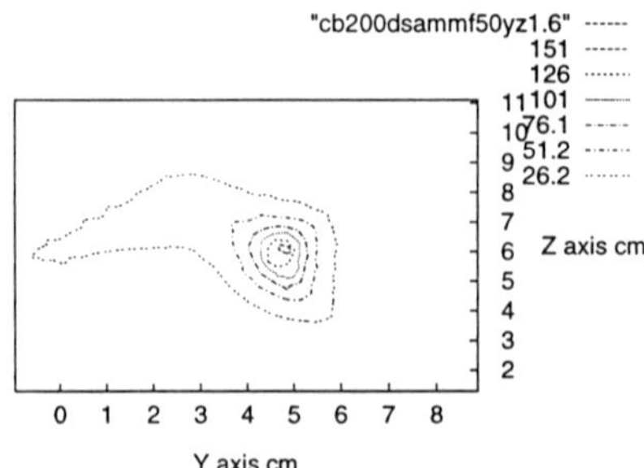

Figure 2: Contourplot for a two-dimensional slice (constant x) of the reciprocal costfunction. The yz-plane shown is the plane where the reciprocal costfunction has its maximum value for one of the sources.

In the latency range of the MMF we find two sources. Fig.2 and Fig.3 show the contourplots in the yz-planes where the reciprocal costfunction has its maxima for the two sources respectively. For comparison, Fig.4 shows the contourplot using 50 time slices around the M100 signal for the same A-averages. The source shown in Fig.3 is therefore located in the auditive cortex and according to its latency, it can not be excluded, that it is the off-response to the stimulus. The other source, however, is found outside the auditive cortex, as shown in Fig.2. Its location is in good agreement with our results reported earlier [4].

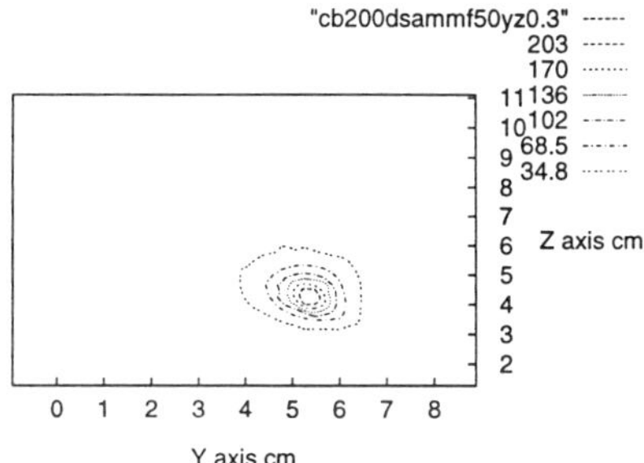

Figure 3: Contourplot for a two-dimensional slice (constant x) of the reciprocal costfunction. The yz-plane shown is the plane where the reciprocal costfunction has its maximum value for the other source.

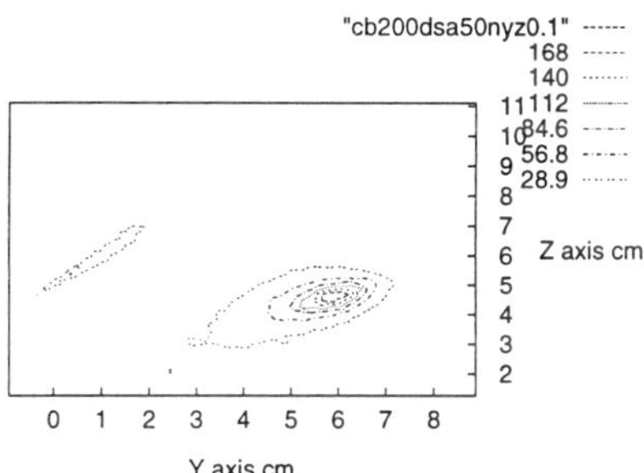

Figure 4: Contourplot for a two-dimensional slice (constant x) of the reciprocal costfunction. The yz-plane shown is the plane where the reciprocal costfunction has its maximum value the for M100 source.

Four subjects have participated in the experiment. For each subject D1 has been 500ms and D2 = 200ms or D2 = 250ms. The interstimulus interval has been 1.25 - 2.75s or 2.5 - 5.5s for each D2. For each subject we therefore have 4 runs. Making grand average, the total number of localizations for one source is 16.

In Fig.5 the grand averages over the 4 subjects is shown. Each source location is measured relative to the position of the M100 source as determined from the same A-average. In the figur the zero point corresponds to the location of the M100 source. Around the locations for the two sources the standard deviations are plotted.

Conclusion

Using the MUSIC algorithm to the MMF signal two souces are found to be involved, when the discrimination is based on tone duration. One of the sources is located very near the M100 source and the other is located about 1 cm anterior, 1.8 cm superior and 0.5cm inferior relative to the M100 source.

Relative source locations

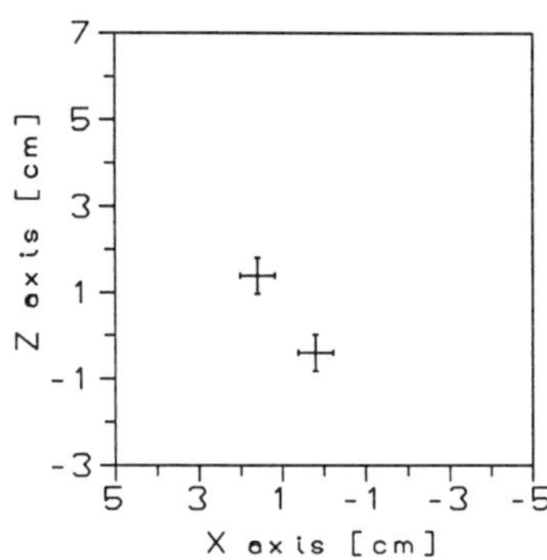
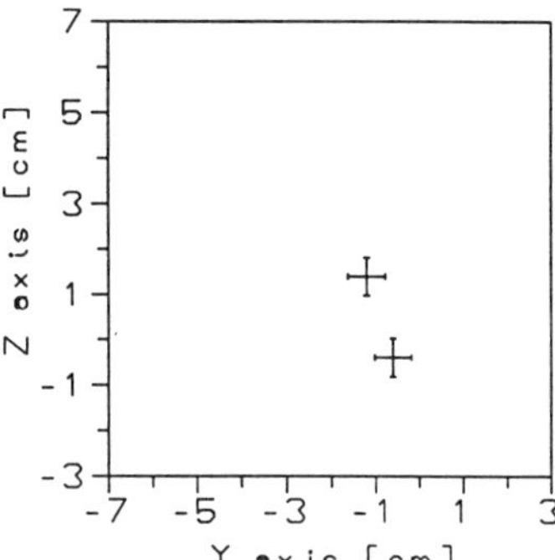

Figure 5: The xz- and yz-location of the two sources found. The locations are plotted relative to the position of the M100 source. The figure shows the grand averages for all subjects and all runs. The figur also shows the error bars for the standard deviation in the measurements

References

[1] Sams M., Kaukoranta E., Hamalainen M., Naatanen R. Cotical activity elicited by changes in the auditory stimuli: different sources for the magnetic N100m and the mismatch responses. Psychophiology. 1991. 28. p21-29.

[2] Csepe V., Pantev C., Hoke M., Hampson S. and Ross B. Evoked magnetic responses of the human auditory cortex to minor pitch changes: localization of the mismatch field Electroencaph. Clin. Neurophys.1992, 84, p538-548.

[3] Mosher J.C., Liwis P.S. and Leahy R.M. Multiple dipole modeling and localization from spatio-temporal MEG data. IEEE Transactions on Biomedical Engineering, 1992, 39, p541-557.

[4] Kofoed B., Bak C.K., Rahn E. and Saermark K. Auditory event-related magnetic fields in a tone-duration discrimination task. Source localization for the mismatch field and for a new component M2". Acta Neurol. Scandinav. 1995. 91, 362-371.

Acknowledgements

We are very grateful to the Danish Natural Science Research Council for financial support.

Measurement of MEG Evoked by Visual Discrimination Task

Kumagai, T., Takeda, T. and Endo, H.

National Institute of Bioscience and Human-Technology, Tsukuba, Japan

Introduction

The oddball paradigm task, a simple discrimination task using infrequent and frequent stimuli, evokes late components after the latency of 300 msec in EEG measurements. It is known that they are composed of many components which are called P3a, P3b and slow wave. Responses to infrequent stimuli are larger than those to the frequent stimuli. They are more sensitive to the cognitive parameters than to the stimulus parameters of the task. Hence late components are defined as endogenous components. It is important to investigate the functions and their sources in time domain in order to understand the style of cerebral information processing.

The late components have been investigated by EEG, PET, implanted electrodes and MEG. However, sources and functions of those components have not been clarified. MEG has a high time resolution like EEG, and it also has a good spatial resolution if the activated areas are fairly concentrated. Although an increasing number of auditory P3 studies with MEG have been reported, there are only few reports on the visual evoked P3[1][2]. Those studies have been carried out using systems covering only a part of the cortex, hence the whole brain activity has not been reported yet.

In this study we examined the magnetoencephalographic (MEG) signals evoked by a visual discrimination task using a 64-channel whole-cortex MEG system. Since P3 may involve a widespread activity of the brain, it is important to measure the entire cortex simultaneously to understand the underlying neural activities. The activated area was estimated with multiple equivalent current dipoles (ECDs) using a simple spherical brain model.

Methods

Equipment:

The measurements were performed using a whole-cortex type MEG system (CTF, Inc.) in which 64 sensors cover the whole brain. The sensors are first order gradiometers whose diameters and base lines are 2 cm and 5 cm respectively. The system was set up in a magnetically shielded room with a reclining chair equipped with a pneumatic lifting mechanism(Fig.1). The visual stimuli were shown to the subjects using a liquid crystal projector system. A translucent screen was set in front of the subject seat. The projector, positioned in front of the magnetically shielded room, projected the stimuli on the screen through the doorway of the magnetic shielded room. In previous experiments we checked that the system can cancel the environmental noises from the doorway through the software noise reduction using reference sensors [3]. The distance between subject's eyes and the screen was 1.5 m.

Stimulus and Paradigm:

The oddball paradigm was composed of yellow and purple color onset stimuli. The frequent stimulus was either yellow or purple while the infrequent stimulus was another color. Frequent and infrequent stimuli were showed with a probability one of 0.8 and 0.2, respectively. The subject had to count silently the number of times in which the infrequent stimulus appeared and to report this number after each trial. The infrequent stimulus appeared between 40 and 45 times in each trial, while the frequent stimulus appeared about four times as often. Fig.2 shows the color onset stimuli used in this experiment. When the stimulus was turned off, a simple gray circle was shown to the subject. While the stimulus was turned on, a purple or a yellow concentric circle was presented on the gray circle. The visual angle of color circles and the gray circle were 3.5 degrees. The luminance of the gray circle, the purple annulus and the yellow annulus were almost the same. A small black fixation point was put at the center of the stimulus. The duration of the stimulus was 200 msec. The inter stimulus interval was chosen from 2 to 2.2 s at random.

Measurment:and Analysis:

The subjects were four right handed normal volunteers ranging from 20 to 40 years old, indicated in the following as S1, S2, S3 and S4. The MEG signals were measured from 200 msec before to 500 msec after the onset of stimulus. The

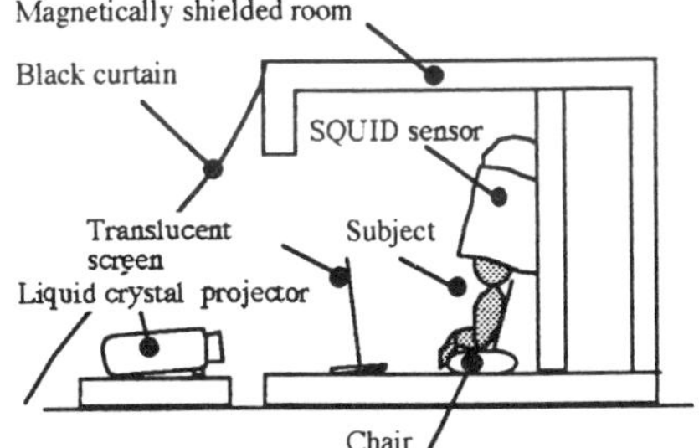

Fig.1 The experimental system

sampling rate was 325 Hz. The measurements that included a magnetic change larger than 900 fT were rejected, since it might be contaminated by artifacts. The responses to infrequent and frequent stimuli were averaged respectively in every trials.

Averaged data were analyzed as follows. DC offset was removed from each channel so that the integration evaluated from 200 msec before to the stimulus onset. After the software noise reduction, we filtered the signal of each sensors with a low-pass filter of 30 Hz, 20th order. Source estimations were performed by a simplex method using Equivalent Current Dipole (ECD) model. Estimation error (E) was defined as $\Sigma \{ (B_{ti} - B_{mi})^2 / B_{mi}^2 \}$. Where B_{ti} is a magnetic field that is calculated from estimated sources at channel i, B_{ti} is a magnetic field measured at channel i. B_{ti} were calculated by the Grynszpan-Gesolowitz equation using a sphere model. In the sphere model we used, the surface of the sphere touched internally the scalp according to the medial sagittal MRI slice of each subject.

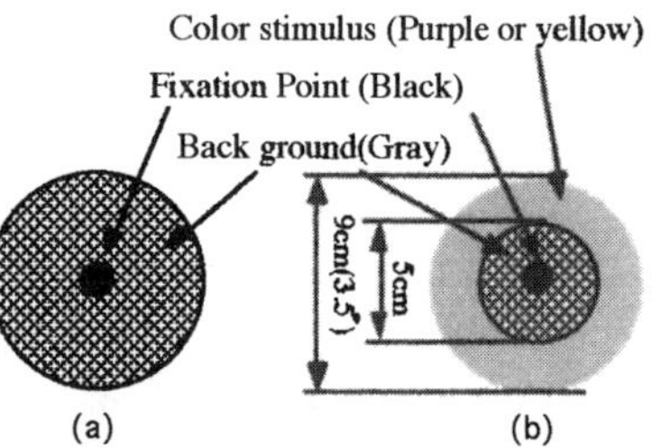

Fig.2 The visual stimulus. (a) The stimulus is turned on. (b) The stimulus is turned off.

Results

Superimpose waveforms:

The fig.3 shows the superimposed waveforms of responses to the purple frequent and the yellow infrequent stimuli. There were no remarkable responses to both waveforms until about the latency of 150 msec. Noise level of responses to infrequent responses was larger than noise level of responses to frequent responses, since the averaging number of infrequent responses was small. Visual evoked responses of the latency of almost 100 msec were not clear since stimuli had little luminance changes.

The responses to frequent stimuli had peaks around 220 and 380 msec. The component around 220 msec was not clear in S3. The component around 380 msec was not clear in S1 and late in S2. S4 had another component at 150 msec. The infrequent stimulus evoked large activities from 150 msec to the end of measurments in S1 and S2. In this response many waves inverted their polarity at the latency of 300 msec. In S3 and S4, the infrequent stimuli evoked activities from 300 msec to the end of measurements. According to above results responses to the infrequent stimulus were able to be classified into two components at least. The first component was from 150 to 300 msec and the second component was from 300 msec to the end of recordings.

Fig.4 Shows the responses of a typical sensor (the right lateral sensor) that recorded the activities to the infrequent stimulus after 300 msec. All subjects showed a large difference between responses to frequent and infrequent stimuli. The rise time of the responses after 300 msec changed slightly for each subject. Therefore it was supposed that the responses after 300 msec were the late complex responses which included P3a, P3b and slow waves, and the responses from 150 msec to 300 msec included N2 and P2 components which were in advance of P3.

Above mentioned characteristics were also shown when we exchanged infrequent and frequent stimuli, i.e. the yellow stimulus was frequent and the purple stimuli was infrequent. Hence, It is believable that the subjects performed the oddball paradigm well and the responses after 300 msec included the P3 components.

Morphology of isocontour maps:

Fig.5 Shows the isocontour maps of the responses to the infrequent stimulus at the latency of 350 and 450 msec of S1. The isocontour maps at 450 msec had sources of magnetic flux at the left frontal and right occipital areas, and sinks of magnetic flux at the right frontal and left occipital areas in all subjects. According to these maps, it seems that there was only one localized activity area in each hemisphere. The isocontour maps of the responses to the infrequent stimulus at 350 msec had more complex patterns than the maps at 450 msec. Therefore it reasonable that there were many active areas. The isocontour maps at 350 msec were similar to the maps at 450 msec except that the maps at 350 msec had a sink of magnetic flux at the left frontal area. The above characteristics of responses to the infrequent stimulus were also found in S3 and S4. However, in subject S2, the maps at 350 msec was similar to the maps at 450 msec.

Source Estimation:

Next we estimated the sources of the late components using two-ECD model. The sources of responses to the infrequent stimulus at the latency of 450 msec were estimated near the bilateral parahippocampal or occipitotemporal gyri. This estimation was consistent in all subjects. In each subject, the E(the estimation error) was lower than 10%.

The strengths of ECDs were from 200 to 400 nAm. The ECDs pointed to the medial direction. Fig.6 shows an example of the estimated source locations.

Sources of the responses to the infrequent stimulus at the latency of 350 msec could not be estimated through two-ECD model with a low estimation error. Contrary the sources of responses to the frequent stimulus at 350 msec were estimated around the deep part of occipital lobe with E less than 15 %. It might mean that the area activated by the frequent stimulus was more localized than the area activated by the infrequent stimulus. However estimated positions were not consistent in all subjects.

Discussion

The sources of the infrequent response at 450 msec were estimated near the bilateral parahippocampal or occipitotemporal gyri. These results are plausible since similar results were obtained using implanted electrodes by Halgren et al.[4]. The results of Halgren suggested that the sources were broad activities and the dipole strength was rather large in our results. Hence, our estimation may be the center of gravity of these broad active areas.

The sources of the responses to the infrequent stimulus at 350 msec could not be estimated by the two-ECD model with a low error. It suggests that these components contained many source generators. These responses probably contain some of P3a, P3b and slow wave. Rogers et al.[2] estimated sources of these late complex responses to the infrequent stimulus on the assumption that ECDs are spatially separated enough to have little effect on each other. They estimated each source separately using single-ECD model. They estimated sources in the medial temporal and occipital lobe. However, in some case it is thought that this assumption is not correct. Hence we tried to estimate the signal source from MEG whole cortex using a four-dipole model. In a few cases, the sources were estimated near the bilateral parahippocampal gyri and the insula with Es less than 10 %. Those results agree with the study of Halgren et al.[4] which reported the activities of the hippocampus and the superior temporal sulcus at 380 msec. Since estimation result with multi-ECD model could have many local solutions, we cannot accept this estimation results as it is. We believe that new evaluation standards through mathematical analysis and anatomical and physiological knowledge as a constraint of estimation should be developed to multiple source estimation.

References

[1]Okada T.C., Kaufman L. and Williamson S.J. The hippocampal formation as a source of the slow endogenous potentials. Electroenceph. clin. Neurophysiol., 1983, 55:417-426.

[2]Rogers R.L., Baumann S.B., Papanicolaou A.C., Bourbon T.W., Alagarsamy S. and Eisenberg, H.M. Localization of the P3 sources using magnetoencephalography and magnetic resonance imaging. Electroenceph. clin. Neurophysiol., 1991, 79:308-321.

[3]Takeda T., Morabito M., Kumagai T. and Endo H., use of CRT as a visual stimulator in MEG measurements, Proc.XI Int. Conf. on Event-related Potentials of the Brain Y.Koga, ed., Elsevier Int Congr Ser

[4]Halgren E., Baudena P., Clarke J.M., Heit G. Liegeois C., Chauvel P. and Musolino A., Intracerebral potentials to rare target and distractor auditory and visual stimuli. I. Superior temporal plane and parietal lobe, Electroencephalor. Clin. Neurophysiol., 1995, 94:191-220.

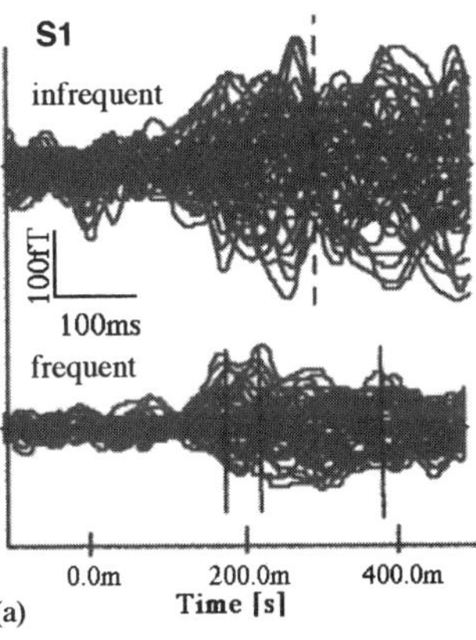

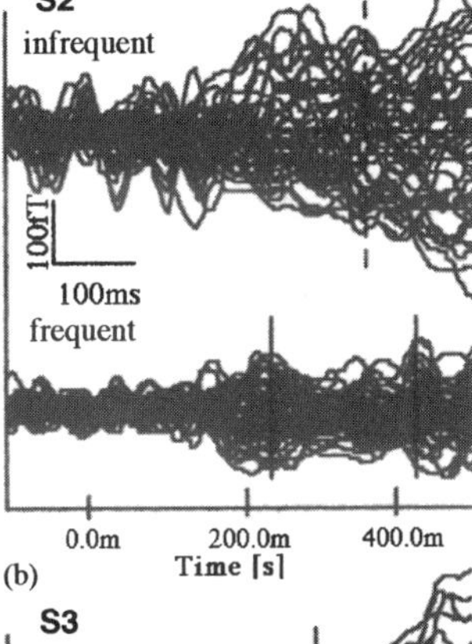

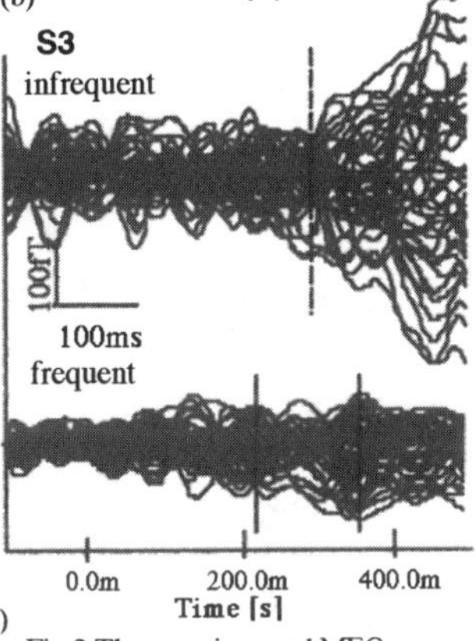

Fig.3 The superimposed MEG waveforms of S1, S2 and S3

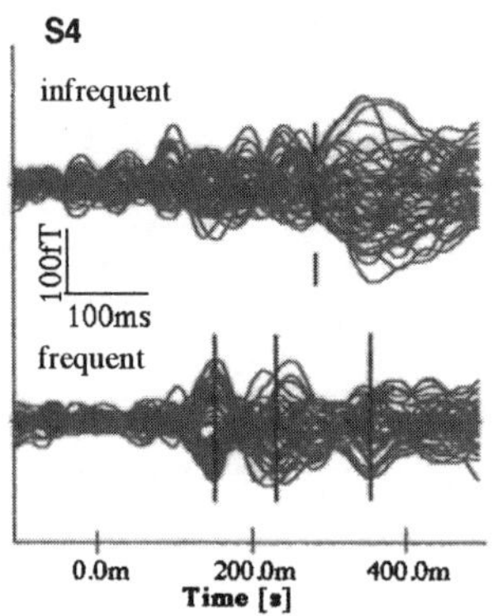

Fig.3(D) The superimposed MEG waveforms of S4

Fig.4 The MEG response of right lateral sensor

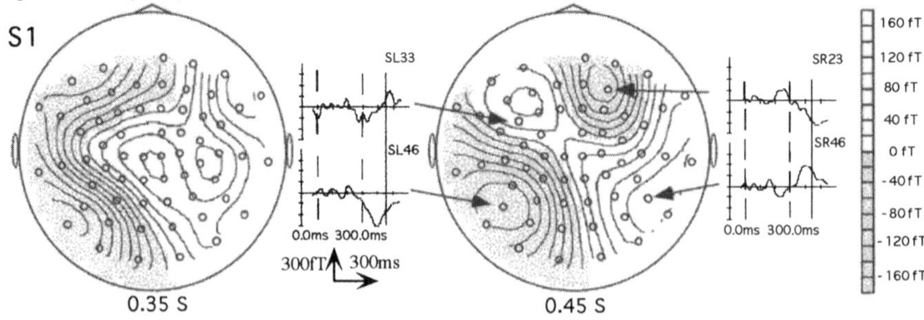

Fig.5 The isocontour maps. The reponses to the infrequent stimulus at the latency of 350 and 450 msec

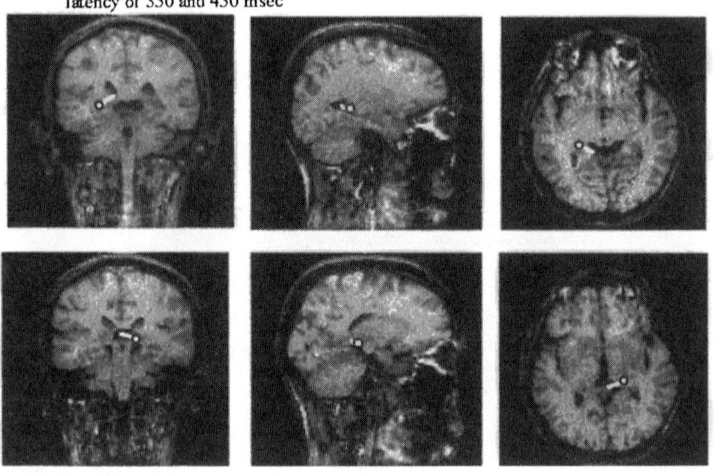

Fig.6 Estimated generator sources at the latency of 460 ms for S2. E was 7 %. The source strength was 210 nAm for the left source and 230 nAm for the right source.

Cerebral Activities in the Occipital and Temporal Regions When Subjects Perform a Matching Task of Visual Stimuli

Kuriki, S.[1,2] and Hirata, Y.[1]

Hokkaido University[1,2], Sapporo, Japan
National Institute for Physiological Sciences[2], Okazaki, Japan

Introduction

One of the most exciting possibilities in the MEG study is that the time course of cerebral activation can be obtained within the brain anatomy of individual subjects during various information processing from sensory to higher functions. Previous MEG studies describe the brain activities associated with visual and higher processing such as object viewing [1] and naming [2]. The main objective of this study is to visualize the time course of the cerebral activation during the recognition of unreadable pseudo-characters in terms of graphical structure. This work would be contrasted with our previous studies on the visual and phonological processing of readable real-characters [3,4].

Method

Five right- and one left-handed male subjects (21-41 ys) participated in the experiments. In each of the 100 trials in a measurement session, the subject was presented with sequential stimuli of three characters (sample), a single character (probe), and a white circle (mark) (Fig.1). The visual stimuli were displayed on a screen in front of the subject's eyes with a visual angle of 0.9 deg of each character at a luminance of 4.3 cd/m^2 in the darkened magnetically shielded room. The three samples and the single probe were displayed symmetrically around the fixation point and at the left lower quadrant of 0.8 deg eccentricity, respectively. The stimulus characters were unreadable, which had been synthesized from Japanese Katakana scripts (readable characters) by shifting the line elements in the vertical or horizontal direction. These pseudo-characters were used as the sample and probe. The subject was instructed to fixate his eyes during the recording period after the probe onset before the mark, and to judge whether the shape of the probe agreed with that of the samples. He replied the match or mismatch by pressing one of two keys after the mark. After pilot measurements it was determined to use identical single pseudo-characters as the three samples to assure high performance of the task. The matching probability between the sample and the probe was adjusted to about 50 %.

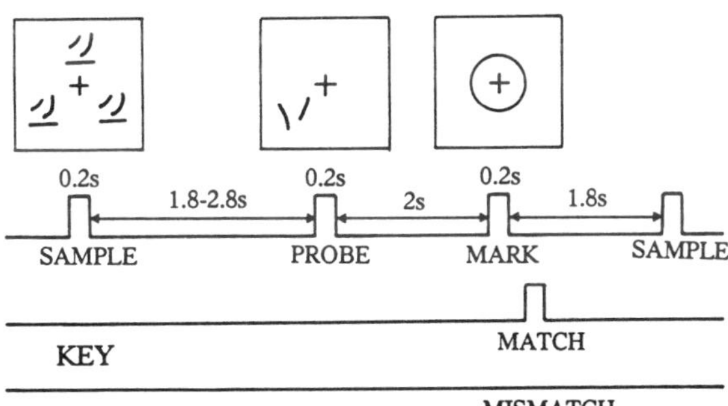

Fig. 1. Visual stimuli of pseudo-characters and the sequence of stimuli for the delayed matching task between the probe and sample characters.

MEG recordings were carried out in the magnetically-shielded room using a dual 37-channel SQUID with upper and lower 37-channel coil sets. The right and left sides of the head were simultaneously measured, varying slightly the recording area from anterior to posterior side in 6-8 sessions. We measured the responses to the single probe character in a 2.0 s recording period after the probe onset. The averaged responses over 100 trials were digitally filtered to 1-40 Hz, and the average in a 0.4 s prestimulus period served as the baseline. Match and mismatch responses were separately averaged. In the waveforms of these responses no significant difference was observed, and therefore we averaged the match and mismatch responses off-line.

Localization of single equivalent current dipoles (ECDs) was made, using a spherical conductor model, separately for the right and left cerebral hemisphere from the 37-channel fields recorded over each hemisphere. The localization was performed every 3.5 ms during the recording period, from which the ECDs having goodness-of-fit values more than 90 % in about 15 ms around the root-mean-squared (rms) peak were selected and averaged. Further averaging of the selected ECDs across recording sessions were made in individual subjects. In all subjects magnetic resonance (MR) tomography images, and surface images in some subjects, were obtained. The MR coordinates were transformed into the head coordinates, and the ECD location was registered in the MR images.

Results

The average of the reaction time for key press, which was measured separately for all subjects was 606 ms. The correct answer of the task recorded during the MEG measurements was above 90 % in most subjects, which indicates that the subjects correctly performed the task. Although the amplitude of the responses depended on the subject, three main peak components were observed in all subjects. The range of the rms peak latency of the 1st, 2nd, and 3rd component was 125-180, 180-250, and 250-460 ms, respectively. Fig. 2 illustrates the responses in duplicated recordings at selected coil locations of high-amplitude positive and negative fields for the three components. Isofield maps for the 2nd component, showing dipolar patterns, are also illustrated.

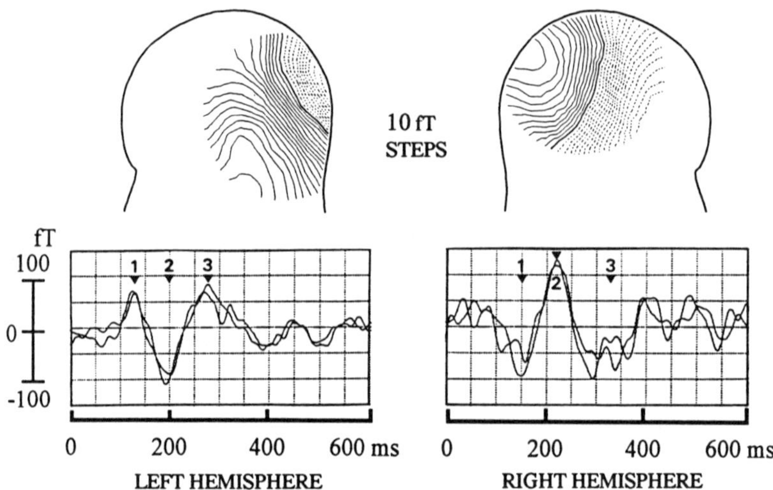

Fig. 2. Lower : waveforms of the responses in duplicated recordings at selected coil locations, where three main components are indicated by numbers. Upper: isofield maps for the 2nd component.

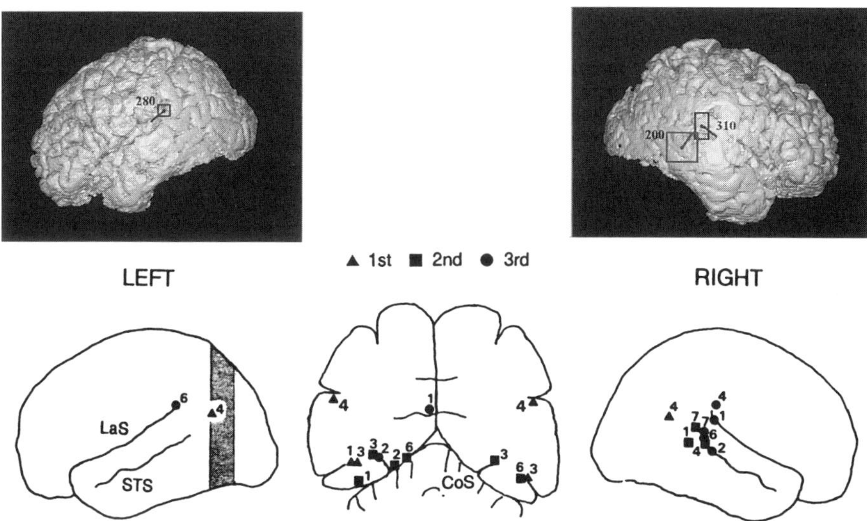

Fig. 3. ECD sources projected on the surface MR images and on the schematic views of the brain, where the sources were classified into three components and numbered to indicate subjects. The locus of the coronal sections is indicated by the hatched area shown on the left lateral surface.

The computed ECD sources were found at specific locations in, or close to, sulci, reflecting the fact that cortical currents tangential to the skull exclusively contribute to the neuromagnetic field. From the projection on the MR images the source location in the individual brain structure was determined in reference to the main sulci. The obtained brain sites of the three sources in all subjects are plotted on the schematic maps shown in Fig.3, together with representative source location on the brain surface MR images. These results indicate that the 1st and some of the 2nd component sources are located at the inferior part of the temporo-occipital region around the collateral sulcus (CoS). This area is included in the posterior part of the fusiform gyrus (FG) and the inferior temporal gyrus (ITG). In the lateral temporal region the 2nd and 3rd component sources are located at the posterior part of the superior temporal sulcus (STS) and the posterior end of the lateral sulcus which is circumscribed by the supramarginal gyrus (SmG). The right-hemisphere preponderance of these sources is a marked contrast to the left-hemisphere dominant activities observed in the readable character task [3,4].

Discussion

From the average reaction time of 606 ms and plausible transmission time from the motor cortex to the muscle for the key press, the judgment of the matching of the probe with the samples should be completed before about 500 ms. Therefore, the observed 1st to 3rd components before 460 ms would reflect the processes involved in the visual perception, recognition, and comparison.

The density of the 1st and 2nd component sources in the FG and ITG in the inferior temporo-occipital extrastriate cortex was left hemisphere dominant, suggesting stronger neural activity in the left side than the right. Sakurai et al. [5] reported the activation of the regional cerebral blood flow in the extrastriate cortex in the posterior inferior temporal region, in addition to that in the medial and lateral occipital region, during a reading task of Japanese characters   (words and non-words).   The activation in the inferior temporal region is significant only on the left side for non-words, and on both sides for real-words but stronger on the left than the right. On the other hand, in visual attention during a shape-matching task of squares and rectangles, significant activation is observed in the occipital inferior region in the lingual gyrus and in the occipitotemporal FG, with stronger amplitude on the left side than the right [6]. Enhancement of the activity in the FG by selective attention is also reported [7]. Considering these results of PET studies, it is suggested that the early responses in the FG and ITG in our pseudo-character task reflect the early graphical processing of the visual stimuli, but are not specific to the scripts or characters. The pseudo-characters are not familiar to the subjects, so that there is strong demand on the analysis of the structure and visual attention.

The sources of the 2nd and 3rd components occurring in 180-460 ms were found in the posterior temporal region including the STS and SmG. This temporal activity was exclusively unilateral on the right side. A few PET studies have delineated the cortical activities in the lateral temporal region. In a shape attention task the activity was significantly elevated along the right STS (Broadmann's area (BA) of 21/22) [6], while in an object discrimination task the left inferior and middle temporal gyri (BA 20, 21) were more strongly activated than those on the right, but the SmG was bilaterally activated [8]. In macaques, areas in the inferior temporal cortex are involved in the object recognition and discrimination [9]. In our previous study of the delayed matching task using Japanese readable characters [3,4] the posterior STS and SmG in the left temporal region are suggested to be related to the phonological process, which is required in the recognition of real characters. These active sites are homologous to the location of the 2nd and 3rd component sources in the right hemisphere in the present study. Thus it may be suggested that the left and right human temporal cortices are specialized into the phonological (more generally, linguistic) and graphical/structural processing. Further studies on the specific functional role of the activities in the right posterior STS and SmG are needed.

In conclusion, the MEG study combined with cognitive task of delayed matching of visually-presented unreadable pseudo-characters revealed the early neural activities in the inferior temporo-occipital extrastriate cortex and the later activities in the right posterior temporal cortex. These activities may be the neural substrate which subserves the visual and subsequent structural processing of the pseudo-characters.

References

[1] Lu ST, Hämäläinen MS, Hari R, Ilmoniemi RJ, Lounasmaa OV, Sams M, Vilkman V, Seeing faces activates three separate areas outside the occipital visual cortex in man. Neurosci. 1991, 43:287-290.

[2] Salmelin R, Hari R, Lounasmaa OV, Sams M, Dynamics of brain activation during picture naming. Nature, 1994, 368:463-465.

[3] Kuriki S, Hirata Y, Fujimaki N, Tsuchiya N, Kobayashi T, Neuromagnetic study of human cortical areas related to visual/auditory information processing. Human Brain Mapping, 1995, Suppl.1: 27.

[4] Kuriki S, Hirata Y, Fujimaki N, Kobayashi T. Cerebral activities of the human brain in a delayed matching task of visual characters. Electroenceph. clin. Neurophysiol. (to be published).

[5] Sakurai R, Momose T, Iwata M, Watanabe T, Ishikawa T, Kanazawa I, Semantic process in Kana word reading: activation studies with positron emission tomography. NeuroRep. 1993, 4:327-330.

[6] Corbetta M, Miezin F, Dobmeyer S, Shulman GL, Petersen SE, Attentional modulation of neural processing of shape, color, and velocity in humans. Science 1990, 248:1556-1559.

[7] Heintz HJ, Mangun GR, Burchert W, Hinrichs H, Scholz M, Münte TF, Gös A, Scherg M, Johannes S, Hundeshargen H, Gazzaniga MS, Hillyard SA, Combined spatial and temporal imaging of brain activity during visual selective attention in humans. Nature, 1994, 372:543-546.

[8] Sergent J, Ohta S, Macdonald B, Functional neuroanatomy of face and object processing. Brain, 1992, 115:15-36.

[9] Gross CG, Rocha-Miranda CE, Bender DB, Visual properties of neurons in inferior temporal cortex of the macaque. J. Neurophysiol. 1972, 35:96-111.

Magnetoencephalographic study on cerebral cortical activities related to speech

Kyuhou, S., Sasaki, K., Nambu, A., Matsuzaki, R., Tsujimoto, T. and Gemba, H.

National Institute for Physiological Sciences, Okazaki, Japan

Introduction

Human motor speech function has been investigated neuropsychologically [1]. Recently by using non-invasive techniques measuring regional cerebral blood flows such as the functional magnetic resonance imaging (fMRI) [3] and the positron emission tomography (PET) [5], it was demonstrated that several cortical areas not only in the dominant hemisphere but also in the non-dominant hemisphere were activated during speech tasks. As these methods, however, have poor time resolution, it is difficult to get temporal information of the cortical activities. The magnetoencephalography (MEG) is suitable for investigating the temporal sequence of the activation of cortical areas on speech, since MEG has good temporal and spatial resolution. The present study is a trial to identify functionally the motor speech center in the frontal lobe of human cerebral cortex by using multichannel SQUID gradiometers. Parts of the results have been reported shortly [8].

Methods

Subjects were six healthy adults of 29-64 years old, five males (four of right handedness, one of left handedness) and one female (right handedness). Handedness was assessed according to Edinburgh Handedness Inventory by Oldfield [4]. An examinee lay on a bed in a magnetically shielded room and two sets of 37-channel SQUID gradiometers (Magnes Model 700, BTi) were attached to both sides of the head for simultaneous magnetic recordings from both the left and right hemispheres. EEG electrodes were also placed over the frontal-parietal part of the head according to the international 10-20 method. Light stimuli of two different colors were delivered in front of a subject for 500 ms in irregular order and random time intervals of 6.0-9.0 s. The subject should respond to either of the stimuli by uttering a short word (noun), e.g., 'pan' [pan] ("bread" in Japanese), 'en' [en] ("round" in Japanese) or the other by vocalizing a short simple voice, e.g., 'pa' [pa], 'e' [e], respectively. The initial sounds of both voices are to be the same, i.e., 'pa' [pa], 'e' [e] in these examples. The subject was asked to recall the meaning of the word every time it was uttered, or to think nothing while uttering the simple voice. Magnetic fields were respectively averaged by the onset pulses of light stimuli 100 times. EMGs were recorded from perioral muscles with silver surface electrodes, rectified and averaged by the pulse. The onset of voice was also monitored with a microphone and was found to be considerably later than that of perioral EMGs.

Results

Evoked magnetic fields (EMFs) preceding utterance of a short word and vocalization of a simple voice are shown respectively in Fig. 1 A and B. The subject is right-handed. Two components of EMFs preceding the utterance were observed over the lateral frontal parts on both sides. The early component at a peak latency of 120-165 ms from the onset of light stimulus was prominent on utterance of the word but this component was hardly seen on that of the simple voice. The late component of magnetic changes at a peak latency of 160-190 ms was observed similarly on both the utterances. The late component appeared about 20-40 ms before the start of perioral EMGs. Perioral EMGs started about 170-220 ms after the onset pulse of light stimuli and differences were hardly observed between latencies of the EMGs on the word and simple voice utterances. These findings suggest that the early activity is important for speech of the word. On the other hand, the late component is thought to be related to the execution of the articulatory movements because latencies between this component and the onset of the articulatory movements were constant. By using the dual 37-channel system, magnetic fields were simultaneously recorded from both the left and right hemispheres (Fig. 2). The time courses of some of the EMFs were similar in both the left and right hemispheres as exemplified in Fig. 2 A and B. There were no detectable differences between respective latencies of the EMFs on both sides in these examples. Locations of the current dipoles responsible for

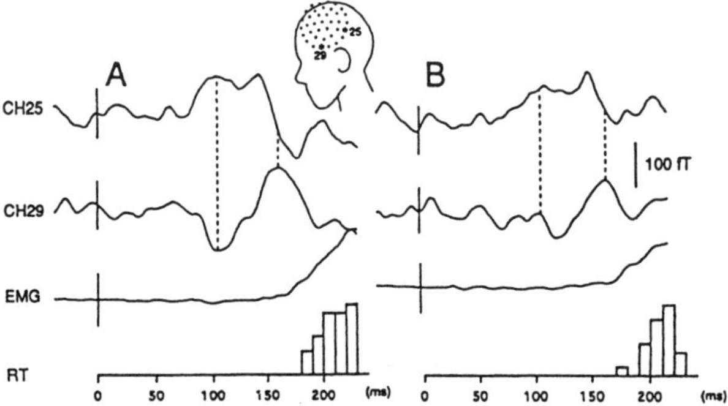

Fig. 1: Evoked magnetic fields on utterance of a short word (A) and on that of a simple voice (B) recorded from the left dorsolateral part of the head of a right-handed subject. Two sites of specimen records (CH25, 29) were indicated by asterisks among dots for 37-channel sites on the head. Reaction times (RT) were measured from the onset of the visual stimuli to that of the EMGs of perioral muscles.

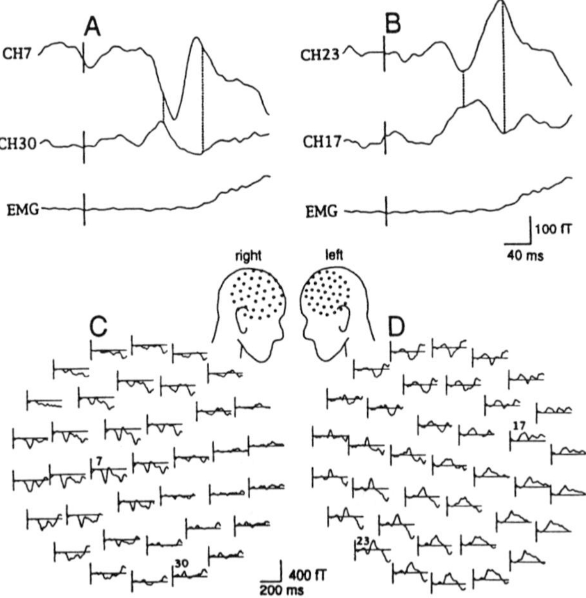

Fig. 2: Simultaneous recordings of magnetic fields from both the right (A, C) and left (B, D) hemispheres on utterance of a short word by using a dual 37-channel system. Records of the channels 7 and 30 in C and the channels 23 and 17 in D are respectively enlarged in A and B with EMGs of perioral muscles rectified. All records were aligned 100 times by the onset pulse of light stimuli. Positions of sensor coils are shown by dots on the right and left side view of the head shapes in the middle. The subject is the same as in Fig. 1.

the two components of EMFs were estimated and were plotted on MRI pictures of the subject (Fig. 3). The estimated current dipoles for the early component were located in the inferior lateral part of the prefrontal cortex (Fig. 3 A1, A2, B1, B2). Those for the late component were positioned on the lateral part of the frontal-parietal cortical areas around the central sulcus (Fig. 3 C1, C2, D1, D2). The estimated current dipoles appeared in the quite symmetrical places in the left and right hemispheres. These results indicate that the inferior lateral prefrontal and motor-sensory cortices are sequentially activated prior to speech in both the hemispheres.

Four other subjects including one of left handedness also showed findings similar to those described above. The remaining right-handed subject revealed current dipoles at latencies of 151-165 ms only in the right prefrontal cortex. Similar activities were sometimes noted also in the left hemisphere but not so conspicuous as in the right hemisphere. This subject showed the frontal mental theta wave on calculation more predominantly in the right frontal lobe than in the left one [7] and the dominant hemisphere would be presumed on the right side.

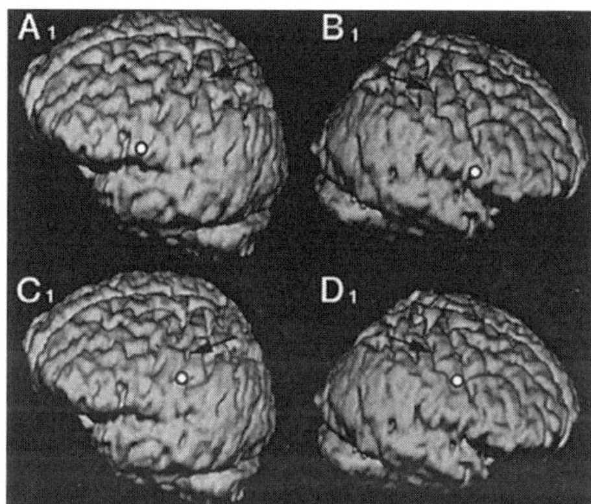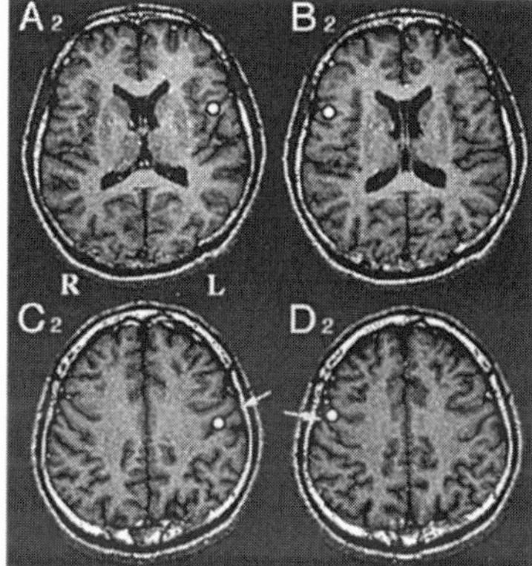

Fig. 3: Locations of the equivalent current dipoles (ECDs) were plotted on the 3-dimensionally reconstructed MRI pictures of the brain (A1-D1). The same results were also plotted on the horizontal sections of MRI pictures (A2-D2). Open circles in A and B indicate the locations of the ECDs responsible for the early component (138 ms after the onset of visual stimulus on both the left and right sides) while those in C and D show the ECDs responsible for the late component (167 and 186 ms after the visual stimuli respectively on the left and right side). The capital letters (R and L) in A2 respectively indicate the right and left side of the head. The arrows indicate the central sulcus. The same subject in Fig. 1 and Fig. 2.

Discussion

It was found in the present study that the inferior lateral prefrontal cortex and the motor-sensory cortex were sequentially activated prior to utterance of a word while only the motor-sensory cortex was activated in preceding vocalization of a simple voice. Pronounced activation of the inferior lateral prefrontal cortex before uttering a word suggests that this cortical area is the motor speech center related to organizing words. The area appears to correspond to the Broca's speech area. In fact, the equivalent current dipoles were located in the frontal opercular region (Broadmann's area 44-45) that is thought to be the Broca's speech area. The motor-sensory cortex activated both on utterance of a word and a simple voice probably corresponds to the face area because it was located in the cortex of the lateral part of the central sulcus. Activation of the face area may induce the articulatory movements for production of words and simple voices.

The present study demonstrated that the bilateral frontal cortices were activated prior to the utterance. Recent MEG [6] and EEG [2] studies also reported the activation of the bilateral frontal cortices on speech tasks such as the picture naming and self-paced utterance. These results indicate that not only the dominant hemisphere but also the non-dominant hemisphere plays important roles on utterance in the normal speech, although the speech center in the dominant hemisphere would be indispensable and crucial.

References:

[1] Broca, P. Remarques sur le siège de la faculté du langage articulé, suivies d'une observation d'aphémie, Bull. Soc. Anat., 1861, 36: 330-357.
[2] Deecke, L., Engel, M., Lang, W. and Kornhuber, W.W. Bereitschaftspotential preceding speech after holding breath, Exp. Brain Res., 1986, 65: 219-223.
[3] Hinke, R.M., Hu, X., Stillman, A.E., Kim S.G., Merkel, H., Salmi, R. and Ugurbil, K. Functional magnetic resonance imaging of Broca's area during internal speech, Neuroreport, 1993, 4: 675-678.
[4] Oldfield, R.C. The assessment and analysis of handedness: the Edinburgh Inventory., Neuropsychologia, 1971, 9: 97-113.
[5] Roland, P. Brain Activation, New York, Wiley-Liss, 1993.
[6] Salmelin, R., Hari, R., Lounasmaa, O.V. and Sams, M. Dynamics of brain activation during picture naming, Nature, 1994, 368: 463-465.
[7] Sasaki, K., Tsujimoto, T., Nambu, A., Matsuzaki, R. and Kyuhou, S. Dynamic activities of the frontal association cortex in calculating and thinking, Neurosci. Res., 1994, 19: 229-233.
[8] Sasaki, K., Kyuhou, S., Nambu, A., Matsuzaki, R., Tsujimoto, T. and Gemba, H. Motor speech centres in the frontal cortex, Neurosci. Res., 1995, 22: 245-248.

Acknowledgements

We thank Mr. O. Nagata and Mr. Y. Takeshima for their technical help.

Present Address

Sasaki, K. and Tsujimoto, T. : Faculty of Medicine, Kyoto University, Kyoto 606-01, Japan.
Gemba, H., Kyuhou, S. and Matsuzaki, R. : Kansai Medical University, Moriguchi 570, Japan.
Nambu, A. : Tokyo Metropolitan Institute for Neuroscience, Fuchu 183, Japan.

Magnetic Recording of the Propagation of Motor Unit Action Potentials in the Human Leg Muscles

Masuda, T., Endo, H., Kumagai, T., and Takeda, T.

National Institute of Bioscience and Human-Technology, Tsukuba, Japan

Introduction

We have developed multichannel electromyography (EMG) for detecting the propagation of motor unit action potentials [1]. With this non-invasive technique it becomes possible to define motor endplate regions based on the source of the propagation. The measurement of muscle fiber conduction velocity has also become possible (Fig. 1). The purpose of the present study was to clarify the possibility of magnetic recording technique for detecting the activity in human skeletal muscles and to develop another new non-invasive method for analyzing and diagnosing human motor functions. With the magnetic recording we can expect better localization of current sources compared with the electrical recording.

Methods

We used a 64-channel SQUID system developed by CTF Systems (Canada). Since this system was designed to detect the magnetic field originating from the human brain, the detecting coils were arranged on a surface of hemisphere. The inter-coil distance was about 40 mm.

The subject were three normal males. They put the knee in the hollow of the Dewar vessel with fully flexing

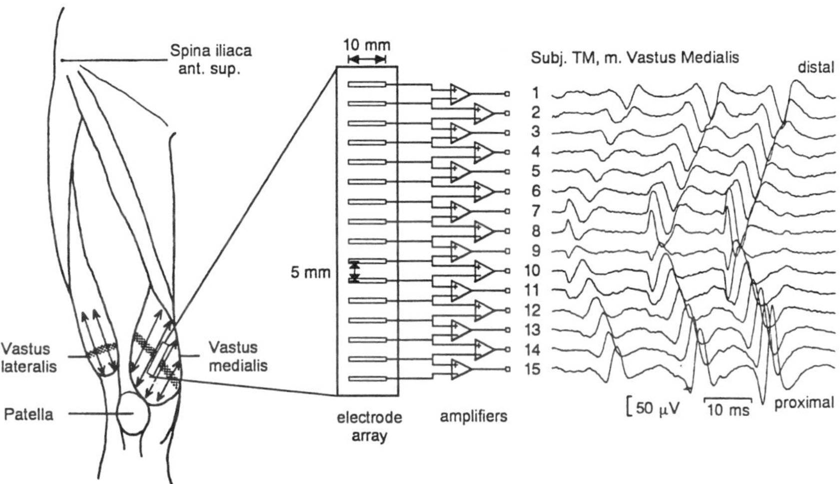

Fig. 1 Propagation of motor unit action potentials in the vastus medialis and the vastus lateralis detected with a linear array of surface electrodes. The shaded area indicates the motor endplate regions defined by the source of the propagation.

the knee joint (Fig. 2). The subject isometrically exerted the force of knee extension. We recorded the magnetic field for 10 s with a sampling frequency of 1.25 kHz. After the recording the sampled signals were high-pass filtered above 5 Hz with a Butterworth filter to eliminate motion artifact. In this setup the motor endplate regions of the vastus medialis and the vastus lateralis were located near the edge of the recording area and the proximal side of the muscles was outside of the SQUID sensors.

To detect discharges of single motor units we used a technique of EMG feedback. Bipolar electrodes were attached on the belly of the vastus medialis. One of the electrodes was placed on the motor endplate regions. The onset of the EMG therefore indicates the start of the depolarization on muscle fibers (Fig. 1). The EMG signal was amplified and presented to the subject as an amplitude modulated sound. The subject controlled the contraction force to produce spike trains that appear at regular intervals with the same intensity. After the recording we averaged the signals from SQUID sensors by using the zero-crossings of the EMG as triggers.

Results

The amplitude of the magnetomyographic (MMG) field recorded during moderate voluntary contraction was 10-20 pT peak-to-peak (Fig. 3), which was 10-100 times greater than that of magnetoencephalogram. There was no difficulty in recording MMG signals. MMG signals were sensitive to motion artifact, which was however removed by high-pass filtering above 5 Hz.

The isofield maps in Fig. 4 showed magnetic field spreading from the proximal side of the vastus medialis to the direction of the patella. The similar spreading was observed for some of the peaks and may correspond to the propagation of motor unit action potentials shown in Fig. 1. However most of the fields showed more complex distribution that was difficult to be explained by a simple moving dipole.

To clarify the components constituting MMG field, we analyzed MMG of a single motor unit. The averaged MMG and EMG signals are shown in Fig. 5. The number of averaging was 84 for this record. MMG of a single motor unit had smaller amplitude of 1-2 pT peak-to-peak. The isofield map of the averaged MMG still showed complex distribution preceding a clear and strong dipole pattern of reversed polarity (Fig. 6). The first map was at the onset of the depolarization.

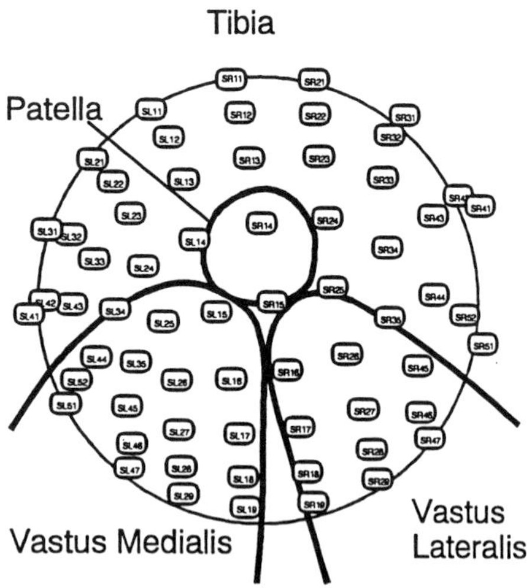

Fig. 2 Configuration of the SQUID sensors and the location of the knee extensor muscles in the recording area.

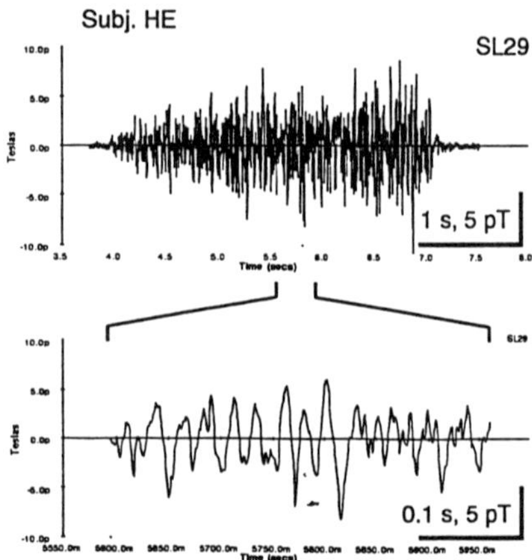

Fig. 3 Raw waveform of MMG detected under moderate voluntary contraction. A portion of the waveform is extended in time (below).

826

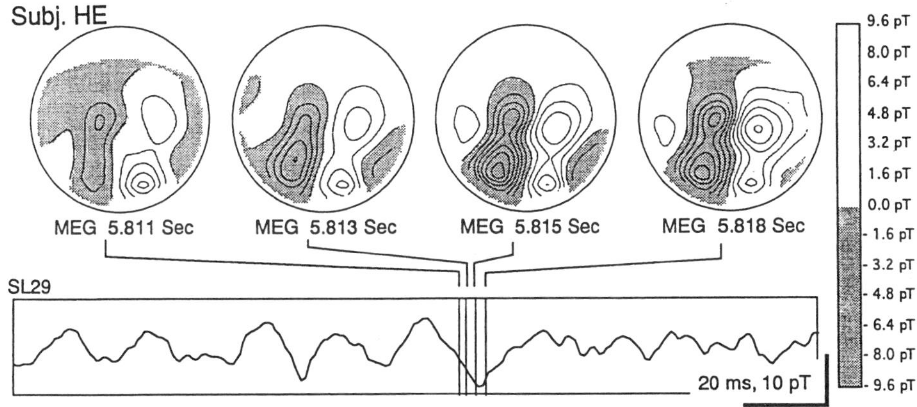

Fig. 4   Isofield maps of the MMG shown in Fig. 3.

Discussion

The magnetic field produced by discharges of single motor unit is explained by a pair of tripoles that arise at the motor endplate regions and propagate in opposite directions along the muscle fibers (Fig. 7) [2]. The currents in each tripole flow in opposite directions and therefore the magnetic field produced by the tripole is cancelled by each other and weak. The resulting magnetic field is equivalent to that produced by a dipole of a current flowing in the direction of the propagation. When the depolarization reaches the tendon, the first current of the tripole diminishes and the second current produces a strong magnetic field with an opposite direction. In the field shown in Fig. 6 the tripole travelling in the proximal direction reaches the tendon earlier than the distal tripole.

If the SQUID sensors are closer to the muscle fibers and have a higher resolution of, for example, 5 mm, the magnetic field of the tripole may become evident. With the resolution of 40 mm, only a subtracted current of the tripole is detected as a dipole. The low resolution causes the complex magnetic field for single motor unit action potentials and more complex field for interference MMG.

As a conclusion we must detect discharges of single motor unit to analyze the magnetic field originating from skeletal muscles or improve the spatial resolution of the SQUID system. If these conditions are satisfied, the magnetic recording technique could have an advantage in localizing the position and in estimating the absolute strength of current sources compared with the conventional EMG recordings.

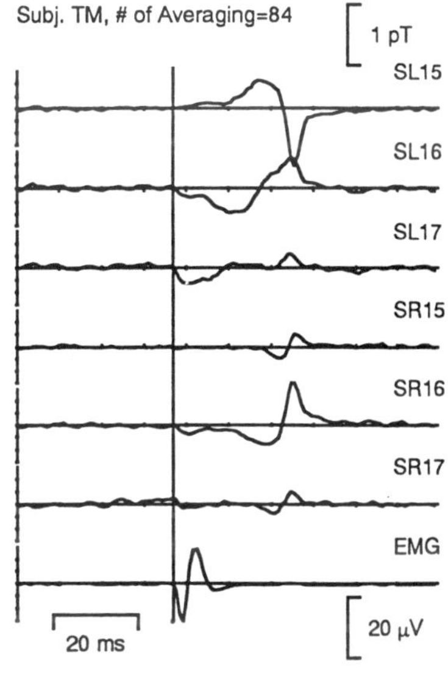

Fig. 5   MMG and EMG of a single motor unit after averaging. The positive-going zero-crossings of EMG were used as triggers.

References:

[1] Masuda, T., and Sadoyama, T. Skeletal muscles from which the propagation of motor unit action potentials is detectable with a surface electrode array, Electroencephalography and Clinical Neurophysiology, 1987, 67: 421-427.

[2] McGill, K. C., and Huynh, A. A model of the surface-recorded motor-unit action potential, In: Harris, G., and Walker, C. Proceedings of the Annual International Conference of the IEEE Engineering in Medicine and Biology Society, 1988, 10: 1697-1699.

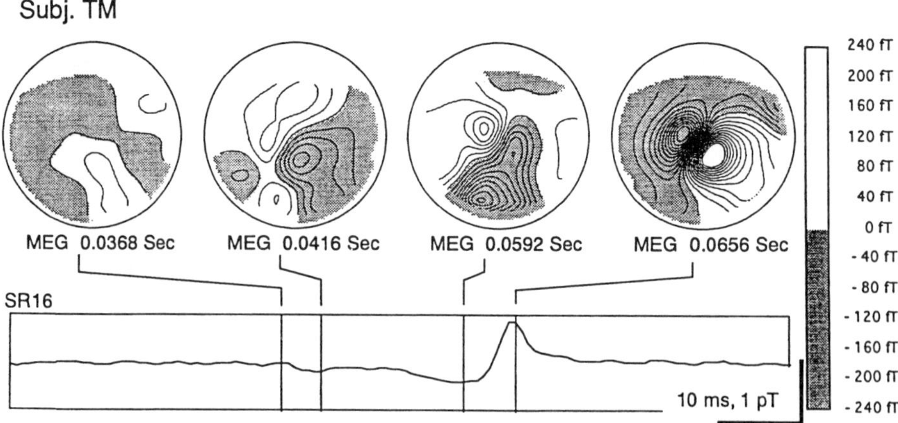

Fig. 6 Four typical isofield maps of MMG produced by a single motor unit shown in Fig. 5.

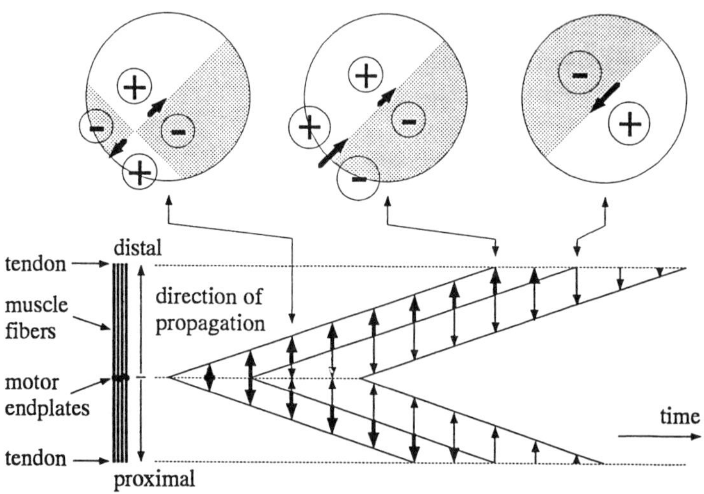

Fig. 7 Schematic diagram explaining the magnetic field produced by a single motor unit. The current in muscle fibers is modeled as a pair of tripoles that propagate in opposite directions to the tendons.

Effect of Luminosity Change on Visual Evoked Field

Matsuhashi, M.[1], Oyanagi, H.[1], Kumagai, T.[2], Endo, H.[2] and Takeda, T.[2]

[1]*Department of Ophthalmology, Toho University School of Medicine, Tokyo, Japan;*
[2]*National Institute of Bioscience and Human-Technology, Tsukuba, Japan*

Introduction

The study of vision in an objective manner is important for a better understanding of the factors which influence the normal visual system [1] [2]. By using the visual evoked potential (VEP), efforts have been made to measure the vision objectively [3]. Responses elicited by flash stimulation are called flash or luminance VEPs. Increase of luminosity in flash stimulation shows larger responses in the flash VEPs. This study is to show the effects of flash intensity on the flash visual evoked brain magnetic field (VEF).

Methods

VEFs were recorded by 64 channel whole head type magnetoencephalogram (MEG) systems by CTF Systems Inc. of Canada. The dewar was designed especially for Japanese people. Other than MEG system, reference SQUIDs for enviromental noise reduction, head positioning system were also included in the whole systems. Measurements were carried out in an AK3b magnetically shielded room by Vacuumschmelze GmbH of Germany.

Setup of hardware, recording and processing of data were made by the MEG software of CTF Systems. Butterworth low pass filter at 40 Hz and notch filters at 50, 100 and 150 Hz were used for recording. Artifact rejection was set on for noise produced by blinking.

Responses were recorded 100 msec for pretrigger and 600 msec for posttrigger. In this paper, however, results are shown only from -100 msec to 400 msec. Signals were averaged 32 times. For data processing, offsets were removed, digital filter of low pass at 40 Hz was applied and third gradient formation was done with the software for the noise reduction [4]. Head positioning data was recorded each time.

A red light emission diode (LED) was used for the flash stimulation. Drive unit for LED was hand made and it was controlled by a personal computer. Stimulus frequency was randomized and its mean was 0.7 Hz, stimulus duration was 10 msec. An LED for fixation was placed 50 cm in front of the eyes and one for stimulus was put 2.9 degrees to the left. The intensity of stimulus was changed by use of neutral density filters to reduce the relative intensity to 10^{-1}, 10^{-2}, and 10^{-3}.

Responses were recorded from three normal subjects. Responses by each stimulus intensity were compared and analyzed. All subjects showed major peaks at around 100 msec and 160 msec. They were called P100-N100*m* and P160-N160*m* respectively in this paper. Amplitude was measured peak-to-peak, from P100-N100*m* to P160-N160*m*. Peak time latency was measured from the onset of the stimulus to the each peak.

Results

With the increase of stimulus intensity, larger responses were recorded. Fig. 1 shows responses at sensor SR17 in subject 1. Amplitude of P160-N160*m* became larger with the increase of stimulus intensity. As for the peak time, they did not differ significantly both at P100-N100*m* and P160-N160*m* in these recordings. Maps at P100-N100*m* in each stimulus intensity are shown in Fig.2, and the location of sensor SR17 is indicated.

Responses at sensor SR27 in subject 2 are shown in Fig.3. Similar rersults in amplitude were recorded. The peak time of P160-N160*m* became shorter in stronger stimulus intensities in these recordings.

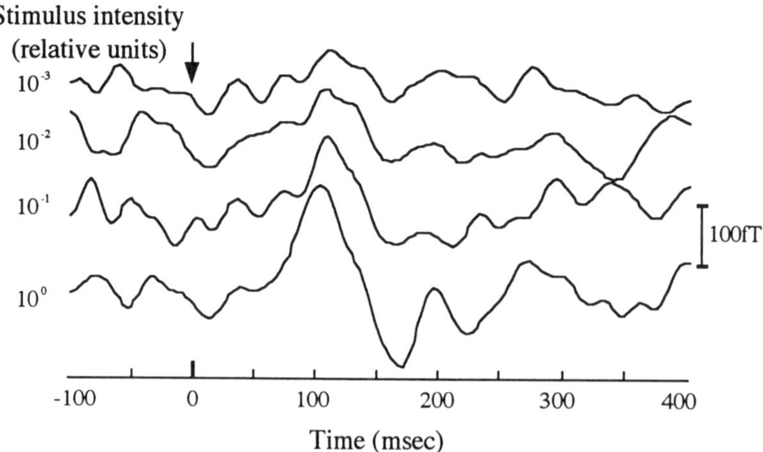

Fig. 1 Visual evoked magnetic fields recorded at sensor SR17 in subject 1. Note that the stronger the stimulus intensity was, the larger the responses became at P100-N100*m* and P160-N160*m*. This phenomenon has been recognized in all subjects. Peak time did not differ significantly both at P100-N100*m* and P160-N160*m* in these recordings. Arrow indicates stimulus onset in this and following figures.

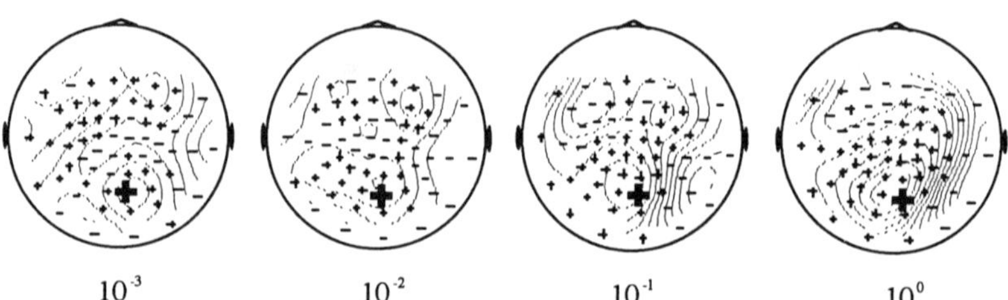

Fig. 2 Maps at P100-N100*m* in subject 1. These maps showed similar topography, suggesting that responses of P100-N100*m* by each stimulus intensity represented similar magnetic field changes. Large bold plus designates sensor SR17, from where responses are shown in Fig. 1.

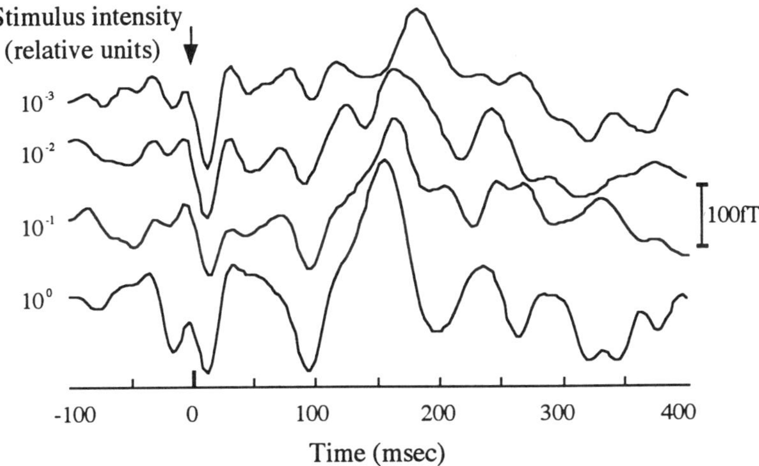

Fig. 3 Visual evoked magnetic fields recorded at sensor SR27 in subject 2. Note that the stronger the stimulus intensity was, the larger the responses became at P100-N100*m* and P160-N160*m* as found in Fig. 1. As for the peak time, peak time delay was recognized at P160-N160*m* in accordance with the diminution of stimulus intensity in these recordings.

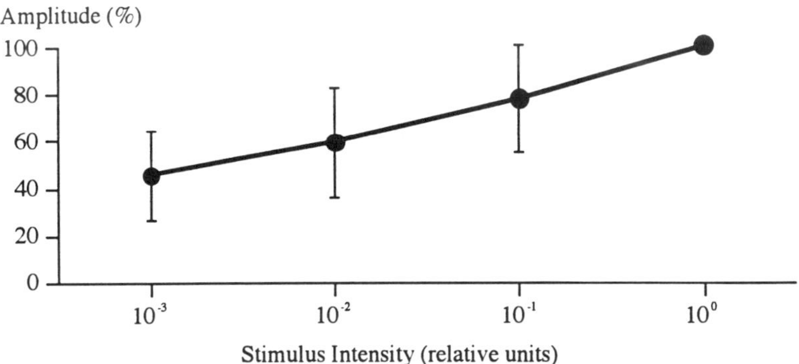

Fig. 4 The amplitude (peak-to-peak) of the averaged evoked magnetic field of P160-N160*m* as a function of the stimulus intensity. The mean amplitute got larger in accordance with the increase of stimulus intensity. By the present stimulus intensity, the amplitude kept rising and there was no steady or decrease. The vertical lines indicate one standard deviation.

In all subjects, with the increase of stimulus intensity, larger responses were recorded. Amplitude of P160-N160m was compared in each subject. Amplitude with the stimulus intensity of 10^0 was made 100 per cent and relative amplitude was calculated. The mean and one standard deviation with each stimulus intensity was 77.9 ± 22.5 at 10^{-1}, 59.5 ± 23.5 at 10^{-2}, and 45.7 ± 19.1 at 10^{-3} (Fig. 4).

Discussion

This study showed the effect of stimulus intensity change on the flash VEFs. The stronger the luminosity was, the larger the responses became as found in VEPs. By the present condition, the amplitude kept rising and there was no steady or decrease. This suggested that the intensity of LED used in this study was not too strong. As for the latency, with the increase of the flash intensity, shortening of peak time was found in most recordings as shown in Fig. 2, however, peak time did not differ significantly in some recordings as shown in Fig.1. Because peak time change was not always consistent, it was not analyzed and illustrated as in Fig.4 in this study. Further study would be needed for analyzing the characteristics in peak time.

Maps shown in Fig. 2 showed similar topography. This suggested responses compared represented the similar ones. However, topograms were not exactly the same configuration. And the source modeling was done in each record and the location of the equivalent current dipole was estimated. Also, they did not show exactly the same position. Therefore, comparing the size of the equivalent current dipole was not applied in this study. However, when those results could.show exactly the same configuration and position, it would be worth comparing the size of the equivalent current dipole. In this study, definite results were found by comparing the amplitude of responses.

References

[1] Matsuhashi, M., Kumagai, T., Endo, H., Nishimura, C., and Takeda, T. Visual evoked multichannel magnetoencephalography, Japanese Journal of Clinical Ophthalmology, 1995, 49:403-406.

[2] Matsuhashi, M.,Oyanagi, H., Kumagai, T., Endo, H., and Takeda, T. Binocular summation in visual evoked brain magnetic fields by pattern stimulation, The Journal of Japan Biomagnetism and Bioelectromagnetics Society, 1995, 8:80-83.

[3] Matsuhashi, M., Oguchi, Y., and Uemura, Y. Clinical ERG and VECP from unsedated children aged under one year, Japanese Journal of Ophthalmology, 1979, 23(supp.):279-285.

[4] Takeda, T., Endo, H., and Kumagai, T. On the measurement reliability of the 64-channel whole-cortex MEG system, The Journal of Japan Biomagnetism and Bioelectromagnetics Society, 1994, 7:224-227.

Evoked Magnetic Fields for Gustatory Stimuli on the Tongue

Murayama, N.[1], Nakasato, N.[2], Hatanaka, K.[3], Fujita, S.[4], Igasaki, T.[1], Kanno, A.[5] and Yoshimoto, T.[2]

[1]Department of Electrical Engineering & Computer Science, Kumamoto University, Kumamoto, Japan; [2]Department of Neurosurgery, Tohoku University School of Medicine, Sendai, Japan; [3]KRI, Kyoto, Japan; [4]R&D Center, Osaka Gas Co. Ltd., Osaka Japan; [5]MEG Laboratory, Kohnan Hospital, Sendai, Japan

Introduction

Animal studies have suggested that gustatory information primarily projects to the bilateral frontal opercula and dorsal insulae [1, 2, 3]. However, clinical studies showed that lesions at the postcentral gyrus [4, 5] or anterior insula [6] caused gustatory disturbances. Scalp topography of gustatory evoked potentials (GEPs) has identified the largest amplitude (about 450 ms in latency) at the central vertex [7]. However, it is difficult to localize the primary gustatory area by non-invasive scalp GEP recording, due to inhomogeneous head conductivity and possible overlaps of bilateral activities. Recently, magnetoencephalography (MEG) has been introduced to localize the sources of various evoked responses. MEG is known to be less influenced by the inhomogeneous head conductivity than electroencephalography. For example, the bilateral cortical sources of auditory evoked magnetic fields were successfully separated and localized precisely on the primary auditory cortex [8]. In the present study, we measured evoked magnetic fields for gustatory stimuli on the tongue in an attempt to identify the location of the primary gustatory area in the human cerebral cortex.

Methods

Five normal subjects participated in this experiment. The subject was seated in a magnetically shielded room with the chin resting on a frame. Taste stimuli were delivered through a tube to the anterior region of the tongue which was protruded slightly out of the mouth. The starting point of the stimulation was detected by an optical sensor mounted at a distance of 5 mm from the nozzle of the tube. As the taste stimuli, 10% glucose and 0.3 mol NaCl were used. Responses obtained by these taste stimuli were compared with that of distilled water (non-taste control). The duration of taste stimuli was about 10 sec. During the interstimulus periods, the tongue was rinsed with distilled water delivered for about 10 s. During the interstimulus period, the tongue was rinsed with distilled water for about 10 s. MEG recording used a helmet-shaped MEG system with 64 channels (CTF Systems - Osaka Gas). Evoked magnetic fields were measured over the entire head except lower part of the face. Signals were averaged over 40 stimulus presentations and two separate sessions were performed for each stimulus condition. A two-dipole model was employed for source estimation of evoked responses, using the best fit sphere for each subject's head surface based on MRI data. Estimated dipoles were superimposed on MRI. Somatosensory evoked fields (SEFs) for median nerve stimuli were measured as physiological landmarks of the central sulcus.

Results

Figure 1 shows the typical waveforms of evoked fields due to the stimulus of 10% glucose on the tongue. Bilateral responses were found over the scalp for glucose, NaCl and distilled water. The most prominent peak, M175, indicated a bilateral two-dipole pattern with a latency of about 130-210 ms (Fig. 2). The amplitude of M175 was larger for 10% glucose than for distilled water.

Sources of the bilateral M175 responses were estimated using a two-dipole model. When the signal/noise ratio was good enough, generators of the M175 were estimated on the most medial parts of the bilateral opercula near the insulae (Fig. 3). These sources were approximately on the border of the frontal and parietal operculum (Fig. 3). The dipole-fit error was lowest for 10% glucose, possibly due to the higher signal/noise ratio than for the other taste stimuli.

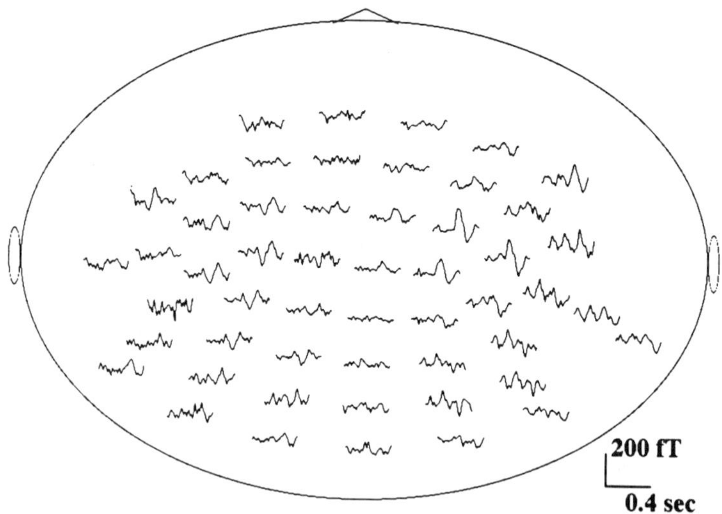

Fig.1 Wave forms of evoked magnetic fields for 10% glucose stimuli on the tongue in subject 1. All 64 MEG channels over the entire head are shown from above.

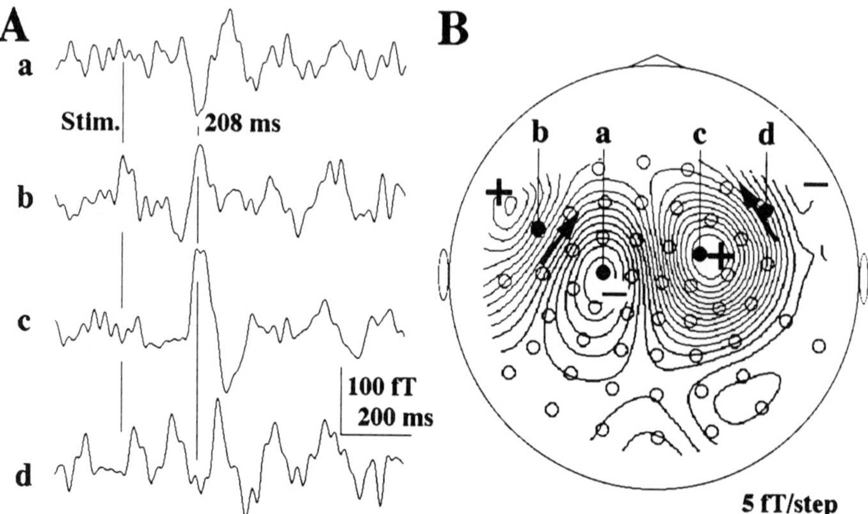

Fig.2 Evoked magnetic fields for 10% glucose stimuli on the tongue in subject 1. A: The first peak (M175) of the evoked fields appeared at a latency of about 200 ms. B: Isofield map indicates a bilateral two-dipole pattern at the M175 peak. Arrows indicate the M175 currents approximately.

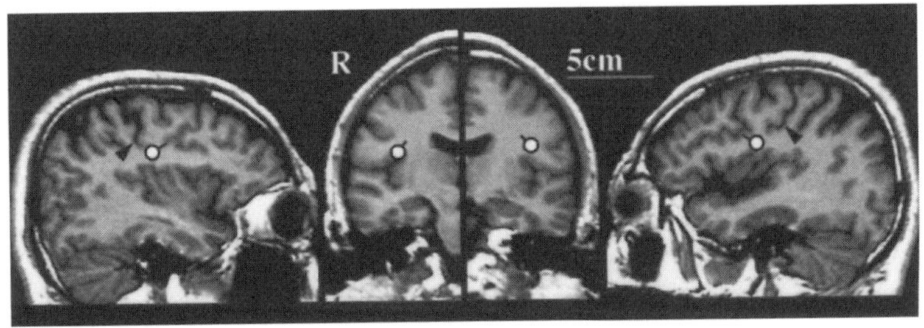

Fig.3   Source estimation of the evoked magnetic fields for 10% glucose stimuli on the tongue in subject 1. Circles and bars indicate the M175 dipole positions and orientations, respectively. Arrowheads indicate the central sulcus identified by the somatosensory evoked fields.

Table 1 M175 component in evoked magnetic fields for taste and non-taste stimuli on the tongue in 5 subjects.

Subjects Age/Sex	Stimulus on Tongue	M175 Response		
		Latency	Isofield Map Pattern	Dipole Estimation
1) 44/M	glucose	208 ms	two-dipole	bilateral opercula
	water	211 ms	two-dipole	n/a*
2) 36/M	glucose	147 ms	two-dipole	bilateral opercula
	water	152 ms	two-dipole	n/a*
3) 23/M	glucose	152 ms	two-dipole	left: operculum right: n/a**
	NaCl	126 ms	two-dipole	left: operculum right: n/a**
	water	131 ms	two-dipole	left: operculum right: n/a**
4) 40/M	glucose	163 ms	two-dipole	left: operculum right: n/a**
5) 32/M	glucose	160 ms	two-dipole	n/a*

glucose = 10% glucose; NaCl = 0.3 mol NaCl; water = distilled water; n/a* not available, dipole fit error > 20%; n/a** = not available, small dipole in the right hemisphere

Table 1 shows a summary of the M175 responses in all five subjects. Although the peak latencies varied , the sources for the gustatory stimuli were estimated on the bilateral opercula in two subjects and unilateral operculum in the two subjects. The latter two subjects indicated an asymmetrical two-dipole pattern with smaller signal amplitude on the right hemisphere. We also found two a dipole-pattern for the distilled water stimuli, but their sources were not localized. We observed some later wave peaks following M175, but the reproducibility of the signals was not good enough for dipole estimation.

Discussion

The M175 response observed in the present experiment might be the response to tactile stimulation of the tongue. However, we believe that gustatory information also contributes to M175 for the following three reasons. First, the two gustatory stimuli evoked larger responses than distilled water. Second, the M175 dipole positions were localized in the most medial part of the operculum near the insula, suggesting areas deeper than the tongue somatosensory areas. Third, the M175 latency (130-210 ms) seems longer than the response from the primary somatosensory area (SI). Karhu et al. [9] recorded SEFs to electrical stimulation of the tongue and observed two peaks (about 50 ms and 140 ms in latency). They speculated that the earlier source originates in the primary somatosensory area (SI) on the contralateral side to the tongue stimulation and that the later source originates in the secondary somatosensory area (SII) in the bilateral hemispheres. More late components were reported in a scalp GEP study. Funakoshi and Kawamura [10] obtained two potential components. They interpreted the earlier one (about 150 ms in latency) as the tactile response, and the later ones (500 ms -1500 ms in latency) as the gustatory response. However, the later potential might be related to higher brain functions such as gustatory recognition.

Although we successfully measured evoked magnetic fields for gustatory stimuli to the tongue, further study is necessary to confirm the location of the human gustatory cortex.

References:

[1] Benjamin. R.M., and Burton, H. Projection of tongue nerve afferents to anterior opercular-insular cortex in squirrel monkey (*Saimiri sciureus*), Brain Research, 1968, 7: 221-231.

[2] Ogawa, H. Gustatory cortex of primates: anatomy and physiology, Neuroscience Research, 1994, 20: 1-13.

[3] Scott, T.R., Yaxley, S., Siekiewicz, Z.J., and Rolls, E.T. Gustatory responses in the frontal opercular cortex of the alert cynomulgus monkey, Journal of Neurophysiology, 1986, 56: 676-890.

[4] Bornstein, W.S. Cortical representation of taste in man and monkey. I. Functional and anatomical relations of taste, olfaction and somatic sensibility. Yale J. Biol. Med., 1940a, 12: 719-736.

[5] Bornstein, W.S. Cortical representation of taste in man and monkey. II. The localization of the cortical taste area in man and a method of measuring impairment of taste in man. Yale J. Biol. Med., 1940b, 13: 133-156.

[6] Motta, G. I centri corticali del gusto. Boll. Sci. Med., 1959, 131: 480-493.

[7] Kobal, G. Gustatory evoked potentials in man. Electroencephalography and Clinical Neurophysiology, 1985, 62: 449-454.

[8] Nakasato, N., Fujita, S., Seki, K., Kawamura, T., Matani, A., Tamura, I., Fujiwara, S., and Yoshimoto, T. Functional localization of bilateral auditory cortices using an MRI-linked whole head magnetoencephalography (MEG) system, Electroencephalography and Clinical Neurophysiology, 1995, 94: 183-190

[9] Karhu, J., Hari, R., Lu, S.T., Raetau, R., and Rif, J. Cerebral magnetic fields to lingual stimulation, Electroencephalography and Clinical Neurophysiology, 1991, 80: 459-468.

[10] Funakoshi, M., and Kawamura, Y. Summated cerebral evoked responses to taste stimuli in man, Electroencephalography and Clinical Neurophysiology, 1971, 30: 205-209.

Acknowledgments

This work has been supported in part by Grants-in-Aid for Scientific Research No. 03404042 and No. 06404050 from the Ministry of Education, Science and Culture of Japan; and Magnetic Health Science Foundation.

Different Effects of Stimulus Rate on the Somatosensory Evoked Magnetic Fields and Potentials

Nagamine, T.[1], Mäkelä, J.P.[2], Mikuni, N.[1,3], Mima, T.[1], Nishitani, N.[1], Sato, T.[1], Ikeda, A.[1], Kimura, J.[4] and Shibasaki, H.[1]

[1]Departments of Brain Pathophysiology, [3]Neurosurgery, [4]Neurology, Kyoto University School of Medicine, Kyoto, Japan; [2]Central Military Hospital, Helsinki, Finland

Introduction

In order to execute the processing of continuing information without delay, sensory cortices require rapid recovery from the preceding stimuli. Depending on the received stimuli, every information is analyzed serially and conveyed parallel through several brain areas. Although the brain areas responsible for these processes have been roughly estimated, exact location remains still unclear. Both somatosensory evoked potentials (SEPs) [1-3] and magnetic fields (SEFs) [4,5] with different interstimulus intervals (ISIs) have been used to characterize the recovery function of the somatosensory cortex. Magnetoencephalography (MEG) is sensitive to intracellular currents flowing tangential to the head surface whereas electroencephalography (EEG) reveals both tangential and radial currents. We have employed simultaneous recordings of electroencephalographic and magnetoencephalographic somatosensory evoked responses (SERs) to elucidate the temporal behavior of the sensory cortex by utilizing recovery function.

Methods

Nine normal volunteers gave informed written consent for the study. The subject was sitting in a chair in a magnetically shielded room (Kyoto University School of Medicine), with the forearm supinated. The subjects were requested to keep their eyes open and to avoid frequent blinking. Unilateral median nerve was stimulated at the wrist with the constant current of electric shock lasting 0.2 ms. Five out of nine subjects were tested on both sides. The intensity of the electric shock was adjusted to invoke weak thumb twitch. Stimulus rate of once every 0.9 s was applied first and then that of 4 s was tested.

A 122-channel DC-SQUID neuromagnetometer, covering the whole scalp [6], was utilized to record the magnetic fields produced by the brain. 61 pairs of two orthogonally oriented figure-of-eight planar first-order gradiometers was designed to detect the two orthogonal derivatives of the radial magnetic field at each pair location. This configuration of the gradiometers allowed us to catch the source current just below the largest signal. The locations of three head position indicator (HPI) coils placed on the scalp were measured with respect to the three anatomical landmarks (bilateral preauricular points and nasion) using a 3-D digitizer (Isotrak 3S1002, Polhemus Navigation Sciences, Colchester, Vermont, U.S.A.).

EEG signals were collected from cup electrodes placed at locations C3, CP3, P3, C4, CP4, P4 and Oz (according to the modified International 10-20 System), referenced to the linked earlobe electrodes. Two electro-oculograms (EOGs) linking the right infraorbital with either the right earlobe or the right supraorbital position were adopted to monitor the eye movements. The impedance of all electrodes was kept below 5 kΩ.

Signals of all EEG, EOG channels and eight selected MEG channels from bilateral central and occipital areas were plotted continuously on the paper so that the vigilance of the subject could be checked.

The signals were digitized at 860 Hz after filtered with analog passband of 0.03–280 Hz for MEG and that of 0.07-280 Hz for EEG and EOG. On-line averaging was done with respect to the electric shock with the time window of 600 ms including 100-ms pretrigger interval. Averaging started after giving five electric shocks. Big artifacts exceeding the preset rejection level of 3 pT/cm for MEG and 150 μV for EEG and EOG were excluded from the averaging automatically.

Averages of around 200 sweeps were obtained for one session, and at least two sessions were employed for 0.9-s stimulus interval condition to confirm the reproducibility. For the responses to 4-s stimulus interval, the first subaverages of 50 responses and the following 50 subaverages were obtained and compared.   Responses with reproducible wave forms were group-averaged across sessions.

Amplitude of the responses was measured by taking the time window from 100 to 5 ms prior to the electric shock as a baseline.   Component configurations were determined at the contralateral somatomotor area.

Results

In all subjects, N20 as well as magnetic counterpart, N20m, was clearly demonstrated over the contralateral somatomotor area.   Two typical subjects are shown in Fig. 1.   SEF responses showed P30m, P40m and P60m, among which P60m demonstrated the biggest amplitude, and some intervening smaller peaks. Although some subjects had subsequent deflections around 80 - 200 ms, they were much smaller than those at 20 - 60 ms.   SEP recordings showed subsequent deflections, labeled as P30, N35, P45, N75, P100, N150 and P250.   Individual variation was evident in the amplitudes, which were clearly larger for stimulus interval at 4 s than that at 0.9 s (Table).   Both SEF and SEP components later than 40 ms were readily suppressed by the 4-s stimulus interval condition (Fig. 2).   Although SEF responses became almost stable later than 150 ms, SEP demonstrated bigger amplitude than those of early latencies. Especially P250 was strongly enhanced with 4-s stimulus interval (Fig. 2).

Table   Latency and absolute amplitude of each component for 0.9-s and 4-s stimulus interval condition. Measurements were performed at the channels showing biggest N20m or N20 component.   Responses following the right median nerve stimulation and those with the left median nerve are put together.

	stimulus interval							
	0.9 s					4 s			
	Latency (ms)		Amplitude (fT/cm)		n	Latency (ms)		Amplitude (fT/cm)		n
	mean	SD	mean	SD		mean	SD	mean	SD	
N20m	21.6	1.6	61.5	28.7	14	21.6	1.6	60.2	27.7	14
P30m	28.5	2.1	22.0	18.5	10	28.5	2.1	21.3	18.1	10
N35m	31.6	2.0	26.8	18.3	5	31.6	2.0	27.2	17.6	5
P40m	38.0	3.5	67.9	46.5	11	38.1	3.2	87.8	60.7	11
N50m	43.7	4.7	43.8	36.0	7	43.9	4.5	54.5	44.8	8
P60m	60.4	8.9	89.7	40.4	12	59.2	8.4	102.9	47.8	13
P100m	91.8	12.6	50.3	39.8	8	91.8	12.6	81.7	57.6	8
N150m	130.5	26.4	20.3	12.8	9	130.5	26.4	38.4	24.3	9

	Latency (ms)		Amplitude (μV)		n	Latency (ms)		Amplitude (μV)		n
	mean	SD	mean	SD		mean	SD	mean	SD	
N20	20.0	1.6	2.06	0.90	14	20.0	1.6	1.59	0.74	14
P30	26.1	1.4	1.11	0.52	8	26.1	1.4	0.87	0.40	8
N35	32.0	3.1	1.42	0.75	8	31.0	1.6	0.97	0.75	8
P45	40.7	3.4	2.11	1.37	13	42.2	5.1	3.43	2.12	6
N75	65.3	10.9	1.78	1.53	9	64.7	10.5	2.85	2.51	10
P100	97.7	7.9	2.23	1.59	5	96.6	6.8	4.25	2.85	7
N150	151.2	23.4	2.46	1.74	8	135.6	20.0	4.31	1.85	12
P250	276.4	16.7	4.04	2.24	4	291.3	19.4	11.59	4.75	12

Lt Median N. Stimulation (Subj. 1)

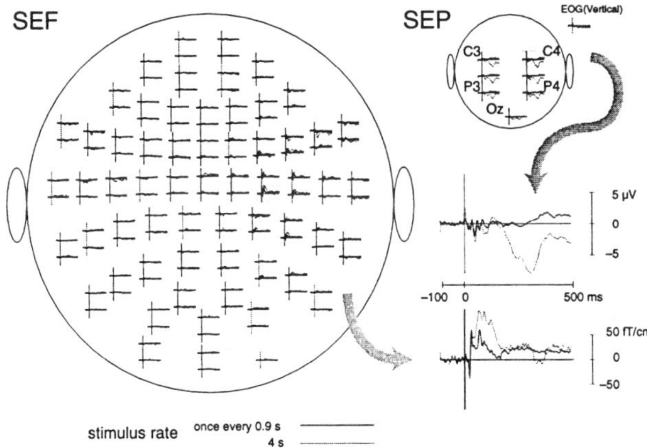

Lt Median N. Stimulation (Subj. 4)

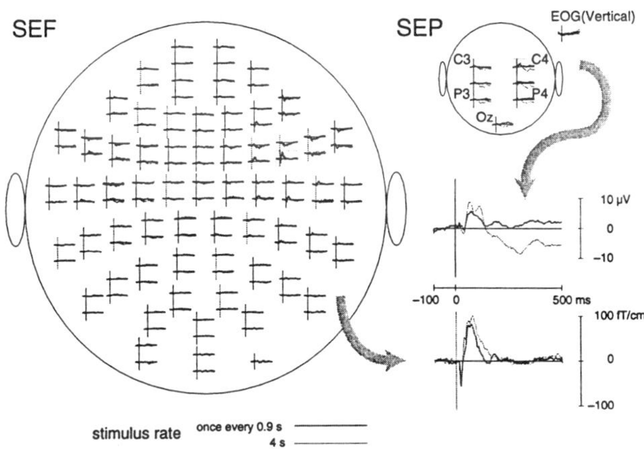

Fig. 1

Simultaneous recording of SEF and SEP in two subjects.

Responses with stimulus rate of once every 0.9 s and 4 s conditions are superimposed. The biggest responses marked by shading are magnified in the right lower corner. One of the MEG channels was not available because of the machine trouble.

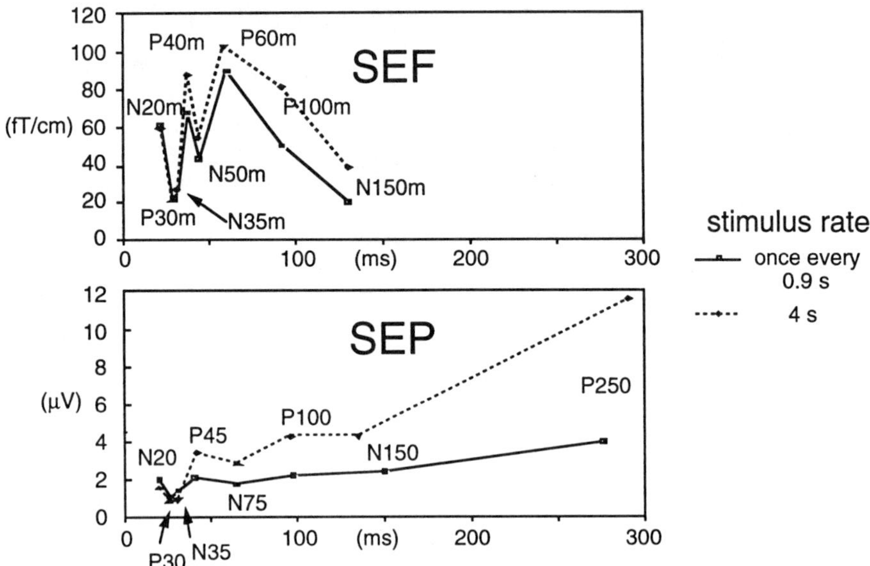

Fig. 2

Mean absolute amplitude of SEF and SEP as a function of latency for different stimulus rate.

Discussion

Clear difference in the behavior with respect to the stimulus rate between SEF and SEP gives us some hint for the sensory processing. Since the prominent responses in SEF have already ended within 100 ms and less affected by the longer stimulus interval, it is suggested that the corresponding generators are responsible mainly in the early stage of sensory processing. Therefore, tangential components of the sensory cortex seem to play a main role during the first 100 ms and be replaced by the radially oriented generators.

References

[1] Nakamura, M., Shibasaki, H., Nishida, S., Neshige, R. 1990, IEEE Trans. Biomed. Eng., 37: 738-740

[2] Emori, T., Yamada, T., Seki, Y., Yasuhara, A., Ando, K., Honda, Y., Leis, A.A., Vachatimanont, P. Electroencephalography and clinical Neurophysiology, 1991, 78:116-123.

[3] Romani, A., Bergamaschi, R., Versino, M., Callieco, R., Calabrese, G., Cosi, V. Electroencephalography and clinical Neurophysiology, 1995, 96:475-478.

[4] Huttunen, J., Ahlfors, S., Hari, R. Electroencephalography and clinical Neurophysiology, 1992, 82:176-181.

[5] Hari, R., Karhu, J., Hämäläinen, M., Knuutila, J., Salonen, O., Sams, M., Vilkman, V. 1993 European Journal of Neuroscience, 1993, 5: 724-34.

[6] Ahonen, A.I., Hämäläinen, M.S., Kajola, M.J., Knuutila, J.E.T., Laine, P.P., Lounasmaa, O.V., Parkkonen, L.T., Simola, J.T. and Tesche, C.D. 122- channel SQUID instrument for investigating the magnetic signals from the human brain. Physica Scripta, 1993, T49: 198-205.

Binaural Fusion of Virtual Pitch in the Human Auditory Cortex

Pantev, C.[1], Elbert, T.[2], Ross, B.[1], Eulitz, C.[1] and Terhardt, E.[3]

[1]Institut for Experimental Audiology, Münster, Germany; [2]Department of Psychology, Konstanz, Germany; [3]Institute of Electroacoustics, Munich, Germany

The vowels of human speech and the sounds produced by many musical instruments are harmonic complex tones. These are comprised of pure tones that are integer multiples of a fundamental frequency. The perceived pitch of complex tonal stimuli is qualitatively different from the one of pure tones. The pitch of pure tones is a basic auditory percept which is monotonically related to its frequency, a phenomenon known as spectral pitch. The pitch of complex tones is termed virtual pitch. It can be assessed by auditory matching to pure tones. The virtual pitch of harmonic complex tones corresponds to a frequency approximating the frequency difference between the components, i.e. to the fundamental frequency, even if the fundamental frequency is missing. The fact that the first six to eight harmonics of a complex tone can be perceived as separate spectral pitches led Terhardt [1] to conclude that the virtual pitch is extracted by a type of "Gestalt" recognition, i.e. virtual pitch is an auditory attribute that is established on a higher level of processing than spectral pitch. The distinction between spectral pitch and virtual pitch can be likened to the distinction between primary visual contours and virtual ("illusory") contours in vision [2]. Just as a visual virtual contour can be regarded as being "inferred" from a set of primary contours, virtual pitch is regarded as the result of a two-stage process which involves peripheral frequency analysis and virtual-pitch formation by a pertinent "pitch processor"[3]. It is believed that the pitch processor is located in more central regions of the auditory system. An important feature is binaural fusion, i.e. the ability to combine spectral components regardless of the ear to which they are presented. It was therefore hypothesized that neural activity ascending from both ears would fuse into one percept [4].

Here we provide first evidence from magnetic source imaging that a pitch processor fuses the information coming from separate ears to produce a virtual pitch represented in the same cortical area as the spectral pitch.

Methods

In the present investigation, the magnetic responses of pure-tones and complex-tones were measured in eight normal hearing, right handed subjects. Four different burst stimuli (duration 500 ms) were presented with a 10 ms cosine slope. The four stimuli differed in their spectral contents (Fig. 1b) as follows: 1) a 250 Hz pure tone; 2) 1500 Hz pure tone; 3) a complex tone composed of eight components, these being the 4th to 11th harmonics of the missing fundamental frequency of 250 Hz (presented monaurally); and 4) two complex tones composed of the even and odd harmonics of the eight-component complex tone (presented binaurally). Stimulus intensity was 60 dB relative to the individual hearing threshold. The interstimulus interval was randomized between 2.5 and 3.5 s. The eight-component complex tone shows a 250 Hz periodicity and although it does not include spectral power below 1000 Hz, 250 Hz periodicity was audible. All subjects were able to match this virtual pitch successfully with a spectral pitch of or close to 250 Hz, that was simultaneously presented to the opposite ear (Fig. 1c, top). In the same way the complex tones containing the even or odd harmonics of the eight-component complex tone (periodicity of 500 Hz) were matched with a spectral pitch of or close to 500 Hz (Fig. 1c, bottom).

Two magnetic sensor arrays (BTi) were used to record the neuromagnetic brain activity (bandwidth 0.1-100 Hz) simultaneously from both auditory cortices. Stimulation was presented in 16 blocks of 150 stimuli each and the block sequence was varied across subjects in a randomized order. Thus each AEF was an average of 600 stimulus related epochs. In order to reduce possible variations caused by changes in attention and vigilance, subjects watched cartoons during the session. Due to the head-sensor constellation the sensing coils of the bottom sensor were further away from the scalp as compared with the upper sensor. This fact causes a reduction of the AEF amplitudes recorded from the right hemisphere (c.f. Fig. 1a).

For each AEF distribution corresponding to the four stimulus conditions a single equivalent current dipole model located in a best fitting local sphere was used to estimate the moment and the space coordinates of a dipole, representing the cortical activity and the center of the activated cortical area.

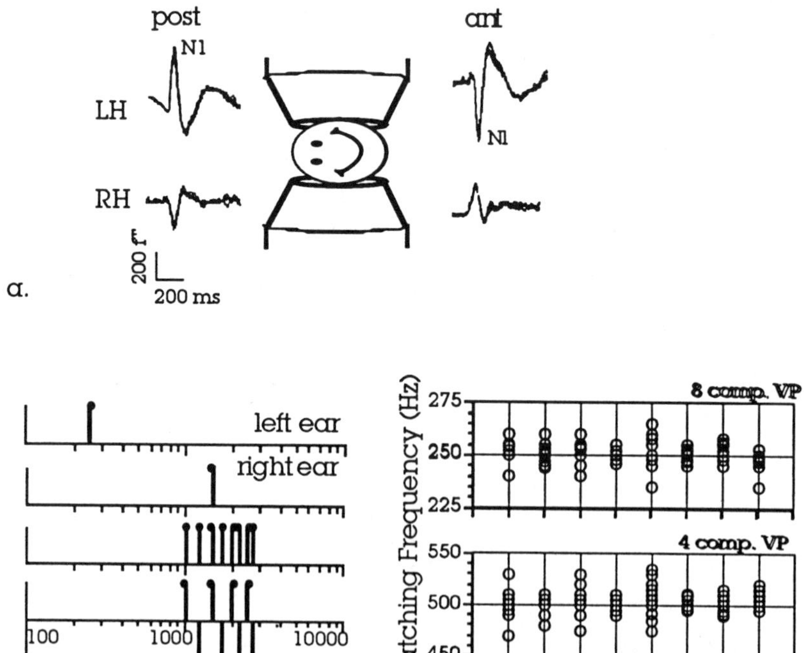

Fig. 1: **a)** Simultaneous magnetoencephalographic recording of auditory evoked activity from the left and the right hemispheres. The waveforms at the posterior and anterior field extremes are superimposed for four repeated blocks of identical stimulation. **b)** Scheme of the spectral content of the stimuli: **First** and **second rows** show spectral pitches of 250 and 1500 Hz presented to the left ear (upward pointing pin). **Third row**: virtual pitch consisting of the 4th to 11th harmonics of the missing fundamental frequency of 250 Hz presented to the left ear. **Fourth row**: two complex tones composed of the even and the odd harmonics of the eight components complex tone presented to the left and right ears, correspondingly. **c)** Ten repeated matching results for each subject. The eight components virtual pitch was matched with a spectral pitch of or close to 250 Hz, presented to the opposite ear (**top row**). The even or odd harmonics of the eight-component complex tone (periodicity of 500 Hz) were matched with a spectral pitch of or close to 500 Hz presented to the opposite ear (**bottom row**).

The averaged value of the source coordinates of the data points around the N1 peak, having the highest two percents of goodness of fit was taken as a measure of the source locations for each of the four different stimulus conditions.. Only data with a goodness of fit greater than 95% were accepted. The data qualified only four subjects suitable for the analyses of the right hemisphere. Therefore, one-way analyses of variance (ANOVA) were calculated separately for the left and the right hemispheres, using STIMULUS CLASS (the four different stimuli) as a repeated measures factor. Probabilities were Greenhouse-Geisser adjusted. Post-hoc Scheffé tests were used to resolve significant ANOVA effects.

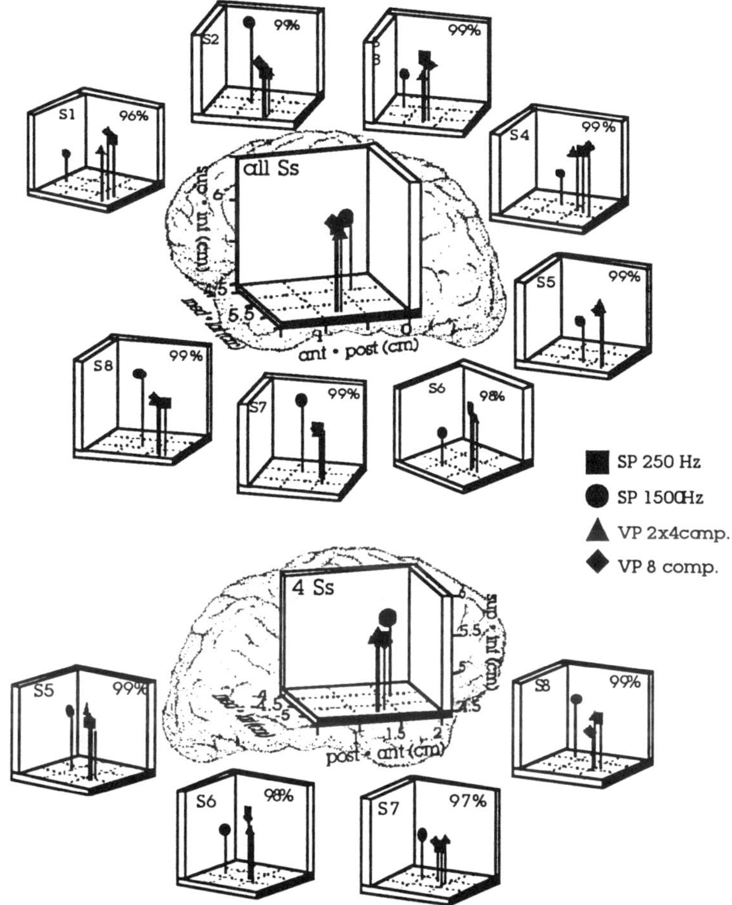

Fig. 2: **(top)** Source locations presented individually for each subject around a sketched brain (subject's number and goodness of fit are listed). The enlarged view at the center represents the spatial grand average across subjects. Squares denote spectral pitch of 250 Hz, circles spectral pitch of 1500 Hz, diamonds virtual pitch of the eight components complex tone (*all presented monaurally*), triangles virtual pitch of the even and odd harmonics of the eight components complex tone (*presented binaurally*). The anterior-posterior axis is denoted by a filled bar, the dimension of each cube is 1.5 x 1.5 x 1.5 cm.
(bottom) Source locations in the right hemisphere as obtained in four out of eight subjects.

Results

The estimated cortical source locations are presented separately for each subject and also as spatial grand averages across subjects in Fig. 2. Goodness of fit equalled 99% in the majority of the cases, confirming the adequacy of a single dipole model. The spatial source coordinates were separately tested for stimulus-related differences.. The separation of the representation of the 1500 Hz source from a location around which the other three sources cluster (Fig. 2) was corroborated by statistical analysis (ANOVA). In the medio-lateral direction the effect was significant for the *left* ($F(3, 21) = 16.39$; $\varepsilon = 0.45$; $p < 0.003$) and the *right* hemisphere ($F(3, 9) = 31.62$; $\varepsilon = 0.72$; $p < 0.001$). Post-hoc Scheffé testing revealed that this effect was exclusively due to the 1500 Hz spectral pitch condition which produced a location different from each of the other three conditions (all $p<0.001$). On the average, this difference amounted to 6.3 mm for the left hemisphere and to 5.5 mm for the right hemisphere. A similar effect was also evident for the posterior-anterior direction over the *left* hemisphere ($F(3,21) = 9.35$; $e = 0.44$; $p < 0.02$). The 1500 Hz spectral pitch source was located 2.9 mm more posteriorly to the sources evoked by the other three stimulus conditions. No significant difference in posterior-anterior direction was found for the *right* hemisphere.

Discussion

If the psychoacoustic prediction of a "binaural fusion" is correct, then the activated brain regions corresponding to the eight components virtual pitch, presented monaurally or binaurally, should activate the same cortical region as the one activated by a 250 Hz pure tone. If there were no binaural fusion, then the brain sources for the binaurally presented virtual pitch would be located consistently more medially than the representation of the 250 Hz-tone due to their matching frequencies of 500 Hz [5]. However, the statistical evaluation of our data clearly show that this is not the case. Given the absence of a difference in the location of the cortical excitation produced by the eight components complex tone when presented either monaurally or binaurally, the results can be considered the first neurophysiological confirmation of binaural fusion in hearing. Furthermore, the data suggest that binaural fusion produces equal results in both hemispheres. Neural excitations travelling along auditory pathways are melded into one pattern at or before the cortical level. At the cortex, we observe the nonlinear addition of information from both ears, having a quality that pertains to a virtual pitch composed of the binaurally presented complex tone. In conclusion, virtual pitch is not determined separately for the input to each ear, but is the result of binaural fusion in much the same way as the perception of monaurally presented complex tones. The finding that the same cortical region is activated by the 250 Hz spectral pitch and the 250 Hz virtual pitch (the complex tone composed of eight harmonics of the missing fundamental frequency of 250 Hz) confirms that the perceived pitch and not simply the spectral content of the auditory stimulus determines the representation within the tonotopic cortical map. This implies that the virtual pitch has been determined at a stage prior to the AEF N1-peak, i.e. the hypothesized pitch processor should be realized in a neuronal network of the primary auditory cortex or in subcortical structures.

References

1. Terhardt, E. Zur Tonhöhenwahrnehmung von Klängen. I. Psychoakustische Grundlagen. Akustica, 1972, 26: 173-186.
2. Terhardt, E. Pitch, consonance, and harmony. J. Acoust. Soc. Am., 1974, 55: 1061-1069.
3. Terhardt, E. Warum hören wir Sinustöne? Naturwissenschaften, 1989, 76: 496-504.
4. van den Brink, G. Monotic and dichotic pitch matchings with complex sounds. In: E. Zwicker and T. E. (Eds.), Facts and models in hearing. Springer, Heidelberg, 1974: 178-189.
5. Pantev, C., Bertrand, O., Eulitz, C., Verkindt, C., Hampson, S., Schuirer, G. and Elbert, T. Specific tonotopic organizations of different areas of the human auditory cortex revealed by simultaneous magnetic and electric recordings. EEG and clin. Neurophysiol., 1995, 94: 26-40.

Acknowledgement

This work has been supported by a grant from the Deutsche Forschungsgemeinschaft (Klinische Forschergruppe Biomagnetismus and Biosignalanalyse). The authors thank Drs. Jablonski, Buchanon and Lawson (BTi) for technical support.

Left and Right Supratemporal Auditory Cortices are Differentially Modulated by Linguistic-Attentional Task Demands

Poeppel, D.[1], Yellin, E.[2], Phillips, C.[2], Roberts, T.P.L.[1], Rowley, H.[1], Wexler, K.[2] and Marantz, A.[2]

[1]*University of California San Francisco, San Francisco, USA;* [2]*Massachusetts Institute of Technology, Cambridge, USA*

Introduction

An important goal in the study of the neural basis of speech perception is to understand the physiological basis of speech processing as a sensory input process and relate it to the well-known anatomical asymmetry for language representation. One advantage gained by focusing on speech perception as a subroutine of language comprehension is that one can use stimuli the acoustic properties of which are explicitly defined and easy to manipulate parametrically.

Beyond the interest in questions of speech and language lateralization, there is an extensive literature devoted to the study of auditory attention, both using event-related potentials and using MEG. Two classes of experimental protocols are commonly employed: tone oddball paradigms, in which one measures the mismatch response to an oddball stimulus in a sequence of standards [2], and 'alternating ear' or dichotic listening paradigms, in which one is asked to selectively attend to the input delivered to the left or right ears [1,2,3].

It is established that 'ear-specific' attention can modulate the M100 auditory evoked field [3]. When subjects were presented dichotically with tones of different frequencies, the M100 generated by a stimulus presented to the attended ear was larger than the M100 generated by the same stimulus presented to the unattended ear. The present study investigated whether 'content-specific, ear-independent' attention also modulates aspects of the M100, and whether there is a *differential* hemispheric effect. This study thus evaluates the possibility of an *attention-dependent gating* of activity that affects the two hemispheres differentially. The same speech stimuli are presented in two experimental conditions, passive listening and performing a simple discrimination task. The prediction was an increased M100 amplitude in the left temporal cortex in the attended condition. There was no prediction about the response of the right hemisphere.

Methods

We investigated the M100 auditory evoked response and its lateralization by recording from the left and right supratemporal planes of six subjects (3 women, 3 men). Four syllables were presented to the contralateral ear 100 times each in pseudorandom order. The four syllables (/bæ/, /pæ/, /dæ/, and /tæ/) were synthesized using a Klatt formant synthesizer. The syllables were 300ms in duration and the voice onset time was 20ms for the two voiced syllables (/bæ/ and /dæ/) and 80ms for the voiceless syllables, reflecting typical values used in natural speech. In one condition, subjects passively listened to the stimuli. In a second condition, subjects were required to make an overt phonetic classification by pressing a response key. Subjects had to group the voiced versus the voiceless syllables. The task was performed with ease, with 5 subjects performing at ceiling and one subject performing at 94%. Neuromagnetic fields were acquired in a magnetically shielded room using a 37-channel system (Magnes, BTi, San Diego, CA). Fields were recorded with a 1.0Hz high-pass cutoff and a sampling rate of 1041.7 Hz (bandwidth 400Hz). The evoked field data were signal-averaged and band-pass filtered (1-20Hz) for further analysis.

Single equivalent current dipole estimates of the M100 field were coregistered with MR images. For each subject, high resolution volumetric magnetic resonance images (SPGR sequence, 128x128x124 matrix, resolution ~ 1x1x1.5mm, TR=36ms, TE=8ms, flip=70deg) were acquired using a 1.5 Tesla SIGNA magnetic resonance scanner (GE Medical Systems, Milwaukee, WI).

Results

The latency of the M100 RMS peak ranged from 82ms to 129ms post stimulus onset. Analyses of variance of the latency data for both conditions revealed no significant main effects or interactions. In particular, there was no effect of the variable hemisphere. Figure 1 plots the M100 amplitude differences. The amplitude data showed the following patterns: (a) In the four right-handed subjects (2 women and 2 men) the task *increased* the mean RMS value in the left hemisphere on 14 of 16 measurements (4 subjects x 4 syllables), with the mean increases over the passive condition ranging from 1.4 femtoTesla (fT) to 54.8fT. (b) In the right-hemisphere measurements from the same four subjects, 15 of 16 measurements showed a *decrease* of RMS peak compared to the passive condition, with

Figure 1 M100 rms difference (task-passive) for each subject and each syllable

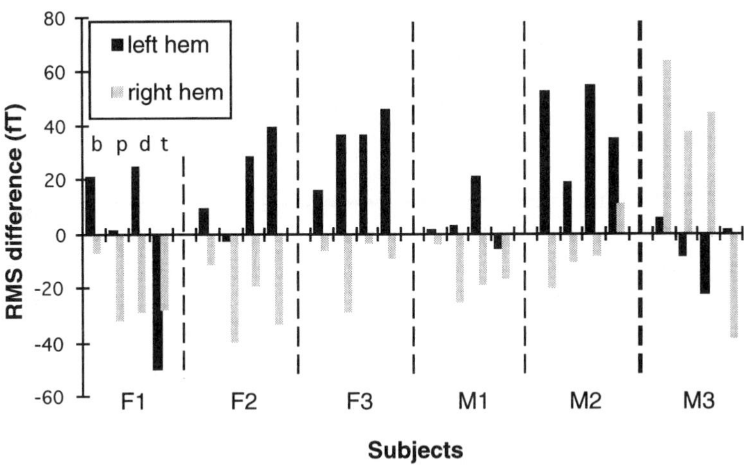

Figure 1 Task - passive M100 rms difference values. Each bar represents the task-passive difference of the M100 rms peak. Positive values indicate that the response amplitude in the task condition was larger than the corresponding measurements in the passive condition. Negative values indicate that the activation in the passive condition was greater than the discrimination condition for that hemisphere. Black bars represent the comparison for the left hemisphere, grey bars represent the comparison for the right hemisphere. Plotted are the mean amplitude changes for the four syllables for each subject. F1-F3, female subjects; M1-M3, male subjects.

Figure 2 Single dipole location for syllables (passive & discriminate) for subject M1

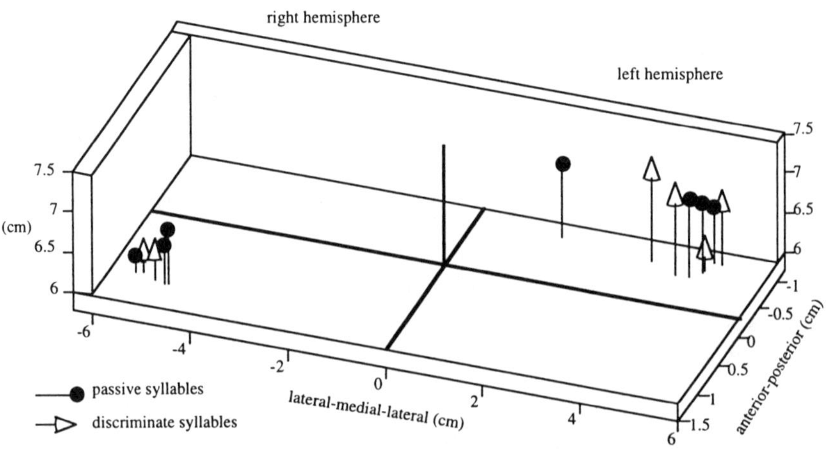

Figure 2 Dipoles elicited by syllable stimuli for one subject. The
filled circles are dipoles derived for the passive condition, the triangles are dipoles
for the VOT discrimination condition. MRI shows that all dipoles localize to
supratemporal auditory cortex. The left-right asymmetry observed for the cluster
of dipoles in this subject is observed by other investigators as well. This
localization asymmetry is typically attributed to the size asymmetry between the
plana, with the left planum temporale being larger in many subjects.

the mean decreases ranging from 3.1fT to 39.5fT. (c) The two left-handed subjects patterned in two ways: One
subject's data showed the same pattern described for the right-handers. The other subject patterned in the opposite
way, with the task decreasing the RMS peak in the left and increasing it in the right hemisphere recordings. With
few exceptions, the task-related changes in RMS were very large when compared with the baseline (passive) RMS
peak values. Typical changes were on the order of tens of percent. Analyses of variance of the RMS data show no
main effect of hemisphere in the passive condition, $\underline{F}(1, 5)=0.089$, $\underline{p}=0.78$, but a significant main effect of
hemisphere in the attention condition, $\underline{F}(1, 5)=11.282$, $\underline{p}=0.02$, despite the low power of the study. Also, the
hemisphere (left/right) x task (passive/attended) interaction was highly significant, $\underline{F}(1, 5)=55.30$, $\underline{p}=0.001$.
Together, these findings suggest that *the task induced the lateralization*. Single equivalent current dipole modeling
(Figure 2) and coregistration of the dipoles with structural magnetic resonance images of the individual subjects
showed that both conditions activated the same small area in supratemporal auditory cortex, thus excluding the
explanation that the presence of the task simply led to the recruitment of different cortical areas.

Discussion

A number of recent results derived from hemodynamic functional brain imaging methods (PET, fMRI)
suggest that speech perception is at least partially mediated bilaterally in the superior temporal gyri. However, when
subjects execute certain task demands, typically 'meta-linguistic' tasks such as phoneme categorization or semantic
classification, the observed neural activity is more highly lateralized. The findings reported here are similar. Passive
listening to syllables revealed complex but bilaterally symmetric responses. In contrast, the execution of a
categorization task with the same stimuli showed that the same region in left temporal cortex increased its response
amplitude (large M100 amplitude increases) while the the region in right temporal cortex showed large M100
amplitude decreases. Attention to speech sounds is thus reflected in lateralized supratemporal cortical responses
possibly concordant with hemispheric language dominance. Taken together, the data suggest that lateralization can be
induced by the attentional demands of a given experiment rather than necessarily reflecting underlying hemispheric
specialization.

The effect of attention needs to be clarified in future experiments to ascertain precisely what kind of
attentional modulation is occurring. The issue that requires clarification is what kind of attentional manipulation
leads to this response attenuation or enhancement. Several possibilities exist. (1) Only attending to the *particular
stimulus set* changes the outcome. The effect is therefore due to subjects having to make overt linguistic
discriminations on the presented material. (2) Attending to *any speech or language stimuli* during the task will result
in this effect. It is not crucial to attend to the stimuli proper; attention to any linguistic material, even though it is
not part of the task, will result in the change. (3) Finally, it is possible that *any* attentional task will lead to such a
modulation, including, for example, the execution of a visual task.

References

[1] Arthur, D., Lewis, P., Medvick, P., and Flynn, E. A neuromagnetic study of selective auditory attention.
Electroenceph. clin. Neurophysiol., 1991, 78: 348-360.
[2] Rif, J., Hari, R., Hämäläinen, M., and Sams, M. Auditory attention affects two different areas in the human
supratemporal cortex. Electroenceph. clin. Neurophysiol., 1991, 79: 464-472.
[3] Woldorff, M.G., Gallen, C.C., Hampson, S.A., Hillyard, S.A., Pantev, C., Sobel, D., Bloom, F.E. Modulation
of early sensory processing in human auditory cortex during auditory selective attention. Proc. Natl. Acad. Sci.
USA, 1993, 90: 8722-8726.

Acknowledgments

We are grateful to Biomagnetic Technologies Inc. for equipment support and to Susanne Honma and Mary Mantle
for their help in running experiments.

Extensive Somatosensory Stimulation Alters Somatosensory Evoked Fields

Rockstroh, B.[1], Vanni, S.[2], Elbert, T.[1], and Hari R.[2]

Department of Psychology[1], University of Konstanz, Germany;
Low Temperature Laboratory[2], Helsinki University of Technology, Espoo, Finland

Introduction

Changes in afferent input lead to functional cortical modifications, i.e., to cortical reorganization resulting in alterations in the cortical responses to stimuli. Animal studies by Jenkins, Merzenich and coworkers [1, 2] demonstrated that amputation of the digits result in an altered representation of the hand in area 3b. Such plastic changes can also be observed in humans using noninvasive measurement techniques. For instance, magnetoencephalographic studies [3,4] demonstrate that after amputation the sensory receptive fields of the face may invade areas which formerly had represented the hand and the digits. On the other hand, increased afferent input, e.g., by behavioral training may produce an enlargement of the cortical area that is activated by a particular task [1,2,5]. Prolonged increase in tactile stimulation to the distal pad of one or two phalanxes has been reported to result in a greatly increased cortical representation specific to that circumscribed area of the hand in owl monkeys [1]. In extreme cases, extensions spanning an area of several millimeters have been observed in the animal brains [6]. In patients suffering from syndactyly, Mogilner et al. [7] reported reorganization of the somatosensory cortex within weeks after surgical separation of the fingers. Training-induced somatosensory plasticity in humans has been demonstrated in blind persons suggesting increased cortical representation of the index fingers used in Braille reading [8]. With MEG-based source imaging Elbert et al. [9] found that in violinists who regularly played and practised their instrumentd the representation of the digits of the left hand in the somatosensory cortex was substantially expanded compared to the digits of the left hand in non-musician controls. The expansion was less pronounced for the left hand, which has the less active task of holding the neck of the instrument, compared to the digits engaged in fingering. The center for cortical responsivity for tactile stimulation of the digits for the left hand was shifted medially in the violinists as compared to age-matched control subjects. The amount of cortical reorganization for the digits' receptive field was significantly correlated with age of onset of musical training and years of training.

On this background the present study was designed to assess the variability of activity in the primary somatosensory cortex SI (area 3b) across time and to determine to what extent intense somatosensory stimulation might change the cortical somatosensory representation. For this purpose repeated magnetoencephalographic measurements were used. The somatotopical organisation of SI is clearly reflected in the distribution of somatosensory evoked magnetic fields (SEFs), and sources of SEFs generally correspond with the somatosensory homunculus derived from intracranial stimulation studies. Separate representations have been described for thumb and little finger, both for early somatosensory evoked fields (assumed to be generated in SI) and for long-latency SEFs, possibly representing SII-activity [10,11], as also documented, for instance, for lingual, trigeminal, facial, tibial, and median nerve SEFs. Despite of the observed interindividual variability in locations of the equivalent current dipoles (ECDs) for early SEFs (SI activity), ECDs of each of the 10 subjects examined by Hari and colleagues [10,11] were concentrated to strips of less than 2 cm in length, 5 to 9 cm lateral to the midline.

Methods

One right-handed adult was trained to read Braille text with the three middle fingers of the left hand. Following the training the same three fingers were kept immobilized and deprived from sensory stimulation by applying local anesthetics (lidocain-prilocain creme) and bandaging. The experimental protocol included 8 baseline measurements, one measurement after 16 hours of Braille training and one after 18 hours, one measurement following 27 hours of sensory deprivation, and 3 additional measurements distributed across 3 weeks following sensory deprivation. The first three baseline sessions comprised stimulations of the left hand only (see Fig. 2).

Somatosensory evoked magnetic fields (SEFs) were measured with a Neuromag–122™ whole-scalp neuromagnetometer which contains 122 planar SQUID gradiometers in a helmet shaped array. Head position indicators (HPI) were attached to the subject's head to give exact information of the location and orientation of the sensor array with respect to the head. Preauricular points and the nasion were marked in order to allow replication of HPI positioning across measurements. During the recordings, the subject was sitting in a magnetically shielded room, with the head supported against the helmet-shaped magnetometer, keeping her head immobile and fixating a point in front of her. Cortical representations of adjacent fingers were determined by monitoring SEFs to electrical stimulation of the of the thumb (D1) and little finger (D5) of both hands. Two Medelec ST 10 stimulators were used for delivering 0.3-ms constant current pulse stimuli at constant intervals of 1 s at intensities of 7-8 mA. Within each session, two series of 500 stimuli were delivered, 250 to the left thumb and 250 to the right thumb in alternating order followed by 500 stimuli alternating between the left and the right little finger.

Signals were bandpass filtered (0.03-300 Hz), digitized at 997 Hz, and 500 responses were averaged on-line for each finger. Averages included a 50-ms prestimulus baseline and 150 ms following stimulation. Trials contaminated by ocular artefacts or large-scale MEG shifts were excluded on-line from the average. Data of two baseline sessions (one prior to and one following experimental manipulations) had to be discarded because of obvious artifacts. Source currents were identified by modelling the signal distribution by equivalent current dipoles (ECDs) during discrete time periods; for each time period the ECD best explaining the magnetic field variance was determined by a least-squares search. For source analysis, a subset of 20 channels over the respective cortex (left for right hand stimulation, right for left hand stimulation) was selected. The goodness-of-fit [12] of the model was calculated and only ECDs explaining more than 90% of the field variance were considered. After localizing estimated sources at different time points a time-varying multidipole model was computed using these sources to explain the entire time period. The sites and orientation of the dipoles were kept fixed but their amplitudes were allowed to change as a function of time. Source localization was further accomplished by superimposing ECDs to the individual magnetic resonance (MR) images. The MRI scan was obtained using a 1.5-T Siemens Magnetom system.

Results

SEFs consisted of prominent deflections at 25-27 ms (M27) and 41-43 ms (M42). The locations of the ECDs during these peaks demonstrated good replicability across baseline measurements (Fig. 1). For the *left D1*, the absolute positions varied in mediolateral, inferosuperior and anteroposterior directions by 4.3 mm, 7.6 mm, and 6.1 mm, respectively (standard deviations across 5 measurements). The corresponding values were 5.5/8.8/7.4 mm for the *left D5*, 3.8/3.2/8.6 mm for the *right D1*, and 3.9/2.9/2.7 mm for the *right D5*. The respective variability in the left D1-D5 distance was 4.3/4.4/4.2 mm. Sources of M27 agreed with somatotopic organization of the SI cortex (with a mean difference of 25.4 ± 4.1 mm for the Euclidean distance between D1- and D5-representations in the left hemisphere and 24.0 ± 4.2 for the right hemisphere/left hand stimulation). Locations of M42 did not systematically differ for the two digits.

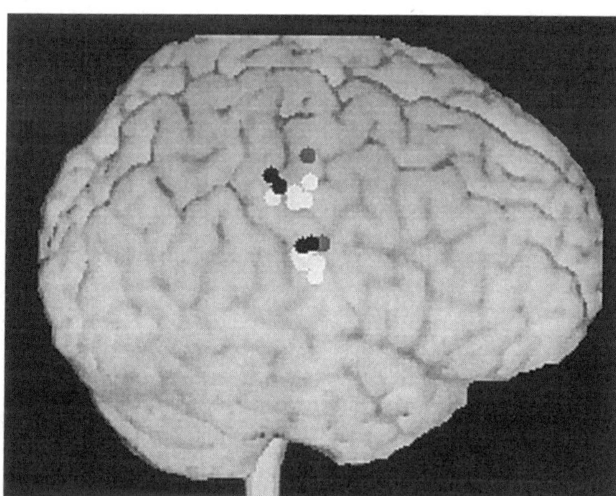

Fig.1. The cortical surface has been reconstructed from the individual MRI. The locations of the ECDs at 25 (D1) and 27 ms (D5) after stimulus onset were superimposed onto this cortical surface. The upper cluster of dots represents the locations for D5, the lower one those for D1 (left hand, right hemisphere). White dots: locations during pre- and post-baseline sessions, black dots: locations following Braille training, grey dots: locations following anesthesia.

Braille training altered representations of the stimulated left (non-dominant), but not of the right hand: The M27 dipole moments for D5, and after the 2nd training period also for D1, were stronger after the training than during any of the baseline or control sessions. (Left D5 baseline: 7.6 ± 1.7 nAm, Braille: $10.5 \pm .3$ nAm, p< .05; right D5 baseline: 11.4 ± 1.8 nAm, Braille $8.9 \pm .6$ nAm; Fig. 2). After training, the distance of left D1-D5 along the curved surface of area 3b tended to be enlarged (Fig. 2). Superposition of M27-ECDs onto the individual magnetic resonance image suggested a shift of the left D5 in medial direction along the central sulcus (Fig. 1).

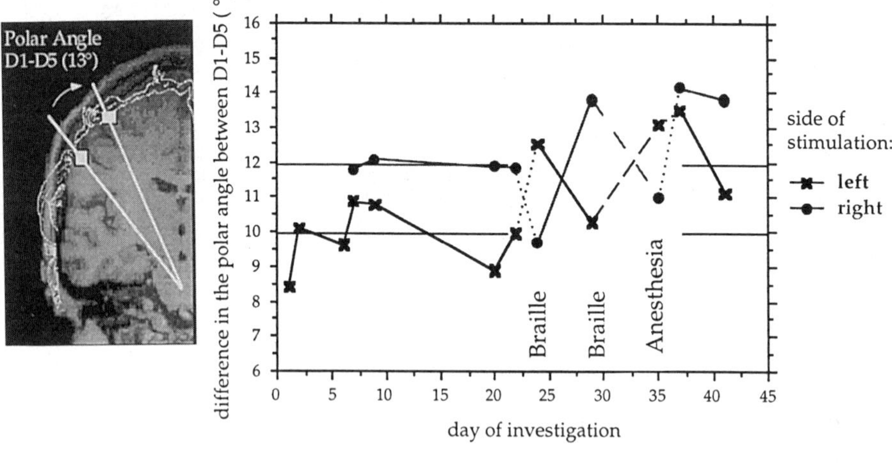

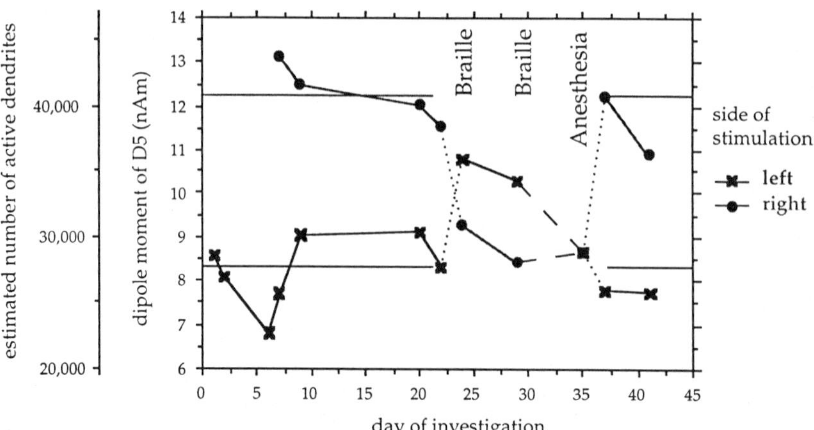

Fig.2. Change of dipole parameters across time; the abscissa indicates the days relative to the start of the investigation. The symbols mark days during which MEG measurements were obtained. The upper panel illustrates the change in the distance between the location of the ECDs for D1 and D5. The distance was measured as the declination (polar angle) between the locations of the represenations of the two digits as viewed from the center of the head, the latter being defined as the midpoint between the two meatuses. The overlay of D1 and D5 locations onto the corresponding coronal MRI section in the upper left corner illustrates this measure. During baseline measurements the polar angle is smaller for left than for right hand stimulation. This difference vanishes following Braille training and anesthesia. The lower panel shows the alteration of the dipole moment following Braille training in nAm. The ordinate on the left is a rough estimate for the number of neurons that may be involved during the cortical response [13]. Again, during baseline the right hand response is larger than the one for the left hand. Following Braille training, this "handedness" effect is reversed.

Discussion

Animal studies have consistently demonstrated that cortical plasticity can be observed during learning if a heavy schedule of training is designed to ensure continuous performance improvements. Thereby the training has to be conducted under conditions of a high motivational drive [1,2,5]. The results of the present study coincide with these considerations. Nevertheless, they must be considered with caution as the study was based on one subject only. To date, repeated measures with whole-head systems are rare. From the variation across baseline sessions it can be concluded that long-term within-subject replicability of MEG recordings should be further improved by advancing methods for head fixation, in order to allow the measurement of small variations in cortical organization. There is considerable evidence that intense stimulation of fingers during a meaningful behavioral tasks seems to be related to changes in cortical representation. Increased use of a limb leads to an expansion of the cortical representation and to a reduction of receptive fields. But alterations are more complex. The present results indicate that changes in representations are not directly related to the enhanced stimulation. Disuse is thought to cause invasion of the representational zone from nearby sites. If this was the case for the present study, then a reduced and not an enhanced neuronal response to stimulation of left D5 should have been seen. While the shift in cortical representation of D5 following training is weak compared to the variations between sessions, the dipole strenghts is clearly enhanced after training. Increased neuronal interconnectivity which is triggered by the trainining may be responsible for the observed enhancement. Similarly, the decrease in the dipole moment of right D5 cannot be explained by simplistic relationships. The laws of interhemispheric transfer and the those underlying cortical reorganization in general have to be studied further.

References

[1] Jenkins, W. M., Merzenich, M. M., Ochs, M. T., Allard, T., Guíc-Robles, E. Functional reorganisation of primary somatosensory cortex in adult owl monkeys after behaviorally controlled tactile stimulation, Journal of Neurophysiology, 1990, 63: 82-104

[2] Jenkins,W.M., Merzenich, M.M. Cortical representational plasticity: Some implications for the bases of recovery from brain damage. In: von Steinbüchel, N., von Cramon, D.Y.,, Pöppel, E. (Eds.) Neuropsychological Rehabilitaion. Heidelberg, Springer, 1992 (pp.20-35.)

[3] Yang, T.T., Gallen, C.C., Schwartz, B., Bloom, F.E., Ramachandran, V.S., Cobb., S. Sensory maps in the human brain. Nature, 1994, 368.

[4] Elbert T., Flor H., Birbaumer N., Knecht S., Hampson S.,Taub E. Evidence for Extensive Reorganization of the Somatosensory Cortex in Adult Humans after Nervous System Injury. Neuroreport, 1994, 5: 2593-2597.

[5] Recanzone, G. H., Merzenich, M. M., Jenkins, W. M., Grajski, A., Dinse, H. R., J. Plasticity in the frequency representation of primary auditory cortex following discrimination training in adult owl monkeys. Journal of Neuroscience, 1992, 67: 1031-1056.

[6] Pons, T., Garraghty, P. E., Ommaya, A. K., Kaas J. H., Taub, E., Mishkin M. Massive Cortical Reorganisation After Sensory Deafferation in Adult Macaques. Science, 1991, 252: 1857-1860.

[7] Mogilner, A., Grossmann, J.A.I., Ribary, U., Joliot, M., Volkmann, J., Rapaport, D., Beasley, R.W., Llinas, R.R. Somatosensory cortical plasticity in adult humans revealed by magnetoencephalography. Proceedings of the National Academy of Science USA, , 1993, 90: 3593-3597.

[8] Pascual-Leone A., Torres, F. Plasticity of the sensorimotor cortex representation of the reading finger in Braille readers. Brain, 1993, 116: 39-52.

[9] Elbert T, Wienbruch C, Rockstroh,B., Pantev C, Taub E. Increased use of the left hand in string players associated with increased cortical representation of the fingers. Science, 1995, 220: 305-307

[10] Hari, R., Reinikainen, K., et al. Somatosensory evoked cerebral magnetic fields from SI and SII in man. Electroencephalography and Clinical Neurophysiology , 1984, 57: 254-263.

[11] Hari, R., Karhu, J., Hämäläinen, M., Knuutila, J., Salonen, O., Sams, M., Vikman, V. Functional Organization of the human first and second somatosensory cortices: a neuromagnetic study. European Journal of Neuroscience, 1993, 5: 724-734.

[12] Kaukoranta, E., Hari, R., et al. Cerebral magnetic fields evoked by peroneal nerve stimulation. Somatosensory Research , 1986, 3: 309-321.

[13] Williamson, S.J., Kaufmann, L. Theory of neuroelectric and neuromagnetic fields. In: Grandori, F., Hoke, M., Romani,G.L. (Eds.) Auditory Evoked Magnetic Fields and Electric Potentials. Basel, Karger, 1990 (pp. 1-39).

Acknowledgement

Research was sponsored by the EU Human Capital and Mobility Programme through the BIRCH Installation in the Low Temperature Laboratory of Helsinki University of Technology and by the DFG.

Laterality of Hippocampal Responses to Infrequent and Unpredictable Omissions of Visual Stimuli

Rogers, R.L., Taylor, S.A., Akhtari, M., and Sutherling, W.W.

Neuromagnetism Laboratory, Epilepsy & Brain Mapping Center, Hospital of the Good Samaritan, Los Angeles, CA 90017 USA

Information obtained in the context of lesion studies and invasive electrical recordings as well as non-invasive event related potential (ERP) and MEG studies points to the possibility that several anatomically distinct sources contribute to the generation of the P3 component (Halgren et al., 1989; Smith et al., 1990; Yamaguchi and Knight, 1991; Rogers et al., 1992; 1993a; 1993b). In our previous work, we have identified two main clusters of sources of P3s irrespective of the stimulus modality: One in the sensory cortex specific to the particular modality of the stimulus and a second, subcortical cluster the precise location of which was not easy to discern. It appeared to be, however, in the general vicinity of the hippocampus and/or the lateral thalamus. In this study we attempted to identify the location of the subcortical cluster of sources, in the context of a visual P3 task. Moreover, in view of suggestions that P3s associated with spatial as opposed to auditory tasks may be generated mostly by right hemisphere sources (e.g. Johnson 1989a; 1989b), we recorded magnetic fields over both hemispheres during the same task. We identified sources bilaterally and subsequently we compared the relative strength of these two clusters of subcortical sources.

Materials and Methods

Ten neurologically normal, right-handed, adult subjects ranging in age from 23 to 53 years (mean = 29.1) were recruited from the university setting.

Magnetic fields and electrical potentials were recorded during a standard oddball paradigm in which a visual stimulus was presented for 50 msec. every 600 msec. and on 10% of the trials was omitted. Omissions were randomly dispersed with restrictions that no fewer than 3 and no more than 15 frequent stimuli were presented. Stimuli consisted of four points of light arranged as a 6 cm square. Light sources were presented by an IBM PC computer with an internal digital-to-analog card. The 5 volt analog output channels were used to power four light emitting diodes outside of the magnetically shielded chamber. The light from the diodes were then transmitted into the room via fiber optic cables that terminated on a panel in front of the subject. Subjects were instructed to fixate on a mark in the center of the square so that all four visual quadrants would be simultaneously stimulated. Subjects were also instructed to count the number of omissions so that their attention was directed toward the infrequent missing stimuli. One hundred such epochs were averaged for each trial, however, subjects were actually presented with between 100 and 120 missing stimulus trials. This was done so that they would not have prior knowledge of the number of omitted stimuli. Trials were repeated for each placement of the neuromagnetic probe.

Magnetic fields were measured using a seven channel probe system (Biomagnetic Technologies, Inc.) which utilized second-order gradiometers with a 4 cm baseline and 1.8 cm diameter of the superconductive sensor coils. A complete trial was performed separately for each probe position. The order of left-right trials was counterbalanced. The number of dewar placements required to cover the lateral aspects of each hemisphere ranged from 5 to 7 (35 to 49 sensor locations). The location of the magnetic sensors in relation to the subjects' heads was determined using three low frequency magnetic field sensors placed on the head with a transmitter mounted on the cryogenic container. Error tolerance was set so that maximum location error would not exceed 2 mm.

Averaged electrical potentials were obtained simultaneously with all MEG recordings. The active electrode sites were Oz, Pz, and Fz referenced to linked mastoids with the common ground on the forehead. Additionally, eye movements in the horizontal and vertical directions were also obtained. Both magnetic field and electric potential measurements were band passed on-line between .1 and 50 Hz.

To ensure the integrity of the average wave forms, each individual epoch was inspected for eye movements as well as variability and normalcy of the peaks. Single trials with obvious artifacts were removed from the averages, but these never exceeded 4% of the total number of sweeps. Variances of the magnetic wave forms were comparable to their electrical counterparts. Frequency histograms of field and potential values were normally distributed about the means.

Dipole parameters including location (Cartesian coordinates), orientation and strength were calculated for each peak using a finite difference version of the Levenberg-Marquardt algorithm for a spherical head model. The size and origin of the sphere was determined separately for each hemisphere by a least-squares fit from the digitization of the temporal areas. Since a single dipole source will produce a distinctive external field pattern with two extrema around which the field strength diminishes in all directions, isofield contour maps were created for each peak and inspected in order to determine that they were consistent with a dipole source. In addition, a minimum least-squares fit which accounted for at least 80% of the total variance between the measured fields and the calculated forward solution was required.

Magnetic resonance images were obtained for projecting the calculated sources on to the appropriate brain image. The procedures and details of coregistration of MEG and MRI data are described in detail elsewhere (Rogers, et al., 1991).

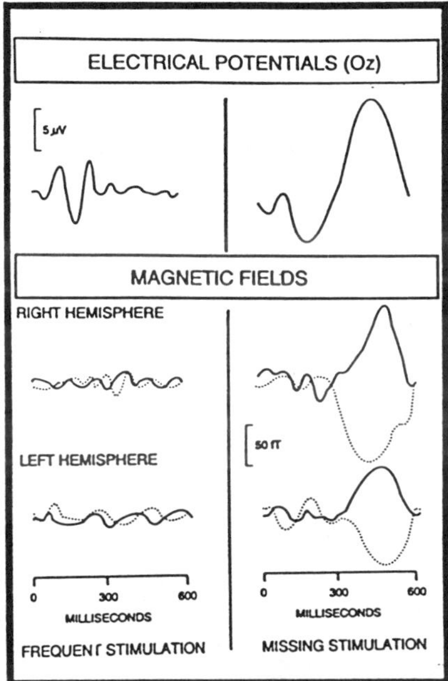

Figure 1: Shows grand average waveforms for both potentials (Cz) and field extrema for the frequent and missing stimulus conditions.

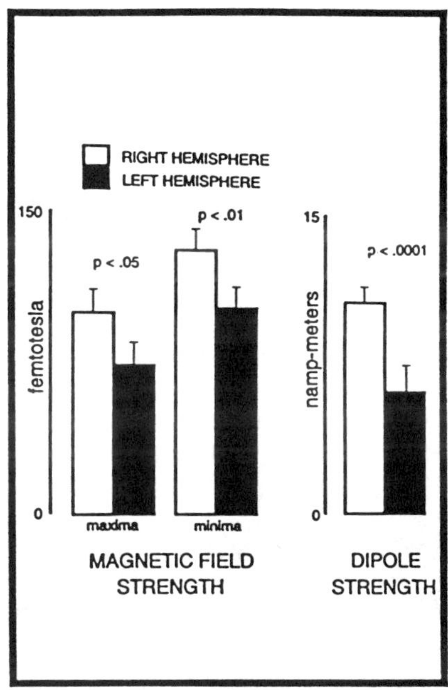

Figure 2: Demonstrates the difference between the two groups for both magnetic field extrema and dipole strength.

Results

Electrical potentials at Oz produced very consistent and predictable results with the frequent stimulation producing the common three peaks as seen in the top-left portion of Figure 1, which represents the average wave forms across all trials for all ten subjects. The omitted stimulus produced two peaks including a negative peak at 221.0 ±19.2 msec. followed by a positive peak seen at 420.1 ±23.1 msec.

Grand average magnetic flux wave forms over the right and left cerebral hemispheres are illustrated in the bottom portions of Figure 1. Magnetic flux to frequent stimulation (left portion of figure) did not show any discernible peaks above noise level. On the other hand, magnetic flux to infrequent omissions of visual

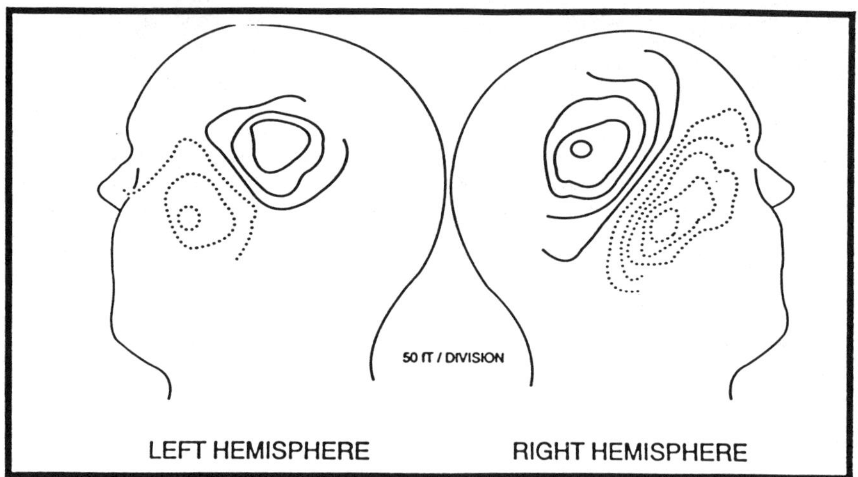

Figure 3: Isocontour maps over the right and left hemisphere during the missing stimulus condition at 455 msec following stimulation. Note the greater strength on the right side.

stimulation produced large distinctive dipole patterns over both the right and left hemispheres in all subjects at latency ranges from 380 to 450 msec.

Figure 2 shows the magnetic flux pattern over the right versus left hemisphere in a single representative subject at a latency of 435 msec. Each contour line in these isofield maps represents an increment of 50 fT, thus, the greater number of contour lines in the right hemisphere indicates that the extrema and gradients are substantially greater over the right hemisphere. Since the location and spacing of

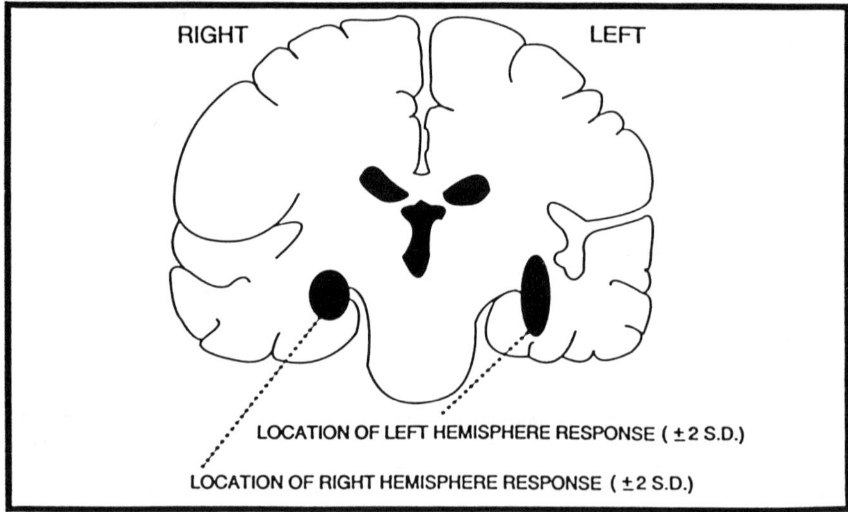

Figure 4 shows the distribution across subjects of the location of right and left hemisphere activity and their 95% confidence intervals.

the extrema are similar in both hemispheres, this would indicate that the generator source emanates from similar locations and orientations in both cases with the major difference being the strength of the sources.

Univarite F-test were performed on the differences between right and left hemisphere responses on the peak amplitudes of both extrema and on the calculated dipole strength. Figure 3 summarizes these results indicating that significantly greater responses occurred on all three of these measures (maxima: $p < .01$; minima: $p < .05$; dipole strength: $p < .0001$). Dipole strength was greater over the right hemisphere for all ten subjects without exception. Since physical laws dictate that magnetic field strength attenuates 50% per centimeter, it is not conceivable that the fields over the two hemispheres are from the same source. Figure 4 shows the distribution across subjects of the location of right and left hemisphere activity and their 95% confidence intervals.

Discussion

Previous studies have demonstrated that externally measured magnetic field patterns are indicative of activity in the vicinity of the right hippocampal formation during infrequent and unpredictable intrusions or omissions of visual stimuli in an oddball evoked response paradigm. These fields occur coincident with late endogenous evoked potential components that are consistently recorded in similar situations. In the present study, magnetic fields temporally corresponding to the late P300 component of simultaneously recorded evoked potentials were accounted for by sources in the vicinity of the left and right hippocampus in addition to previously reported sources in the vicinity of the primary visual cortex. Projection of these sources onto MRIs suggested that both hippocampal structures are simultaneously active and that there is an amplitude and strength-related dominance of the right hippocampal sources to visual stimulation.

References

Halgren, E., Squires, N.K., Wilson, C.L., Rohrbaugh, J.W., Babb, T.L. and Crandell, P.H. Endogenous potentials generated in the human hippocampal formation and amygdala by infrequent events. Science, 1980, 210: 803-805.

Johnson, Jr., R. Auditory and visual P300s in temporal lobectomy patients: evidence for modality-dependent generators. Psychophysiology, 1989a, 26: 633-650.

Johnson, Jr., R. Developmental evidence for modality-dependent P300 generators: a normative study. Psychophysiology, 1989b, 26: 651-667.

Rogers, R.L., Papanicolaou, A.C., Baumann, S.B. and Eisenberg, H.M. Late magnetic fields and positive evoked potentials following infrequent and unpredictable omissions of visual stimuli. Electroenceph. clin. Neurophysiol., 1992, 83: 146-152.

Rogers, R.L., Basile, L.F.H., Papanicolaou, A.C. and Eisenberg HM. Magnetoencephalography reveals two distinct sources associated with late positive evoked potentials during visual oddball task. Cerebral Cortex, 1993a, 3: 163-169.

Rogers, R.L., Basile, L.F.H., Papanicolaou, A.C., Bourbon, T.W. and Eisenberg, H.M. Visual evoked magnetic fields reveal activity in the superior temporal sulcus. Electroenceph. clin. Neurophysiol., 1993b, 85(5): 344-347.

Smith, M.E., Halgren, E., Sokolik, M., Baudena, P., Musolino, A., Liegeois-Chauvel, C. and Chauvel, P. The intracranial topography of the P3 event related potential elicited during auditory oddball. Electroenceph. clin. Neurophysiol., 1990, 76: 235-248.

Yamaguchi, S. and Knight, R.T. Anterior and posterior association cortex contributions to the somatosensory P300. J. Neurosci., 1991, 11: 2039-2054.

MEG Measurement of Auditory Sensory Memory Persistence via the M100 in Children and Adults

Rojas, D. C. ,[1] Sheeder, J. L. ,[1] Teale, P. ,[1] Walker, J. R.,[1] Robertson, B. A. ,[1] Selig, S. ,[1] and Reite, M.[1,2]

[1]University of Colorado Health Sciences Center, Denver, USA; [2]Veterans Administration Medical Center, Denver, USA

Introduction:

There have been several papers linking the auditory evoked field (AEF) component M100 to the auditory sensory memory trace, or echoic memory [1,2,3]. Recently, a preliminary effort to describe the AEF in children has been reported [4]. It is apparent that children do not show the same component structure as adults when presented with the same stimuli. The most striking finding of this report was the apparent lack of M100 response in the younger children, a finding which we have also replicated (see below). One possible explanation, proposed by Paetau, et al. [4], is that the M100 refractory period is longer in children than adults. Thus, the field strength of the M100 would be much weaker than adults at similar interstimulus intervals. Also, if the refractory period is longer in children than adults, then the echoic memory trace duration would also be longer, assuming that M100 indexes the trace duration, as proposed by Lü, et al. [1]. As this possibility is theoretically important to both psychology and MEG research, two experiments were undertaken in children and adults to explore the development of the AEF M100 component.

Our first attempt to describe the M100 in children was a standard AEF mapping experiment to pure tones delivered at one interstimulus interval (ISI), similar to the procedure we typically employ when mapping the M100 in adults. The results of this experiment led to a second experiment in which we positioned the intrument over a single, putative M100 extremum, and measured the AEF while systematically varying the ISI between tones. Both experiments are described in this paper.

Methods:

A BTI Model 607 7-channel, second order gradiometer with 1.8 cm coil diameters and a 4.0 cm baseline was used for MEG recordings in both experiments. Subjects were recorded reclining in a custom non-magnetic bed. All recordings were made in a magnetically-shielded room constructed from 6.3 cm mu-metal and aluminum and lined with 0.76 mm thick mu-metal. The room provides 35 dB noise reduction at DC. MEG data were acquired via BTI amplifiers and anti-aliasing low pass filters, and digitized at 1 kHz with a 12-bit, 64-channel Scientific Solutions analog-to-digital converted. Neuroscan SCAN software running on an 80386 computer was used for signal acquisition and processing. MEG was written out on a Grass Model 78 ink-writer unit to facilitate visual inspection of the raw signal for magnetic artifact, eye movement, subject movement and other potential artifacts. Coil locations and orientations relative to the head were determined using a sonic digitizer, as described in Reite et al. [5]. Custom software permitted co-localization in space of the subject's head (referenced to nasion and the left and right pre-auricular points), each recording site, and each gradiometer coil (and orientation). The accuracy of the digitizer system is ± 1 mm.

In the first experiment, we recorded data from left and right hemispheres in 17 right-handed children (6 girls and 11 boys) ranging in age from 6-9 years old. To map the distribution of the AEF, we recorded auditory evoked responses using 25 msec (3 msec rise/fall), 1 kHz toneburst stimuli. Tones were delivered at a constant rate every 2 seconds, and were presented at 85 dB SPL over insert earphones. Auditory stimuli were generated using a SoundBlaster SB16, 16-bit sound card housed in an 80286 PC. Responses to 100 stimuli, recorded from 35 sites per hemisphere (5 dewar positions), were averaged together. Six of these children (4 males and 2 females) were recorded again one year after their first recording session to ascertain any changes over time.

Five children (4 males and 1 female ages 8-10 years old) and 12 adults (6 males and 6 females ages 22 to 47 years old) participated in the second, variable ISI experiment. Subjects listened to 1 kHz tones presented monaurally at 85 dB SPL over insert earphones. Interstimulus interval was varied in randomly presented sessions of 1, 1.5, 2, 3, 4, and 6 seconds. MEG recordings were made from the contralateral hemisphere with a BTI Model 607-21 7 channel gradiometer, positioned over the putative anterior M100 extremum. Recordings were made from both hemispheres at all 6 ISIs while subjects watched silent movies of their choice. MEG data were bandpass filtered between 0.1 and 300 Hz, digitized at 2 kHz, low pass filtered off-line at 40 Hz, and baseline corrected using a 200 msec pre-stimulus window for further analyses.

The time constant of decay of the echoic trace, or τ, was computed by curve fitting the measured M100 amplitude data from the 6 ISIs to the equation

$$F(t) = A\left(1 - e^{-(t-t_0)/\tau}\right)$$

where $F(t)$ is the field amplitude in fT at time t, A represents the asymptotic amplitude, t_0 is the time axis intercept, and τ is the time constant, or trace lifetime of echoic memory [1]. Curves were fit to the data using a non-linear regression procedure (Marquardt-Levenberg algorithm) to minimize the sums of squares as A, t_0, and τ were varied. The trace lifetime τ represents the time necessary for 2/3rds of the trace to decay. Two constraints were placed on the fits a priori: a) A was not allowed to be higher than 1000 fT which is reasonable given both the theoretical constraints upon source generator strength and our own practical experience with the M100, and b) t_0 was not allowed to be less than 0.0 seconds, reasonable in light of the fact that negative values of the time intercept would imply memory trace formation prior to the presentation of the stimulus.

Results:

In the first experiment, all children exhibited identifiable auditory evoked magnetic fields, with several dipolar-like components. However, the latency and polarity of the components varied considerably. Fig. 1 shows overlapping AEF data from both hemispheres of children and adults to the same stimulus paradigm. Several components are evident in both children and adults, but differences between them are also clearly identifiable. The children's AEF morphology consistently differed from that typically seen in adults both in the number of components within a 500 msec window and the variability between children. Table 1 lists the number of

Table 1. Number of Dipolar Components Evident in 35 Channel Recordings

Subj. ID	Sex	Age	Year 1			Year 2		
			Left	Right	M100?	Left	Right	M100?
1	male	9	5	4	R(2)	4	5	R(2)
2	female	8	5	4	R(2)	5	4	R(2)
3	male	9	6	3	-	-	-	-
4	male	8	4	3	-	3	4	-
5	male	6	4	2	-	3	5	L(1)
6	male	9	4	4	R(2)	-	-	R(2)
7	male	6	3	5	R(2)	-	-	R(2)
8	male	9	3	4	-	4	4	R(1)
9	female	8	6	5	R(2)	-	-	R(2)
10	female	9	5	5	R(2)	3	5	R(2)
11	female	9	5	3	-	-	-	-
12	female	7	5	5	R(2)	-	-	R(2)
13	male	9	4	4	-	-	-	-
14	male	7	6	4	-	-	-	-
15	male	9	5	7	R(2)	-	-	R(2)
16	female	7	5	5	R(2)	-	-	R(2)
17	male	9	4	4	R(2)	-	-	R(2)
mean	n/a	8.12	4.65	4.18	n/a	3.67	4.50	n/a
SD	n/a	1.11	0.93	1.13	n/a	0.82	0.55	n/a

Note to Table 1. Left and right refer to the two hemispheres. Numbers in right and left hemisphere columns reflect the number of dipolar components evident in the 35 channel AEF recordings. R = Right, and L= Left. Number in parentheses reflects serial position within AEF of component with similar polarity and latency to an adult M100.

components seen in both hemispheres for all children, including those recorded a second time 1 year apart. There were no significant correlations between age and number of components in either left or right hemisphere, r = .10 and r = .08, p > .05, respectively. Left and right hemisphere topographic maps constructed from a representative female child and adult are also illustrated in Fig. 1.

In the second, variable ISI experiment, the adults' mean τ was 1.56 ± .30 (SEM) for the left hemisphere, and 1.01 ± .20 for the right. The left hemisphere mean for the children was 3.44 ± 1.5 and the right hemisphere mean was 1.91 ± .93. Fig. 2 illustrates the curve fits for each group separately.

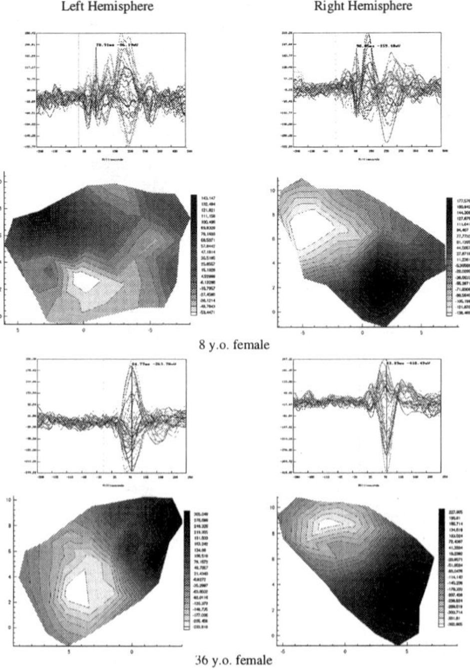

Fig. 1. Waveforms and topographic maps for 35 channel recordings. Scale for both in fT. Axes on maps are in cm, and reflect headframe coordinates. Positive y is superior, and positive x is anterior.

Discussion:

Most of the children did not show adult-like morphology and latency. Specifically, the M100 ingoing and outgoing extremum were not seen at comparable latencies to adults in almost all of the child recordings. At this point, we are not confident that any of the components seen are analogs to the adult M100, since in most of the children a large component of the correct polarity was not seen until 200 msec post-stimulus. Of the children who did show adult-like morphology and latency, it was only present in the right hemisphere recordings. We interpret this finding to mean that there is independent development of the right hemisphere from the left with respect to the auditory system. The preliminary data from the second experiment support the contention that children have longer M100 refractory periods than adults, and perhaps also longer echoic trace lifetimes. This would explain why the children in the first experiment did not exhibit M100-like responses, since at the same ISI an adult would have a much stronger response than a child. Interestingly, the left hemisphere appears to have slightly longer trace lifetimes than the right in the children. As with the mapping experiment, which revealed more adult-like responses in the right

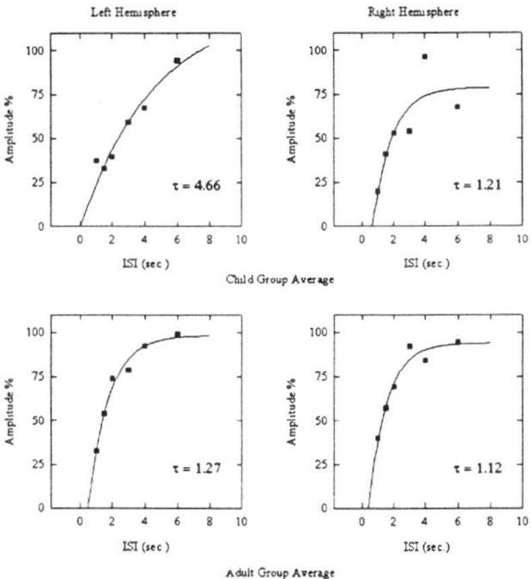

Fig. 2. Amplitude - ISI relationship for putative M100 component in adults and children. Amplitude is represented as normalized response, expressed as percentage of largest response in fT. Taus are curve-fit to mean responses.

hemisphere, this suggests that the auditory cortex of the two hemispheres may develop independently.

More data will need to be collected to further delineate these developmental changes. First, although there is clear evidence of a change between ages 6-9 and adulthood, the timecourse of that maturation will need to be explored in greater detail. There is also a pressing need to demonstrate that the children's putative M100 response is related to echoic memory, as has been done before in adults [1] with the use of psychoacoustic methods. If the relationship holds, these data suggest that echoic trace lifetime is longer in children than adults.

References:

[1] Lü, Z. -L., Williamson, S. J., and Kaufman, L. Science, 1992, 258: 1668-1670.
[2] Lü, Z. -L., Williamson, S. J., and Kaufman, L. Brain Research, 1992, 572: 236-241.
[3] Sams, M., Hari, R., Rif, J., and Knuutila, J. Journal of Cognitive Neuroscience, 1993, 5: 363-370.
[4] Paetau, R., Ahonen, A., Salonen, O., and Sams, M. Journal of Clinical Neurophysiology, 1995, 12: 177-185.
[5] Reite, M., Adams, M., Simon, J., Teale, P., Sheeder, J., Richardson, D. and Grabbe, R. Cognitive Brain Research, 1994, 2: 13-20.

Acknowledgments:

This work was supported by a grants from the Developmental Psychobiology Research Group at the UCHSC School of Medicine, #250, USPHS MH47476, and a grant from the Alzheimer's Association, PRG-93-143. Salary support for D. C. Rojas was provided by USPHS 5T32 MH15442 and M. Reite salary support was provided by USPHS grant 5K02 MH46335.

The Relation Between Gustatory-Evoked Magnetic Fields and Reaction Times to Different Taste Qualities

Saito, S.S.[1], Kobayakawa, T.K.[1], Ayabe-Kanamura, S.A.[2], Endo, H.E.[1], Yamaguchi, Y.Y.[1], Kikuchi, Y.K.[3], Ogawa, H.0.[4] and Takeda, T.T.[1]

[1]National Institute of Bioscience and Human Technology, MITI, Tsukuba, Japan; [2]University of Tsukuba, Tsukuba, Japan; [3]Tokyo Medical and Dental University, Tokyo, Japan; [4]Kumamoto University, Kumamoto, Japan

Introduction

Kobayakawa et al. reported the development of a new gustatory stimulation apparatus for which the rise-time was less than 20 ms for a measurement of EEG [1]. They also modified the apparatus for a measurement of MEG, and reported the estimation of the source of the gustatory-evoked magnetic fields (GEMs) in the human cerebral cortex [2][3]. In the present study we measured the reaction times (RTs) to the different taste qualities and compared them with the GEMs of different taste qualities to find the mutual relation between the reaction times and the onset of GEMs.

Methods

In separate experiments we measured the RTs and the GEMs, using the same sharp stimulation method. Six

Table 1. Results of RTs and GEMs onset

Subj.	Stimulus	RTs ave. (ms)	RTs Diff.* (ms)	RTs SD (ms)	n	subject's introspection	GEMs onset (ms)	GEMs Diff.*	subject's introspection
AS	NaCl	350	81	111	37	salty	40	60	at first sour and salty
	saccharine	431		83	40	sour > sweet	100		at first sour and sweet
YY	NaCl	354	117	76	40	salty	70	110	salty, moderately weak
	saccharine	471		150	36	moderate sweet	180		fairly sweet
KT	NaCl	372	66	107	37	salty	30	50***	salty
							210		salty
	saccharine	438		106	34	sour > sweet	340		sweet
							80		sour and sweet
TA	NaCl	629	247	110	39	salty	130	150	salty
	saccharine	876		244	36	sweet and a little sour	280		sweet and a little sour
MK	NaCl	730	81	184	40	salty	60	80	salty
	saccharine	811		186	37	sweet	140		sweet
EH	NaCl	369	-11	95	36	distinct salty	140	-20	clearly salty
	saccharine	350		68	37	strange sweet tooth paste with salt taste (not just salt)	120		strange sweet tooth paste with salt (not just salt)
Ave.	NaCl	467	96	114	38		93**	72	
	saccharine	563		140	37		172**		

* difference between NaCl and Saccharine

** the data were averaged for a subject having two GEMs onsets

*** 80ms (sour and sweet) - 30ms (first onset)

subjects (3 females) participated in this experiment; their ages ranged from 21 to 32 years old. A subject was seated on a wooden chair in a magnetically shielded room, and his or her jaw was supported by a chin rest attached to the chair. A Teflon tube through which a taste solution flowed was placed on the subject's tongue using a guide attached to the chair.

A solution of 1 M NaCl was used as a salty taste and a solution of 3 mM saccharin was used as a sweet taste. We used saccharin instead of sucrose because saccharin presents a sweet taste in low concentration, which would not contaminate the apparatus and not wound the membranes of taste cells. However, saccharin caused other taste qualities (sour, bitter, or salty) depending on the location of the placement of a hole of the tube on the tongue. We instructed subjects to position the hole of the Teflon tube on the part of their tongue where they perceived only or primarily a sweet taste. The duration of each stimulus was 400 ms. During the inter-stimulus interval, deionized water flowed to rinse the tongue for about 30 s. The temperature of both the solution and the deionized water was maintained at the temperature of the tongue's surface.

Results

Table 1 shows for each subject, RTs, the latency until the GEMs onset, and subjects' introspection for NaCl and saccharin.

Subject's introspection: No subject reported a tactile or temperature differences between the taste solution and deionized water. All subjects perceived NaCl as being salty. Two subjects reported an "only sweet taste" for saccharin. Subject EH reported a "strange sweet-like tooth paste with salt taste" for saccharin. Two other subjects reported "primarily sour with some sweet", and one subject reported "primarily sweet with some sour".

Reaction Time: For each subject, an averaged reaction time was calculated for two tastants. The saccharin showed significantly longer RTs than NaCl except for subject EH.

GEMs: When treating the GEMs data, the trials which contained eye movement were rejected and the remainder of the trials were averaged. The number of trials for averaging was about 35 per subject. The sampling rate was 250 Hz. Averaged GEMs of the 64 channels were superimposed on the same graph, aligned according to the time after the raising of the stimulus. We estimated the onset of GEMs of each tastant for each subject.

The onset of GEMs was significantly later for saccharin than NaCl, for five of the six subjects. Fig. 1 and 2 show the GEMs of NaCl or saccharin for subject YY; the pattern is typical of all subjects.

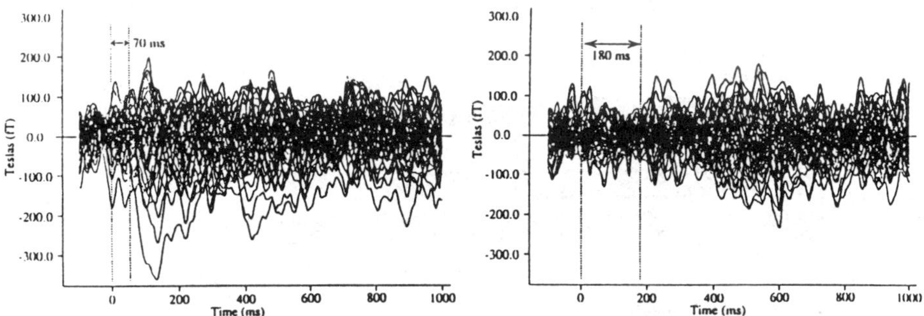

Fig. 1 GEMs onset of NaCl (subject YY)   Fig. 2 GEMs onset to saccharin (subject YY)

In the case of subject KT, we measured two GEMs' experiments of 40 trials for each stimuli and estimated each onset for each GEMs' experiment; the GEMs onsets were very different in both taste qualities (Table 1). For saccharin solution the perception of "only sweet" for saccharin showed the later GEMs onset (Fig. 3) than the perception of "sour and sweet" (Fig. 4). For NaCl solutions perceived as salty in both experiments, one had a GEMs onset of 30 ms after stimulation, another had a GEMs onset of 210 ms after stimulation. The GEMs containing the faster onset of GEMs showed a second onset at the same time as the later GEM onset. In the cases of other subjects, subject AS, MK, and YY showed the GEMs onset at 40, 70 or 60 ms after the stimulation like to the faster onset of subject KT, and two of them had the second onset at 190 ms (subject AS) or 260 ms (subject YY) after the stimulation. Subject TA and EH showed the GEMs onset 130 or 140 ms after the onset of stimulation like to the later onset of subject KT.

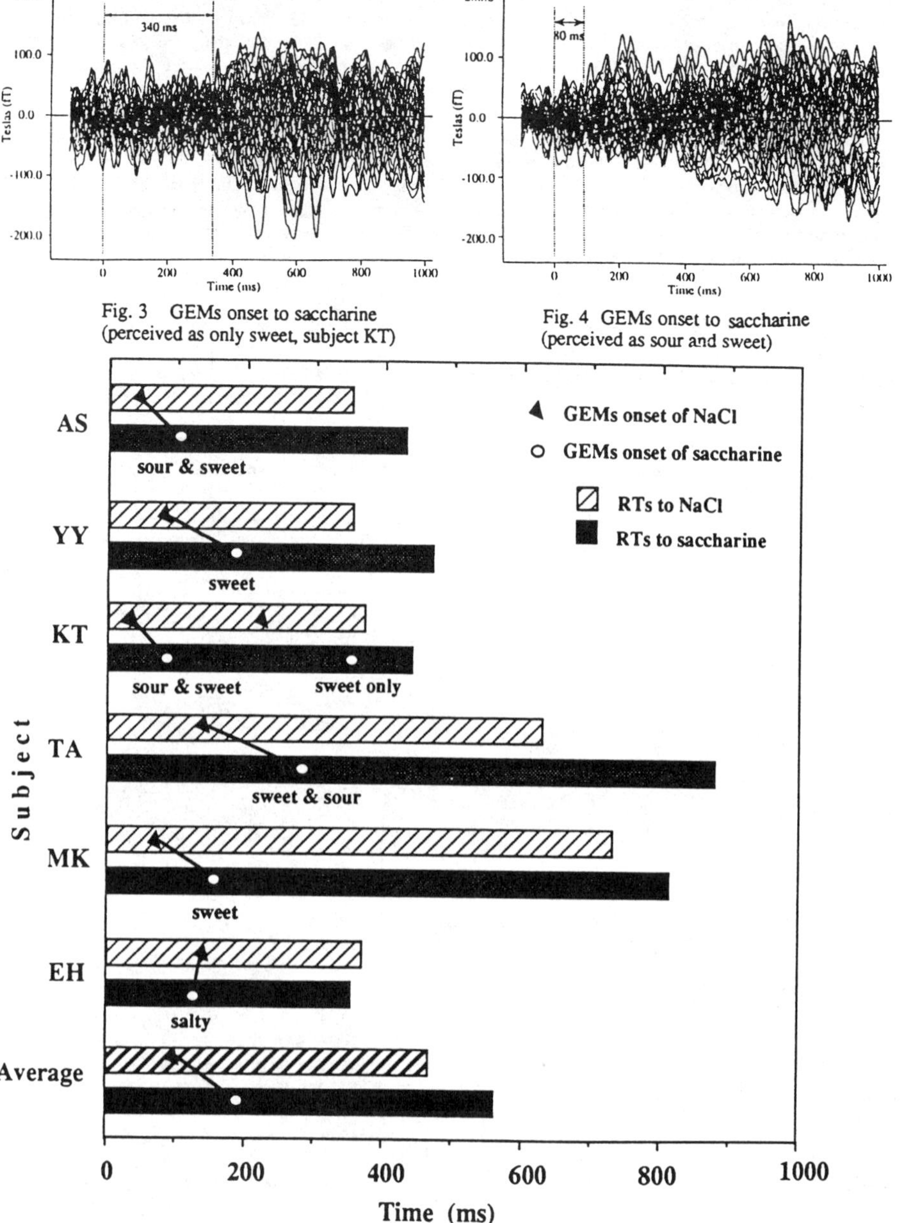

Fig. 3　GEMs onset to saccharine
(perceived as only sweet, subject KT)

Fig. 4　GEMs onset to saccharine
(perceived as sour and sweet)

Fig. 5 The relation between GEMs onsets and reaction times to NaCl and saccharine

YY showed that the GEMs of NaCl began to respond 70 ms after the stimulation, whereas the GEMs of saccharin began to respond about 180 ms after the onset of the stimulus. The difference of RTs to the two tastants 117 ms, which was similar to the difference (110 ms) of the GEMs onsets for the two tastants. Similarly, subject AS showed that the GEMs of NaCl began to rise 40 ms after the stimulation, and the GEMs of saccharin began to rise about 100 ms after the onset of the stimulus. Also, the difference between the RTs to the two tastants was 81 ms, which was similar to the difference (60 ms) between the GEMs onsets for the two tastants.

Discussion

Reaction time to 1 M NaCl was reported as 433 ms that was shorter than RT to sweet taste of sucrose in Yamamoto & Kawamura [4]. The averaged RTs to 1 M NaCl measured in this study was 467 ms that was consistent with the finding of Yamamoto & Kawamura. Reaction times measured in this study showed longer RTs to saccharine than NaCl except for subject EH. In the case of subject EH, there was no difference between two tastants for the RTs and the latency until the GEMs onset. Subject EH reported a salty taste in saccharin, which may account for the same RTs to NaCl. This result confirms the GEMs findings by Kobayakawa [2][3].

Besides, in every subject, a difference between the RTs to NaCl and the RTs to saccharin was nearly identical with a difference between the latency until the GEMs onset of NACl and that of saccharin. It means the RT subtracted from the latency until the GEMs onset is constant in spite of tastant. It's supposed that this definite time consists of processes of cognition and reaction to button except for sensation and perception, and that the difference between tastants in RTs or GEMs onset is caused by the difference of receptor mechanisms in sensation and perception process. In fact, the receptor mechanism of a salt by ion activity is assumed to be simpler than that of a sweetener through receptor site of protein.

In the case of subject KT, the perception of "only sweet" for saccharin showed the longer GEMs onset than the perception of "sour and sweet". This suggests that the onset of GEMs is related to perceived phenomena. Otherwise in the case of KT's NaCl, the GEMs data containing the faster GEMs onset showed a second onset at the same time as the later onset. Besides, other subjects showed the fast GEMs onset, the late GEMs onset or both GEMs onsets. More experiments are needed to explain exactly these results, but one speculation is that there are two peaks within 1000 ms of the onset on NaCl stimulation.

Conclusion

For all subjects, the latency until the GEMs onset is correlated to the RTs; this result confirms the measurement of GEMs by Kobayakawa [3][4]. Then, the difference between RTs for two tastants and the difference between the onset times of GEMs for two tastants were virtually identical.

Acknowledgment

We are grateful to Dr. T. Yamamoto, Dr. B. P. Halpern and Dr. B. Challis for their valuable advice and comments.

References

[1] Kobayakawa, T., Saito, S., Ayabe-Kanamura, S., and Yoshimura, S. (1995) Gustatory-Evoked potentials (GEPs) varied according to the concentration, Proceedings of the 17th meeting of Association for Chemo-reception Sciences: 263.

[2] Kobayakawa, T., Endo, H., Saito, S., Ayabe-Kanamura, S., Kikuchi, Y., Yamaguchi, Y., Ogawa, H., and Takeda, T. (1996) Trial measurement of gustatory-evoked magnetic fields. Electroencephalography and clinical Neurophysiology (in press).

[3] Kobayakawa, T., Endo, H., Saito, S., Ayabe-Kanamura, S., Kikuchi, Y., Yamaguchi, Y., Ogawa, H., and Takeda, T. (1996) Measurement of gustatory-evoked magnetic fields with sharp stimulation. Biomag96.

[4] Yamamoto, T., and Kawamura, Y. (1981) Gustatory reaction time in human adults. Physiology and Behavior, 26: 715-719.

SQUIDs and Stereopsis

Shahani, U.[1], Weir, A.I.[1], Lang, G.[1], Mansfield, D.C.[2], Halliday, D.M.[3], Hutson, D.[1], Maas, P.M.[1] and Donaldson, G.B.[1]

[1]Wellcome Biomagnetism Unit & Department of Physics and Applied Physics, University of Strathclyde; [2]Department of Ophthalmology, Southern General Hospital; [3]Department of Physiology, University of Glasgow, Glasgow, Scotland

INTRODUCTION

Stereopsis implies the perception of solid form and is generally used to denote the perception of solid objects by the human visual system. Although monocular cues are important for depth perception at a distance, for objects less than a 100 feet away, this process is also mediated by stereoscopic vision. Fixation at a point with both eyes renders the image of that point falling on the centre of the retina in each eye. However, because the two eyes are about six centimetres apart, three-dimensional objects produce slightly different images on different parts of the 2 retinae. The degree of this binocular disparity depends on the distance of the object from the plane of fixation of the 2 eyes. Thus, points on a three-dimensional object just outside the fixation plane stimulate different points on each eye and multiple disparities provide cues for stereopsis - which results in the perception of depth. The hypothesis that binocular disparity is the effective stimulus for depth perception is attributed to Kepler, the astronomer but the demonstration of that disparity alone can produce depth sensation was first provided by Wheatstone in 1838 with his invention of the stereoscope [1]. Bela Julesz[2] discovered that apparent three-dimensionality of objects can be perceived by the human visual system in random-dot stereograms which contain hidden images. This fact suggests that the main clue required for the perception of stereopsis is retinal disparity which excites the "cyclopean" group of neurones in the brain. These neurones force fusion of the discrete retinal images to give the perception of apparent third dimension in an essentially two-dimensional image.

More than three decades of research has gone into the study of the human and primate visual system. Physiological evidence of the existence of binocular cells in the primate brain comes from several neurophysiological studies of disparity sensitive single neurones in the primary visual and visual association cortices [see 1 for review].

Human stereopsis was studied using techniques which primarily measured retinal disparity by presenting spatially disparate images to each eye and asking human subjects to either fuse these images or have displaced images in each eye as a result of diplopia. Several hypothesis were used to describe the nature of these disparity sensitive neurones in the cortex the discussion of which is beyond the scope of this paper [1].

EEG and EP studies of stereoscopic depth processing providing physiological correlates to the psychophysical studies in man have been few [1]. However, several recent studies have concentrated not so much on mapping depth processing as on other related areas of the brain involved in verbal tasks [3,4], visual imagery and visual representation using multichannel EEG and MEG as well as functional imaging using either PET or MRI.

In this study we report the response of the cortical neurones involved in stereopsis recorded from over the occipital and parietal cortices using a conventional single channel axial SQUID neuromagnetometer. We have used frequency analysis to study the resonance phenomena that have been described as a characteristic of higher order processing in the brain [5].

METHODS

In our eddy current shielded room, free running magnetoencephalograms were recorded from over the occipital and parietal cortices in humans in order to study the frequency responses of cortical neurones during stereoscopic vision. Points $O_{1/2}$ and $P_{3/4}$ of the 10-20 electrode system were used for the location of the SQUID neuromagnetometer. The measurement points cover areas V_1, V_2, V_3 and V_4 of the visual system all of which are involved in processing depth information.

Bandpass settings were 0.3 to 100 Hz and signals were amplified using a Grass Model 12 Neurodata Acquisition System. Sequential recordings were made from subjects with eyes closed, eyes open and when perceiving the cyclopean *wallpaper* pattern under conditions of uniform luminance. Recordings of at least one minute were made of the subjects' response to each of the experimental conditions. For the first minute of recording subjects were asked to keep their eyes closed. Another minute recording was made of the subjects' with their eyes open but perceiving the 2-d wallpaper pattern. Finally subjects were asked to focus on the pattern in an attempt to perceive the hidden 3-D object. Subjects indicated perception of the 3-D object and its loss. At least one minute of sustained 3-D perception was recorded over each point. The subjects were then asked to close their eyes and relax and a further resting recording was made.

The results were recorded on a DAT tape recorder and analysed off-line in the frequency domain with a Hewlett-Packard 3561A Dynamic Signal Analyzer using the exponential window function and rms averaging. The exponential window function uses a slow overlapping temporal slice to average the data so that the oldest sample is given the least weight. The segments of data were then subjected to statistical analysis by the discrete Fourier transform and parameters derived from it, all of which were estimated by dividing the records into a number of disjoint sections of equal duration and estimating spectra by averaging across these discrete sections. Following standard time series practice a log transformation was used to obtain confidence limits for frequency spectra allowing distinct periodicities in the MEG recordings to be characterised [6].

RESULTS

MEG signals from all of our subjects showed increased activity at frequencies between 8 and 12 Hz in the alpha range in the eyes closed condition which attenuated on eye opening. Examples of each of these taken from Subject DMcK from over the right parieto-occipital cortex at point P_4 of the 10-20 system are shown in Fig 1. These data are consistent with a recent report in the literature which suggests that MEG activity at 8-12 Hz attenuates on eye opening and visual imagery recorded from the parieto-occipital cortex [4]. On perception of the three-dimensional image, this alpha rhythm was further attenuated and there was a marked increase in the spectral component at several frequencies between 35 and 75 Hz. There was substantial inter-individual variation of peak frequency but for constant location intra-individual variation was small. Figure 2 shows the activity of Subject DMcK on perception of the hidden 3-D image in the wallpaper pattern. There is an increase in general background activity. This is most marked at frequencies of 42-43 Hz and 67-68 Hz. Recordings over point P_3 on the left parieto-occipital hemisphere showed increases in power at similar frequencies on perception of the hidden 3-D image. While recording over point P_4 , the subject lost the image and then regained it. Spectral content at frequencies associated with stereopsis decreased dramatically on loss of perception of the 3-D image. A decrease was seen in the spectral content at all frequencies. Regaining perception of the hidden 3-D image resulted in an increase in the spectral content at 42-43 Hz and 65-70Hz . There was an increase in similar frequency ranges seen when he first perceived the 3-D image. All data presented in the figures was also subjected to statistical analysis by the discrete Fourier transform and parameters derived from it, all of which were estimated by dividing the records into a number of disjoint sections of equal duration and estimating spectra by averaging across these discrete sections and a log transformation was used to obtain 95% confidence limits for frequency spectra which were statistically significant.

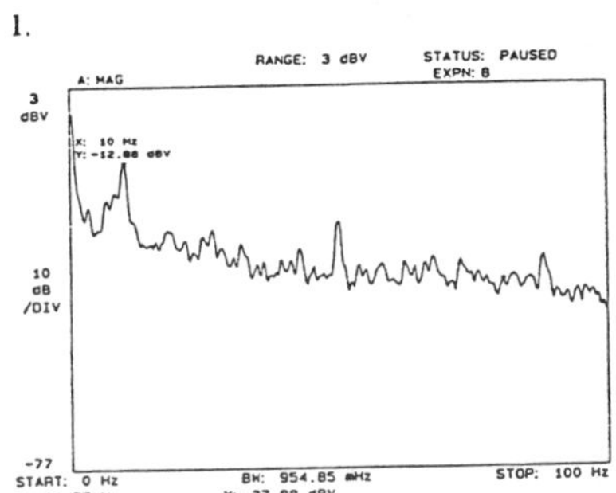

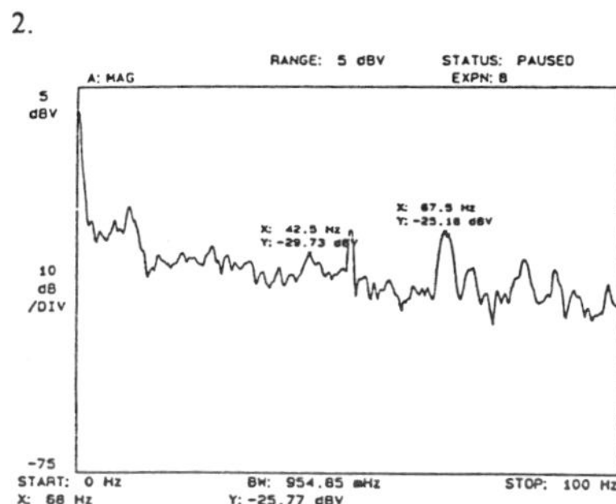

Fig.1 Eyes closed (point P_4, Subject DMcK). Increase in alpha rhythm. Fig.2 Perception of 3D (Subject DMcK). Increase in frequencies of 40-50 Hz and 60-70 Hz. Each figure is a RMS exponential weighting average of MEG. In addition, the data was found to be statistically significant.

In order to test whether the increase in spectral content of frequencies of 40-50 Hz and 60-70 Hz was directly associated with stereoscopic vision, we performed a series of control experiments.

The first set of control experiments were devised to test whether the increase in the spectral content of the frequency ranges associated with stereopsis was due to perception of 3-D per se and not the sudden recognition of an identifiable object. Subjects were presented with a non-cyclopean wallpaper pattern without a hidden image. The picture itself could be seen in 3-D. Fig. 3 shows results obtained from Subject GL in response to one such non-cyclopean wallpaper pattern. There was an increase in spectral content at frequencies of 40-50 Hz and at 69Hz. Fig 4 shows results from the same Subject GL in response to a cyclopean wallpaper pattern in which a hidden image is perceived in 3-D. There was an increase in spectral content at 58-59 Hz in addition to that seen in the previous Fig suggesting that object recognition could influence the activity of cortical neurones during stereopsis.

3.

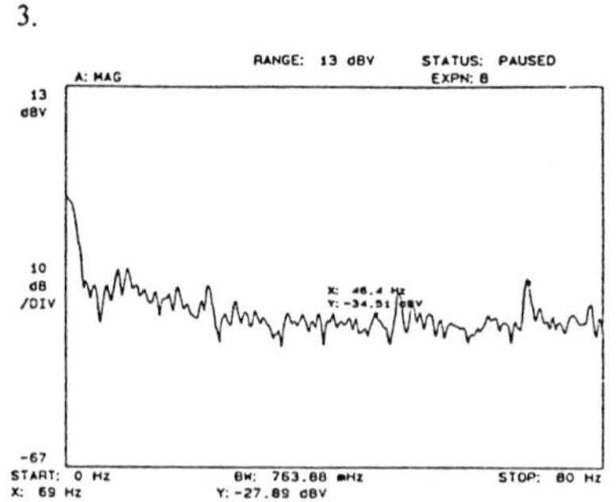

4.

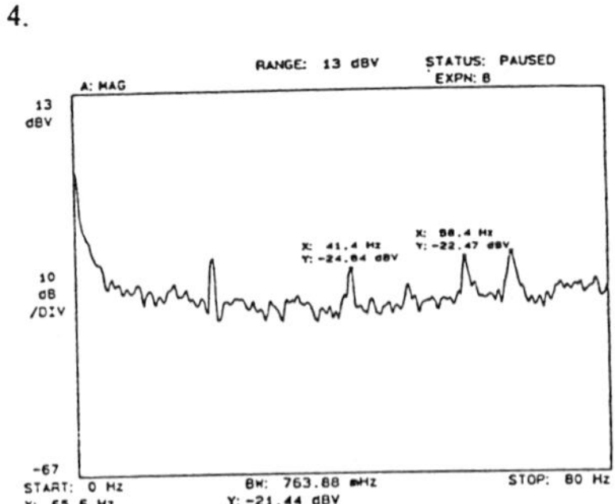

Fig.3 Perception of 3D. Perception of non-cyclopean wallpaper pattern. Fig.4 Perception of 3D (Subject GL). Each figure is a RMS exponential weighting average of MEG. In addition, the data was found to be statistically significant.

The second set of control experiments were devised in order to test whether the increase in activity at frequencies associated with the perception of depth was not a result of converging or diverging the eyes. Convergence or divergence of the eyes through the 2-D plane of focus is the technique most commonly used to focus on a wallpaper pattern or a random dot stereogram in order to perceive depth. To control for convergence or divergence we used a 6-dioptre prism in front of one eye. In addition, the wallpaper pattern was rotated through 90 degrees in order to prevent perception of depth. Results from subjects suggested that convergence or divergence alone without stereoscopic vision did not produce an increase in the spectral content at the frequencies associated with stereopsis. Figs.5 and 6 show the results of this experiment in Subject DCM.

5.

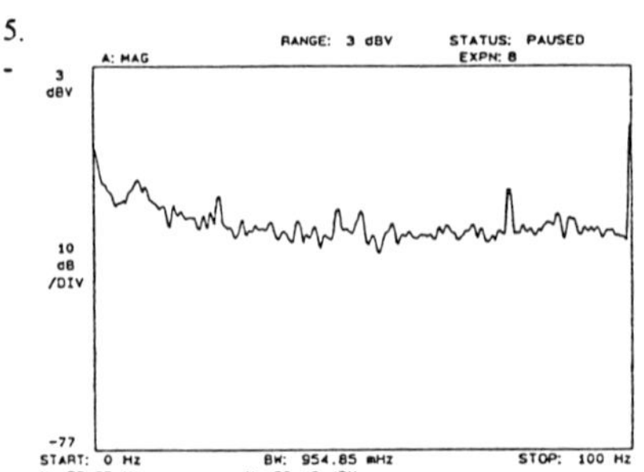

6.

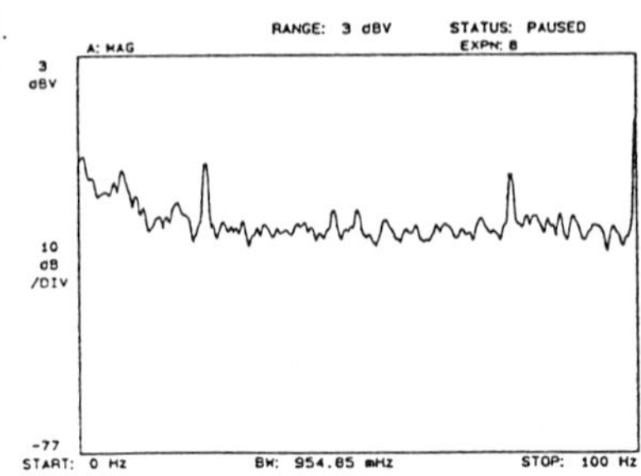

Fig.5 Convergence and Fig.6 Divergence (DCM). Each figure is a RMS exponential weighting average of MEG. In addition, the data was found to be statistically significant.

SUMMARY

We have found that the spectral content of frequencies in the range of the alpha rhythm recorded from the parieto-occipital cortex was reduced on eye opening and perception of a visual image. Perception of 3-D during stereoscopic vision resulted in a further attenuation of this alpha-rhythm. These results are in agreement with reports in the literature on the reduction of the alpha rhythm during tasks such as visual imagery and verbal association [4].

We have seen increased cortical activity at frequencies specific to stereopsis and the perception of depth. This increase at specific frequencies may be influenced by object recognition. The increased activity at frequencies when perceiving 3-d in a stereogram is not the same when convergence or divergence occurs without stereopsis.

Our results are consistent with reports in the literature which suggests that frequencies above 40 Hz are involved in focused attention, pattern recognition and higher order visual activity [see 5].

REFERENCES:

[1] Regan, D. Human Brain Electrophysiology. Evoked potentials and evoked magnetic fields in science and medicine. 1989 Elsevier.

[2] Julesz, B., Vision Research, 1986, 26: 1601-1612.

[3] Gevins, A., Leong, H, Smith, M.E., Le, J and Du, R. Mapping cognitive brain function with modern high-resolution electroencephalography. TINS, 1995, 18(10) : 429-436.

[4] Salenius, S., Kajola, M., Thompson, W.L., Kosslyn, S. and Hari, R. Reactivity of magnetic parieto-occipital rhythm during visual imagery.   EEG Journal 1995, 95 : 453-462.

[5] Basar, E., Basar-Eroglu, Demiralp,T. and Schurmann, M. Time and frequency analysis of the Brain's Distributed Gamma-band system. IEEE 1995 July/August : 400-410.

[6] Conway, B.A., Halliday, D.M., Farmer, S.F., Shahani, U., Maas, P., Weir, A.I. and Rosenberg, J.R. Synchronization between motor cortex and spinal motonuronal pool during the performance of a maintained motor task in man. J.Physiol. 1995, 489 (3) :917-924.

Acknowledgements : This work was supported by the Wellcome Trust and EPSERC.

Source Localization of a N400-like Negativity in a Sentence-Reading Paradigm Using Evoked Magnetic Fields and Magnetic Resonance Imaging

Simos, P.G., Basile, L.F.H., Zouridakis, G., and Papanicolaou, A.C.

Department of Neurosurgery, University of Texas -Houston Health Science Center, Houston, Texas, USA

Introduction

The N400 component of the Event-Related Potential (ERP) corresponds to a negative-going deflection that peaks at approximately 400 ms after the onset of the final word in a sentence [1]. N400 amplitude is believed to be an index of the degree that a given concept is deemed "appropriate" within a particular linguistic context [2]. Little is known regarding the sources of neural activity that give rise to N400-like negativities measured from the scalp. Identifying the neural substrates of N400 activity could provide important information regarding the brain structures that participate in language comprehension. Intracranial recordings have revealed sources of electrical activity, that become active at the same latency as the scalp-recorded N400 response, in paleo- and archeocortical regions as well as in neocortical areas. The former included the hippocampus and the parahippocampal gyrus [3, 4]. Potential neocortical sources may include the inferior parietal cortex, temporal areas surrounding the collateral sulcus [3], and the middle temporal gyrus [5]. Magnetoencephalographic (MEG) measurements show excellent temporal resolution and combined with magnetic resonance imaging (MRI) have enabled researchers to localize the intracranial sources of several ERP components within neurologically plausible anatomical regions [6, 7, 8]. However, the only reported attempt to localize the sources of the magnetic equivalent of the N400 component (N400m) in the brain [9] has produced inconclusive results.

The present investigation employed a standard paradigm for eliciting the N400 response [1, 10], in which the subjects are asked to read sentences presented one word at a time in order to be able to recognize them after the testing session. Half of the sentences ended with a semantically inappropriate word (e.g., *He was stung by a towel*). Using standard MEG analysis techniques, we attempted to identify brain areas that give rise to the magnetic equivalent of the N400 response.

Methods

Eight right-handed adults, native speakers of English were tested in this study. The subjects were between 18 and 45 years old, had normal or corrected-to-normal vision, and no history of neurological disorders. They were required to sign consent forms and they were paid for their participation. Eight hundred five-to-eleven word sentences with highly predictable final words served as stimuli in the recording sessions. Sentence completion norms were available for all of the sentences. Half of these sentences were transformed into semantically meaningless sentences by substituting an inappropriate word for the original ending. Congruous and incongruous final words were matched for word frequency and length. Word endings in the incongruous sentences were always grammatically and syntactically appropriate. The stimuli were randomly assigned to seven lists of 80 sentences each. The sentences were presented one word at a time in the center of a computer screen. Each word was flashed for 200 ms with an inter-word interval (onset to onset) of 800 ms. The inter-sentence interval was randomly varied between 2 and 4 s. The stimuli were drawn white on a black background and the final words subtended between 0.8° and 2.0° of horizontal visual angle. To ensure that subjects were engaged in the task, they were given after each session a list of 16 sentence bodies and were asked to mark the ones that they had seen in the previous block (sentence recognition task).

The ERPs were recorded from Fz and Pz referenced to the right mastoid, using a Nicolet Compact Four system. The electro-oculogram (EOG) was monitored and recorded to identify epochs contaminated by eye movements. The EEG data were analog-filtered from 0.01 to 30 Hz. Further analysis of the electrical signals was done in common with the magnetic signals. MEG was simultaneously recorded with a 7-channel Neuromagnetometer (Biomagnetic Technologies, Inc., Model 607), using second-order gradiometers with a 4 cm baseline and 1.8 cm coil diameters. Recordings were done in a magnetically shielded room. The position of the magnetometer relative to the subject's head was automatically recorded with a BTi probe-position indicator which had an error tolerance of less than 3 mm and an accuracy of 0.1 mm. Each subject's fiduciary points -- i.e., the nasion and the two preauricular points -- served as the basis for the MEG Cartesian coordinate system. The Y axis in that system was defined by the two preauricular points, the X axis by the nasion and the center of the head, while the Z axis was normal to the X-Y plane and passing through the X-Y intersection (i.e., the center of the head). The MEG analog filters were set at 0 to 50 Hz. The signal was digitized at 200 Hz starting 100 ms before the onset of the final word and continuing for 1 s after the onset. All the MEG epochs were filtered off-line with a bandpass of 0.01 to 20Hz (Butterworth 4 pole filter, 24dB/octave rolloff) in order to improve signal quality. Before averaging, trials on which an eye movement or blink had occurred (as indicated by a peak-to-peak amplitude in the EOG channel in excess of 70 μV) were rejected. Artifacts that were produced by the magnetometer itself typically consisted of high amplitude noise bursts confined to a single MEG channel. Those artifacts were easily recognizable by visual inspection of the MEG epochs and excluded from subsequent analyses. The data were then baseline adjusted relative to the mean amplitude in the 100-ms prestimulus period, subjected to a computerized linear detrending procedure, and averaged separately for each condition (i.e., Congruous, Incongruous) and recording session. A minimum of 30 single-trial MEG records were used to calculate each averaged waveform. Neuromagnetic measurements were obtained from 49 locations on the left side of the scalp corresponding to seven consecutive placements of the recording device. The intracranial sources of the observed evoked magnetic fields (EMFs) were modeled as equivalent current dipoles, by using a finite-difference version of the Levenberg-Marquardt algorithm. The dipole fitting method was applied at every 5 ms, from 200 to 800 ms after stimulus onset. Dipole parameters (spatial coordinates, orientation [Psi] and strength [Q]) were computed with reference to the best-fitting sphere, the curvature of which was defined by the surface of each subject's scalp over the region of the head containing the field extrema. The criterion we used for accepting a dipole solution as satisfactory was a correlation of at least .85 between the measured and the predicted fields at each time point. MRI scans from three subjects were used to identify the brain regions that corresponded to the estimated locations of the equivalent current dipoles. For each of the remaining five subjects, spatial dipole coordinates were projected on MRI scans from three other neurologically normal individuals with closely matching head size. A computer algorithm was used to convert the MEG dipole coordinates into corresponding MRI coordinates [6].

Results

All subjects reached the 85% correct criterion in the sentence recognition task in every recording session. The amplitude of the N400 response, that was measured between 320 and 420 ms post-stimulus (240 and 320 ms in S#3), was more negative in the Incongruous than in the Congruous condition, $F(1,156)=20.88$, $p=.0001$. N400 amplitude varied in the same direction across conditions in every subject. No systematic pattern of amplitude or latency variation emerged between recording sessions. Figure 1A shows representative average ERPs from two subjects.

In the Incongruous condition EMF waveforms displayed a prominent deflection during the course of the electrical N400 wave. In every case the peak-to-peak field strength (i.e., the algebraic sum of outward and inward flux) was smaller in the Congruous condition. Figure 1B displays representative magnetic waveforms from the two subjects. Individual contour maps (Figure 1C) contained a single pair of extrema. Seven subjects produced satisfactory dipole solutions. The vertical lines drawn in Figures 1A and 1B indicate the range of time points associated with satisfactory dipole solutions during the N400 latency window. In every subject, for whom reliable dipole parameters were computed in both conditions (n = 5), the entire range of Q values computed during the N400 latency window was larger in response to incongruous words than in response to congruous words. These findings provide strong evidence that the N400 response and its magnetic equivalent (N400m) belonged to the family of negativities which are presumably sensitive to linguistic context [2]. After superimposing the dipole coordinates onto the appropriate horizontal MRI slice, we used a standard stereotaxic atlas of the human brain [11] to identify the cerebral structures (e.g., sulci, and gyri) and cytoarchitectonic areas where dipoles were localized. MRI slices from the three subjects whose MRI scans were available are presented in the upper row in Figure 1E. The bottom row in Figure 1E displays the estimated locations of the sources of the N400m response for the five subjects for whom MRIs were not available. Given that there was close agreement in the dipole regions across the three different MRIs used in each case we present only one slice per subject. Five subjects demonstrated dipole-like activity localized in posterior temporal neocortex near the temporo-parietal junction, typically in the middle temporal gyrus. These areas lie near the banks of the superior temporal sulcus. In one subject (#6) dipole regions extended more posteriorly into the temporo-occipital association cortex (area 37). In two subjects (#1 & #7) solutions fell in the medial temporal lobe, in the vicinity of the hippocampus and the parahippocampal gyrus.

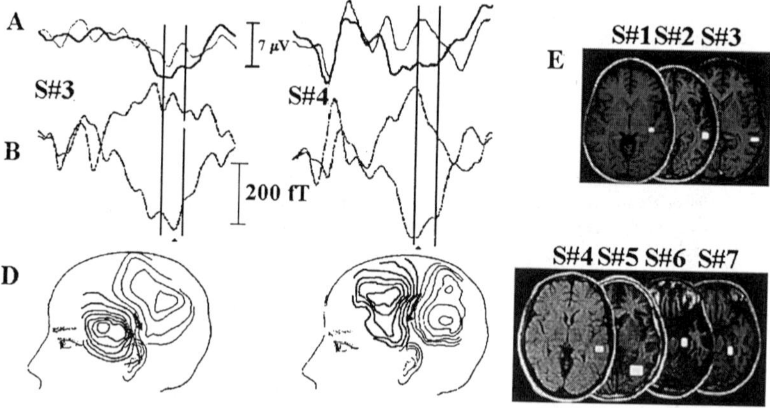

Discussion

The present study was successful at recording the magnetic equivalent of the N400 response elicited by the final words in sentences. In every case dipole solutions were localized in temporal lobe structures, including neocortical areas on the lateral surface of the left hemisphere as well as more medially located paleo- and archeocortical regions. The finding of dipole activity in posterior temporal regions near the temporo-parietal junction is in agreement with the results of a recent study that combined ERPs and metabolic imaging using Positron Emission Tomography (PET) [12], and with the outcome of intracranial recordings [5]. On the basis of our data it seems likely that the sources of the magnetic equivalent of the N400 response reflect focal activity in areas specialized for those aspects of language which are presumably indexed by that ERP component. These processes are involved in lexical access that occurs within a previously established semantic context and leads to word recognition. In agreement with this claim, several lines of converging evidence have established a firm link between temporal and temporo-parietal cortex and language processes such as word recognition [13] and semantic comprehension of written material, including lesion [14] and PET studies [15, 16].

In addition to evidence for superficial sources, our data showed dipole solutions in more medial aspects of the temporal lobe, in the vicinity of the hippocampus and the parahippocampal gyrus. This finding is consistent with reports of intracranially-recorded electrical potentials from medial temporal lobe structures which shared many of the characteristics of the scalp-recorded N400 response [3, 4]. The functional anatomy of the hippocampal formation makes this structure ideal for mediating some of the neural processes involved in recent memory [17]. Inasmuch as the N400 is a sign of certain operations involved in word recognition and it is affected by experimental manipulations that influence the course of this process (such as semantic context, subjective expectancy, etc.), it seems likely that the intracranial generation of this component is closely linked to neural events in the medial temporal lobe.

References

[1] Kutas, M., & Hillyard, S.A. (1980). *Biol. Psych.*, 11, 99-116.

[2] Halgren, E. (1990). Insights from evoked potentials into the neuropsychological mechanisms of reading. In A.B. Scheibel & A.F. Wechsler (Eds.), *Neurobiology of Higher Cognitive Function.* New York, NY: The Guilford Press.

[3] McCarthy, G., Nobre, A.C., Bentin, S., & Spencer, D.D. (1995). Language-related field potentials in the anterior temporal lobe: I. Intracranial distribution and neural generators. *J. Neurosci., 15,* 1080-1089

[4] Smith, M.E., Stapleton, J.M., & Halgren, E. (1986). *Electroencephal. Clin. Neurophys., 63,* 145-159.

[5] Guillem, F., N'Kaoua, B., Rougier, A., & Claverie, B. (1995). Intracranial topography of event-related potentials (N400/P600) elicited during a continuous recognition memory task. *Psychophysiol. 32,* 382-392

[6] Rogers, R.L., Baumann, S.B., Papanicolaou, A.C., Bourbon, T.W., Alagarsamy, S., & Eisenberg, H.M. (1991). Localization of the P3 sources using magnetoencephalography and magnetic resonance imaging. *Electroencephal. Clin. Neurophys., 79,* 308-321.

[7] Papanicolaou, A.C., Baumann, S.B., Rogers, R.L., Saydjari, C., Amparo, & E.G., Eisenberg, H.M. (1990). Localization of auditory response sources using MEG and MRI. *Arch. Neurol., 47,* 33-37.

[8] Basile, L.F.H., Rogers, R.L., Bourbon, W.T., & Papanicolaou, A.C. (1994). Slow magnetic fields from human frontal cortex. *Electroencephal. Clin. Neurophys., 90,* 157-165.

[9] Schmidt, A.L., Arthur, D.L., Kutas, M., & Flynn, E. (1989). Neuromagnetic responses during reading meaningful and nonmeaningful sentences. *Psychophys. Abstr., 26,* S6.

[10] Polich, J. (1985). Semantic categorization and event-related potentials. *Brain Lang., 26,* 304-321.

[11] Talairach, J. & Tournoux, P. (1988). *Co-planar stereotaxic Atlas of the human brain.* New York, NY: Thieme Medical Publishers.

[12] Nenov, V.I., Halgren, E., Smith, M.E., Badier, J-M., Ropchan, J., Blahd, W.H., et al. (1991). Localized brain metabolic response correlated with potentials evoked by words. *Behav. Brain Res.,*

[13] Damasio, A.R. & Damasio, H. (1983). The anatomic basis of pure alexia. *Neurol. (Cleveland), 33,* 1573-1583.

[14] Hart, J. & Gordon, B. (1990). Delineation of single-word semantic comprehension deficits in aphasia, with anatomical correlation. *Ann. Neurol., 27,* 226-231.

[15] Price, C.J., Wise, R.J.S., Watson, J.D.G., Patterson, K., Howard, D., & Frackowiak, R.S.J. (1994). Brain activity during reading: The effects of exposure duration and task. *Brain, 117,* 1255-1269.

[16] Howard, D., Patterson, K., Wise, R., Brown, W.D., Friston, K., Weiller, C., et al. (1992). The cortical localization of the lexicons. *Brain, 115,* 1769-1782.

[17] Squire, L.R. (1986). Mechanisms of memory. *Science, 232,* 1612-1619.

Information Theory Predicts the Magnitude of an MEG Response Related to the Operation of the Auditory Sensory Memory

Sinkkonen, J.[1,2], Tiitinen, H.[1], May, P.[3], Ilmoniemi, R. J.[4], and Näätänen, R.[1]

Cognitive Psychophysiology Research Unit[1], Department of Psychology, University of Helsinki, Finland; Department of Radiology[2], Helsinki University Central Hospital, Finland; Department of Mathematics[3], King's College, London, UK; BioMag Laboratory[4], Medical Engineering Centre, Helsinki University Central Hospital, Finland

Introduction

An electric field generated by the primate brain, MMN, and its magnetic counterpart, MMNm, have established their status as physiological correlates of a transient auditory memory[5].

Small changes in the energy consumption of a limited area of the cortex should be linearly proportional to amplitude changes of the scalp-recorded ElectroMagnetic Fields: EMFs are mainly due to the summation of post-synaptic potentials of cortical neurons which together with their cause, the action potentials, consume most of the energy available for the nervous system[1,3,4].

Information, as defined by Shannon[6], is proportional to $\log p$. In the absence of any *a priori* meaning of the stimuli to the organism under observation, Shannon information can be used as a measure of the importance of the stimuli. *Optimal resource sharing* should then express itself in stimulus-evoked changes of the EMF being directly proportional to the information of the stimulus, $-p \log p$, where p is the probability of the stimulus event: this way the same amount of processing capacity becomes allocated for each incoming unit of information.

Methods

We studied 8 subjects with a whole-scalp magnetometer[2] in an auditory environment containing an unpredictable sequence of two mutually distinguishable sinusoidal tones (stimulus duration 60 ms including 10-ms rise and fall times, ISI 200 ms, frequencies 1000 Hz and 1150 Hz; the subject received the stimuli to her left ear while reading self-selected texts). The probability of the 1150-Hz tone was varied from 0.02 to 0.40 in eight steps; the presentation order was interleaved and counter-balanced. For illustration, see Fig. 1.

For each subject, one channel over the right auditory cortex with good S/N ratio was selected, and temporal average (100-200ms; shaded in the Fig. 2) of the signal from the selected channel was used as a measure of the resource usage of the cortex. The signal consisted almost exclusively of MMNm, as N1m and other EMF components are attenuated in nonattending subjects hearing fast stimulus sequences with small deviations[7]. See Fig. 2.

Results

Results, averaged over subjects, are presented in Fig. 3. They closely matched our prediction based on the principle of optimal resource sharing.

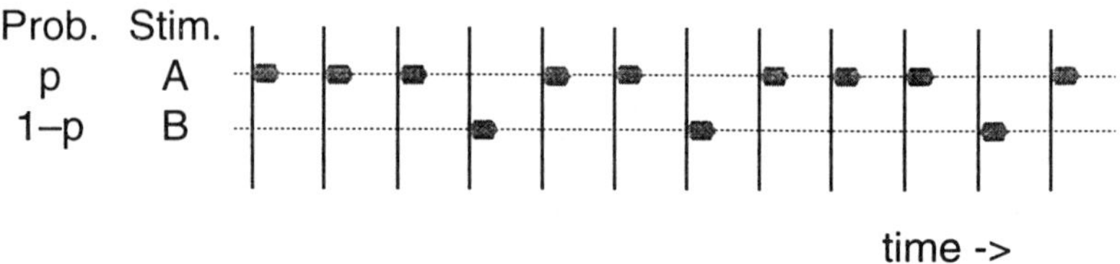

Figure 1: The stimulus paradigm consisted of independent but regularly occurring sinusoidal tones of two frequency. The probabilities of the tones were varied (see text for details).

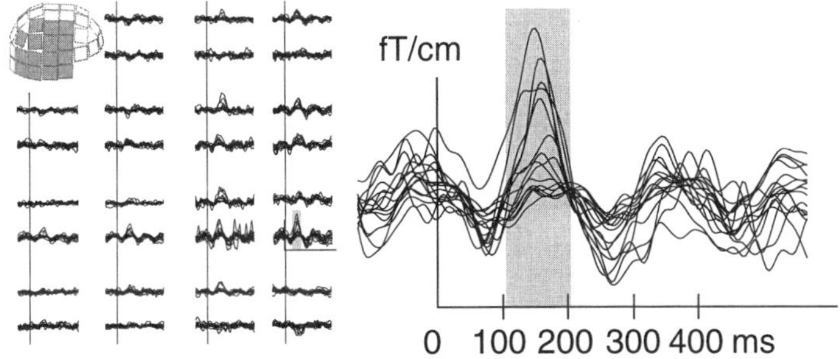

Figure 2: Data of a representative subject for each stimulus probability, measured over the right auditory cortex. Location of the illustrated sensors relative to the whole-head SQUID array is shown in the inset (left). One channel with good S/N ratio was selected for each subject, and the data was integrated from 100 ms to 200 ms after stimulus onset to get a quantification of the MMNm amplitude (right).

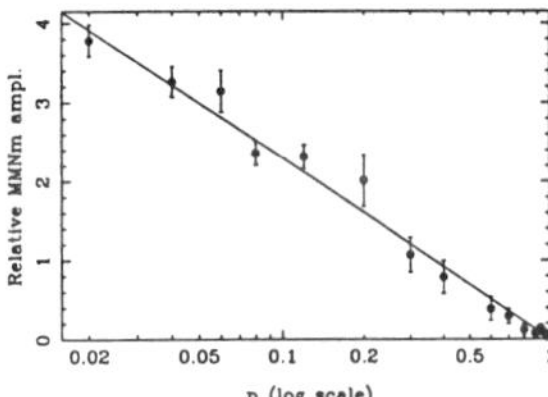

Figure 3: The predicted linearity of MMNm amplitude (line) and the averaged response amplitude of eight subjects (dots with bars) as a function of logarithmic stimulus probability. Because of subject-wise variation of the overall amplitude of the responses, normalization of the data over subjects was necessary before averaging. Bars denote the standard error of the mean. Fluctuations from linearity are within random variation.

Discussion

Energy invested for the processing of unpredictable stimuli seems to be proportional to their information content. The present result suggests that optimal resource allocation is one of the principles of information processing of the cortex. The characterization of the behaviour of the MMNm amplitude as a function of probability also allows optimization of the auditory odd-ball paradigm for applications where fast data collection is important.

References

[1] Creutzfeld, O. & Houchin, J. in *Handbook of Electroencephalography and Clinical Neurophysiology* (ed A. Remond) Vol. 2C, 5-55 (Elsevier, Amsterdam, 1974).
[2] Knuutila, J. *et al. IEEE Trans. Magn.* **29,** 3315-3320 (1993).
[3] Lorente de No, R. *A Study of Nerve Physiology* **132,** part 2 (Rockefeller Institute, New York, 1947).
[4] Martin, J. H. in *Principles of Neural Science* 3rd ed. (eds Kandel, E. R., Schwartz, J. H. & Jessell, T. M.) 779 (Elsevier, New York, 1991).
[5] Ritter, W. *et al. Ear and Hearing* **16,** 52-67 (1995).
[6] Shannon, C. E. *Bell Syst. Tech. J.* **27,** 379-423 (1948).
[7] Tiitinen, H. *et al. Nature* **372,** 90-92 (1994).

Acknowledgements

We thank M. Huotilainen, K. Kaila, S. Kaski, J. Lavikainen, and G. Nyman for their help in conducting the experiments and in preparation of this manuscript and the associated poster. This work was supported by the Academy of Finland.

Authors' contact information: e-mail: janne@iki.fi; fax: +358 0 471 5781; phone: +358 40 5464798.

Determination of the Nature of the Connection Between the Auditory and Visual Cortices

Stephen, J.M.[1], Broadhurst, J.H.[1], Knuth, K.H.[1*] and Schwartz, B.J.[2]

[1]Physics Department, University of Minnesota, Minneapolis, USA; [2]Scripps Research Institute, La Jolla, USA

Introduction

Studies confirming the influence of visual information on the response of the auditory cortex to speech sounds [1] lead us to question the nature of the connection between the visual and auditory cortices. The superior colliculus exhibits visual-sensitive neurons that can be modulated (both inhibited and excited) by auditory stimuli [2]. Since the superior colliculus is one of the nuclei that the optic nerve passes through as it travels to the visual cortex, this suggests that one might expect a modulation of the visual response with a simultaneous auditory stimuli.

However, the superior colliculus is not traversed by the main auditory nerve as it passes through the brain stem to the auditory cortex. Therefore, it is first important to consider if there is any influence of the auditory response with a corresponding visual stimulus and if there is to determine the possible region of the convergence of information. This led us to study the response of the auditory cortex to a click and unrelated light flash. In animal studies some interesting results that show that some neurons located in the auditory cortex are excited by a visual stimulus [3]. Amongst conflicting results in electroencephalographic studies, one study claimed that the influence of a visual stimulus to the auditory cortex was a linear superposition of the individual responses [4]. In the present study the results show that the response to the light flash and click is not a linear summation of the responses to the individual stimuli.

Methods

Seven normal subjects were presented with three different stimuli. The response of the auditory cortex to the three different stimuli was measured using a 37-SQUID biomagnetometer. First, the subject was presented with a red light point flash of duration 0.5 msec at a distance of approximately one meter. The second stimulus was a click of duration 0.5 msec at sound pressure level between 60-100 dB depending on the subject. Third, the subject was presented with the light flash and click simultaneously. Each stimulus was presented many times and the response was obtained by on-line signal averaging.

To determine if there is any interaction between the incoming visual and auditory information we determined if the response to the combined stimulus was a simple linear superposition of the individual responses. This was done by adding the response of the auditory cortex to the light flash (Fig. 1a) to the response of the auditory cortex to the click (Fig. 1b) and subtracting off the combined response (Fig. 1c). These responses were from one and the same channel, that was

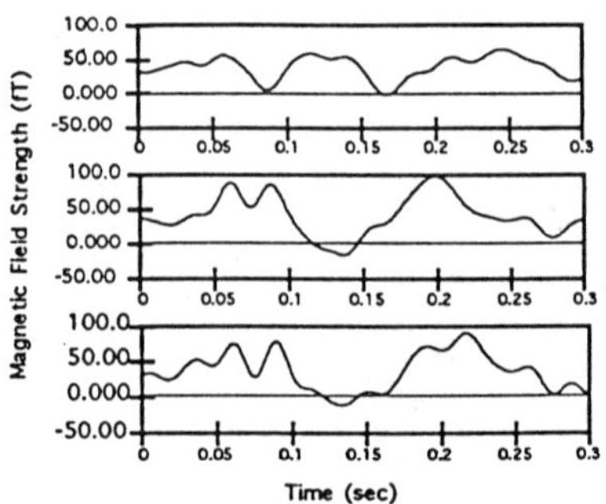

fig. 1 Magnetic field responses of subject PQ for one channel (a) to flash alone (b) to click alone (c) to simultaneous flash and click

chosen based on the high signal-to-noise ratio in that channel. This difference response was then compared across subjects to determine if the pattern persisted among different individuals. The comparison was done by taking a correlation between two subjects' difference plots and determining if the correlation was significant (based on the t-test).

* Now at Kresge Hearing Institute of the South, New Orleans, USA

Results

If there were no interaction between the two modalities we would expect the difference plot to be at the level of the random noise. For example, if the magnetic fields measured were the result of different areas of the brain being activated by the different stimuli without any merging of information, when the two stimuli are presented together the total magnetic field should be a simple summation of the magnetic fields to the individual responses. As can be seen in the difference plot of subject PQ (Fig. 2), the result is well above the expected noise level. The difference plots for all seven subjects were consistent with this result.

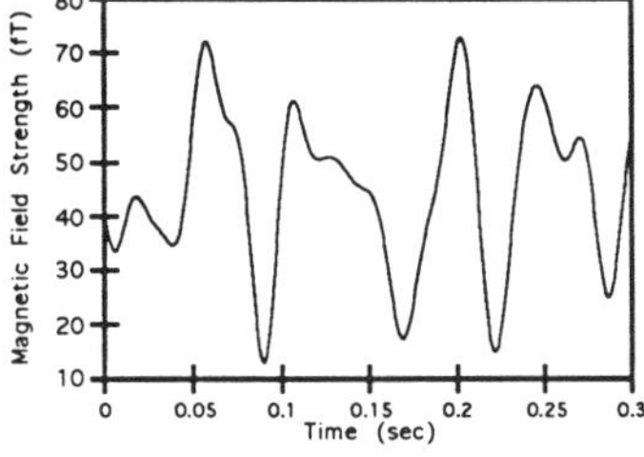

fig. 2 Difference plot for subject PQ

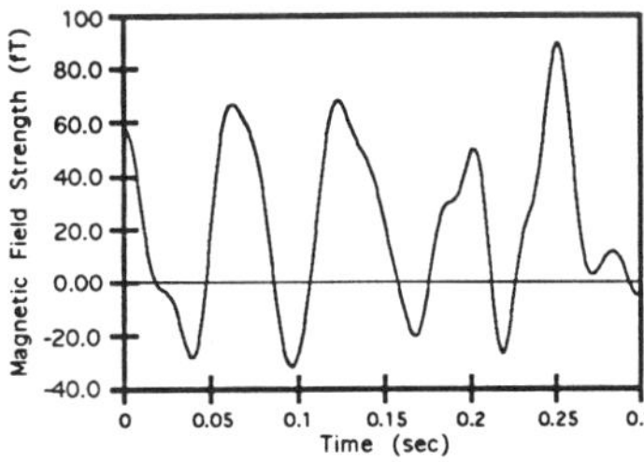

fig. 3 Difference plot for subject EH

By comparing the difference plots of Fig. 2 and Fig. 3, one can see that not only did all seven subjects have a nonzero result, but there is also a similar pattern across subjects. The apparent similarities were confirmed by the finding that there is a significant correlation between Fig. 2 and Fig. 3 with an offset of 2 msec of Fig. 2 with respect to Fig. 3 as can be seen in Fig. 4. The correlation at lag time 2 msec was 0.26±0.08. Similar results can be seen in Fig. 6, which is the correlation of two other difference plots, Fig. 3 and Fig. 5. Although there is a slightly different offset for the maximum correlation, this plot also has a significant correlation at lag time 12 msec of 0.30±0.08.

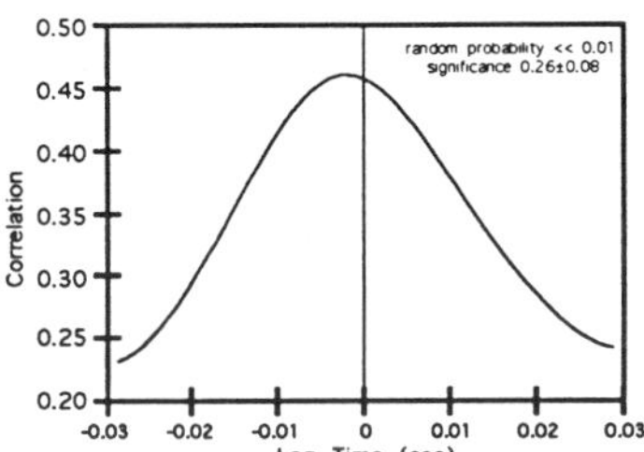

fig. 4 Correlation plot of figures 2 and 3

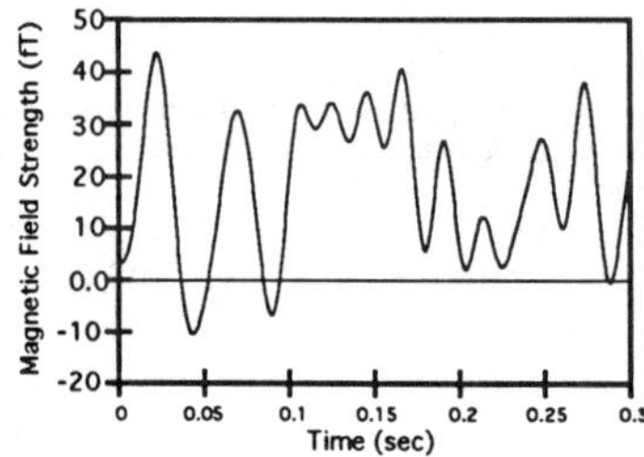
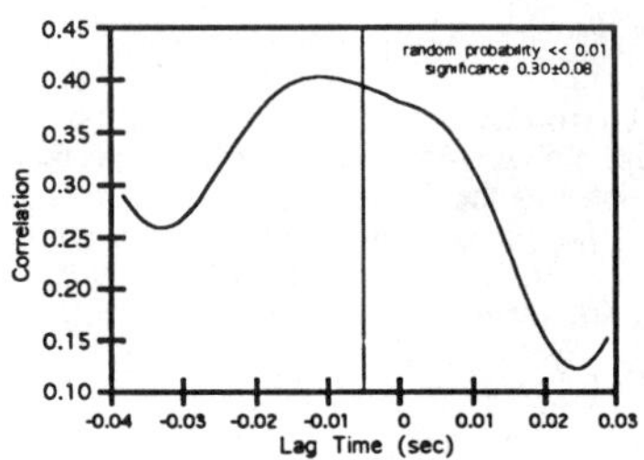

fig. 5 Difference plot for subject DO fig. 6 Correlation plot of figures 3 and 5

Discussion

The significant correlation between two different subjects for multiple pairings leads us to believe that there is a definite interaction between the incoming information from the separate modalities. The early appearance of this interaction implies that there is a convergence of information prior to any cortical processing. The question then is where this interaction may be occurring. Based on anatomical studies we know that the auditory pathway passes through the inferior colliculus on its way to the auditory cortex. The optic nerve only enters the brainstem at the level of the superior colliculus. Some studies have shown that there are auditory sensitive neurons in the superior colliculus, despite the fact that conventional knowledge does not point to the auditory pathway passing through this portion of the colliculus. This implies that there might be some connection between the inferior and superior colliculus that not only allows some auditory information to pass to the superior colliculus but also allows some information on the visual stimulus to pass to the inferior colliculus to affect the auditory signal.

To thoroughly cover the study of the connection between the auditory and visual cortices we plan on testing the influence of an auditory stimulus on the visual response.

References:

[1] Sams, M., Aulanko, R., Hämäläinen, M., Hari, R., Lounasmaa, O.V., Lu, S.-T., Simola, J. Seeing speech: visual information from lip movements modifies activity in the human auditory cortex, Neuroscience Letters, 1991, 127: 141-145.
[2] Meredith, M.A., and Stein, B.E. Visual, auditory, somatosensory convergence on cells in superior colliculus results in multisensory integration, Journal of Neurophysiology, 1986, 56: 640-662.
[3] Watanabe, J., Iwai, E. Neuronal activity in visual, auditory and polysensory areas in the monkey temporal cortex during visual fixation tasks, Brain Research Bulletin, 1991, 26: 583-592.
[4] Ciganek, L. Evoked potential in man: interaction of sound and light, Electroenceph. Clin. Neurophysiol., 1966, 21: 28-33.

Acknowledgments

This study was supported in part by the University of Minnesota, Biomagnetic Technologies Inc. and the Scripps Research Institute. We thank Lacy Kurelowich, Joslyne Foley, and Patti Quint, the technicians at Scripps, for helping collect the data, Steve Cobb at BTI for the technical support, and Dr. Benjamin Bayman for our discussions on the analysis of the data.

Visualization of Tonotopy of Auditory Cortex Using a Newly Developed 129 Channel Vector Magnetoencephalography

Takanashi, Y.[1], Iwamoto, K.[1], Yoshikawa, K.[1], Makino, M.[1], Tomita, T.[2], Tomita, S.[2], Kajihara, S.[2], Kondo, Y.[2], Okamura, S.[2], Yoshida, Y.[2], Ueda, M.[2,3] and Nakajima, K.[1]

[1]Department of Neurology, Research Institute for Neurological Diseases and Geriatrics, Kyoto Prefectural University of Medicine, Kyoto, Japan; [2]Technology Research Laboratory, Shimadzu Corporation, Kyoto, Japan; [3]Superconducting Sensor Laboratory, Chiba, Japan

Introduction

The existence of the human tonotopic organization has been revealed using the neuromagnetic measurement. These early studies on the tonotopic organization in the human were performed by measuring sequentially at many sites using the magnetoencephalography (MEG) with a few channels[1,2,3], so these studies needed a vast effort and time to fulfill a study. This meant that we could not use mangentoencephalography (MEG) to investigate the auditory function of patients in clinical setting. Now, many MEG systems, which are commercially available, consist of multi-channel first-order axial gradiometers or planar gradiometers. These have some advantages; specially, axial gradiometer is sensitive to the radial component of magnetic fields, and planar gradiometer is favorable to obtain their tangential component. Even using these two systems, however, we could not obtain three dimensional components about magnetic fields. Consequently, we have developed a new 129 channel vector magnetometer system[5].

The purpose of the present study is to evaluate the reliability of source localization in our new MEG system, and to visualize the human tonotopic organization on MR imaging.

Subjects and methods

Nine normal-hearing subjects (9 males, all right-handed), with no history of otological or audiological disorders and aged among 20 and 46 years, participated in this study after giving us their informed consent. Five subjects were investigated in both hemispheres; two subjects, the left one; remaining two subjects, the right one. Tone-burst with 100 ms duration (rising time, 10 ms; falling time, 10 ms) and an interstimulus interval 1.00-1.86 msec were at random delivered to subject's contralateral ear, and white noise with 65 dB strength was delivered to ipsilateral ear to the examined hemisphere through a plastic tube. The carrier frequencies were 500, 1000, 2000 Hz with 85 dB, and 4000 Hz with 95 dB strength.

Magnetic auditory evoked fields (AEFs) were recorded in a magnetically shielded room using a novel 129-channel vector neuromagnetic imaging system[5]. In brief, our MEG system consisted of 43 units of three orthogonally-oriented ellipse-shaped pick-up coils. Each sensor coil was positioned at the same center point and the three planes of coils intersected each other. The effective diameter of the recording area was 190 mm. The baseline was 50 mm. The distance between the centers of the sensor coils was 25 mm. Our system could cover half of the cerebral hemisphere. Our MEG system can measure changes of the magnetic fields at 43 sites, and can determine the three dimensional direction of magnetic fields at each site. Each subject calmly lay on his side on a bed in a magnetically shielded room. We determined the exact location of the subject's head in relation to each pick-up coil by measuring the magnetic signals of three positioning coils placed at known sites on the head. We ensured reliable positions by averaging the signals 50 times prior to the experiments. If two extrema were not included in the obtained magnetic fields, the position of sensors was changed to cover the two extrema. The signals were filtered with a band-pass filter (0.7-100 Hz), and were digitized at 1024 Hz. Two hundred single responses were averaged off-line, including signals for 50 ms before (pre stimulus baseline) and 300 ms after the stimulus. To identify sources of the equivalent current dipoles (ECDs) with high correlation values were found by a least-squares algorithm on the basis of a one-dipole model for 300-ms period after stimulation. Generator sites of the ECDs were superimposed on appropriate axial, coronal and sagittal MRIs. Magnetic resonance images were obtained for all subjects with 0.5T MAGNEX-a (Shimadzu, Japan) after MEG measurement. In order to create the MEG 3D coordinate system in the MR images three hollow disks filled with NiCl solution (0.5%) were affixed at sites where three positioning coils were placed in each subject before MRI measurement.

Results

Fig. 1 shows two sets of data; one is raw data, and the other is calculated data from one of our subjects obtained with 500 Hz tone-burst. Three dimensional components derived from the raw data were clearly shown, and enables us easily recognize the direction of the vector of magnetic field obtained at each recording site.

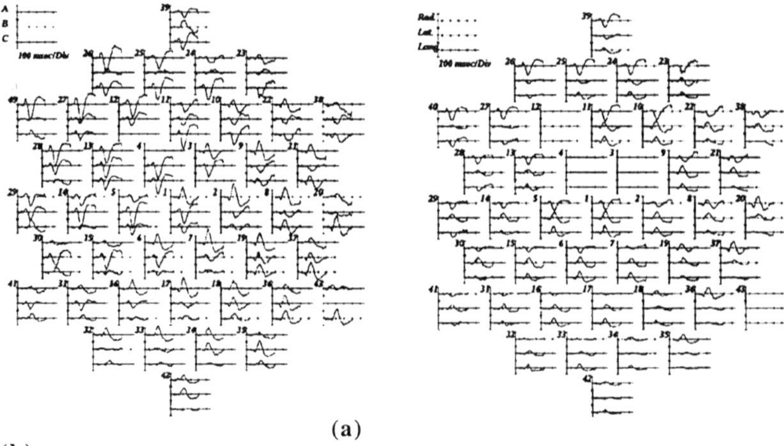

(a)

(b)

Fig. 1   Two sets of 129 averaged waves evoked by 500 Hz tone-burst stimulation.   (a) show the raw data set obtained from the right hemisphere,   (b) is the set of converted data to the radial, longitudinal, and latitudinal component on the scalp surface.   Note that 4 sets of data were not calculated due to the lack of one of three data from A, B, and C channel.

Table 1 shows the latencies, the dipole moment (the strength of current), and the goodness of fit of N100m, which were obtained from both hemispheres with each carrier frequency in 5 subjects. We could not find any significant difference between the right and the left hemisphere on the latency of N100m, however, the dipole moment of ECD in the right hemisphere is stronger than that in the left one. The values of goodness of fit of N100m in the right hemisphere were generally larger than that in the left.

878

Table 1:   The parameters of N100m in both hemisphere

	Latency (msec)		Dipole moment (nAm)		Goodness of fit (%)	
	Rt.	Lt.	Rt.	Lt.	Rt.	Lt.
500 Hz	103 ± 9.8	112.6 ± 7.0	42.8 ± 7.5	26.6 ± 11.2	97.8 ± 1.1	94.6 ± 3.0
1000 Hz	98.2 ± 10	107.1 ± 14	43.6 ± 9.7	27.9 ± 8.8	97.0 ± 2.0	94.0 ± 3.2
2000 Hz	99.0 ± 11.2	105.5 ± 6.0	32.2 ± 13.1	24.5 ± 8.3	96.6 ± 1.5	92.3 ± 8.6
4000 Hz	103.2 ± 11	107.8 ± 4.8	36.0 ± 22.6	16.9 ± 7.8	95.8 ± 2.2	90.1 ± 6.1

n = 5

Fig. 2 shows the estimated distance of the ECDs of N100m from the center of a sphere model.   These graphs indicate that the location of N100m ECDs is found to vary depending upon the tone-burst frequency.   However, their locations at 4000 Hz tone-burst varied among subjects.

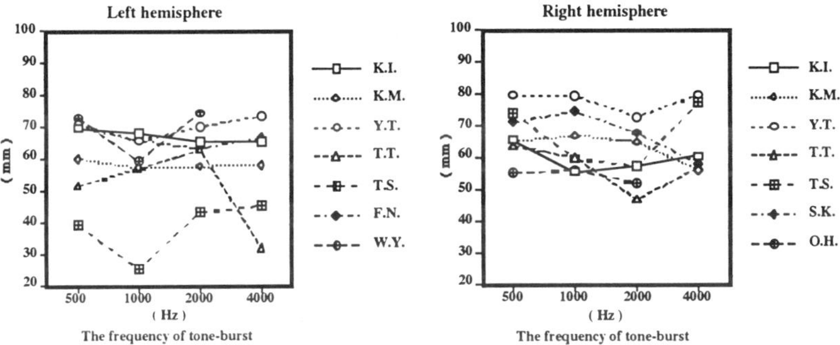

Fig. 2. The estimated distance of the equivalent current dipole of N100m from the center of a sphere model in each subject.   The left column shows data obtained from the left hemisphere, the right column from the right hemisphere.

Fig. 3 shows the locations of the estimated N100m dipoles evoked by 4 different carrier frequency. Although the locations of estimated N100m dipoles with each career frequency were different each other, we superimposed each dipole of N100m on an axial and a sagittal slice of MRI obtained from subject Y.T. in order to compare four different dipoles each other on the same anatomical location.

Discussions and conclusions

We have developed a new 129 channel vector magnetometer system linked with MRI.   The location of each ECD dipole was estimated using a single-dipole model in a spherical model determined individually according to the MRI images.   There was found to be the tendency that the location of N100m ECD was dependent upon the tone-burst frequency.   However, the locations at 4000 Hz tone-burst varied among subjects.

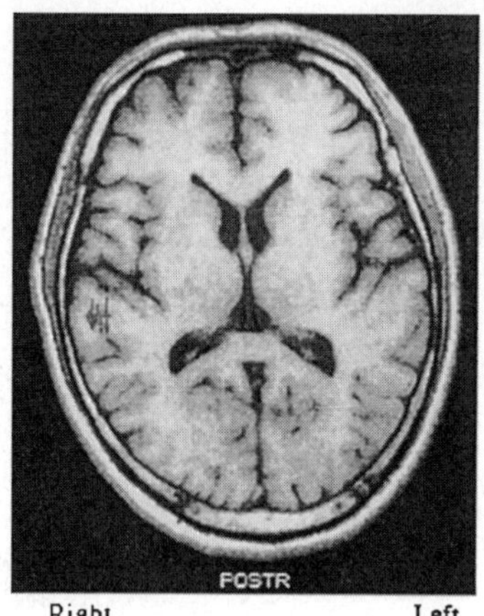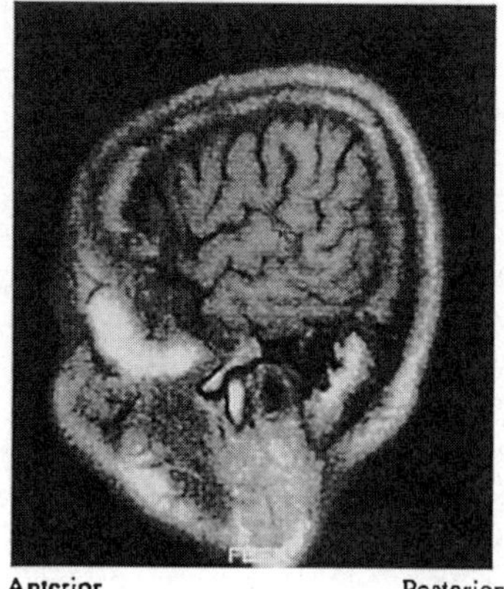

Right   Left   Anterior   Posterior

Fig. 3. The estimated dipoles of N100m on axial and sagittal MRIs obtained from subject Y.T. Four estimated dipoles were superimposed on each MRI. Note that all dipoles are located in the superior temporal gyrus

Previous studies on the visualization of the human tonotopy have always discussed based on the values of coordinate system[1, 2, 3]. Recently, Pantev et al. superimposed the estimated N100m source locations on individual MRI images[4]. Our results are in good agreement with their data. This study clearly demonstrates that a newly developed vector magnetometer system has the good reliability of source localization when visualizing the human tonotopy.

References

[1] Romani, G.L., Williamson, S.J. and Kaufman, L. Tonotopic organization of the human auditory cortex. Science, 1982, 216: 1339-1340.

[2] Elberling, C., Bak, C., Kofoed, B., Lebeck, J. and Saermark, K. Audiotorymagnetic fields. Source location and 'tonotopical organization' in the right hemisphere of the human brain. Scand Audiol., 1982, 11:61-65.

[3] Pantev, C., Hoke, M., Lehnertz, K., Lutkenhoner, B., Anogianakis, G., and Wittkowski, W. Tonotopic organization of the human auditory cortex revealed by transient auditory evoked fields. Electroenceph. clin. Neurophysiol., 1988, 69: 160-170.

[4] Pantev, C., Bertrand, O., Eulitz, C., Verkindt, C., Hampson, S., Schuierer, G., and Elbert, T. Specific tonotopic organizations of different areas of the human auditory cortex reveald by simultaneous magnetic and electric recordings. Electroenceph. clin. Neurophysiol., 1995, 94: 26-40.

[5] Yoshida, Y., Arakawa, A., Kondo, Y., Kajihara, S., Tomita, S., Tomiata, T., Takanashi, Y., and Matsuda, N. A 129-channel vector neuromagnetic imaging system. in This volume.

Measurement of Accommodation Related MEG Responses

Takeda, T., Hashimoto, K., Kumagai, T. and Endo, H.

*Human Informatics Department, National Institute of Bioscience and Human-Technology
1-1 Higashi, Tsukuba City, Ibaraki, 305, JAPAN*

Introduction

We have been developing optometers to measure accommodation, eye movement and pupil diameter simultaneously under real working conditions[1,2,3]. By using them, we have found that the accommodation is sensibly controlled by apparent depth perception judged by brains[4]. Yet, there is much to be clarified on the control mechanism of the accommodation. Hence, the study of accommodation both from ocular responses and brain activities has high possibility to light a new insight into brain and visual functions.

A special relay lens system has been developed to impose visual stimuli and at the same time to measure accommodation responses with a dynamic refractometer[1]. As for the brain response, a 64-channel Whole-cortex MEG systems (CTF Inc., Canada) was employed. Using those systems, MEG and accommodation responses were measured simultaneously for the first time (Fig. 1).

Methods

Fig.1 Measurement System.

One of the most prominent characteristics of the MEG system is a software noise reduction (SNR) technique[5], which uses noise data sensed by reference coils and subtract them from the measured data. The efficiency of the software noise reduction technique was examined and confirmed by a phantom head examination and a method, which allowed measurement with opening the door of a magnetic shielded room, was established[6].

In order to measure the accommodation, we have made a special relay lens system (see Fig.1), which optically transfers the subject's eye image to the outside of the magnetically shielded room. It consists of 4 identical spherical lenses with 400 mm focal lengths, forming a focal system. Thereby it moves the eye by 3200 mm to form a real image just in front of the dynamic refractometer where the subject's eye is placed in normal measurement. The accommodation measurement is done with the real image of the eye.

The dynamic changes of the focus point of the eye were measured by a servo-controlled optical system, using the retinal reflection of injected infra-red beams. A target in the system was moved by a personal computer (NEC, PC-9800). The resolution of the measurement is about 0.1 D (Diopter; it is defined by the reciprocal of a focal point) and the speed of measurement is 6.4 Hz. Details of the optical system of the dynamic refractometer are explained elsewhere [2].

Subjects were 3 right-handed volunteers with adequate accommodative power (HK, 36, male with 4D of Accommodative Power (AP); EH, 32, male with 4D of AD; KC, 26, female with 6D of AP) and no ocular problems other than myopia.

The AP was measured with the dynamic refractometer by moving a target from a position located farther than the far point (the farthest point which the subject can focus) to another position nearer than the near point (analogously defined) [1]. The target was changed stepwise with a random time interval of 5 ± 0.5s. The amplitude of the stimulus was set to be about 60 % of each subject's AP to prevent excessive visual fatigue. Subjects were instructed to try to watch the target as clearly as possible.

The measurements were done on the right eye, occluding the left eye with an eye patch. The light in the magnetically shielded room was turned off to dilate the pupil diameter. The experiments consisted of more than 64 recordings; the collected data were averaged after discarding those contaminated by eye blinks. The sampling rate was set to 250 Hz. The recording time started 1s before the onset of the stimulus and lasted 2.5s. The evoked fields, the accommodation responses and the target positions were simultaneously recorded by a personal computer. The trigger signal for MEG measurement was made by a predetermined threshold of the target position. The MEG signals were removed its offset using 1s pre-recording and filtered with a zero phase shift, 1-40 Hz band-pass filter.

Results

Only the responses of subject HK are shown in this paper to save space. The responses of the other subjects were similar with the presented data.

Fig. 2 depicts a control response of HK when he gazed at the target in the refractometer which was positioned at the farthest (-5D) without moving. MEG recordings of all the 64 channels are superimposed. It shows noise-like response with intensity of about 50 fT all the time.

Then the target was moved between -5 and -8 D, because HK was -5 D myopic. Fig. 3 depicts whole channels' responses for the target movement of far to near, shown on a circular projection of the head. Clear responses are found on the sensors located over occipital and lateral lobes.

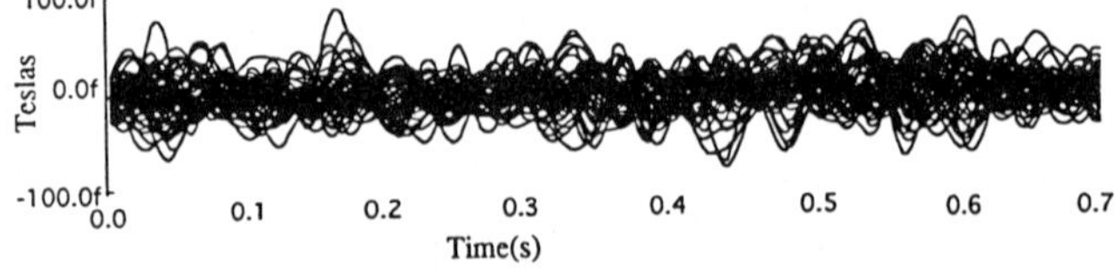

Fig.2 Superimposed MEG responses of HK when the target was not moved.

Fig. 4 shows the accommodation and MEG responses for the same stimulus together with the target movement. The time lag of the accommodative response was 288 ± 50 ms. It was calculated by averaging the time lags of the 64 records determined by inspection. The start of accommodative response is indicated by an arrow in the figure. The averaged time lag was stable within each subject. Though the amplitudes do not match with the stimuli, it is natural and is called as the accommodation lag[7]. The mean time lag and the shape of the accommodation responses were in good agreement with the literature[8]. Superimposed MEG responses indicated well synchronized activation of the brain around 100 and 200 ms. Both peaks showed the phase reversal phenomenon. On the other hand, the responses were rather quiet around 290 ms when the average accommodative response began to rise.

Fig. 5 shows the accommodation and MEG responses when the target was moved from

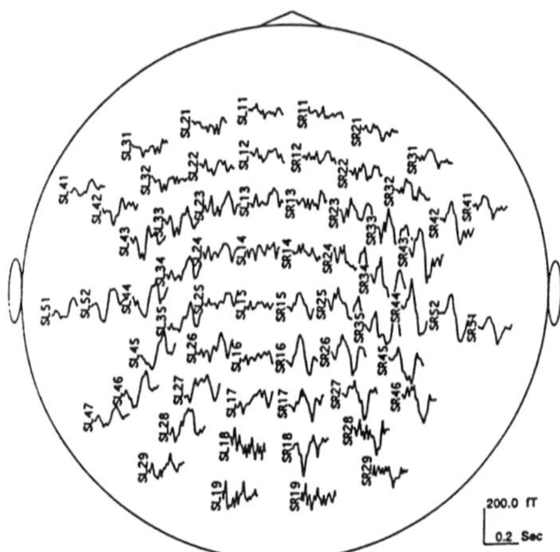

Fig.3 Responses of all the 64 channels which show large signals on the channels located over occipital and lateral lobes.

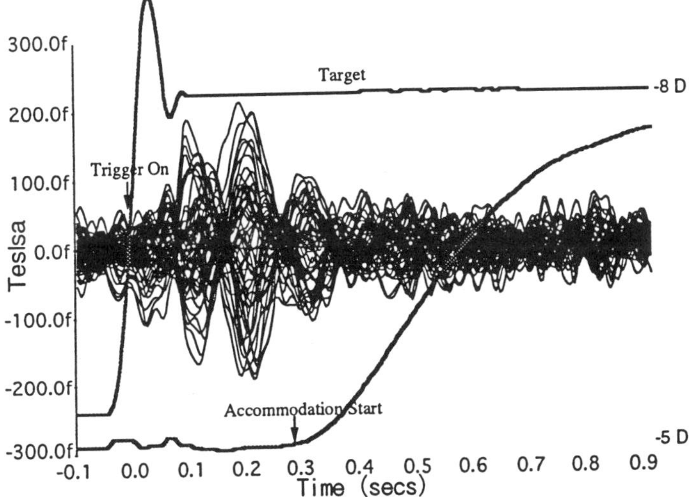

Fig.4 Superimposed MEG and accommodation responses together with target change which was abruptly moved from far (-5 D) to near (-8 D).

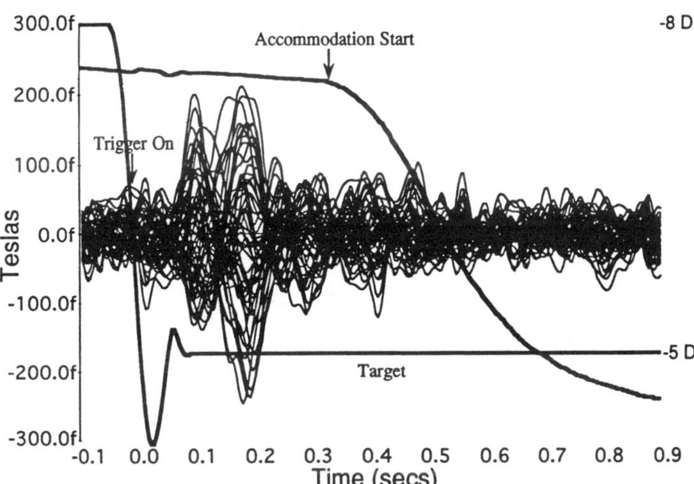

Fig.5 Superimposed MEG and accommodation responses together with target change which was abruptly moved from near (-8 D) to far (-5 D).

near to far, together with the target movement. The time lag of the accommodative response was 341 ± 56 ms. Superimposed MEG responses showed a tendency similar to the ones in Fig. 4. There were at least two phase reversal peaks around 100 and 190 ms and a relatively low activation when the accommodation started.

Discussion

There are no previous reports on the accommodation related MEG measurement. On the other hand, there are a few studies concerning the location of the control center of the accommodation, employing implanted electrodes in animals[9]. However no basic relationship between cerebral activity and accommodation response was found nor was identified the control center of the accommodation.

In the present study, the superimposed MEG responses and the accommodation responses (Fig. 4,5) clearly suggest that perception of blur and determination of some control signals for accommodation are made at around 100 and 200 ms after the onset of the accommodative stimuli. It can be confirmed by comparing the control responses shown in Fig.2, where the average of 64 responses shows virtually no synchronized responses except about 50 fT of external and brain noises. Hence, the two peaks in Fig. 4 and 5 should at least have strong correlation with blur perception and accommodation control.

However, there might be many factors relating to these MEG responses. The left eye should be moved up to 10° by accommodative vergence, because right eye position is fixed to the optical axis and the left eye should be directed from -5D (20 cm) to -8 D (12.5 cm). The sensors located near the left eye recorded several hundreds fT drifts. Though the slow components of the eye movement were removed by the band-pass filter, the fast components might be involved in Fig. 4 and 5

Fig.6 shows superimposed MEG responses when HK gazed at a spot presented on a CRT, which was moved up to 10° for the same period of the accommodative stimuli. The MEG data filtered with the same band showed much smaller responses compared with the responses shown in Fig. 4 and 5. Hence eye movement should not have major importance in those responses.

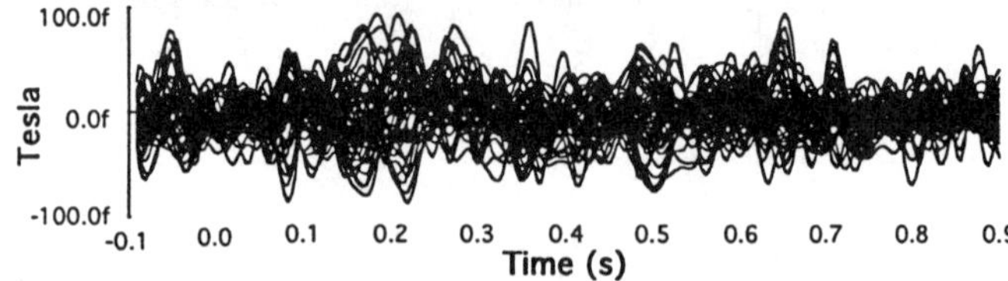

Fig.6 Superimposed MEG responses of HK when target's direction was moved slowly.

Near accommodation drives pupil constriction in general. As the change of pupil diameter is not so large judging from TV monitoring, it must have not great influence on the MEG response in those conditions. Anyway, it will be checked soon.

It is noteworthy that the MEG data in Fig.4 shows another activation after the onset of the accommodation and data in Fig. 5 dose not show. This might be related to the fact that the ciliary muscle is activated for the far to near accommodation but is relaxed for the near to far accommodation [7].

In concluding, the present experiment does show the MEG responses which should correlate with the accommodation control. Because the accommodation must have close correlation with stereoscopic perception[8], the present study on accommodation should have tremendous importance in developing 3-D TVs in the future. However, more detailed experiments are needed to clarify the timing, location and characteristics of the accommodation-related brain activities.

References

[1]Takeda, T., Ostberg O., Fukui Y. and Iida, T., Dynamic Accommodation Measurements for Objective
    Assessment of Eyestrain and Visual Fatigue, J. of Human Ergology, 17, 1, 21-30, 1988

[2] Takeda, T., Fukui, Y. and Iida, T., Three-dimensional Optometer, Applied Optics, 27, 12, 2595-2602, 1988.

[3] Takeda, T., Fukui, Y. and Iida, T., Three-dimensional Optometer Ⅲ, Applied Optics, 32, 22, 4155-4168, 1993.

[4] Takeda, T., Iida, T. and Fukui, Y., Dynamic eye accommodation evoked by apparent distances, Optometry and
    Vision Science, 67, 6, 450-455, 1990.

[5] Vrba J. et al., 1982, Spatial discrimination in SQUID gradiometers and 3rd order gradiometer performance,
    Can. J. Phys., 60, 1060-1073.

[6] Takeda, T., Morabito, M., Kumagai, T. and Endo, H., Use of CRT as a visual stimulator in MEG
    measurements, Event-related potentials of the brain edited by Y. Koga, Elsevier, 1995 (in press).

[7] Stark, L., 1987, Presbyopia in light of accommodation, in Presbyopia edited by L. Stark and G. Obrecht,
    Fairchild Publications New York, 264-274.

[8] Campbell, F.W. and Westheimer, G., 1959, Factors influencing accommodation responses of the human eye, J.
    Opt . Soc. Am., 49, 568-571.

[9] Bando, T. and Toda, H., 1991, Cerebral cortical and brainstem areas related to the central control of lens
    accommodation in cat and monkey, Comp. Biochem. Physiol. 98C, 1, 229-237.

Electromagnetic source localization of cognitive processes related to attention to single or omitted tones

Tarkka, I. M., and Stokić, D. S.

Baylor College of Medicine, Division of Restorative Neurology, Houston, Tx, USA

Introduction

Magnetic and electric recordings have confirmed that the P300, a late positive component of the human event-related potentials (ERP), can be evoked not only by a target discrimination task but also by the omission of targets [1,2,3]. Recently it was shown that also single target stimuli, delivered without distractors, can evoke a P300 response [4]. In this case the task is simpler than the oddball paradigm which may render it more useful for clinical applications. However, the analogy of the single stimulus P300 to the classical P300 needs to be shown.

The purpose of the present study was to extend our previous work on the identification of the brain regions generating the P300 potential elicited by an auditory oddball paradigm. Our goal was to find out if the P300 responses elicited by rare or single tones or the omission of tone differ in their morphology and/or in the underlying cerebral structures responsible for generating the P300. Noninvasive electric source localization technique was employed by using the Brain Electrical Source Analysis (BESA) software in order to develop spatio-temporal equivalent electric dipole models for the scalp-recorded electric data. Individual data was modeled using individual head radius for the spherical head model.

Methods

Subjects

Seven healthy subjects (1 woman, 6 men, mean age 32 y, from 25 y to 39 y) volunteered for the study and gave an informed consent prior to the experiment. The experimental protocol was approved by the local committee for human subject research. None of the subjects had a history of neurological symptoms and they were not under any medical treatment.

Experimental protocols

Three different paradigms using auditory stimuli constituted the experiment. First, a classical oddball paradigm was used delivering tones at the rate of 1.1 Hz. The frequent tones were low pitched (1000 Hz) and the target tones, occurring randomly 18 % of the time, were high pitched (2000 Hz). The plateau duration of the tone was 50 ms with rise and fall times of 10 ms. Tones were delivered binaurally through ear-phones at an 85 dB intensity level with 60 dB masking with white noise. Subjects were asked to silently count the number of rare, high pitched tones. Three to four short sessions were recorded, each consisting of about 40 rare tones. Short breaks were given to the subjects between sessions.

Second, no frequent tones were delivered, however the target tones (2000 Hz) remained delivered as in the previous oddball condition. Thus, at irregular intervals single tones were presented and the subjects were asked to silently count them. Again each short session consisted of about 40 tones.

In the third condition, the frequent tones were delivered as in the rare tone condition but this time target tones were omitted. Therefore the irregular absence of the tone was to be counted. Data collection was performed in a same way for the rare tone, single tone, and omitted tone conditions. All subjects performed the three tasks with better than 96% target detection rate.

ERP recording

ERPs were recorded with 30 silver/silverchloride cup electrodes (diameter 10 mm) evenly spaced on the scalp. During the recording linked earlobes served as a reference, for further analysis average reference was used. Electrode impedances were kept below 3 kΩ. The EEG bandpass was from 0.07 Hz to 80 Hz. The 700 ms recording window consisted of a 100 ms pre-stimulus baseline and 600 ms for the post-stimulus epoch. During recording the frequent and rare wave forms corresponding to each condition were sorted into two memory buffers and averaged on-line. Automatic artefact rejection was used to control the quality of the data.

Source analysis

Prior to source analysis all the wave forms were base-line corrected and filtered at 30 Hz high cut-off frequency. The Brain Electric Source Analysis (BESA) program, version 2.0, [5,6,7] was used to develop equivalent electric dipole models to suggest locations of the active brain regions. Individual head radius was used for each subject's 4-shell spherical head model. In addition to descriptive measures and topographic mapping of the scalp recorded data, the program allowed iterative simultaneous or sequential fitting of the location and orientation of equivalent electric dipoles. The development of the models for the rare, single and omitted tone conditions was done independently from each other. The strategy for model development was to first develop models for grand average data of each condition. This was done by dividing the 600 ms window into two parts; one corresponding to the auditory evoked potential (0-180 ms) and one to the cognitive potential (180-600 ms). These two models were then combined and fitted in order to model the 600 ms window of the grand average data. The models explaining grand average wave forms were used as starting parameters for developing the individual models. The locations, orientations and strengths of the dipoles were compared between the three conditions for each of the subjects.

Results

A late positive component, P300, was evoked in all studied conditions. The peak latency of the P300 evoked by single stimuli was shorter than that evoked by the rare (p < .05) or the omission (p < .01) condition. The P300 amplitudes of the rare and single tone conditions were similar, in the order of 15 μV, and the P300 evoked by the omitted tone had a smooth and widespread morphology with a peak amplitude of about 10 μV. The spatio-temporal models for rare and single tone conditions both consisted of 6 dipoles and the model for omitted tones consisted of 3 dipoles. The grand average models for each of the conditions are shown in Fig. 1. The residual variances (RV %) of the models of the individual subjects, indicating the goodness of fit, ranged from 5.1 % to 11.9 % in the rare tone condition, from 5.4 % to 14.1 % in the single tone condition and from 3.1 % to 25.7 % in the omitted tone condition. The differences in dipole locations of the individual models between rare and single tone conditions were less than 15 mm for all three axis in 95.3 % of the data. As the equivalent electric dipole presumably accounts for neural activity of an area of 10 mm in diameter, it makes the differences in dipole locations between the rare and single tone conditions unremarkable.

When the parameters of the average dipole locations were transferred into a brain atlas [8], a fairly lateral region of the superior temporal gyrus bilaterally was implicated by the dipoles number 1 and 2. The source potentials of dipoles 1 and 2 indicated that their major contribution was at the time of the late auditory evoked potential in both rare and single tone conditions (see Fig. 1.). A deeper region, bilaterally, in the superior temporal gyrus was implicated by dipoles 3 and 4. The source potentials of these dipoles implied that these source were active both at the time of the late auditory evoked potential and the cognitive potential. The dipoles 5 and 6, bilaterally, located deeper in the medial temporal lobe, more than 10 mm posterior to dipoles 3 and 4, suggesting hippocampal and parahippocampal regions as active sources. Dipoles 5 and 6 had the majority of their strength during the P300 component. The 3-dipole model explaining the P300 evoked by omitted tones consisted of a right unilateral dipole deep in the superior temporal gyrus showing most of its strength at the time of the long latency auditory evoked potential as seen in the rare or single tone conditions. The P300 component appeared to be explained by two deep dipoles, one in temporal lobe region and one in the posterior frontal lobe.

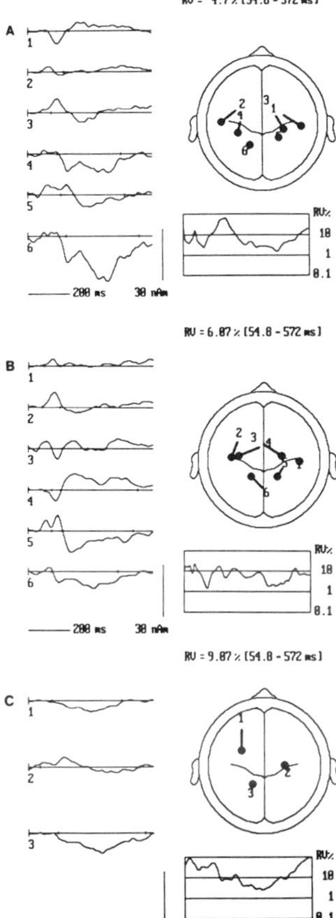

Fig. 1. Spatio-temporal multiple dipole source models developed for the grand average ERP data (7 subjects) of rare tone (A), single tone (B) and omitted tone (C) conditions. On the left are the source potentials for each of the dipoles and the RV % in the middle shows the residual variance throughout the analyzed window. The overall RV % is shown at the top for the entire window. Note, that the models for rare and single tone conditions are fairly similar.

Discussion

The target tones delivered in a classical oddball paradigm and the single tone paradigm produced similar P300 components in their peak amplitude and scalp topography. The peak latency of the P300 produced by single tone was shorter than that of the rare tone or the P300 evoked by the omission of tone. In this aspect the single tone evoked P300 resembles the novel P300. The comparison of the dipole locations of the individual models revealed close similarity of the models which implies that the rare and single tone tasks activated adjacent anatomical regions. These two models support the notion that bilateral superior temporal lobe areas, primary and secondary auditory cortices, are active especially at the time of the long latency auditory evoked potential and furthermore they suggest that the deeper medial temporal lobe sources, in part in the deep superior and transverse temporal gyri and in part in the hippocampal/parahippocampal area, contribute to the late cognitive component. The parameters of the models for the rare tone in the present study were also compared with the parameters of the multiple dipole models we presented previously for the same type of auditory evoked P300 response but recorded in a different group of subjects [9]. In spite of the difference in radiuses of the head models used in these two studies, a close correspondence of the models was found. The similarity of the two independent models adds credibility to the modeling approach. At the same time these results imply the same active neural generators for auditory evoked P300 potential. Also, source localization performed on magnetoencephalographic auditory evoked long latency fields and P300 components has suggested similar areas for the generators of the human P300 as the present study [10,11,12,13]. The similarity found in the models for rare and single tone evoked P300 appears to support the suggestion of Polich et al. 1994 [4] that the underlying neural and cognitive mechanisms may be the same for the rare and single tone conditions.

References

[1]   Sutton, S., Braren, M., Zubin, J., and John, E. R. Science, 1965, 150:1187-1188.

[2]   Papanicolaou, A.C., Baumann, S.B., and Rogers, R.L. Source estimation of late components of omitted tone evoked magnetic fields. In: Hoke, M. et al. Biomagnetism: Clinical aspects, Amsterdam, Elsevier, 1992, pp. 177-180.

[3]   Rogers, R.L., Papanicolaou, A.C., Baumann, S.B., Eisenberg, H.M. Late magnetic fields and positive evoked potentials following infrequent and unpredictable omissions of visual stimuli. Electroencephalography and clinical Neurophysiology, 1992, 83:146-152.

[4]   Polich, J., Eischen, S.E., and Collins, G.E. P300 from a single auditory stimulus. Electroencephalography and clinical Neurophysiology, 1994, 92:253-261.

[5]   Scherg, M., and von Cramon, D. Evoked dipole source potentials of the human auditory cortex. Electroencephalography and clinical Neurophysiology, 1986, 65:344-360.

[6]   Scherg, M., Vajsar, J., and Picton, T.W. A source analysis of the late human auditory evoked potentials. Journal of Cognitive Neuroscience, 1989, 1:336-355.

[7]   Scherg, M., and Berg, P. Brain Electrical Source Analysis Handbook, 1994, V. 2.0., Herndon, Virginia.

[8]   Talairach, J., and Tournoux, P. Co-Planar Stereotaxic Atlas of the Human Brain. Thieme Medical Publishers, Inc., New York, 1988.

[9]   Tarkka, I.M., Stokić, D.S., Basile, L.F.M., and Papanicolaou, A.C. Electric source localization of the auditory P300 agrees with magnetic source localization. Electroencephalography and clinical Neurophysiology, 1995, 96:538-545.

[10]   Hari, R., Rif, J., Tiihonen, J., and Sams, M. Neuromagnetic mismatch fields to single and paired tones. Electroencephalography and clinical Neurophysiology, 1992, 82:152-154.

[11]   Okada, Y.C., Kaufman, L. and Williamson, S.J. The hippocampal formation as a source of the slow endogenous potentials. Electroencephalography and clinical Neurophysiology, 1983, 55:417-426.

[12]   Rif, J., Hari, R., Hamalainen, M.S., Sams, M. Auditory attention affects two different areas in the human supratemporal cortex. Electroencephalography and clinical Neurophysiology, 1991, 79:464-472.

[13]   Rogers R.L., Papanicolaou, A.C., Baumann, S.B., Bourbon, W.T., Alagarsamy, S., Eisenberg, H.M. Localization of P3 sources using magnetoencephalography and magnetic resonance imaging. Electroencephalography and clinical Neurophysiology, 1991, 79:308-321.

Acknowledgements

This work was supported by the Vivian L. Smith Foundation for Restorative Neurology, Houston, Tx, USA.

A Cortical Substrate For Hand Skill: An MEG Evaluation

Thoma, R. J., Lewine, J.D., Davis, J.T., and Orrison, W.W., Jr.

The New Mexico Institute of Neuroimaging, The New Mexico Regional Federal Medical Center, Albuquerque, New Mexico, USA.

INTRODUCTION

It is well established that, independent of culture, approximately 90% of the human population demonstrates a right hand preference on most tasks. This population asymmetry is especially noteworthy because of its magnitude and because it is uniquely human. The current study is an attempt to relate differential hand skill directly to a measure of cortical functional anatomy. Specifically, it was hypothesized that more sensorimotor cortex would be dedicated to the dominant versus non-dominant hand. To assess this, we utilized magnetic source imaging, a method that combines magnetoencephalographic (MEG) data on brain function with magnetic resonance (MR) data on brain structure. Current flow associated with neuronal activity produces a magnetic field, just as does the current flow in a wire. The magnetic field that can be measured outside the head selectively reflects current flow within the dendrites of pyramidal cells oriented parallel to the skull surface. Using signal averaging techniques it is possible to identify the magnetic signals time-locked to particular motor or sensory events.

Evoked fields associated with self-paced movements are characterized by a slow field shift beginning several hundred milliseconds prior to the movement, and a large magnetic signal during/after the movement. The field prior to movement is believed to reflect predominantly activities in motor cortex, whereas the sharp signal at the time of movement reflects mostly feedforward/feedback activation of somatosensory cortex. The somatosensory component is typically of larger amplitude and more easily localized. By combining simplifying assumptions about the conductivity properties of the brain, skull, and scalp with assumptions about the configuration and number of neural generators, it is often possible to infer the spatial location of those cells that generate specific magnetic field components. In particular, the head is assumed to be well modeled by a spherical volume conductor, and it is assumed that the magnetic field pattern at any particular instant in time can be modeled as though it were generated by an equivalent current dipole intended as a mathematical representation of the neuronal currents of a focal population of cells. Available data indicate that the dipole model is adequate for characterizing the primary motor field and the associated somatosensory field, with sources localizing to within 5 mm of the central sulcus.

To infer the amount of cortex devoted to sensorimotor processing for the right versus left hand, we took advantage of the somatotopic organization of sensorimotor cortex. Specifically we localized sources associated with movements of the thumb and little finger and took the distance between these sources to be indicative of the amount of cortex devoted to the hand.

METHODS

<u>Subjects</u>: Data were successfully collected from 9 volunteers drawn from a population comprised of students and faculty of the Psychology Department at the University of New Mexico and staff from the Department of Radiology at the Albuquerque VA Medical Center (VAMC). The sample consisted of 4 right-handed females, 3 right-handed males and 2 left-handed males. Subjects ranged in age from 18 to 41 years old. The MEG data collection often lasted several hours and subjects were compensated for their time with pizza.

<u>MEG Data Collection</u>: All MEG data were collected at the Albuquerque VAMC using a large array biomagnetometer (BTi) equipped with 37 first-order, axial gradiometers. Prior to the experiment, a head-centered coordinate frame was defined by touching the stylus of a 3D digitizer to right and left preauricular points and to the nasion. The stylus was also used to digitize the entire headshape in order to define a best-fitting sphere for the source modeling calculations. During each experimental test block, subjects were required to make 150 self-paced movements of a specified digit (right thumb, right little finger, left thumb, left little finger) at a rate of about 1 per 3 seconds. The movements caused the closing of a photo-optic switch that provided a time stamp to the data acquisition system. The magnetometer was initially positioned over the parietal/frontal region contralateral to the hand being tested. The unit was repositioned during the course of the experiment, as needed, to ensure that data were recorded over both the positive and negative field extrema (region of maximal exiting and entering flux, respectively). Fifteen hundred msec of data were acquired prior to each movement, and 500 msec were acquired after each movement. Data were filtered on-line with a bandpass of 0.1 - 50 Hz, and they were digitized at a rate of approximately 200 Hz. Each block of trials lasted 20-30 minutes.

MEG Data Analysis: Prior to signal averaging, trials where the magnetic field on any sensor exceeded 10 picoTesla were rejected. Typically, less than 5% of trials were contaminated. The data were signal averaged, baseline corrected, and subjected to a 20Hz low-pass digital filter. For this project, analyses focused on the large component coincident with the movement, because this was recorded more reliably in all conditions. Also, this component is better characterized by the dipole-in-a-sphere model. The dipole-in-a-sphere modeling algorithm provides the location, orientation and strength of that current dipole which best accounts for the magnetic field pattern. A dipole fit was considered valid if, and only if, it accounted for more than 90% of the variance in the field map and the confidence volume was less than one cubic centimeter. The spatial parameters of the dipole were taken as indicative of the coordinates of the relevant sensorimotor cortex. To assess the extent of sensorimotor cortex devoted to the hand, a linear distance between dipole sources for the first and fifth digit was calculated. Ideally, distances should be measured along the actual cortical surface, but prior experience indicated that this is a very time-consuming process, and that the linear distance correlates with the cortical distance.

Magnetic Resonance Imaging (MRI): A 3D sagittal volumetric data set (1.5 mm slice thickness, no gap) was obtained using a Seimens 1.5T whole-body imaging system. CEMAX VIP software running on a SUN workstation was used to create a volume rendering of the data. By identification of the MEG head-frame fiducials (preauricular points and nasion) on the MR data set, it was possible to plot MEG dipoles on the brain volume. The data were then inspected to confirm that dipole sources localized to the sensorimotor cortex.

Handedness Measures: During the MEG data collection phase of the project, experimenters were blind as to the handedness of subjects. After the MEG data were analyzed, all subjects were given the Waterloo Handedness Questionnaire as a measure of hand preference. The questionnaire allows subjects to circle answers (left always, left usually, equal, right always, right usually) to questions about which hand they use for certain tasks, for example, holding a toothbrush. Hand skill was measured with a pegboard task [1]. Each trial on the task requires subjects to move 10 pegs from one holder and fit them sequentially into holes in another holder approximately 3 inches away. Five trials with each hand are completed by each subject.

RESULTS

Signals were recorded successfully from the four test conditions for nine subjects. In all cases, dipole sources localized to the sensorimotor cortex of the contralateral hemisphere. In every case, the linear distance between localized source points for the first and fifth finger was larger contralateral to the subject's dominant hand than the corresponding measure in the opposite hemisphere [Fig 1].

For right-handers, the average left hemisphere distance was 14.5 mm, and average right hemisphere distance was 11.7 mm. In left-handers, the averages were 10.3 mm and 14.5 mm, respectively. Differences between cortical distances for the two hemispheres (right-hemisphere thumb to fifth finger distance minus left-hemisphere thumb to fifth finger distance) were significantly correlated with differences in total times for right and left hands on the pegboard task (right-hand score minus left-hand score) r= .81, p=.008. That is, the more sensorimotor cortex devoted to a hand, the faster the performance of the contralateral hand. Additional analyses showed a significant correlation between Waterloo hand preference total scores and cortical distance difference scores (r=.7153, p=.046). The Waterloo questionnaire also correlated with pegboard difference scores (r=-.762, p=.028).

DISCUSSION

Through the use of MSI, we were able to demonstrate that the linear distance between finger representations on the post-central gyrus is greater contralateral to the dominant hand than the respective distance contralateral to the non-dominant hand. In addition, differences calculated by subtracting subjects' left-hemisphere from right-hemisphere cortical measurements were highly correlated with differences (R-L) in pegboard scores. Subjects' reported hand preference was also correlated with cortical distance scores.

Unfortunately, the present analysis does not conclusively define the causal relationship between handedness and cortical representation. There are two, not necessarily mutually exclusive, possible causal chains. Given that there is evidence for right thumb sucking preference by 10 weeks gestational age [2], one possibility is that a larger hand representation in sensory-motor cortex is established very early in life, and that this is the causal substrate of hand preference. Alternatively, increased hand representation in the dominant hemisphere may be an effect of preferential useage of the contralateral hand. Recent research with animals has shown changes in somato-sensory cortex occur as a result of dynamic interaction between cortical tissue and behavioral stimulation [3]. Differential hand use, early in life, may cause a cascade of environment/brain feedback events that results in the pattern of hand-use and cortical measures in the current dataset.

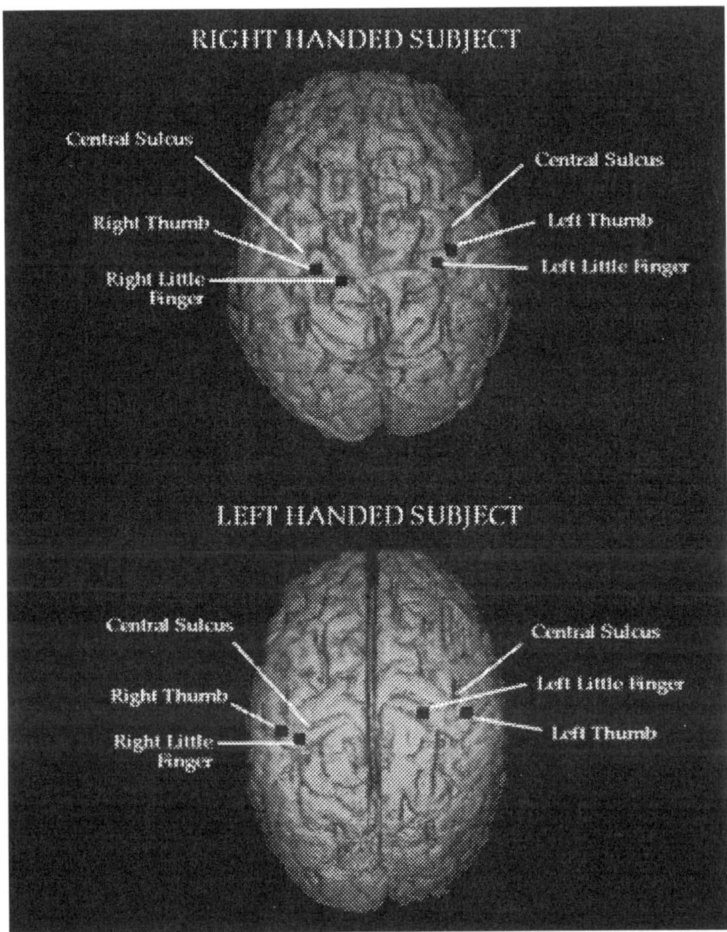

Fig 1: MEG results for a right versus left handed patient. For the right hander the distance between thumb and little finger responses is greater over the left hemisphere. For the left hander, the difference is greater over the right hemisphere.

REFERENCES

[1] Annett, M. Handedness in families, Annals of Human Genetics, 1973, 37: 93-105.

[2] Hepper, P.G., Shahidullah, S., and White, R., Handedness in the human fetus. Neuropsychologia, 1991, 28: 1107-1111.

[3] Rencanzone, G., Merzenich, M., & Schreiner, C., Changes in the distributed temporal response properties of SI cortical neurons reflect improvements in performance on a temporally based tactile discrimination task. Journal of Neurophysiology, 1992, 67.

Separation of Overlapping Sources Induced by Mixed Stimulations Using 3-Dimensional MEG Mappings

Uchikawa, Y.[1], Kobayashi, K.[1], Sakawa, K.[1] and Kotani, M.[2]

[1]Faculty of Science and Engineering, Tokyo Denki University, Saitama 350-03 Japan; [2]Faculty of Engineering, Tokyo Denki University, Chiyodaku, Tokyo 101 Japan

INTRODUCTION

There is a problem of separating multi-sources overlapping in time when many distinct areas of the cortex are active. Several methods for evaluating source localization from spatio-temporal MEG data have been reported [1,2,3,4]. Almost all SQUID magnetometers for MEG study are for measuring magnetic field perpendicular to the scalp. There is a limitation in spatial resolution, using flux-detection coils with finite radius in regular devices, to separate multi-sources [5]. To get more accurate information about source configuration, we have proposed a new 3-D measurement of MEG, namely vector magnetoencephalogram [6].

This paper describes the separation of two sources overlapping in time, located in the brain, by using a 3-D second-order gradiometer [6]. First, to discriminate and to separate multi-sources, we used 3-D MEG data from magnetic fields produced by two current dipoles implanted in a cranium model. Second, to apply and to confirm for human MEGs with the results of model study, we carried out 3-D MEG measurements for mixed AEFs and SEFs overlapping in time.

METHODS

The cranium model shown in Fig. 2 and 3 was based on a skull sample for medical education (The circumference of nasion-inion measured 32.3 cm, T3-T4 in the international 10-20 system measured 21.8 cm) and was made with colorless epoxy resin filled with salt water. The electrode for a current dipole (6 µAm) was made with an insulated copper wire (0.5 mm diameter, 6 mm length) and was set in the head model with a depth of 1 cm from the inner surface and parallel to the scalp. The 3-D MEG measurement was made by a 3-D second-order gradiometer. In the case of two current dipole sources, the magnetic fields were generated by two different frequencies of 30 and 70 Hz with constant current, and measured by separating the frequencies. As shown in Fig.1, a 3-D second-order gradiometer, which can simultaneously measure magnetic fields both of perpendicular to and tangential to the scalp, consists of three separate second-order gradiometers. Each coil is orthogonally wound with Nb-Ti wire on the rectangular solid of 3 x 3 x 6 cm. These three coils are connected to three rf-SQUIDs. The base line of the detecting coil for Br, which is the magnetic field perpendicular to the scalp, is 2.8 cm and that of the coils for Bθ and Bϕ, which are the magnetic field tangential to the scalp, are 1.4 cm

Fig.1 Coordinate system for measuring 3-D magnetoencephalogram using a 3-D second-order gradiometer.

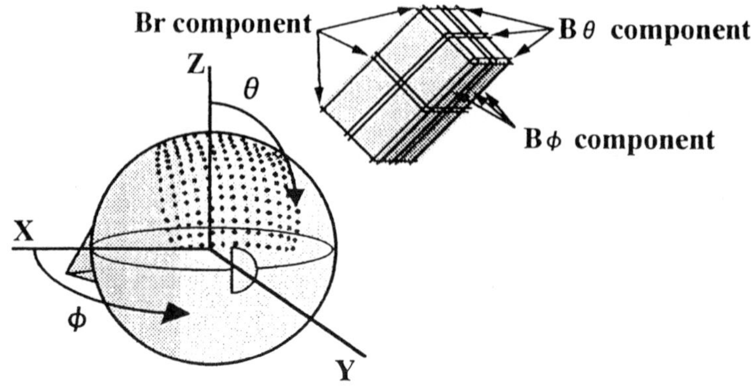

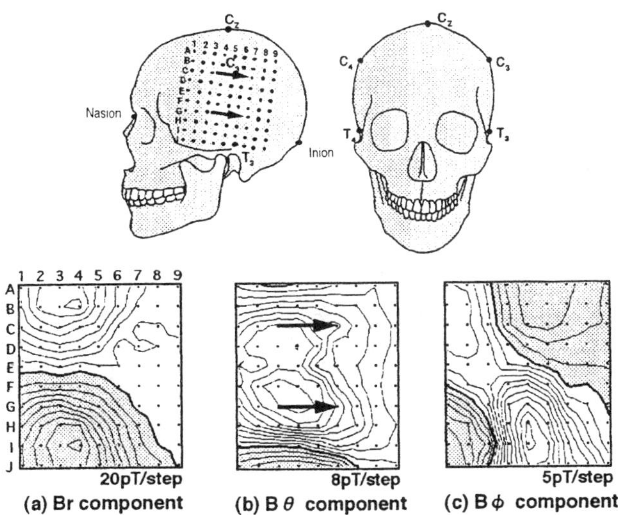

20pT/step	8pT/step	5pT/step
(a) Br component	**(b) B θ component**	**(c) B φ component**

Fig. 2 Isofield contour maps for two cuurent dipoles, oriented in the same direction, implanted at a depth of 1 cm from the inner surface of a human cranium model. The separation of the dipoles is 4 cm. Arrows indicate the cuurent dipoles.

respectively. Coordinate system for measuring 3-D MEG is shown in Fig. 1. Gradiometers for Bθ and Bφ, in particular, can provide the maximum and/or minimum field with a sign just above the location of a single current source [6,7]. This is a merit for separating multi-sources and is maybe the same thing with other coil configurations such as the 2-D coil [8].

RESULTS

Fig. 2 shows example of isofield contour maps for two current dipoles with the same direction, at C5 and G5, implanted at a depth of 1 cm from the inner surface of a cranium model. The separation of the dipoles is 4 cm.

In the Br component (Fig. 2(a)), a dipole distribution with the maximum of 140 pT (magnetic flux outflow) at B4 and the minimum of -126 pT (magnetic flux inflow) at I4 can be seen. It is possible to see that the magnetic field distribution is organized by a single current dipole located at E3, because two current dipoles go in the same direction. However, the Bθ component (Fig. 2(b)) shows two extremes of about 80 pT with same polarity at both of D4 and F4. In this case, by referring to the map of Bθ, we can clearly say that this map consists of "two sources with same polarity" underlying each extreme corresponded to the measurement points of D4 and F4. The direction and the location of two sources estimated by visual inspection were represented by arrows on the scalp as shown in Fig. 2(b).

Fig. 3 shows example of isofield contour maps for two current dipoles with the opposite direction, at C5 and G5, implanted at a depth of 1 cm from the inner surface. The separation of the dipoles is 4 cm. This source configuration is important for MEG measurement since electric potentials are attenuated at the scalp while magnetic fields are not. As shown in Fig. 3(a), the map of Br significantly misses the dipole pattern and shows one extreme of - 160 pT at E5. However, the map of Bθ clearly shows two extremes of -96 pT at C5 and 112 pT at G5 with opposite polarity. Positions of C5 and G5 correspond to the places implanted each current dipole. We can clearly say that this map consists of "two sources with opposite polarity" underlying each extreme corresponded to the measurement points of C5 and G5. The direction and the location of two sources estimated by visual inspection were represented by arrows on the scalp as shown in Fig. 3(b)

893

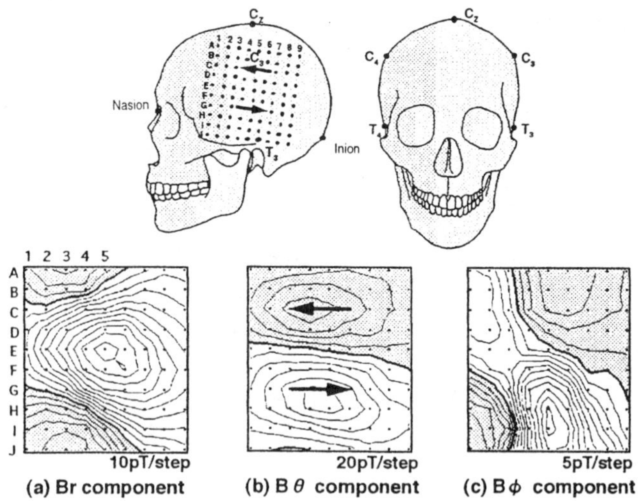

10pT/step	**20pT/step**	**5pT/step**
(a) Br component	**(b) B θ component**	**(c) B φ component**

Fig. 3 Isofield contour maps for two current dipoles, oriented in the opposite directions, implanted at a depth of 1 cm from the inner surface of a human cranium model. The separation of the dipoles is 4 cm. Arrow indicate the current dipoles.

APPLICATION FOR MIXED AEF AND SEF

To separate multi-sources for human MEG with the results of model study, we carried out 3-D MEG measurement of mixed AEFs and SEFs overlapping in time. AEFs were elicited by 1 kHz tone bursts of 50 ms duration to the right ear and SEFs were elicited by electric pulses of 0.2 ms duration with 6 to 8 mA to the median nerve of the right wrist. The interval of stimulation was 500 ms. Stimulation to the median nerve was delivered 40 ms later than to that of the tone burst. There were two evoked field, namely mixed AEF and SEF, overlapping in time. All magnetic data were average by 300 to 400 measurements. The sampling interval was 0.5 ms. A band-pass filter was used in the range of 0.5 to 40 Hz.

Fig. 4 shows example of isofield contour maps on the left hemisphere at 120 ms latency of AEF and at 80 ms latency of SEF. As shown in Fig. 4(a), map of Br shows one extreme with about 950 fT at measurement position C3, but it is a quite similar to that of Fig. 3(a). Map of Br, in this case, was not helpful to estimate the location and the number of sources due to lack of a dipole pattern. By referring to maps of Bθ and Bφ shown in Fig. 4(b) and Fig. 4(c), which are tangential magnetic field, we can clearly see two extremes with opposite polarity in both map. The locations and the directions of sources estimated by visual inspection from these maps were indicated by arrows on the scalp. This field pattern is almost the same as that of Bθ map in Fig. 3(b). In this case, we can clearly predict that these maps consist of the magnetic fields generated by "two sources with opposite polarity" underlying each extreme corresponded to the locations on the scalp indicated by arrows. These results from the maps of Bθ and Bφ, relating to the estimation of source configuration, were confirmed by comparison with coronal MRI images of a subjet's head showing the estimated source locations as shown in Fig. 4(d).

CONCLUSION

We have developed a 3-D second-order gradiometer, which can simultaneously detect field components perpendicular to and tangential to the scalp. It was found that this gradiometer becomes a useful tool for

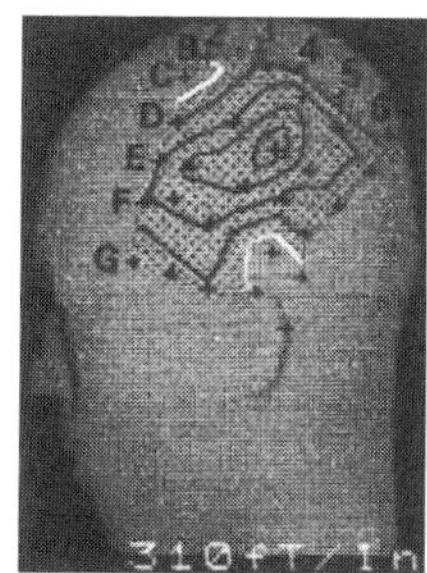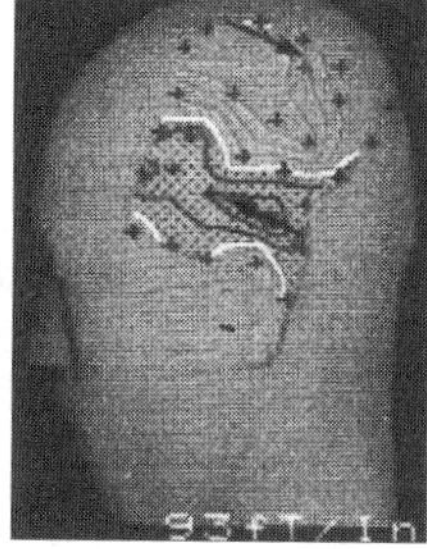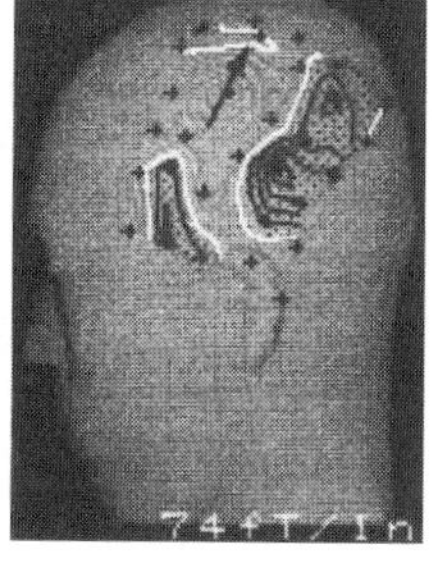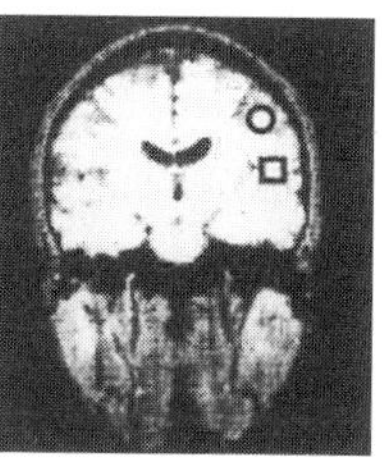

| (a) Br component | (b) Bθ component | (c) Bφ component | (d) Coronal MRI |

Fig. 4 Isofield contour maps consisting of mixed AEFs and SEFs. The latency of an AEF is 120 ms. The latency of a SEF for median nerve stimulation in the right wrist is 80 ms. Coronal MRI of a subject's head showing the source localization of AEFs (square) and SEFs (circle).

separating multi-sources overlapping in time, in the cortex. Two gradiometers for detecting tangential field component, in particular, work as spatial band-pass filter having the characteristic of the second derivative tangential to the scalp. Namely, we conclude that this 3-D second-order gradiometer can provide the information of the constraint conditions by visual inspection for calculating the inverse problem with multi-sources, which is difficult in so far as we have used the MEG data of magnetic field perpendicular to the scalp.

REFERENCES
[1]Smith, W. E., Estimation of the Spatio-Temporal Correlations of Biological Electrical Sources from Their Magnetic Fields, IEEE Trans. Biomed. Engng., 1992, 39 : 997-1004.

[2]Baumgartner, C., Sutherling, W. W., Shi Di and Barth, D.S., Spatiotemporal modeling of cerebral evoked magnetic fields to median nerve stimulation, Electroencephalography and clinical Neurophysiology, 1991, 79 : 27-35.

[3]Achim, A., Richer, F. and Saint-Hilaire, J.M., Methodological considerations for the evaluation of spatio-temporal source models, Electroencephalography and clinical Neurophysiology, 1991, 79 : 227-240.

[4]Achim, A., and Richer, F., Methods for Separating Temporally Overlapping Sources of Neuroelectric Data, Brain Topography, 1988, 1 : 22-28.

[5]Okada, Y., Discrimination of localized and distributed current dipole sources and localized single and multiple sources, Biomagnetism:Applications and Theory, ed. by H. Weinberg, G. Strolnk, T. Katila, 1985 : 266-272.

[6]Uchikawa, Y., Matsumura, F., Kobayashi, K. and Kotani, M., Application to Discrimination of Multi-Sources to Somatosensory Evoked field using 3-Dimensional second-order gradiometer, In : Deecke, L., Baumgartner, C., Stroink, G. and Williamson, S.J (eds.), Biomagnetism :Fundamental research and clinical application, 1993 : 219-220.

[7]Ahonen, A., Hamalainen, M., Kajola, M., Knuutila, J., Lounasmaa, O., Simola, J., Tesche, C., and Vilkman, V., Multichannel SQUID systems for brain research, IEEE Trans. Magn., 1991, 27 : 2786-2792.

[8]Cohen, D., Proc. 12th Intern. Conf. on Med. and Bio. Eng., Jerusalem, Israel, 1979.

ACKNOWLEDGMENT
This study has been supported by the grant from the Center for Research, Tokyo Denki University.

Measurements of Visual Evoked MEG Responses Associated with Color Discrimination

Ueno, K.[1], Ueno, S.[1] and Weinberg, H.[2]

[1]*Institute of Medical Electronics, Faculty of Medicine, University of Tokyo, Tokyo, Japan;*
[2]*Brain Behaviour Laboratory, Simon Fraser University, Burnaby, B.C., Canada*

Introduction

Recent rapid progress of SQUID techniques made it possible to measure MEG signals over the whole cortex of the brain simultaneously. It is extremely important for evoked MEG research in which the signals have to be averaged. With auditory, visual and somatosensory stimulation, the electrical activities evoked by stimuli can be recorded as a response of the brain. Recently, investigations on higher brain function such as memory and cognition have been progressing. For example, P300 has been introduced as a neuro-electro-magnetic phenomenon associated with the higher brain function. The P300 is an electric response of the brain to target stimuli around 300msec after the onset of stimulation. Investigations of P300 have mainly focused on auditory evoked MEG[1]-[4]. We studied the P300 evoked by visual stimulation using a discrimination task consisting of two different colours and estimated the source of the P300 response using a 2-dipole model.

Method

The experiments were carried out on 3 healthy adults (A: male, 32 years, B: male, 32 years, C: female, 28 years). The visual evoked MEG was recorded using a multi-channel 3rd-order gradiometer system (CTF Systems Inc.). The paradigm to measure the P300 component was an oddball paradigm. Visual stimuli were displayed in the center of a monitor screen. The distance from the subject's eyes to the center of the monitor screen was 3.64m. The stimulus was a rectangle (27.5x18.5cm) filled with a single colour. The colours used as stimulation were red and blue.

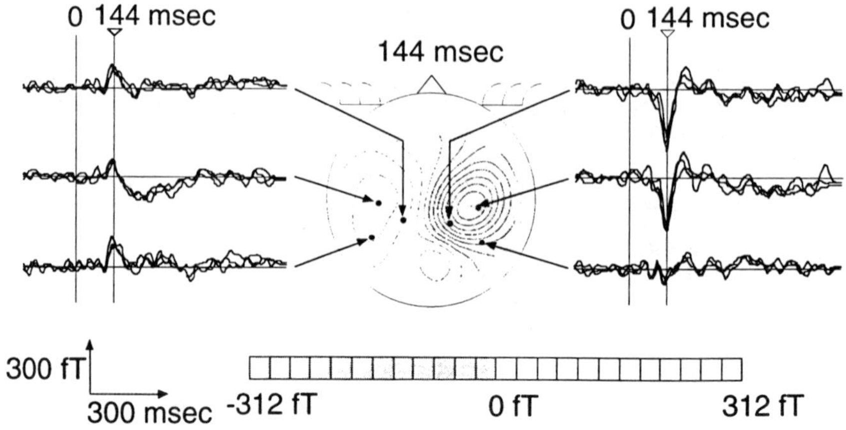

Fig.1 Signal traces and a topography of visual evoked MEG in control experiments for subject A. Each trace is the average of 100 response to repeated stimuli (red).

Each stimulus was randomly presented for 1 sec at irregular intervals of 1 to 2 sec. Low frequency colour, red, was the target. It was presented 100 times at a frequency of 25%. The subject's task was to count the number of targets. For comparison, we also stimulated 300 times using a single colour red. The data were collected at a digitizing rate of 250Hz and averaged after being filtered through a 30Hz lowpass filter. In the P300 experiment, we averaged only the data associated with target colour.

Results and Discussion

Fig. 1 shows the visual evoked MEGs with repeated single colours as the stimulus for subject A. Each trace is the average of 100 responses. Three waves were very similar and had clear peaks at latencies of about 150 and 200 msec. The responses are associated with the visual processing of colour in the brain. The sources at a latency of about 150 msec were estimated with a 2-dipole model for all three subjects. The results were shown in Fig.2. The two dipole were located in the occipital areas of each of the right and left hemisphere, corresponding to the visual cortex. The dipole moments were between about 15 to 45 nAm.

Fig. 3 shows the superposition of two MEG waveforms for subject C. One is the response to red stimulus in the control experiment, the other is the response to red, the target stimulus, in the P300 experiment. A clear difference could be found at the latency from 250msec where there is a reversal between posterior and anterior, and also between right and left. The peak of the difference was at a latency of about 400 msec. The difference appeared to be a P300 response.

Fig. 4 shows the topographies of the differences at a latency of 396 msec for all three subjects. The topographies have similar patterns which are reversed between anterior and posterior, and also between right and left hemispheres, so it is possible to assume the sources of these topographies are also the same. The sources with a 2-dipole model was estimated as shown in Fig. 5. The estimated sources for all three subjects was located near the center of the

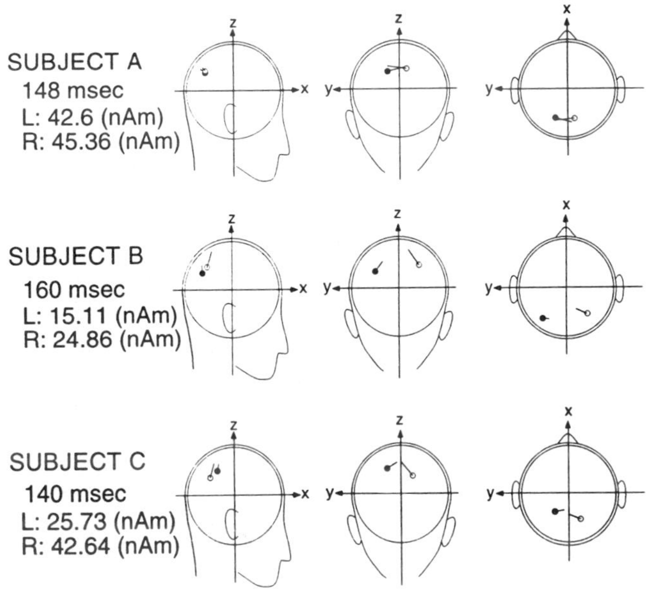

Fig.2 The estimated sources of the visual evoked MEG at latency about 150 msec in control experiment for three subjects.

head model. All the P300 responses recorded in these three subjects can be produced by similar sources in the brain. However, each dipole has a much larger moment than the estimated sources on control wave at about the latency of 150 msec in Fig.2. This means that the dipole model cannot be used in source estimation. Because a dipole doesn't have any spread, its moment should be adequately small. The sources of the P300 component might be a distributed system or might be composed of several partial sources. It is necessary to estimate the sources by taking into account anatomical and functional brain structures.

Acknowledgements

This research was supported in part by the Grants from the Ministry of Education, Science and Culture in Japan, and the Collaborative Research And Development Grant of the Natural Sciences and Engineering Council of Canada.

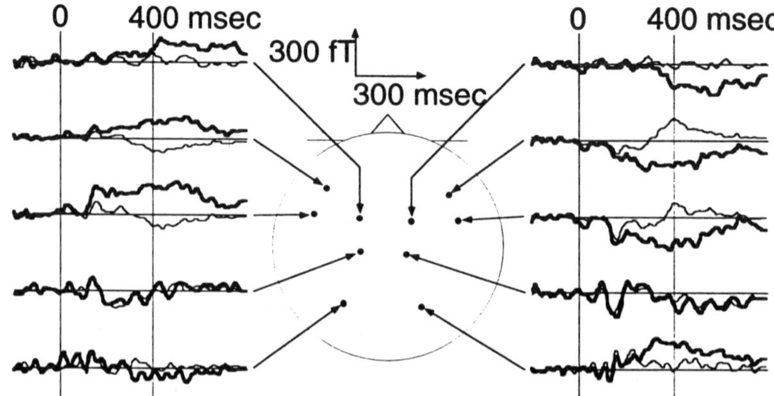

Fig.3 The superposition of two MEG waveforms for subject C. The bold trace is the response to red stimulus in the control experiment, the thin trace is the response to red, the target stimulus, in the P300 experiment.

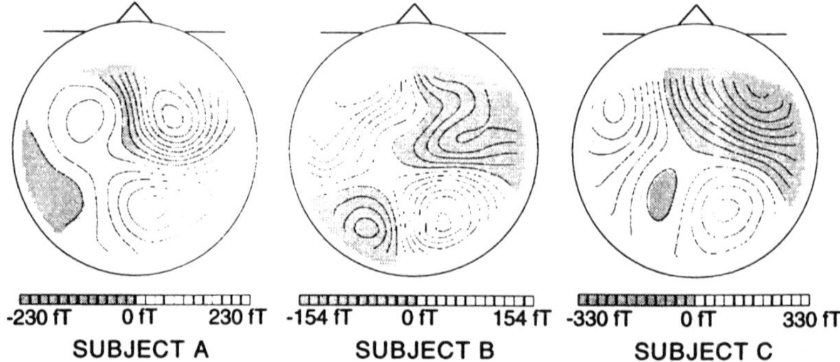

Fig.4 The topographies of the differences at a latency of 396 msec for three subjects. The topographies have similar patterns which are reversed between anterior and posterior, and also between right and left hemispheres.

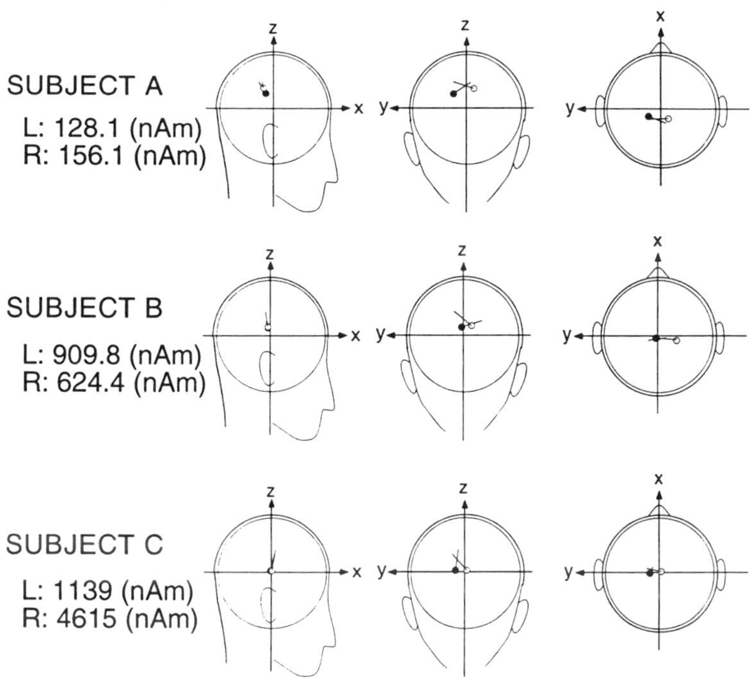

Fig.5. The estimated sources of the P300 component at latency of 396 msec for three subjects.

References

[1] E. Gordon et al., Magnetoencephalography: locating the source of P300 via magnetic field recording., Clinical & Experimental Neurology. 23:101-10, 1987.

[2] E. Gordon, C. Rennie and L. Collins, Magnetoencephalography and late component ERPs., Clinical & Experimental Neurology. 27:113-20, 1990.

[3] R. L. Rogers, A. C. Papanicolaou, S. B. Baumann and H. M. Eisenberg, Recording of event-related potentials (P300) from human cortex., Electroencephalography & Clinical Neurophysiology. 83(2):146-52, 1992 Aug.

[4] R. Neshige and H. Luders, Recording of event-related potentials (P300) from human cortex. , Journal of Clinical Neurophysiology. 9(2):294-8, 1992 Apr.

Measurement and Source Analysis of MEG Activities Associated with Short-Term Memory and Identification Processes

Yoshida, H.[1], Ueno, S.[2] and Weinberg, H.[3]

[1]*Department of Computer Science and Communication Engineering, Faculty of Engineering, Kyushu University, Fukuoka 812, Japan;* [2]*Institute of Medical Electronics, University of Tokyo, Tokyo 113, Japan;* [3]*Brain Behaviour Laboratory, Discovery Park at Simon Fraser University, Burnaby, B.C., Canada V5A 1S6*

Introduction

Sternberg's paradigm [1] has been very useful in examining the short-term memory process, and many EEG studies have been undertaken using this paradigm. Some studies reported slow wave [2] or DC potential shifts [3] in memory tasks. We thought that MEG was more applicable than EEG in measuring low frequeny signals. In this study, subjects memorized the order of four tones and compared it with that of another set of four tones. They had 1.6 seconds to memorize the order of the first set of tones, 1.4 sec as an inter-stimulus interval, and 1.6 sec to compare the order of the last set of tones. Extremely slow magnetic fields were observed by using this paradigm. We estimated the current dipole sources underlying short-term memory for tones.

Methods

The whole-cortex 64 channels of MEG data were digitized with a sampling frequency of 125 Hz, and filtered using a zero phase shift, low-pass digital filter with a cut-off at 20 Hz. Fig.1 shows the sequence of events in each trial. Two kinds of tones (two octaves apart: the 1046.5 Hz tone and the 261.6 Hz tone) were used. Three right-handed male subjects (age 26, 28, 50 years old) participated. The subjects memorized an order of four tones, which was chosen randomly. These four tones were called the memory set. After 1.4 sec, the other four tones called the probe set were presented. The subjects discriminated between the order of the first set of tones from that of the last set of tones. When another long (1 sec) tone was presented 2.4 sec later, the subjects were to push either a true or a false button. Pushing the true button meant that the order of the first four tones was the same as that of the last in the forward order. Pushing the false button meant that the order was not the same. In a control task, subjects had to hear the memory and probe set without treating. Data were averaged twenty times. The power spectra was calculated using the 2.56 sec data from the start of the trial. Additionally, it were also calculated using the data from 3.0 sec to 5.56 sec. In the estimation of the equivalent current dipolar sources, the head was assumed to be a homogeneous spherical volume conductor with a radius of 100mm and the Grynszpan-Geselowitz equation [4] was solved.

Results

Fig.2 shows the MEG recordings at C4 of the 10-20 international system. In comparison with the control magnetic fields (a), the magnetic fields during the task (b) clearly included very low frequency components. Fig.3 shows the power spectra at C4 of the 10-20 international system. Fig.3(a) is the control power spectra. Fig.3(b) is the power spectra during the memory task, and Fig.3(c) is the power spectra during the identification task. The intensity of the control power spectral components at 0.5 Hz was less than 15 fT(rms)/√Hz. However, the intensity of the power spectral component at 0.5 Hz during the memory task was 25 fT(rms)/√Hz, and that during the identification task was 45 fT(rms)/√Hz. Because the memory processes affected the very low frequency components of the MEG, the current dipole source models were made from the data filtered from below 0.5 Hz. Fig.4 shows the MEG topographies for three subjects. The control topographies showed very weak and complicated patterns. However, all the topographies during the task showed one dipole pattern in the frontal region of the head. The polarity of the MEG topographies was opposite for the time the subjects were listening to the tones (a), (c) and afterward (b), (d). Fig.5 shows the two-dipole models. Fig.5(a),(b) are the two-dipole models during the memory task, and Fig.5(c),(d) are the two-dipole models during the identification task. Three pairs of two-dipole sources were superimposed. The two-dipole sources were very close and their directions were contrary to each other. Compared with the source models during the memory task, the locations of the two-dipole sources during the identification task were slightly anterior to the head. We suggested that the location of the estimated sources should be in the cingulate

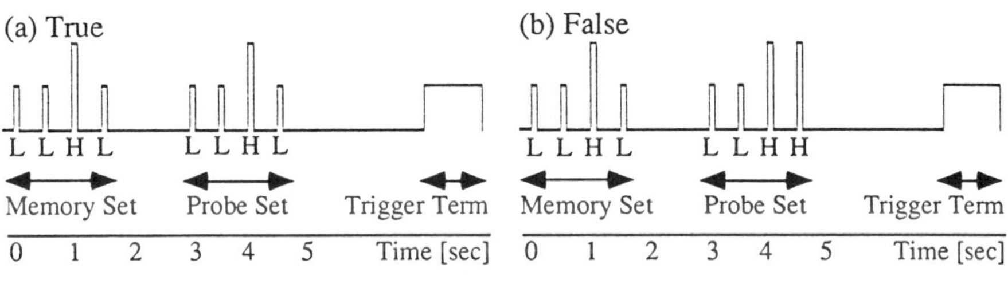

Fig.1 The time chart of the short-term memory and identification experiment.

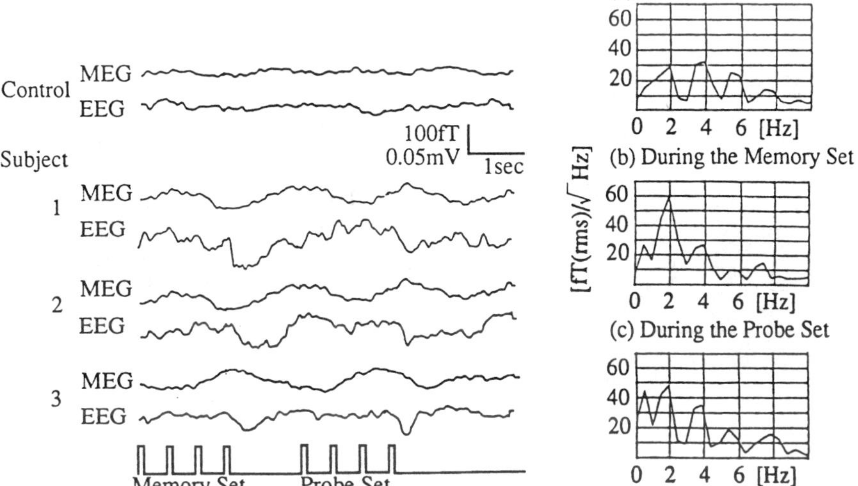

Fig.2 The recordings of MEG and EEG at C4 of the 10-20 international system (0-5.0Hz).

Fig.3 The power spectra at C4 of MEG data of the subject 1.

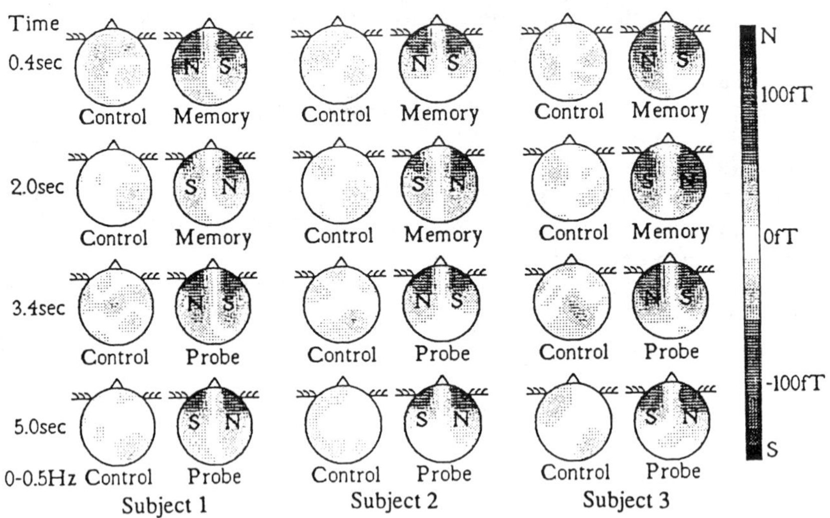

Fig.4 The MEG topographies (0-0.5 Hz) for three subjects.

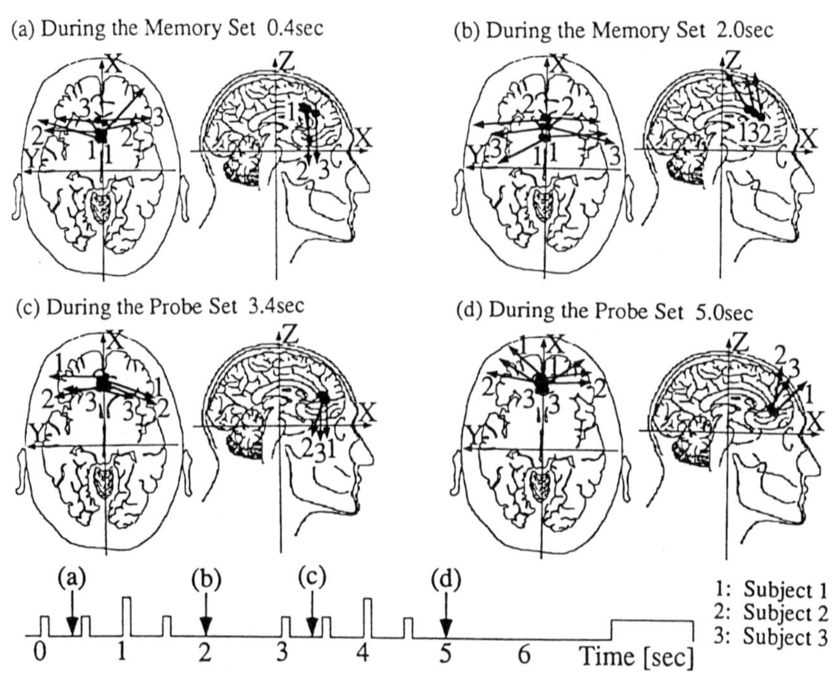

Fig.5 The two-dipole source models (0-0.5 Hz) for three subjects.

902

gyrus. Because many cases have been reported in which illness of the cingulate gyrus caused amnesia, we thought that the estimated locations as proper[5].

Discussion

The Sternberg paradigm employed many kinds of tones, and it was difficult to discriminate a certain tone from others. That was why we employed only two kinds of tones. The correction ratio was increased (more than 95%) in every trial. Information was given to subjects in a serial format, and the analysis term was prolonged. As a result, we could observe extremely slow magnetic fields.

We also tried using a single dipole estimation before estimating with two dipoles. However, the estimated location of the single dipole was in the longitudinal cerebral fissure. It wasn't proper. Based on the results, we supposed that there were two current dipole sources in the brain.

The polarity of the MEG topographies was opposite for the time during which the subject was listening to the tones and the time afterward. We suggested that the mass current in the neurons turned over because of the feedback during both the memorizing process and the identification process.

Conclusions

We investigated the influence of a short-term memory task using a whole-cortex MEG sensor. We observed that MEG activities during the task included extremely slow waves. Based on the results, we proposed a two-dipole source model. Two current dipole sources were estimated in the cingulate gyrus. They were very close and their directions were contrary to each other.

Acknowledgements

This study was supported in part by grants of the ministry of education, culture, and science on Japan, and collaborative research and development grant by natural sciences and engineering council of Canada.

References:

[1] Sternberg, S., High-speed scanning in human memory. Science, 1966, 153: 652-654.

[2] Ruchkin, D.S., Sutton, S., Kietzman, M.L. and Silver, K. Slow wave and P300 in signal detection. Electroenceph. Clin. Neurophysiol., 1980, 50: 35-47.

[3] Lang, M., Lang, W., Uhl, F., Kornhuber, A., Deecke, L. and Kornhuber, H.H. Slow negative potential shifts indicating verbal cognitive learning in a concept formation task. Hum.Neurobiol., 1987, 6: 183-190.

[4] Geselowitz, D.B. On the magnetic field generated outside an inhomogeneous volume conductor by internal sources. IEEE Trans. on Magn. MAG-6, 1970, No.2: 346-347.

[5] Iversen, S.D. Do hippocampal lesions produce amnesia in animal? Int. Rev. Neurobiol. 1976, 19: 1-49.

Oscillatory Brain Activities

Lopes da Silva, F. (Organizer)[1], Basar, E.[2], van Dijk, B.[3], Narici, L.[4], Pantev, C.[5], Ribary, U.[6], and Salenius, S.[7]

[1]Institute of Neurobiology, Graduate School Neurosciences Amsterdam, University of Amsterdam, The Netherlands; [2]Institute of Physiology, Medical University Lubeck, Germany; [3]Institute of Opthalmological Research, Graduate School Neurosciences Amsterdam, The Netherlands; [4]Dipartamento di Fisica, Università di Roma "Tor Vergata", Italy; [5]Center of Biomagnetism, Institute of Experimental Audiology, University of Münster, Germany; [6]Center for Neuromagnetism, Department of Physiology and Biophysics, New York University Medical Center, New York, USA; [7]Low Temperature Laboratory, Helsinki University of Technology,Espoo, Finland.

Introduction

The workshop was an attempt to present an overview of new lines of research with respect to the problems of how to detect, analyze and interpret signals generated by neuronal populations particularly with an oscillatory character. Indeed in the last decade there was a resurgence of interest for the recording and the analysis of oscillatory activities of the brain. This can be ascribed to three main developments; at the cellular level, the unraveling of the conditions under which neurons can display membrane oscillations even *in vitro* [18]; at the level of neuronal populations, the finding that cortical neurons can display enhanced correlated oscillations within a frequency range between 20 and 60 Hz (beta and gamma band) in response to specific stimuli [13][9] and, at the macroscopic level, the possibility of analyzing similar types of oscillations in the human brain during cognitive tasks using the new technology of Magnetic/Electric functional brain mapping. The most important issues related to these aspects that were discussed in the Workshop are summarized here.

Nature of Oscillatory Activities

Intrinsic membrane processes and synaptic circuit properties

The basic mechanisms responsible for the generation of oscillations in the neuronal networks of the brain reflect both intrinsic and synaptic, i.e. local circuit, properties, as it has been clearly shown in thalamocortical circuits [44]. Indeed in the intact brain, both synaptic interactions and input signals play an essential role in setting the conditions that are necessary for oscillations to occur. The frequency of the latter depends on the intrinsic (non-linear) properties of the neurons and on the strength and time constants of the synaptic interactions [21]. The existence of feedback circuits shapes the dynamics of the neuronal networks creating the conditions for the occurrence of different types of oscillatory behaviour. In this respect an essential property is the capacity of the neurons, within certain populations, of being active in synchrony, due to mutual synaptic interactions both excitatory and inhibitory.

Types of oscillations: linear vs. non-linear

Under certain conditions there are classes of brain oscillations that may be described as resonance phenomena of neuronal networks that behave in a linear regimen submitted to relatively weak (random) inputs. Nevertheless, in general, these networks exhibit nonlinear behaviour. This behaviour becomes evident when spectral analysis of the corresponding field potentials reveals the presence of significant relationships between harmonic components. This has been shown for some alpha rhythms of the visual cortex and beta/gamma activities of the olfactory bulb and cortex (review in [22]). The nonlinear character of the dynamic behaviour of, for example, the neuronal networks of the olfactory cortex is also evident from a number of physiological changes that occur during processing of olfactory information. One of these changes is that the median frequency of the gamma activity increases during odour sampling [11], while the subharmonic component is strongly enhanced [6]. Particularly interesting is the fact that this increase in subharrnonics, revealing strong nonlinear dynamics, is most conspicuous when the odour being sampled has acquired a special meaning through conditioning (Fig 1). This highlights the fact that neuronal networks exhibit nonlinear dynamic behaviour while analyzing significant information.

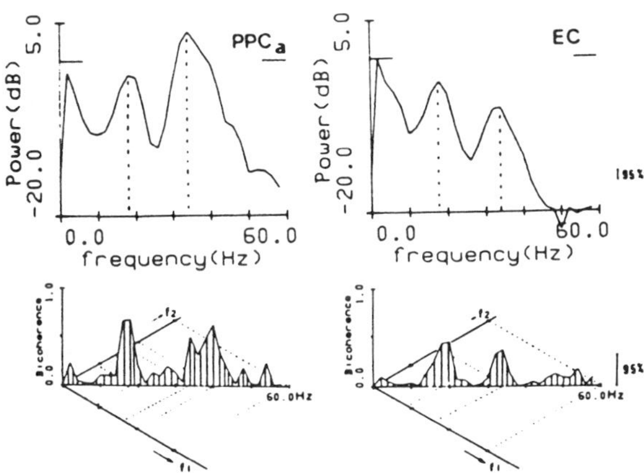

Fig. 1: Evidence for non-linear spectral properties of EEG signals recorded from the depth of two areas of the temporal cortex of the cat: PPCa = anterior prepyriform cortex, and EC = entorhinal cortex. Above: power spectra of signals recorded simultaneously from the two cortical areas. Below: the corresponding bi-coherence that shows the phase relationship between different frequency components of each signal. Note the existence of two main frequency peaks around 36 and 18 Hz, with the former dominating in the PPCa and the latter in the EC. In addition the bi-coherence shows two main peaks at the intersection between 18 and 18 Hz, indicating a harmonic relation between 18 Hz and 18+18 Hz, and another around the intersection 36 Hz and 36 Hz indicating a harmonic relation between 36 and 36+36 Hz, that cannot be distinguished in the power spectra.

Transfer Between Synaptic Activity and Neuronal Firing in Neuronal Populations (Correlation Images)

To estimate the functional connectivity of the brain, in vivo, requires recording of the activity at many sites in the cortex simultaneously and a model of the signal transfer. The first requirement has to be met because bivariate measurements of correlations suffer from the well known problems of a common source resulting in an overestimate of coherence, or of a missing source resulting in an underestimate of coherence. The second requirement is purely mathematical.

The data set used [43] consisted of pairs of radial cortical activity profiles. We recorded the low frequency (3-200Hz) content of the intra-cortical EEG at 14 sites at varying depths with respect to the visual cortical surface simultaneously. We called this profile the *synaptic activity profile*. To do this we applied spatial filtering with a 3 point 2nd-order derivative along the array to ensure that our recordings were from a restricted area. The other profile, called the *spiking activity profile,* was the envelope of the high frequency (1000-SOOOHz) content of the intra-cortical EEG at 16 sites at varying depths. These two profiles were recorded simultaneously from area 18 of the lightly anaesthetized cat. We used a linear non-parametric predictive model, and tested wether we could predict these profiles from each other. In these experiments we observed that a linear model quite accurately predicted even the individual features in each recording from the simultaneously recorded second data set. This held true in general. We found multiple coherencies in the range 0.7 to 0.9 using this straightforward linear model. The prediction of the synaptic activity profile from the spiking activity profile was a little more accurate than the reverse prediction. An obvious explanation would be that every spike results in some form of synaptic activity, while not every synaptic activity results in spikes.

The compound (multi-in/multi-out) transfer function of the complete predictive filter describes how the signals from one of the profiles influence the signals of the other profile. We therefore used these compound transfer functions to estimate the functional connectivity within the cortical column from which we recorded. This estimate is probably not precise, given that the cortex is certainly not a linear system, and given that we probably undersampled spatially. However, we wanted to test the *hypothesis* that *the functional connectivity in the visual cortex changes upon visual stimulation.* Thus we do not need a precise estimate as long as the estimate is consistent.

Figure 2 shows data from an experiment where we used three different stimuli: a homogeneous nonmodulated grey TV screen (no-stimulus), a homogeneous flashing TV screen (flash stimuli), and a sinusoidally moving rectangular black bar on a white TV screen (moving bar stimulus) in random alternation. In this figure we depict the impulse response functions of the filter model that predicts the synaptic activity profile from the spiking activity profile. Specifically in each panel we show the synaptic activity profile that would result if there would be a unit pulse change in the spiking activities at the electrodes nrs.8 through 11, corresponding to the input layers of the cortex (bottom half of layer IV). The predicted synaptic activity at the surface of the cortex is depicted at the top (electrode nr. 1) each subsequent electrode lies 150 mm deeper in the cortex. From the figure we see that both the response wave shape and the spatial pattern of the compound impulse response function, depend on the type of stimulus used. We also see that for each stimulus the impulse response functions show oscillatory components, typically in the 50-70Hz band, which are most outspoken for the moving bar stimulus. These oscillations can not be due to an oscillatory common driving signal, since these would be nulled out by the analysis method used (any signal that is common to more than 2 electrode sites is removed by the non-parametric predictive model). The conclusion drawn from this experiment is that indeed the functional connectivity of the visual cortex changes upon visual stimulation.

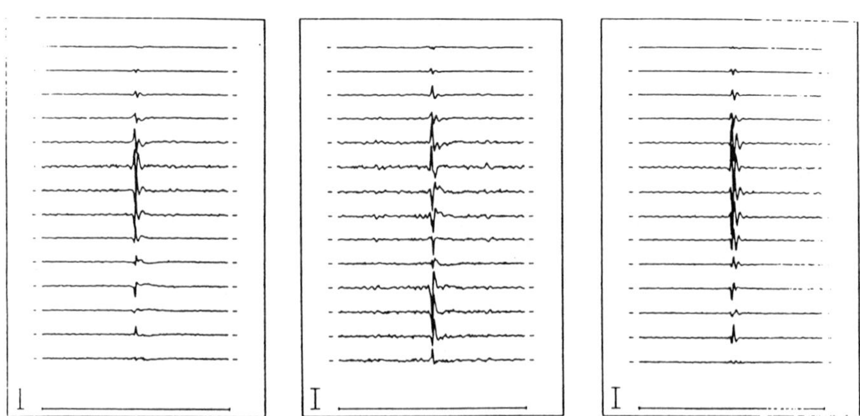

Fig. 2: Compound impulse response functions as calculated for three different stimulus conditions: Non-stimulated (left column), Flash-stimuli (middle column), and Moving-bar-stimulus (right column). The traces within each column depict the predicted synaptic activities that would result at the intra-cortical electrode array, if the spiking activity would be a unit pulse of I sample duration (1/1024 s) at electrodes close to the bottom of layer IVC and identically zero at all other electrode sites. The pulsing electrodes are marked at the right side of each column. The traces at the top of each column represent activity from superficial electrodes, the traces at the bottom from deep electrodes. The horizontal scale bars correspond to 1024ms, the vertical scale bar corresponds to 50 uV.

To better understand the interplay between changes in functional cortical connectivity and changes in the dynamics of cortical activity, we studied much simpler "toy systems of artificial neurons, 50% excitatory and 50% inhibitory, with spiking dynamics and synaptic plasticity resembling long term potentiation (LTP) and long term depression (LTD), that have been characterized in different brain areas notably in the hippocampus. A full account of this simulation studies can be found in Kalitzin and van Dijk [15].

For such a simple network, we can calculate the dependence of the average connectivity strength on the average activity strength from the plasticity dynamics, and we can also calculate the dependence of the average activity strength on the average connectivity strength from the spiking dynamics. Where these two relations cross the network is in equilibrium. We found that there is a parameter range where the system has multiple equilibrium states of which two (or more) can be stable. This implies that when the system is started at a random initial connectivity, the system goes to one of the stable equilibrium states. For instance the network could go to either a state of low average connectivity and low average activity or to a state of high average connectivity and high average activity. If the network is at a high connectivity equilibrium the neurons of the network are more synchronized.

If we supply the network with a diffuse external driving signal (affecting either all the neurons in the network or only a randomly drawn subset of the neurons) we can force the network to go into another equilibrium state. For example when the network is in a low connectivity state, then a sufficiently strong correlated input will drive the network to the high connectivity state. When the external input is switched off the network will stay in the new state, with more strongly coherent ongoing activity either forever, or until internal noise, or a noisy input, drives it out of that equilibrium again. Which of these two evolutions occur depends on the actual parameter settings; in this context the parameters that determine the rate of the plasticity changes are the most crucial. The response of the network to an external stimulus has two components: one that is time- and phase-locked (evoked), governed by the spiking dynamics of the network, and one that is time locked but not phase locked (induced), governed by the plasticity dynamics and driven by the synchronicity of the external stimulus. Anti-synchronous input can force the network to lower connectivity equilibria, and less coherent ongoing activity. The network functions as a *synchrony enhancer,* where the synchrony of the ongoing activity reflects the synchrony of the last external stimulus, i.e. as a *memory mechanism.* We postulate a similar role for synchrony and plasticity in the sensory cortex. Our experimental data have shown that the fast *plasticity* of the functional connectivity is present in the visual cortex.

These experimental results and model studies allow establishment of the dynamic relationships between the activity at the level of single neurons and that of neuronal populations.

Spatio-temporal Distributions of Oscillations in the Brain: The Role of MEG/MFT

Among all non-invasive functional brain imaging techniques available today, MEG/EEG have the unique potential of a high time resolution, offering the possibility to analyze oscillatory activities in the human brain. MEG offers, in addition, a better accuracy in localizing generator sources within the brain. The proper choice of MEG recordings permits to select small time segments of raw data and to extract several frequency components, or rhythms, in the human brain, which are present simultaneously at deeper and higher levels [35]. In addition, the high temporal resolution of the technique further enables to monitor the three-dimensional distribution of these individual oscillations in the human brain in time steps of less than 1 msec. In our laboratory a multi-channel MEG system was used to record the magnetic activity from human subjects, the raw signals were filtered to extract the oscillatory activity at around 40Hz [19]. By analyzing the *spatio-temporal distribution of this oscillatory activity* (Fig. 3), we could demonstrate a possible correlation to some sort of human cognitive brain functions [14].

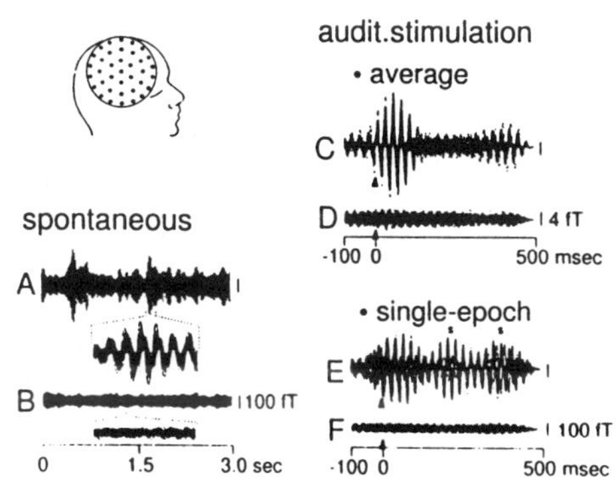

Fig. 3: Synchronized and coherent 40-Hz oscillations recorded form the right hemisphere of the human brain. A: spontaneous 40-Hz oscillations in the absence of any sensory stimuli, compared with the system noise (B). Reset of spontaneous 40-Hz activity by auditory stimulation (300 msec tones), and induced synchronized activity with respect to auditory onset, as seen in average data (C) and in single epochs (E), compared to the appropriate system noise (D,F). (modified from Llinas and Ribary 1993 [19]).

Earlier findings indicate that the oscillatory activity at around 40-Hz in humans is continuously generated by the central nervous system, and is altered during different states of consciousness [19]. We demonstrated large spontaneous 40-Hz coherent magnetic activity in the awake and in REM sleep states but very reduced such activity during delta sleep. Because spontaneous 40-Hz oscillation was seen in wakefulness and in dreaming, we proposed it as a correlate of cognition. In addition, spontaneous 40-Hz oscillation has been shown to be reset or modified by sensory stimuli in the awake state but this was not observed during REM or delta sleep [19]. These data indicate that the 40Hz oscillations could be involved in integrating outside sensory information with the ongoing cognitive activity of the individual, probably resulting from coherent 40-Hz resonance between thalamocortical specific and nonspecific loops [19].

Using Magnetic Field Tomography (MFT) on humans [35], we earlier demonstrated that such 40-Hz coherence displayed well defined cortico-subcortical correlations with a time shift that is consistent with thalamo-cortical conduction times. In particular, the MFT results indicated that the onset of activity at the thalamic level was followed by widespread activation of the thalamo-cortical system which resulted in large coherent thalamo-cottical 40-Hz oscillations organized in space and time, which was altered in pathological states such as Alzheimer's disease [34][35]. These studies provided, in conjunction with intracellular recordings from cortical interneurons [20] and thalamic neurons [45], further support of the hypothesis that recurrent thalamo-cortico-thalamic activity is involved in organizing and supporting coherent 40-Hz oscillations as a major synchronized event during sensory or cognitive processing [19]. As such, consciousness may arise by the resonant 40-Hz coactivation of at least two systems, which would temporarily conjoin cerebral cortical sites specifically activated at or around 40-Hz frequency. In this manner the specific system would provide the content, and the nonspecific system the temporal binding of such content into a single cognitive experience [19].

Spontaneous Oscillations vs. Event-Related Activities: Phase-Locked and Time-Locked (Evoked) vs. Time-Locked but not Phase-Locked (Induced) Activities

Basic mechanisms

As indicated above two types of changes in the electrical activity of the cortex may occur upon sensory stimulation: one change is *time-locked and phase-locked (evoked)* and can be extracted from the ongoing activity by simple linear methods such as *averaging;* the other is *time-locked but not phase-locked (induced)* and can only be extracted trough some non-linear method such as *envelope detection.* Which mechanisms underlie these types of responses? The time- and phase-locked response can easily be understood in terms of the response of a stationary system to the external stimulus, the result of the existing wiring of the cortex. The second response component can not be understood in such terms. We demonstrated that the non-phase locked component can be understood as a changed ongoing activity, resulting from changes in the functional connectivity within the cortex. We based this conclusion on the results using two data sets: intra-cortical recordings from lightly anaesthetized cat visual cortical area 18, and simulated responses from artificial neural systems with well defined plasticity.

The anatomical connectivity is the pattern of "wires" that interconnect neurons, or local ensembles of neurons, such as cortical areas or laminas, i.e. a spatial and directed set of possible signal transfer pathways; the functional connectivity reflects the ways in which the activity of neurons, or of local ensembles of neurons, influences the activity of other neurons or local ensembles, i.e. a spatio-temporal directed set of actual signal transfer. In artificial neural systems the anatomical connectivity would be the general layout of the network, the functional connectivity would be the complete set of synaptic weights. Both the anatomical and the functional connectivity of the cortex change continuously, the changes in functional connectivity occur probably faster, and as explained above (section 3) can account for the emergence of *time-locked but not phase-locked oscillations,* i.e. of *induced activity,* elicited by a sensory event.

Analysis methodologies: FRP

The reactivity of underlying brain rhythmicity to stimulation in the same frequency band of the rhythms themselves, was studied using the recently introduced Frequency Responsiveness Procedure (FRP) [23]. This procedure allows to discriminate on the sole basis of responsiveness, the brain activities that share the same frequency band and that often feature a widely overlapping topography, while it significantly increases the inter- and intra-individual reproducibility with respect to simple spontaneous activity measurements. The different ways rhythmical activities are affected by simple sensory stimulations can be compared to results of recent cognitive investigations that suggest similar discriminations [16][17]. The combined information relative to the same activity (cognitive correlates plus sensory responsiveness) may improve our ability of building functional models. Moreover FRP can be used as a simple rhythm classifier. In consideration of the good reproducibility of FRP responses, and of the quite short measurement time (<10 min), it may also be used to build functional/pathological indicators.

Alzheimer's disease, for instance, has been recently described to produce effects on the frequency responsiveness of the cortex [32]. The first FRP tests, performed with the 28 channel neuromagnetometer (in use at CNR-ESS - Rome) were able to discriminate activities in the lower and higher alpha band and to measure stimulation-rate-dependent alpha depression as well as a slowing down of the alpha rhythm following rhythmical stimulation [23]. Further experiments, performed with the Neuromag 122 channel neuromagnetometer (LTL laboratory - Helsinki University of Technology) permitted to thoroughly investigate the interaction between sensory stimulation and cortical spontaneous activities [24].

The FRP stimulation consists in delivering intermittent bursts of stimuli at a given rate intermingled by a pause of constant length. The entire stimulation epoch (burst plus pause) is repeated several times for each rate, in order to perform off-line averaging of the response spectra. The rates are scanned within the extended alpha band (from 6.0 Hz to 14.0 Hz at step of 0.5 Hz). Preliminary studies set the number of repetitions to 10 (but probably 5 could have been enough to achieve comparable results) to rule out possible influence of the scanning order of the stimulation. Two of the FRP findings were described in more detail: (i) the co-existence of two rhythms within the alpha band with strongly overlapping topography but different responsiveness to rhythmical stimuli; and (ii) the induction of more "Organized States" following FRP stimulation.

Co-existence of activities in the alpha band (Fig. 4):
The magnetic field associated with the bioelectrical activity evoked by the stimulation was measured with the 28 channel superconductive system in use at the CNR-ESS laboratory in Rome [10]. The data were acquired at a rate of 1.0 kHz (recording bandwidth 0.48 Hz - 250 Hz). The visual stimuli were short light pulses delivered by two red LEDS, mounted on a blackened scuba-mask (about 5 cm from the eyes, at an eccentricity of about 5 degrees) in the left hem)field. The subject was asked to maintain fixation on two green LEDS, constantly 'on', positioned in the mask with zero eccentricity.

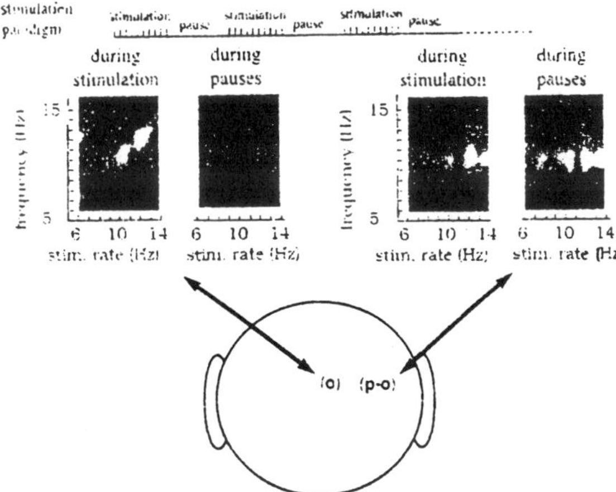

Fig. 4:   Different activities are present in the same frequency band (10-12 Hz) with overlapping topography and can be discriminated by the FRP procedure. Measurements following left hemifield visual stimulation (led short flashes) from two sites (out of 28 neuromagnetic sensors) are shown. Each image is a collection of 17 vertical spectra, each relative to the responses to rhythmic stimulation at rates ranging from 6.0 Hz to 14.0 Hz, step 0.5 Hz. Spectral responses during the stimulation and in the pause immediately following it are shown. The spectral amplitude is gray coded, white being maximal. One activity, peaking at '10 Hz, present mostly in the parieto-occipital area (p-o), is not influenced by the stimulation train, even when "resonant" (i.e. at 10 Hz). This is also the most evident activity seen in spontaneous measurements. Another activity peaking at '10 Hz and .12 Hz, stronger in the occipital region (o), is enhanced by the stimulation and is neither present in the pauses nor in the spontaneous measurements (not shown). Further correlation studies confirm that the former is not locked by the stimulating train while the latter is. For further details see the original work [23].

The FRP investigation managed to discriminate two activities within the 9 Hz-12 Hz band featuring a strongly overlapped topography, on the sole basis of their responsiveness. One activity, mostly on the low end of the band (<10 Hz) was quite undisturbed by the stimuli: it was present indifferently during the stimulating burst and during the pauses, and did not show improvements in the correlation across channels due to the stimulating bursts, even when these were delivered at 10 Hz (the peak of this on-going, or spontaneous, activity). The second one was, instead, locked to the stimululi as shown by a high amplitude measured during the burst while this activity decreased during the pauses, as well as by the sign)ficant improvement of cross-channels ccrrelation during bursts.

This second activity featured a wider spectrum covering the band 10 Hz - 12 Hz, however peaking at about 12 Hz. The phase-locked activity was localized slightly more occipital than the latter that was stronger in the parieto-occipital area.

These findings support the hypothesis of a lower alpha band reflecting "attentional" processes, constantly "on" during the test, not affected by the stimulation, as opposed to a higher alpha band reflecting stimulus related (cognitive) processes [16][17]. An innovative, non-linear, analysis technique, able to extract the time behaviour of oscillating activities buried in noise, permitted also to confirm this discrimination on the basis of the different dynamics of the two types of oscillatory activities with respect to the stimulating bursts [12].

Induction of organized states

Preliminary analysis from measurements performed with the 122 channel Neuromag whole-head neuromagnetometer (LTL/ Helsinki University of Technology) shows the induction of activity depression following FRP stimulating protocol [24]. In this study FRP was performed using the stimulation of three sensory modalities (for a total of 21 measurement sessions): somatosensory, visual and auditory. The data were acquired at 0.5 kHz with a passband 0.03 Hz - 160 Hz. Somatosensory stimuli consisted of short current pulses delivered at the median nene of the left wrist, with an intensity aufficient to elicit a painless thumb twitch. Visual stimuli, 30 ms long, were delivered by a matrix of 16 red leds (4 x 4 array, 10 mm x 10 mm) at an eccentricity of about 7 degrees in the left hem)field. Auditory stimuli were noise bursts 20 ms long, generated with a 44 KHz sampling rate.

Regardless of stimulating modality and rate, a depressive effect on the cortical activity outside the major activity peaks, in the band between 5 and 25 Hz, was obsened following the bursts of stimuli. The same effect was not measured when stimulating with a single stimulus. The depression of MEG activities is a whole cortex effect (appearing as a sort of more "Organized" State of the cortex: OS), lasting a few tens of seconds. A number of stimuli larger than three appears to be needed to induce it. The intra- and inter-individual reproducibility of the induction of OS is remarkable. The OS induction can be described by a new indicator: the "Global Activity Factor" (GAF). GAF is calculated for each channel as the difference between percentages of "active" frequency bins measured during the pauses immediately following the stimulating bursts, and the baseline (spontaneous) condition. A channel is defined "active" when its amplitude is larger than a threshold (choosing a threshold between 2 and 10 fT produce the same results). GAF grand average over subjects clearly shows the independence of OS induction from stimulation modality and cortical region.

The depression of activities over all cortical sites strongly suggests thalamic involvement in inducing the OS. OS could be described as an "alert" short-term cortical condition following trains of stimuli. The FRP appears a useful discriminating tool to investigate overlapping (in time-frequency and space) rhythmical activity and to study their functional correlates.

Brain "Resonance" Responses

The group of Basar has carried out a long series of experiments both in cat and human, where the relationships between evoked and spontaneous activities, particularly of the auditory and visual systems, were analyzed in detail [3]. Evoked potentials were studied in the frequency domain by means of the Fourier transform and adaptive digital filtering was applied. They found resonances mainly at 5, 10, 20, and 40Hz frequency bands, depending on the sites of measurement and the stimulus modality [1]. They concluded that evoked activities have components that are superpositions of induced rhythms caused by resonance phenomena in neuronal networks. They used the expression "gamma responses" for the 40Hz potentials that were shown to be widely distributed in the cat and human brain, not only upon sensory stimulation, but also after cognitive stimulation [3][21]. Brain gamma oscillations appear to have a broad spectrum of correlates with functional states, since they are widespread in a large variety of species (from human to snails) and are observed in multiple brain areas. Therefore one can speak of distributed gamma systems. Altough the research related to rhythmic activities in the beta/gamma frequency band have been very popular in recent years, one should also realize that other induced rhythms, particularly in the alpha frequency range, manifest functional correlates, namely with primary sensory and cross-modality processing [2].

Psychophysiological Correlates: Evoked and Induced Activities

The finding that after transient sensory stimulation two types of activity in gamma-band range can be differentiated: *evoked (time and phase-locked) and induced (time but not phase-locked) is* illustrated in Fig. 5. The evoked gamma-band activity is stimulus time- and phase-locked and it can be extracted from the noise in time-domain using the averaging procedure. In contrast, the induced gamma-band activity is not phase-locked to the stimulus or it strongly jitters, and thus it can not be extracted in the time domain.

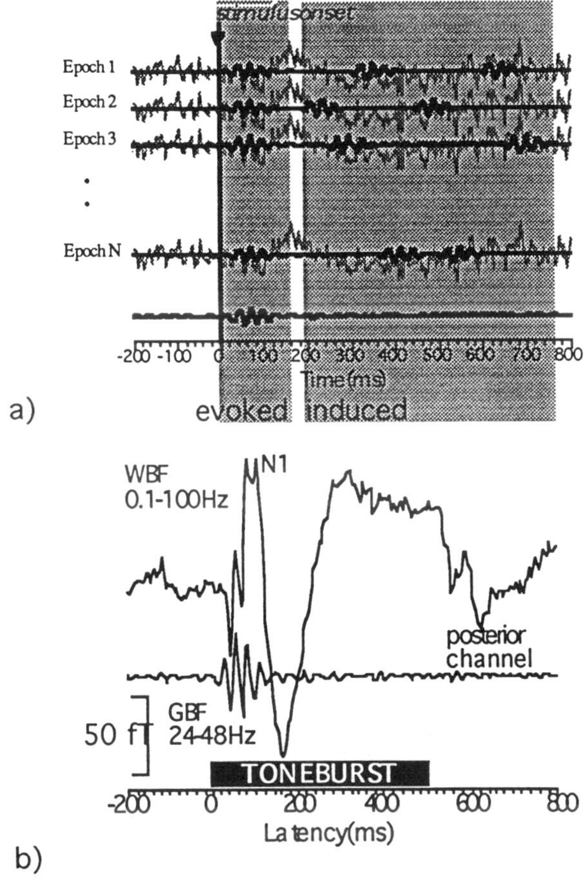

Fig. 5: (a) Schematic representation of evoked and induced gamma-band activity (black line) as superimposed on noise (grey line).
(b) Wide-band (top) and gamma-band evoked field due to tone-burst stimulation (posterior field maximum, N=384).

A transient auditory stimulus elicits not only the well known large slow wave components with the most prominent N1 peak, but also a noticeable gamma-band activity during the first 100 ms after the stimulus onset with spectral energy between 30 and 40 Hz (c.f. Fig. Sb). The equivalent source of the transient gammaband field (GBF) was estimated to be located on the supratemporal plane, few millimetres more medial and anterior as compared with the equivalent N1 source [29]. With series of functional studies we tried to increase our understanding about the transient gamma-band response as well as the function it may serve for sensory processing in the human auditory cortex. The functional behaviour of the gamma-band field was investigated with respect to changes in the stimulus parameters: 1) different kinds of auditory stimuli (in order to investigate the morphology of the gamma-band response); 2) different stimulus repetitions rates (in order to examine the recovery function of the gamma-band response); 3) different stimulus rise times (in order to differentiate between the middle latency and the gamma-band evoked responses as well as to test the gamma-band response originality and independence); 4) the dependence upon the carrier frequency of the stimulus (in order to test whether the gamma-band response represents the tonotopy of the underlying cortical equivalent sources) and 5) the dependence of the evoked gamma-band response upon attentional variations (in order to study how far the gamma-band response demonstrates an electrophysiological correlate of attention). In all these studies we compared the functional behaviour of the gamma band response in time-, frequency and "source space"-domain with the corresponding behaviour of the best investigated N1 peak of the slow evoked response.

We compared the morphology of the evoked transient GBFs elicited by different type of simuli, such as click, tone-pip and tone-bursts, that differed in duration (500 or 2000ms) or in quality (pure tone or complex tone burst). The different stimulus configurations did not produce sign)ficant changes in the gamma-band response morphology. For all different stimuli we were able to verify consistently an oscillatory response, configured by four to five waves with a global duration between 100 and 200ms.

Using tone pips with carrier frequency of lOOOHz and stimulus rates of 1/8, 1/4, 1/2 and lHz the recovery functions of the evoked GBF and the N1 component of the slow auditory evoked field (sAEF) were compared within four different subjects [27]. Both the GBF and N1 amplitudes decreased at high stimulus rates, however N1 amplitude decreased faster. The normalized rate-amplitude characteristics, i.e. the dependence of the root mean square field amplitude (RMS, calculated over the 37 recording channels) upon the stimulus rate, showed different slopes, indicating different recovery patterns. This finding was confirmed by the results in source space, demonstrating two separate groups of source locations in the supratemporal cortex with the GBF sources medial and anterior to N1. Thus, these results suggested that the GBF and the N1 have different origins and may arise from different processes in the auditory pathway.

A major problem when trying to separate the evoked GBF, is the possible involvement of the middle latency components, that are in the same frequency range. Therefore in ten subjects the waveforms of the evoked GBFs, elicited by 50 ms tone burst with short rise/decay time of 2 ms and long rise/decay time of 25 ms were compared. The first stimulus possesses the ability to elicit the middle latency reponse, due to its high degree of synchronization, but the second one does not. The GBF, when using a short stimulus, is accompanied by a characteristic time course in the initial part, that coincides with Na and Pa waves of the middle latency response. This effect could not be observed for the GBF elicited by the long stimulus. In contrast no obvious difference was observed in the final part of the GBFs, evoked by short and long stimuli.

The tonotopy is one of the general principles of functional organization of the auditory system. The depth of the N1 source of the sAEF indicates this tonotopic organization: the higher the stimulus frequency, the deeper the equivalent source. Moreover, in most subjects the depth of N1 increases linearly with the logarithm of the stimulus frequency [28][25]. Twelve subjects were investigated at 3 different stimulus frequencies: 500, 1000 and 4000 Hz (short-tone bursts of 50 ms duration, 3 ms rise/decay time, intensity of 60 dB nHL and interstimulus interval randomized between 600 and 800 ms) and EEG and MEG were recorded simultaneously [5]. The N1 scalp potential maps demonstrated the well known topographic changes with increasing stimulus frequency: an increase of the postero-lateral potential extremum was observed. In contrast, on the gamma-band scalp potential maps no clear effects were discernible. The magnetic data also showed obvious results. The estimated gamma-band and the N1 source locations, that were obtained in each single subject were averaged across subjects and presented as a function of the stimulus frequency. Significant dependence on frequency was observed only for the medial-lateral source coordinate of N1, but not for the transient GBF.

Functional Relations Between Brain Oscillations, Perception, Attention, and Other Cognitive Processes

Gamma-band MEG fields

The effects of attention and processing on the transient GBF were studied in six subjects by Pantev and Elbert [26]. In a classical odd-ball paradigm rare stimuli (low pitch of 667 Hz, probability of 33%) were randomly interspersed in a sequence of frequent stimuli (high pitch of 1500 Hz; 67% probability). Stimuli were Gaussian tone-pulses of the above mentioned frequencies with half-value time of 6 ms and an intensity of 60 dB nHL. The interstimulus interval varied randomly from 1.5 to 2.5 s. The stimuli were presented in twelve blocks, with one block having a random length between 150 and 250 stimuli. The subjects' task was either to attend to the oddball stimuli or to process a completely different task in order to distract from the oddball stimuli. During the "attend condition", the subject was instructed to concentrate on the appearance of rare stimuli and to count them. During the "distract condition", the subject had to perform a numbers-to-letter task involving mental arithmetic with considerable demand on memory. With the begin of the stimulus block, the subject started to calculate the sum of the numbers corresponding to the rank of the letters of a certain well known sentence. The RMS field values of the wide-band and gamma-band evoked responses were calculated separately for the rare and frequent stimuli of the "attend" and "distract" conditions and then averaged across subjects. A clear difference was obtained within the time-range of P300 for the "attend" and "distract" conditions, indicating that the subjects complied well with the experimental task. Higher amplitudes of Nl were obtained for the rare stimuli in both "attend" and "distract" conditions. This effect was explained by the well-known short term habituation effect in the auditory system. No substantial difference was obtained for the GBF, neither between the rare and frequent stimuli, nor between the "attend" and "distract" conditions. In addition, no sign)ficant results were observed for the "attend" and the "distract" condition in source space.

The results that we obtained in different studies of auditory evoked transient GBF, led us to conclude that the transient evoked gamma-band activity (GBF) represents an early distinctive event of the human auditory cortex. This effect reflects the synchronization of a neuronal network, involving primary auditory cortical areas and probably some part of the primary integration region, when an adequate stimulus enters the auditory system. This rather global cortical synchronization mirrors the auditory cortical perception and probably serves as a dynamical link between cortical subregions that are involved in the evaluation of the physical parameters of the auditory stimulus.

"40-Hz" coherent MEG activity

The specific "40Hz" activity was also shown to be related to auditory perception, i.e. coherent 40Hz activity appears to correlate to the temporal binding of incoming auditory stimuli [14]. Ribary's group tested, in particular, the hypothesis that the minimal interval required to identify separate auditory stimuli correlates with the time needed for the reset of the 40Hz MEG signal.

In control subjects, experimental and modeling results indicated a stimulus-interval-dependent response with a critical interval of 12-15 msec. At shorter intervals only one 40-Hz response, to the first stimulus, was observed. With longer intervals, a second 40-Hz response appeamd, which coincided with the subject's perception of a second distinct auditory stiumulus. These results indicated that oscillatory activity near 40-Hz, recorded during the first 100 msec post stimulus, represent a neurophysiological correlate to the early temporal processing of auditory stimuli underlying cognition [14]. It is important to note that this finding, the 12-15 msec interval required for perceptual separation, does not simply reflect the conduction time of the auditory system, which is in the order of 15-20 msec. Neuropsychological and psychophysical observations indicate that the brain is capable to distinguish two auditory stimuli (separated by more than 2-3 msec) and to identify two distinct stimuli (separated by more than 15-20 msec). In addition, it is well known that the auditory system is able to detect delays orders of magnitude smaller, in order to localize sound in space. Indeed, the auditory system is known for its excellent time resolution, as opposed to the visual system which has a delay that is deeply embedded in the integration time for the retinal system. Therefore, our data suggest that the 40-Hz activity does not only correlate with the primary processing of auditory information per se but with temporal processing of auditory events, which happens at a critical time period of 12-15 msec, in order to bind single elements as one cognitive event.

These findings are demonstrating one example, that we can use an MEG system, record the magnetic activity from a human brain, extract the oscillatory activity at around 40Hz, analyze its spatio-temporal distribution and demonstrate a coding-mechanism which starts at 25-30 msec post stimulus and relates to the perception or the awareness of auditory input [14], and more importantly, a coding-mechanism which is altered during cognitive impairment such as dyslexia (see in this book: Ribary et al.). This sign)ficant evidence indicates that the 40-Hz response could represent a major mechanism underlying quantum cognitive brain function. In addition, it also indicates that the MEG technology has many more potential applications than it is commonly used today, i.e., that the MEG technique can do better and be an important competitive tool to other functional brain imaging techniques.

Sensorimotor rhythms, mu and "40-Hz"

MEG easily detects the sensorimotor mu rhythm over the central sulcus in practically all human subjects. This rhythm typically consists of two distinct frequency components, one around 10- and one around 20-Hz, which may be transiently phase-locked giving rise to the classical comb-shaped mu rhythm. Detailed source analysis shows that the 10-Hz component arises mainly posterior to the central sulcus, whereas the 20Hz component arises slightly more anteriorly and medially [40]. Both components are attenuated during movement; however, after the movement, a transient increase above the pre-movement baseline, a rebound, typically occurs. The post-movement rebound for the 20-Hz component systematically occurs 300-500 ms earlier than the 10-Hz component [39]. These results, led to the hypothesis that the 20-Hz component is somatomotor whereas the 10-Hz component is mainly somatosensory, and that the components may have different functional roles. The hypothesis gains further support from a recent study showing that the 20-Hz rhythm, in contrast to the 10-Hz rhythm, displays what one could call "motorotopy": for the 20-Hz rhythm the attenuation during movement and the post-movement rebound follow the "motor homunculus", whereas for the 10-Hz rhythm the rebound centers in the somatosensory hand area regardless of which body part is moved [41]. Analysis and quantification of reactivity may be facilitated by focusing on rhythms modulated by peripheral stimulation; these reactions may be more robust and easier to replicate and quantify than the reactivity of ongoing or spontaneous rhythms possibly reflecting lower degrees of the induced rhythms. An example is provided by the transient mu rhythm enhancement occurring after median nerve stimulation (and after movements). This enhancement is suppressed by movements and sensory stimulation and the suppression is easily quantified [38]. Interestingly the enhancement is suppressed also by motor imagery, indicating involvement of primary sensorimotor cortex during imagery [42][38][36].

Although movement-related increase of rhythmic activity around 40-Hz has been reported repeatedly [8][30], most subjects do not show sign)ficant sensorimotor activity in the frequency bands above 20 Hz. Nonetheless, we have recently studied such activity in detail in a subject known from EEG recordings to show enhancement of 40-Hz activity in association with finger movements. The movement-related 40-Hz rhythm was found to increase before and during self-paced finger movements and to originate in the precentral motor cortex [37]. Furthermore, the cortical 40-Hz rhythm was found to be coherent with the rhythmicity of the electromyogram simultaneously recorded from the contracting muscle (Fig 6).

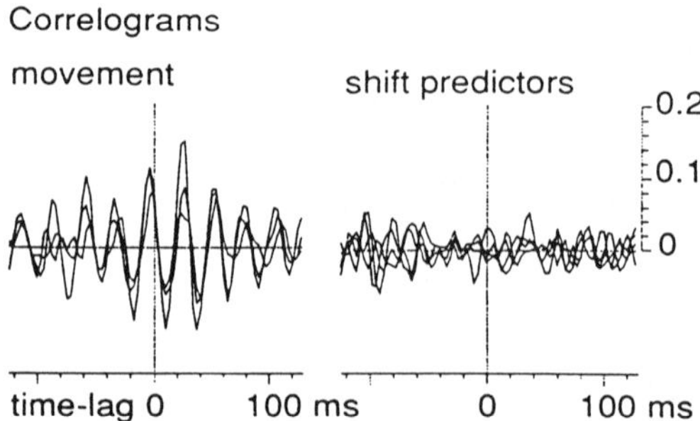

Fig. 6: Averaged cross-correlograms between rolandic MEG and extensor EMG, computed during contralateral index-finger movements, and compared with shift predictors, i.e. cross-correlograms computed using MEG and EMG signals from different movements. Adapted from Salenius et al. 1996 [37].

Conway et al. [7] were the first to report on coherence between MEG and EMG, although in association with sustained contraction and not movements. They found coherence between 15 and 30 Hz in different subjects. Based on the above findings one may speculate that the 40-Hz activity obsened in association with movement planning and execution in our subject could represent an exception, and that the functionally relevant enhancement of movement-related rhythmic activity in most subjects occurs at frequencies between 15 and 35 Hz. The coherence between MEG and EMG obviously reflects communication between the cortex and the periphery. The cortical rhythms may reflect "binding" in the motor cortex, that is, synchronization of control of different muscles, by means of a common frequency and phase, although it is still possible to argue that they are epiphenomenal [38].

The sensorimotor mu rhythm can thus be decomposed in partly independent frequency components. Further studies may confirm already suggested subdivisions of, for example, the parieto-occipital alpha rhythm, based on frequency or main source areas, with partly independent reactivity and possibly functional differences (see above 5.2). However, functionally corresponding frequency components may sometimes show sign)ficant interindividual variability, as suggested by the coherence between motor cortex and the motor units, which varies at least between 15 and 40 Hz in different subjects [7][37]. Furthermore, even though characterization of the reactivity of distinct frequency components may yield important information about the functional states of a neural network, a functional state can indeed be associated with oscillatory activity at several frequencies, or with aperiodic behavior. In this case an approach based on analysis of narrow frequency bands may yield a simplified or, in the worst case, misleading picture of the oscillatory dynamics of the network under study. Interpreting MEG results in terms of activity of underlying neuronal populations requires careful consideration of data from invasive measurements. Source analysis with multi-channel MEG enables a tight coupling of macroscopic rhythms to the underlying functional anatomy, and may thus, in combination with analysis of temporal behavior and coherence, help in elucidating the functional sign)ficance of cortical rhythms.

General Points of Discussion and "Take Home Messages"

(i) The question of which analysis methods are most suitable to extract and to quantify oscillatory activities from the MEG, was generally discussed in the Workshop. A general conclusion was that it is important to apply wide frequency band spectral analysis before any narrow-band pass filtering, in order to be able to determine whether different, mutually related, frequency components, i.e. different harmonics , contribute to the oscillatory phenomenon of interest. If this is not done one may be led to overemphasize some very specific frequency components. Of course, the latter may be prominently present such as an alpha rhythm between 10 -12Hz, or a beta/gamma rhythm of 38-42Hz, but it is always indicated to start the analysis of the correlations between MEG/EEG signals and cognitive processes using wide-band spectral analysis. After the relevant frequency bands, within a given context, are identified, the latter may be extracted for further analysis, for example, to obtain MEG/EEG brain maps of their spatio-temporal distributions.

(ii) Furthermore there is an increased awareness that neuronal networks have intrinsic nonlinear dynamic properties, that are reflected in the characteristics of the oscillations they generate, namely in their harmonic composition. Changes in the neuronal network activity may be reflected in several properties of an oscillation: - an increase/decrease in amplitude of a given oscillatory component, representing an increase/decrease in the proportion of synchronously active neurons within a population; - a change in frequency, and/or in the relative amplitude of different harmonic components, representing a change in the operating point of the underlying neuronal network that lies in its nonlinear range of operation. The latter can take place due to a change in the strength of the input to the network and/or of in that of modulating signals.

(iii) In this respect, but also in general terms, it is important to emphasize the need to make bridges between studies of oscillatory brain dynamics in experimental animals and in human subjects. An interesting research line into the sign)ficance of human MEG/EEG oscillations with respect to brain functional states, is to analyze the dynamics of these oscillations under well chosen behavioural and cognitive conditions as examplif~ed above (section 6.1 and 6.2). In this respect a particularly fruitful approach is to calculate brain maps and MFT distributions, of the changes in the MEG/EEG relevant parameters, representing transitions between dynamic states that correspond to different stages of cognitive/perceptual or motor tasks. This poses a number of technical problems since it involves analysing nonstationary processes that may have complex dynamics but this topic goes beyond the scope of this overview [31][33][46]. Nevertheless the use of *timevarying spectral analysis* offers a practical possibility for estimating spectral changes in MEG/EEG relative to well defined events.

Finally, the following "take home messages" were formulated:

- **Urs Ribary**: use temporal information of the MEG to study global brain networks, their interactions and the correlation of their activities with human brain cognitive functions.

- **Christo Pantev**: the transient auditory evoked gamma-band activity mirrors global cortical synchronization, reflects cortical perception and probably serves as a dynamic link between cortical subregions involved in the evaluation of stimulus parameters.

- **Erol Basar**: the brain's elementary oscillations, from delta to gamma frequencies, are distributed and have multiple functional correlates.

- **Stephan Salenius**: source localization, in combination with analysis of reactivity and coherence, reveals functional subdivisions of macroscopic cortical rhythms.

- **Livio Narici**: use the frequency responsivity procedure (FRP) as a rhythm discriminating method. In this way "lockable" and "unlockable" rhythms within the same frequency band can be readily distinguished.

- **Bob van Dijk**: sensory (visual) stimuli can change the functional connectivity of the cortex. These changes, in turn, lead to changes in the dynamics of ongoing activity. the latter are time-locked but not phase-locked response components.

- **Fernando Lopes da Silva**: MEG/EEG signals indicate the dynamical states of distributed neuronal networks of the brain. These networks have essential nonlinear properties. The analysis of the characteristics of the corresponding dynamical states using multi-channel recordings and both linear and nonlinear methods of quantification (e.g. time-varying (bi)spectra), can help to formulate specific hypothesis about the dynamics of functional states of brain systems with respect to cognitive functions.

References

[1] Basar, E., Basar-Eroglu, C., Rahn, E,. and Schdrmann, M. Sensory and cognitive components of bMin resonance responses. An analysis of responsiveness in human and cat brain upon visual and auditory stimulation. Acta Otolaryngol. Suppl., 1991, 491: 25-34.

[2] Basar, E., and Schurmann, M. Functional aspects of evoked alpha and theta responses in humans and cats. Occipital recordings in "cross modality" experiments. Biol. Cybern., 1994, 72: 175-183.

[3] Basar, E. and Bullock, Th. Induced rhythms in the brain, Boston, Birkhauser, 1992.

[4] Basar, E. and Schurmann, M. Alpha rhythms in the brain; functional correlates. News in Physiol. Sciences, 1996. In Press.

[5] Bertrand, O. and Pantev, C. Stimulus frequency dependence of the transient oscillatory auditory evoked responses (40 Hz) studied by electric and magnetic recordings in human. In: Pantev, C., Elbert, T. and Lutkenhoner, B. Oscillatory Event Related Brain Dynamics, New York, Plenum, 1994: 231-24.

[6] Boeijinga, P.H. and Lopes da Silva, F.H. Modulations of EEG activity in the entorhinal cortex and forebrain olfactory areas during odour sampling. Brain Res., 1989, 478: 257-268.

[7] Conway, B.A., Halliday, D.M., Farmer, S.F., Shahani, U., Maas, P., Weir, A.I., and Rosenberg, J.R. Synchronization between motor cortex and spinal motoneural pool during the performance of a maintained motor task in man. J. Physiol., 1995, 489: 917-924.

[8] DeFrance, J. and Sheer, D. Focused arousal, 40-Hz EEG, and motor programming. In: Giannitrapani, Murri The EEG of Mental Activities. (Karger, Basel) (1988) pp. 153-168.

[9] Eckhorn, R., Bauer, R., Jordan, W., Brosch, M., Kruse, W., Munk, M. Reitboeck. Coherent oscillations, a mechanism of feature linking in the visual cortex? Biol. Cybern., 1988, 60: 121-130.

[10] Foglietti, V., Del Gratta, C., Pasquarelli, A., Pizzella, V., Torrioli, G., Romani, G.L., Gallagher, W.J., Ketchen, M.B., Kleinsasser, A.W., Sandstrom, R.L. 28-channel hybrid system for neuromagnetic measurements. IEEE Trans-MAG, 1991, 27: 2959-2962.

[11] Freeman, W.J. and Viana di Prisco, G. Relation of olfactory EEG to behavior: time series analysis. Behav. Neurosci., 1986, 100: 753-763.

[12] Gaeta, R., Benzi, R., Peresson, M. and Narici, L. Singular Spectrum analysis applied to neuromagnetic signals. Biomag 1996, this volume.

[13] Gray, C.M., Koenig, P., Engel, K.A., Singer, W. Oscillatory responses in cat visual cortex exhibit intercolumnar synchronization which reflects global stimulus properties. Nature, 1988, 338: 334-337.

[14] Joliot, M., Ribary, U. and Llinas, R. Neuromagnetic coherent oscillatory activity in the vicinity of 40-Hz coexists with cognitive temporal binding in the human brain. Proc. Natl. Acad. Sci. USA, 1994, 91: 1174811751.

[15] Kalitzin, S., van Dijk, B.W., Spekreijse, H. and Van Leeuwen, W.A. Coherency and connectivity in oscillating neural networks: linear partialization analysis. Biological Cybernetics, in press.

[16] Klimesch, W., Schimke, H., Ladurner, G., and Pfurtscheller, G. J. Alpha frequency and memory performance, Psychophysiol., 1990, 4: 381-390

[17] Klimesch, W., Schimke, H. and Pfurtscheller, G. Alpha frequency cognitive load and memory performance, Brain Topography, 1993, 5: 241-251.

[18] Llinas, R.R. The intrinsic electrophysiological properties of mammalian neurons: insights into central nervous function. Science,1988, 242: 1654-1664.

[19] Llinas, R. and Ribary, U. Coherent 40-Hz oscillation characterizes dream state in humans. Proc. Natl. Acad. Sci. USA 1993, 90: 2078-2081.

[20] Llinas, R. Grace, A.A. and Yarom, Y. I vitro neurons inmammalian cortical layer 4 exhibit intrinsic activity in the 10 to 50Hz frequency range. Proc. Natl. Acad. Sci. USA, 1991, 88: 897-901.

[21] Lopes da Silva, F.H. Neural mechanisms underlying brain waves: from neural membranes to networks. Electroenceph. clin. Neurophysiol., 1991, 79: 81-93.

[22] Lopes da Silva, F.H. The generation of electric and magnetic signals of the brain by local networks. In: Greger, R. and Windhorst, U. Comprehensive Human Physiology, Berlin-Heidelberg, Springer-Verlag, 1996.

[23] Narici, L. and Peresson, M. Discrimination and study of rhythmical brain activities in the alpha band: a neuromagnetic frequency responsiveness test. Brain Res., 1995, 703: 3144.

[24] Narici, L., Portin, K., Salmelin, R. and Hari R. In preparation.

[25] Pantev, C., Bertrand, O., Eulitz, C., Verkindt, C., Hampson, S., Schuirer, G. and Elbert, T. Specific tonotopic organizations of different areas of the human auditory cortex revealed by simultaneous magnetic and electric recordings. EEG and clin. Neurophysiol., 1995, 94: 26-40.

[25] Pantev, C. and Elbert, T. The transient auditory evoked gamma-band field. In: Pantev, C., Elbert, T. and Lutkenhoner, B. Oscillatory event related brain dynamics, New York, Plenum, 1994: 219-230.

[27] Pantev, C., Elbert, T., Makeig, S., Hampson, S., Eulitz, C. and Hoke, M. Relationship of transient and steady-state auditory evoked fields. Electroenceph. clin. Neurophysiol.,1993, 88: 389-396.

[28] Pantev, C., Eulitz, C., Elbert, T. and Hoke, M. The auditory evoked sustained field: origin and frequency dependence. EEG and clin. Electrophysiology, 1994, 90: 82-90.

[29] Pantev, C., Makeig, S., Hoke, M., Galambos, R., Hampson, S. and Gallen, C. Human auditory evoked gamma-band magnetic fields. Proc Natl Acad Sci USA, 1991, 88: 8996-9000.

[30] Pfurtscheller, G. and Neuper, C. Simultaneous EEG 10 Hz desynchronization and 40 Hz synchronization during finger movements. NeuroReport, 1992, 3: 1057-1060.

[31] Pijn, J.P., van Neenen, J, Noest, A. and Lopes da Sillva, F.H. Chaos or noise in EEG signals: dependence on state and on brain site. Electroenceph. clin. Neurophysiol., 1991, 79: 371-381.

[32] Politoff, A.L., Monson, N., Hass, P., and StadLer, R. Decreased alpha bandwidth responsiveness to photic driving in Alzheimer disease. Electroencephalogr. Clin. Neurophysiol., 1992, 82: 45-52.

[33] Pritchard, W.S. and Duke, D.W. Measuring chaos in the bMin: a tutorial review of nonlinear dynamical analysis. Int. J. Neurosci., 1992, 67: 31-80.

[34] Ribary, U., Llinas. R., Kluger, A., Suk, J. and Ferris, S.H. Neurophathological dynamics of magnetic, auditory, steady-state responses in Alzheimer's disease. In: Williamson, S., Hoke, M., Stroink, G. and Kotani, M. Advances in Biomagnetism, New York, Plenum Press, 1989, pp. 311-314.

[35] Ribary, U., Ioannides, A.A., Singh, K.D., Hasson, R., Bolton, J.P.R., Lado, F., Mogilner, A. and Llinas, R. Magnetic field tomography (~;T) of coherent thalamo-cortical 40-Hz oscillation in humans. Proc. Natl. Acad. Sci. USA, 1991, 88: 11037-11041.

[36] Salenius, S., Kajola, M.,Thompson, W.L., Kosslyn, S. and Hari, R. Reactivity of magnetic parietooccipital alpha rhythm during visual imagery. Electroenceph clin Neurophysiol., 1995, 95: 453-462.

[37] Salenius, S., Salmelin, R., Neuper, C., Pfurtscheller, G. and Hari, R. Human cortical 40-Hz rhythm is closely related to EMG rhythmicity. Neurosci. Lett., 1996, In Press.

[38] Salenius, S., Schnitzler, A., Salmelin, R., Jousmaki, V., Hari, R. Modulation of human rolandic rhythms during natural sensorimotor tasks. Submitted.

[39] Salmelin, R. and Hari, R. Characterization of spontaneous MEG rhythms in healthy adults. Electroenceph clin Neurophysiol., 1994a, 91: 237-248.

[40] Salmelin, R. and Hari, R. Spatiotemporal chaMcteristics of sensorimotor MEG rhythms related to thumb movement. Neurosci., 1994b, 60: 537-550.

[41] Salmelin, R., Hamalainen, M., Kajola, M. and Hari, R. Functional segregation of movement-related rhythmic activity in the human brain. Neuroimage, 1996, 2: 237-243.

[42] Schnitzler, A., Salenius, S., Salmelin, R., Jousmaki, V. and Hari, R. Involvement of primary sensorimotor cortex in motor imagery: a neuromagnetic study. Soc. Neurosci. Abstr., 1995, 21: 558.

[43] Spekreijse, H., van Dijk, B.W., Kalitzin, S.N. and Vijn, P.C. Non-linear dynamics of columns of cat visual cortex revealed by simulation and experiment. In: Ciba Foundation Symposium, 1994, 184: 88-99.

[44] Steriade, M. and Llinas, R R. The functional states of the thalamus and the associated neuronal interplay. Physiol. Rev., 1988, 68: 649-742.

[45] Steriade, M., Curro Dossi, R., Pare, D. and Oakson, G., Fast oscillations (20-40Hz) in thalamocortical systems and their potentiation by mesopontine cholinergic nuclei in the cat. Proc. Natl. Acad. Sci. USA, 1991, 88: 4396-4400.

[46] Theiler, J. and Rapp, P.E. Re-examination of the evidence for low-dimensional, nonlinear srtucture in the human electroencephalogram. Electroenceph. clin. Neurophysiol., 1996, 98: 213-222.

Acknowledgements

We would like to acknowledge the excellent secretarial assistance of T. Ngo-Ha.

Time-Frequency Analysis of Oscillatory Gamma-Band Activity: Wavelet Approach and Phase-Locking Estimation

Bertrand, O., Tallon-Baudry, C., and Pernier, J.

Brain Signals and Processes Laboratory, INSERM U280, Lyon, France

Introduction

High-frequency oscillatory activity above 20 Hz (gamma-band activity) has been reported from both animal and human recordings (see [1, 2] for a review). Theoretical works have proposed a specific role for oscillatory synchronization of distributed neural assemblies in feature binding and sensory information processing [3]. Most human EEG and MEG studies have been considering the transient 40 Hz activity occurring in the first 100 ms following the stimulus onset in the auditory modality [4]. However, although non phase-locked 40 Hz activity was observed in animal data after visual [5], auditory [6, 7] or somatosensory [8] stimulations, only few human studies reported stimulus induced gamma-band response [9, 10], suggesting a role in focused arousal and/or cognitive information processing. In recent studies in the visual modality, we have shown that a 40 Hz non phase-locked activity occurs 200-300 ms after the stimulus onset and that it is specifically related to coherent visual stimuli (illusory triangles) compared to low-frequency evoked components [11, 12].

Detecting phase-locked and non phase-locked gamma-band activity in single trials and averaged responses requires some specific tools and related statistical tests. We present here time-frequency analysis methods allowing to detect and quantify both evoked and induced activities. The sensibility of the methods will be presented in simulation studies, and some experimental results will be briefly discussed in the visual and auditory modalities.

Methods

Our goal was to detect and characterize stimulus-related oscillatory activities in the gamma-band (ranging from 20 to 100 Hz). As both latency and frequency of oscillatory bursts are a priori not known, a time-frequency representation of the signal is required. For that purpose, we used a wavelet transform of the signals. This approach, compared to the more classical moving short-time Fourier transform, provides a better compromise between time and frequency resolution [13].

The signal to analyze (EEG or MEG, single-trial epoch or averaged response) is convoluted by complex Morlet's wavelets (sometimes called Gabor wavelets) $w(t,f_0)$, having a gaussian shape both in the time domain (standard deviation σ_t) and in the frequency domain (standard deviation σ_f) around its central frequency f_0:

$$w(t,f0) = A.\exp(-t^2 / 2\sigma_t^2).\exp(2i\pi f_0 t), \text{ with } \sigma_f = 1/2\pi\sigma_t.$$

Wavelets are normalized so that their total energy is equal to 1 ($A = 1/(\sigma_t \sqrt{\pi})^{1/2}$). A wavelet family, as opposed to classical windowed short-time Fourier transform, is defined by a constant ratio f_0/σ_f, which should be chosen in practice greater than about 5, i.e., the same number of cycles under the gaussian envelopes for any f_0. The time-frequency resolution, defined by $2\sigma_t$ and $2\sigma_f$, is thus varying with frequency (with $f_0/\sigma_f = 7$, this means 111 ms / 6 Hz in the 20 Hz band, and 22 ms / 29 Hz in the 100 Hz band). Below 20 Hz, wavelet duration is of hundreds of ms, thus leading to a poor time resolution. The time-varying energy $E(t,f_0)$ of the signal in a frequency band around f_0 is the squared norm of the result of the convolution of the complex wavelet $w(t,f_0)$ with the signal $s(t)$: $E(t,f_0) = |w(t,f_0)*s(t)|^2$. Repeating this calculation for a family of wavelets will provide a time-frequency representation of the energy of the signal (TF energy).

This analysis can be applied on the averaged evoked response to get the oscillatory activity phase-locked to the stimulus onset which remains after averaging. It can also be applied to each single trials and then all TF energies can be averaged. In this case, both phase-locked and non phase-locked activities are summed up, and bursts of high signal-to-noise ratio will emerge as long as their time-jitter does not exceed too much the wavelet time-duration. The TF energy of the post-stimulus epoch will be referred to a baseline level define as the mean value of the pre-stimulus TF energy in each frequency band.

To statistically test the phase-locking of a particular time-frequency component, we adapted the "phase-averaging" methods previously proposed in the frequency domain [14]. We consider the normalized complex time-varying energy of each single trial i associated to each wavelet $w(t,f_0)$: $P_i(t,f_0) = w(t,f_0)*s_i(t)/|w(t,f_0)*s_i(t)|$.

For all single-trials, these normalized complex values are distributed on a unit circle in the complex plane. When a given time-frequency component is rather phase-locked to the stimulus, all complex points have roughly the same orientation on this circle. Conversely, when it is absolutely non phase-locked to the stimulus, the points are randomly distributed on the unit circle. The vector average of these points gives a representation of the phase distribution in a given time-frequency region: its length will further be called the phase-ordering factor. It is independent of the signal amplitude, and it is ranging from 0 (purely non phase-locked activity) to 1 (strictly phase locked activity). With this method, even very low amplitude signals can be identified provided that they are rather strictly phase-locked. Statistical approach (Rayleigh test) has been proposed [15] to test the uniformity of the angle distribution on the unit circle. Critical values for the Rayleigh test are known exactly and are widely tabulated [15]: for 200 single-trials, for instance, the 0.01 error level corresponds to phase-ordering factors greater than 0.152.

To statistically detect a non-phase-locked oscillatory burst as emerging from the baseline activity, we use a non-parametric sign test applied to each time-frequency region. It consists in comparing, for all single trials, each post-stimulus time-frequency energy to its corresponding pre-stimulus baseline value in the same frequency band. Due to the fact that TF energy distribution across trials are generally not gaussian, we choose the median value of the pre-stimulus period as a representative of the baseline energy. A Z value can be computed at each time-frequency point: $Z(t, f_0) = (N^+(t, f_0) - 1/2N)/1/2\sqrt{N}$, where $N^+(t, f_0)$ denotes the number of post-stimulus TF energy values $E(t, f_0)$ above the baseline median value, and N is the total number of single trials. Tabulated probability values show that p<0.01 when $|Z(t, f_0)|$ is greater than 2.33. Positive and negative values can be obtained for Z, corresponding to increased and decreased activity compared to baseline.

Four types of time-frequency representation of the data is thus proposed : TF energy of the averaged response and TF phase-ordering factor for phase-locked activity, average of the single-trial TF energies and TF sign test Z values for non phase-locked activity.

Simulation study

To evaluate how these TF representations varies with the time-jitter and the amplitude of the oscillations, several simulations were performed with oscillatory bursts embedded in real EEG noise. Sets of data were made of 200 single-trial EEG epochs on which a wavelet-like signal was added. Gamma-band bursts of frequency ranging from 15 to 60 Hz were considered. Fig. 1 shows the four TF representations in case of 40 Hz signals peaking at 200 ms with different time-jitters and amplitudes (2ms/1μV, 20 ms/2 μV, and 40 ms/4 μV).

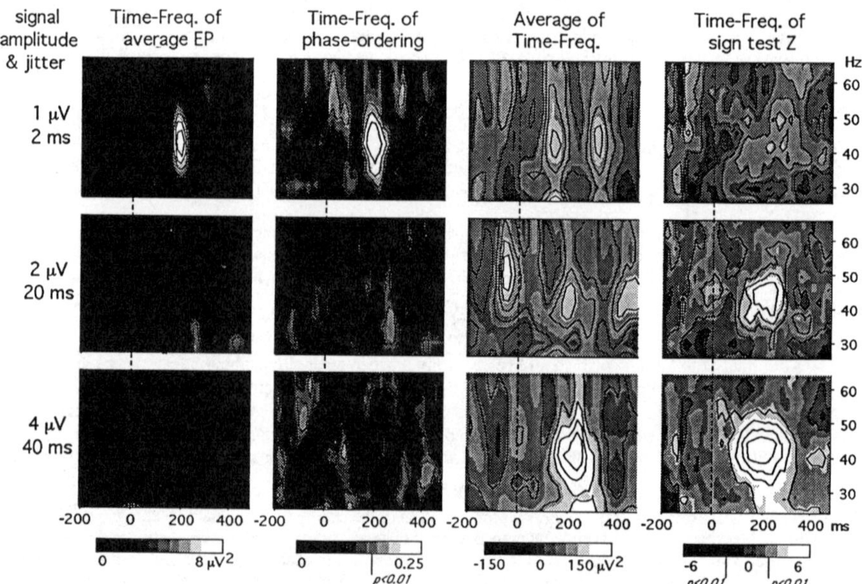

Fig.1. Time-frequency representations obtained on simulated data with different time-jitters and amplitudes combinations (see text for more details).

It appears that signal having a small time-jitter (2ms) are clearly discernible on the TF energy of the averaged response and the TF phase-ordering factor, even with amplitude as low as 1μV. However, it is not discernible on the average of the single-trial TF energies. Conversely, signal having an important time-jitter is no longer visible on phase-locked TF representations while it appears on the average of the single-trial TF energies as long as its signal-to-noise ratio is high enough (amplitude 4 µV). Nevertheless, for signal of lower amplitude (2 µV), while the 200ms/40Hz peak is not very clearly detectable on the average of the single-trial TF energies, the TF sign test Z value improves the detection of this non phase-locked activity, with a rather good elimination of noisy peaks.

A more systematic evaluation with time-jitters ranging from 0 to 150 ms, and signal amplitudes ranging from 0 to 5 µV, was conducted on wavelet-like oscillatory bursts (frequency ranging from 15 to 60 Hz). By means of the statistical tests based on the phase-ordering factor and the sign test Z value, limits (p=0.01) have been computed in the time-jitter/amplitude plane to characterize the detectability of phase-locked and non phase-locked signals (Fig. 2). Due to the low level of EEG background noise in the upper part of the gamma-range (40 Hz and above), phase-locked signals as low as 0.5-1 µV can be detected provided that their time-jitter does not exceed 5 ms. Detectability of non-phase locked signals is less sensitive to time-jitter in the lower part of the gamma-band (below 40 Hz), but requires higher amplitudes of about 2-3 µV.

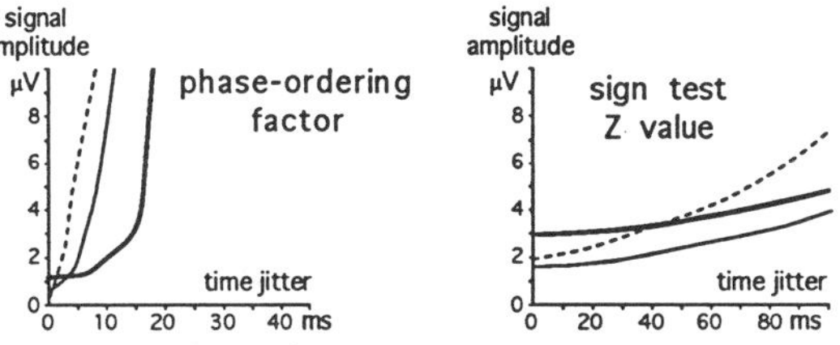

Fig. 2. Limits of statistical detection of oscillatory bursts (thick line: 15 Hz, thin line: 40 Hz, dotted line: 60 Hz). All signals having a combination of time-jitter and amplitude above those lines are detectable with the phase ordering factor (rather phase-locked activity) or the sign test Z value (induced non phase-locked activity).

Experimental studies

It has been suggested that feature binding could be achieved in the visual cortex by synchronization of spatially distributed neurons in the gamma-band. We have recently tested this hypothesis in human [18], using three types of visual stimuli: two coherent stimuli (a Kanizsa illusory and a real triangle) and a non-coherent one ("no-triangle"). The time-frequency analysis of averaged as well as single-trial EEG epochs (8 subjects) was performed to characterize phase-locked and non phase-locked high-frequency activities. An early phase-locked 40 Hz component was found maximal at electrodes Cz-C4, and does not vary with stimulation type. We described a second 40 Hz component, emerging from pre-stimulus baseline level around 280 ms, not phase-locked to stimulus onset. This component shows a higher energy in response to a coherent triangle, no matter whether real or illusory: it could thus reflect a mechanism of feature binding based on high-frequency synchronization. At the same latencies, the low-frequency evoked components behave differently, suggesting that low- and induced high-frequency activities have different functional roles.

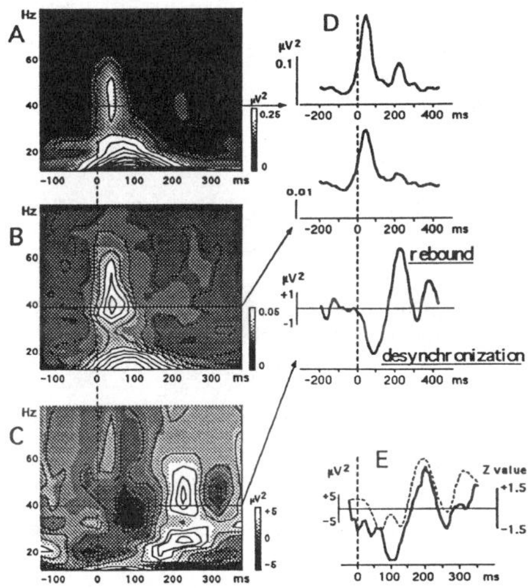

Fig.3. 40 Hz activity acoustic tone-burst presentation. Time-frequency representations (grand average over 10 subjects) of (A) the averaged evoked potential, (B) the phase-ordering factor, and (C) the average of the single-trial energies. (D) shows more precisely the time-varying energy in the 40 Hz band. (E) shows for one single subject the average of the single-trial energies (dotted line) and the related sign test Z value (thin line) around 40 Hz.

921

In the auditory modality, recent animal studies [7] have reported spontaneous 40 Hz activity recorded on an ECoG epipial electrode array placed over the auditory cortex of a rat. Single-trial spectral analysis of click responses indicated an inhibition of gamma-band oscillations during 150 ms, with subsequent enhanced gamma-activity between 250 and 350 ms. The epipial distributions of pre-stimulus and enhanced post-stimulus gamma-oscillations were the same. We present here preliminary time-frequency results in human EEG (grand-average of 10 subjects) of 40 Hz activity related to tone-burst presentation (1000 Hz, 50 ms duration) (fig. 3). A clear transient 40 Hz response peaking around 50 ms and phase-locked to the stimulus onset was observed (fig. 3A,B) as reported in previous papers [2]. The representations of the average of the single-trial TF energies and of the TF sign test Z values show a clear 40 Hz activity peaking around 200 ms, emerging from pre-stimulus baseline level (fig. 3C). Since it almost does not appear in the averaged data and phase-ordering factor, it most probably corresponds to a non phase-locked induced activity. This is consistent with previous findings in human recording reacting to click stimuli [9]. However, we also observed a significant reduction of 40 Hz activity immediately after the stimulus onset, with a minimum around 100 ms (fig. 3C,E). These 40 Hz de-synchronization and rebound phenomena are consistent with those observed on rat auditory cortex [6]. The transient 40 Hz response is clearly visible in the averaged phase-locked response even though it has a very small amplitude. However, it is very small in the single-trials compared to the spontaneous and induced activities, and thus does not show up in the average of TF energies.

Discussion

We have presented here methods to analyze EEG or MEG signals in the time-frequency domain for the detection and characterization of event-related gamma-band oscillatory activity. These methods have been evaluated with EEG simulations and applied in EEG experimental studies. Transient phase-locked and induced non phase-locked 40 Hz bursts have been observed in auditory and visual modalities. The visual data provides some evidence for a specific functional role of induced gamma-band in feature binding. Additional experimental studies should be performed to further clarify the respective functional roles of evoked and induced gamma-band, compared to low-frequency components.

References :

[1] Basar, E., and Bullock, T.H. Induced Rythms in the Brain, Boston, Birkhauser, 1992.

[2] Pantev, C., Elbert, T., and Lütkenhöner, B. Oscillatory Event-Related Brain Dynamics, New York, Plenum Press, 1994.

[3] von der Malsburg, C., and Schneider, W. A neural coktail-party processor, Biol. Cybern., 1986, 54: 29-40.

[4] Pantev, C., Makeig, S., Hoke, M., Galambos, R., Hampson, S. and Gallen, C. Human auditory evoked gamma-band magnetic fields, Proc. Nat. Acad. Sci. USA, 1991, 88: 8996-9000.

[5] Eckhorn, R., Schanze, T., Brosch, M., Salem, W., and Bauer, R. Stimulus-specific synchronization in cat visual cortex: multiple microelectrode and correlation studies from several cortical areas. In: Induced Rythms in the Brain, Basar, E., and Bullock, T.H. Boston, Birkhauser, 1992: 47-82.

[6] Basar-Eroglu, C. and Basar, E. A compound P300-40 Hz response of the cat hippocampus. Int. J. Neuroscience, 1991, 60: 227-237.

[7] Franowicz, M.N., and Barth, D.S. Comparison of evoked potentials and high-frequency (gamma-band) oscillating potentials in rat auditory cortex, J. Neurophysiol., 1995, 74, 1: 96-112.

[8] MacDonald, K.D., and Barth, D.S. High-frequency (gamma-band) oscillating potentials in rat somatosensory and auditory cortex, Brain Research, 1995, 694: 1-12.

[9] Jokeit, H. and Makeig, S. Different event-related patterns of gamma-band power in brain waves of fast- and slow-reacting subjects. Proc. Nat. Acad. Sci. USA, 1994, 91: 6339-6343.

[10] Lutzenberger, W., Pulvermüller, F., and Birbaumer, N. Words and pseudo-words elicit distinct patterns of 30 Hz EEG responses in humans. Neurosci. Letters, 1994, 176: 115-118.

[11] Tallon, C. Bertrand, O., Bouchet, P., and Pernier, J. Gamma-range activity evoked by coherent visual stimuli in humans, Eur. J. Neurosci., 1995, 7: 1285-1291.

[12] Tallon-Baudry, C., Bertrand, O., Bouchet, P., and Pernier, J. Variation of gamma-range oscillations correlates with stimulus coherency in human. Human Brain Mapping, 1995, suppl. 1: 41.

[13] Sinkkonen, J., Tiitinen, H. and Näätänen, R. Gabor filters: an informative way for analysing event-related brain activity, J. Neurosci. Methods, 1995, 56: 99-104.

[14] Jervis, B.W., Nichols, M.J., Johnoson, T.E., Allen, E., and Hudson, N.R. A fundamental investigation of the composition of auditory evoked potentials, IEEE Trans. Biomed. Eng., 1983, 30: 43-50.

[15] Mardia, K.V. Statistics of directional data, London, Academic, 1972.

[16] Tallon-Baudry, C., Bertrand, O., Delpuech, C., and Pernier, J. Stimulus specificity of phase-locked and non phase-locked 40 Hz visual responses in human. Submitted.

Inverse Imaging of Distributed Oscillatory Activity

Bradshaw, L.A. and Wikswo, J.P., Jr.

Vanderbilt University, Nashville, TN, USA

Introduction

Several types of bioelectric current source produce external electric and magnetic fields with oscillatory characteristics. Often, these sources are not well localized and may be distributed over a wide spatial range or may consist of several simultaneously-active localized sources distributed over a wide area. Additionally, the oscillation associated with such activity may not be synchronized over space -- activity in different regions may oscillate at different frequencies. The brain's alpha rhythm is one such example, with activity spread over a large region of the brain oscillating at a characteristic frequency between 8 and 12 Hz [1]. The basic electrical activity of the gastrointestinal system is another example. Frequencies of oscillation in gastrointestinal electrical activity include a 3 cycle-per-minute (cpm) slow-wave in the stomach and slow-waves that range from 8-12 cpm in the small intestine, with more distal segments of the small bowel typically having lower frequencies than more proximal segments [2,3]. We present a technique to address the problem of estimating the source current distribution from external measurements of the electrical activity.

Methods

Distributed oscillatory activity may be considered as collections of localized oscillators. We will represent these localized oscillators as current dipoles for purposes of evaluating the algorithm. In this formulation, the positions of the oscillators that comprise the source current distribution are considered fixed and the magnitudes of the dipole strengths vary according to the electrical activity. The current dipole strength oscillates according to a sine wave with a characteristic amplitude and phase. In this manner, we can treat bioelectric activity for which the oscillation is sinusoidal in nature. Other more complicated oscillatory patterns may also be treated by this technique if the oscillation can be characterized according to certain parameters.

Two situations arise with multiple oscillators characterized in this way. Each oscillator may have a different characteristic frequency, or two or more oscillators may have the same frequency with different phases. The former case arises with activity such as the gastrointestinal basic electrical rhythm (BER) where pacemaking oscillators appear along the bowel at progressively lower frequencies with more distal locations. The latter situation occurs when any type of propagating electrical activity is present.

The phenomenon of propagation can be characterized by phase shifts in the signal at different time instants. Uniform propagation implies that the phase shifts linearly with time for an oscillator at a given location. Since the frequency of the propagation oscillator remains the same as that of nearby oscillators, one may utilize the phase spectrum of the oscillation pattern to determine the nature of the propagation [4].

External multichannel measurements of the magnetic fields from bioelectric current sources record the magnetic fields adjacent to a region that typically includes the source activity. Several methods are available for estimating the source current distribution from external measurements. A review of some of these methods is provided in reference [5]. These methods all have varying degrees of usefulness for tracking the time variation in the signals, but none explicitly addresses oscillatory characteristics.

We describe a method for localizing sources of oscillatory activity by decomposing the frequency content of the magnetic field pattern. By exploiting both magnitude and phase information that results from a frequency-domain analysis, we effectively restrict the number of possible solutions for the inverse procedure.

Before the algorithm may be applied, the nature of the oscillatory activity and its characterizing parameters must be determined. The parameters define a basis function that represents the oscillatory activity. For the purposes of demonstrating the technique, we will use sinusoidal oscillations that are characterized by their amplitude and phase. The basis function in this case is thus a sine wave. More complex waveforms may be necessary for precise studies of activity in which the oscillation cannot be easily characterized as sinusoidal, but as long as oscillation parameters can be defined, the algorithm still applies with minor adjustments.

Once the oscillation has been characterized, it is necessary to determine the frequency or frequencies present in the recorded magnetic field pattern. Frequency decomposition techniques are applied to the entire magnetic field pattern and the results are averaged to obtain an idea of which frequencies are most prominently represented in the magnetic data. The magnitude spectrum of the Fourier transform is one such technique for identifying multiple frequencies of oscillation present in the signal.

The next step is to determine the parameters of the basis function that most closely describe the recorded data for each frequency present in the recording. Using the sine wave as a basis function, we find the amplitude and phase of the sine wave that best fits the observed data in each channel for each frequency present in the signal. This process is basically a complicated form of Fourier decomposition.

Let $B_i(t)$ represent the data recorded in channel i of an n-channel magnetometer array, and nf represent the total number of frequencies present in the signal as determined by the frequency decomposition. For each channel and for each frequency present in the recording, we wish to determine the amplitude and phase of the sine wave that best matches the recorded data. In other words, we wish to know $A'_{i,j}$ and $\phi_{i,j}$ given that

$$B_i(t) = A'_{i,j}\sin(\omega_j t + \phi_{i,j}) \,;\, i = 1, ..., n \,;\, j = 1, ..., nf \,, \tag{1}$$

where ω_j is one of the signal frequencies. Once the amplitudes and phases are determined, the phase information in $\phi_{i,j}$ may be evaluated to determine the presence or absence of propagation in the magnetic field pattern. Linear increases in best-fit phase indicates uniform propagation. If propagation is not indicated by the phase spectrum, single dipole fits to the sine wave amplitude pattern $A'_{i,j}$ isolate oscillators at different frequencies. If propagation is indicated, multiple dipole fits determine the location of the several oscillators in the propagation pattern whose frequencies are the same, but whose phases differ.

To illustrate this algorithm, we apply it to two hypothetical situations, one in which the underlying current distribution results in a magnetic field pattern with different frequency components, and another in which all oscillators have the same frequency but different phases simulating uniform propagation in a "V" pattern.

Results

The first example uses six oscillators whose distinct frequencies are the odd harmonics of a square wave. The amplitude and phases of the oscillators are chosen with their locations to result in a square wave magnetic field in the middle channel of the magnetometer array. The magnetic fields in the array are calculated at points that would coincide with the coils in the 37 channel Biomagnetic Technologies, Inc. Magnes™ system. Figure 1 shows the dipole distribution and resulting magnetic field pattern used for evaluation of the Frequency-Domain Inverse Algorithm. Fast Fourier Transformation (FFT) of the data in each channel was performed and the results of the 37 FFTs were averaged to obtain the average magnitude spectrum. The six frequencies corresponding to each of the dipolar oscillators were apparent in the average spectrum. For each of the six frequencies present, amplitudes and phases of sine waves with the given frequency were determined. Single dipole fits to the six sine wave patterns resulted in the six derived dipoles indicated in Figure 2A. The actual dipole distribution is also plotted, although it is not apparent since the actual and derived dipole distributions are so nearly equivalent. For comparison, moving dipole solutions to

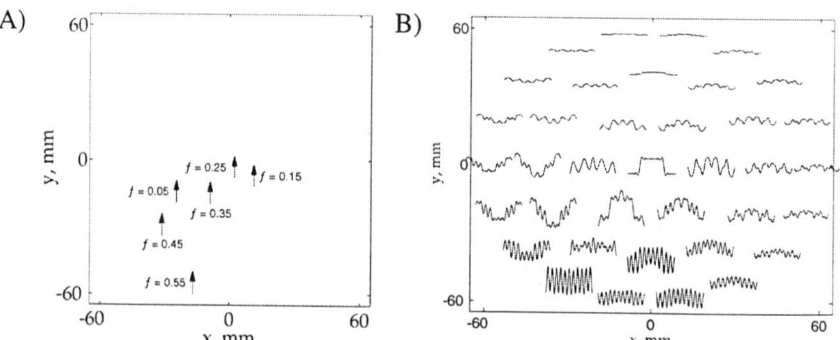

Figure 1. (A) Six dipoles with different oscillation frequencies indicated are used to evaluate the Frequency-Domain Inverse Algorithm. (B) The magnetic field pattern calculated for the 37 channel system from the six dipoles in (A).

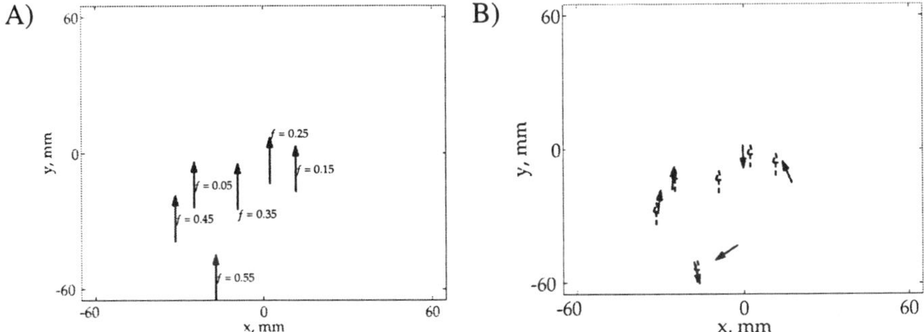

Figure 2. (A) Frequency-Domain Inverse Algorithm solutions to the magnetic field pattern in Figure 1B. (B) A six-dipole fit to a single time instant of the magnetic field data in Fig. 1B. In each case, the true dipole distribution is indicated by dashed arrows and the derived distribution by solid arrows.

the magnetic field patterns at successive time instants as well as a six-dipole fit to the magnetic data at single time instants were calculated. The results of the six-dipole fit are shown in Figure 2B.

The Frequency-Domain Inverse Algorithm requires minor modification when applied to data in which propagating oscillatory activity is present. Since propagation is characterized by several oscillators with a particular frequency and different phases, the single dipole fits to sine waves of one frequency that were used in the previous example will not be sufficient to fully describe patterns of propagation, and multiple dipole fits are necessary.

We simulate the propagation of electrical activity by assigning five dipoles arranged in a "V" pattern identical frequencies and phases that vary linearly according to the position of the dipoles, as shown in Figure 3. In this case, the same frequency decomposition that was applied to the square wave magnetic field example yields only one peak in the average magnitude spectrum at the frequency of the five dipoles. However, the phase spectrum of each dipole suggests the presence of propagating activity. The amplitude and phase of the best-fit sine wave are again calculated. We know that multiple dipole fits are necessary to adequately describe propagating activity, but the number of dipoles to fit is unclear; in fact, it is somewhat arbitrary. Since propagation is in essence a distributed activity, it follows that the more dipoles used in the fit, the better agreement with the actual data. However, including more dipoles in an inverse procedure makes the procedure more computationally intensive, and an optimal number of dipoles to fit

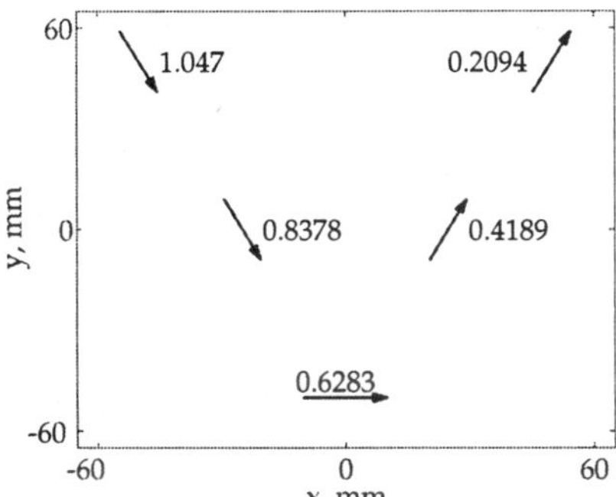

Figure 3. Five dipoles arranged in a "V" pattern. All five dipoles have the same frequency. The phases vary in a linear manner with dipole position to simulate uniform propagation of electrical activity. The phase values (in radians) are listed next to the arrows representing the dipoles.

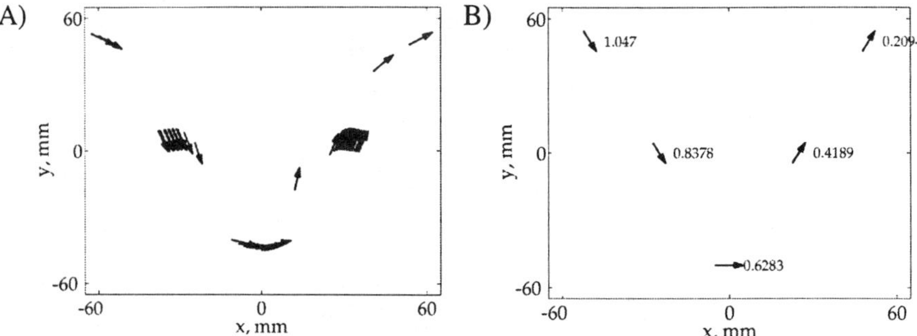

Figure 4. (A) Moving single dipole solutions to the magnetic field pattern from the propagating dipole distribution of Figure 3. (B) Five dipoles localized by the Frequency-Domain Inverse Algorithm. The actual dipole distribution is also displayed but cannot be seen since it corresponds so nearly to the inverse solution.

is desired. To this end, we perform moving single dipole fits to the magnetic field in successive time instants and cluster the results of the moving dipole analysis to determine the minimum number of dipoles required to properly describe the propagation pattern. Five clusters are apparent in the results of the moving dipole approach shown in Figure 4A. The geometric centers of these five clusters are used as starting parameters for the Levenberg-Marquardt fit of five dipoles to the magnetic field pattern. The results of the multiple dipole fit are shown in Figure 4B.

Discussion

The Frequency-Domain Inverse Algorithm was applied to two simulations of distributed oscillatory activity. In both cases, the dipole oscillators were correctly identified from distant magnetic field calculations. Using the goodness-of-fit parameter [5] commonly used in biomagnetism, it was found that the Frequency-Domain Inverse Algorithm solutions accounted for 99.8% of the square wave magnetic field pattern and 99.9% of the propagating "V" magnetic field pattern. In contrast, a multiple-dipole approach to the square wave magnetic field pattern provided a goodness-of-fit of only 65.0% while the average moving-dipole solution accounted for only 2.0% of the magnetic field pattern. This algorithm could be applied with little or no modification to many of the problems in which characterization of periodic or oscillatory activity is required from external electric or magnetic measurements.

References

[1] Nunez, P. L. *Electric Fields of the Brain*. New York, Oxford University Press, 1981.
[2] Christensen, J. D., Schedl, H. P., and Clifton, J. A. The small intestinal basic electrical rhythm (slow wave) frequency gradient in normal men and in patients with a variety of diseases, *Gastro*, 1966, 50: 309-315.
[3] Smout, A. J. P. M., Van Der Schee, E. J., and Grashuis, J. L. What is measured in electrogastrography? *Dig. Dis. and Sci.*, 1980, 25: 179-187.
[4] Bradshaw, L. A. *Measurement and Modeling of Gastrointestinal Bioelectric and Biomagnetic Fields*. Ph.D. Dissertation, Vanderbilt University, 1995.
[5] Hämäläinen, M., Hari. R., Ilmoniemi, R. J., Knuutila, J. and Lounasmaa, O. V. Magnetoencephalography - theory, instrumentation, and applications to noninvasive studies of the working human brain. *Rev. Mod. Phys, 1993*, 65(2): 413-497.

Acknowledgments

This work was supported in part by grants from the Veterans Administration and by an NIH SBIR grant to Conductus, Inc.

MEG Measurements of 40 Hz Auditory Evoked Response in Human Brain

Ciulla, C., Takeda, T., Morabito, M., Kumagai, T. and Endo, H.

Behaviour Control Laboratory, National Institute of Bioscience and Human-Technology, 1-1, Higashi, Tsukuba, Ibaraki 305 Japan

Introduction

Steady-state responses (SSRs) evoked by auditory stimulation in the brain were discovered by Galambos using EEG [1]. The amplitude of the response was found to be dependent on repetition rate, frequency and intensity of the stimulus, and it was maximum at repetition rates near 40 per second. Another EEG study conducted by Galambos [2] described intermodality cancellation of tactile and auditory 40 Hz potentials evoked by steady-state stimulations. A subcortical structure of the brain (possibly the rostral reticular formation) was assumed to be the generation site of the 40 Hz evoked activity. A model of the 40 Hz auditory-event related potentials was proposed and evaluated by EEG in Spydell et al., [3]. Comparison of the phase lag of the EEG response at the stimulation rate of 40 Hz between normal and thalamic and midbrain lesioned subjects, suggested again subcortical origin of the 40 Hz response.

More recently MEG was successfully employed in studies concerned with 40 Hz auditory evoked activity in the human brain. Makela et al., [4] reported that the presumed generators of the 40 Hz magnetic response were located at the primary auditory cortex deep within the sylvian fissure, close to the sources of the components of the transient response. The generation mechanism of the 40 Hz steady-state response was explored by Hari et al., [5]. SSRs were obtained in response to clicks presented between 10 and 70 Hz. Assuming that each click evokes a similar response at all repetition rates, it was proposed that the amplitude enhancement at 40 Hz was due to the superposition of the responses evoked by single clicks. The results of those studies were achieved by measurements carried out on only one hemisphere.

The present research aimed to study the origin in the human brain of the auditory SSR at 40 Hz. The use of a whole-cortex MEG system allowed simultaneous detection of the neuromagnetic responses from both hemispheres and estimation of the equivalent dipolar sources.

Methods

(a) Subjects, stimuli and neuromagnetic recordings

Five normal-hearing subjects (3 females and 2 males) aged between 27 and 48 years participated in the experiments. All of the subjects were right handed. Subjects were instructed to rest in relaxed wakefulness while eyes closed. Stimuli were 500 Hz tone bursts repeated at the rate of 40 per sec (6 ms duration, 1 ms rise and fall time). They were delivered through a 4.75 m plastic tube to the right ear at the intensity of 60 dB. Neuromagnetic fields were recorded by a whole-cortex 64-channels MEG system (CTF System Inc.) placed in a magnetically shielded room. Detection coils were 1st order axial gradiometers, 20 mm in diameter, and the distance between coils was 42 mm center to center. The pulse generator, the amplifier, and the transducer were placed outside the shielded room in order to avoid contamination of the neuromagnetic signal with the stimuli. Three coils placed at the preauricular points (tragions) and the nasion, defined the right-handed MEG head coordinate system with the origin located at the mid point between the tragions. The same fiducial points marked with pills having a 4 mm diameter were used in order to reproduce the head coordinate system on magnetic resonance images. MRI were determined individually for each of the subjects. The thickness of the MRI slices was 2 mm.

(b) Experimental session

Prior to the experiments, the estimation of the MEG system instrumental noise, the determination of the baseline spontaneous brain activity, and a control experiment were conducted. MEG measurements were lowpass filtered on-line (0-100 Hz) with a sampling rate of 1250 Hz. Six hundred epochs of 1500 ms analysis period were recorded and averaged. The use of on-line notch filters removed the powerline noise (50 Hz). Trials containing artifacts produced by eye blinks were eliminated automatically by software. Both baseline spontaneous brain activity and the control experiment were determined in one subject instructed to rest while eyes closed in relaxed wakefulness. The aim of the control experiment was to ascertain whether or not 40 Hz activity could be evoked when the stimulus was a 500 Hz continuous tone.

In order to check the replicability of source estimation, the experiment was repeated twice on different days (subjects KA and AY). In an additional experiment with two subjects, 40 Hz SSRs were obtained delivering the stimuli to the left ear (subjects KA and KI).

(c) Data analysis

The signal was extracted by off-line bandpass filtering (35-45 Hz). An equivalent current dipole (ECD) was used to represent the source of the brain activity, and its location was estimated in a spherical homogeneous volume conductor. The sphere gave the best fit to the curvature of the brain at the auditory cortices of both hemispheres, and it was determined for each of the subjects. The radius and the position of its center with respect to the origin of the MEG system were determined on the basis of magnetic resonance images.

An iterative least square minimization algorithm based on the Grynszpan-Geselowitz formulae provided the location and the orientation of the estimated dipolar sources in respect to the MEG head coordinate system. Only estimates with goodness of fit greater than 92% were considered. The depth of the sources was computed as radial distance from the surface of the spherical head model. The locations of the sources on MRI were determined applying the rotation/translation matrix of the MRI and the MEG head coordinate systems. Correction matrices were determined individually for each of the subjects.

Results

The instrumental noise of the MEG system (less than 20 fT for single epoch) was reduced to less than 1 fT (20 fT/$\sqrt{600}$) by averaging (n=600). The frequency spectra of the MEG sensors had an amplitude level lower than 1 fT rms/$\sqrt{Hz}$ (Fig. 1). The baseline of the averaged (n=600) spontaneous brain activity had a frequency spectra around 40 Hz comparable to that of the instrumental noise. The frequency spectra of the averaged brain activity had no peak around 40 Hz when the stimulus was a 500 Hz continuous tone (control experiment). A clear and distinct peak around 40 Hz was seen when the stimuli (500 Hz tone bursts) were delivered at the rate of 40 per sec (Fig. 1). Forty-Hertz magnetic SSRs were recorded with relatively high signal to noise ratio (about 10). Fig. 2 shows a time window of the raw MEG signals (0-100 Hz) recorded, the waves present a clear and consistent rhythm at 40 Hz. The neuromagnetic isocontours had for each hemisphere two extrema of opposite polarity located over the fronto-temporal and the temporo-parietal areas of the head (see Fig. 3). Considerable stability during the analysis period (1500 ms) was observed in the location of the fields extrema, suggesting the activation of a restricted area of the brain. Variability in the amplitude of the response between subjects was noticed. The maximum amplitude recorded was about 40 fT (subject KA). Comparisons between the ECDs estimated in the two hemispheres (left and right) was conducted on the basis of the dipole moment and the depth. With both right and left ear stimulation, the source located in the left hemisphere was estimated deeper (20-36 mm) and with slightly bigger dipolar moment (1.2-5.7 nAm) with respect to that of the right hemisphere (7-24 mm and 1.2-4 nAm). Table I summarizes for 5 subjects and for each hemisphere the maximum amplitude recorded by the MEG sensors, and the results of dipole estimation (location: x, y, and z; declination: d; dipolar moment: q; depth: D; and goodness of fit: g). The declination was computed with respect to the z axis of the MEG head coordinate system.

For all of the subjects 3-dimensional MRI locations of both contralateral and ipsilateral ECDs agreed with the activation of the supratemporal plane (SP) deep within the sylvian fissure. ECDs declination varied within 20-50 deg and 23-43 deg respectively for the left and the right hemisphere (ranges across subjects), compatibly with the anatomy of the supratemporal plane. Fig. 4 shows dipolar sources estimated for subject KA superimposed on MRI. Coronal and sagittal sections and top views are given for both left and right hemisphere.

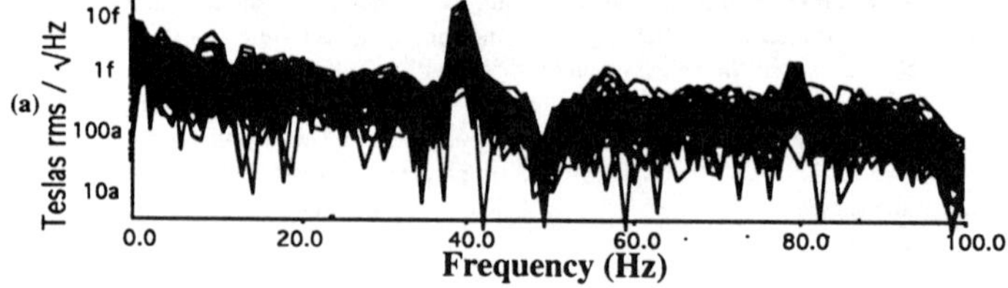

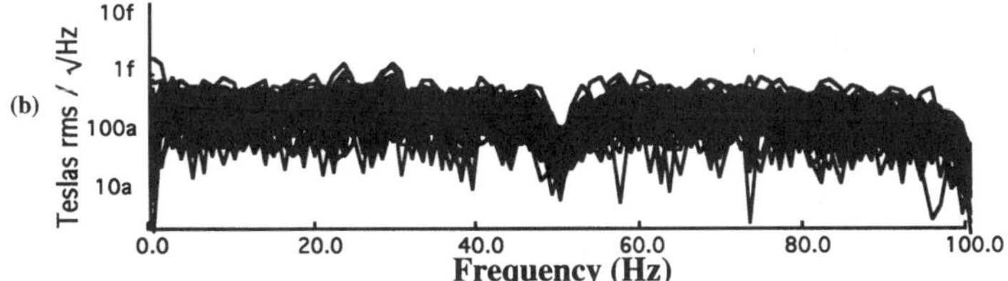

Fig. 1. Frequency spectra of the MEG signals. (a) Typical MEG response to an auditory steady-state stimulation at 40 Hz. Signals from all sensors of the MEG system are superimposed (0-100 Hz), units are in Tesla rms/√Hz. The brain response is peaked around 40 Hz. (b) The noise level of the MEG system. The instrumental noise was below 1 fT rms/√Hz (n=600).

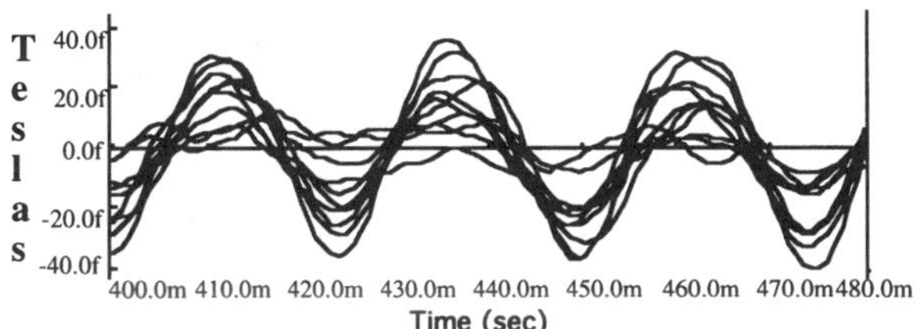

Fig. 2. Time window of the MEG signals recorded in response to auditory steady-state stimulation at 40 Hz.

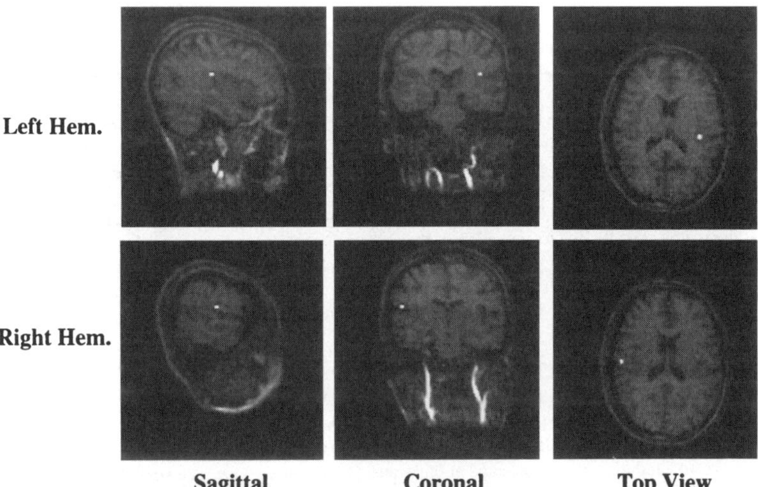

Fig. 3. Equivalent current dipoles superimposed on MRI (subject KA). The location of both contralateral and ipsilateral ECDs was in agreement with the activation of the supratemporal plane deep within the sylvian fissure. Sources of the left hemisphere were estimated deeper than those of the right hemisphere both with left and right ear stimulation. Sagittal section, coronal section and top view are shown.

	KA		AY	
	Right Hem.	Left Hem.	Right Hem.	Left Hem.
A (fT)	22	15	18	12
	20	20	20	12
x (mm)	1 05	-1 83	-5 09	-9 056
	-2 49	-8 3	5 691	0 66
y (mm)	-56 56	34 18	-48 36	31 73
	-52 44	38 2	-41 81	29 99
z (mm)	51 12	56 93	47 1	46 13
	61 79	63 37	56 12	59 32
d (mm)	42	50	26	23
	26	25	35	30
q (nAm)	2 85	4 06	3 7	4 9
	2 07	4 66	3 99	5 68
D (mm)	7 25	31 36	20 38	33 71
	11 63	28 77	24 6	36 02
g (%)	96 8		98 1	
	97		96 2	

	KA (left ear stim.)		KI (left ear stim.)	
	Right Hem.	Left Hem.	Right Hem.	Left Hem.
A (fT)	20	38	20	10
x (mm)	-2 356	-2 77	4 7	-2 11
y (mm)	-53 96	47 58	-58 82	49 76
z (mm)	57 01	58 13	49 79	52 96
d (mm)	30	20	43	28
q (nAm)	3 45	4 73	1 67	2 15
D (mm)	10 28	19 6	12	26 22
g (%)	93 4		96 2	

	KI	
	Right Hem.	Left Hem.
A (fT)	20	18
x (mm)	7 1	-3 86
y (mm)	-53 53	47 21
z (mm)	50 65	53 44
d (mm)	37	30
q (nAm)	1 87	2 83
D (mm)	16 89	28 2
g (%)	95 4	

	NH	
	Right Hem.	Left Hem.
A (fT)	8	12
x (mm)	1 11	-1 58
y (mm)	-53 2	52 83
z (mm)	45 78	45
d (mm)	23	20
q (nAm)	1 18	1 83
D (mm)	20 52	25 34
g (%)	95 9	

	TA	
	Right Hem.	Left Hem.
A (fT)	12	8
x (mm)	12 47	6 22
y (mm)	-50 52	46 63
z (mm)	62 03	59 66
d (mm)	40	29
q (nAm)	1 85	1 17
D (mm)	11 46	23 65
g (%)	96 2	

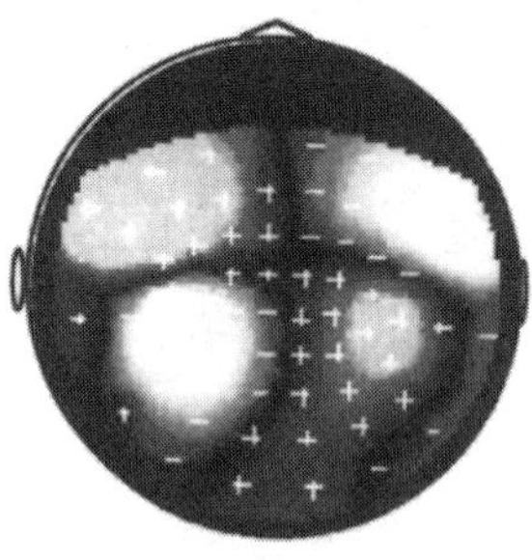

Fig. 3. Typical neuromagnetic isocontour of the 40 Hz response.

Table I

Maximum amplitude of the 40 Hz response, localization, strength, depth and goodness of fit of the equivalent current dipoles (5 subjects).

Discussion

The present research provides evidence for the anatomical origin of the 40 Hz steady-state response to ipsilateral stimuli and can be discussed on the basis of the former research. While subcortical generators were proposed [2], [3], evidence for at least one cortical generator (the contralateral) were presented by [4], rendering improbable the hypothesis of a single midbrain generator. The cases studied by this research extended the findings that were reported earlier by Makela et al., [4]. MRI confirmed that the ECDs of 40 Hz steady-state response to both contralateral and ipsilateral stimuli were located in the primary auditory cortex at the supratemporal plane. Sources of the right hemisphere were found to have depths remarkably shallower and smaller dipolar moment than those of the left hemisphere. To attribute this fact to a predominance of the contralateral subcortical activity as suggested by Hashimoto et al., [6], and Moller et al., [7], seems questionable since the difference in both depth and dipolar moment remained the same also with left ear stimulation.

On the basis of the results presented here, it is likely to assume that the 40 Hz magnetic steady-state response recorded at the scalp is largely due to activation of the supratemporal plane. Thus an exclusive midbrain origin seems at least doubtful. However as proposed earlier by Galambos [2], waves might originate in the reticular formation and then generate a similar response at cortical level by conduction of impulses. If this was true then would be likely to expect that subcortical structures might contribute to the generation of the neuromagnetic response recorded at the scalp.

References:

[1] Galambos, R., Makeig, S., and Talmachoff, P.T. A 40-Hz auditory potential recorded from the human scalp. Proc. Natl. Acad. Sci. USA., 1981, 78: 2643-2647.
[2] Galambos, R. Tactile and auditory stimuli repeated at high rates (30-50 per sec) produce similar event related potentials. Ann. N.Y. Acad. Sci., 1982, 338: 722-728.
[3] Spydell, J.D., Pattee, G., and Goldie, W.D. The 40 Hz auditory event-related potential: normal values and effects of lesions. Electroenceph. clin. Neurophysiol., 1985, 62: 193-202.
[4] Makela, J.P., and Hari, R. Evidence for cortical origin of the 40 Hz auditory evoked response in man. Electroenceph. clin. Neurophysiol., 1987, 66: 539-546.
[5] Hari, R., Hamalainen, M., and Joutsiniemi, S.L. Neuromagnetic steady-state responses to auditory stimuli. J. Acoust. Soc. Am., 1989, 86: 1033-1039.
[6] Hashimoto, I., Mashiko, T., Yoshikawa, K., Mizuta, T., Imada, T., and Hayashi, M. Neuromagnetic measurements of the human primary auditory response. Electroenceph. clin. Neurophysiol., 1995, 96: 348-356.
[7] Moller, A.R., and Jannetta, P.J. Evoked potentials from the inferior culliculus in man. Electroenceph. clin. Neurophysiol., 1982, 53: 612-620.

Oscillatory Neuromagnetic Activity Induced by Verbal and Non-Verbal Stimuli Presented in Visual and Auditory Modalities

Eulitz, C.[1,3], Maess, B.[2], Pantev, C.[3], Friederici, A.[2], Feige, B.[1] and Elbert, T.[1]

[1]University of Konstanz, Germany; [2]Max-Plank-Institute for Cognitive Neuroscience, Leipzig, Germany; [3]University of Münster, Germany

Introduction

The evaluation of brain rhythms has been used as an alternative method for quantifying functional changes in brain activity [1, 2, 3]. Recently it was applied successfully to the investigation of processes during speech perception and production [4, 5].

The neurophysiological plausibility regarding the evaluation of brain rhythms in different frequency bands is given by a number of experimental findings reviewed by Lopes da Silva [6]. As to the question of whether oscillations in neuronal networks, as reflected in brain waves, are merely epiphenomena, or reflect general functions, it has been suggested that higher frequency oscillations (above 20 Hz) may mediate the formation of assemblies of neurons that represent a given stimulus pattern. The activity of such an assembly of neurons is characterized by the coherent firing of large groups of neurons. Experimental evidence for coherent neuronal activity induced by changes in essential features of sensory stimuli comes from local slow wave field-, single- and multi-unit recordings from the olfactory system and the visual system of animals (for review see [7]). If these coherently activated neurons represent some part of a cortical cell assembly, then cell assemblies can be assumed to be the units by which elementary cortical functions are realized. Thus, one way of exploring these distinct cognitive processes in the human brain could be to measure coherent oscillations with non-invasive techniques such as EEG or MEG.

For the detectability of cell assembly activity using MEG recordings, the number of coherently activated neurons and their spatial distribution is critical. Previously, only simulation studies were available to estimate the number of neurons belonging to one cell assembly. Aertsen and coworkers [8] recently reported sizes of cell assemblies ranging from a few thousand to several ten thousand neurons, with a fraction of 1/10 to 1/5 of the assembly activated at any given moment (within a 2-4 ms time-window). This means in the ideal case, that about twenty thousand neurons of one cell assembly could be activated at a time. As pointed out by Wikswo [9], the combined activity of ten to one-hundred thousand neurons is necessary to generate a magnetic field detectable outside the skull. From these considerations it is clear that only large, not too widely distributed cell assemblies can be detected noninvasively. Given that the estimated number of neurons belonging to one cell assembly is correct, MEG recordings of cell assembly activation may operate near the boundary of detectability.

The first goal of this study was to quantify stimulus induced changes in oscillatory brain activity and to analyze whether or not these changes were different for the perception of language compared to non-verbal stimuli. Secondly, the study examined the dependence of oscillatory brain activity on the modality of stimulus presentation, and analyzed topographical aspects, in particular, hemispheric lateralization. A third question was whether or not the changes in oscillatory brain activity could be explained by an activity of word-specific cell assemblies, as suggested by the "neurological theory of language" [10]. This model states that the activation of word specific cortical cell assemblies can be measured as an enhancement of the normalized spectral power in frequency bands above 20 Hz across activated brain regions. If this is the case, there should be an enhancement of spectral power in the gamma band in response to the auditory and visual presentation of word stimuli across brain regions relevant for language processing, but one should detect topographically different alterations in gamma band activity when non-verbal stimuli are presented.

Methods

MEG (37-channel system, Magnes®, BTi) was recorded over the left and right hemispheres in twelve right-handed subjects. During the session, words two syllables in length and physically similar non-verbal stimuli were presented with equal probability in a randomized order in either the visual or auditory modality. Approximately 15% of these stimuli served as targets and each subject responded to them by deflecting a beam of light. The experiment consisted of 4 recording blocks (auditory and visual stimulation respectively, for the left and the right hemisphere). The order of blocks was balanced across subjects with the constraint that half of the subjects were measured over the left hemisphere first and that half of these subjects started with the auditory modality. For each subject, a total of 252 stimuli per experimental condition were delivered.

Power spectral estimates were obtained for subsequent 107 ms Welch-windowed segments (overlapping by one half) during a pre-stimulus baseline of 600 ms and 1200 ms post-stimulus onset. The geometric mean was calculated across trials. The mean time-dependent spectra were transformed with respect to baseline-related changes in two different ways. First, normalized mean power spectra were calculated by baseline dividing. A second parameter was the portion of single trial spectral power estimates above or below the mean single trial baseline spectral power (relative gain). Effects were considered only if both the normalized spectral power and the relative gain indicated statistically significant differences between experimental conditions.

Results

Fig. 1 illustrates changes of normalized spectral power values in time and frequency. These spectral power values represent the verbal condition in the auditory modality. Spectral power was grand averaged across all channels over the left hemisphere and then across all subjects. There are two major time/frequency regions where the spectral power is enhanced compared to the baseline and one time/frequency region showing a reduction of spectral power in most of the experimental conditions. Regions of enhanced spectral power occur around 129 ms in the 9.3–18.6 Hz bands, and around 883 ms in the 13.9–32.5 Hz bands. A reduction of normalized spectral power is seen around 345 ms in the 13.9–41.8 Hz range. When comparing these changes between experimental conditions, no modality independent and hemisphere specific reduction or enhancement of spectral power was found.

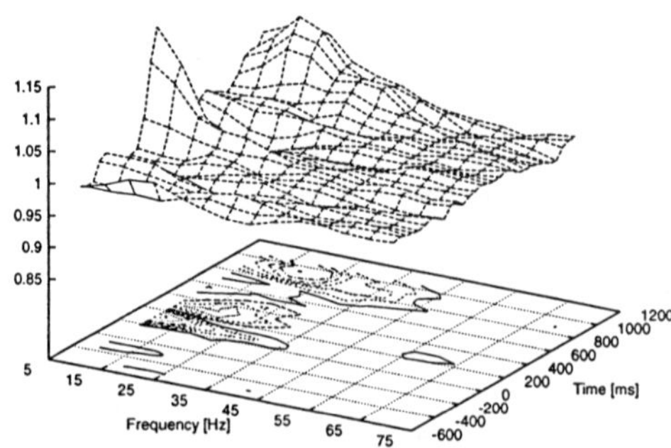

Fig. 1: Plot shows changes of normalized spectral power over the left hemisphere for the verbal condition in the auditory modality grand averaged across all subjects. The x-axis represents the frequency, the y-axis the latency, and the z-axis the normalized spectral power. The surface plot indicates the changes in normalized spectral power in the time-frequency plane. The same information is illustrated in the bottom half using an isocontour plot with a contour step of 1.5%. A normalized spectral power of 1 corresponds to the mean spectral power across nine pre-stimulus time windows (separately calculated for each frequency band). A normalized spectral power of 1.15 indicates an enhancement by 15% above baseline level.

In the higher frequency range, Fig. 1 shows a slight enhancement of the mean normalized spectral power in the time window centered at 237 ms in the 60.3 Hz and 65.0 Hz bands. A three-way univariate analyses of variance (ANOVA) for a 2 MODALITIES (auditory vs. visual) x 2 STIMULUS CLASSES (verbal vs. non-verbal) x 2 HEMISPHERES (left vs. right) design (repeated measures) delivered a statistically significant STIMULUS CLASS by HEMISPHERE interaction ($F(1/9)=7.42$; $p<0.03$; only results for the 60.3 Hz band will be reported in detail). There was no influence of MODALITY on this effect. For a more detailed topographical analysis of this effect, subaverages of the normalized spectral power from channels over the frontotemporal, the centroparietal and the temporooccipital brain regions were calculated. Statistical analyses using the BRAIN REGIONS (3 levels) as an additional repeated measures factor revealed on top of the STIMULUS CLASS by HEMISPHERE interaction ($F(1/9)=9.51$; $p<0.02$) a significant main effect of the BRAIN REGIONS ($F(2/14)=4.44$; $p<0.04$). The BRAIN REGIONS did not, however, significantly interact with other repeated measures factors. Post-hoc Scheffé comparisons suggested that the main effect was mostly caused by differences between centroparietal and frontotemporal brain regions ($p<0.04$). Mean values of the normalized spectral power showed a well expressed enhancement over the right hemispheric centroparietal and temporooccipital brain regions for the non-verbal condition, whereas for the verbal condition the enhancement of spectral power was largest over the corresponding left hemispheric brain regions for both the auditory and the visual modality. An extended view of the normalized spectral power changes for the 60.3 Hz frequency band as a function of time is presented in Fig. 2.

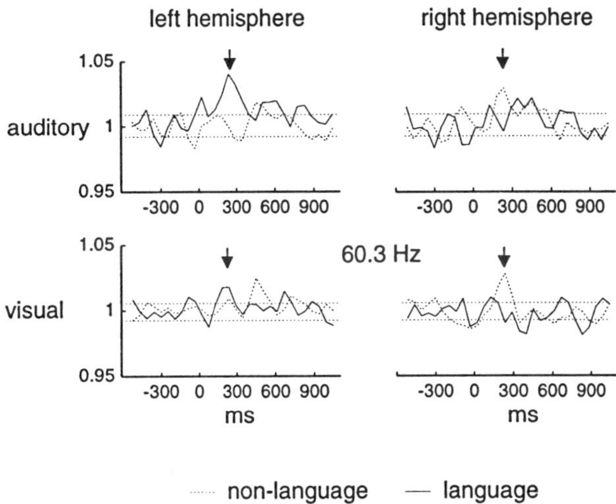

Fig. 2: The waveforms show the changes of normalized spectral power as a function of time for the frequency band around 60.3 Hz. They are averaged across channels over centroparietal brain regions for the auditory modality and across temporooccipital brain regions for the visual modality (and then across all subjects). The dotted line represents the non-verbal condition; the solid line indicates the verbal condition.

Discussion

The results of this study show that an enhancement of the normalized spectral power compared to a baseline level in the frequency band around 60 Hz and the latency range around 300 ms might be interpreted as a correlate of specific cognitive processing. In an experimental situation where simple lexical processing took place, the normalized spectral power was predominantly enhanced over temporooccipital and centroparietal brain regions of the language-dominant left hemisphere. In contrast, during the perception of visually and acoustically presented non-verbal patterns, the enhancement of the normalized spectral power was largest over centroparietal areas of the right hemisphere. This result might be taken as evidence for the activation of those cortical cell assemblies responsible for language processing located for the most part in the left hemisphere, thus confirming Braitenberg and Pulvermüller's [5, 10] cell assembly theory of language processing, i.e. word presentation should lead to a cell assembly ignition, because an adequate stimulus has been perceived, while presentation of non-verbal stimuli should fail to ignite specific assemblies in the same brain regions. Gamma-band responses (above 25 Hz) to words should be stronger than responses to non-verbal stimuli over brain regions processing the words. This hold true for the present results. As all subjects were right-handed one could expect that speech would be predominantly processed in this region [11]. The results for the non-verbal stimuli indicated a predominant activation of brain regions in the right hemisphere, as assumed for the processing of visual-spatial information [12] or complex acoustic pattern and prosodic information. This might be taken as evidence for the activation of other cortical cell assemblies situated for the most part in the right hemisphere thus accounting for the processing of non-verbal patterns. However, the percentage of change in the normalized spectral power was quantitatively small. Given that a). cell assemblies exist, b). specific cell assemblies were activated during the experimental procedure and c). the applied methods of data analyses were adequate, one can conclude that MEG recordings presumably of cell assembly activation were slightly above the threshold of detectability.

Changes of the normalized spectral power in other time/frequency bins indicated differences in the processing of stimulus classes not directly related to linguistic processing. The reduction in normalized spectral power (15-45 Hz) mirrored task specificity rather than stimulus specificity alone, thus not confirming previously published results [5]. In contrast, the left hemispheric enhancement of spectral power present around 240 ms in the 60-65 Hz band seems to reflect oscillatory patterns specific to the processing of words.

References

[1] Kaufman, L., Curtis, S., Wang, J.-Z., and Williamson, S. J. Changes in cortical activity when subjects scan memory for tones, Electroenceph clin Neurophysiol, 1991, 82: 266-284.

[2] Pfurtscheller, G. Event-related synchronization (ERS): an electrophysiological correlate of cortical areas at rest, Electroenceph. clin. Neurophysiol., 1992, 83: 62-69.

[3] Salmelin, R., and Hari, R. Characterization of spontaneous MEG rhythms in healthy adults, Electroenceph. clin. Neurophysiol., 1994, 91: 237-248.

[4] Klimesch, W., Pfurtscheller, G., Mohl, W., and Schimke, H. Event-related desynchronisation, ERD-mapping and hemispheric differences for words and numbers, Int J Psychophysiol., 1990, 8: 297-308.

[5] Pulvermüller, F., et al. Brain rhythms, cell assemblies and cognition: evidence from the processing of words and pseudowords, PSYCOLOQUY, 1994, 5(48): brain-rythms.1.Pulvermueller.

[6] Lopes da Silva, F.H. Neural mechanisms underlying brain waves: from neural membranes to networks, Electroenceph. clin. Neurophysiol., 1991, 79: 81-93.

[7] Singer, W. Synchronization of cortical activity and its putative role in information processing and learning, Annu. Rev. Physiol., 1993, 55, 349-374.

[8] Aertsen, A., Erb, M., Palm, G., and Schütz, A. Coherent assembly dynamics in the cortex: multi-neuron recordings, network simulations and anatomical considerations, In: Pantev, C., Elbert, T., and Lütkenhöner, B. Oscillatory Event Related Brain Dynamics, New York, London, Plenum Press, 1994, pp. 59-83.

[9] Wikswo, J.P. Biomagnetic sources and their models, In: Williamson, S.J., Hoke, M., Stroink, G., and Kotani, M. Advances in Biomagnetism, New York, Plenum Press, 1989, pp. 1-18.

[10] Braitenberg, V., and Pulvermüller, F. Entwurf einer neurologischen Theorie der Sprache, Naturwissenschaften, 1992, 79: 103-117.

[11] Rasmussen, T., and Milner, B. The role of early left-brain injury in determining lateralisation of cerebral speech functions, In: Dimond, S., and Blizzard, D., Evolution and lateralization of the brain, New York, New York Academy of Sciences, 1977.

[12] Rösler, F., Heil, M., and Henninghausen, E. Distinct cortical activation patterns during long-term memory retrieval of verbal, spatial and colour information, Journal of Cognitive Neuroscience, 1995, 7: 51-65.

Acknowledgement

Research was supported by grants from the Deutsche Forschungsgemeinschaft

Singular Spectrum Analysis applied to brain signals

Gaeta, R.[1], Benzi, R.[1], Peresson, M.[2] and Narici, L.[1]

[1]Dipartimento di Fisica - Università di Roma "Tor Vergata", Roma, Italy; [2] Divisione di Neurologia, Ospedale Fatebenefratelli - S. Pietro, Roma, Italy

Introduction

A large number of concurrently active processes are usually responsible for the spontaneous or evoked Electro/Magneto Encephalogram (EEG & MEG). Rhythmical activity, often referred to as "oscillating" activity, is most often measurable in many areas of the cortex. Interest in this kind of activities is increasing fast, due to the recent evidences of their correlation with functional, cognitive and perceptual activities. It has been shown that these are linked to rhythms within narrow frequency bands and that several times activities with different functional correlates share the same frequency window and often approximately the same source location.

The investigation of these "oscillations", their behavior during or following sensory stimulation or performance of cognitive tasks, is, therefore, a most interesting study in order to discriminate among similar activities with different correlates, as well as to approach the undestanding of processes such as learning and habituation. However most of the analysis tools to measure the dynamics of the many rhythmical activities sharing the same frequency band present important disadvantages. Time studies of narrow band filtered signals require prior knowledge of the band and risks to distort the real signal cutting important parts of it. Tools such as running FFT techniques [1] rely on same sort of time smoothing and therefore partially loose the time information, others (for example Single Sweep Analysis [2]) rely on pre-determined models of the oscillation and may force non-existent features in the dynamics.

In this paper we propose a novel technique (Singular Spectrum Analysis, SSA[3,4]) that is able to discriminate oscillatory activities in noisy signals, and to follow their dynamics. As a first test we apply this technique to neuromagnetic measurements of the activity (spontaneous as well as evoked by bursted rhythmical visual stimulation) over the occipital cortex of a normal subject.

The SSA technique has been recently developed and applied to studies of oscillatory components in the oceanographic dynamics or to the investigation of slow oscillation in the atmospheric dynamics. In these studies this novel technique proved to be quite useful. However in such cases the Signal to Noise Ratio (SNR) was favorable. The poor SNR proper of brain signals required a revision of the SSA procedure to identify oscillations in order to achieve meaningful results. SSA technique combined with our novel oscillation-retrieving procedure permitted to succesfully identify several oscillations in the extended α band (6.0 Hz - 14.0 Hz) on a relaxed subject, over the occipital cortex , and to compare them to the oscillation measured following repetitive rhythmic visual stimulation.

In the following we present a description of the SSA technique and of our oscillation-retrieving procedure. The reader interested in more technical details about SSA is addressed to the original papers [3,4]. The results from the first test of this technique, performed on neuromagnetic signals, are discussed and compared to those obtained, on the same set of data, with Fourier-based analyses [5,6]. We mention also first results of the dynamical studies relative to the found rhythmical activities.

Methods

The SSA is an application of the Karhunen-Loève expansion theorem. It is based on a Principal Component Analysis (PCA), performed in the vector space of delay coordinates for a single time series. It expands a process $X(t_i)$ as a sum of M components where M is an arbitrarily chosen 'embedding dimension', in contraddistinction with the fixed dimension of the data vectors in PCA. Vautard and Ghil (VG) [3,4] have shown that whenever in the $X(t_i)$ process a periodic oscillation is present, then a couple of "nearly equal" eigenvalues exists in the spectrum of the autocovariance matrix and the associated eigenvectors are orthogonal. In more detail, let be $\mathbf{X} = (X_1, X_2, \dots ,X_n)$ a stationary process with zero average. We construct the M x (N-M+1) matrix $\mathbf{Y}$ which has as i^{th} row the vector $(X_i, X_{i+1}, \dots ,X_{i+N-M})$ [$1\leq i \leq M$]. Let $\mathbf{T}$ be the autocovariance matrix of the process $\mathbf{X}$ (estimated by $\mathbf{Y \cdot Y^T}/N$). We define $\mathbf{S}$ as the MxM matrix which has the autocovariance eigenvectors (Φ) as rows. We also define

$\Theta = S \cdot Y$ the M x (N-M+1) matrix featuring as rows the M 'principal components'. It has been shown that the process X can be succesfully reconstructed as a sum of M Reconstructed Components (RCs) and that the sum of the RCs relative to a couple of "nearly equal" eigenvalues λ of T describe the time behavior of the oscillation with no phase shift with respect to the original oscillation [3,4].

To maintain the maximum possible resolution when scanning for existent oscillations the highest possible M value should be used. External information (neurophysiological, models) should then be used to re-assemble those oscillations relative to the same physiological source.

Using the oscillation-retrieving procedure suggested by VG only oscillations that could account for a consistent part of the signal (SNR >> 0.5) were succesfully retrieved. Such procedure was infact tested with signals with quite low noise levels. We performed further tests (simulated and with the real neuromagnetic signals): we can show that this procedure does not work for brain signals, where the SNR may be quite lower than 0.5 for long time intervals [7]. The VG procedure was based on the comparison between couples of eigenvectors featuring similar eigenvalues. The spectra of these eigenvalues often featured more than one peak. The choice of the peak to be considered in the frequency comparison strongly depended on the used algorithm. This was, of course, an unwanted feature. To solve such problem, we propose to perform the comparison between the frequencies using the two RCs instead of the eigenvectors themselves. In this way it is possible to show that the results does not depend by the algorithm used and is quite robust, leading to a percentage of retrieving success several times better (more than 10 times) than with the VG procedure. Briefly: once all the Φ are calculated, as well as the RCs relative to similar λ and their respective Fourier spectra, we define "oscillation" of frequency

$$f = \frac{f_a + f_b}{2} \pm \frac{n}{2N}$$

the sum of the two "a" and "b" RCs ($\lambda_a = \lambda_b$), if the condition

$$|f_a - f_b| \leq \frac{n}{N}$$

is satisfied (n=acquisition rate; N=total number of points).

In most cases, when the frequency window where oscillations are looked for is centered on values $f_0 << 1/n$ it might be useful to reduce the number of points per second averaging K consecutive points. In this case the useful frequency window is $\approx [0, n/K]$ Hz

The neuromagnetic signals used to test the SSA technique and our novel retrieving procedure are measured with the 28 channel neuromagnetometer in use at the IESS-CNR in Rome [8] over the occipital cortex of a normal subject. We measured spontaneous activity and activity following intermittent bursted visual stimulation at rates r finely scanned ($\Delta r = 0.5$ Hz) within the extended a band (6 Hz - 14 Hz) [5,6]. Visual stimuli were delivered via two red leds positioned within a blackened mask, at an eccentricity of ≈ 5 degrees, in the left hemifield (fixation point was constantly indicated by two centered green leds). The magnetic field was measured over the right occipital region. Each stimulating burst contained 10 stimuli, the non stimulated periods between bursts was 1.3 s. Consequently the length of the entire epoch variated between 2.8 s (for r=6.0Hz) and 1.9 ms (for r=14.0 Hz). A complete SSA analysis was performed on the signal measured by the center channel of the neuromagnetometer. To check whether or not the found oscillation were localized on such channel we performed the same analysis, for a single stimulation rate (r=12.5 Hz), on the first Principal Component - PC - (relative to a PCA over all the channels), which explained 92% of the total variance.

Results

In Fig. 1 we show the retrieved oscillation from the center channel during the entire spontaneous and stimulated session. In the abscissa we present the stimulation rates and in the ordinate the frequency of the magnetic response. The dots correspond to the found oscillations. The diameters of the dots are linearly proportional to the rms amplitude of the reconstructed oscillations. The continuous horizontal line represent the largest oscillation found in the spontaneous measurement. The frequencies of the oscillations relative to the RCs of the first PC (r=12.5 Hz) are reported in the plot with asterisks.

The frequency values found for the existent oscillations in the signal appear to cluster around three regions indicated in the figure with dashed lines: i) on the f=r line, ii) on the f=2r line and iii) on the f≈9.5 Hz line. There is an appreciable slowing between the spontaneous main activity (≈ 9.8 Hz) and the oscillations apparently unaffected by the stimulation and clustered about 9.5 Hz [(iii) region]. This is more so farther from those r that elicit the strongest actvities on the (i) and (ii) regions (≈ 6 Hz, ≈ 10 Hz, ≈ 12 Hz).

Fig. 2 illustrates the behaviour of the sum of the two RCs with a frequency ≈ 12 Hz, as well as the sum of the RCs at ≈ 10 Hz (both relative to r=12.0 Hz). The dynamics of these two rhythms are compared with the the stimulation bursts, shown at the bottom. Zooms of the plots for a shorter time span are shown on the right.

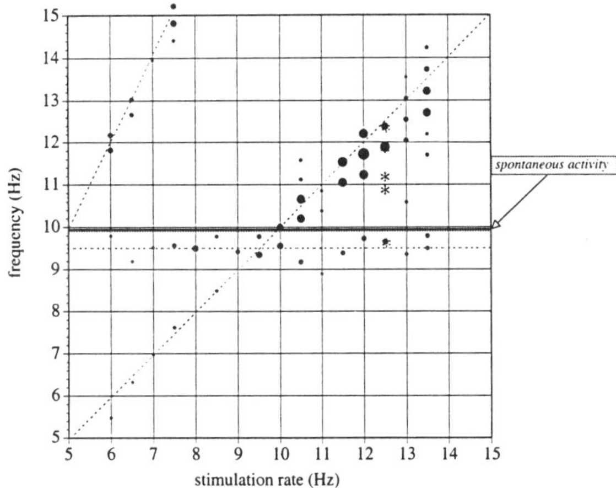

FIGURE 1   Plot of all the found oscillations in function of their frequency and rate of stimulation. Each RC is represented by a dot. The dot diameter is proportional to the rms amplitude of the RC. The horizontal solid line corresponds to the frequency of the RC found in the analysis of the spontaneous measurement. The asterisks correspond to the RCs relative to the first PC (obtained from all the channels) of the r=12.5 Hz measurement.

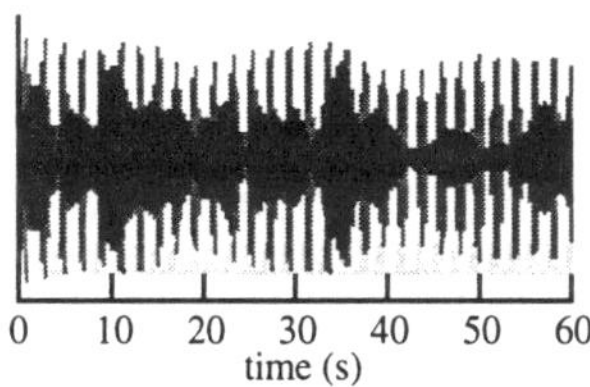
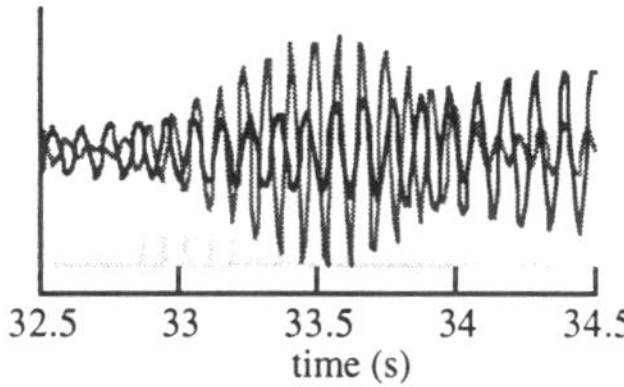

FIGURE 2   Time dynamics of the two oscillations (r=12 Hz) at $\approx$ 12 Hz (gray) and at $\approx$ 9.7 Hz (black) compared with the stimulating bursts (at the bottom, in gray). Note that only the former one is locked by the stimulation.

Discussion

The found oscillations, in the three 'regions', (i), (ii) and (iii) have been already evidenced with Fourier techniques [5,6]. The (i) region is the "driving region": cortical activity follows the driving of the stimulation. This driving effect is quite smaller at rates slower than the "spontaneous" a rhythm, that in this subject is at about 9.8 Hz. Also the (ii) region can be interpreted as "driving": it corresponds to the second harmonic of the stimuli, that, being sharp pulses, contains all the harmonics of the fundamental rate. On the basis of the inter channel cross correlations and of the topographical distribution of the activities [6], rhythms in the (iii) region have been interpreted as either unaffected by the stimulation rate, or locked by the stimulating bursts, as a component of the activity featuring its major peak at about 12 Hz.

It is interesting to note that most of the found oscillations appear in coupled frequencies with differences that are equal to the epoch rate (Figs. 1 and 2). This is due to the combined presence in the stimulation of the burst rate as well as of the epoch rate. This is not so in the case (iii) where the activities cluster around a constant $\approx$ 9.5 Hz frequency, indicating a lack of correlation with the stimulation rate. This is clearly evident in Fig. 2 where we present the results relative to r=12 Hz. At this rate the inter epoch interval is 2.05 s corresponding to 0.5 Hz. This corresponds to the difference between the center frequencies of the two added RCs (relative to the $\approx$12 Hz response).

A clear correlation with the stimulating bursts can be appreciated for the 12 Hz response while the $\approx$10 Hz one appears to be quite independent from the presentation of the stimuli.

These findings appear to confirm most of those obtained with Fourier analysis on the same data, while providing improvements in frequency resolution, and offering new insights. In this single channel test it is already possible to appreciate the presence of at least two distinct actvities within the same frequency band: one that follows the stimulating bursts, one undisturbed by them.

In this case the two discriminated activities feature different frequencies and different time dynamics. The first one ($\approx$10 Hz), not linked to the stimuli, is probably generated by some of the sources also responsible for the α rhythm, as it is similar in frequency (and time behaviour) to the spontaneous α rhythm. The second ($\approx$ 12 Hz) is strongly stimulus correlated. This response might generate from the cortical/thalamic pathways. Its dynamics (building up during the burst, and decaying after the end of the stimuli) may suggest that under visual stimulation the circuits between thalamo-cortical and reticular-thalamic cells first oscillates, under the stimulus driving, and then, at the end of the external driving, are dampened by the powerful inhibitory post synaptic potentials generated by neurons of the reticular nucleus in the thalamo-cortical pathways.

The SSA test on the first PC relative to r=12.5 Hz (asterisks in Fig. 1) shows clearly that the effect is not confined to the single channel studied but that the same activities are present over most of the channels. It also suggests a new activity at $\approx$ 11 Hz (also locked by the stimulation), not present in the studied channel, but evidently detected by most of the others.

The proposed technique allows to study the dynamics of the activities that will not show in averaged evoked-MEG measurements (such as our stimulus uncorrelated $\approx$10 Hz rhythm), and to investigate in more detail the effect of the length of the non stimulated period. However, to fully exploit it, it is mandatory to obtain separate information, possibly neurophysiological, to perform the correct reconstruction of the rhythms. As an example we had reasons to separate the $\approx$ 10 Hz activity from the $\approx$ 12 Hz activity in Fig. 2, as they responded differently to stimulation. But in general we may not have enough information to decide which of the RCs describe the same activity and therefore should be combined together.

In summary this quite powerful technique may be used to retrieve dynamical information on cortical oscillations in many situations and, when used in conjunction with other neurophysiological information may allow a complete reconstruction of rhythms dynamics.

References

[1] Jones, D.L. and Parks, T.W., A resolution comparison of several time-frequency representations, IEEE-Trans. Sign. Proc. 40(2).

[2] Liberati, D., Narici, L., Santoni, A., Cerutti, S., The dynamic behavior of the evoked Magneto-EncephaloGram detected through parametric identification, J. Biomed. Eng., 1992, 14:57-63

[3] Vautard, R., Ghil, M.,Singular spectrum analysis in nonlinear dynamics, with applications to paleoclimatic time series, Physica D, 1989, 35:395

[4] Vautard, R., Yiou, P., Ghil, M., Singular-spectrum analysis: a toolkit for short, noisy chaotic signals, Physica D, 1992, 58:95

[5] Narici, L, Peresson, M., A new procedure to discriminate and study different rhythmical cortical activities on the basis of their responsiveness to simple sensory stimulation, 1996, this volume.

[6] Narici, L, Peresson, M., Discrimination and study of rhythmical brain activities in the α band: a neuromagnetic frequency responsiveness test, Brain Res., 1995, 703:31-44.

[7] Gaeta R., Singular Spectrum Analysis applicata a segnali cerebrali, Thesis (in italian), 1995, Physics Department - University of Rome "Tor Vergata".

[8] Foglietti, V., Del Gratta, C., Pasquarelli, A., Pizzella, V., Torrioli, G., Romani, G.L., Gallagher, W.J., Ketchen, M.B., Kleinsasser, A.W., Sandstrom, R.L., 28-channel hybrid system for neuromagnetic measurements, IEEE Trans-MAG, 1991, 27:2959-2962.

Acknowledgements

We are grateful to V. Pizzella and F. Tecchio for the help in the measuring session.

Characterization of Photosensitive Discharges by Oscillatory Responses to Driving Stimuli

Karhu, J.[1] and Tesche, C.D.[2]

[1]*Department of Clinical Neurophysiology, Kuopio University Hospital, Kuopio, Finland;*
[2]*Low Temperature Laboratory, Helsinki University of Technology, Espoo, Finland*

Introduction

Photosensitivity has drawn an increasing amount of interest since the first report of epileptic fits induced by television [1], mainly because of the dramatically increased exposure of children and adolescents to videoscreens and videogames. Two stimulus-dependent mechanisms seem to be involved in photosensitive reactions. In some individuals electroencephalographic (EEG) discharges and/or seizures are induced by the flicker of the screen, which is scanned at the same frequency as that of the A.C. mains supply. In others, seizures may be induced by the raster pattern, which presents a grating pattern alternating at half the mains supply [2]. In addition to pure sensory mechanisms, there is growing evidence of seizure induction by cognitive effort during videogames.

The sign of photosensitivity in EEG is a widespread discharge following intermittent photic or flickering stimulus. This paroxysmal discharge is called a photoconvulsive response (PCR) and it consists of sharp waves and/or atypical (poly)spikes and waves. The PCR may be accompanied by jerking movements of the muscles, i.e., myoclonus, by absence, or by no visible clinical signs at all. The most common combination is a generalized spike and wave discharge with slow wave frequency at 3-3.5 Hz and an associated absence. Several determinants of experimental stimuli which trigger PCRs are well known (for a review, see [2]). The frequency of intermittent photic simulation (IPS) is optimally 15-20 Hz. Longer duration and higher stimulus intensities result in a higher incidence of photosensitive reactions. Adding patterns to intermittent visual stimulation markedly increases the probability of evoking abnormal responses as compared with diffuse flashes. Patterns of small squares formed by narrow lines (raster patterns) and closely spaced parallel lines seem to be most effective.

Although numerous studies have investigated the stimuli which trigger photosensitive seizures, the brain structures and neuronal networks which mediate the sensitizing input and elicit PCRs are poorly known. EEG methods have met with limited success because of the "generalized" (or distributed) nature of the PCR. Other functional imaging tools such as PET and fMRI do not have sufficient temporal resolution to evaluate the dynamics of rapidly spreading cerebral discharges. In addition, several pathophysiological and methodological problems may hamper the study of triggered epileptogenic brain activity. The epileptogenic activity may be triggered by a diffuse, widespread brain structure such as the reticular system, it may initiate in subcortical structures, or it may be multifocal. Finally, the experimentally triggered discharges may not represent the cerebral activity which causes clinical symptoms.

Magnetoencephalographic (MEG) arrays provide both sufficient temporal resolution and spatial information to enable the identification of components of very rapidly spreading brain activity. The millisecond-scale temporal resolution is particularly valuable for the characterization of changes of responses to the fast-frequency stimuli which are an essential component of photosensitivity. This study was directed towards the evaluation of distributed brain activity during PCRs utilizing novel signal analysis tools developed for the exploitation of topographic MEG information [3,4]. We first characterized neuronal populations in the frequency domain by their topographic field patterns and resonant behaviors during intermittent photic stimulation (IPS). The abnormal brain activity during PCRs was then spatially filtered with respect to these topographies to reveal waveforms and interactions identified with activation of cach topography.

Methods

A 122-channel whole-head MEG array (Neuromag-122™) was used to record ongoing spontaneous activity and magnetic responses evoked by a bright flash (Grass stroboscope; distance from the eyes 2 m). Results are reported here for data recorded from a 9-year old non-medicated subject. The photosensitivity of the patient had been previously verified in several paper-EEGs and in a 24-hour videotelemetry recording. Photic stimuli were delivered in 4-s trains at 1, 3, 6, 10, 20, and 35 Hz. The subject was seated in a magnetically shielded room (Euroshield Ltd.) during the recordings (passband 0.03-100 Hz, sampling rate 397 Hz). The accurate position of the subject's head inside the helmet-shaped magnetometer was detected by measuring the magnetic field produced by three coils attached to the scalp.

In time-domain analysis of electromagnetic signals recorded on the scalp, a model with one or more equivalent current dipoles (ECDs) is frequently used to approximate localized sources of synchronized post-synaptic activity in the brain [5, 6, 7]. In MEG, magnetic field patterns are calculated as a function of the amplitude, location and orientation of the ECDs. These parameters are varied until a reasonable fit (usually above 90%) is obtained between the calculated and measured field values (for a review see [8]). Information about underlying sources may also be obtained from data which has been transformed from the time domain to the frequency domain by taking the fast Fourier transform (FFT) [3, 9]. In particular, any inversion algorithm which is applicable to time domain data can also be formally applied to FFT transformed data. ,

Time-domain ECD models are often not efficient in the case of distributed or complex brain sources such as those eliciting PCRs and other "generalized" epileptic discharges. Spatially distributed neuronal networks have oscillatory properties with characteristic physiologic frequencies for example at 6 Hz, 10 Hz, and 20 Hz [10]. These networks may be "driven" by external stimuli around these frequencies [11]. Since photosensitivity clearly depends on stimulation frequency and most patients are sensitive to stimulation within the physiological frequency range, we characterized the neuronal sources activated during photic stimulation by scanning the spectral behavior of the recorded brain signals. Topographies for prominent responses at the frequency of stimulation and its (sub)harmonics were selected from the FFT of each of the 122 channels of evoked data. Spectral "peaks" generated during ongoing spontaneous activity were also used to define MEG topographies. The FFT of a single epoch of a data generated a complex number characterized by an amplitude A_k and complex phase $\emptyset_k$. Each topography was characterized by the specific choice of frequency and complex phase.

Evoked and spontaneous FFT field patterns which were clearly dipolar (see Fig. 1) were characterized by ECDs. These frequency-domain ECDs represent activity of neuronal populations with specific oscillatory content (see Fig. 2) [3]. A different strategy was used for distributed or complex field patterns which were not adequately modeled by a limited number of frequency-domain ECDs. The most general description of activity in the brain from MEG or EEG data is provided by the values recorded by sensors. Signal-space projection (SSP) utilizes the pattern of responses in the detector array to characterize components of the signal which may be generated by either distributed or localized brain sources [4]. Although specific distributions of current flow in the brain determine unique vectors s_i in a 122-dimensional "signal-space", it is impossible to extract a unique current distribution from a set of sensor readings. However, waveforms for the ongoing activity of a set of SSP components may be extracted from the data [4]. In this study, we used the 122 numbers which generated the ith frequency-domain topography to define unit vectors s_i in signal-space. Thus the FFT field patterns generated during photic stimulation and spontaneous brain activity were used to construct signal-space "filters" which were applied subsequently to data recorded during PCRs. The resultant SSP waveforms described the evolution of discharge in terms of the responses of distributed oscillatory networks in brain.

Results

Fig. 1 shows topographic FFT field patterns in the photosensitive patient for spontaneous oscillatory activity during relaxed wakefulness with eyes closed, for 3-Hz activity during 3-Hz photic "driving" stimulation, and for 10-Hz activity during 10-Hz "driving" stimulation. The field pattern which was elicited by the about 10.5-Hz spontaneous activity was dipolar with an ECD location consistent with those of occipital MEG alpha rhythm. The oscillatory activity during the 10-Hz IPS produces a complex topographic field pattern dominated by responses over occipital and parieto-occipital areas. This response was modeled adequately by two ECDs (goodness-of-fit over 90%). The oscillatory activity driven by the 3-Hz photic stimulation was most complex with some dipolarity over frontal brain areas and could not be modeled adequately by dipoles.

The field patterns for the frequency-domain 10-Hz ECDs are shown in Fig. 2. Although these generators have similar oscillatory content, the topograpbies are consistent with distinct ECD source locations and orientations in parieto-occipital and in occipital areas (called parieto-occipital and occipital SSP components). The parieto-occipital field pattern closely resembles that elicited by the spontaneous 9-Hz activity.

We used the FFT field patterns shown in Figs. 1 and 2 to define signal-space components for the decomposition of PCRs. Fig. 3 shows the temporal waveforms in parieto-occipital and occipital SSP components during 3-Hz, 6-Hz, 10-Hz, and 20-Hz photic stimulation *(left)*. The frontal SSP component was derived from the FFT topography produced by the 3-Hz IPS (see Fig. 1). Frontal, parieto-occipital and occipital SSP components were all utilized in the decomposition of PCRs *(right)*. The amplitude of responses for each SSP waveform is scaled to the values derived from the field patterns [4]. The occipital SSP component shows the least activation, whereas prominent activation of frontal and parieto-occipital SSP components shows synchronized activity during the discharge at around 4-5 Hz.

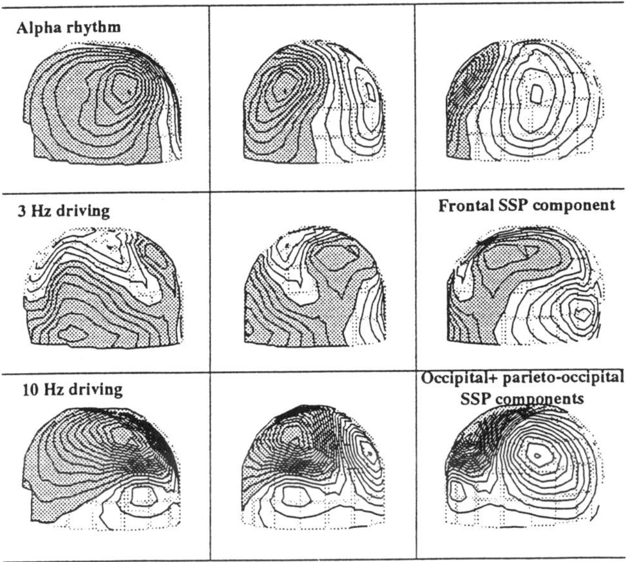

Figure 1 shows the FFT field patterns elicited by spontaneous activity and 3-Hz and 10-Hz photic stimulation.

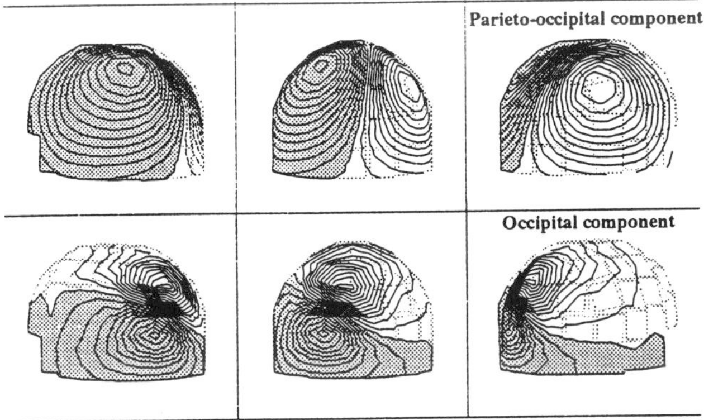

Figure 2 shows the simulated topographic field patterns for the two frequency-domain 10-Hz ECDs.

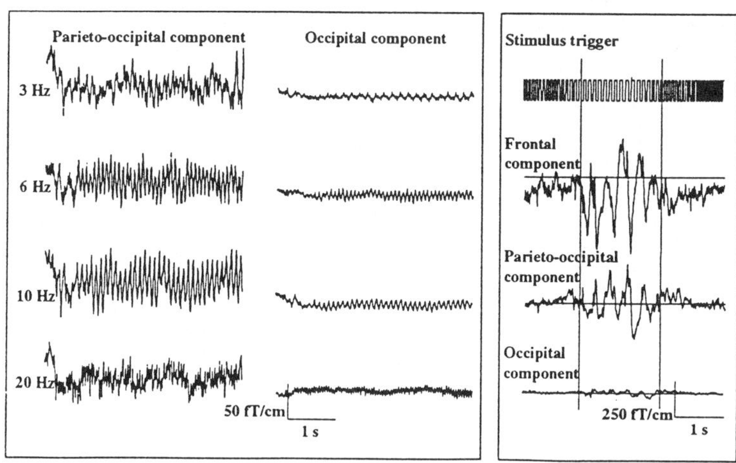

Figure 3 shows the spatially filtered temporal waveforms in SSP components during photic stimulation at various frequencies *(left)* and during a PCR (right).

Discussion

This case report from an ongoing study of photosensitivity demonstrates that complex oscillatory responses could be characterized by their topographies and resonant properties during driving utilizing FFT field patterns and frequency-domain ECDs. When a distributed photoconvulsive response was spatially filtered with respect to the SSP components defined in frequency-domain, activation in frontal and parieto-occipital components showed a time-locked, synchronized oscillatory behaviour during the discharge. The smaller amplitude of the response in the occipital SSP component suggests that either this component was not equally active during the PCR or that its activation was enhanced during driving stimuli relative to the activation during the PCR.

The oscillatory behaviour of spatially distributed neuronal networks in human brain is quite well defined, even if the overall function of these oscillations is not fully clear. The network oscillations are almost always present and can be manipulated by external stimuli. Characterization of these multiple networks in the frequency-domain and evaluation of their behavior during complex brain activation utilizing spatial filtering in signal-space seems to be a very promising and physiologically relevant technique for the study of "generalized" epileptic discharges and other distributed cerebral responses.

References

[1] Livingston, S. Am. J. Dis. Children, 1952, 83, 409.
[2] Hardmg, G. and Jeavons, P. Photosensitive epilepsy, 1994.
[3] Tesche, C. and Kajola, M. Electroenceph. Clin. Neurophysiol., 1993, 87:408-416.
[4] Ilmoniemi, R.J. and Williamson, S.J. Soc. Neurosci. Abstr., 1987, 13:46. Tesche, C.D., Uusitalo, M.A. Ilmoniemi, RJ., Huotilainen, M., Kajola, M. and Salonen, O. Electroenceph. Clm. Neurophysiol., 1995, 95: 189-200.
[5] Brenner, D., Lipton, J., Kaufman, L. and Williamson, S.J. Science, 1978, 199: 81-83.
[6] Scherg, M. and von Cramon, D. Electroenceph. Clin. Neurophysiol., 1986, 65: 344-360.
[7] Mosher, J.C., Lewis, P.S. and Leaby, R.M.. IEEE Trans. on Biomed. Engin., 1992, 39: 541-557.
[8] Hämäläinen, M., Hari, R, Ilmoniemi, R., Knuutila, J. and Lounasmaa, O.V. Rev. Mod. Phys., 1993, 65: 413-497.
[9] Lütkenhöner, B. In: 5th International Congress: International Society for Brain Electromagnetic Topography. Westfälische Wilhelms-Universität, Münster, Germany, 1994: pp. 54.

Equivalent Current Sources of the Auditory Evoked Gamma Band and Steady State Fields

Knuth, K.H.[1,2], Broadhurst, J.H.[2] and Schwartz, B.J.[3]

[1]Kresge Hearing Research Laboratory, New Orleans, USA; [2]University of Minnesota, Minneapolis, USA; [3]Scripps Clinic, La Jolla, USA

Introduction

Neural activity in the auditory cortex of the human brain can be evoked by presenting transient acoustic stimuli to the ear. Using the Magnes Biomagnetometer System at the Scripps Clinic in La Jolla CA, we measured the magnetic fields generated by this neural activity. The auditory steady-state response (SSR), first studied by Galambos et al.[1], is a sinusoidal response elicited by periodic trains of auditory click transients presented at rates ranging from 15-70 Hz. Galambos et al. proposed that the SSR was generated by a superposition of the individual transient middle-latency responses.

By presenting single and multiple auditory click transients at 25 ms intervals, we found that the response evoked by a long train of clicks (SSR) is generated by a nonlinear superposition of transient responses, and verified that several peaks in the transient response (middle-latency peaks also known as the auditory evoked gamma-band response, AEGBR), which occur at approximately 25 ms intervals, were responsible for the SSR. [2]

We used the sampled field data from the SQUID array to locate the positions of the source currents representing the activity. The response to a single click was found to be due to two sources, the mPa-mNb and the mNb'-mPb, named after their corresponding EEG peaks. One main source of the response to multiple click stimuli was found, the mPab-mNbc, which is probably the best fit current dipole to the field generated by the simultaneous activity of the mPa-mNb and the mNb'-mPb.

Methods

Three healthy right-handed subjects, two women and one man, with normal hearing and no history of otological or neurological disorders, aged between 25 and 40 years participated in the study. Magnetoencephalographic (MEG) measurements were made of the radial component of the evoked magnetic field using the Biomagnetic Technologies 37-channel biomagnetometer at the Scripps Research Institute. The subjects were placed in a magnetically shielded room to reduce background magnetic noise. To improve the signal to noise ratio, the responses from 256 to 1024 stimuli were averaged.

The sound stimuli were generated digitally by an Amiga 3000 computer and were delivered into the room via a plastic tube coupled to an earplug. The sound stimuli consisted of sets of single and multiple clicks, 0.5 ms square wave condensation clicks, numbering from 1-5 per set, with a separation of 24 ms between the individual clicks. This way, as the number of clicks in the stimulus are increased, the onset of the SSR could be studied. The interstimulus intervals between successive stimuli were randomly varied 0.8334 +- 0.1667 sec (the two-click stimulus had an ISI of 0.8584 +- 0.1667 s). The separate sets of stimuli (one click, two click, etc.) were not randomly mixed in their presentation, but each set was presented the designated number of times before proceeding to the next set.

The sounds were delivered to the right ear and the magnetic fields were measured over the contralateral hemisphere. Responses were digitized at a sampling rate of 1041.7 Hz and conditioned between 0.1 and 400 Hz using digital high and low pass filters. Further analysis was performed off-line. To perform source localizations of the 40 Hz activity, the magnetic field data was bandpass filtered about 40 Hz with a half-bandwidth of 12 Hz. Head shape data were taken to define the headframe coordinate system, and to locally fit a sphere to the measurement area. The system used the Levenberg-Marquardt nonlinear search algorithm[3] to solve the inverse problem. The results of the dipole fits were examined and only the dipoles with goodness of fit parameters greater than 0.95 were considered acceptable. Good responses were obtained in two of the three subjects, one man and one woman. Reproducibility of the dipole locations and magnitudes across experiments for a given subject allowed averages and standard deviations of the positions and moments to be calculated. Two dipoles were considered to be distinct as long as their two sigma volumes did not intersect. Due to the large variability in the dipole locations between subjects, the equivalent current dipoles were not averaged over subjects. In one subject the locations of these sources were superimposed on previously obtained MRI images.

Results

Acceptable dipole fits were obtained for two subjects using single and multiple click stimuli. The locations of these dipoles were extremely reproducible for both subjects across experiments allowing the averages and standard deviations of the dipole positions and magnitudes to be calculated. Fig. 1 shows the left anterior MEG recorded from Subject 1 in response to a single auditory click. The lower waveform represents the 40 Hz activity obtained by bandpass filtering the upper waveform. The vertical lines represent the times at which successful equivalent current dipole, ECD, fits were obtained. The dipoles were named according to the EEG peaks with which they coincide. Four distinct ECDs were found, mPa, mNb', mNb, and mPb. The ECD positions and magnitudes showed that these four sources represent two dipole pairs. Essentially two distinct true sources were detected, the mPa-mNb', and the mNb-mPb, for the single click stimulus.

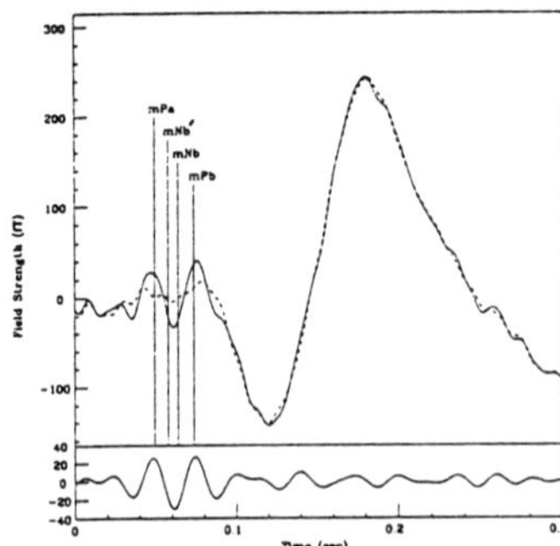
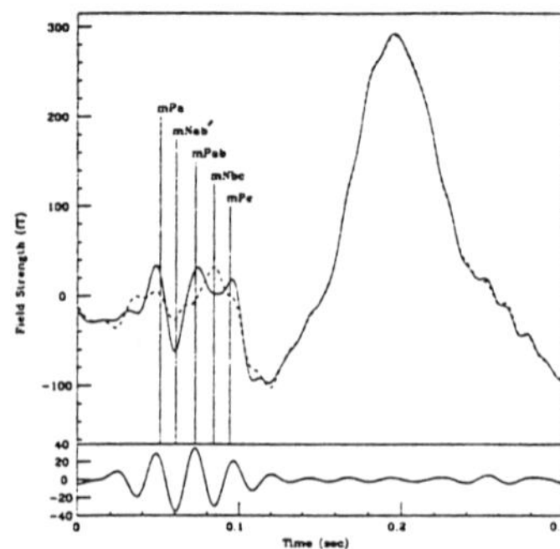

Fig. 1   The left anterior MEG in Subject 1 in response to a single auditory click (left) and two auditory clicks presented with an interval of 24 ms (right).

To demonstrate the effect of removing the DC component of the field, Fig. 2 shows the magnetic field contour plots at an interval of 12.4 ms, corresponding to the ECDs mPa (48.6 ms) and mNb (64.0 ms). The fields, after filtering, are of approximately the same form, but with opposite polarity. The arrows represent the approximate ECD location corresponding to the field. Most sources obtained were found to be a member of a source pair. This result lends credibility to the belief that the evoked 40 Hz activity is comprised of distinct sources.

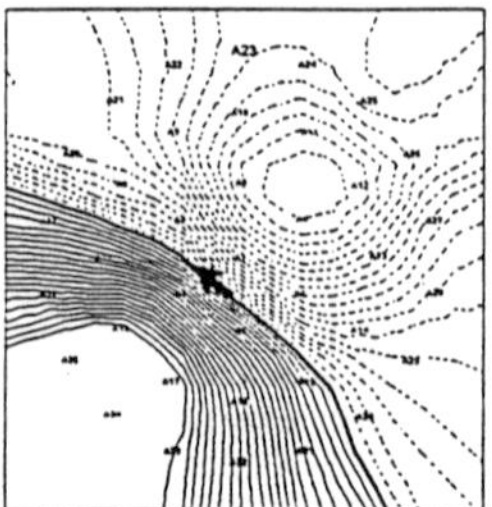
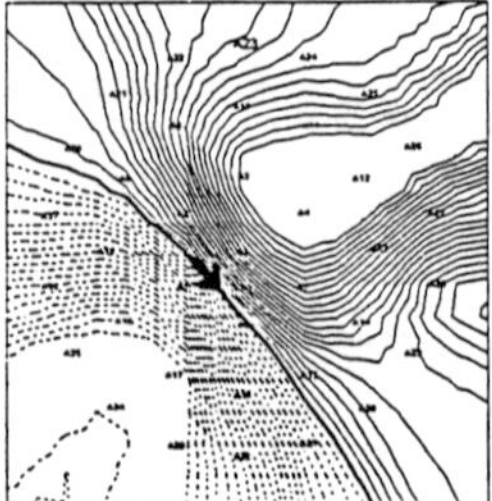

Fig. 2   A contour plot of the radial component of the magnetic field from Subject 1 found by interpolating the data sampled at the detector locations A1 through A37, at latencies 48.6 ms corresponding to mPa (left) and 64.0 ms corresponding to mNb (right). The approximate locations of the ECDs have been drawn in. Solid lines represent flux emerging from the head and dashed lines represent flux entering the head. The dark line represents zero field strength. The contour step is 1 fT.

Examining the responses of Subject 1 elicited by a two click stimulus, we found five distinct ECDs (Fig. 1 right). Similar ECDs were obtained for the other multiple (3-5) click responses. They are named according to the EEG peaks with which they coincide, but in this case the peaks from the two stimuli are overlapping. For example, mPab occurs at a time where the Pa and Pb peaks should be overlapping. The ECDs, mPab and mNbc, constitute a dipole pair. The source, mPab-mNbc, probably represents the best fit current dipole to the simultaneous activation of mPa-mNb and mNb'-mPb. We would have expected mPa to have its dipole pair mNb, but mNab' was found to be distinct from mNb at the two sigma level. This is interesting in that it suggests that the source mPa-mNb was modified by the simultaneous activity of mNb'-mPb during its activation. The final ECD, mPe, was found to be distinct from mPb and mPab. Similar sources were obtained for subject 2, but they were found to be more medial and anterior to those of subject 1.

The locations of the dipoles found above are displayed in Fig. 3 below. The left figure shows the ECDs due to a single click stimulus and the right figure shows the ECDs due to the two click stimulus. These figures show only the (x,y) coordinates of the ECDs in the headframe coordinate system, (i.e. the view is oriented so as to look down on the top of the head with the x-axis protruding from the nasion, and the y-axis protruding from the left ear). The boxes surrounding the ECDs represent two standard deviations in the positions of the ECDs.

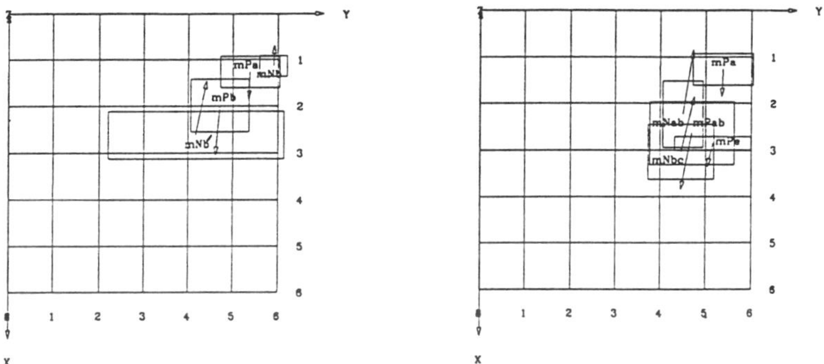

Fig. 3 The locations and the magnitudes of the ECDs obtained in Subject 1 elicited by a single click stimulus (left) and a two click stimulus (right). The view is looking down on the z-axis. The boxes surrounding the ECDs represent two standard deviations in their positions.

The ECD positions obtained for Subject 1 were superimposed on the subject's previously obtained MRI images, Fig. 4. These images show that the dipoles are within or fairly close to the Sylvian fissure.

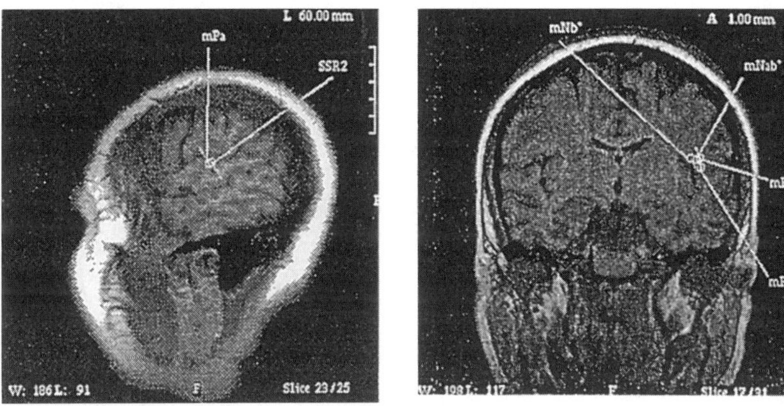

Fig. 4 The MRIs from Subject 1 superimposed with the locations of some of the obtained ECDs.

945

Discussion

By presenting stimuli consisting of small numbers of clicks (numbering from one to five), we were able to study both the AEGBR in response to single and multiple click stimuli. This allowed us to examine the relationship between the AEGBR and the onset of the SSR. By examining the amplitudes and durations of the gamma band activity we determined that the steady state response is generated by a nonlinear superposition of the AEGBR, also known as the 40 Hz early middle latency responses [2]. Using the magnetic field data we were able to locate the equivalent current dipoles for the sources responsible for the AEGBR, and the multiple click evoked GBR, which represents the onset of the SSR.

To locate the current sources responsible for the 40 Hz evoked activity, we bandpass filtered the data about 40 Hz with a 12 Hz half-bandwidth. This filtering removes the DC component of the field. This is significant in that a field that is simply turning on and off will appear as being generated by two alternating equivalent current dipoles at the same location, but with opposite dipole moments appearing at one half cycle intervals. Thus any current source will appear as a "dipole pair". The true current source, however, consists of the sum of these dipoles and the current source that generated the original field DC offset. This effect was observed with most of the sources we obtained which lends credibility to the finding that the 40 Hz evoked activity is the result of a few distinct sources.

We found that the single click evoked gamma band response was due to the activation of two sources, mPa-mNb and mNb'-mPb. The fact that the AEGBR consists of two independent sources which are not simultaneously active explains why the phase of the evoked gamma band field is observed to "sweep" across the cortex. Studies of the multiple click evoked gamma band field revealed a main source, the mPab-mNbc, which is probably the best fit current dipole to the two sources mPa-mNb and mNb'-mPb. In addition to this main source, the mPa was always observed; however, its "pair" which should have been mNab' was not in the same location. This implies that the true source mPa-mNb was affected by the second click during its activation. This may provide some insight into the nonlinear interactions occurring between mPa-mNb and mNb'-mPb during their simultaneous activation.

Similar results were obtained with Subject 2 for the multiple click stimuli (no acceptable fits were obtained for the single click stimulus in this subject). For the two click stimulus, five ECDs were obtained at similar latencies as Subject 1. In addition, these ECDs had the same relationships as the ECDs obtained for Subject 1. The relative locations of these ECDs also were comparable to those found for Subject 1; however, the absolute positions were found to be more anterior and medial than in Subject 1. Acceptable M100s were not obtained in this study. Other experiments performed examining the behavior of the SSR yielded too few acceptable dipoles for comparison, which is probably because the studies were concerned with amplitude and phase of the SSR and too few averages were taken to reduce noise.

These findings support our earlier results [2] that the multiple click GBR and the SSR are generated by a nonlinear superposition of 40 Hz middle latency responses (the AEGBR). The fact that the superposition was found to be nonlinear implies an interaction between the sources mPa-mNb and mNb'-mPb of the AEGBR, which was observed in the multiple click ECDs as the difference between mPa's pair in the single click (mNb) and multiple click (mNab) stimuli.

References

[1] Galambos, R., Makeig, S., and Talmachoff, P.J. A 40 Hz auditory potential recorded from the human scalp., Proc. Natl. Acad. Sci. USA, 1981, 78:2643-2647.
[2] Knuth, K.H., Broadhurst, J.H., Schwartz, B.J. Dynamical behavior of the generation of the auditory evoked 40 Hz steady state response. In preparation.
[3] Marquardt, D. An algorithm for least-squares estimation of nonlinear parameter. SIAM Journal of Applied Mathematics, 1963, 11:431-441.

Acknowledgements

We would like to thank the following people for their technical assistance: Steve Cobb, Joslyne Foley, Lacey Kurelowich, and Patti Quint. This work was supported in part by the University of Minnesota, Biomagnetic Technologies Inc., Scripps Clinic, Kresge Hearing Research Laboratory, and NIH NIDCD 5 T32 DC00007.

Local Modulation of MEG Rhythmic Activities Related to the Alternation of Binocular Rivalry and Binocular Fusion

Kobayashi, T., Kato, K. and Kuriki, S.

Research Inst. for Electronic Science, Hokkaido University, Sapporo, Japan

INTRODUCTION

Binocular rivalry (BR) is the process where dissimilar images presented to the left and right eyes at the same time are perceived alternately at any moment in time. This phenomenon has been studied primarily through psychophysical experiments [1, 2]. Recently, BR is considered to be a clue for understanding aspects of visual awareness [3]. Several physiological studies on cats and monkeys have been carried out in the past in an attempt to demonstrate a neural basis of BR [4, 5]. The next step in understanding BR is to elucidate the neural activities underlying this phenomenon in humans. However, experiments on human subjects require noninvasive techniques for observing the processes.

In previous papers [6, 7] we reported that altered neurophysiological processes with respect to the states of BR and binocular fusion (BF) may be reflected as variations in the attenuation of EEG power in the 8-13 Hz band in humans. In the present study, we measured and analyzed the spatio-temporal modulation of MEG power in the 8-13 Hz band related to the alternation from BF to BR and vice versa to study cortical activities involved in BR and/or BF.

METHOD

Five healthy subjects (males, 24-25 years old) with normal or corrected-to-normal visual acuity participated in the experiments. Two pairs of vertical and/or horizontal gratings used in the experiments are shown in Fig.1 with the time course of the stimuli's alternation and MEG data acquisition in the experiment. Each grating of the two pairs, 1° x 1° in visual angle, white, 10 cd/m^2, were projected on a screen (0.2 cd/m^2) in a shielded room, and presented to the left and right eyes separately through a mirror stereoscope. One of the two gratings was replaced with a grating which had an orientation orthogonal to the previous grating in intervals of 5 to 8 s. During the course of the experiment, subjects perceived BF and BR alternately by the replacement of the grating. We repeated this procedure until the selective average of the MEGs of 200 trials was completed. The subjects perceived a fused grating in BF when identical gratings were presented. On the other hand, when orthogonal gratings were presented, the subjects perceived one of the two gratings exclusively at any given time and then perception switched to the other grating in BR. This subjective switching occurred repeatedly during BR.

A SQUID magnetometer, with 19 second-order gradiometers covering a circular area of 146 mm in diameter on a spherical surface with a radius of 120 mm, was used for the measurements. The MEG data obtained were

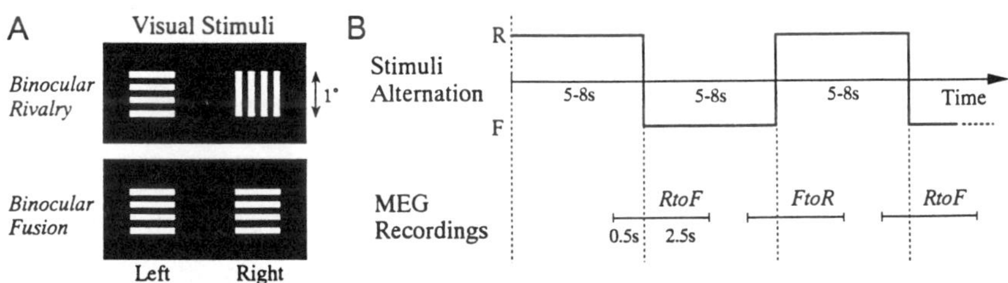

Fig. 1 A: Two pairs of gratings, 1° x 1° in a visual angle, white, 10 cd/m^2, on a black screen (0.2 cd/m^2), used as the visual stimuli of binocular rivalry (R) and binocular fusion (F). B: Time course of the stimuli alternation of R and F and MEG data acquisition.

digitally filtered in the 8-13 Hz band. To determine how the power in the 8-13 Hz band varies with time, the variance in 200 trials in the band was obtained with the time of alternation of the images as a reference trigger. Furthermore, we calculated the variance normalized by the averaged variance between -500 and 0 ms prior to the alternation from BF to BR. MEG data were obtained from two areas, where the center channel was located in the middle between P_z and P_3, and P_z and P_4 of the international 10-20 EEG electrode system. The measurement area covered the parietal, occipital and posterior-superior temporal regions in left and right hemispheres.

For two subjects (KK and TN), EEGs of 100 trials were measured and analyzed for the comparison with MEGs from 20 electrodes (F_{p1}, F_{p2}, F_7, F_3, F_z, F_4, F_8, T_3, C_3, C_z, C_4, T_4, T_5, P_3, T_6, P_z, P_4, O_1, O_z and O_2) placed on the scalp referenced to linked ears by the same procedure described above.

RESULTS

The sample plots of the normalized MEG variances in the 8-13 Hz band as a function of latency from the subject KK are shown in Fig. 2. A clear local modulation of the variances in the band was found after the alternation of visual stimuli. A similar modulation was observed in all subjects. The changes in normalized MEG variance in all subjects and their average at the channel located in the vicinity of P_3, where the largest variation was observed, are presented in Fig.3. The same modulation was observed at the corresponding locations in the right hemisphere. The variance was greatly attenuated until 700 ms after switching from BF to BR and then slightly recovered. On the other hand, no or only slight attenuation was observed in the variance until 700 ms and a subsequent rapid recovery to a stable level was observed in the alternation from BR to BF. The variance in BF, which averaged between -500 to 0 ms prior to the alternation of visual stimuli, was significantly larger (n=5, p<0.005; paired t-test) than that in BR.

The normalized variance in BR, which averaged between -500 to 0 ms, for both EEGs and MEGs in the 8-13 Hz band for two subjects, are shown in Fig.4 as topographies. The topographies were obtained by linear interpolation of the values at 20 electrode positions for EEGs and 19 sensor positions for MEGs. The top of each topography is the rostral direction. The normalized variance of EEGs was small in the occipital, parietal and posterior temporal regions of the two subjects. These spatial features are consistent with these of our previous results [6, 7]. On the other hand, large local variations were observed in the topographies of MEGs. Although there

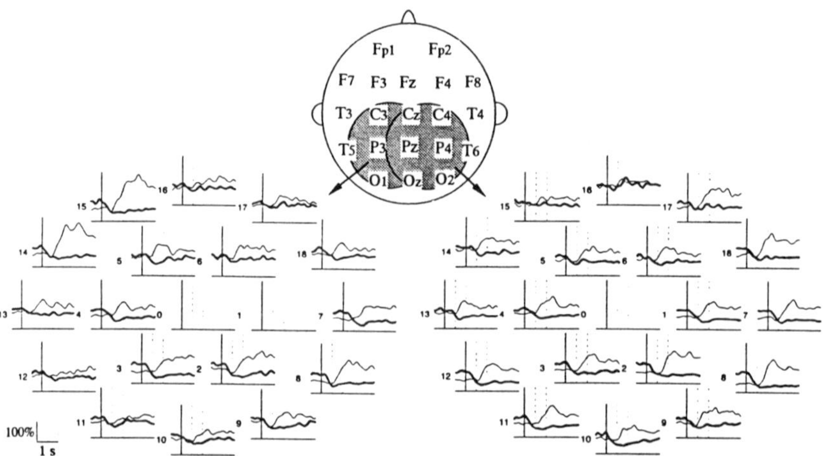

Fig.2 Changes in normalized MEG variance in the 8-13 Hz band (from Sub.KK) elicited by the alternation from BF to BR (thick line) and BR to BF (thin line) at two areas where the center channel #0 (unusable temporarily in these measurements) is located in the middle between P_z and P_3, and P_z and P_4. A part of two measurement areas is overlapped as illustrated in the top figure. The top of the channel layout is the rostral direction. The solid vertical line in each graph indicates the time of alternation of visual stimuli.

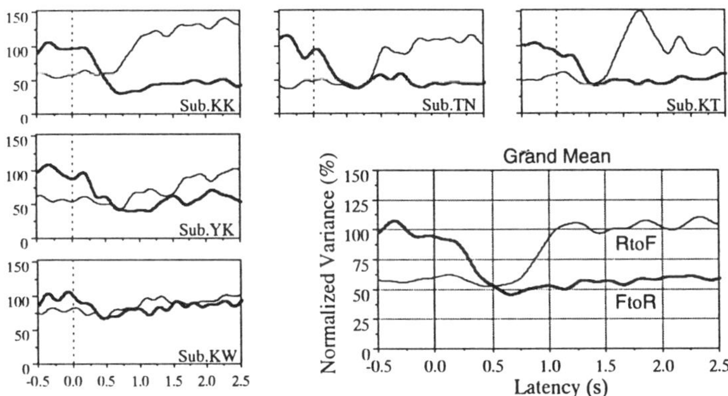

Fig.3 Changes in normalized MEG variance in the 8-13 Hz band elicited by the alternation from BF to BR (FtoR; thick line) and BR to BF (RtoF; thin line) in all subjects and their grand mean at the channel located in the vicinity of P_3, where the largest variation was observed in each subject.

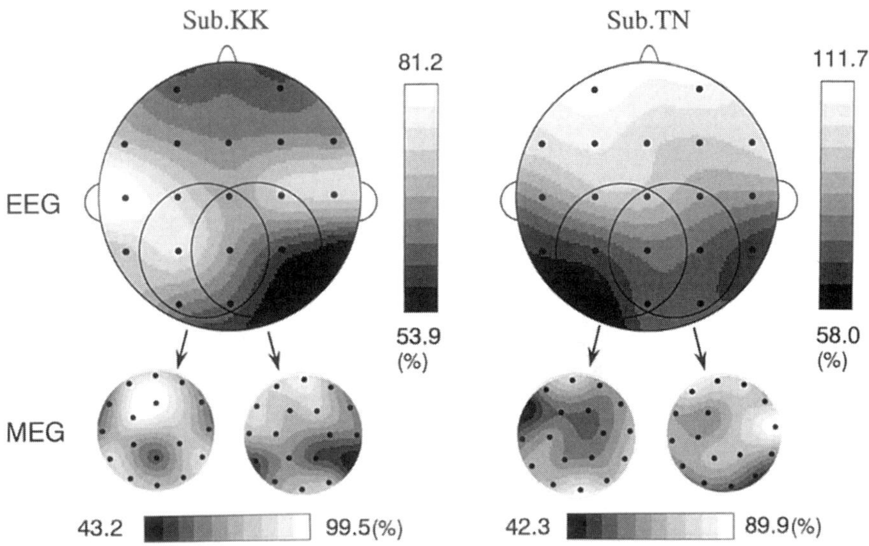

Fig.4 Topographical maps of the normalized EEG and MEG variance in BR, which averaged between -500 to 0 ms prior to the alternation of visual stimuli in the 8-13 Hz band, for two subjects (KK and TN). The top of the channel layout is the rostral direction. The EEG and MEG topographies were drawn, maintaining the relative size of their measurement areas. The 20 electrode positions for EEGs and the 19 sensor positions for MEGs are shown by dots in the figures. The scale bar was divided into 10 levels between the maximum and minimum values of variance in each map.

were individual differences in the MEG topographical distributions, a common feature was that they showed the smallest value in the parietal region.

DISCUSSION

As shown in Fig.3, the attenuation of MEG variance in the 8-13 Hz band in BR detected over the parietal region was significantly larger than that in BF, although the stimulus strength, i.e., luminance, size, shape and contrast, given to both eyes in the two states were physically identical. The mechanism of the attenuation of MEG power in the 8-13 Hz band, which is thought to be the same phenomenon as that of EEG, has not been clarified yet. However, it is widely accepted that the attenuation represents a change in the nature of excitation from basically synchronous to more asynchronous, occurring in a large number of cortical neurons during the period of the attenuation [8,9,10]. Furthermore, recent studies have revealed that cortical neurophysiological processing related to mental activity is reflected in the modulation of EEG power, and this phenomenon is known as event-related desynchronization and event-related synchronization [11]. Therefore, it is possible to interpret the modulation of MEG variance in the 8-13 Hz band observed in this study as follows. The attenuation until 700 ms and subsequent recovery after the alternation of visual stimuli reflects transient neural activities related to the alternation since the modulation was commonly observed in both cases of stimuli alternation between BF and BR. On the other hand, differences in the variance in the 8-13 Hz band at -500 to 0 ms prior to the stimuli alternation, where the variance seemed stable, reflects the altered cortical activities involved in BR and BF.

It is difficult to understand exactly what these altered activities are from only the results of the present experiments. Blake and Overton [12] tried to discover the site of binocular suppression by pshycophysical experiments, and proposed that the suppression occurs at a site in the human visual system after the locus of grating adaptation and, hence, after the striate cortex. Physiological and anatomical findings in the primate visual system suggest that different components of visual information processing are segregated into largely independent parallel pathways [13]. Livingstone and Hubel [14] have suggested, based on results of their psychophysical experiments, that BR results from activity in the magnocellular pathway. Although we are not certain whether the human visual system is similar to that of primates due to the lack of neuroanatomical information for humans, the idea of parallel pathways may explain the altered neurophysiological activities between BR and BF. The parvocellular pathway may be involved both in BR and BF because the form of visual stimuli may be processed commonly in the two state, whereas the magnocelullar pathway may be involved only in BR. Further study is needed to estimate where in the cortex and which pathway is involved in BR and BF.

REFERENCES:

[1] Blake, R., A neural theory of binocular rivalry, Psychol. Rev., 1989, 96:145-167.
[2] Wolfe, J.M. , Stereopsis and binocular rivalry, Psychol. Rev., 1986, 93:269-282.
[3] Crick, F. and Koch, C., Are we aware of neuronal activity in primary visual cortex?, Nature, 1995, 375:121-123.
[4] Leopold, D.A. and Logothetis, N.K., Activity changes in early visual cortex reflect monkeys' percepts during binocular rivalry, Nature, 1996, 379:549-553.
[5] Sengpiel, F., Blakemore C. and Harrad, R., Interocular suppression in the primary visual cortex: a possible neural basis of binocilar rivalry, Vision Res., 1995, 35:179-195.
[6] Kobayashi, T., Kato, K., Owada, T. and Kuriki, S., Difference of EEG spectral powers observed between binocular rivalry and binocular fusion, Fornt. Med. Biol. Engng., 1995, 7:11-19.
[7] Kobayashi, T., Kato, K. and Kuriki, S., Scalp distribution of the attenuated EEG rhythmic activities for dissimilar and/or moving images in binocular vision, Fornt. Med. Biol. Engng., in press.
[8] Kaufman, L., Curtis, S., Wang, J.-Z. and Williamson, S.J., Electroenceph. clin. Neurophysiol., 1991, 79:211-226.
[9] Markand, O.N., Alpha rhythm. Journal of Clinical Neurophysiology, 1990, 7:163-189.
[10] Steriade, M., Gloor, P., Llinás, R.R., Lopes da Silva, F.H. and Mesulam, M.-M., Basic mechanisms of cerebral rhythmic activities. Electroenceph. clin. Neurophysiol., 1990, 76:481-508.
[11] Pfurtscheller, G., Event-related synchronization (ERS): an electrophysiological correlate of cortical areas at rest. Electroenceph. clin. Neurophysiol., 1992, 83:62-69.
[12] Blake, R. and Overton, R., The site of binocular rivalry suppression. Perception, 1979, 8:143-152.
[13] Felleman DF and Van Essen DC. Distributed hierarchical processing in the primate cerebral cortex. Cerebral Cortex. 1991;1:1-47.
[14] Livingstone, M.S. and Hubel, D.H., Psychological evidence for separate channels for the perception of form, color, movement, and depth. The Journal of Neuroscience, 1987, 7:3416-3468.

Magnetic Oscillatory Components of Human Flash-VEP: Luminance and Eccentricity Functions and Effect of Acute Scopolamine.

Lopez, L.[1], Narici, L.[2], Conforto, S.[1] , Romani, G.L.[1] and Sannita, W.G.[3,4]

[1]Institute of Advanced Biomedical Technologies, University of Chieti, Italy; [2]Department of Physics, University of Tor Vergata, Rome, Italy; [3]Center for Neuropsychoactive Drugs, University of Genova, Genova, Italy; [4]Department of Psychiatry, SUNY, Stony Brook, NY, USA

A full-field flash stimulation elicits a series of oscillatory potentials (100- 120 Hz) that are superimposed on retinal electroretinogram (ERG) [1] and are enhanced by highpass filtering at about 60-80 Hz. Comparable potentials were recorded from scalp locations [2,3] with longer latency (+25-30 ms). There is evidence [4] that a cortical/subcortical origin of these potentials is conceivable.

The functional significance of the oscillatory potentials has not been clarified yet, although there seem to be marked similarities between these potentials and the gamma band responses obtained with patterned stimulation [5], which would point to a stimulus-dependent synchronization that involves portions of the thalamo-cortical pathway and is instrumental in stimulus processing [6].

In the present work, a characterization of these potentials has been attempted, by means of different stimulus luminance and eccentricities and by cholinergic blockade.

Methods

Eight healthy subjects were recruited for a pharmacological study, and signed an informed consent. The study protocol involved a baseline recording with flash stimuli full field (at intensities ranging from 0.9 to 3.5 c.s.m^2) and in spots of varying diameter and eccentricity (1-3 cm; 5-30 deg from fixation indicated by a red LED point). In all cases, stimulus was monocular with duration of 20 μs and repetition rate of 0.8 Hz; white noise at 60 dB over threshold was delivered to mask strobe artefacts. The same sequence of stimuli was repeated at 30' from cholinergic muscarinic antagonist scopolamine in acute 0.5 mg i.m. doses).

Recordings at retinal level were obtained with conventional techniques (dermal electrodes on inferior orbital rim referred to linked mastoids); scalp recordings were made electrically (O1 and O2 vs Fpz) and magnetically (CNR-IESS 28-channel neuromagnetometer, 16 gradiometers, 9 magnetometers, 3 cancellation coils). MEG recordings were contralateral to the stimulated eye. After continuous acquisition, 65-75 artefact-free epochs were averaged and digitally filtered with 80-150 Hz bandpass.

Results

Oscillatory potentials (OPs) were recorded at retinal and scalp locations, in response to both full field and spot stimuli and were phase locked to stimuli and superimposed in latency to the ERG and the earlier waves of flash-VEP respectively. When the spot stimuli were used, there were no ERG, or retinal OPs, Scalp MEG recordings displayed definite oscillatory and broad band responses, as long as the stimuli were presented in the first 5° of eccentricity (Fig.1). An increase in latency of about 10-20 ms was observed as compared to full-field stimuli, but both latency and amplitude of the oscillatory responses were almost unaffected by the different spot diameters in the 5° position. When stimuli where presented in the 15° or 25° there was no oscillatory response, while the broad band VEP could still be recorded with graded increase in latency and decrease in amplitude.

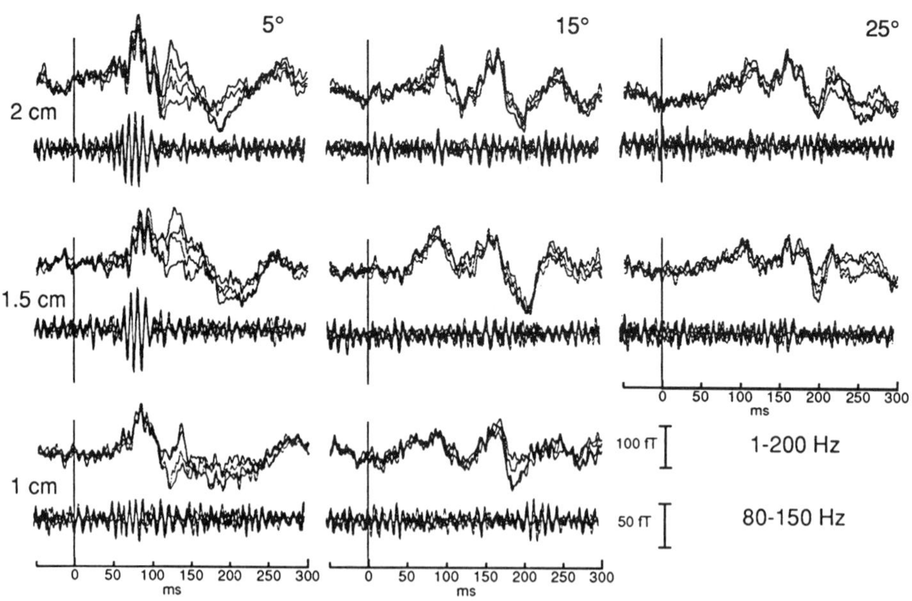

Fig. 1. MEG recordings in response to spot stimulation. Upper tracings
Rows: increasing eccentricities; columns: increasing spot diameter. All spots were
presented in the superior left half field

After injection of acute 0.5 mg scopolamine a marked reduction of the broad band late components
was observed as described earlier [7], while the oscillatory responses were still recordable in all subjects. The
oscillatory activity after scopolamine seemed to be enhanced. (Fig 2). These data were also confirmed by the analysis
of the "locking indicator" L, which is a normalized version of a parameter recently introduced by other investigators
[8,9] and describes the degree of locking of a narrow band response to an external stimulus. It is obtained by filtering
of the raw data and calculation of i) the square of the averaged signal (where the average is performed with respect to
the stimuli; ii) average of the squared signal. An integration is then performed and the square root of the ratio i/ii is
taken. If the oscillating activity is locked by the stimulus L goes to 1. If the activity is mostly unlocked L goes to
smaller values.

When we applied this analysis to our pre- and post-drug data, we observeed, in all subjects, an
increase in locking of the 80-150 Hz activity in response to full field, after administration of scopolamine. The
locking activity was also organized in two peaks of locked activity which occurred at about 50 ms and 80 ms from
the stimulus, the second peak being generally more prominent and having longer duration (50 ms) (Fig. 3). When
the same analysis was applied to the responses to spot stimuli we observed a reduction of locked activity after
scopolamine.

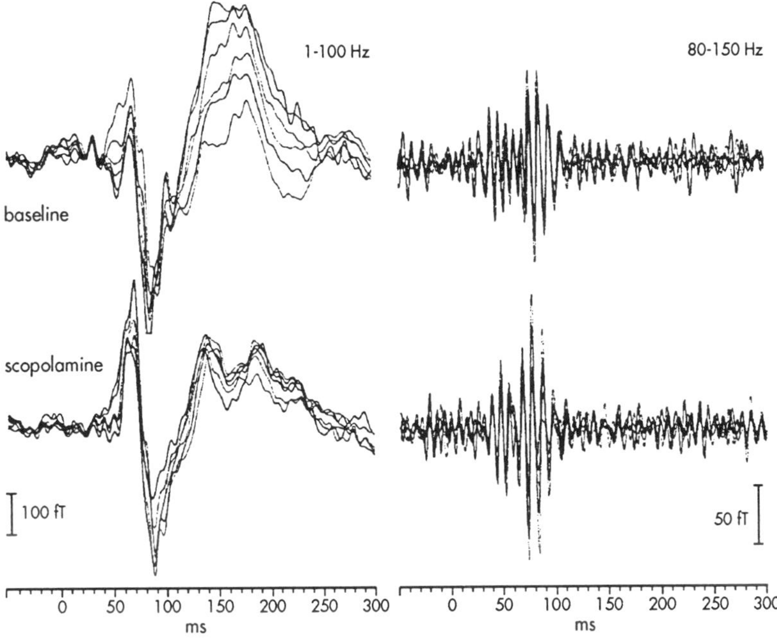

Fig. 2. MEG broad-band and oscillatory responses to full-field flash stimulation, before and after cholinergic blockade.

Discussion

These observations are consistent with experimental studies, in which *in vivo* recordings from latero-geniculate nucleus lamina 6 [10] indicate negligible contributions from volume-conducted pre-cortical phenomena to scalp-recorded oscillatory events.

In the present study, we observed the different behavior of the oscillatory activity in response to spot stimulus as compared to the broad band response. Namely, the oscillatory activity falls off as the stimulus is moved away from the central 5-10 degrees of visual filed, which would suggest independent cortical generators for broad-band VEP and oscillatory activity. The latter may reflect a predominance of events triggered by foveal/parafoveal stimulation in the generation of cortical. In this sense they would be the result of a parallel and partially independent stimulus processing with respect to the broad-band responses. The dichotomy between the effects of scopolamine on the

953

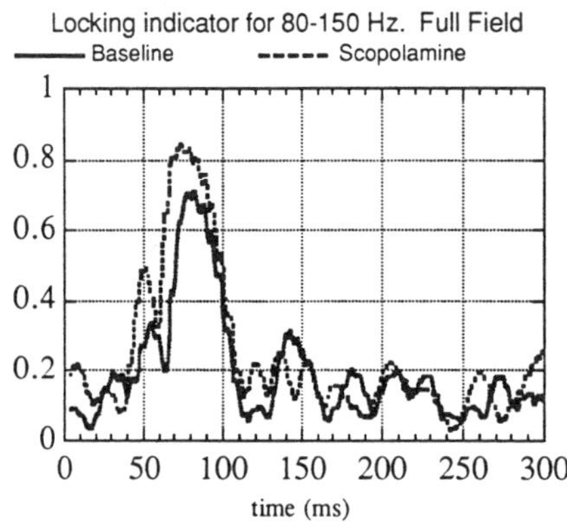

Fig. 3. Locking indicator diagram for the 80-150 Hz activity, in response to full-field

oscillatory responses to full-field and spot stimulation may also suggest a role of acetylcholine in the stimulus-related synchronization that occurs in visual cortex upon sensory input, and is suggested to depend on the physical characteristics of the stimulus [6] and is probably involved in feature binding processing.

References

[1] Karwoski, C. and Kawasaki, K. Oscillatory potentials. In: Heckenlively, J.R. and J.B. Arden Principles and practice of clinical electrophysiology of vision, St. Louis, Mosby-Year Book, 1991, pp 125-131.

[2] Harding, G.F.A. and Rubinstein, M.P. The scalp topography of the human visually evoked subcortical potential. Invest. Ophtalmol. Vis. Sci., 1980, 19: 318-321.

[3] Cracco, R.Q. and Cracco, J.B. Visual evoked potentials in man: early oscillatory potentials. Electroencephalography and clinical Neurophysiology, 1978, 45: 731-739.

[4] Lopez, L., Pasquarelli, A., Romani, G.L., Torrioli, G. and Sannita, W.G. Magnetic recording of cortical oscillatory potentials in response to flash stimulation in man. In: Baumgartner, C., L. Deecke, . Stroink and S.J. Williamson Biomagnetism: fundamental research and clinical applications, 7, Amsterdam, Elsevier, 1995, pp 170-173.

[5] Sannita, W.G., Lopez, L., Piras, C. and Bon, G.D. Scalp-recorded oscillatory potentials evoked by transient pattern-reversal visual stimulation in man. Electroenceph. clin. Neurophysiol., 1995, 96: 206-218.

[6] Gray, C.M., Engel, A.K., König, P. and Singer, W. Stimulus-Dependent Neuronal Oscillations in Cat Visual Cortex: Receptive Field Properties and Feature Dependence. European Journal of Neuroscience, 1990, 2: 607-619.

[7] Sannita, W.G., Balestra, V., Bon, G.D., Marotta, V. and Rosadini, G. Human flash-VEP and quantitative EEG are independently affected by acute scopolamine. Electroenceph. clin. Neurophysiol., 1993, 86: 275.

[8] Salmelin, R. and Hari, R. Spatiotemporal characteristics of sensorymotor MEG rhythms related to thumb movement. Journal of Neuroscience, 1994, 60: 537-550.

[9] Pfurtscheller, G. Event-related synchronization (ERS): an electrophysiological correlate of cortical areas at rest. Electroenceph. clin. Neurophysiol., 1992, 83: 62_69.

[10] Schroeder, C.E., Tenke, C.E. and Givre, S.J. Subcortical contributions to the surface recorded flash-VEP in the awake macaque. Electroenceph. clin. Neurophysiol., 1992, 84: 219-231.

Increased variability of the period duration for waveforms in Epochs of MEG data as a function of arousal state

Moran, J. E.[1], Tepley, N.[1,2]

Henry Ford Hospital[1], Detroit MI, USA; Oakland University[2], Rochester MI, USA

Introduction

In clinical evaluation of spontaneous magnetoencephalography (MEG) or EEG time series data, the state of arousal is assigned of one of 5 NREM scores (awake, stages 1-4) based on visual examination if the data for the presence of specific waveforms and the general amplitudes and frequencies of these waveforms[1]. However, this method of quantifying the state of arousal is not linearly related to changes in the spectrum of the data and totally insensitive to small shifts of arousal state. These deficiencies required us to develop a new system to quantify changes of arousal for our long duration MEG studies. Our arousal measurement parameters are based on a spectrum of period duration versus power, generated by our DPA (dynamic period analysis) technique. DPA spectra are not complete spectra, such as produced by FFT methods. We found FFT spectra to be ill suited for quantifying arousal state because of large subject to subject spectral differences dominated arousal spectral change. In our DPA spectra, spectral differences between individuals are relatively small compared to spectral changes associated with the state of arousal. Further, a DPA spectrum for any state of arousal can be divided into a linear combination of just two unique DPA spectra. One of these unique spectra is identical to a typical awake DPA spectrum and the other we have been able to identify as the time averaged power spectrum of a highly random component of the MEG data which is only present during states of arousal other than awake. We established these two unique DPA spectra by applying a principal component analysis of a large set of DPA period spectra and obtaining the two unique principal component spectra[2]. During shifts from awake to stages 2, 3, and 4 arousal states the fraction of the awake principal component DPA spectrum is significantly decreased while the proportion of the drowsy/sleep principal component DPA spectrum increases from the zero amplitude awake value. We have found the drowsy/sleep principal component DPA spectrum to exhibit changes in amplitude and amplitude variability which make it especially useful for distinguishing fluctuations of the awake state from a transition state of arousal that signifies the onset of stage 2 sleep. Also, we have established a relationship of our arousal parameters to the traditional classification system such that the results of our system can be related to this familiar and widely used scale of arousal state.

Methods

MEG recordings of spontaneous cortical activity obtained from six subjects were used to develop the two principal component DPA arousal state spectra. The MEG data were acquired with a 7 channel Neuromagnetometer (BTi model 607) equipped with second order gradiometers. The center channel was positioned at P3 of the international 10-20 EEG measurement system and 45 minutes of data was acquired from each subject lying on his/her side. For these subjects, the data was partitioned into 100 second epochs and DPA spectra generated for each epoch. For two additional subjects, included in this report, simultaneous EEG recordings were made with electrode placements, P3-A2 and Cz-F2. The 45 minutes of EEG and MEG data were partitioned into 25 second epochs. DPA spectra were produced for the center channel of the sensor array and the epochs of EEG data were scored for arousal state (awake, stages 1 - 4). For each DPA spectra, the amplitudes for the awake principal component spectrum and the drowsy/sleep principal component spectrum were generated.

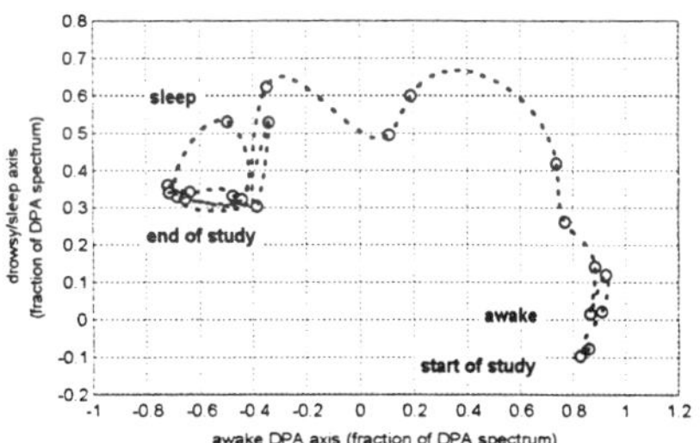

Figure 1 DPA spectral composition during 45 minute study, awake to sleep.

With these amplitudes plotted as X-Y coordinates, the angle from the horizontal (awake) axis was used to relate our two new parameters for quantifying arousal state to the existing system (see Fig. 1).

During subsequent analysis, these two arousal coordinate amplitudes were generated for DPA produced from epochs of durations 4, 8, 16, 32, 64, and 128 seconds. Since DPA spectra are time averaged spectra, these data were used to investigate changes in both the amplitude and amplitude variability of our principal component arousal parameters as a function of data epoch length and arousal state. The standard deviation of the amplitude of both the awake and drowsy/sleep components during extended data segments of a specific state of arousal were used to quantify the amplitude variability associated with the state of arousal and DPA epoch duration. The average amplitude for these same data segments was used to quantify the amplitude of the awake and drowsy/sleep components of the composite DPA spectra.

Results

The DPA spectra of the two principal components extracted from the DPA data using a principal component analysis are shown in Fig. 2. Our data indicates the onset of drowsiness gives rise to a shift in cortical function which introduces progressively greater durations of activity characterized by high variability in both frequency and amplitude. In Fig. 3, we demonstrate that the drowsy/sleep principal component is the time average spectrum of this random cortical process. For this Figure, 17 epochs (25 second epochs used to generate DPA spectra), recorded during stage 1 and 2 arousal states were averaged after subtracting the awake DPA spectral component. This averaged spectrum is nearly identical to the drowsy/awake spectrum in Fig. 3. The single epoch DPA spectrum plotted in Fig. 3 reflects the low rate of occurrence and random amplitude of these waveforms.

In Fig. 4, the amplitude and amplitude variability of the fraction of the a DPA spectrum corresponding to the awake DPA principal component is graphed as a function of both arousal state and epoch duration used to generate the DPA spectrum. For both awake and stage 2 arousal states this fractional amplitude of the awake principal component requires a relatively short epoch of data before reaching a stable value with very low variability. In Fig. 4, both the amplitude and amplitude variability are nearly constant, especially for DPA epochs longer than 16 seconds. Relative to the awake state, the amplitude of the awake principal component of a stage 2 DPA spectrum is less and amplitude variability increased. These amplitude differences are nearly the same for all epoch durations, such that, the graphs of awake and stage 2 are almost parallel.

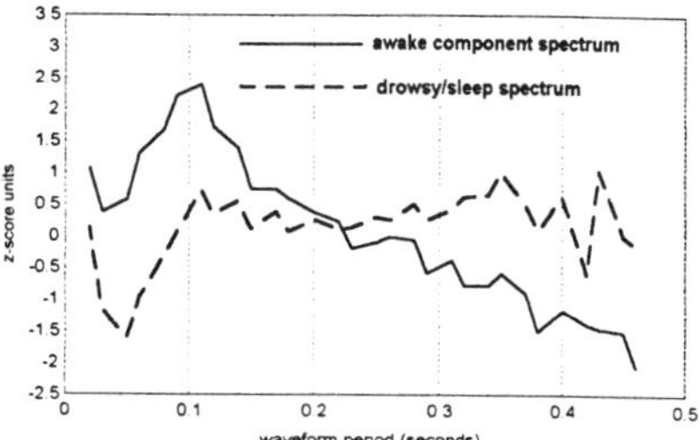

Figure 2 Awake and drowsy/sleep DPA spectra.

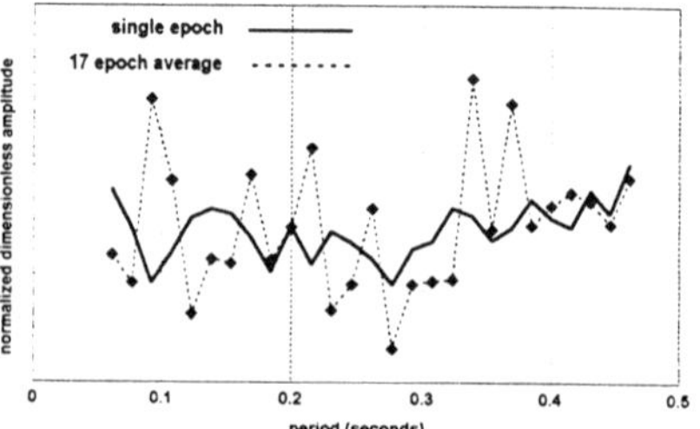

Figure 3 Single epoch DPA spectrum and 17 epoch average spectrum, after removal of awake spectral component.

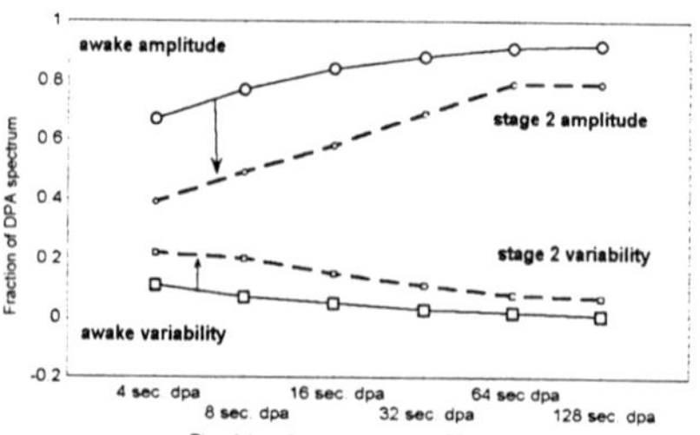

Figure 4 Change in the awake component of the DPA spectrum during transition from awake to stage 2 arousal state.

However, the amplitude and amplitude variability of the drowsy/sleep component content of awake and stage 2 DPA spectra demonstrate a more complex relationship to the state of arousal and the epoch duration used to generate the DPA spectra. As a result, this component is especially useful for recognizing the onset of sleep. This relationship allows the DPA epoch duration to be optimized for recognizing and quantifying changes of arousal. In Fig. 5, the drowsy/sleep content of the awake DPA spectra is nearly zero for all DPA epoch durations. However, for stage 2 DPA spectra, the amplitude of the drowsy/sleep fraction of the DPA spectrum is significantly increased when a longer epoch duration is used to generate these DPA spectra. Therefore, a larger amplitude difference of the drowsy/sleep component between awake and stage 2 arousal states is created when a longer DPA epoch duration is used. Even more interesting and potentially useful is the amplitude variability of the drowsy/sleep component as a function of the duration used to generate the DPA spectra. An adequate sample of the waveforms contributing to the awake DPA spectrum is obtained with a short epoch of data. Thus, random variability of the awake DPA spectrum is reduced when a longer epoch duration is used to generate DPA spectra. However, a very long epoch of data must be used to acquire all waveforms contributing to the stage 2 arousal state DPA spectrum. As a result, the variability of the drowsy/sleep component remains nearly constant until greater than 60 second epochs are used (see Fig. 6). Therefore, the difference in amplitude variability of the drowsy/sleep component between awake and stage 2 arousal states also increases when longer epoch durations are used. During sleep onset, the epoch duration effect on amplitude variability of the drowsy/sleep DPA component can be more dramatic than changes of the mean amplitude. In Fig. 7, a large increase in the variability of the drowsy/sleep component is easy to detect during the stage 1 sleep recorded just prior to the transition to stage 2 (1400 seconds to 1700 seconds). Notice, during this interval, the amplitude variability of the drowsy/sleep component is the nearly same as after the transition to stage 2 (1700 seconds to 2100 seconds). However, during this stage 1 epoch, the amplitude of the drowsy sleep component has an average amplitude identical to the awake state and abrupt increases with stage 2 onset. This sustained increase in amplitude variability is absent during other epochs of stage 1 sleep which alternated with epochs of awake. For this graph, a 32 second epoch duration optimized the compromise between maximizing the drowsy/sleep component amplitude difference between awake and sleep arousal states and the ability to resolve temporal changes of the arousal state. In actual application of this technique, the time series of drowsy/sleep amplitudes for more than one DPA epoch duration are examined for arousal state changes.

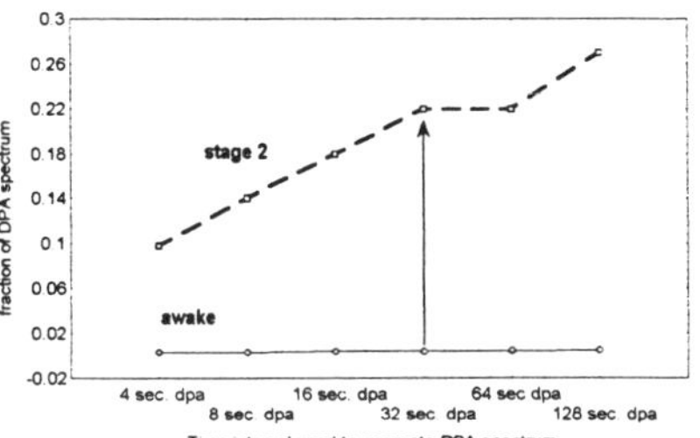

Figure 5   Amplitude of drowsy/sleep component of the DPA spectrum. Awake versus stage 2 arousal.

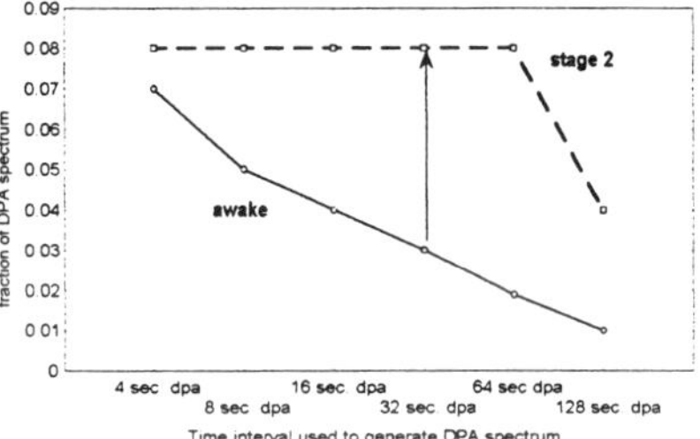

Figure 6 Variability of the drowsy/sleep component of the DPA spectrum. Awake versus stage 2 arousal.

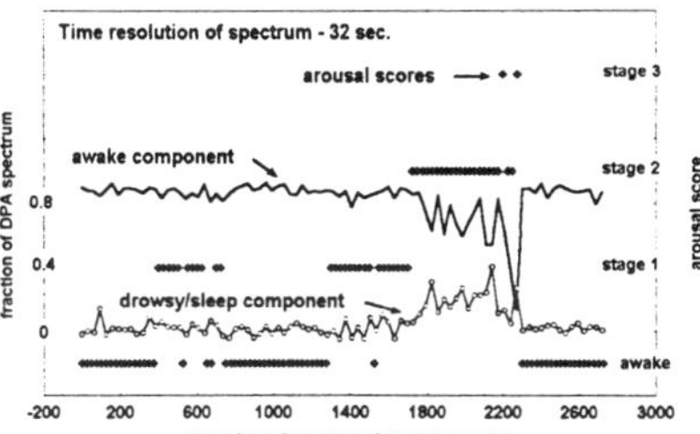

Figure 7 Time series of clinical arousal scores versus amplitudes of awake and drowsy/sleep fraction of the DPA spectra.

Discussion

Our results indicate that during the transition from awake to sleep some portion of the cortex begins to function in a mode distinctly different than the awake state cortex. The time average spectrum of the MEG waveforms arising from this drowsy/sleep mode of cortical function is identical to the DPA drowsy/sleep principal component we extracted. This spectrum has the property that the same time average power occurs at all period durations less than 0.46 seconds, included in our DPA spectra (see Fig. 2). DPA spectra are formed by counting the number of waveforms with each specific period duration and the corresponding waveform power. Therefore, during this mode of cortical function, for each period duration, the number of waveforms multiplied by the time average power of the waveform is equal to a constant for all period durations. If the power and duration of a waveform are on average proportional to the area of cortex generating the recorded activity, then this result implies that the number of waveforms arising from cortex of specific area is inversely proportional to the size of that area for this mode of cortical function.

Compared to awake state waveforms, drowsy/sleep waveforms occur infrequently and are characterized by large amplitude variability. Thus, a very long epoch duration is required to accumulate enough waveforms of all periods to produce a completely stable DPA spectra during sleep arousal levels. As a result, both the amplitude of the drowsy/sleep DPA component and the amplitude variability increase when longer epochs are used for DPA spectra generated from less than 60 seconds of data. Therefore, the amplitude variability of the drowsy/sleep fraction of DPA spectra can be used to detect the switch from awake to sleep cortical function (see Figs. 5, 6 and 7).

Also, we have developed a mathematical formula which relates our new arousal parameters to the present clinical scores. This was done to facilitate

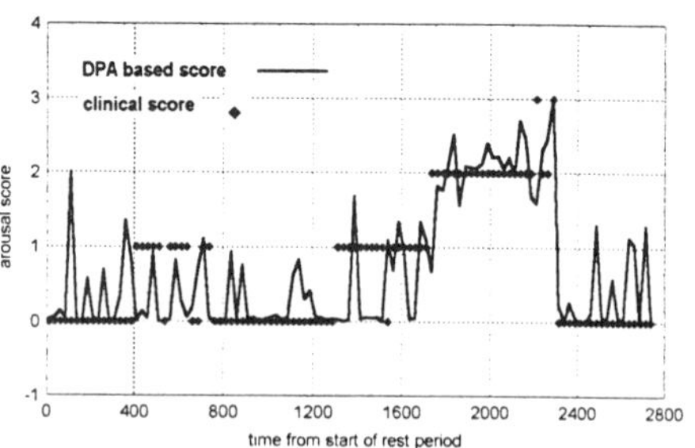

Fig. 8 The time series of DPA based and clinical arousal scores for 45 minute study of single subject.

the use of our new parameters which provide a continuous parameter of changes in the spectrum of MEG data associated with arousal state. As previously described, the amplitude of the awake and drowsy/sleep content of each DPA spectra can be used as X, Y coordinates (see Fig. 1). The angle from the awake coordinate axis was used to develop the relationship. We found the relationship is best described by;

log(DPA angle / 140) = 1.15 (clinical arousal score) - 3.6 angle in degrees

Using this formula on a time series of DPA spectra the same trend in arousal change and approximately the same arousal level is predicted as found in the actual time series of clinical scores (see Fig 8).

References

[1]   Rechtchaffen, A., Kales, A., eds. A manual of standardized terminology, techniques and scoring system for sleep stages of human subjects. Public Health Service, U. S. Government Printing Office, Washington D.C., 1968.

[2]   Norusis, M.J., Factor analysis in: SPSS-X advanced statistics guide., McGraw-Hill, 1985, pp. 125-163.

Acknowledgements

Research supported by NIH/NINDS Grant No. 1R01 NS30914
Dr. Timothy Roehrs, Sleep Disorders Center, Henry Ford Hospital, who provided arousal scores and useful suggestions.

Suitability of FFT Spectra versus DPA Period
Spectra for Identification of Arousal States

Moran, J. E.[1], Tepley, N.[1,2]

Henry Ford Hospital[1], Detroit MI, USA; Oakland University[2], Rochester MI, USA

Introduction

To detect changes in arousal which often occurred during long duration DC MEG studies in our study of migraine, we found the present clinical method inadequate. Small fluctuations of arousal can not be quantified using the 5 stage scores available for NREM arousal states [1]. In addition, this scale is not linearly related to changes in the spectral activity observed in the MEG data. Therefore, we developed a zero crossing period measurement technique [2] with a built-in frequency selection filter (dynamic period analysis, DPA). DPA produces a period based power spectrum quantifying the dominant waveform activity recorded in a segment of data. For all subjects, changes of arousal level resulted in characteristic changes in the corresponding DPA spectra. This motivated us to apply a principal component analysis [3] to extract a set of principal component spectra for quantifying the small changes of arousal which we needed to detect. A set of DPA spectra were used in this analysis. Two principal spectral components were extracted and used to construct a single arousal parameter. Unlike clinical arousal scores (awake, stage 1, etc.), this arousal parameter is continuous and allows us to detect even small changes of arousal and quantify the associated DPA spectral changes. Because FFT methods of spectral analysis are more familiar, we attempted to construct a similar set of arousal parameters based on FFT spectra generated from sequential epochs of MEG data for these subjects. Unfortunately, the principal component analysis of these FFT spectra did not extract component spectra which described a common set of arousal changes shared by all individuals. Rather than corresponding to arousal states, the two FFT principal components primarily formed a basis for differentiating between individual subjects.

Methods

MEG recordings of spontaneous cortical activity obtained from six subjects were used to develop the two principal component DPA arousal state spectra. The MEG data were acquired with a 7 channel Neuromagnetometer (BTi model 607) equipped with second order gradiometers. The center channel was positioned at P3 of the international 10-20 EEG measurement system and 45 minutes of data was acquired from each subject lying on his/her side. This site was accessible to our MEG system and data recordings contained less muscle and eye movement artifacts. For these subjects, the data was partitioned into 100 second epochs and a DPA spectrum was generated for each epoch. For the principal component analysis, it was necessary to limit the frequency band to frequencies greater than 0.5 Hz. Characteristic amplitude changes of the DPA power spectra for these frequencies are sufficient to distinguish all states of arousal. FFT spectra were generated for these same epochs of data. These FFT spectra were constructed from the same 100 seconds data epochs by averaging seven FFT spectra obtained from data within 40 second time windows which overlapped by 75%. This is a commonly employed technique for constructing an average FFT spectrum for semi-stationary data when the FFT software incorporates a window function. Window functions are used to minimize significant leakage of spectral power from the true frequency component into other nearby frequency components [4, 5]. However, if the spectral properties of the data are not completely stationary throughout the data segment then window functions may weight the data in a manner in which the final FFT does not represent the average spectrum of the data. We used FFT software which incorporated a Hanning window to weight the data within each 40 second window prior to the FFT transformation [6]. The FFT frequency spectra were transformed to FFT period duration spectra such that they could be compared to the corresponding DPA spectra and also to consolidate the power of the high frequency components into fewer components such that shifts in power to lower frequencies occurring with drowsiness and sleep are given more weight in the mathematical analysis.

For two additional subjects, included in this study, simultaneous EEG recordings were made with electrode placements, P3-A2 and Cz-F2. The 45 minutes of EEG and MEG data were partitioned into 25 second epochs. DPA spectra were produced for the center channel of the sensor array and the epochs of EEG data were scored for

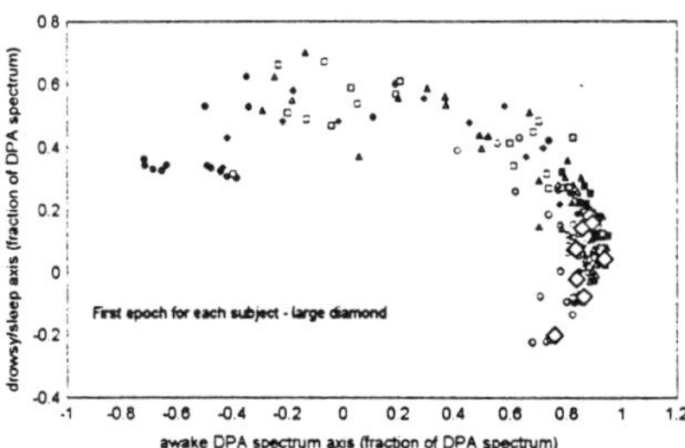

Fig. 1 Plot of β_{awake} and β_{drowsy} amplitudes for DPA spectra of eight subjects. Awake state nearly the same.

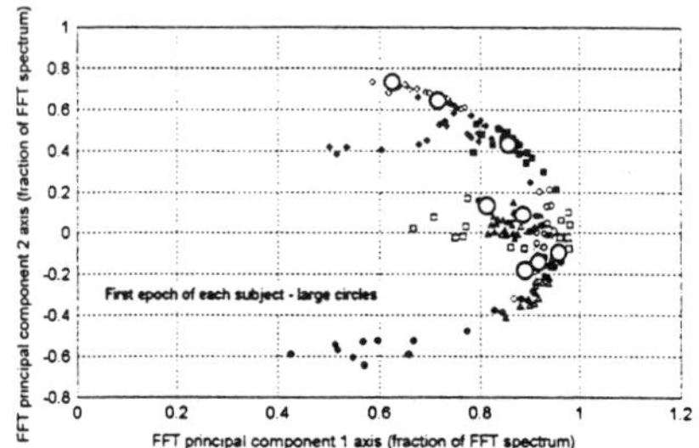

Fig. 2 FFT spectra principal component amplitudes for eight subjects. Awake state co-ordinates very different.

arousal state (awake, stages 1 - 4).

A principal component analysis was performed on the set of 162 DPA spectra from six individuals. Similarly, a principal component analysis was applied to the set of FFT spectra and two principal component spectra were generated.

Results

For the DPA spectra, the principal component analysis generated two principal component spectra. One spectrum was identical to the average awake spectrum of these subjects. The other component was found to be the time averaged power spectrum of waveforms which are present in the MEG data only in states of arousal other than awake. For any subject, a linear combination of the two principal component DPA spectra can be used to obtain a DPA spectra corresponding to the arousal state of the subject. Mathematically this relationship is:

$$DPA_{arousal\ state} = \beta_{awake}\ DPA_{awake} + \beta_{drowsy}\ DPA_{drowsy} + Random$$

with: $DPA_{arousal\ state}$ = DPA spectrum of subject during an epoch of the study
DPA_{awake} = principal component DPA spectrum corresponding to awake state
DPA_{drowsy} = principal component DPA spectrum of waveforms occurring if not awake
β_{awake} = amplitude of the fraction of $DPA_{arousal\ state}$ which is associated with DPA_{awake}
β_{drowsy} = amplitude of the fraction of $DPA_{arousal\ state}$ which is associated with DPA_{drowsy}
Random is the spectrum of random spectral fluctuations that remain in the DPA spectrum

Using β_{awake} as an X coordinate amplitude and β_{drowsy} as a Y coordinate amplitude, the set of 162 DPA spectra for these six subjects are plotted in Fig. 1. In Fig. 1, the awake DPA spectrum generated from the first 100 seconds of data for each subject is closely aligned with the horizontal positive awake axis. Sleep state DPA spectra from subjects that were found asleep at the end of the study were at the other end of the arc of data points which defines an arousal state path through this two component vector space.

The results of the principal component analysis on the FFT spectra can be described by an equation similar to the DPA relationship above, except the awake state spectra for each of the six subjects were located very far apart in the X,Y coordinate system plot in Fig. 2. The awake spectra from the first epoch can not be used to define a set of awake state coordinates. While arousal changes are easily seen for each subject, there is no common state of arousal path. These FFT principal component spectra are seen to primarily differentiate the spectra according to properties unique to the individual rather than change of arousal state.

During the first 100 second epoch of the study each subject was awake. Therefore, in order to quantify individual spectral differences, we cross-correlated each DPA spectrum from this first epoch with the first epoch

DPA spectra of the other subjects. We did the same cross-correlation analysis for the FFT spectra. For the first 100 second DPA spectra the average cross-correlation for all subjects was 0.81. The corresponding FFT cross-correlation average was only 0.74. To quantify changes of spectrum due to a shift in the state of arousal we cross correlated the DPA from the first 100 second epoch to the DPA spectra generated from the 26 subsequent 100 second epochs of data for each subject. The same cross-correlations were performed on the FFT spectra. The average arousal state change cross-correlation for the DPA spectra was 0.70 and 0.91 for the FFT spectra. Thus, FFT spectra exhibit relatively large subject to subject differences and relatively small spectral change with shifts of the state of arousal. The DPA spectra have the opposite characteristics. Changes in the DPA spectrum caused by a shift in the arousal state are greater than the differences in spectra between individuals. This was investigated in more detail using the two subjects for which simultaneously recorded EEG data was scored for arousal state. In Fig. 3, the sequence of cross-correlation amplitudes of the first DPA to the subsequent 26 DPA spectra is plotted with the corresponding FFT cross correlations and the time series of arousal scores. In Fig. 3, the DPA spectra are characterized by large spectral changes with a shift in arousal state and the corresponding FFT spectra change much less. These differences in sensitivity to arousal state are easily seen in plots of individual DPA and FFT spectra. In Fig. 4, the awake state DPA and FFT spectra for this subject are over-plotted. In Fig. 5, the stage 2 arousal state DPA spectrum and the corresponding FFT spectra are plotted. Comparing Fig. 4 and Fig. 5, the DPA spectra exhibits greater shifts of spectral power between arousal states than the corresponding FFT spectra.

In Fig. 1, the state of arousal is observed to be defined by the angular location away from the horizontal awake spectrum axis and the maximum angle is approximately 140 degrees. Therefore, we defined a new arousal parameter to be equal to the angle of DPA spectrum plotted in these coordinates divided by 140. With this scaling, zero corresponds to the awake state and 1.0 would correspond to the stage 4 arousal state. However, stage 1, 2 and 3 are not located at equal intervals along this scale. The relationship of our arousal state score (0.0 to 1.0) to the present clinical scores is not linear.

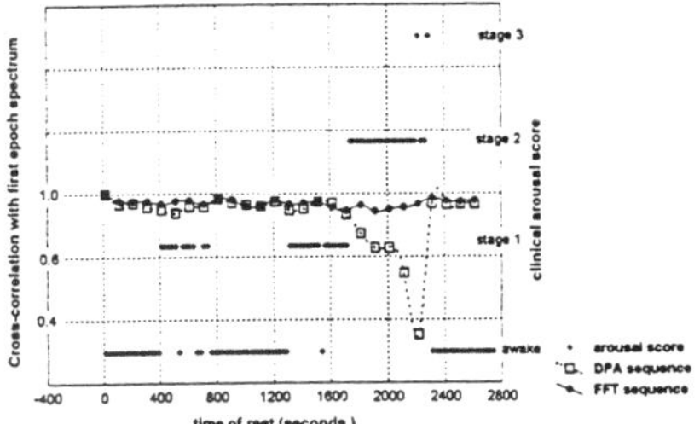

Fig. 3 The time series of DPA and FFT spectral change correlation amplitudes with the corresponding arousal scores.

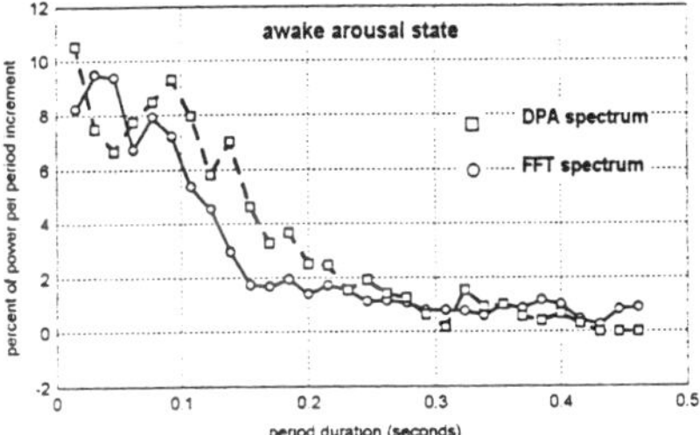

Fig. 4 The corresponding awake DPA and FFT period spectra.

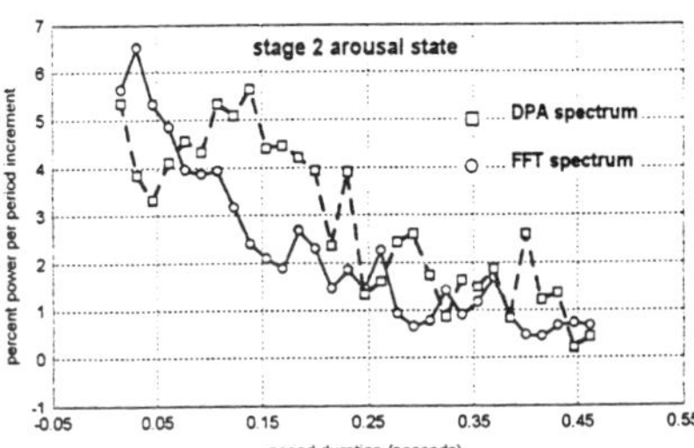

Fig. 5 The stage 2 DPA and FFT spectra.

For our data, mathematically,

$$\log(\text{DPA arousal score}) = 1.15 \, (\text{clinical arousal score}) - 3.6$$

We used this relationship to predict the clinical arousal score from our DPA arousal score. In Fig. 6, the time series of predicted and actual scores are plotted. Both measures of arousal state display the same trends but do not completely agree at each epoch. While this plot is useful to those more familiar with the clinical scale, we find time series plots of the β_{awake} and β_{drowsy} amplitudes of the DPA spectra to be more useful for assessment of changes of arousal state.

Discussion

Assessment of the frequency content of data is primarily performed by either visual inspection or advanced mathematical techniques, such as FFT based power spectra. In this study, we demonstrated that the continuum of arousal state changes of MEG data can be quantified by applying a principal component analysis to DPA spectra and generating a set of unique arousal spectra. All DPA spectral changes occur in the corresponding FFT spectra. However, within the complete FFT spectrum the arousal related changes are small relative to subject to subject

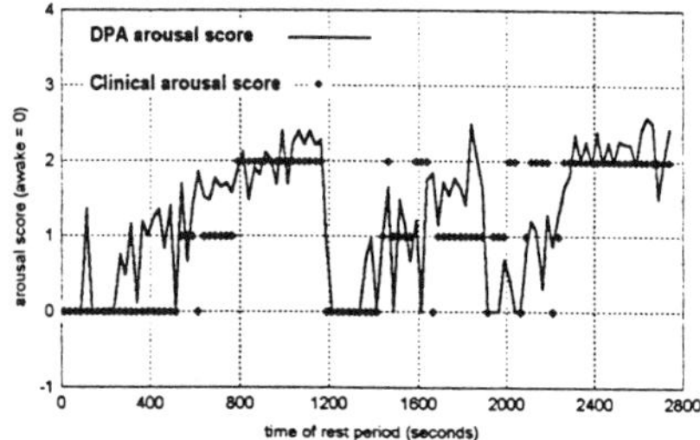

Fig. 6   Predicted arousal state score based on DPA parameters versus standard clinical scores.

spectral differences. Thus FFT spectra are not suitable as a basis for development of arousal state parameters using a principal component analysis.

We use the arousal parameters, β_{awake} and β_{drowsy}, to identify and quantify amplitude and spectral changes of MEG data associated with shifts of the arousal state. However, these parameters may also be useful in assessing the physiological state of arousal based on changes observed in MEG or EEG data. The variability of the β_{drowsy} component may be especially useful for determining sleep onset.

References

[1] Rechtchaffen, A., Kales, A.. eds. A manual of standardized terminology, techniques and scoring system for sleep stages of human subjects. Public Health Service, U. S. Government Printing Office, Washington D.C., 1968.

[2] Frost J.D. Jr. Mimetic techniques. In: Gevins, A.S., Remond, A. eds., Methods of analysis of brain electrical and magnetic signals. EEG handbook, Elsevier, Amsterdam 1987 (revised series, Vol 1)

[3] Norusis, M.J., Factor analysis in: SPSS-X advanced statistics guide., McGraw-Hill, 1985, pp. 125-163.

[4] Lopes da Silva F., EEG analysis: theory and practice. In: Niedermeyer E., Lopes da Silva F., eds. Electroencephalography: basic principles, clinical applications, and related fields., 3rd. ed. Baltimore, USA: Williams & Wilkins 1993:1097-1123

[5] Dumermuth G., Molinari L., Spectral analysis of EEG background activity. In: Givens A.S., Remond A. eds., Methods of analysis of brain electrical and magnetic signals. EEG handbook, Elsevier, Amsterdam 1987 (revised series, Vol. 1)

[6] BTi Neuromagnetometer: Programmer's Reference Manual. Biomagnetic Technologies, Inc., 1987

Acknowledgements

Research supported by NIH/NINDS Grant No. 1R01 NS30914

Dr. Timothy Roehrs, Sleep Disorders Center, Henry Ford Hospital, who provided arousal scores and useful suggestions.

Chaos in the Brain: A New Strategy to Discriminate Deterministic Low Dimensional Dynamics in the Spontaneous Activity of the Human Cortex

Narici, L.[1], Boccaletti, S.[2], Giaquinta, A.[2] and Arecchi, T.[2]

[1]*Dipartimento di Fisica - Università di Roma "Tor Vergata", Roma, Italy;* [2]*Istituto Nazionale di Ottica, Arcetri, Firenze, Italy*

Introduction

Spontaneous activity of the human brain in normal subjects is not likely to be "low dimensional" being the result of superpositions of many correlated and uncorrelated activities. These activities are not constantly present, neither detectable uniformly over the whole scalp. It is likely that this difficulty in defining the "measured system" and the large time variability of the Signal to Noise Ratio (SNR) is one of the reasons for the wide spectrum of results that can be found in the literature on chaotic estimators especially when investigating spontaneous activity. As an example the Correlation Dimension (CD) for the waking stage (mostly α rhythm) ranges from 2.4 to 11 (see, for example, [1]). While some agreement can be found on the dimension decrease under epileptic seizures or other specific pathologies the attempts for a dimensional classification of α rhythm have been mostly unsuccessful.

The difficulties in obtaining reliable estimates of the correlation dimension and of the other dynamical indicators in these cases are often recognized (see for example [1-3]). However in several instances some authors suggest that even though absolute estimate were questionable, comparative results were significant. While the comparative use of these parameters appear to be 'safer', we object that the values of the chaotical evaluators given by the algorithms may be strongly influenced by the noise (including the brain concurrent activities that are, most often, outside the control of the experimenters). Thus, differences between two chaotic estimators, obtained in two different conditions may be due not only to the differences in the studied underlying dynamics, but also to the different way in which 'noise' and experimental conditions affect the calculation. Therefore we believe it is important to set precise limits for acceptance of the estimators and to propose 'standardized' procedures to calculate them, featuring suitable tests able to check their validity.

The α rhythm, largest cortical activity, appears in a narrow frequency band (α band $\approx$8 Hz to 12 Hz) and in *spindle* at irregular time intervals. Outside these *spindles* the SNR<<1. Passband filtering would sharply cut the amount of independent points. A very simple minded calculation shows this number to be $N \approx 2T\Delta f$, where T is the data length in time and Δf the filter bandwidth. As it may be plausible to define a sort of 'maximum stationarity time window' during sleep ($< 10^3$ s) and awake stages (α rhythm)($< 10^2$ s) this impose an upper bound to N ($\approx 10^3$ for α rhythm, passband filtered). This translates in an upper limit for the calculus of CD: $\leq$6 [4], or even $\leq$4 [5].

In this work we refer, for simplicity, to all activities present in the α band in the occipital region as 'α rhythm', even if it has been recently shown that different activities are sharing the same frequency window and often the same approximate source location (see, for example, [6,7]). We also assume that the reader is familiar with the most general procedures used to calculate chaotic estimators from experimental data (the interested reader is addressed to the many reviews available, for example: [1,8-10]).

Methods

Our contribution is two-fold: we propose a procedure to look for determinism and chaotic estimators (CD) in EEG/MEG data, using recent tests described in the literature and we propose a novel filtering technique [11], based on the local features of the wavelet tranformation. The procedure is briefly sketched in the following.

i) - We first calculate the autocorrelation function and Mutual Average Information [12] by which embedding delay time is chosen. ii) - The suitable embedding dimension m is then provided by the false nearest neighbor algorithm [10,13] that calculates the lowest dimension needed to eliminate all false self crossings of the orbit. The number of false crossing must go to zero for increasing m for a signal to be deterministic. This algorithm can therefore be applied to discriminate the original signal from a surrogate (a particular realization of a coloured noise which has the same distribution and autocorrelation function of the signal but all the phases of the Fourier component randomly chosen): if this discrimination is not possible or if the found embedding dimension is not sufficiently low the data must be filtered first. iii) - Test for determinism is carried out (β-test [14]) and chaotic estimators are calculated (Grassberger and Procaccia algorhithm, GP [15]), both for the original and for the surrogate data. iv) - If needed [see (ii)] the signal is filtered in wavelet space (see below); (ii) and (iii) are therefore repeated with the filtered signal.

The procedure is tested on spontaneous occipital neuromagnetic activity measured in a normal subject. The magnetic field over the left occipital cortex of a healthy subject is measured with a 28-channel superconductive system in use at the CNR-IESS laboratory in Rome [16]. The subject was lying prone in a bed wearing a blackened rubber mask, watching fixedly two centered green LEDs to maintain steady eyeball position. We measured 1 minute of occipital activity at an acquisition frequency f=1 KHz, with a band pass filter 0.4 Hz-250 Hz. The raw data were stored in an optical disc for the off-line analysis.

The advantage of filtering in the wavelet space, is that wavelets represent a localized transform. As a consequence, a signal s picked at a given frequency, that gives rise in the Fourier space to very few channels different than zero in which information on the signal is stored, in the WT space preserves many components as a consequence of the fact that a single frequency is delocalized in time. In this work we choose as the analysing wavelet the second derivative of the Gaussian function. To determine the best filtering thresholds for the WT we propose to use a strategy recently introduced [11,14] based on the use of an adaptive recognition of the dynamics of the signal [14]. This strategy, here thereafter called βWT, permits to fix these thresholds in such a way to achieve the extraction of the deterministic component of the original signal. The details of this procedure can be found in the original papers. Let us mention that for each embedding dimension m a coefficient β can be calculated that scales as $\exp(-k_1 m)$ for a deterministic signal whereas as $\exp(k_2 m)$ ($k_1,k_2 > 0$) for noise. In the case of the surrogate of a deterministic system, one half of the total number of degrees of freedom (the phases of each Fourier component) scales like noise, while the other half (the amplitudes of each Fourier component) scales like the deterministic signal. As a result, we expect an intermediate global scaling law. Other kinds of colored noise should also feature an intermediate result. An adaptive algorithm based on the minimization of the slope of $\beta(m)$ permits to implement a new type of filter for the noisy background in a temporal series. This criterion can, of course, be applied also to filters acting in other transform spaces (Fourier, Laplace, etc..), but, as mentioned, the localized feature of the WT may offer an important advantage when dealing with brain signals. The study of the $\log \beta(m)$ can thus be used for two purposes: as a criterion for fixing the optimal WT filtering thresholds, and as a test for the presence of determinism in the analysed signals. We will show that the βWT filter leaves mostly unaltered those frequencies featuring the largest interesting signal power (≈ 6 Hz to ≈ 40 Hz).

Results

The results presented here are relative to one channel of the neuromagnetometer, where the presence of activity in the α band was most prominent, in the perieto-occipital region. Studies on other channels also featuring strong α activity produced quite similar results.

Both mutual average information algorithm and evaluation of the the the zeroes of the autocorrelation function provided the same estimate for the embedding delay time which permits to fully reconstruct the invariant set of points of the attractor: τ=27 ms. The false nearest neighbors algorithm showed no saturation, probably as a consequence of the high level of additive noise. Therefore, the signal was filtered using the βWT filter. The frequency window containing the most important part of the signal is unaffected by the βWT filter. This can be easily appreciated from Fig. 1 where we report the βWT filter results. We show for comparson the Fourier spectra, as well as a 0.1 s long period of the time behavior, of the raw and βWT filtered data. Very little differences can be appreciated by eyes.

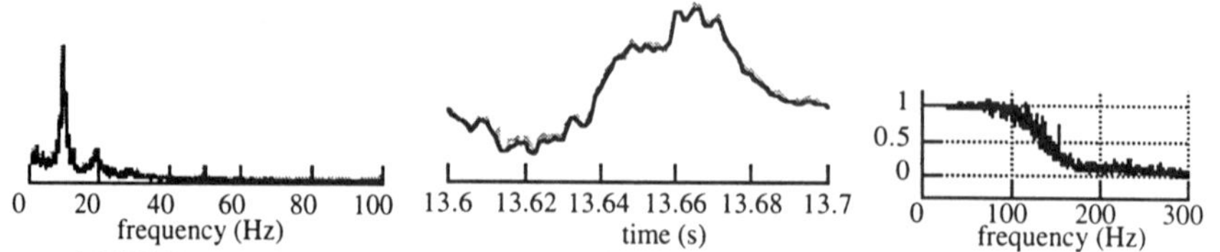

FIGURE 1   Raw data (gray) compared with βWT filtered data (black) in the frequency and time domain. On the right the resulting filter transfer function. Very little differences can be appreciated by eyes.

The study of $\log\beta(m)$ is presented in Fig. 2. We show $\log\beta(m)$ for the raw, for the βWT filtered signals and for the surrogate of the βWT filtered signal.

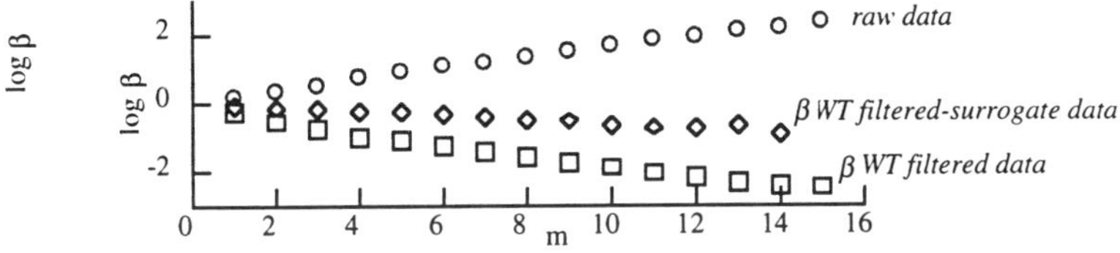

FIGURE 3   Plot of the log of the β parameter (see text) vs. embedding dimension m.

The βWT technique worked only partially on our data as it was not possible to determine the minimum slope of $\log\beta(m)$. We found increasingly negative slopes when increasing the filtering action: a compromise had to be done in order to maintain a sufficiently large number of independent data. Therefore in choosing the filtering thresholds, a certain degree of subjectivity remained. However, for those choices of thresholds for which the filtered signal showed negative slopes of the $\log\beta(m)$, the corresponding surrogate was clearly discriminated as shown in Fig. 3. The filtered signal keeps about 16% of the information originally contained in the original signal (about 10,000 independent points). Data has been filtered also in Fourier space, keeping the same amount of independent points (corresponding to $\Delta f \approx 70$ Hz), centered on the α band. In this case the slope of $\log\beta(m)$ was again negative, but it was not possible to discriminate the surrogate.

The false nearest neighbors algorithm applied on the βWT filtered data provided $m \approx 7$ as embedding dimension while when applied on the surrogate we found no saturation.

Finally we calculated the correlation integral C_2 (GP algorhithm, [15]) after having properly reconstructed the invariant set in a space of dimension 7: no reliable plateau indicates that CD > 7. As a check we performed the GP algorhithm, reconstructing the data set in dimensions $2 \leq m \leq 12$, and produced the same results (Fig. 4).

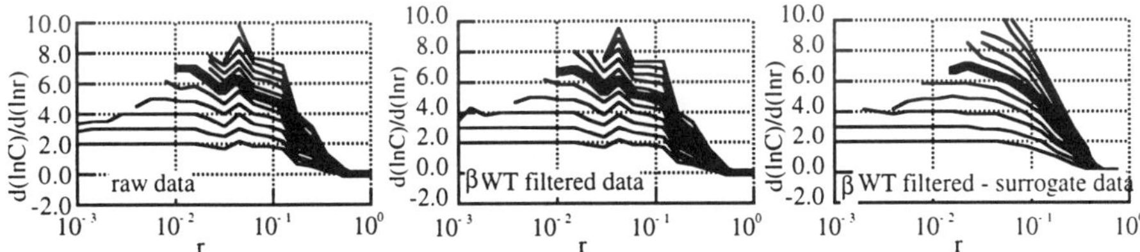

FIGURE 4   GP algorhithm results (local scaling exponents dlnC/dlnr vs logr for different embedding dimension ($2 \leq m \leq 12$); C: correlation integral; r: box size). The thicker line refer to m=7, the embedding dimension found with the false nearest neighbour algorhithm. No plateau is found.

Discussion

With the use of the βWT filter we increased N and thus the CD limits to 6 [5] or 8 [4]. Even with these larger limits CD could not be calculated. However our procedure allows to claim the deterministic nature of some of the waking cortical activity. The β-test did in fact show that our signal is mostly deterministic and that a colored noise, as a surrogate extracted from our signal, can be discriminated. This result opens a new perspective in studying the brain activity in other situations (as e.g. in stimulated cases) where the signal to noise ratio is more favorable and hence the influence of the filter is weaker. We will address such question elsewhere.

The unsuccess in the determination of the minimum of the $\log\beta$ slope is probably linked to the high level of noise in the original signal as well as to the possibly many processes concurrently present. However, as shown in our results, a compromise can be reached extracting from the signal a mostly deterministic component, that can be discriminated from the stocastic component, and that is actually containing all the frequency band where are the most significant signals.

The localized action of the Wavelet Transform permits a time selective filtering that is most useful in those cases, as cortical rhythms, where the signal is concentrated in erratic, short, time windows, such as spindles. The proposed criteria for the choice of the filter parameters, the minimization of the slope of $\log\beta$ vs. m , appears of larger applicability, whenever the process of signal extraction can be assumed to be equivalent to the discrimination of the deterministic components in the raw data. Even when, as in our case, minimization cannot be performed, the combined requirements, negative $\log\beta$ slope and discrimination of the surrogate, appear to be sufficient for a reliable filtering. Finally we would like to stress that a passband filter maintains a much smaller number of independent points and fails the false nearest neighbors test. Even using the same number of independent points passband filtering produces results that cannot be discriminate from surrogate.

References

[1] Basar E., Chaotic dynamics and resonance phenomena in brain function: progress perspectives and thoughts, in: Basar, E. ed. Chaos in Brain Functions Berlin-Heidelberg-New York, Springer, 1990.

[2] Pijn J.P., Van Neerven J., Noest A. and Lopes da Silva F.H., Chaos or noise in EEG signals; dependence on state and brain site, Electroencephal. Clin. Neurophysiol., 1991, 79: 371-381.

[3] Achermann P., Hartmann R., Gunzinger A., Guggenbühl W and Borbély A.A., All-night sleep EEG and artificial stochastic control signals have similar correlation dimensions, Electroencephal. Clin. Neurophysiol., 1994, 90: 384-387.

[4] Ruelle, D., Deterministic chaos: the science and the fiction, Proc. R. Soc. Lond. A, 1990, 427:241.

[5] Kantz, H. and Schreiber, T., Dimension estimates and physiological data, Chaos, 1995, 5:143-154.

[6] Salmelin, R., Hari, R., Characterization of spontaneous MEG rhythms in healthy adults, Electroencephal. Clin. Neurophysiol., 1994, 91: 237-248.

[7] Narici, L., Peresson, M., Discrimination and study of rhythmical brain activities in the α band: a neuromagnetic frequency responsiveness test, Brain Res., 1995, 703:31-44.

[8] Rapp P.E., Bashore T.R., Martinerie J.M., Albano A.M., Zimmerman I.D., Mees A.I., Dynamics of brain electrical activity, Brain Topography, 1989, 2: 99-118.

[9] Pritchard W.S. and Duke D.W., Measuring chaos in the brain: a tutorial review of nonlinear dynamical EEG analysis, Intern. J. Neuroscience, 1992, 67: 31-80.

[10] Abarbanel, H.D.I., The analysis of observed chaotic data in physical systems, Rev.Mod.Phys., 1993, 65:1331.

[11] Perrone, A.L., Boccaletti, S., Basti, G., Arecchi, T.F., A mutually recursive method to detect and remove noise in chaotic dynamics, SPIE, 1994, 2242:130-139.

[12] Fraser, Swinney, Independent coordinates for strange attractors from mutual information, Phys. Rev. A, 1986, 33:1134.

[13] Kennel M.B., Brown R. and Abarbanel, H.D.I. Phys. Rev., 1992, A45:3403

[14] Arecchi, F.T., Basti, G., Boccaletti, S., Perrone A.L., Adaptive recognition of a chaotic dynamics, Europhys. Lett., 1994, 26:327-332.

[15] Grassberger, P. and Procaccia, I.,Measuring the strangeness of strange attractors, Physica D, 1983, 9:189

[16] Foglietti, V., Del Gratta, C., Pasquarelli, A., Pizzella, V., Torrioli, G., Romani, G.L., Gallagher, W.J., Ketchen, M.B., Kleinsasser, A.W., Sandstrom, R.L., 28-channel hybrid system for neuromagnetic measurements, IEEE Trans-MAG, 1991, 27:2959-2962

Acknowledgements

We are grateful to M. Peresson, V. Pizzella and F. Tecchio for the help in the measurement session, and to H. Kantz for providing the software for the CD calculation.

A New Procedure to Discriminate and Study Different Rhythmical Cortical Activities on the Basis of their Responsiveness to Simple Sensory Stimulation

Narici, L.[1] and Peresson, M.[2]

[1]Dipartimento di Fisica - Università di Roma "Tor Vergata", Roma, Italy; [2]Divisione di Neurologia, Ospedale Fatebenefratelli -S. Pietro, Roma, Italy

Introduction

The investigation of the spontaneous as well as stimulus related rhythmical brain activities has been receiving increasing attention in the recent years. Cognitive functions as well as arise of pathologies, have been correlated with time and frequency behaviour of cortical rhythms [1-9]. However the understanding of such correlations and of the genesis of the rhythms is still in a primitive stage. The erratic behaviour of the power of most of the spontaneous rhythms, appearing in spindles at random instants, and the quite large inter-individual variability caused difficulties when investigating such correlations or looking for functional and/or pathological indicators. Moving from the observation that those brain structures generating spontaneous oscillations have higher responsiveness to sensory stimulation in the same frequency band where such oscillations are detected [10], we propose to classify the activities on the basis of their degree of responsiveness to rhythmic sensory stimulation. Here we present a procedure able to discriminate these activities within the α band on such basis.

The measurement procedure is fast: less then 10 minutes. In the following we describe the stimulation paradigm, data analysis, data presentation that allow for studying frequency selective enhancements or suppressions of the cortical responses in selected frequency windows. This Frequency Responsiveness Procedure (FRP) combines, in one single test, multichannel detection, a finely rate scanned intermittent stimulation that permits to study its effect on the underlying activity and a compact presentation to appreciate the whole rate/frequency/topographical dependency of the spectral responses. This is a promising new tool for the study of cortical responsiveness to sensory stimuli. The classification of the underlying activity, on the basis of responsiveness indicators may become useful in the diagnosis of pathologies that have frequency selective effects on the reactivity of spontaneous rhythms to burst of stimuli such as the Alzheimer's disease [8].

The FRP is tested (under separate visual and somatosensory stimulation) on neuromagnetic measurements on healthy normal subjects: a first non-invasive study of the frequency responsiveness within the extended alpha band (6 Hz to 14 Hz) [11]. The degree of irregularity of the rhythms, as well as the intra- and inter-individual variability is quite reduced. Robustness of the test and stationarity of the responses is improved. Several of the results presented in this work suggest that we are eliciting similar responses than those obtained with complex cognitive experiment. This would help in the understanding of the linking mechanism between cognitive functions - sensory stimuli - rhythmical activity.

Methods

The stimulation consists in intermittent burst of stimuli (10 stimuli each burst) at a given rate. The pause between the burst is fixed to 1.3 s in this test. The entire stimulation epoch (burst plus non stimulated period) is repeated several times for each rate, in order to perform off-line averaging of the response spectra. The rates are scanned from 6.0 Hz to 14.0 Hz at step of 0.5 Hz. Preliminary studies are used to set the number of repetition to 10 (but probably 5 could have been enough to achieve comparable results), and to ruled out possible influence of the scanning order of the stimulation.

The FRP was tested on five normal subjects. The visual stimuli were short light pulses delivered by two red leds, mounted on a blackened scuba-mask (about 5 cm from the eyes, at an eccentricity of about 5 degrees) in the left hemifield. The subject was asked to maintain fixation on two green leds, constantly 'on', positioned in the mask with zero eccentricity. The somatosensory stimuli were current pulses ($\approx$0.1 ms long) delivered to the right median nerve at wrist with an amplitude just above motor threshold. In the somatosensory modality we tested two different subject conditions (open and closed eyes) to investigate possible cross-modality interactions between somatosensory stimulation and α activity. One minute of spontaneous activity was measured from the same cortical region at the beginning of each session to be used as baseline.

The magnetic field associated with the bioelectrical activity evoked by the stimulation was measured with the 28 channel superconductive system in use at the CNR-IESS laboratory in Rome[1] [12]. The data were acquired at a rate of 1.0 kHz (recording bandwidth 0.48 Hz - 250 Hz), and stored on optical disk for the off line analysis.

Analysis is performed separately in two segments in each epoch. The first during the stimulation, and the second following it, during the pauses (starting one inter-stimulus interval after the delivery of the last stimulus in the previous burst). Each segment is 1024 ms long. In each segment we calculate the spectrum, and then average them over the 10 repetitions relative to the same stimulation rate. The procedure is repeated for all stimulation rates. The spectra of the spontaneous activity is calculated during fifteen 4 s long periods.

For each channel seventeen averaged spectra for the stimulated and non stimulated periods (corresponding to the seventeen rates of stimulation), as well as fifteen spectra for the spontaneous activity (corresponding to the fifteen 4 s long time intervals) have been calculated. These spectra are collected in three interpolated maps, where the amplitude of the frequency response is gray mapped and the two independent variables are the frequency of the response and either the rates of stimulation (for the stimulated and non stimulated period maps) or the time (for the spontaneous maps). For each analysis (spontaneous, during stimulation, during pauses) we collected all the interpolated maps in a composite image, where the maps relative to each channel are positioned at the approximate scalp location of the channel.

Results

Complete results are shown for subject S1 in Fig. 1. In both modalities we show the composite images relative to the time behavior of the spontaneous activity (left) as well as to the rate behavior of the responses during stimulation and pauses (center and right). For the somatosensory test both open and closed eyes sets of results are presented. In Fig. 2 we present the results for three subjects (only somatosensory test, stimulated period, open eyes) to show the good inter-individual reproducibility.

<u>visual test</u>   In the spontaneous activity we observe a single, broad, irregularly intermittent peak at about 10 Hz. Responses tho the FRP procedure show at least two active frequencies: $\approx$ 10 Hz and $\approx$ 12 Hz. Both of them present during stimulation ('driven'), closer to the calcarine fissure, while only the $\approx$10 Hz activity is present in the non stimulated period, confined to the parieto-occipital cortical region. When stimulating with r<10 Hz all the activities, during and following stimulation, are strongly attenuated. A hint for a negative shift in frequency of the major 10 Hz peak between the spontaneous measurement and the measurements during pauses is apparent after careful investigation. This is evidenced in Fig. 3 where we show the ratio between 'spontaneous' and 'pause' mean frequency of the 10 Hz peak for the 7 parieto-occipital channels (see also [11,13]).

<u>somatosensory test</u>   In the somatosensory test it is strikingly evident the reduction of the parietal $\approx$ 10 Hz activity when opening the eyes. This is the known behavior of the α rhythm. The spontaneous activity is almost absent with the eyes open. With the eyes closed it is centered at about 10 Hz, and it is very strong in the parietal region, weaker and more irregular in the rolandic region. The driven activity during stimulation is present for most of the stimulation rates (slightly stronger at $\approx$10 Hz) and it is confined in the rolandic region, across the rolandic fissure. A much weaker $\approx$ 10 Hz activity can be found also during pauses, both in the rolandic and in the parietal regions, even in the open eyes condition, where very little spontaneous activity was found. Note the attenuation for r < 10 Hz similar to what found during visual test; this is not as evident, however, for all the studied subjects.

Discussion

Recent investigations have suggested that activities in the α band may have different functional correlates. Many authors prefer to divide the α band into a 'lower' sub-band (6-10 Hz) and a 'higher' sub-band (10-14 Hz), the former recently related to activity linked mainly to attentional processes, the latter to activity linked to stimulus correlated cognitive processes [5]. The results obtained with the FRP, show different 'reactivities' within the α band, and support this subdivision idea providing information about the interaction of these rhythms (and their functional correlates) with simple sensorial stimulation. For example, we can speculate that the apparent frequency shift (Fig. 3) could be explained referring to the recent work by Klimesh and coworkers [5] where it was demonstrated a decrease in α frequency when attentional demands and/or memory loading was increased. In our case we can speculate that the rate 'decoding' process (needed to allow for the rate-dependent response behavior) might be linked to more attentional

---

[1] At the time of the visual measurements, this system featured 16 vertical gradiometers plus 12 outer planar gradiometers. To simplify data interpretation, we are presenting only the results coming from the inner 16 channels. One of the inner 16 channels was not working properly during the visual measurements, therefore we present results from a total of 15 channels. The somatosensory measurements were performed with a modified version of the system, where the 12 outer planar gradiometers were substituted by 12 magnetometer, three of which used for electronic noise cancellation. These results are therefore presented from a total of 25 channels.

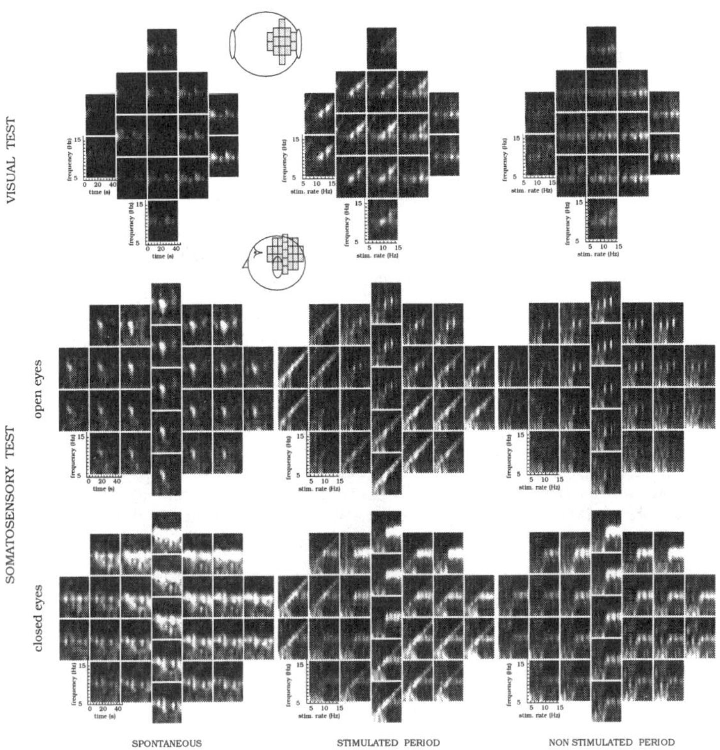

FIGURE 1 Composite images relative to subject S1. Each interpolated, gray scaled map (white: largest amplitude) is the collection of all the spectra measured by the same channel, and represent the amplitude of the responses in function of time and frequency (left, spontaneous measurements), or in function of the stimulation rate and frequency. The positions of the channels are indicated in the head silhouettes.

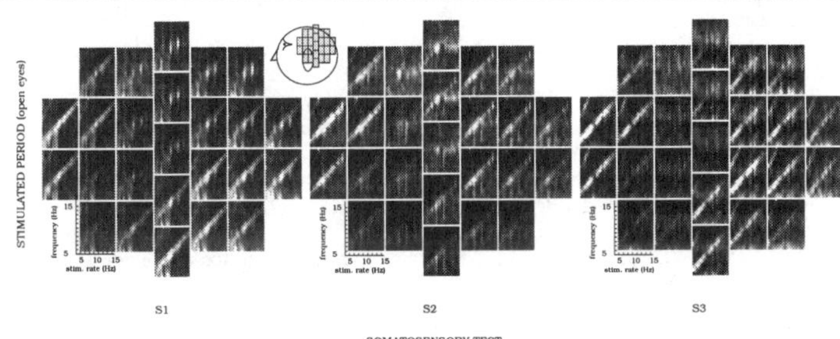

FIGURE 2 Composite images collecting all the spectra relative to three subjects (Somatosensory stimulation, open eyes, during the stimulating burst). The positions of the channels are indicated in the head silhouette.

demands and, therefore, to a decrease in α frequency.

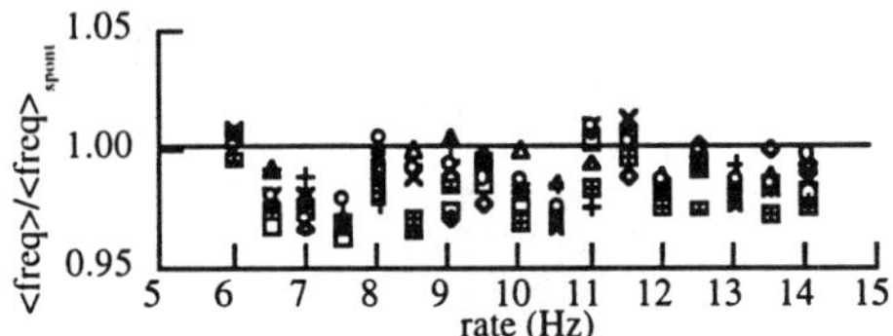

FIGURE 3   Ratio between the mean frequency of the $\approx$ 10 Hz activity during the pauses of the burst stimulation and during the spontaneous activity measurements. Different symbols correspond to seven different parietal channels, where the activity was most prominent.

The rate dependent alpha suppression appears to be modality independent. Studies on the activity in the cortical regions close to the hand and foot areas [6,7] have demonstrated the increase in amplitude of occipital α and concurrent decrease of μ rhythm, following voluntary movements. Our paradigm is somewhat different (the thumb twitch is not voluntary), but for r<10 Hz, the occipital α rhythm (closed eyes) appears strongly attenuated, however for r>10 Hz the rhythm seems unaffected by the stimulation. The same behavior can be observed to a lesser extent in the open eyes condition. Other studies described α suppression effects and related them to memory scanning [3]. In our case we suggest that memory is probably involved for rate decodification, and speculate a different effect on the rhythm whether the rate is faster or slower than the rhythm itself. This effect seems not to be linked to the specific sensory pathway: only the rate decodification process appears to matter. Recent neuromagnetic studies [9] showed activities in the occipital and parieto-occipital cortex. Their frequency content and source location agreed with what found here. The present work evidences that these two rhythms have also distinct sensorial 'reactivity'.

References

[1] Basar, E., Lopes, F., Hari, R., Schürmann, M. Eds. Alpha Processes: Sensory and Cognitive Behavior, Boston, Basel, Berlin, Birkäuser *in press*
[2] Basar, E. and Schürmann, M., Functional aspects of evoked alpha and theta responses in humans and cats, *Biol. Cybern. in press.*
[3] Kaufman, L., Curtis, S., Wang, J.Z., and Williamson, S.J. Electroencephalogr., Changes in cortical activity when subjects scan memory for tones, Clin. Neurophysiol., 1991, 82: 266-284
[4] Klimesch, W., Schimke, H., Ladurner, G., and Pfurtscheller, G. J., Alpha frequency and memory performance, Psychophysiol., 1990, 4: 381-390
[5] Klimesch, W., Schimke, H. and Pfurtscheller, G., Alpha frequency cognitive load and memory performance, *Brain Topography*, 1993, 5:241-251.
[6] Pfurtscheller, G., Event-related synchronization (ERS): an electrophysiological correlate of cortical areas at rest, *Electroencephalogr. Clin. Neurophysiol.*, 1992, 83:62-69.
[7] Pfurtscheller, G., Neuper, C., Event-related synchronization of mu rhythm in the EEG over the cortical hand area in man, *Neurosci. Lett.*, 1994, 174:93-96.
[8] Politoff, A.L., Monson, N., Hass, P., and Stadter, R., Decreased alpha bandwidth responsiveness to photic driving in Alzheimer disease, *Electroencephalogr. Clin. Neurophysiol.*, 1992, 82:45-52.
[9] Salenius, S., Kajola, M., Thompson, W.L., Kosslyn, S., Hari, R., Reactivity of magnetic parieto-occipital alpha rhythm during visual imagery, Electroencephal. Clin., Neurophysiol., 1995 in press.
[10] Basar, E., EEG-Brain Dynamic, Amsterdam, Elsevier, 1980
[11] Narici, L., Peresson, M., Discrimination and study of rhythmical brain activities in the α band: a neuromagnetic frequency responsiveness test, Brain Res., 1995, 703:31-44
[12] Foglietti, V., Del Gratta, C., Pasquarelli, A., Pizzella, V., Torrioli, G., Romani, G.L., Gallagher, W.J., Ketchen, M.B., Kleinsasser, A.W., Sandstrom, R.L., 28-channel hybrid system for neuromagnetic measurements, IEEE Trans-MAG, 1991, 27:2959-2962..
[13] Gaeta, R., Benzi, R., Peresson, M. and Narici, L., Singular Spectrum Analysis applied to neuromagnetic signals, 1996, this volume.

Acknowledgements

The authors express gratitude to V. Pizzella and F. Tecchio for the help during the measurement sessions.

Cognitive Temporal Binding and its Relation to 40Hz Activity in Humans: Alteration During Dyslexia

Ribary, U.[1], Joliot, M.[2], Miller, S.L.[3], Kronberg, E.[1], Cappell, J.[1], Tallal, P.[3] and Llinás, R.[1]

[1]Center for Neuromagnetism, Dept. of Physiology and Neuroscience, New York University Medical Center, New York, USA; [2]SHFJ, DRIPP-CEA, Orsay, France; [3]Center for Molecular and Behavioral Neuroscience, Rutgers University, Newark, USA

Introduction

Spontaneous oscillatory electrical activity in the human brain at a frequency near 40-Hz and its reset within thalamo-cortical systems by sensory stimulation has been proposed earlier [1-3] and has been suggested to be related to cognitive processing and to the temporal binding of sensory stimuli [3,4]. In particular, we demonstrated spontaneous 40-Hz coherent magnetic activity in the awake and in REM sleep states that is reduced during delta sleep [3]. This 40-Hz magnetic oscillation has been shown to be reset by sensory stimuli only in the awake state but not during REM or delta sleep. Because spontaneous 40-Hz oscillation was seen in wakefulness and in dreaming, we proposed it as a correlate of cognition, probably resultant from coherent 40-Hz resonance between thalamocortical specific and nonspecific loops [3].

Using Magnetic Field Tomography (MFT) on humans, we earlier demonstrated that such 40-Hz coherence displayed well defined cortico-subcortical correlations with a time shift that is consistent with thalamo-cortical conduction times [2]. In particular, the MFT results indicated that the onset of activity at the thalamic level was followed by widespread activation of the thalamo-cortical system which resulted in large coherent thalamo-cortical 40-Hz oscillation, organized in space and time, which is altered in pathological states such as Alzheimer's disease [2].

40-Hz oscillatory activity has been suggested to subserve cognition by conforming relevant sensory information into specific time segments, that implement cognitive binding [3]. It was also proposed that 40-Hz binding abnormalities—a dyschronia [5]—could be one of the neurophysiological correlates of the temporal deficits recorded in reading disabilities [6]. In particular, we were interested 1) in testing this hypothesis and in determining specifically whether the minimal interval required to identify separate auditory stimuli correlates with the reset of the 40-Hz magnetic signal, and 2) in analyzing any alterations in the 40Hz magnetic activity in subjects with language-based learning disabilities (dyslexia).

Methods

We used a 37-channel MEG system (BTi), in order to record magnetic activity from nine awake human subjects in response to one or a pair of auditory clicks [4], presented at interstimulus intervals of 3, 6, 9, 12, 15, 18, 24, and 30 msec. The subjects were asked to lie on a bed with their eyes closed and to stay awake and attentive. The magnetic-sensor array was positioned over the auditory area of the right hemisphere. Stimulus pairs were presented with interstimulus-pair intervals of 130 ± 15 msec. MEG activity was recorded from 10 msec before to 100 msec after the onset of the first stimulus (bandpass 1-400 Hz, sample rate 1041 Hz). Transient responses were averaged using the onset of the first stimulus as a trigger.

In order to analyze differences in magnetic recordings from subjects with language-based learning disabilities (diagnosed at Rutgers University), we further modified the technical approach within the paradigm, and implemented a new set of recordings on 9 healthy controls and on 6 dyslexic subjects. In particular, we increased the recording epoch to 140 msec (-10, 0, 130 msec), and changed the presentation of interstimulus intervals to 6, 12, 15, 18, 21, 24, 30 and 36 msec. The Stimulus pairs were presented with interstimulus-pair intervals of 160 ± 15 msec. Our laboratory is presently equipped with a dual 37-channel MEG system (BTi) which allowed us to record, for the first time, bilateral magnetic activity from 4 control subjects and 2 dyslexic subjects, taking into account the responsiveness of both brain hemispheres.

Two distinct models were utilized in the analysis of MEG data. Data-simulation-procedures were implemented in this study in order to account for the MEG data [4]. Following magnetic recordings, a perceptual task was performed on each subject [4], using 6 sets of 2 click presentations with the same interstimulus intervals, and the perceptual threshold was computed.

Results and Discussion

A power spectral analysis of the raw data revealed a significant component near 40 Hz, indicating the presence of a synchronized 40-Hz event in the human brain [4]. The data was filtered at 20-50 Hz for further analysis to remove a 10-Hz component, which modified the baseline in the raw data in each of the 37 channels differently, depending on the 10-Hz component in that particular channel. The high frequency rhythm near 40 Hz in the raw data was well correlated with the 40-Hz response in the filtered data.

Data showed a clear reset of 40-Hz oscillatory activity in control subjects, followed by a stable synchronized and coherent time-locked response with respect to one auditory click [4]. We define the 40-Hz response produced by a single stimulus as a 2.5 oscillatory cycle, demonstrating two and a half oscillations at 40-Hz. Using the dual probe MEG system in control subjects, preliminary data indicate a similar response in both hemispheres. In dyslexics, preliminary data suggest an unstable reset and a prolonged period in synchronized activity in response to auditory stimuli.

Our evidence supports the hypothesis that the minimal interval required to identify separate auditory stimuli correlates with the reset of the 40-Hz magnetic signal (Fig.1). In particular, perceptual tests in control subjects indicated that 15-20 msec are required for the perceptual separation of two stimuli, as reported earlier in the literature. Neuropsychological and psychophysical observations indicated that the auditory system is capable of tonality discrimination of two stimuli that are separated by only 1-2 msec [7], while 15-20 msec are required for the perceptual separation of two stimuli [8]. Experimental and modeling results, based on MEG data, further indicated a stimulus-interval-dependent response with a critical interval of 12-15 msec. At shorter intervals only one 40-Hz response, to the first stimulus, was observed. With longer intervals, a second 40-Hz wave appeared, which coincided with the subject's perception of a second distinct auditory stimulus [4]. These results indicate that oscillatory activity near 40-Hz, recorded during the first 100 msec post stimulus, represents a neurophysiological correlate to the early temporal processing of auditory stimuli underlying cognition.

Using the modified paradigm, perceptual tests in control subjects confirmed our previous findings, that 15-20 msec are required for the perceptual separation of two stimuli. In subjects with language-based learning disabilities, preliminary results indicated a delayed perceptual threshold of 20-30 msec for the identification of a second auditory stimulus (Fig.2), in agreement with earlier findings [9]. In addition, our findings indicate a different time interval necessary for the appearance of a second 40-Hz wave (Fig.3). In particular, data may suggest an increased time period for the processing of incoming stimuli, resetting to the first stimulus, or resetting to the second stimulus and deleting, at least parts, of the first stimulus. This would indicate that, by

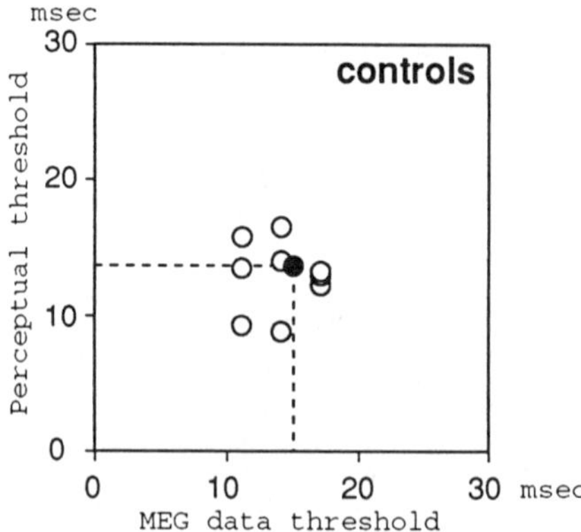

Fig.1: Relationship between perceptual-threshold (average: 13.7 ± 1.32 msec) and MEG-threshold (average: 15.0 ± 2.6 msec) for identifying two clicks with a minimum interstimulus interval for control subjects. The graph plots the thresholds for each of the 9 subjects (hollow circles) and the mean across all subjects (filled circle). (Modified from [3]).

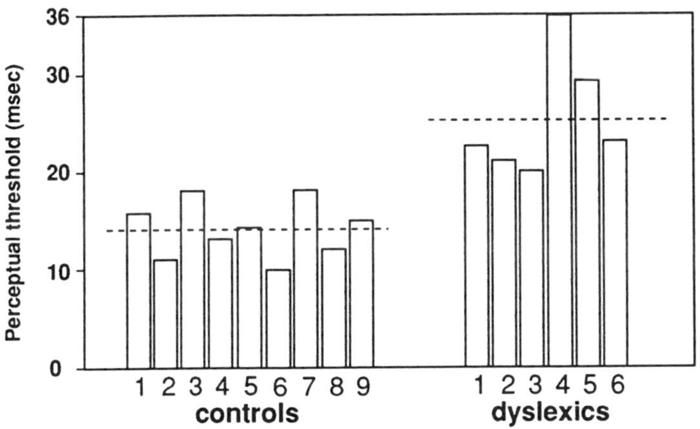

Fig.2: Perceptual-thresholds (minimum interstimulus intervals) required for identifying two separate clicks for 9 individual control subjects and for 6 subjects with language-based learning disabilities. Dashed lines indicate the average values for controls (14.06 ± 2.95 msec) and dyslexics (25.29 ± 6.16 msec).

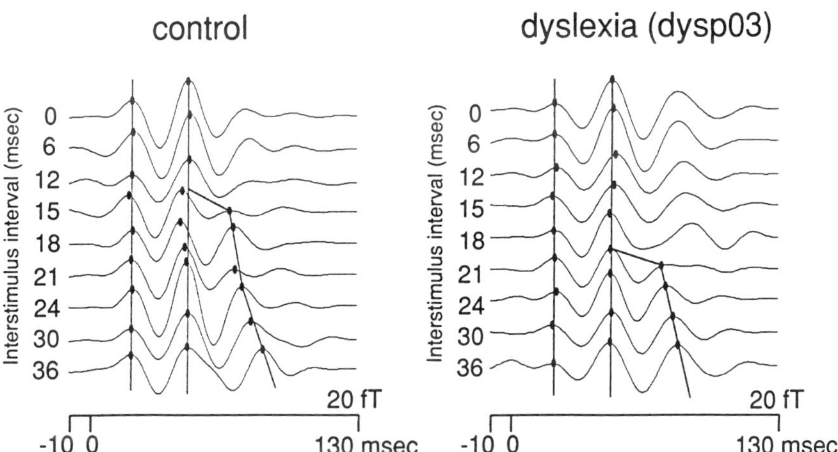

Fig.3: Effect of increasing the interstimulus interval on MEG activity in one control and in one dyslexic subject. The largest of the 37 responses, obtained by the single-click experiment and filtered at 20-50 Hz, was selected and that same channel recording is the actual one illustrated for different interstimulus intervals (6-36 msec). The peak latencies of the responses are marked and the solid lines indicate the reset to one or two clicks. Note that the minimal interval required for a second reset in 40Hz activity is increased in the dyslexic subject (around 21 msec) compared to the control (around 15 msec).

using our recording and analysis procedures, it may be possible to distinguish different levels of severity or different subgroups of dyslexia and to find a neurophysiological marker underlying sensory information processing of auditory stimuli in subjects with language-based learning disabilities, as a powerful new methodology to understand and detect dyslexia for further therapeutical procedures [10]. However, a large population of subjects will be necessary before this phenomena can be characterized more completely.

References

[1] Llinás, R. Intrinsic electrical properties of mammalian neurons and CNS function, In: Fidia Research Foundation Neuroscience Award Lectures, Raven Press, New York, 1990, 4: 175-194.

[2] Ribary, U., Ioannides, A.A., Singh, K.D., Hasson, R., Bolton, J.P.R., Lado, F., Mogilner, A., and Llinás, R. Magnetic field tomography (MFT) of coherent thalamo-cortical 40-Hz oscillation in humans, Proc. Natl. Acad. Sci. USA, 1991, 88: 11037-11041.

[3] Llinás, R., and Ribary, U. Coherent 40-Hz oscillation characterizes dream state in humans, Proc. Natl. Acad. Sci. USA, 1993, 90: 2078-2081.

[4] Joliot, M., Ribary, U., and Llinás, R. Neuromagnetic coherent oscillatory activity in the vicinity of 40-Hz coexists with cognitive temporal binding in the human brain, Proc. Natl. Acad. Sci. USA, 1994, 91: 11748-11751.

[5] Llinás, R. Is dyslexia a dyschronia? Annal NY Acad. Sci., 1993, 682: 48-56.

[6] Tallal, P., Miller, S. and Fitch, R.H. Neurobiological basis of speech: A case for the preeminence of temporal processing, Annal NY Acad. Sci., 1993, 682: 27-47.

[7] Miller, G.A., and Taylor, W.G. The perception of repeated bursts of noise, J. Acoust. Soc. Am., 1948, 20: 171-182.

[8] Hirsh, I.J. Auditory perception of temporal order, J. Acoust. Soc. Am., 1959, 31: 759-767.

[9] Tallal, P. Auditory temporal perception, phonics, and reading disabilities in children, Brain and Language, 1980, 9: 182-198.

[10] Tallal, P., Miller, S.L., Bedi, G., Byma, G., Wang, X., Nagarajan, S.S., Schreiner, C., Jenkins, W.M., and Merzenich, M.M. Language comprehension in language-learning impaired children improved with acoustically modified speech, Science, 1996, 271: 81-84.

Acknowledgements

The support of BTi and the excellent technical assistance by Richard Jagow is greatly acknowledged. This study was supported by The Charles A. Dana Foundation.

A Technique for the Identification of Hippocampal Theta from MEG Data

Tesche, C. D.

Low Temperature Laboratory, Helsinki University of Technology, 02150 Espoo, Finland

Introduction

Rhythmic macroscopic field potentials in hippocampus between 4–12 Hz (theta) are prominent during attention and exploratory behavior in rats and other small mammals [1]. Theta oscillations modulate the firing of individual hippocampal and entorhinal cortical neurons and are believed to be essential for normal memory trace formation [2]. Observations of hippocampal theta in primates are scarce. There are only two reports of human hippocampal theta, both from depth electrode recordings in patients [3], and no reports of the observation of ongoing spontaneous activity in normal human hippocampus. Population responses in human hippocampal formation have been inferred from averaged MEG responses evoked by attention to oddball targets embedded in a sequence of standard stimuli [4,5]. This paper describes a method utilized to determine ongoing spontaneous activity in normal human hippocampus from MEG data recorded with a whole-head array [6].

Methods

MEG data was recorded with a 122-channel array of planar gradiometer detectors covering the whole scalp (Neuromag-122^{TM}). Seven healthy right-handed adults performed two tasks after providing informed consent. In a mental calculation task, subjects silently added and subtracted numbers corresponding to the position in the alphabet of printed letters. For example, c a m g ... was the calculation 3 + 1 - 13 + 7 .... The second task was passively viewing a picture. Data was recorded in 60 s epochs (bandpass filter 0.03–330 Hz, sampling rate 1 kHz). A vertical electro-oculogram was used to identify eye movements and blinks.

Current flow at a set of source locations in the neighborhood of the hippocampus was modelled as a vector sum of three orthogonal equivalent current dipole (ECD) sources. Parietal-occipital alpha and rolandic mu, if prominently represented in the data, were modelled by equivalent current dipole (ECD) sources determined from a least-squares fit between the FFT of the data at specific frequencies and simulated magnetic field patterns [7]. Signal-space projection (SSP) [8,9] was used to derive waveforms for cortical and hippocampal sources from the raw data. Magnetic field patterns were calculated for a set of N ECDs using a spherical head model for superficial sources and a single-compartment boundary-element model for the conducting volume of the head for deeper sources [10]. The 122 sensor readings computed for the ith ECD defined a unit vector s_i in a 122–dimensional space. The measured MEG signal M(t) defined a time-dependent vector in the same space, where

$$\mathbf{M_T}(t) = \sum_{i=1}^{N} a_i(t)\mathbf{s}_i + \mathbf{n}(t). \tag{1}$$

The signal-space waveform $a_i(t)$ describes the strength of the component $\mathbf{s}_i$ as a function of time. The time-dependent component $\mathbf{n}(t)$ includes both intrinsic system noise and current flow within the brain which are not explicitly modelled by the vectors $\mathbf{s}_i$. A pseudoinverse of the source matrix $\mathbf{S} = (\mathbf{s}_1, \mathbf{s}_2, \ldots, \mathbf{s}_N)$ determines SSP waveforms for the N sources $\mathbf{a}(t) = \{a_1(t), a_2(t), \ldots, a_N(t)\}^T$,

$$\mathbf{a}(t) = \mathbf{S}^+ \mathbf{M_T}(t). \tag{2}$$

where $\mathbf{S}^+ = [\mathbf{S}^T\mathbf{S}]^{-1}\mathbf{S}^T$. This approximation yields an unbiased estimate for $\mathbf{a}(t)$ when $\mathbf{n}(t)$ is spatially uncorrelated, normally distributed noise [9].

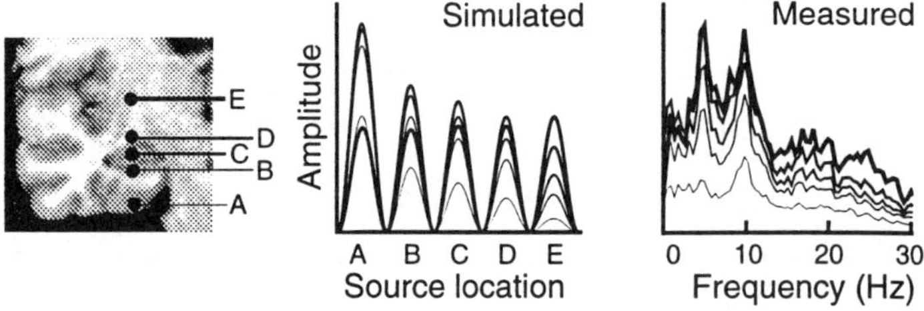

Fig. 1. SSP waveforms computed for a family of "probe" ECD sources in the neighborhood of left anterior hippocampus (left). The orientations of the ECDs are perpendicular to the plane of the MR image. The simulated data contains an isolated source at one of the locations A–E. An example of SSP spectra computed from the data for the same set of "probe" ECDs is shown on the right. The task was mental calculation.

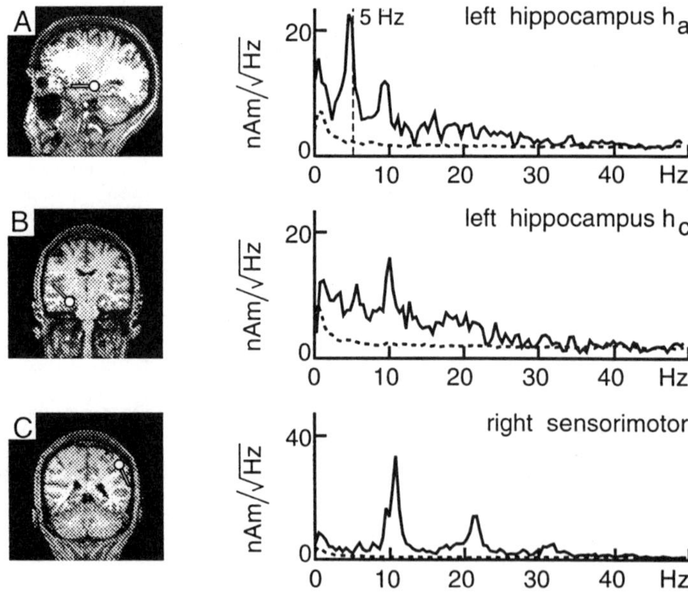

Fig. 2. MR images of the locations and orientations of ECDs in (A–B) left anterior hippocampus and (C) right sensorimotor cortex. Spectral densities as a function of frequency for 5.6 s epochs of mental calculation for each source are shown on the right. The dashed lines indicate the system noise for each source.

There is no unique solution to the inverse problem in MEG. However, waveform scaling has been applied to the study of evoked hippocampal responses to attended auditory oddballs to check the consistency between the SSP results and the MEG data [5]. In this study, waveform scaling was extended to the frequency domain. Averaged spectra were computed from the SSP waveforms for a set of "probe" ECD components with locations in the vicinity of the hippocampus. Results were obtained for the same set of "probe" ECDs for data generated by sources located in the neighborhood of the hippocampus. Fig. 1 shows that the dependence of the spectral components on the location of the "probes" is consistent with that observed from simulations for results generated by an ECD source at location D.

Fig. 2 shows results for subject #4. Both spectral peaks at 5 and 9 Hz and considerable broadband spectral components are visible for activity in hippocampus. In contrast, in sensorimotor cortex the spectral peaks at 11 and 22 Hz (rolandic mu) are much larger than the broadband spectral components.

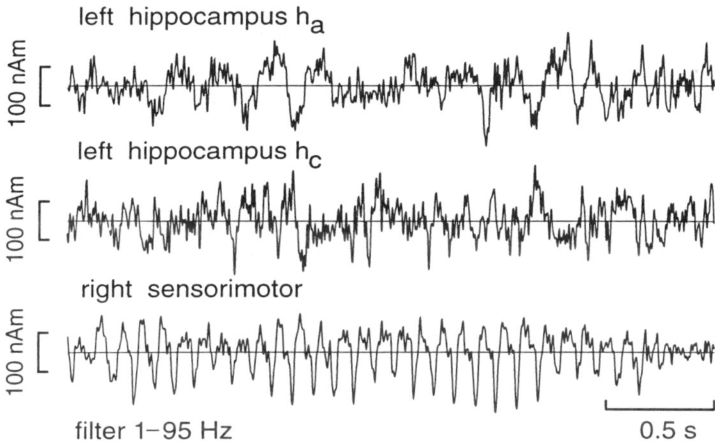

Fig. 3. Waveforms for hippocampal and sensorimotor activity for a 3 s interval of the epoch used to generate the spectra shown in Fig. 2.

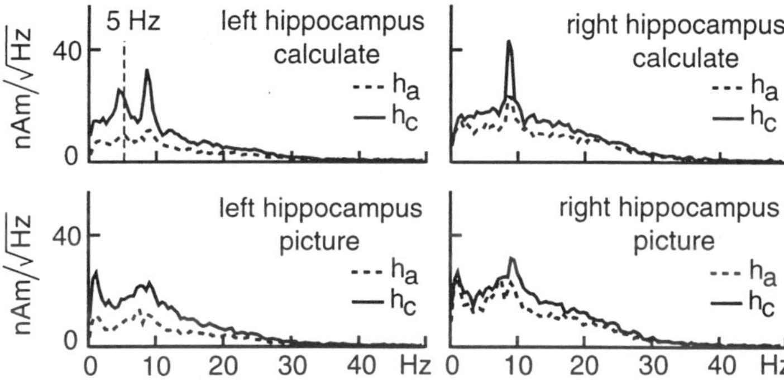

Fig. 4. Spectra for hippocampal activity during the performance of two different tasks; mental calculation (calculate) and passive viewing of a picture (picture). Additive system noise has been subtracted from these spectra.

Short bursts of oscillatory activity were only occasionally seen in hippocampal waveforms (Fig. 3) In contrast, the relatively smaller broadband component and the harmonic structure of the sensorimotor spectra is reflected in the well-defined comb-shaped waveform typical of rolandic mu.

Task-dependent features of hippocampal activity are difficult to extract directly from the time-domain data. As a result, the ability to compute detailed averaged spectra for specific components of the signal [9] is an important tool for the study of human hippocampal population activity. Fig. 4 shows task-dependent features in spectra averaged over 60 s epochs (subject #1). A spectral peak at 5 Hz is present in component h_c during mental calculation, but not during passive viewing of a picture. Spectral peaks at 9–10 Hz also show task dependent features.

Discussion

Spectra consistent with population activity in anterior hippocampus were derived from whole-head MEG data. Broadband spectral components were present both during mental calculation and passive viewing of a picture, with spectral peaks between 4–7 Hz particularly apparent during mental calculation. The application of new analysis tools for MEG data, such as signal-space projection and frequency-domain waveform scaling, opens up the possibility of investigating the function of population activity in normal human hippocampus.

References

[1] Grastyan, E., Lissak, K., Madarasz, I. and Donhoffer, H. Hippocampal electrical activity during the development of conditioned reflexes. Electroenceph. Clin. Neurophysiol., 1959, 11: 409–430. Vanderwolf, C.H. Hippocampal electrical activity and voluntary movement in the rat. Electroenceph. Clin. Neurophysiol., 1969, 26: 407–418.

[2] Buzsáki. G., Leung. L.S. and Vanderwolf, C.H. Cellular basis of hippocampal EEG in the behaving rat. Brain Res., 1983. 6: 139–171.

[3] Arnolds. D.E.A.T., Lopes da Silva, F.H., Aitink, J.W., Kamp, A. and Boeijinga, P. The spectral properties of hippocampal EEG related to behaviour in man. Electroenceph. Clin. Neurophysiol., 1980, 50: 321–328. Halgren, E., Babb, T.L. and Crandall, P.H. Human hippocampal formation EEG desynchronizes during attentiveness and movement. Electroenceph. Clin. Neurophysiol., 1978, 44: 778–781.

[4] Okada, Y.C., Kaufman, L. and Williamson, S.J. The hippocampal formation as a source of the slow endogenous potentials. Electroenceph. Clin. Neurophysiol., 1983, 55: 417–426.

[5] Tesche, C., Karhu, J. and Tissari, S. Non-invasive detection of neuronal population activity in human hippocampus. Cog. Brain Res., in press.

[6] Tesche, C.D., Non-invasive detection of theta activity in human hippocampal formation with a whole-head magnetoencephalographic array. Soc. Neurosci. Abstr., 1996, 22 (1): 276.

[7] Tesche, C.D. and Kajola, M. A comparison of the localization of spontaneous neuromagnetic activity in the frequency and time domains. Electroenceph. Clin. Neurophysiol., 1993, 87: 408–416.

[8] Ilmoniemi, R.J. and Williamson, S.J. Analysis for the magnetic alpha rhythm in signal space. Soc. Neurosci. Abstr., 1987. 13 (2): 46.

[9] Tesche, C.D., Uusitalo, M.A., Ilmoniemi, R.J., Huotilainen, M., Kajola, M. and Salonen, O. Signal-space projections of MEG data characterize both distributed and well-localized neuronal sources. Electroenceph. Clin. Neurophysiol., 1995, 95: 189–200.

[10] Hämäläinen, M. and Sarvas, J. Realistic conductivity geometry model of the human head for interpretation of neuromagnetic data. IEEE Trans. Biomed Eng., 1989, 36: 165–171.

Acknowledgements

I thank J. Karhu for useful discussions and R. Hari for comments on the manuscript. This work was supported by NIH National Research Service Award 1 F33 NS09623-01H from the NINDS. The MRI scans were obtained at the Radiology Department of the Helsinki University Central Hospital.

VI. Magnetoencephalography (MEG): Clinical Studies

MEG: Clinical Applications

[1]Mäkelä, J.P., [2]Elbert, T., [3]Kakigi, R., [4]Lewine, J., [5]Lopez, L., [6]Nagamine, T., [7]Nakasato, N., [8]Reite, M., and [9]Sannita, W.

[1]*Low Temperature Laboratory, Helsinki University of Technology, Espoo; Central Military Hospital, Helsinki, Finland,* [2]*University of Konstanz, Konstanz, Germany,* [3]*Department of Integrative Physiology, National Institute for Physiological Sciences, Okazaki, Japan,* [4]*The New Mexico Institute of Neuroimaging, Albuquerque, New Mexico, U.S.A.,* [5]*Institute for Advanced Biomedical Technologies, Chieti, Italy,* [6]*Department of Brain Pathophysiology, Kyoto University School of Medicine, Kyoto, Japan,* [7]*Tohoku University, Department of Neurosurgery, Sendai, Japan,* [8]*University of Colorado, Health Sciences Center, Denver, Colorado, U.S.A.,* [9]*Center of Neuropsychoactive Drugs, University of Genova, Genova, Italy; State University of New York, New York, U.S.A.*

During the last two decades, magnetoencephalography (MEG) has become available for monitoring human cortical activity. Although impressive results have been obtained with instruments having smaller coverage, new whole-scalp neuromagnetometers [1;2] began a new era in the neuromagnetic patient studies. The simultaneous activity of different widely spaced cortical areas can now be observed. This is an advantage in studying propagation of activity, *e.g.* in epilepsy, and also in studying functional alterations after lesions, which may occur far from the site of the lesion. Furthermore, individual variation is remarkable even in healthy subjects' MEG responses to simple stimuli. For instance, some control subjects may show gross hemispheric differences in auditory evoked fields (AEFs) to simple tones [3;4]. Increased recording speed allows collection of large normative data bases, necessary for clinical applications.

Non-invasive MEG measurements may help in clinical decision making in individual patients. There are examples of epileptic surgery guided, in part, by MEG information. MEG source localization may also aid in locating functionally irretrievable areas before neurosurgery. In addition, MEG may give information of mechanisms underlying diseases and recovery. Furthermore, one can also view lesions in neurologic diseases as "experiments of nature" and record MEG in patients to obtain indirect information about normal brain functions.

Search for MEG clinical applications should be directed to areas where no good objective tools exist at present, or where the nowadays applied methods are inaccurate, expensive or risky for the patients. The non-invasiveness and excellent time resolution of MEG should be taken into account in search of such alternatives. The following chapters describe approaches which are potentially interesting in this respect.

Functional localization before neurosurgery

MEG source localization may aid in pinpointing functionally irretrievable areas before neurosurgery. Tumours or vascular malformations may distort brain anatomy so as to render localization of. *e.g.*, motor areas impossible on the basis of anatomic landmarks. Functional localization of the eloquent cortical areas may prevent postoperative hemiparesis, or to encourage neurosurgeon to operate in an otherwise unclear situation.

MEG sources are now routinely superimposed on MR images using co-registration techniques. An MRI-linked whole head MEG system [2] has been applied for more than 400 subjects including normal volunteers and patients with various types of neurological diseases. The sources of the somatosensory evoked fields (SEFs) to median nerve stimuli are localized on the posterior wall of the central sulcus [5;6]; this localization can be applied especially usefully in neurosurgery since primary motor cortex, to be protected during operations, lies on its anterior wall. Although MRI scans may identify "anatomically" the central sulcus, it is controversial as to whether the "functional" central sulcus is shifted due to structural lesions such as arteriovenous malformations (AVMs). AVMs are congenital and can lead to central nervous system (CNS) reorganzation; anatomical localization does not take into account the plasticity of brain function. In patients with AVM adjacent to the central sulcus, the N20m sources for the median nerve SEF in the affected hemispheres were localized on the anatomical central sulcus with a 4-mm standard deviation [7]. This was so even in two cases with a large and diffuse AVM occupying almost the entire frontal lobe, suggesting that AVM may not shift the localization of the functional central sulcus.

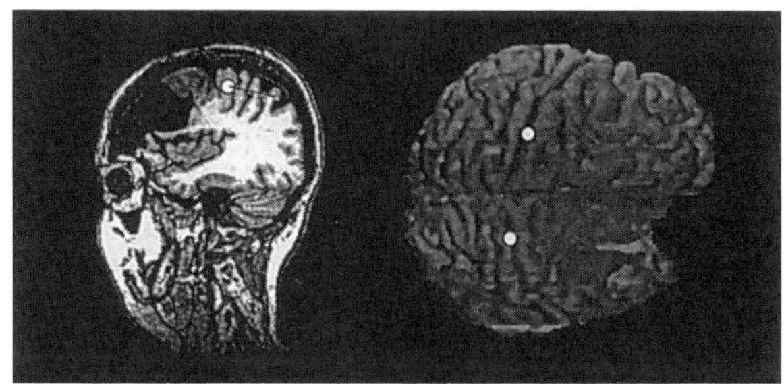

Fig. 1. Sources of SEF P30m deflection overlied on 3-D MR image of a patient after removal of right frontal glioma. Two years after operation, the patient's symptoms reappeared and reoperation was planned. Sagittal section shows that tumor infiltrates the primary motor cortex, immediately anterior to the SEF sources. Because of high probability of postoperative hemiparesis, reoperation was cancelled. Unpublished data, courtesy to Dr. Forss.

The N100m sources of the auditory evoked magnetic fields have been localized in the vicinity of the primary auditory cortex bilaterally on superior temporal surfaces [3;8;9]. In AEFs measured with the MRI-linked whole head MEG system [4], averaged position of the estimated N100m dipoles was precisely on the upper surface of temporal lobes with a vertical 2-mm standard deviation. Disappearance of N100m response [10;11] and delay of the N100m latency [12] have been reported in patients with temporal lobe lesions. Current MEG techniques cannot localize speech-related cortical areas directly. However, functional localization of the auditory cortex is useful for surgery especially on the left temporal lobe, because the left auditory cortex is surrounded by the language-related cortex.

Visual evoked fields (VEFs) are equally applicable in the functional localization of eloquent cortical regions. In recordings of pattern reversal VEF in a patient with homonymous left hemianopsia due to right occipital AVM, a two-dipole pattern of P100m appeared over the occipital area bilaterally. In this patient, the medial part of the calcarine fissure was occupied by the AVM whereas the lateral part was spared [13]. The results suggest good correspondence between functional localization and cortical anatomy. Further MEG studies, especially in patients with brain lesions, are warranted to provide more information about cortical sources of evoked responses.

The parameters of function used in presurgical evaluation need to be stable across populations of patients and age-matched controls. SEF N20m stability has been evaluated in detail [14].The median nerve, thumb and little finger of the two hands were stimulated, and N20m localization, distance between homologous equivalent current dipoles (ECDs) in the two hemispheres, distance between thumb and little finger ECDs, and the difference in this distance in the two hemispheres were calculated. In patients 2-7 mo after hemispheric stroke, at least one of these parameters was aberrant, implying functional reorganization [15] In the study of plasticity (see below) it is also important to have predictive parameters that describe the intra-individual relationship between homologous areas in the two hemispheres.

Functional MR imaging (FMRI) has recently been applied to location of sensory, motor, visual, and auditory activities. Possibilities of making functional measurements with 1.5 T instruments, already available in hospitals, are specially alluring. At present, no clear understanding exists of the spatial relations of neural activity and altered blood flow and changes of oxygenation. Comparison of FMRI and MEG localization of unilateral motor activity resulted in 17 mm difference between the centers of activity estimated with these methods [16]. Methodological improvements of FMRI are progressing rapidly. MEG measurements may give one reference point for this development.

AVMs, containig blood vessels, probably provide difficulties in FMRI localization because of aberrant blood flow and vessel permeability. In addition to activations in the regions expected for the paradigms employed in a FMRI study of AVM patients, qualitatively reproducible activations were observed in unexpected regions. This was attributed to plasticity due to AVM [17]. Comparison with MEG measurements could be beneficial in selecting functionally important activations from those due to AVM.

Detection of abnormal slow-wave activity

Brain lesions are often accompanied by increased EEG activity in the theta and delta bands. Whereas this increase in slowing has been reported in the EEG literature for years, the MEG counterpart of the phenomenon has been characterized only recently. In EEG, slowing is typically diffuse, often encompassing all of the electrodes over one or both hemispheres. In contrast, slowing in the MEG is often focal, especially in patients with circumscribed lesions. MEG field patterns associated with slow waves are typically complex, implying contributions from multiple, simultaneously active brain regions. Nevertheless, for some slow waves, an equivalent current dipole model provides an excellent account of the field pattern, presumably because a relatively focal area of dysfunctional tissue is responsible for its generation. Whereas it would be of interest to disentangle the multiple sources contributing to complex slow waves, presently available analysis algorithms preclude this on a routine basis. It is thus fortunate that clinical information can be ascertained by focusing analysis on the slow waves with a dipolar field configuration.

A challenge in the evaluation of slow waves is distinguishing pathological slowing from normal one, such as that associated with sleep. This is particularly problematic in patients, because they often become drowsy during an examination. To deal with this problem, several investigators have developed specific criteria for identification of pathological slowing [19-22]. One of the most critical of these criteria relates to spatial clustering of sources. In healthy subjects, it is rare to find co-localization of the dipolar sources from separate slow wave events within a $1cm^3$ volume. In patient groups, clustering of sources is common. For example, when clustering analyses are used [19], clusters of slow wave sources are found in less than 10% of normal subjects, invariably at occipital or high parietal locations, but in more than 70% of patients with gross brain lesions, *e.g.*, neoplasms or encephalomalacia associated with cerebrovascular accidents. Fig. 2 shows data from a patient with a clear MRI lesion. Delta activity is obvious within the raw data, and as shown by the isofield contour map, the indicated slow wave has a highly dipolar field pattern. In this case, sources localize all along the lateral margin of the lesion, as shown on the MR section.

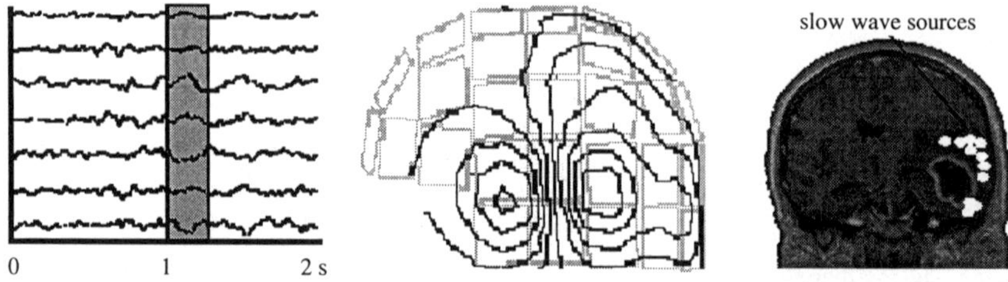

Fig. 2. Left: slow wave data from a patient with a left temporal lesion with cystic cavity, ring enhancement and surrounding edema. Middle: Field pattern of a slow wave, overlied on the sensor helmet of a 122-channel instrument, seen from the left. Right: slow wave sources, overlied on the patient's MRI, localize to the lateral margins of the lesion.

In stroke patients, clustering of slow wave sources is common along the margins of the lesion, in the ischemic penumbra. The data from patients with neoplasms and strokes lends "face validity" to the notion that cluster analysis of slow wave activity provides for good identification of areas of cortical dysfunction. However, a less intense mirror focus of slowing is found in the healthy hemisphere in 20-30% of patients. Also, it is not uncommon to find slow wave sources localizing deep into the white matter. This probably reflects erroneous modelling of an extended cortical sheet with a single dipole. Given this situation, slow wave activity must be interpreted within the entire clinical context, with acknowledgment that source localizations of slow waves are probably accurate only to within 1-2 cm, especially in the depth direction.

One interesting clinical development in the assessment of slow waves has been the identification of significant dipolar slowing in patients that have neurobiological or psychiatric dysfunction, but no gross anatomical lesions. For example, slowing is found in 80% of patients with epilepsy, including those without obvious anatomical changes. In patients with both interictal spikes and slowing, spikes and slow wave sources co-localize. Dipolar slowing is also found in 65% of patients with post-concussive syndromes subsequent to mild head trauma [20]. This is true even in cases with normal MRI, normal CT, and even normal clinical EEG. Other conditions that show slowing include transient ischemia [22], schizophrenia, dementia, and chronic alcohol abuse.

Although the presence of dipolar slow wave clusters is believed to be a strong indication of underlying pathology, the converse is not necessarily true; an absence of slow wave clusters does not always indicate that all is well. Some patients have clearly abnormal delta and theta activity in EEG and MEG power spectra, but no dipolar slow wave clusters. This is believed to reflect the fact that the relevant pathology is not focal. The precise relationship between the location of slow waves and underlying primary pathology must be considered carefully. For example, thalamic lesions generate cortical slow waves [18]. That is, the slowing localizes cortically, even though the critical lesion is subcortical.

Despite the care that must be given to the clinical interpretation of slowing, available data suggest that MEG characterization of slow waves is of clinical utility. MEG slowing provides objective evidence of brain dysfunction that is lacking through other neuroradiological and neurophysiological procedures.

Pain representation in CNS

Pain is a common clinical problem. To date, no objective pain measurement system is available; subjective psychophysical pain assessment is the only alternative. Objective pain indexes would be of considerable interest. To develop them it would be useful to understand CNS pain representation.

Painful electric or CO_2-laser stimulation elicits a complex pattern of activation in MEG [23]. In addition to primary somatosensory activation, bilateral second somatosensory areas (SII) [24;25] and cingulate gyrus [26] are activated. However, the signals about 200 ms onwards are complex, show large intraindividual variation and cannot easily be attributed to dipolar sources [26]. Probably, attentional networks are strongly activated by painful stimuli. Therefore, more experimentation is needed to subtract pain-specific components from those related to general attention.

Cortical plasticity in adult humans

Structure and function of various brain areas are modified following injury or as a result of experience. The follow-up of this mutability by non-invasive methods may be interesting in evaluation of rcovery and results of rehabilitation. One can, in some cases, see objective signs of recovery and reorganization after the lesion in the CNS by MEG. The cortical representation of the sensory perception relates in an orderly way to the spatial arrangement of peripheral receptors. Changing the relative weighting of connections by increased stimulation or by inactivating a pathway may alter these maps.

Somatosensory plasticity

In animal studies, increased use of a limb expands its cortical representation and reduces receptive field size, whereas deafferentation or disuse generates an invasion in the representational zone from nearby sites. Synchronous, behaviorally relevant stimulation at nearby sites (*e.g.,* different fingers) results in the fusion of representations, whereas asynchronous stimulation to two different sites segregates the representations [27;28]. Such a reorganization in adult humans can be detected by MEG. The somatotopic map that spans across a relatively large distance along the posterior wall of the central sulcus can be studied particularly well with this technique.

Patients with unilateral upper extremity amputation have altered topographic representation of the cortical face area and the cortical upper arm (stump) areas which lie adjacent to the former receptive field of the amputated limb. Sources of SEFs elicited by facial stimulation were shifted 1-2 cm towards the receptive field which would normally receive input from the now amputated hand and fingers [29;30]. A similar tendency was observed for the receptive field for the stump of the upper arm. These alterations imply extensive plastic cortical reorganization in adult humans following injury. Furthermore, in human upper extremity amputees a strong correlation exists between extent of cortical reorganization and the amount of experienced phantom limb pain [31] as well as with the amount of body surface from which painful stimuli evoked sensations that were perceived to be emanating from the the phantom limb [32]. Although phantom pain is a disadvantageous consequence of injury, it nevertheless suggests that cortical reorganization can have an outcome of perceptual significance for the individual. These studies also demonstrated fluctuation of cortical reorganization, phantom pain and referred sensations [33]. Furthermore, a pilot investigation, using MEG, suggests that the persistent, incoming stimulation of chronic pain may alter the central representation of body parts related to the pain process. Patients who experience chronic noxious stimulation over extended periods of time seem to produce more extensive activation of neural assemblies than controls, when phasic somatosensory stimuli are applied to body sites related to their chronic pain [34]. It is of interest to try to analyze why alterations related to SI area relate to pain; as seen above, pain-specific responses ocur primarily in the SII-cortices.

The phenomena leading to phantom pain may be aberrant derivatives of the mechanism permitting the processes associated with growth, learning, and adjustment to environmental demands. For instance, the cortical representation of the left hand digits of string players, which carry out the demanding task of fingering the strings, is larger than that in controls [35]. No such differences were observed for the representations of the right hand digits. Furthermore, extensive training of Braille reading enlarges the representation of the reading fingers within days [36]. In patients with syndactyly, where the fingers are congenitally fused, the SEF sources of finger stimulation are not organized somatotopically. When the fingers were surgically separated, they became individually represented in the cortex within weeks [37]. Obviously, the representation of different parts of the body in the human primary somatosensory cortex changes in a use-dependent manner. In amputees, the presumed increase of afferent input from the intact hand, based on the requirement that the hand should be used more to compensate for the loss of the contralateral one, should consequently result in enlargements of the intact finger representatios. This was shown to be so in a study comparing the cortical organisation of the representation of the intact arm and the amputation zone in upper extremity amputees with the one in controls [38].

On cellular level, receptive fields are shaped, in part, by lateral inhibition. A small denervation causes disinhibition of viable, but normally suppressed inputs from areas adjacent to the affected region and leads to the expression of new or expanded receptive fields. These processes may be related to phantom pain. Cortical alterations have, in part, been explained by arborization of thalamocortical neurons that had retained their normal sources of input and that were adjacent to the affected cortical areas that had lost their afferents: however, this mechanism can account for only a small fraction of the massive reorganization covering up to a few cm, observed after the arm amputation in man [29;30]. In addition to cortical alterations, plasticity in the lower levels of somatosensory pathways probably contribute to the effects observed in the cortex. Small changes in the local circuits of subcortical maps can be relayed to cortex and appear as major reorganizations [39].

Auditory plasticity

In a similar vein, MEG can be used to detect alterations of tonotopic organization in the auditory cortex. Several tonotopic maps have been described and their plasticity is currently under investigation: The sources of N100m for higher frequencies lie in progressively deeper locations, whereas the sources of the middle-latency P30m for higher frequencies are located progressively more superficial as frequency increases [40] . Tinnitus, a ringing in the ears, may be an auditory phantom perception. Patients with tinnitus have deviance in their tonotopic map; the amount of deviation is correlated with the subjective tinnitus strength [41]. However, tonotopy of N100m is not always apparent in controls (for references, see *e.g.*, [42]). In animal experiments, frequency specificity of single cells in AI is affected by stimulus intensity [43]. Obviously, carefully delineated studies of *e.g.*, stimulus intensity vs. frequency are needed before reliable clinical studies can be performed.

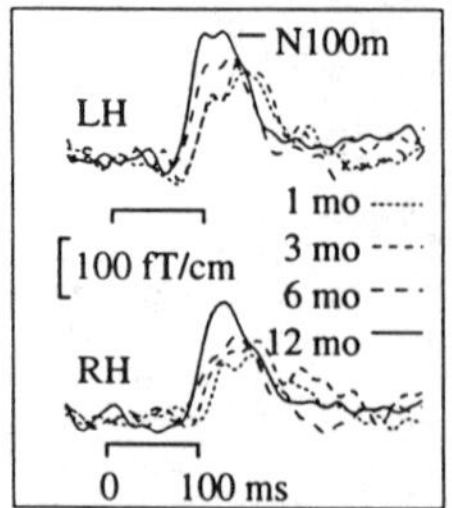

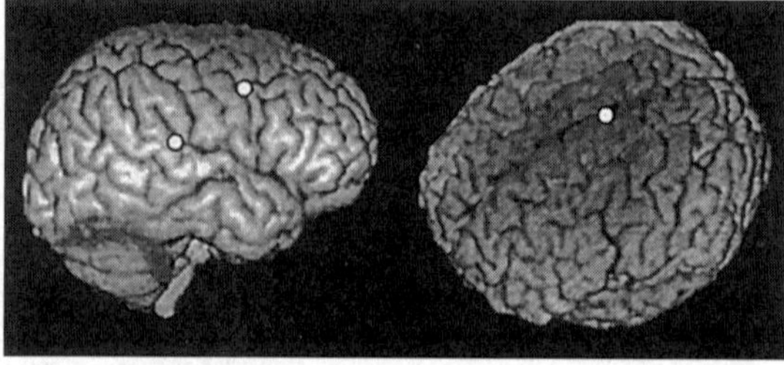

Fig. 3. Left: Development of AEFs to healthy ear stimulation in a patient with left-sided deafness after operation of acoustic neuroma. LH indicates left and RH right hemisphere. Adapted from [44]. Right: Sources of AEFs to non-attended healthy-ear stimulation in a patient after sudden unilateral hearing loss. Adapted from [45].

Other types of auditory reorganization have been observed as well. Figure 3 shows AEFs in a patient deafened in the left ear after an acoustic neuroma operation; the stimuli were delivered to the healthy, right ear. On the basis of plastic somatosensory reorganization, one could expect that the N100m ipsilateral to the stimulated healthy ear would increase in amplitude during the follow-up, because there would be reorganization due to loss of input from the contralateral, operated nerve. This turned out not to be so: the responses were initially dampened in both hemispheres and recovered towards normal values during one year after the operation, even exceeding them in some cases [44]. The activation of both auditory cortices appeared to become more brisk and concise during the recovery. This may be due to enhanced coherence of postsynaptic potentials underlying N100m. No systematic changes were seen in the AEF source locations.

Three out of 8 patients with sudden unilateral hearing loss of unknown origin showed atypical AEFs over head midline. Furthermore, over the right hemisphere, the same three patients had two active cortical areas, an early posterior and a late anterior area. The later, anterior source reflected activation in the frontal association areas ([45]; see Fig.3). It is possible that the increased difficulty in the analysis of auditory stimuli due to unilateral hearing defects has generated a compensatory increase in the processing of stimuli by auditory polysensory pathways and enhanced their cortical source areas so as to make them visible in AEFs.

In animal experiments, cortical and subcortical maps reorganize after restricted damage of the cochlear receptors. The cortical area normally representing frequencies of the lesioned cochlear portion is occupied by neighbouring frequencies. However, this does not mean central compensation of peripheral hearing loss; sensitivity to those frequencies that would normally produce activation in the damaged region of the cochlea is not restored [46]. On the other hand, enhanced mean responses of single cells, and increase of the coherence between activities of different neurons participating in the same processing assembly would lead also to behavioural improvement by means of increased reliability and shortening of required processing time [47]. Little attention has been paid to changes in temporal properties of neural signalling during development of lesion-induced changes, although MEG's superior temporal resolution is optimal for such studies. The results of [44] show that indeed, shortening of the response latencies and increased coherence of neural activity, reflected as larger signals, occur in the human central nervous system during recovery after peripheral nerve lesions.

Diagnostics of motor disorders

Currently available MEG techniques for investigating motor disorders are extensions of approaches previously used in EEG. One possibility to detect abnormality of voluntary movement is to measure movement-related magnetic fields. They are produced, in part, in the primary motor cortex: its location may be shifted by lesions around somatomotor area. Onset of spontaneous movements, detected by electromyography, may be used as a fiduciary point for backward averaging. Jerk-locked back-averaging is an useful tool to explore cortical hyperexcitability in patients with involuntary movements: it can be shown that some myoclonic jerks in patients with progressive myoclonus epilepsy are cortically initiated.

Analysis of magnetic spontaneous activity has been applied to elucidate CNS mechanisms of tremor in parkinsonian patients. If tremor is produced by cortical functions subserving normal movements, somatomotor mu rhythm disruption might occur during tremor as well as during other movements. In patients with hemi-parkinsonism, the spectra of spontaneous activity showed a clear dampening of 10-Hz rhythm during tremor [48]. Apparently, parkinsonian tremor activates cortical mechanisms normally producing rapid, alternating movements. Recently, synchronized and tremor-related MEG activity over the frontoparietal cortical regions was observed in parkinsonian patients. This activity could be attributed to source areas that were previously ascribed for voluntary movements, and was suggested to be generated by rhythmic activity in the diencephalon, generating central motor loop oscillations [49].

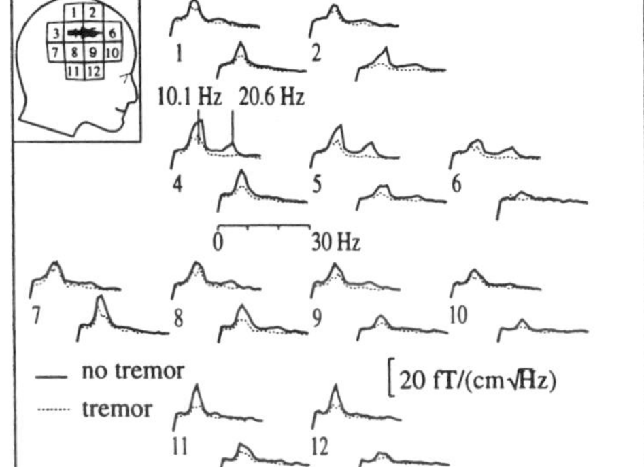

Fig. 4. Dampening of 10-Hz rhythm during parkinsonian tremor. The inset shows the measurement location; the arrow depicts the N20m source for median nerve stimulation. Modified from [48].

MEG may reveal abnormalities in primary sensorimotor cortices and, in special cases, also in premotor and supplementary motor cortices. Activities in basal ganglia can only be estimated indirectly by current MEG techniques. Responses from primary motor areas show extensive variation in normal subjects, exceeding clearly the variability of *e.g.*, SEFs; this should be kept in mind when evaluating signals from patients. However, careful observation of clinical symptoms and carefully planned experimental setups, strenghtened by advances in MEG technique, will undoubtedly clarify the mechanisms of motor disorders.

Pharmaco-MEG

Analysis of alteration of CNS activity by different pharmacological manipulations by MEG is still in its infancy. Complex and interacting subcortical structures modulate cortical activity. It is not possible on the basis of MEG measurements only to conclude at which level the drug effects on cortical/subcortical function take place. Preliminary MEG data, however, suggest that in humans this approach can provide information on drug action that is complementary to that obtained by conventional quantitative EEG methods [50;51]. Chemically specified subsystems exist in sensory systems. A functional dissection of neural circuits by pharmacological manipulation has allowed segregation of evoked response components [52;53], and showed local drug interference at synaptic and non-synaptic level [54]. Cortical and precortical evoked phenomena depend upon characteristics of the information processing in the CNS that *e.g.*, in visual system, can be activated independently and are selectively affected by drugs. Hypotheses exist about the functional roles of dopaminergic, GABAergic and cholinergic transmission in the CNS, and the effects on brain electrophysiology of compounds interfering with these systems have been described [53;55;56]. Interaction between cortical-subcortical acetylcholine modulation and the carrier systems mediating in the transfer of sensory information *e.g.*, visually evoked oscillatory potentials [57;58], is also conceivable. Preliminary human data [59] indicate that scopolamine increases the synchronization of cortical oscillatory response after full-field luminance stimulation while reducing it when stimulation is in spots.

By MEG, it is feasible to study activity within a limited cortical region, to use *e.g.*, dipole modelling to create a "spatial filter" representing activity of a cortical region. Thereafter, it is possible to observe modulations of the activated region when changing different stimulus parameters, or by changing the state of the subject with different pharmacological manipulations.The best example at present is modulation of AEF sources by intravenous lidocaine in patients with tinnitus. In three out of four patients with tinnitus, transient amelioration of the symptoms during lidocaine injection was reflected in shortened and sharpened activation of the N100m source [60].

Anesthesia generates strong effects on brain electric activity. Monitoring these changes is feasible by MEG: This was demonstrated with an experiment, showing burst-suppression pattern of MEG signals generated by enflurane anesthesia in a dog [61]. The result itself is by no means unexpected: however, the experiment demonstrates that it is perfectly feasible to record MEG despite apparatus providing inhalation anesthesia and monitoring the subject state. Selection of epileptic drugs is at present time-consuming. It is possible to monitor epileptic cortical activity by MEG and study *e.g.*, with intravenous injections of how the selected drug alters the activity of the epileptic focus (Knutsson, pers. comm.). The optimal selection of antiepileptic agent may avoid several outpatient visits and additional epileptic seizures during tests of ineffective medication.

Several pharmacologic agents mediate their effects through receptors affecting ion channels in cell membranes. The effects of direct modulation of these ionic channels on MEG signals has been analyzed in vitro in slabs of turtle cerebellum [62]. This type of basic research will undoubtedly increase potential of pharmaco-MEG, depending on the extent to which animal and human data can be comprehensively compared.

In some cases, the topography of spontaneous activity or evoked responses recorded at scalp level matches the available evidence of cortical generators. However, localization may be less important in human neuropharmacology than the definition of relative activation of discrete brain portions and of the dynamics of drug effect, connected to characteristics of neurotransmission and pharmacological mechanisms of action. "Transparency" of low-pass filtering skull, and independence from reference favour MEG application in neuropharmacology.

MEG in psychiatry

The importance of MEG's time resolution, compared with other imaging techniques, may become useful in psychiatry. The combination of MEG to other techniques should enable clinicians to find diagnostic tools for psychiatric diseases, in which the understanding of the brain information processing is crucial. The earliest applications of MEG in psychiatry have studied AEF N100m deflections. In controls, the sources of N100m are more anterior over the right than the left hemisphere. In schizophrenics, this difference is smaller than in controls [63]. Probably this reflects, at least in part, structural differences of temporal lobes evident in CT and MRI as well.

Transient auditory hallucinations delay N100m as does real speech in controls, suggesting that auditory cortex is activated during hallucinations by endogeneous neural firing [64], in good agreement with recent FMRI data [65].

The anomalous asymmetry demonstrated by relative antero-posterior left-right N100m source localization in schizophrenia appears to be sex-specific. In actively psychotic males with schizophrenia, there is less asymmetry of N100m source localization than in control males; schizophrenic males have sources relatively further anterior on the left superior temporal gyrus (STG) and relatively more posterior on the right STG (see Fig. 5). On the other hand, actively psychotic females with schizophrenia demonstrate increased asymmetry, with N100m sources further anterior on the right STG compared with control females [66]. The same group of schizophrenic males also demonstrated smaller left STGs than controls, whereas in females with schizophrenia the STG volume was normal [67].

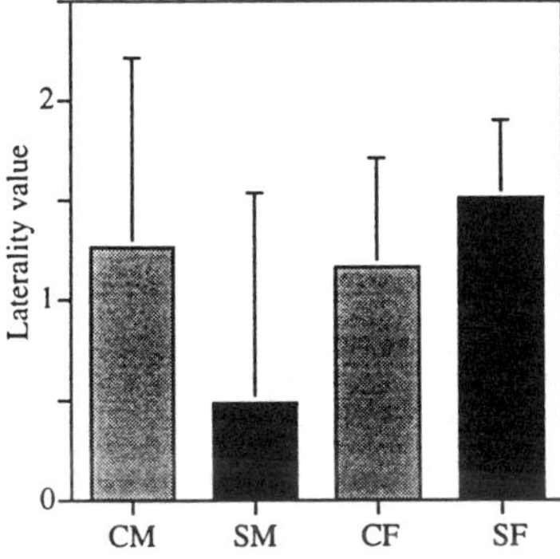

Fig. 5. Laterality value ± S.D. of control males (CM), schizophrenic males (SM), control females (CF) and schizophrenic females (SF). The laterality value was calculated by subtracting the left N100m Y coordinate value from the right Y coordinate value for each group. The antero-posterior Y axis runs through the nasion, perpendicular to the line connecting preauricular points. Modified from [66].

Spontaneous 10-Hz activity in the vicinity of the N100m sources is suppressed when subjects scan memory to recall melodies [68]. This suppression is clear over both left and right STG in controls. In schizophrenic males, however there is a deficient suppression over the left but not over the right STG during such task [69]. In a pilot study with schizophrenic females, normal suppression was also observed over the right hemisphere but not over the left one. Thus, only males with schizophrenia had anatomic abnormalities in the left STG whereas both males and females showed functional deficit over the left hemisphere. This suggests relative independence of of functional and structural abnormalities in schizophrenia.

Electroconvulsive therapy in depression increases slow EEG rhythms. Similar slowing is evident in MEG spontaneous activity: the frequencies below 8 Hz increase in a dose-dependent manner during treatment and disappear about 1 month after the treatment. The alterations appear to occur in the precentral and central areas, in the posterior temporal lobes, and around temporo-pariteo-occipital junctions [70]. The relation of the changes to the effect of the therapy are unknown at present. Obviously, increase in understanding of the basic processes involved in attention, learning, and memory will further the study of psychiatric disorders by MEG.

Conclusions

MEG is useful at the level of single patients in defining the location of functionally irretrievable areas before surgery. MEG studies may reveal the mechanisms of diseases and recovery and also allow conclusions to be drawn regarding how the normal brain functions, on the basis of findings in neurologic patients. The temporal resolution of MEG is useful in studying how signal coherence and response speed improve after CNS lesions. The future will show if more complex methods of signal analysis, *e.g.*, coherence calculations, provide additional benefit from this time resolution in the study of patients. In general, most MEG measurements have been performed in research units distant from hospitals; the studied patient groups are small and selected. Recently, whole-head MEG systems have been installed in medical centers as well, and the near future will clarify the role of MEG in clinical use.

References

[1] Ahonen, A, Hämäläinen, M, Kajola, M, et al. (1993) 122-channel SQUID instrument for investigating the magnetic signals from human brain. Physica Scripta, T40: 198-205.
[2] Vrba, J, Betts, K, Burbank, M, et al. (1993) Whole cortex 64 channel SQUID biomagnetometer system. IEEE Transactions of Applied Superconductors, 3: 1878-1882.
[3] Mäkelä, JP, Ahonen, A, Hämäläinen, M, et al. (1993) Functional differences between auditory cortices of the two hemispheres revealed by whole-head neuromagnetic recordings. Human Brain Mapping, 1: 48-56.
[4] Nakasato, N, Fujita, S, Seki, K, et al. (1995) Functional localization of bilateral auditory cortices using an MRI-linked whole head magnetoencephalography (MEG) system. Electroencephalography and clinical Neurophysiology, 94: 183-190.
[5] Wood, C C, Cohen, D, Cuffin, B N, et al. (1985) Electric sources in the human somatosensory cortex: identification by combined magnetic and potentials field recordings. Science, 227: 1051-1053.
[6] Kawamura, T, Nakasato, N, Seki, K, et al. (1996) Neuromagnetic evidence of pre-and post-central cortical sources of somatosensory evoked responses. Electroencephalography and clinical Neurophysiology, 100: 44-50.
[7] Kawamura, T, Nakasato, N, Ohtomo, S, et al. Neuromagnetic identification of the somatosensory cortex in cases with arteriovenous malformation adjacent to the central sulcus. In this volume.
[8] Hari, R, Aittoniemi, K, Järvinen, M L, et al. (1980) Auditory evoked transient and sustained magnetic fields of the human brain. Localization of neural generators. Experimental Brain Research, 40: 237-240.
[9] Pantev, C, Hoke, M and Lehnertz, K (1990) Identification of source of brain neuronal activity with high spatiotemporal resolution through combination of neuromagnetic source localization (NMSL) and magnetic resonance imaging (MRI). Electroencephalography and clinical Neurophysiology, 75: 173-184.
[10] Mäkelä, JP, Hari, R, Valanne, L, et al. (1991) Auditory evoked magnetic fields after cerebral vascular lesion. Annals of Neurology, 30: 76-82.
[11] Nakasato, N, Fujita, S and Matani, A Clinical application of the whole-head MEG: Auditory evoked response in patients with intracranial structural lesions. In: Baumgartner, C, Deecke, L, Stroink, G, Williamson, SJ (Eds.), Biomagnetism: Fundamental research and clinical applications. Studies in applied electromagnetics and mechanics. Elsevier/IOS press, Amsterdam, 1995: 186-190.
[12] Kanno, A, Nakasato, N, Ohtomo, S, et al. Normalized N100m latency in the auditory evoked fields after surgical removal of temporal lobe gliomas. In this volume.
[13] Seki, K, Nakasato, N, Fujita, S, et al. (1996) Neuromagnetic evidence that the P100m component of the pattern reversal visual evoked response originates in the bottom oc calcarine fissure. Electroencephalography and clinical Neurophysiology, in press.
[14] Rossini, PM, Narici, L, Martino, G, et al. (1994) Analysis of interhemispheric asymmetries of somatosensory evoked magnetic fields to right and left median nerve stimulation. Electroencephalography and clinical Neurophysiology, 91: 476-482.
[15] Rossini, P.M, Pizzella, V, Rossi, S, et al. SEFs to median nerve and finger stimulation: analysis of interhemispheric asymmetries in healthy subjects and following hemispheric lesions. In this volume.
[16] Beisteiner, R, Gomiscek, G, Erdler, M, et al. (1995) Comparing localization of conventional functional magnetic resonance imaging and magnetoencephalography. European Journal of Neuroscience, 7: 1121-1124.
[17] Maldjian, J, Atlas, S, Howard, R, et al. (1996) Functional magnetic resonance imaging of regional brain activity in patients with intracerebral artriovenous malformations before surgical or endovascular therapy. Journal of Neurosurgery, 84: 477-483.
[18] Mäkelä, JP, Salmelin, R, Kotila, M, et al. (1994) Neuromagnetic correlates of memory disturbances caused by infarction in anterior thalamus. Society for Neuroscience Abstracts, 20(1): 810.
[19] Lewine, JD, Orrison, WW, Astur, RS, et al. Explorations of pathophysical spontaneous activity by macnetic source imaging. In: Baumgartner, C, Deecke, L, Stroink, G, Williamson, SJ (Eds.), Biomagnetism:Fundamental research and Clinical applications. Studies in applied electromagnetics and mechanics. Elsevier/IOS Press, Amsterdam, 1995: 55-59.
[20] Lewine, JD, Orrison, WW, Sloan, JH, et al. Neuromagnetic assesment of pathophysiological brain activity induced by minor head trauma. In this volume.
[21] Vieth, J, Kober, H and Grummich, P (1996) Sources of spontaneous slow waves associated with brain lesions, localised by using the MEG. Brain Topography, 8: 215-221.
[22] Vieth, JB, Kober, H, Stippich, C, et al. Time course of abnormal MEG activity associated with transient ischemic attacks. In this volume.
[23] Kakigi, R, Koyama, S, Hoshiyama, M, et al. (1995) Pain-related magnetic fields following painful CO2 laser stimulation in man. Neuroscience Letters, 192: 45-48.

[24] Hari, R, Kaukoranta, E, Reinikainen, K, et al. (1983) Neuromagnetic localization of cortical activity evoked by painful dental stimulation in man. Neuroscience Letters, 42: 77-82.

[25] Huttunen, J, Kobal, G, Kaukoranta, E, et al. (1986) Cortical responses to painful CO_2 stimulation of the nasal mucosa. Electroencephalography and clinical Neurophysiology, 64: 347-349.

[26] Kitamura, Y, Kakigi, R, Hoshiyama, M, et al. (1995) Pain-related somatosensory evoked magnetic fields. Electroencephalography and clinical Neurophysiology, 95: 463-474.

[27] Jenkins, WM, Merzenich, MM and Recanzone, G (1990) Neocortical representational dynamics in adult primates: implications for neuropsychology. Neuropsychologia, 28: 573-584.

[28] Wang, X, Merzenich, MM, Sameshima, K, et al. (1995) Remodeling of hand representation in adult cortex determined by timing of tactile stimulation. Nature, 378: 71-75.

[29] Elbert, T, Flor, H, Birbaumer, N, et al. (1994) Evidence for extensive reorganization of the somatosensory cortex in adult humans after nervous system injury. NeuroReport, 5: 2593-2597.

[30] Yang, TT, Gallen, C, Ramachandran, VS, et al. (1994) Noninvasive detection of cerebral plasticity in adult human somatosensory cortex. NeuroReport, 5: 701-704.

[31] Flor, H, Elbert, T, Knecht, S, et al. (1995) Phantom-limb pain as a perceptual correlate of cortical reorganization following arm amputation. Nature, 375: 482-484.

[32] Knecht, S, Henningsen, H, Elbert, T, et al. (1995) Cortical reorganization in human amputees and mislocalization of painful stimuli to the phantom limb. Neuroscience Letters, 201: 262-264.

[33] Knecht, S, Henningsen, H, Elbert, T, et al. (1996) Reorganizational and perceptual changes after amputation. Brain, In press.

[34] Flor, H, Braun, Ch, Birbaumer, N, et al. Chronic pain enhances the magnitude of the magnetic field evoked at the site of pain. In: Baumgartner, C, Deecke, G, Stroink, G, Williamson, SJ (Eds.), Biomagnetism:Fundamental research and Clinical applications. Studies in applied electromagnetics and mechanics. Elsevier/IOS Press, Amsterdam, 1995: 107-111.

[35] Elbert, T, Pantev, C, Wienbruch, C, et al. (1995) Increased cortical representation of the fingers of the left hand in string players. Science, 270: 305-307.

[36] Rockstroh, B, Vanni, S, Elbert, T, et al. Extensive somatosensory stimulation alters somatosensory evoked fields. In this volume.

[37] Mogilner, A, Grossman, A, Ribary, U, et al. (1993) Somatosensory cortical plasticity in adult humans revealed by magnetoencephalography. Proceedings of the National Academy of Sciences of USA, 90: 3593-3597.

[38] Elbert, T, Sterr, A, Rockstroh, B, et al. Cortical reorganization in amputees: alterations of somatosensory representation in the hemisphere contralateral to the intact side. In this volume.

[39] Florence, SL, Wall, JT and Kaas, JH (1991) Cortical projections from the skin of the hand in squirrel monkeys. Journal of Comparative Neurology, 311: 563-578.

[40] Pantev, C, Bertrand, O, Eulitz, C, et al. (1995) Mirror-image tonotopy of different areas of human auditory cortex revealed by simultaneous magnetic and electric recordings. Electroencephalography and clinical Neurophysiology, 94: 26-60.

[41] Mühlnickel, W, Flor, H, Elbert, T, et al. Deviations from the tonotopic map are correlated with tinnitus strength. In this volume.

[42] Mäkelä, JP and Hari, R Long-latency auditory evoked magnetic fields. In: Sato, S. (Eds.), Magnetoencephalography: Clinical Applications and Comparison of EEG. Raven Press, 1990: 177-191.

[43] Phillips, DP, Semple, MN, Calford, MB, et al. (1994) Level-dependent representation of stimulus frequency in the cat primary auditory cortex. Experimental Brain Research, 102: 210-226.

[44] Vasama, J-P, Mäkelä, JP, Pyykkö, I, et al. (1995) Abrupt unilateral deafness modifies function of human auditory pathways. NeuroReport, 6: 961-964.

[45] Vasama, J-P and Mäkelä, JP (1995) Auditory pathway plasticity in adult humans after idiopathic sudden unilateral sensorineural hearing loss. Hearing Research, 87: 132-140.

[46] Irvine, DRF and Rajan, RR (1993) Plasticity in the frequency organization of auditory cortex of adult mammals with restricted cortical lesions. Biomedical Research, 14: 55-59.

[47] Ahissar, E and Ahissar, M (1994) Plasticity in auditory cortical circuitry. Current Opinion in Neurobiology, 4: 580-587.

[48] Mäkelä, JP, Hari, R, Karhu, J, et al. (1993) Suppression of magnetic mu rhythm during parkinsonian tremor. Brain Research, 617: 189-193.

[49] Volkmann, J, Joliot, M, Mogilner, A, et al. (1996) Central motor loop oscillations in parkinsonian resting tremor revealed by magnetoencephalography. Neurology, 46: 1359-1370.

[50] Sannita, WG, Lewine, JD, Maclin, EL, et al. Quantitative EEG and MEG effects in healthy subjects of acute oral phenobarbital (100 mg). In this volume.

[51] Wiebruch, C, Eulitz, C, Lehnertz, K, et al. Methohexital-induced changes in spectral power of neuromagnetic signals: reduced enhancement of beta-band power over the hemisphere ipsilateral to the epileptogenic focus. In this volume.

[52] Kraut, MA, Arezzo, JC and Vaughan, H (1990) Inhibitory processes in the flash evoked potential of the monkey. Electroencephalography and clinical Neurophysiology, 76: 440-452.

[53] Arakawa, K, Peachey, NS, Celesia, GG, et al. (1993) Component-specific effects of physostigmine on the cat visual evoked potential. Experimental Brain Research, 95: 271-276.

[54] Mavroudakis, N, Brunko, E, Nogueira, MC, et al. (1991) Acute effects of diphenylhydantoin on peripheral and central somatosensory conduction. Electroencephalography and clinical Neurophysiology, 78: 263-266.

[55] Bodis-Wollner, I (1990) Visual deficits related to dopamine deficiency in experimental animals and Parkinson's disease patients. Trends in Neurosciences, 13: 296-302.

[56] Sannita, WG, Balestra, V, Di Bon, G, et al. (1993) Human Flash-VEP and quantitative EEG are independently affected by acute scopolamine. Electroencephalography and clinical Neurophysiology, 86: 275-282.

[57] Singer, W (1993) Synchronization of cortical activity and its putative role in information processing and learning. Annual Review of Physiology, 55: 349-374.

[58] Sannita, WG, Lopez, L, Piras, C, et al. (1995) Scalp-recorded oscillatory potentials evoked by transient pattern-reversal stimulation in man. Electroencephalography and clinical Neurophysiology, 96: 206-218.

[59] Lopez, L, Narici, L, Conforto, S, et al. Oscillatory retinal and cortical responses to luminance stimulation: eccentricity function and effect of acute scopolamine. In this volume.

[60] Shiomi, Y, Fujiki, N, Hirano, S, et al. (1996) Tinnitus remission by lidocaine demonstrated by auditory evoked magnetoencephalogram: preliminary report. Association for Resarch in Otolaryngology, book of abstracts, 19: 13.

[61] Jäntti, V, Baer, G, Yli-Hankala, A, et al. (1995) MEG burst suppression in an anaesthetized dog. Acta Anesthesiologica Scandinavica, 39: 126-128.

[62] Lopez, L, Chan, CY, Okada, YC, et al. (1991) Multimodal characterization of population responses evoked by applied electric field in vitro: extracellular potential, magnetic evoked field, transmembrane potential and current source density analysis. Journal of Neuroscience, 11: 1998-2010.

[63] Reite, M, Teale, P, Goldstein, L, et al. (1989) Late auditory evoked magnetic sources may differ in the left hemisphere of schizophrenic patients. Archives of General Psychiatry, 46: 565-572.

[64] Tiihonen, J, Hari, R, Naukkarinen, H, et al. (1992) Modified activity of the human auditory cortex during auditory hallucinations. American Journal of Psychiatry, 149: 255-257.

[65] David, AS, Woodruff, PWR, Howard, R, et al. (1996) Auditory hallucinations inhibit exogeneous activation of auditory association cortex. NeuroReport, 7: 932-936.

[66] Reite, M, Sheeder, J, Teale, P, et al. (1996) Magnetic source imaging evidence of sex differences in cerebral lateralization in schizophrenia. Archives of General Psychiatry, in press.

[67] Reite, M, Teale, P, Sheeder, J, et al. (1996) Neuropsychiatric applications of MEG. Electroencephalography and clinical Neurophysiology, Suppl. Visualization of Information processing in the human brain; Recent advances in MEG and Functional MRI: in press.

[68] Kaufman, L, Curtis, S, Wang, J-Z, et al. (1991) Changes in cortical activity when subjects scan memory for tones. Electroencephalography and clinical Neurophysiology, 82: 266-284.

[69] Reite, M, Teale, P, Sheeder, J, et al. (1996) MEG evidence of abnormal early auditory function in schizophrenia. Biological Psychiatry, in press.

[70] Salmelin, R, Mäkelä, JP, Heikman, P, et al. (1996) Human brain rhythms and electroconvulsive therapy. Society for Neuroscience Abstracts.

Changes of the Functional Organization of the Somatosensory Cortex in Chronic Pain Patients

Braun, C.[1], Flor, H.[1,2], Birbaumer, N.[1] and Elbert, T.[3]

[1]Institute of Medical Psychology, University of Tübingen, Tübingen, Germany; [2]Department of Psychology, Humbolt University, Berlin, Germany; [3]Department of Psychology, University of Konstanz, Konstanz, Germany

Introduction

Does the persistent noxious stimulation experienced by patients suffering from chronic back pain lead to cortical reorganization? Previous research has shown that changes in sensory inputs lead to changes in the cortical representation of the relevant somatic structures. Cessation of peripheral input leads to the reduction in size of the cortical representation of the corresponding area, while adjacent cortical areas expand [1]. If a body part is used extensively, however, the corresponding representation in somatosensory cortex expands [2]. Chronic pain patients have experienced years of persistent noxious stimulation in the lumbar region. Cortical reogranization should thus be present in an area of the somatosensory cortex that represents the back, but not in unrelated areas of somatosensory cortex, such as those representing the finger.

We investigated the cortical responses of chronic back pain patients and control subjects to electrical stimulation of different magnitudes to either the back or the finger. If the cortical representation of the painful body region is increased, this should be reflected in the magnetic field generated in response to the stimulation. The magnetic fields should exhibit higher amplitudes and a broader topographical distribution when evoked from the affected body part of the chronic pain patients. Furthermore, dipole analysis should reveal a shift in the cortical locations of the back in chronic pain patients.

Methods

Three groups of subjects participated in the experiment. The patients suffering from continuous, long term lower back pain were divided into two groups: Chronic patients (n=5), suffering for an average duration of 18 years from back pain, and Subchronic patients (n=5), suffering for an average of 7 years from back pain. Age- and gender-matched healthy subjects (n=9) comprised the control group.

Electrical bipolar pulses with a duration of 10 ms. and a preexperimentally determined current intensity (max. 5 mA) served as pain stimuli. Stimuli were applied through an intracutaneous gold electrode to two different sites: (1) the left back, in the region of the most intense pain, and (2) the left index finger. Tactile perception and pain threshold were determined in two ascending and descending series using the method of limits. Three stimulus intensities were presented: (1) sub-threshold, which was determined for each subject to be the point midway between the perception and pain thresholds; (2) painful, which was determined to be 50% above the pain threshold, and (3) a standard stimulus of 500 μA. Each stimulus was presented to the finger and back electrodes in a blocked design. After each block of 24 trials, subjects were asked to rate (on a verbal 10 point scale) the stimulus intensity and the current level of clinical pain.

Magnetic fields were recorded from 37 locations centered over the right C4, contralateral to the side of stimulation. In the BTi Neuromagnetometer system, the coils were arranged in a circular array (diameter 14.4 cm) on a spherical surface (with 12.2 radius). The coil diameter was 2.0 cm and the distance between the center of two adjacent coils was 2.2 cm. Data were sampled at a rate of 297 pts/s.

Trials were excluded from analysis if the difference between the maximum and the minimum (computed across a 1 second epoch) exceeded 2.5 pT in any of the MEG channels. This criterion also served to exclude larger eye movements or blinks which would have led to excessive amplitudes in most anterior channels. An average of 230 artifact free trials per subject were analyzed. Responses were averaged for each subject separately for the different experimental conditions. Parameters were extracted from the time course of the RMS at 80 to 125 ms across all MEG channels.

The center of the activated source was determined for all stimulation types using a single moving dipole model. A sphere was fit locally to the digitized head shape in the region proximal to the sensor array for each subject. The location (x,y,z positions), orientation, and amplitude of a best fitting equivalent current dipole were estimated for each point in time. The anterior-posterior, medial-lateral, and inferior-superior coordinates of the dipole location with their confidence volumes were calculated in the head-frame based coordinate system. Only estimates with a correlation between the measured and the expected field of >.95 and a confidence volume of less than 1 cm^3 were used.

Results

Amplitude of the evoked magnetic field at the back (r=5, p<.05), but not at the finger (r=0, n.s.) correlated with the chronicity of the pain problem. With increasing duration of the illness, a higher response of the magnetic field is found.

An ANOVA of the M100 RMS activity revealed a significant difference between the chronic pain patients and the combined subchronic and control subjects. The 100 ms component was significantly enhanced during back stimulation, but not finger stimulation, in the chronic patients compared to subchronic patients and the control subjects (F(1,16)=5.34, p<0.35) (see Figure 1).

The latency window between 70 and 80 msec yielded the most pronounced peak and was chosen for the analysis of the location of the cortical activity in response to painful stimulation. Neither the inferior-superior nor the anterior-posterior dipole coordinates were significantly different between the groups, although they differed between the back and finger (see Figure 2). In the medial-lateral direction of the back stimulation response site, however, the dipole analysis revealed a source location which was shifted 2.5 cm medially for the chronic patients compared to the other groups.

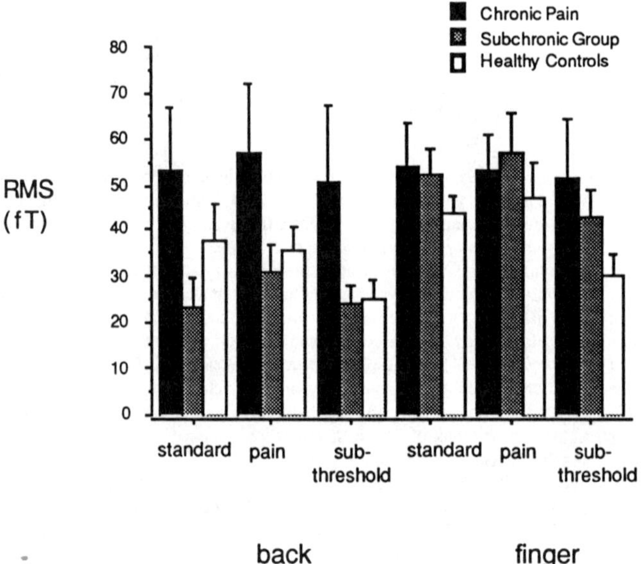

Figure 1. Peak amplitudes of the root mean square (RMS) across all 37 MEG-channels, computed for the range 80-125 ms.

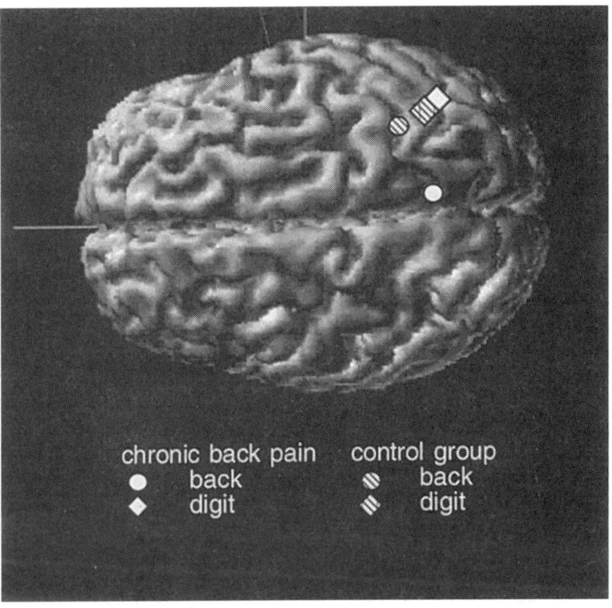

Figure 2. Localization of the dipoles for control subjects and chronic pain patients when stimulated at the back or finger.

Discussion

The results of this study support the hypothesis that chronic pain leads to reorganization of the cortical representation of the affected body region. Those patients who had suffered from chronic noxious stimulation of the lower back region for the longest amount of time were those with the most enhanced MEG response to stimulation of the back. Finger stimulation revealed no differences between patient and control groups: only when the back was stimulated did the differences between patients and controls emerge [3].

The amplitude of the magnetic field evoked at the finger was positively correlated with chronicity of the pain problem, suggesting that long-lasting proprioceptive input led to enhanced cortical responses to tactile stimulation. This result is in accordance with our assumption about stimulation-produced plasticity of the somatosensory cortex. Furthermore, dipole analysis revealed differences in dipole location between the groups for the back, but not the finger stimulation. Taken together, these results suggest that the cortical representation of the chronic pain patients' affected body region is increased. Years of painful stimulation may have led to cortical reorganization that may now maintain the pain without peripheral input.

References

[1] Merzenich, M.M., Nelson, R.J., Stryker, M.P., Cynader, M.S., Schoppman, A., Zook, J.M. Somatosensory cortical map changes following digit amputation in adult monkeys. Journal of Comparative Neurology, 1984, 224: 591-605.
[2] Recanzone, G.H., Merzenich, M.M., Jenkins, W.M., Grajsi, K.A., Dinse, H.R. Topographic reorganization of the hand representation in cortical area 3b of owl monkeys trained in a frequency discrimination task. Journal of Neurophysiology, 1992, 65: 12031-1056.

[3] Flor, H., Braun, C., Birbaumer, N., Elbert, T., Ross, B., Hoke, M. Chronic pain enhances the magnitude of the magnetic field evoked at the site fo pain, In: Baumgartner, C. et al. Biomagnetism: Fundamental Research and Clinical Applications, Elsevier Science, 1995.

Acknowledgement

We thank Laura Helmuth for helpful comments and for checking this manuscript. The study was supported by a grant of the Deutsche Forschungsgemeinschaft (Bi 195/24).

Clinical MEG II: Development Of A Normative Database

Davis, J.T., Lewine, J.D., Edgar, J.C., Hoesing, J., and Orrison,W.W., Jr.

The New Mexico Institute of Neuroimaging, The New Mexico Regional Federal Medical Center, Albuquerque, New Mexico, USA.

INTRODUCTION

As clinical MEG matures, the need for normative data on both evoked responses and spontaneous brain activity is becoming acute. Towards this end, The New Mexico Institute of Neuroimaging has developed a comprehensive series of protocols for examining brain activity in both patient and normal control populations [1]. This protocol, which can be completed in about two hours, evaluates spontaneous brain activity and visual, auditory, and somatosensory, functioning. A major objective of this program is to develop techniques for quantitative MEG that will be similar to those previously developed for quantitative EEG, but hopefully, of superior clinical utility, owing to MEG's exceptional spatial resolution. We are in the process of developing a database of information on 100+ neurologically normal individuals between the ages of 20 and 60.

A wide range of parameters are being included in the database, including information on the latency, amplitude, and source locations for multiple evoked potential components in auditory, visual and somatosensory domains. Spontaneous data are evaluated for spectral composition in both eyes open and eyes closed conditions. We are also beginning to apply algorithms that characterize interhemispheric, interlobar, and intralobar coherence in delta, theta, alpha, and beta frequency bands. Once a normative database is completed, it will be possible to use multivariate statistical procedures to identify abnormalities in patient data sets. Ultimately, it should be possible to develop discriminant functions that allow for rapid patient diagnosis.

METHODS

<u>Subjects</u>: Prior to study, subjects complete a comprehensive questionnaire that includes questions about developmental milestones, head trauma, and psychiatric and neurologic dysfunction. Only subjects with normal developmental histories and no evidence of substance abuse, head trauma, or neurologic or psychiatric dysfunction are included in the normative data base. Subjects must also demonstrate a normal magnetic resonance imaging examination prior to inclusion in the normative data base and they must have a normal EEG.

<u>Evoked Data</u>: Information on evoked potential paradigms are presented in reference [1]. For auditory activity, we are evaluating responses to 500 Hz, 2000 Hz, and 4000 Hz tone pips presented binaurally (duration 50 msecs), and also monaural 2000 Hz tones. Each subject's response is ultimately characterized by the latency and amplitude of M50, M100, and M200 components over right and left hemispheres, plus dipole parameters. For somatosensory activity, responses to electrical stimulation of the median and tibial nerves are being characterized. The latency and amplitude of M20 and M30 components are being assessed, as is information on associated sources. Visual responses to stimuli in right and left upper and lower visual fields are being evaluated with emphasis of the latency and amplitude of M100 and M 130 components.

<u>Spontaneous Data</u>: The details of data collection are described in reference [1]. Spontaneous data are visually inspected for gross abnormalities including focal slowing and epileptiform transients. It should be noted that the presence of these signals per se is not considered an exclusionary criteria from the data base. Exclusion is based on other than MEG factors, as the use of MEG as an exclusionary parameter would be circular to the development of a normative database. Following visual inspection, average power spectra are calculated from artifact free data segments. A Fast Fourier Transform is applied to successive windows of approximately 3.4 seconds in length, which are slid by 1.7 second steps. Epochs with signals in excess of 3 picotesla are excluded from analysis. At each non artifact step, a transform is calculated and included in the average, until a total of one hundred averages is obtained. Despite attempts at automatic artifact rejection, processed data are sometimes contaminated with various forms of low amplitude artifact that adds power to the spectra. Typically, the artifactual contribution can be recognized in the spectral data itself and a subspace projection method is used to remove the artifact signal (e.g. heartbeat artifact). In order to obtain regional measures of spectral values, the 122 sensors are grouped so that the data from each group approximates a traditional head region. The number of sensors in each region is kept similar, except for the midline occipital group, which has only 2 sensors, and the midline frontal group which is excluded from further analyses. Within each regional grouping, a grand-average power spectra is calculated across all encompassed

sensors. Figure 1 shows the sensor groupings. Several parameters are entered into the database for each grouping, including regional power in each band.

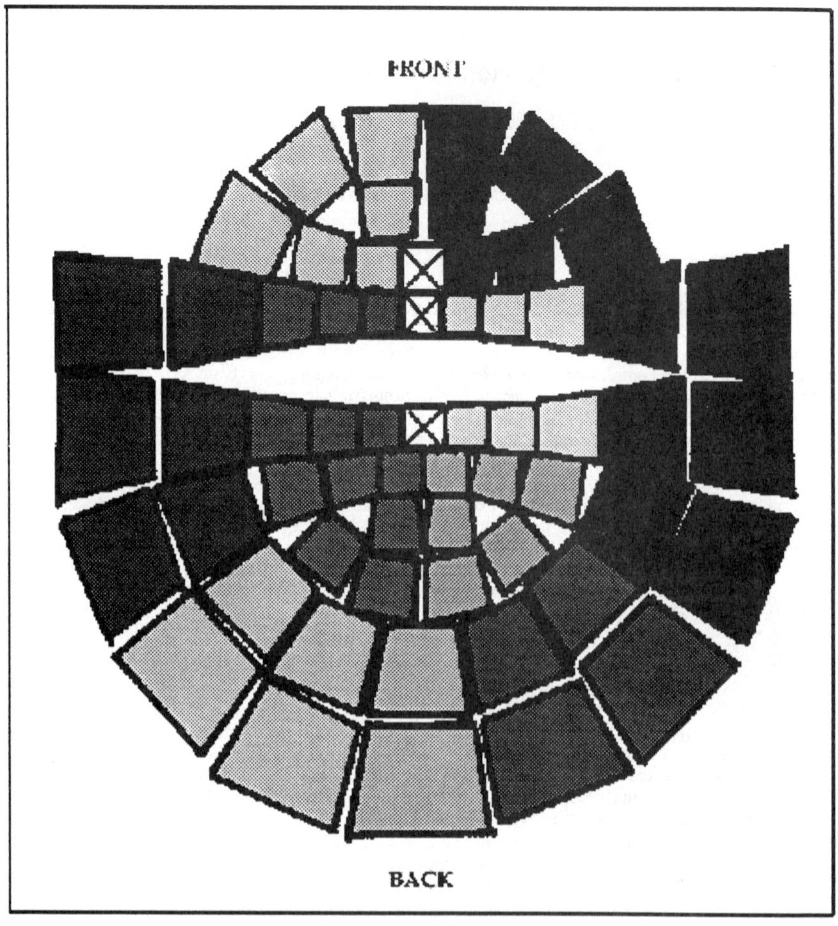

Fig 1: Sensor groupings for spectral analyses. Each box represents two of the 122 channels in our biomagnetometer.

RESULTS

Full exposition of all of the contents of the database is well beyond the scope of this brief report, so only a subset of the data is described here.

Subset 1: At present, auditory data have been completely analyzed for 12 right handed males between the ages of 25 and 40 years old. The mean latency of the M100 response to bilateral 2000 Hz tones is 107 msec with a standard deviation of 10 msec. The magnitude of the associated current dipoles range from 23 to 30 nA, with no significant difference between right and left hemispheres in either latency or dipole magnitude. The orientation of the M100 dipole response invariably has a negative z component (ventrally directed) which is usually the largest component. The y component is usually negative (posteriorly directed), and the x component (right-left) is near zero. The relative location of the right and left hemispheric dipoles is such that the right is on average 10 mm anterior to the left.

This admittedly small database has already proven useful in the identification of abnormalities in several patient populations. For example, figure 2 shows a comparison of auditory evoked data (2000 Hz, binaural tone) from a typical control subject and an individual with a dual diagnosis of bipolar disorder and Parkinson's disease. Clearly the M100 component of the auditory response is delayed for the patient.

Subset 2: Spectral data have also been analyzed from 10 right handed females between the ages of 20 and 40. Figure 3A shows delta and theta regional power for a representative subject. The overall amount of delta and theta power is low, and there are no gross regional asymmetries. The situation for a female patient with a left temporal-parietal tumor is markedly different. Excessive delta and theta is seen, with marked regional asymmetries consistent with the patients pathology.

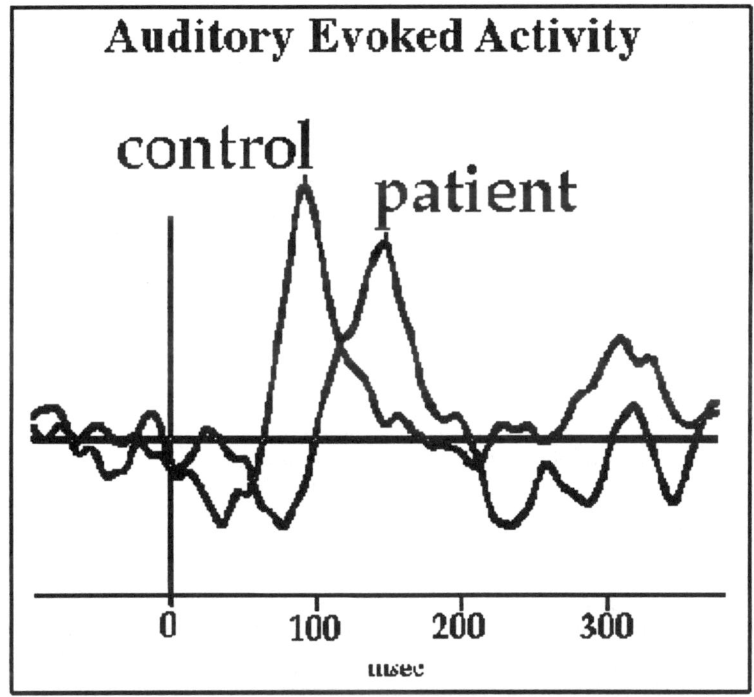

Fig 2: Auditory evoked activity from a control subject and a patient with bipolar disorder and Parkinson's disease. The patient's M100 response is delayed significantly.

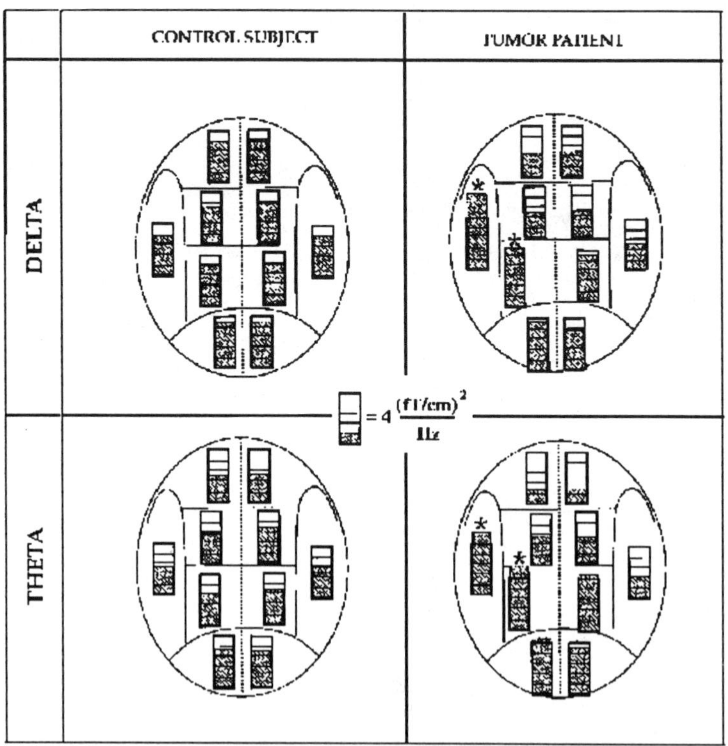

Fig 3: Regional spectral data in delta and theta bands for a control subject
and a patient with a left temporal-parietal tumor.

DISCUSSION & CONCLUSIONS

The normative database that we are starting to compile has already proven useful in the objective identification of abnormal data sets. As the database grows, we anticipate greater sensitivity to subtle pathologies. As has been done for quantitative EEG, it is anticipated that the normative profile can be used in a multivariate approach, in which disorders, such as schizophrenia, can be recognized by the deviance from the overall normative profile. The current task is to continue to accumulate the normative data.

REFERENCES

[1] Lewine, J.D., Davis, J.T., Davis, L.E., et al., Clinical MEG I: Towards a standardized examination, Proceedings, Biomag'96, this volume.

Cortical Reorganization in Arm Amputees: Alterations of the Somatosensory Representation of the Intact Arm

Elbert, T.[1], Sterr, A.[1], Rockstroh, B.[1], Charbonnier, D.[1], Flor, H.[2], Pantev, C.[3], Wienbruch, C.[3], Knecht, S.[3], and Taub, E.[4]

University of Konstanz[1], Germany; Humboldt University[2], Berlin, Germany; University of Münster[3], Germany; University of Alabama[4], Birmingham, U.S.A..

Introduction

Modulation in the flow of somatosensory input is followed by changes in the cortical representation of the body in the somatosensory cortex. Intense, behaviorally relevant stimulation of a set of receptors leads to an expansion in the cortical representation of the stimulated parts of the body [5,7]. Magnetic source imaging (MSI) has provided evidence that use-dependent reorganization also occurs in humans [3,8]. String players, for example, provide a model for the effects of differential afferent input to the two hemispheres. It was found that the representation of the digits of the left hand proved to be larger than that in controls [3]. Disuse of limbs also brings about changes in the representation of the body in the somatosensory cortex. Such phenomenona have been observed after total loss of input. In owl monkeys, Pons and coworkers [6] reported massive reorganization of primary somatosensory cortex subsequent to sensory deafferentation of an entire arm 12 years earlier. These investigators also demonstrated that the adjacent representations from the face had expanded into the hand areas, so that stimulation of the face evoked responses in the area formerly occupied by the hand and digits. Extensive cortical reorganization following somatosensory deafferentation also occurs and persists in humans following upper extremity amputation, as demonstrated by MSI [2,4,9]. It has also been shown that the magnitude of cortical reorganization is strongly associated with the amount of phantom limb pain experienced by the amputees [5].

A loss of sensory input from the amputated limb causes the alterations in the hemisphere contralateral to the amputated side. The question arises if there is a measurable change also in the hemisphere contralateral to the intact side. Two different mechanisms may be involved in this alteration: First, homotopic regions of somatosensory cortex are linked such that plasticity induced in one hemisphere is mirrored in the other hemisphere [1]. Second, the obvious increase of afferent input from the intact hand may result in enlargements of the representational zones of the fingers.

The present study was designed to investigate the nature of the change of somatotopic representation in upper arm amputees contralateral to the intact hand. Using magnetic source imaging, the representation of digits D1 and D5 was compared with that of controls.

Methods

Six upper arm amputees (2 with left, 4 with right arm amputation) and six control subjects participated in the study. The amputation had occurred 2 to 49 years (mean 21 years) prior to the current investigation.

Somatosensory stimulation was delivered separately to the first and fifth digit of the hand (both hands in controls) and to the lower lip at the corners of the mouth. The stimulation consisted of a brief and light, non-painful superficial pressure delivered through a pneumatic device. One thousand stimuli were delivered to each of the stimulation sites (first and fifth digit, left and right corners of the mouth) at an average rate of 0.5 Hz. The interval between stimulus onsets varied between 50 and 500 ms. The sequence of stimulation was balanced across subjects. The sensor array of magnetic detectors was positioned over the hemisphere (centered over C3 or C4) contralateral to the stimulated limb and lip site.

Within the range of 30-75 msec a first major peak was identified in each of the evoked waveforms (Fig.1). For this peak a single equivalent current dipole (ECD) model (best fitting local sphere) was fitted and the medians of the dipole moment and the dipole location were computed from a selection of time points within a 20 msec time segment (eleven sampling points) around the maximal root mean square (RMS) across the 37 channels. Points were selected if they met the following criteria: (1) RMS indicating a signal-to-noise ratio > 3, (2) goodness of fit of the ECD-model to the measured field $> .95$ and (3) a minimal confidence volume of the ECD location < 300 mm^3.

The representation of hand and face in area 3b of controls was compared to the representation of hand and face in the deafferented hemisphere (contralateral to the amputation) and contralateral to the intact side of the amputees. Euclidean distances between lip- and D1-, lip- and D5-, and D1- and D5-locations, were calculated to uncover changes in the cortical representation of the intact hand. In order to evaluate the changes in lip representation, if any, due to amputation ECD-values of both hemispheres were averaged for the controls and compared with (a) parameters of deafferented hemispheres and (b) locations in hemipsheres contralateral to the intact side. Anovas and t-tests were calculated to confirm differences.

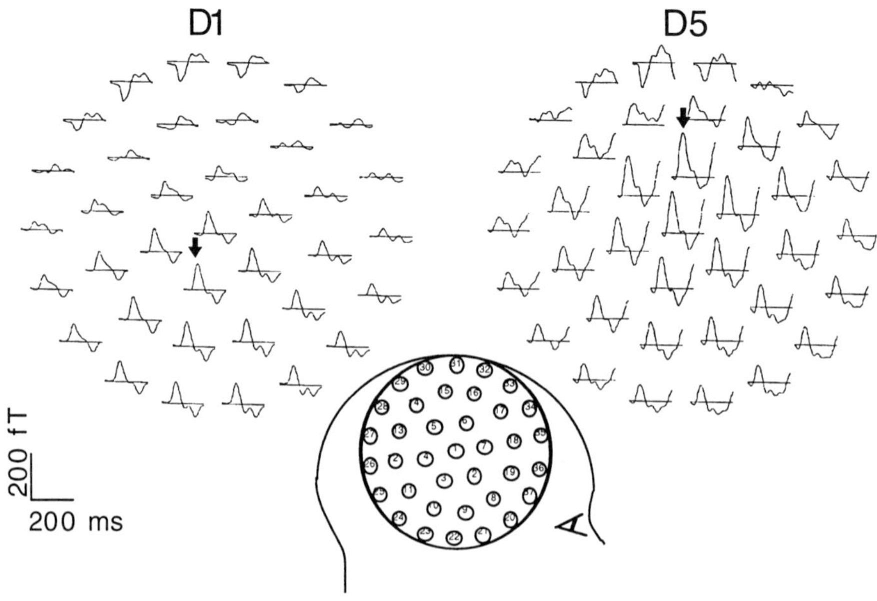

Fig. 1: The lower insert represents the sensor array placement. The set of waveforms indicates the magnetically evoked responses averaged for D1 (left) and D5 (right) respectfully. The black arrow marks the first major peak chosen for dipole modeling . The field maxima (black arrows) suggest that D1 is located more inferiorly than D5. This homuncular organization is confirmed by the dipole analysis.

Results

A typical example for the averaged evoked fields is presented in Fig.1. There was no obvious difference in waveshape between amputees and controls. In all subjects the peak for D1 was inferior to the peak for D5.

In amputees the cortical representations of D1 and D5 of the intact hand encompassed a greater area (Euclidean distance between digits: 12.4 ± 5.5 mm) than in controls (6.9 ±1.6 mm), $t = 2.4$, $p<.05$. The average difference on the inferior-superior plane was 5.5 mm larger in amputees than in controls ($t=3.2$, $p <.01$). When compared to the cortical representations in control subjects, only the D1-representation in the amputees had shifted along the central sulcus. The shift was limited to the lateral direction (7.7 mm, $t=2.1$, $p<.1$) and not significant in the inferior direction (2.6 mm). The distance between the representations of lip and D5 was 5.6 mm greater in amputees than in control subjects ($t=2.6$ $p<.05$). The distance between D1 and lip did not differentiate groups. There was also no group difference in the magnitude of the dipole moments of the digits.

In the amputees' deafferented hemisphere the cortical representation of the lip had shifted in a superior direction (13.9 mm, $t=2.9$, $p<.01$) compared to controls. The location of the lip contralateral to the intact side corresponded to the lip location in control subjects. The locations of the ECDs obtained in this study are schematically summarized in Fig. 2.

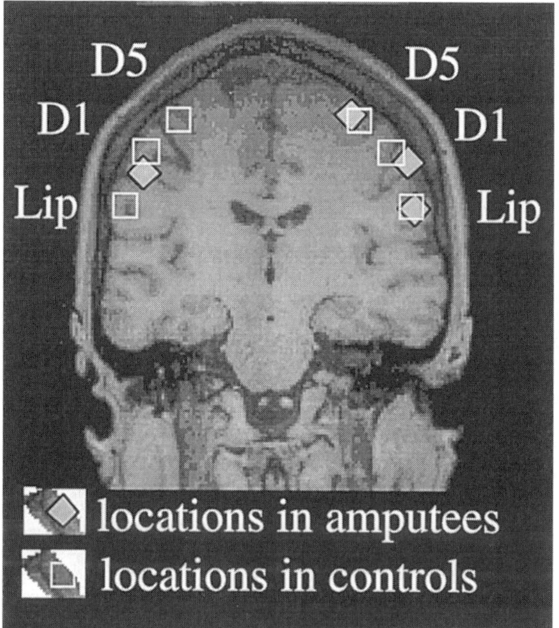

Fig.2. Schematic locations of the ECD for D1-, D5- and Lip-stimulation are seen here as projections onto a coronal MR-slice (of one control subject). Open squares indicate the average locations for controls, filled diamonds represent average locations for arm amputees. For amputees, the left half illustrates representations in the deafferented hemisphere. Compared to controls, the lip location in amputees has shifted contra- but not ipsilateral to the amputated side. Contrallateral to the intact side, the finger area of amputees has expanded, mainly as a result of the shift in D1-location.

Discussion

The results of the present study indicate that cortical reorganization after amputation is not restricted to the hemisphere contralateral to the amputated side. After an upper arm amputation, the representation of the intact fingers become enlarged. The shaping of the receptive field features of a cortical neuron by the input flow constitutes one basic mechanism underlying cortical reorganization [5]. Therefore, the extension of the hand area in amputees could be explained by the increased use of the remaining extremity and the subsequent increase of somatosensory input. The shift in the intact side was most pronounced for D1 and smaller for D5. Differences in the frequency of use of the separate digits may account for this result.

An invasion of the lip representation into the area formerly occupied by the amputated hand has been previously reported. Our present results suggest that such an invasion is not transferred to the hemipshere contralateral to the intact hand, the plastic changes in one hemisphere are not mirrored in the other.

The plastic alterations of the somatosensory cortex in amputees result from the interplay of two principles which govern cortical reorganization: (1) As the organization of the adult human brain seems to be use-depended and continuously "adapts" to the current needs and experiences of the individual the representational zone of the fingers of the remaining hand becomes enlarged [3]. (2) In the hemisphere deprived of input from the fingers, the face and lip representations expand into the now deafferented region [2]. This expansion may be responsible for the asymmetry of the location of the lip-response, observed in amputees. It could be speculated that the expansion encompassed by the digits may not alter the lip represention in the hemisphere contralateral to the intact side. Flor et al. [4] reported a strong relationship between hemispheric asymmetries in the lip representation and phantom limb pain. We can now conclude, that this relationship arises from alterations in the deafferented hemisphere and not from changes contrallateral to the intact side. Future investigations should study the temporal stability and the time course of the observed alterations.

References

[1] Calford, M., and Tweedale, R. Interhemispherictransfer of plasticity in the cerebral cortex. <u>Science</u>, <u>249</u>: 805-807, 1990.
[2] Elbert, T., Flor, H., Birbaumer, N., Knecht, S., Hampson, S., and Taub, E. Evidence for extensive reorganization of the somatosensory cortex in adult humans after nervous system injury. <u>Neuroreport</u>, <u>5</u>, 2593-2597, 1994.
[3] Elbert, T., Pantev, C., Wienbruch, C., Rockstroh, B., and Taub, E. Increased use of the left hand in string players associated with increased cortical representation of the fingers. <u>Science</u>, <u>220</u>: 305-307, 1995.
[4] Flor, H., Elbert, T., Knecht, S., Wienbruch, C., Pantev, C., Birbaumer, N., Larbig, W., and Taub, E. Phantom limb pain as a perceptual correlate of cortical reorganization following arm amputation. <u>Nature</u>, <u>375</u>: 482-484, 1995.
[5] Jenkins, W. M., Merzenich, M. M., Ochs, M. T., Allard, T., and Guíc-Robles, E. Functional reorganisation of primary somatosensory cortex in adult owl monkeys after behaviorally controlled tactile stimulation, <u>Journal of Neurophysiology</u>, <u>63</u>: 82-104, 1990.
[6] Pons, T., Garraghty, P. E., Ommaya, A. K., Kaas, J. H., Taub, E., and Mishkin, M. Massive cortical reorganisation after sensory deafferation in adult macaques. <u>Science</u>, <u>252</u>: 1857-1860, 1991.
[7] Recanzone, G. H., Merzenich, M. M., Jenkins, W. M., Grajski, A., Dinse, H.R. Plasticity in the frequency representation of primary auditory cortex following discrimination training in adult owl monkeys. <u>Journal of Neuroscience</u>, <u>67</u>: 1031-1056, 1992,
[8] Rockstroh, B., Vanni, S., Elbert, T., and Hari, R., Extensive somatosensory stimulation alters somatosensory evoked fields. This volume.
[9] Yang, T.T., Gallen, C., Schwartz, B., Bloom, F.E., Ramachandran, V.S., and Cobb., S. Sensory maps in the human brain. <u>Nature</u>, <u>368</u>: 592-593, 1994.

Acknowledgement

This work was supported by grants from the Deutsche Forschungsgemeinschaft.

Comparison of Magnetoencephalographic and Intraoperative Electrocorticographic Localization of Sensory Response

Greenblatt, R.[1], Bucholz, R.[2], Sobel, D.[3], Schwartz, B.[3,4] and Hirschkoff, E.[3,4]

[1]Source/Signal Imaging, San Diego, CA, USA; [2]Saint Louis University, St. Louis, CA, USA; [3]The Scripps Clinic and Research Foundation, La Jolla, CA, USA; [4]Biomagnetic Technologies, Inc., San Diego, CA, USA

Introduction

Magnetoencephalography (MEG) has been applied to identify regions of eloquent cortex potentially endangered by a planned surgical procedure. The resulting presurgical functional maps serve as an aide for the neurosurgeon during presurgical evaluation and planning. Prior work examined the reliability of this method [1] and assessed its accuracy by comparison with a variety of intraoperative results [2, 3]. Comparison of the three dimensional location of a dipolar source of somatosensory response as determined from MEG data with the corresponding location determined intraoperatively using electrocorticography (ECoG) has been limited by two primary factors: 1) the localization precision of routine intraoperative electric measurements, and 2) the limitation of routine intraoperative measurements to the cortical surface. Efforts at comparison have been further complicated by the effect of the relative orientation of the neurons responding to the stimulus; routine intraoperative electric recording usually makes the simple assumption that the source lies directly below the point of voltage phase reversal as measured on the surface of the exposed cortex[4], an assumption that can lead to substantial error [5]. Prior work attempted to account for this orientation effect in an approximate manner [6].

The present study was undertaken to take into account the fact that the source of sensory response generally lies below the surface of the cortex and may have any spatial orientation to provide a better test of the accuracy of non-invasive MEG somatosensory localization. Four patients who were scheduled to undergo open craniotomy were subject to presurgical mapping of somatosensory cortex using a 37 channel biomagnetometer (Biomagnetic Technologies Magnes). Subsequently, electrocorticographic localization of the response to median nerve stimulation was undertaken during surgery using a grid of EEG electrodes mounted on the exposed surface of the cortex. Single and multiple dipole modeling [7] of the somatosensory response was then undertaken for both the magnetic and electric data and the results compared.

Methods

Patients. 4 patients with tumors near eloquent cortex scheduled to undergo open craniometry were studied presurgically with MEG and intraoperatively with ECoG. Informed consent was obtained from all patients.

MEG. A 37 channel Magnes biomagnetometer (Biomagnetic Technologies, Inc.) in a magnetically shielded room was used to record evoked field data. Fields were evoked by tactile mechanical stimuli to individual fingers or lips.

ECoG. A 4x5 rectangular electrode grid (1 cm spacing) was placed over the exposed cortex. Responses were evoked by electrical stimulation of individual fingers or median nerve, and recorded with a Nicolet Viking system (10-13.5mA, 0.2 msec). 120 to 630 sweeps were used to generate on-line averages.

MRI. A T1-weighted volumetric image set was obtained from each subject using a 1.5T MRI system (General Electric).

Coregistration. 12 indelible scalp marks were placed on each patient, (including nasion and left and right preauricular points). The locations of these fiducial marks were measured prior to MEG acquisition using a Fastrak digitizer (Polhemus). The same marks were measured in the operating room with an LED digitizer (Picsys), as were the locations of the individual electrodes, permitting coregistration of the MEG and ECoG data. The nasion and preauricular points were marked with vitamin E capsules prior to magnetic resonance imaging, and these 3 fiducials were used for coregistration of the ECoG, MEG, and MRI data.

Source localization. Middle latency MEG data were fit with a single equivalent current dipole, using the Magnes software. The dipole at the latency with the largest correlation with the data was used for comparison with the ECoG source location. ECoG data were analyzed with EMSE software (Source/Signal Imaging) using the PRoMS algorithm [7], and a 1 shell sphere model with origin fit to the electrode location curvature. The location of the peak of the resulting source map was selected manually for comparison with the MEG dipole.

Results

Fig. 1 shows representative waveforms for ECoG (top) and MEG (bottom) for one subject. Several points are worth noting. Clearly the ECoG signal is more complex than the MEG signal, simply by inspection. This is consistent with the results of single dipole fitting. Good single dipole fits could always be obtained from the MEG data, but not from the ECoG data, which generally required 2 or more dipoles to account for greater than 95% residual variance. In addition, there are significant differences in time course between the MEG and ECoG signals, which are probably due to the difference in stimulus method (tactile for MEG vs. electrical for ECoG). Lastly, we have observed considerable intersubject differences in the ECoG data, which in part may be accounted for by the parietal lesions in these patients.

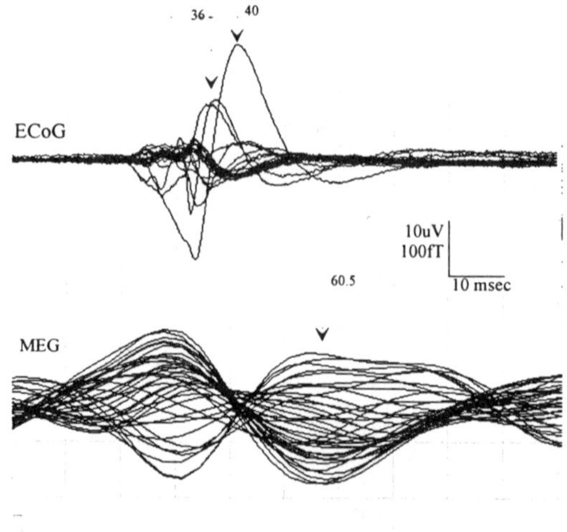

Fig. 1. (A) Averaged electrocorticogram for patient 1489, referenced to contralateral ear. The stimulus to the contralateral thumb was presented at t=0, indicated by the vertical mark. The 20 individual traces have been superimposed to a common origin to emphasize the temporal patterns. The arrows indicated the latencies for which ECoG-derived source maps are displayed in figure 2. (B) Averaged magnetoencephalogram for the same patient. The stimulus was presented at t=0, indicated by the vertical mark. The 37 individual traces have been superimposed to a common origin to emphasize the temporal patterns. The arrow indicates the latency for which the MEG-derived source map is displayed in figure 2.

Source maps from ECoG data for this patient are shown in fig. 2. Similar source maps could be obtained from 3 of the 4 patients who had adequate signal/noise in the ECoG. In 1 patient, data was used from the independent stimulation of 2 digits. For these 4 datasets, the location of the peak of the deeper, more tangential dipole from the ECoG source map was compared with the MEG single equivalent dipole location. The mean discrepancy between the 2 localization methods was 8±3 mm (mean±standard deviation).

The locations of each of the 4 pairs of sources were transferred to the corresponding MRI images, which were read by a trained observer. The location of each of the sources was consistent with a location in the postcentral gyrus.

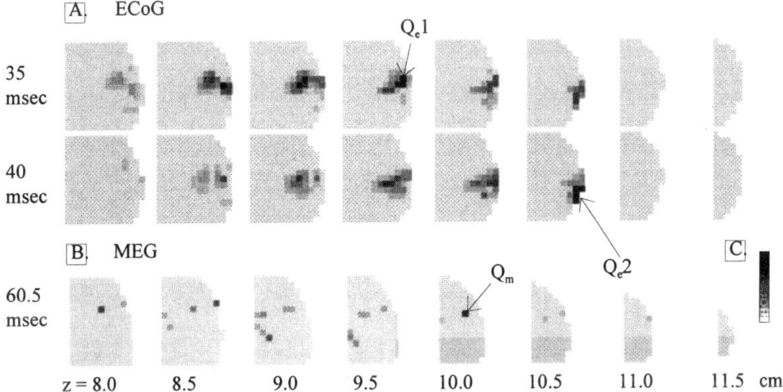

Figure 2. **A**. Source maps from ECoG at 36 (top) and 40 (middle) msec. poststimulus latencies following stimulation of the contralateral thumb. Source maps are displayed as axial slices, 5 mm resolution, with the nose (+X) towards the top of the figure, the right ear (-Y) towards the right. The cortical surface was determined by extrapolation from the measured electrode locations. The 36 msec slice is dominated by the activity of a primarily tangential source (Q_e1). The 40 msec slice is dominated by the activity of a primarily radial source (Q_e2). **B**. Source maps from MEG at 60 msec. poststimulus latencies. Source maps are displayed as in A, but span slightly different cortical extents, the cortical surface was approximated from scalp measurements in B.. The slice is dominated by the activity of a source (Q_m) at the approximate location of Q_e1. C. Source map scale maximum is $3{\times}10^{-10}$ ampere meters for A, $3.5{\times}10^{-8}$ ampere meters for B. The difference in source strength may be attributed to the more distributed source pattern in A.

Conclusion

Middle latency MEG data was adequately modeled as a single equivalent current dipole. Middle latency ECoG data required more complex modeling, usually involving at least two sources. Both MEG and ECoG-derived sources could be located reliably to postcentral gyrus. The location of the ECoG-derived tangential source agreed with that of the MEG-derived dipolar source to within about 8 mm on average.   This discrepancy may be a limitation imposed by the combined imprecision of the MEG method, the ECoG method, and the coregistration process. These data suggest that the use of appropriate modeling techniques may improve the utility of ECoG data, and provide additional support for MEG non-invasive functional mapping of somatosensory cortex.

Literature cited

[1] Gallen, C.C., Schwartz, B.J., Rieke, K., Pantev, C., Sobel, D., Hirschkoff, E., and Bloom, F.E., Intrasubject reliability and validity of somatosensory source localization unsing a large array biomagnetometer. *Electroenceph and Clin Neurophys.*, **1994**, 90:145-156.

[2] Gallen, C.C., Sobel, D.F., Waltz, T., Aung, M., Copeland, B., Schwartz, B.J., Hirschkoff, E.C., and Bloom, F.E., Noninvasive presurgical neuromagnetic mapping of somatosensory cortex. *Neurosurgery*, 1993, **33**:260-268.

[3] Morioka, T., Yamamoto, T., Katsuta, T., Fujii, K., and Fukui, M., Presurgical three-dimensional magnetic source imaging of the somatosensory cortex in a patient with peri-rolandic lesion: tecnical note. *Neurosurgery*, **1994**, 34:930-934.

[4] Allison, T., McCarthy, G., Wood, C.C., and Jones, S.J. Potentials evoked in human and monkey cerebral cortex by stimulation of the median nerve. *Brain*, 1991 **114**:2465-2503.

[5] Suzuki, A., Yasui, N., J. Interoperaticve localization of the central sulcus by cortical somatosensroy evoked potentials in brain tumor. *Neurosurgery*, 1992, **76**:867-870.

[6] Gallen, C., Schwartz, B., Bucholz, R., Malik, G., Barkley, G., Smith, J., Tung, H., Copeland, B., Bruno, L., Assam, S., Hirschkoff, E., Bloom, F., Presurgical localization of functional cofrtex using magnetic soruce imaging. *J Neurosurgery*, 1995, **82**:988-994.

[7] Greenblatt, RE., Probabilistic reconstruction of multiple soruces in the bioelectromagnetic inverse problem. *Inverse Problems*, 1993, **9**:271-284.

Origin of Slow Wave Observed in Cerebrovascular Disease

Hatanaka, K.[1], Nakasato, N.[2], Kanno, A.[3], Ohtomo, S.[2], Seki, K.[2], Fujiwara, S.[2] and Yoshimoto, T.[2]

[1]KRI, 17 Chudoji Minami-machi, Shimogyo-ku, Kyoto, Japan; [2]Department of Neurosurgery, Tohoku University School of Medicine, 1-1 Seiryo-cho, Aoba-ku, Sendai, Japan; [3]MEG Laboratory, Kohnan Hospital, 4-20-1 Nagamachiminami, Taihaku-ku, Sendai, Japan

Introduction

Due to excellent spatio temporal resolution and noninvasive nature of the measurements, magnetoencephalography (MEG) becomes a powerful tool for clinical application. This is especially the case in evoked MEG applied for presurgical diagnosis of the patients who suffered brain diseases. However, in spontaneous activity such as epileptic discharges, applicability is not so simple. This is because, for spontaneous waves, several brain activities or noises were superimposed on the detected signal, and separation for target activity is generally difficult. Small array of sensing coils also restricted the spatial resolution and sometimes brought erroneous results simply because they could not cover the entire brain at once. The appearance of whole-head MEG directly linked to magnetic resonance imaging (MRI) reduces data acquisition time, patient fatigue and localization error. It also provides objective information about patients which is essential for correct diagnosis. Spontaneous MEG studies used to be concentrated to epileptic spike activity. However, slow wave activity, typically observed by electroencephalography (EEG), is quite common in stroke, tumor, epilepsy, cerebrovascular disease, or head injury. Earlier investigations of slow wave activity with MEG [1-3] were restricted to small (1 to 37) channel system. Here we present the results of whole head MEG and source analysis of a slow wave activity.

Methods

A 13-year-old male subject, who had brain infarctions due to a cerebrovascular disease, was seated on a chair in a magnetically shielded room. We employed a 64 channel MEG (CTF Systems Inc., Canada) and a standard international 10-20 system EEG (Nihonkohden Inc., Japan) for measurements and recording. Subject was instructed to stay at rest, and eventually he fell asleep during the recording. By monitoring the EEG chart, we could pick up twenty 10 seconds epochs of prominent slow wave activities from the session lasting for about an hour. Sampling clock of MEG was 250 Hz and the signal filtered by 0.5 to 40 Hz band pass and 50 and 100 Hz notches was stored on a disk for further analysis. Seven channels of EEG, FP1, CZ, F3, F4, P3, O1 and O2, were analog to digital converted and passed through the same filters and stored simultaneously with MEG data for comparison.

Source analysis of MEG data was carried out after measurement using single current dipole model. The effect of volume current was considered using Grynszpan and Geselowitz spherical head shape approximation. Position and radius of ' best sphere' (*i.e.* sphere which fit into subject's head in the least square sense) were determined by magnetic resonance imaging (MRI) of the subject. Conversion of MEG coordinates system to MRI was carried out by reference to three fiducial points (two preauricular points and nasion), each marked by small coils for MEG and oil-containing capsules for MRI. Positions and orientations of the estimated equivalent current dipoles (ECDs) were directly superposed on the MRI of the subject.

Results

We found that slow wave activities included delta wave, theta wave, and sharp (spike) and slow wave complex. These observations naturally contained normal (*e.g.* slow wave sleep) as well as pathological activity. We rejected artifacts such as eye movement or blink by carefully reading the EEG chart. Although magnetic noises from moving vehicles were mostly reduced by the low cut filter (0.5 Hz) of the band pass, we lost dc or lower frequency components [4,5] of slow wave activities at the same time. It is essential to find out pathological activities peculiar to the brain disease in the recording. Examples of a pathological slow wave are shown in Fig. 1. These activities were observed in light sleep (stage 2) epochs.

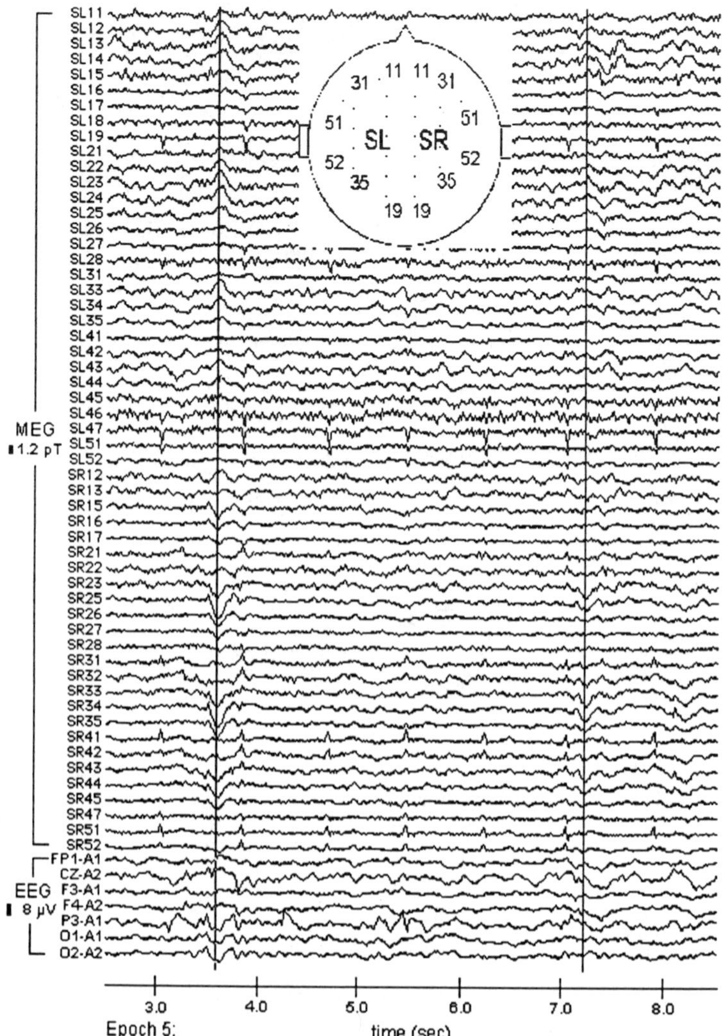

Fig. 1

An example of the simultaneous recording of MEG and EEG during light sleep stage. Bandpass filter was 0.5 to 40 Hz. The positions of the pickup coils of MEG are indicated on schematic head. Some noisy MEG channels are deleted in the above chart, and seven EEG recording stored in the MEG system are shown. Two pathological slow wave activities are observed: one peaked at 3.612 sec and the other at 7.216 sec. Note that slow wave peak at 3.612 sec is observed both in MEG and EEG, while peak at 7.216 sec is only detected by MEG. This is because magnitude of the dipole is about half for the later wave (see Table 1). Spikes of about 0.8 sec period observed in SL19 etc. are magnetocardiogram detected by MEG pick-up coils.

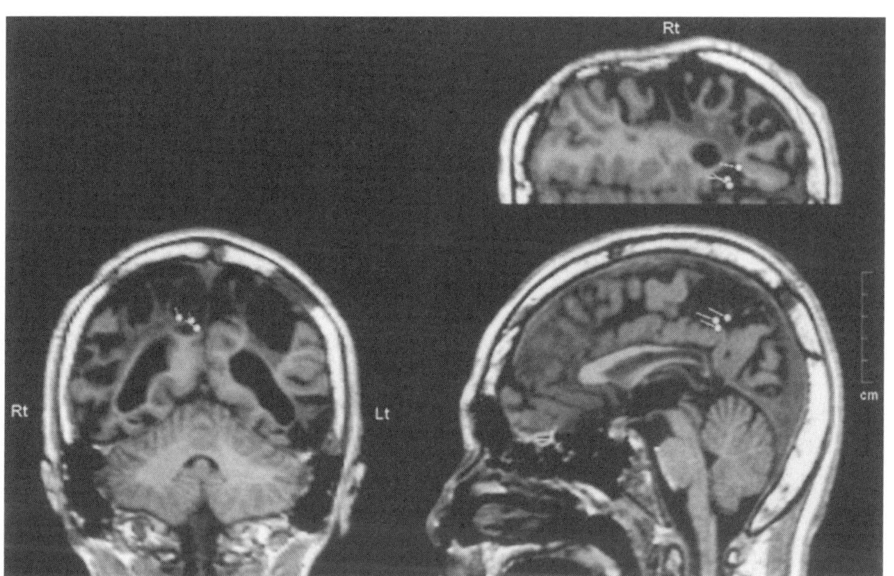

Fig. 2
Positions (circles) and orientations (bars) of the ECDs estimated from the slow waves in a case of cerebral infarction.

Table 1. The ECD parameters of the slow waves in a case of cerebral infarction (x, y, z: anterior, left lateral, and upward coordinates of the MEG system; φ: azimuthal angle of dipole, 0 for+x direction; θ: declination angle of dipole, 0 for +z direction; Q: magnitude of current dipole)

Dipole	Epoch:	time (sec)	x (cm)	y (cm)	z (cm)	φ (deg)	θ (deg)	Q (nAm)	Error (%)
1	5:	3.612	-0.98	-0.32	8.16	332	82.9	650	7.8
2	5:	7.216	-0.78	-0.57	8.47	358	83.7	333	15.2
3	6:	5.816	-1.24	-1.16	8.72	354	79.1	443	5.5

Results of the source analysis of the slow wave activities are shown in Fig. 2, and the parameters of ECDs are summarized in Table. 1. The ECDs were localized at the lower bottom surface of a large infarct on right parietal lobe, and the orientations were anterior-posterior direction.

Discussion

Completed infarct foci are replaced by cerebrovascular fluid or connective tissues and thus electrically silent. Therefore, abnormal slow waves are believed to be originated in the surrounding brain tissues. ECDs might be distributed or extended in space. Surprisingly, our present study demonstrated that some of abnormal slow waves indicated clear single dipole patterns with relatively low fitting errors even for the whole head data. These findings suggest that electrically active tissue exist at least a part of the boundary of infarct foci.

Infarction grows from ischemia, and cerebrovascular diseases generally accompany focal neurological symptoms and signs resulting from blood vessels. MEG studies in animal models have demonstrated that focal cerebral anoxia [4] or ischemia [5] brought slow or dc (direct current) variations of magnetic fields. As Suthering pointed out [6], it is very important to confirm these results because there could be potential diagnostic value of detecting slow wave in human cerebrovascular dieseases. Our results, though not dc-MEG, support this comment.

EEG may detect the peculiar slow wave activity we found in a cerebrovascular disease patient. MEG can, however, detect and localize sources of the activity accurately. Further studies of whole head MEG in cerebrovascular disease patients will reveal the occurrence and development mechanism of these diseases in future.

References

[1] Vieth, J.B. Magntoencephalography in the study of stroke (cerebrovascular accident), In: Sato, S. *Advances in Neurology* 54: 261-269, New York, Ravan Press, 1990.

[2] Vieth, J. B., Kober, H., Weise, E., Daun, A., Moeger, A., Friedrich, S. Functional 3D localization of cerebrovascular accidents by magnetoencephalography (MEG), *Neurological Research* ,1992, 14: S132-S134.

[3] Gallen, C.C., Sobel, D.F., Schwartz, B., Copeland, B., Waltz, T., and Aung, M. Magnetic source imaging Present and future, *Investigative Radiology*, 1993, 28, Supplement 3: S153-S157.

[4] Takanashi, Y., Chopp, M., Levine, S.R., Kim, J., Moran, J.E., Tepley, N., Chen, Q., Barkley, G.L., and Welch, K.M.A. Magnetic fields associated with anoxic depolarization in anesthetized rats, *Brain Research*, 1991, 562:13-16.

[5] Chen, Q., Chop, M., Chen, H., and Tepley, M. Magnetoencephalography of focal cerebral ischemia in rats, *Stroke*, 1992, 23:1299-1303.

[6] Suthering, W.W. Editorial comment, *Stroke*, 1992, 23:1303.

Acknowledgments

This work has been supported in part by Grants-in-Aid for Scientific Research No. 03404042 and No. 064050 from the Ministry of Education, Science and Culture of Japan.

Magnetic Source Imaging in the Evaluation of Chronic Alcohol Abuse

Hill, D., Waldorf, V.A., Lewine, J.D., Provencal, S.L., Moyers, T., Yeo, R.

The New Mexico Institute of Neuroimaging, The New Mexico Regional Federal Medical Center,
Albuquerque, New Mexico, USA

INTRODUCTION

Chronic alcohol abuse is a significant problem for a large number of individuals, especially in the population of U.S. Veterans. Research to date has documented some of the long-term deleterious effects associated with chronic abuse. Cognitive difficulties may include impairments in short-term memory, paired associate learning, visual-spatial skills, complex problem-solving, and psychomotor functioning. Neuroradiographic findings show that cortical atrophy and white matter changes are prominent at late stages of alcoholism, but intermediate effects are less well documented. From a brain functional perspective, evoked potential studies demonstrate that subjects in the late stages of alcohol abuse show changes in P3 amplitude for both the visual and auditory stimuli.

A limitation of previous research has been the focus on patients with severe, chronic conditions such as Wernicke-Korsakoff syndrome. The present study was focused on specification of neurobiological and neuro-psychological deficits that may manifest at intermediate stages of alcohol abuse. In particular, we used magneto-encephalography, neuropsychology, and magnetic resonance imaging to determine if patients with a history of substance abuse demonstrate focal abnormalities.

METHODS

Subjects: Four volunteer subjects enrolled in the out-patient alcohol treatment program at the Albuquerque Veterans Administration Medical Center (VAMC) participated in this project. The subjects were all right-handed. All four subjects had a diagnosis of alcohol dependence, based on the Diagnostic and Statistical Manual of Mental Disorders-Third Edition (APA, 1987). Average daily alcohol intake was 106 ounces (range= 32-226), with an average of 112.3 days elapsed between last drink and testing (range= 26-180). No history of head trauma or psychoactive substance dependence was reported. Case histories were as follows: Case 1 was a 49 year old Caucasian male with 12 years of education and a 27 year drinking history. Case 2 was a 39 year old Native American male with 12 years of education and a 23 year drinking history. Case 3 was a 42 year old Hispanic male with 16 years of education and a 20 year drinking history. Case 4 was a 50 year old Caucasian male with 15 years of education and a 26 year drinking history.

Neuropsychological Testing: Each subject was administered a comprehensive battery of neuropsychological tests assessing the cognitive domains of general intelligence, perceptual-motor skills, motor skills, memory skills, executive planning skills, and language skills. Standard norms were used to evaluate performances of each subject, and age and education appropriate norms were used whenever possible.

Magnetic Resonance Imaging (MRI): Magnetic resonance imaging was performed with a 1.5 Tesla Magnetom system (Siemens). Images were obtained using modified gradient and spin echo sequences with the following parameters: Sagittal - 1.5 mm contiguous slices, TR=40, TE=6; Coronal - 5.0 mm slices with 1 mm gaps, TR=800, TE=15; Axial - 5.0 mm slices with 1 mm gaps, TR=2500, TE=90. The images were examined for pathology by a neuroradiologist who was blind to the clinical status of the subjects. The sagittal volumetric data set was re-formatted and re-sliced into 1.0mm thick coronal slices for the purpose of whole-brain volumetric measurement. Volume analyses were performed using the NIH Image 1.52 program.

Magnetoencephalography (MEG): The studies were performed in a magnetically and radio-frequency shielded room equipped with a large array 37-channel biomagnetometer or a recently installed whole-head system with 122-channels. For each subject, 3 minutes of continuous data was recorded in an eyes closed state from each of multiple positions to provide for a whole-head examination. MEG examinations typically required 2 hours of subject time. MEG data were examined for epileptiform transients, abnormal power spectra, and abnormal low frequency magnetic activity (ALFMA). For the evaluation of ALFMA, data were digitally filtered with a bandpass of 1-4 Hz in order to isolate activity in the delta range. After identification of the latencies of local amplitude peaks where the magnetic field strength was greater than 200 fT, single dipole modelling was performed. Only highly dipolar slow waves were considered in further analyses. A spatial clustering algorithm was used to determine if a particular cortical region was responsible for generation of multiple dipolar low frequency events. This is a critical aspect of the analysis procedure because available data suggest that it is the repetitive generation of dipolar low frequency activity that provides the most reliable sign that a region of tissue is dysfunctional. A 1-cm radius volume element was considered pathological only if it was responsible for generation of five distinct low frequency events for each minute of recorded data. Previous work indicates that these particular parameter values provide for good separation of control and patient populations.

RESULTS

As summarized below, and in table 1, abnormalities were found on one or more tests for three of the four patients.

<u>Case 1</u>: This subject's general intelligence was in the above average range. The skill domains of perceptual-motor, memory, executive planning and language were all found to be within the normal limits. However, this individual showed significant impairment on the motor task. This subject's MRI was read as normal. Whole brain volume was $1298cm^3$, a value within normal limits. MEG examination was abnormal, with extensive ALFMA over left temporal regions. In addition, occasional epileptiform transients were identified bi-parietally, R>L.

<u>Case 2</u>: Although this subject showed a similar drinking history to the other subjects, his performance on the neuropsychological battery was within or above normal limits. MRI was normal. Whole brain volume was $1297cm^3$. MEG examination was normal.

<u>Case 3</u>: Given this individual's educational level, general intelligence scores were lower than would be expected. Relative strengths were noted in perceptual-motor and language skills. However, tests of higher cognitive abilities showed mild deficits, although the scores fell within the low normal range. The motor skills test scores suggested a lateralized deficit, as dominant hand performance was significantly below average, whereas non-dominant fell within the normal range. MRI results indicated diffuse white matter changes. Whole brain volume was $1170cm^3$. This is almost two standard deviations below normal. MEG results indicated abnormal power spectra (excessive delta) over right and left temporal regions, but dipolar sources of ALFMA were not identified. In addition, some left parietal epileptiform transients were identified.

<u>Case 4</u>: This subject's general intelligence, memory, language, and motor skills were in the average range. However, this individual showed impairment on tests of executive functioning and perceptual-motor skills. MRI revealed diffuse white matter changes, and note was made of cerebral volume loss, especially in the frontal lobes. Whole brain volume was $1162cm^3$, a value almost two standard deviations below normal. MEG revealed abnormal power spectra (excessive delta) over left and right temporal lobes and left parietal areas. Much of the left parietal slowing had dipolar fields with sources of ALFMA localizing around the central sulcus. It is noteworthy that this patient was re-evaluated with MEG 14 months after cessation of alcohol abuse. During this follow-up, MEG power spectra were normal and no ALFMA was identified.

TABLE 1. Assessment Summary: Neuropsychological, MRI, and MEG Findings.

	Neuropsych	MRI	Brain Volume	MEG
Case 1	Motor Abnormalities	Normal	$1298cm^3$	LT-A,P-E
Case 2	Normal	Normal	$1297cm^3$	Normal
Case 3	Motor Abnormalities	White Matter Changes	$1170cm^3$	R/LT-S, LP-E
Case 4	Executive & Perceptual-Motor Abnormalities	White Matter Changes & Atrophy	$1162cm^3$	R/LT-S, LP-S, LP-A

MEG Abnormalities

<u>location</u>	<u>type</u>
L: Left	A: ALFMA
R: Right	S: Spectral (high delta)
T: Temporal	E: Epileptiform transients
P: Parietal	

DISCUSSION

In this study, a novel multimodal approach was utilized to further understand the long-term effects of chronic alcohol use. Techniques for assessing cognitive function (neuropsychological evaluation), brain structure (magnetic resonance imaging) and electrophysiology (magnetoencephalography) were all used with a common patient population to explore the extent to which alcohol abuse induces focal abnormalities.

Neuropsychological findings indicated that 3 of the 4 cases demonstrated clinically significant deficits. Findings included motor and perceptual-motor impairments, as well as problem-solving difficulties. These deficits are consistent with those found in other research with this population.

MRI findings revealed white matter changes in two of the patients, a finding consistent with other radiological studies of alcoholics. Two cases had abnormal MRI examinations indicating white matter changes. For both, atrophy was also present, whole brain volumes being almost two standard deviations below normal (1313 cm^3, sd= 89 cm^3). Both of these patients demonstrated neuropsychological deficits and MEG abnormalities.

Overall, MEG identified abnormalities in 3 of the 4 patients. Abnormalities included abnormal low frequency magnetic activity, excessive delta in power spectra, and in some cases, epileptiform transients (Fig 1). The patient without MEG abnormalities was also found to be normal on MRI and all neuropsychological tests. It is also noteworthy that Case 3 had a normal followup examination after 14 months of sobriety. This suggests that MEG assessed brain dysfunction may be partly reversible.

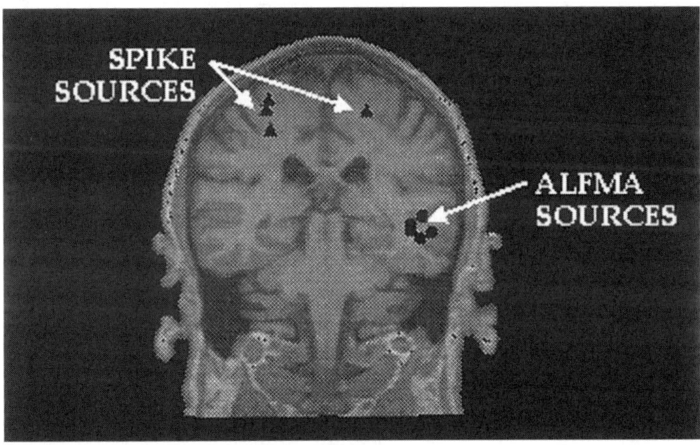

Fig 1: Case 1, MEG shows left temporal ALFMA and bi-parietal epileptiform transients. MRI was normal, as was whole brain volume. Neuropsychological testing revealed multiple deficits including memory and perceptual-motor problems.

CONCLUSIONS

The preliminary results presented here suggest that MEG may be useful for providing objective evidence of brain dysfunction in patients with a history of chronic alcohol abuse. An ability to demonstrate to patients the negative impact of alcohol abuse on their brain function may be beneficial in convincing patients to stay in therapy programs. Also, a better understanding of how alcohol effects brain physiology may ultimately lead to better interventional strategies.

MEG and MCG in a Clinical Environment: BioMag Laboratory, Helsinki

Ilmoniemi, R.J.[1], Ahonen, A.[4], Alho, K.[5], Aronen, H.J.[2], Huttunen, J.[1], Karhu, J.[6], Karp, P.[1], Mäkijärvi, M.[1,3], Näätänen, R.[5], Paetau, R.[7], Pekkonen, E.[5], Standertskjöld-Nordenstam, C.-G.[2], Tiihonen, J.[8], Toivonen, L.[1,3] and Katila, T.[9]

[1]BioMag Laboratory/Medical Engineering Centre; [2]Department of Radiology; [3]1st Department of Medicine, Helsinki University Central Hospital; [4]Neuromag Ltd, Helsinki; [5]Cognitive Brain Research Unit, University of Helsinki; [6]Department of Clinical Neurophysiology, Kuopio University Hospital, Kuopio; [7]Children's Castle Hospital, Helsinki; [8]Department of Forensic Psychiatry, Niuvanniemi Hospital, Kuopio; [9]Laboratory of Biomedical Engineering, Helsinki University of Technology, Espoo, Finland

Introduction

The success of biomagnetism will critically depend on how useful the technique is as a clinical tool and what kinds of applications exist. We have started work at the new BioMag Laboratory of the Helsinki University Central Hospital to bring this technology clinically available and to evaluate its possibilities. The project follows extensive development and pioneering work for over two decades at the Helsinki University of Technology in neuromagnetism [1] and cardiomagnetism [2] done in collaboration with the Helsinki University Central Hospital. While some clinical applications already have been established [1–4] and we are actively working on them, much of our research is conducted on various patient groups and healthy subjects in order to evaluate and assess this new technology and to extend our knowledge of the functioning of the brain and the heart.

Here, we first describe some goals, design criteria, and technical solutions in building our new biomagnetism laboratory. Then, ongoing clinical and technical projects will be summarized.

Methods

The large-array magnetometers made available by Neuromag Ltd. were the starting point for the design of the laboratory. In addition to the whole-scalp 122-channel neuromagnetometer, a 68-channel cardiomagnetometer was developed and built for this project by Neuromag in collaboration with the Medical Engineering Laboratory of the Helsinki University of Technology. Although in an ideal case brain and heart research would have separate shielded rooms, we decided to place both systems in the same room (see Fig. 2), requiring that the switch from brain to heart measurements, and vice versa, is easy and quick and that as much as possible of the electronics and computing facilities can be shared between the systems. Although the magnetometers, together with gantries and counterweights, weigh about 270 kg each, they can be interchanged easily on air cushions by one person.

The site for the laboratory was selected on the basis of careful measurements of the external magnetic noise level and the requirement that the distance from cardiology and neurology departments is not very long. The level of mechanical vibrations at the site is low (Typical rms levels for acceleration, speed, and displacement of the underlying floor in the bandwidth 5–100 Hz were 0.35 mm/s^2, 1.2 μm/s, and 14 nm, respectively; transient levels could be 2 orders of magnitude higher.). Otherwise, mechanical damping for the shielded room would have been considered. The recorded environmental noise spectrum in the selected location is shown in Fig. 1. It was decided that a shielding factor of 30 (at 0.1 Hz) to 80 dB (100 Hz) would be sufficient at this site. Actually, this minimum shielding requirement was exceeded by 10–20 dB (Euroshield Ltd., see [5]).

A ramp mechanism was built to allow easy wheeling of the patient bed into the shielded room [5]. Lighting was arranged via glass fibers to exclude magnetic interference from the lamps. A rectangular hole was built in the wall to allow the subject to look at a computer screen or a video display (up to 15"). Another opening enables cardiac monitoring inside the room during intervention studies. In many experiments, in particular with children, elderly people or demented patients, movies shown to the subjects help them to sit still during the measurement.

Proper use of the biomagnetic technique requires the utilization of other imaging modalities to help solve the inverse problem (MRI, fMRI, EEG/ECG) and to display the results (MRI). We have built in the facility a 64-channel EEG/ECG system [6] to be used simultaneously with magnetic measurements. The laboratory's local area network is connected to the hospital's fast image network to allow easy transfer of MRI images from the Radiology Department to be used in conjunction with data analysis. In addition, we have realized a fast connection to the Helsinki University of Technology so that data can be analyzed directly at the site where MCG software is being developed and where much of the data are analyzed. Software maintenance by Neuromag Ltd. is performed telematically via the networks as well. A special consideration in the setup of these connections is the strict requirement of patient data security. Unauthorized access to information about patients is prevented in two ways. First, patient names are separated from the data. Second, connecting to the computers from the outside is possible by authorized users only.

To maximize effective recording time, the laboratory (about 200 m^2) was designed so that a patient can be prepared (change of clothes, attachment of electrodes and marker coils, digitization of head or chest, etc.) while the previous patient is still being studied. Space was assigned in the laboratory also for data analysis, technical development work, and for cardiac surface-potential measurements. A registered nurse accompanies most patients in the shielded room, making sure that the patient stays still and performs the assigned task. MCG measurements usually take less than 5 minutes; MEG paradigms for patients are designed to last at most 40–60 minutes.

The 122-channel helmet-shaped MEG magnetometer array [7] consists of 61 identical plug-in units, each containing two orthogonal planar gradiometer flux transformers and dc SQUIDs. Although the 16.5-mm baselength of the flux transformers is relatively short and the sensitivity pattern of each sensor is quite narrow, the array as a whole is able to see as deep in the brain as an array of axial sensors. The advantages of the planar design include very good balance and precise calibration; in addition, the raw signals are easy to interpret, because the measured signal is at maximum right above the source. Furthermore, because of the tighter sensitivity pattern collecting brain noise, the signal-to-noise ratio in the SQUID output from planar gradiometers is typically higher than from axial gradiometers [7].

The 68-channel MCG device [8] has 62 planar and 6 axial gradiometers over a slightly curved dewar bottom (diameter 300 mm). Since the advantage of the planar gradiometer geometry in MCG may not be as clear as in MEG, the axial gradiometers were built in order to evaluate the gradiometer types experimentally. The high intrinsic balance and ease of calibration were lost, but the signal level was increased in these channels as expected.

Data from the two magnetometer arrays are collected by a real-time computer, which also controls the head-position indicator electronics and provides stimulus triggers when needed. On-line and off-line data analysis is performed by HP 725/100-series PA-RISC workstations interconnected by a local area network. An X-terminal in the patient preparation room is used for entering patient data, for 3D digitizing, and for data analysis. Additional terminals and workstations can be connected to the local area network directly or via ATM-level networks.

Liquid helium transfer is carried out 3 times per 2 weeks. For quality assurance purposes, an empty room measurement and phantom measurement should be performed at least once a week. A preventive maintenance service is scheduled for four days per system in a year.

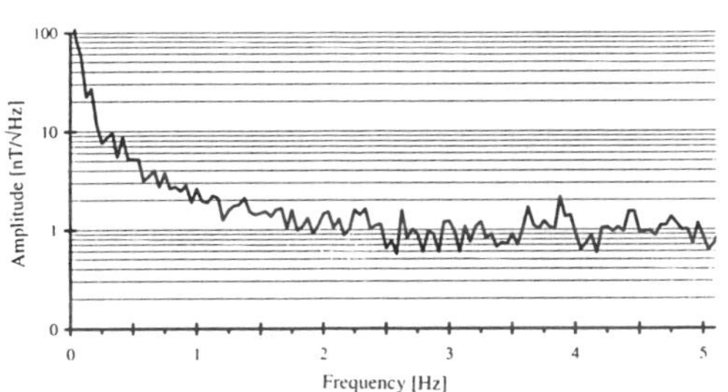

Figure 1. Magnetic field noise level at the site of the laboratory prior to construction of the shielded room. At some other candidate sites at the hospital, 2–5 times higher noise levels were measured.

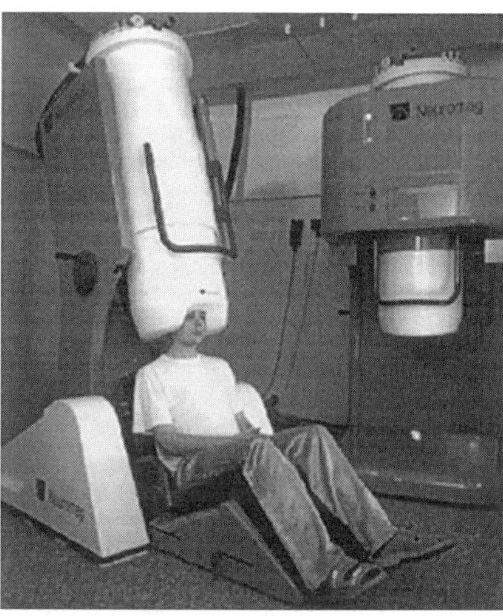

Figure 2. The interior of the shielded room showing the MEG system in measuring position, the cardio-magnetometer being in the background. For a figure showing an MCG experiment, see [8].

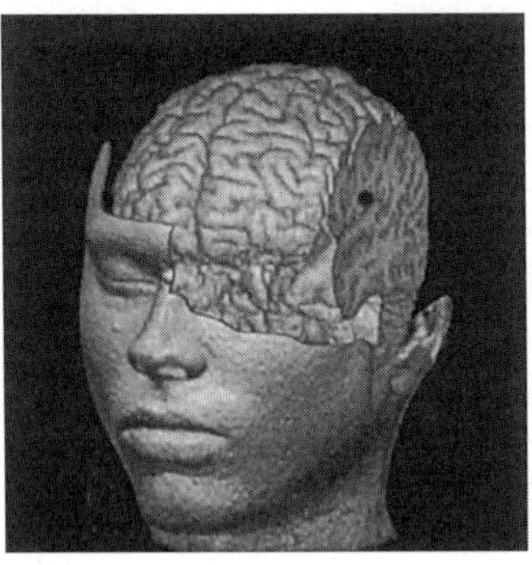

Figure 3. 3D-rendering of a subject's head showing the location of an auditory response.

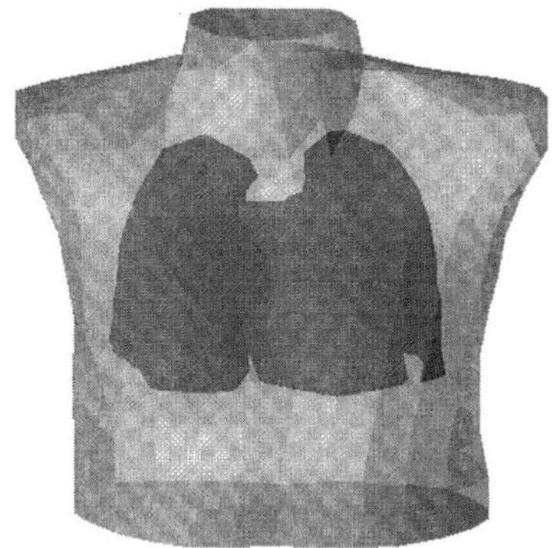

Figure 4. Torso model used in realistic-geometry computations of cardiac magnetic fields.

Results

Our major clinical projects with MEG include the localization of epileptic foci and the determination of the location of the central sulcus in patients going to brain surgery. Extensive work has been initiated to study patients with Alzheimer's and Parkinson's diseases, schizophrenia [9,10], alcoholism, or developmental disorders [11]. We have also used MEG in studies of neuronal changes associated with disorders of central and peripheral auditory processing and after brain infarction. Heart studies [8] include the determination of myocardiac arrhythmogenic foci, the estimation of the risk of sudden cardiac death, repolarization disorders and left-ventricle hypertrophy. During the first year of operation, we completed a total of 767 measurement sessions, of which 288 were on patients.

We have compared MEG and fMRI localization of somatosensory activation [12] in normal subjects and in patients scheduled for surgery. On the basis of these studies, the method of locating the central sulcus with MEG looks very promising, as previously reported by Lewine *et al.* [4]. In some cases, however, artefacts may be a problem. To eliminate the effect of external noise, eye movements, or magnetic particles left inside the head during previous surgery, we have successfully used the Signal-Space Projection method [13].

We have studied somatosensory and motor functions in 20 patients with acute brain infarction in order to detect plastic changes during recovery. While it has become clear that patients with even a moderate paresis can be studied up to 60 minutes, the current plans of manufacturers to allow recording also from supine position is more than welcome. Equally important are efforts to help maintain the head position during data collection.

Our experience with the mismatch paradigm [14,15] shows that MEG studies lasting up to 45 minutes can be performed even in severely demented subjects (Mini-Mental State score down to 14/30). During these experiments, a motion picture is often shown in order to keep the patient vigilant and comfortable.

In patients being evaluated for surgical treatment of epilepsy, low-amplitude fast cortical activity not seen in scalp EEG has been detected in ictal whole-head MEG. In such patients, MEG seems to reduce the need for invasive preoperative recordings. In patients with different clinical epilepsy syndromes, the determination of cortical generators of the underlying activity seems to help in understanding the pathogenesis and in the planning of therapy.

Very young children may fall asleep or are unable to sustain conscious attention to head position even during a brief measurement. The quality of the data was greatly improved in such children by using a special examination seat, allowing exact positioning of the head. For MEG examination of infants, the MCG equipment is used.

Preliminary MEG results obtained in schizophrenics showed a reversed interhemispheric asymmetry of temporal auditory areas [9,10], although the MRIs in these patients were normal. It thus appears that MEG might be capable of revealing subtle changes in the functional anatomy in major psychiatric disorders. This kind of application may be very useful in the screening of high-risk populations.

MCG has been performed on over 100 patients who have the Wolff–Parkinson–White (WPW) syndrome, life-threatening ventricular tachyarrhythmias after myocardial infarction, cardiomyopathy, or congenital long-QT syndrome [8]. A routine measurement takes just 5 minutes, while patient preparation takes about 10 minutes. The preliminary results from our studies are encouraging. The 3-dimensional location of the overt accessory pathway in WPW syndrome can be determined with an accuracy of 1–1.5 cm. The preoperative information has been helpful in the planning of ablative therapy. Patients with malignant ventricular tachyarrhythmias after myocardial infarction have low-amplitude high-frequency electromagnetic activity, so called late fields, at the end of the ventricular depolarization phase. These signals are absent in patients without arrhythmic events after myocardial infarction. Using spectral turbulence analysis, the patients with tachycardias after infarction can be separated from those without.

Similar abnormal ventricular activity can also be found in patients with cardiomyopathy and ventricular tachyarrhythmias. Long-QT patients have an inhomogeneous pattern of repolarization in MCG, clearly distinguishable from the normal pattern of their healthy relatives. In summary, multichannel MCG seems a powerful new clinical tool for identifying risk factors and for locating the foci in patients with life-threatening arrhythmias.

Discussion

In clinical work, an essential requirement is reliability. The hardware and software must work without interruption; in addition, one has to be able to rely on results from the inverse problem algorithms used; operator errors must be minimized and wrong judgements avoided. To reach high reliability with a new complicated technology, effective training of the personnel as well as of the clinicians is necessary. Reliability and efficiency from the human point of view is best achieved by developing standardized experimental and analysis procedures. We have started systematic work in this direction. Standardized procedures will help in collecting normative data banks, which can be used as reference in patient studies. These data banks will facilitate studies with new patient groups and in pharmacological evaluations and support assessment studies for developing clinical practices.

To support the clinical studies and to push the technologies further, work is done at the BioMag Laboratory in modeling, signal and image processing, multimodal functional imaging and in transcranial magnetic stimulation.

References

[1] Hämäläinen, M., Hari, R., Ilmoniemi, R.J., Knuutila, J., and Lounasmaa, O.V. Magnetoencephalography—theory, instrumentation, and applications to noninvasive studies of the working human brain, Rev. Mod. Phys., 1993, 65: 413–497.

[2] Nenonen, J. and Katila, T. Noninvasive functional localization by biomagnetic methods, J. Clin. Eng., 1991, 16: 423–434, 495–503.

[3] Hari, R. Magnetoencephalography as a tool of clinical neurophysiology, In: Niedermeyer, E., Lopes da Silva, F. Electroencephalography: Basic Principles, Clinical Applications, and Related Fields, Baltimore, Williams & Wilkins, 1993.

[4] Lewine, J., Baldwin, N., Sanders, J., Anson, J., Halliday, A., and Orrison, W. Noninvasive localization of somatosensory cortex in neurosurgical patients—A comparison of structural (MRI) and functional (MEG) approaches with validation by intraoperative monitoring, Soc. Neurosci. Abstr., 1995, 21, Part 2: 1419.

[5] Paavola, M., Ilmoniemi, R.J., Sohlström, L., Meinander, T., Penttinen, A., and Katila, T. High performance magnetically shielded room for clinical measurements, these proceedings.

[6] Virtanen, J., Parkkonen, L., Huotilainen, M., Teder, W., Pekkonen, E., Ilmoniemi, R., and Näätänen, R. Facilities for simultaneous EEG and MEG recording, Human Brain Mapping, 1993, Suppl. 1: 111.

[7] Knuutila, J.E.T., Ahonen, A.I., Hämäläinen, M.S., Kajola, M.J., Laine, P.P., Lounasmaa, O.V., Parkkonen, L.T., Simola, J.T., Tesche, C.D. A 122-channel whole-cortex SQUID system for measuring the brain's magnetic fields, IEEE Transactions on Magnetism, 1993, 29: 3315–3320.

[8] Montonen, J., Ahonen, A., Hämäläinen, M., Ilmoniemi, R.J., Laine, P., Nenonen, J., Paavola, M., Simelius, K., Simola, J., and Katila, T. Magnetocardiographic functional imaging studies in BioMag Laboratory, these proceedings.

[9] Tiihonen, J., Katila, H., Pekkonen, E., Aronen, H., Ilmoniemi, R.J., Huotilainen, M., Räsänen, P., Virtanen, J., and Karhu, J. Cerebral asymmetry in schizophrenia, submitted for publication.

[10] Karhu, J., Tiihonen, J., Pekkonen, E., Katila, H., Huotilainen, M., Virtanen, J., and Ilmoniemi, R.J. Reversed hemispheric asymmetry of auditory N100m in schizophrenics, these proceedings.

[11] Sipilä, L., Heikkilä, E., Autti, T., Sainio, K., Huttunen, J., Aronen, H.J., Korvenoja, A., Ilmoniemi, R.J., and Santavuori, P., Somatosensory evoked magnetic fields from primary sensorimotor cortex in juvenile neuronal ceroid lipofuscinosis, submitted for publication.

[12] Korvenoja, A., Wikström, H., Huttunen, J., Virtanen, J., Laine, P., Aronen, H.J., Seppäläinen, A.–M., and Ilmoniemi, R.J. Activation of ipsilateral primary sensorimotor cortex by median nerve stimulation, NeuroReport, 1995, 6: 2589–2593.

[13] Tesche, C.D., Uusitalo, M.A., Ilmoniemi, R.J., Huotilainen, M., Kajola, M., and Salonen, O. Signal-space projections of MEG data characterize both distributed and well-localized neuronal sources, Electroenceph. Clin. Neurophysiol., 1995, 95: 189–200.

[14] Huotilainen, M., Tiitinen, H., Lavikainen, J., Ilmoniemi, R.J., Pekkonen, E., Sinkkonen, J., Laine, P., and Näätänen, R. Sustained fields of tones and glides reflect tonotopy of the auditory cortex, NeuroReport, 1995, 6: 841–844.

[15] Pekkonen, E., Huotilainen, M., Virtanen, J., Sinkkonen, J., Rinne, T., Ilmoniemi, R., and Näätänen, R. Age-related functional differences between auditory cortices: a whole-head MEG-study, NeuroReport, 1995, 6: 1803–1806.

The support of TEKES, SITRA, the Academy of Finland, and the Ministry of Education is gratefully acknowledged.

MEG Traces of Painful Processes

Kakigi, R, Kitamura, Y., Koyama, S., Hoshiyama, M., Shimojo, M., Watanabe, S., Takeshima, Y. and Nagata, O.

Department of Integrative Physiology, National Institute for Physiological Sciences, Okazaki, 444, Japan

Introduction

The cerebral representation of pain perception in man is poorly understood as compared with other modalities of sensations such as touch or vibration. This is mainly because of a lack of the appropriate instrumentation for stimulation and recording. Therefore, we recently studied magnetoencephalographies (MEGs) following painful electrical stimulation [1,2] and CO_2 laser stimulation [3,4] to detect the cerebral areas which are responsible for pain perception in man, and reviewed them.

Methods

Five normal volunteers (1 female and 4 males; mean age 32 years, range 28 to 41 years) were studied. Informed consent was obtained from all participants prior to the study.

The electric stimulus was a constant voltage square-wave pulse delivered transcutaneously to the right index finger. We adopted the three levels of intensities, "weak", "moderately painful" and "very painful", depending on the subjective feeling of each subject. Actually, 2-4 mA, 5-7 mA and 10-13 mA were used for weak, moderately painful and very painful session, respectively. Stimuli was applied at random. The interstimulus interval was between 1500 and 5000 msec and the stimulus duration was 1 msec. Responses were filtered with a 0.1-100 Hz bandpass filter, and digitized at a sampling rate of 1041.7 Hz. The analysis time was 100 msec before and 900 msec after the stimuli, and DC offset was achieved using a pre-stimulus period as the baseline. Each session was comprised of an average of two hundred trials, and at least two averages were obtained to ensure reproducibility. The order of each session was at random.

A special CO_2 laser stimulator (Nippon Infrared Industries, Kawasaki, Japan) was used to record MEGs.. The laser wavelength was 10.6 μm, and the diameter of the irradiation beam was about 2 mm. The stimulus duration was 20 msec. The intensity was approximately 18 mJ/mm^2, which elicited a sharp pain that all subjects described as tolerable, "like a pin- prick". To avoid causing burns or habituation, the irradiated points were moved slightly for each stimulus. The interstimulus interval of the laser beam was at random, between 3 to 10 seconds. To avoid magnetic noises, the stimulator was settled outside a shielded room, and the laser beam was carried through the optical fibers approximately 3.5 m in length, which penetrated the wall of the shielded room. The beam was applied to the bilateral arms and legs in separate sessions. Responses were recorded with a 0.1-50 Hz bandpass filter, followed by that at 1-30 Hz. Responses were digitized at a sampling rate of 512 Hz. Fifty trials were averaged in one session.

MEGs were measured with dual 37-channel biomagnetometers (Magnes, Biomagnetic Technologies Inc., San Diego, CA). The detection coils of the biomagnetometer were arranged in a uniformly distributed array in concentric circles over a spherically concave surface. The device was 144 mm in diameter and its radius was 122 mm. The outer coils were 72.5^0 apart. Each coil was 20 mm in diameter and the distance between the centers of each coil was 22 mm. Each coil was connected to a superconducting quantum interference device (SQUID). We recently reported several papers of MEGs using this magnetometer [5-11]. The measurement matrix was centered at the five different positions; around the C3, C4, Fz, Cz and Pz of the International 10-20 System in each subject. The whole head was mostly covered by these positions of magnetometer placement. A spherical model was fitted to the digitized shape of the head of each subject, and the location, orientation and amplitude of a best-fitted single equivalent current dipoles (ECD) were estimated at each time point.

Magnetic resonance imaging (MRI) were obtained using a GE Signa 1.0 T system. The T1-weighted coronal and axial images with contiguous 3 mm slice thickness were adopted for overlays with ECD sources detected by MEG.

Results

(1) MEGs following painful electrical stimulation

Four main deflections and their counterparts were clearly identified in all subjects in all "weak", "moderately painful" and "very painful" sessions less than 100 msec in latency (Fig .1). The position of ECD of each deflection was located around the hand sensory area of the SI in the left hemisphere which was contralateral to the stimulated finger.

Two main middle-latency deflections, N100m-P100m and P250m-N250m, were identified only in the "moderately painful" and "very painful" sessions, and they were much larger in the latter session (Fig. 1). Their ECDs were located around the superior bank of the Sylvian fissure, probably in the second sensory cortex (SII) in bilateral hemispheres. The following large deflection, P250m-N250m, was clearly identified in 4 of 5 subjects. However, the position of their ECDs was unexpected from the early deflections. When the probe was centered at the C3 or C4, ECDs were located in bilateral SII (Fig. 2) like the N100m-P100m. When the probe was centered at the Cz position, clear polarity-reversal deflections were also identified in 4 subjects and their ECDs were located along the interhemispheric fissure, probably in the cingulate cortex (Fig. 2).

When the probe was centered at the Fz and Pz positions which covered the mid-frontal and mid-parietal areas, no clear polarity-reversed deflections were identified. No consistent ECD considered to be generated in the mid-frontal and mid-parietal areas were identified.

When the lower limb was stimulated, the results were similar to the results following upper limb stimulation in terms of the waveform and ECD location, although the latency was different.

(2) MEGs following painful CO_2 laser stimulation

Consistent, clear magnetic fields were identified in bilateral cerebral hemispheres following the stimulation of each hand and foot in 4 of 5 subjects. Their duration was relatively long, being about 80-100 msec. The peak latencies varied in each subject, being about 180- 210 and 220-250 msec for arm and foot stimulation, respectively. No definite difference in latency was identified in either hemisphere. For example, when the probe was centered around C4, the latencies following the left arm and the right arm stimulation were similar. The amplitude following arm stimulation was 2 to 3 times larger than that following leg stimulation in each subject. The position of ECDs was located along the superior bank of the Sylvian fissure, around the SII in bilateral hemispheres. Their positions were very similar to those of the N100m-P100m and N250m-P250m following painful electrical stimulation (see Fig. 2). However, consistent ECD was identified in neither SI corresponding to the stimulated area nor the cingulate cortex, even when the probe was centered at the Fz, Cz and Pz.

Discussion

To our knowledge, only a few systematic studies of MEG following painful stimulation have been reported. Hari et al. [12] studied the painful electrical stimulation of tooth pulp, Huttunen et al. [13] used CO_2 stimulation of the nasal mucosa, Hawland et al. [14] applied high-intensity painful electric finger stimulation, and recently Laudahn et al. [15] reported MEGs following painful CO_2 laser stimulation applied to the temple. In the present studies, we used 2 sets of a large array of the magnetometer with a small intra-coil distance, and used two different ways of painful stimulation to elucidate the activated areas in the brain following painful stimulation in man in more detail, and found the ECDs of the pain-related components in SII in bilateral hemispheres and in the cingualate cortex. These findings were generally compatible with the previous studies [12-15], in which ECDs were found in SII or the surrounding areas, but activation in the cingulate cortex has not been reported previously. It may be due to the difference of method, because they did not focus on the activities in the cingulate cortex.

Although the functions of SII are not clear, SII probably provides for a very fundamental somatosensory capacity because this area has been found in nearly every mammalian species, including animals lacking extensive manipulative skills [16]. SII is the first cortical stage of the somatosensory system in which neurons process information from both sides of the body [16]. Therefore, it is reasonable to assume that the bilateral SII are activated following painful stimulation.

Cerebral activities following the initial excitation of neurons in SII to painful stimulation should be very complex. Whereas the periods of the large deflections, P250m-N250m, were definitive, we did not consider that they were generated by a single ECD. We considered that multiple areas including bilateral SII and cingulate cortices were activated simultaneously and independently. Therefore, a large inter-individual difference in location of the ECD was recognized, because magnetic fields generated by multiple areas should

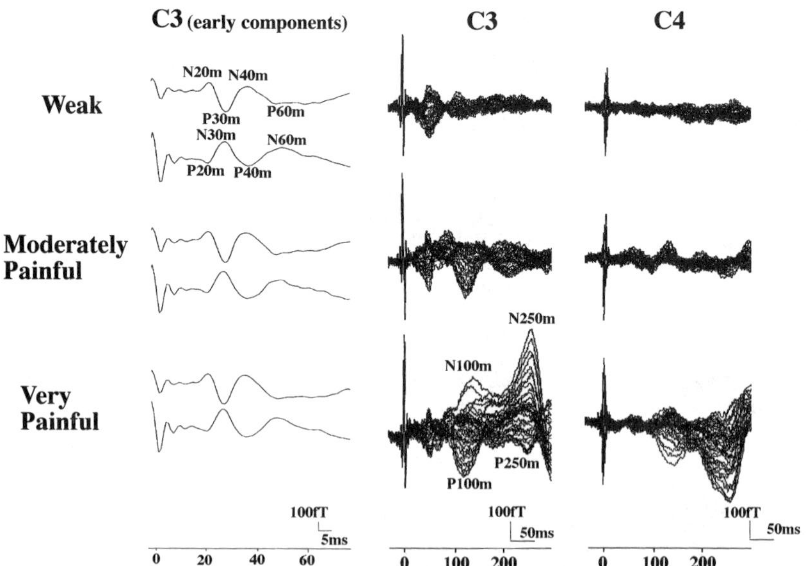

Fig. 1: MEGs waveforms recorded in C3 and C4 positions following "weak", "moderately painful" and "very painful" electrical stimulation applied to the right finger in a subject. The left figure shows the early components recorded at 2 representative channels at the C3 position. No significant difference was found in each condition. The middle and the right figure shows the superimposition of all 37 channels in each condition. The N100m-P100m and M250m-P250m were identified only in "moderately painful" and "very painful" sessions, being much larger in the latter.

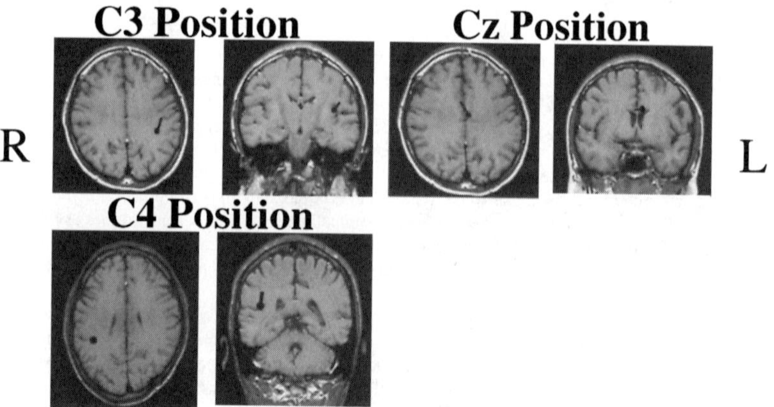

Fig. 2: Localization of representative ECDS of the N250m-P250m following very painful electrical stimulation applied to the right finger recorded at the C3, C4 and Cz positions overlapped on axial and coronal views of MRI. They were located in bilateral SII and the cingulate cortex.

interfere with each other. Positron emission tomography (PET) has demonstrated that painful heat stimuli causes significant activation of SI in the hemisphere contralateral to the stimulated hand, bilateral SII and the cingulate cortex [17], fundamentally compatible with the present study.   The reason why the activation in the cingulate cortex was not found following CO_2 laser stimulation was unknown, but the small S/N ratio may be one reason.

In conclusion, when we receive painful stimulation, bilateral SII are initially activated. It is now not clear whether SII are activated by signals receiving from SI or by signals reaching directly from the thalamus. Then, multiple areas including SII and cingulate cortex are activated simultaneously and independently.

References

[1] Kitamura, Y., Kakigi, R., Hoshiyama, M., Koyama, S., Shimojo, M. and Watanabe, S. Pain-related somatosensory evoked magnetic fields. Electroencephalogr. Clin. Neurophysiol., 1995; 95: 463-474..

[2] Kitamura, Y., Kakigi, R., Hoshiyama, M., Koyama, S., Shimojo, M. and Watanabe, S. Pain-related somatosensory evoked magnetic fields following lower limb stimulation. 1996, submitted.

[3] Kakigi, R., Koyama, S., Hoshiyama, M., Kitamura, Y., Watanabe, S. and Shimojo, M. Pain-related magnetic fields following painful CO_2 laser stimulation in man. Neurosci. Lett., 1995, 192: 45-48.

[4] Kakigi, R., Koyama, S., Hoshiyama, M., Kitamura, Y., Watanabe, S. and Shimojo, M. Pain-related brain responses following CO_2 laser stimulation: magnetoencephalographic studies. In: Hashimoto, I., Visualization of Information Processing in the Human Brain, Amsterdam, Elsevier, 1996, in press.

[5] Kakigi, R. Somatosensory evoked magnetic fields following median nerve stimulation. Neurosci. Res., 1994, 20: 165-174.

[6] Kakigi, R., Koyama, S., Hoshiyama, M., Watanabe, S., Shimojo, M. and Kitamura, Y. Gating of somatosensory evoked responses during active finger movements: magnetoencephalographic studies. J. Neurol. Sci., 1995, 128: 195-204.

[7] Kakigi, R., Hoshiyama, M., Koyama, S., Shimojo, M., Kitamura, Y. and Watanabe, S. Topography of somatosensory evoked magnetic fields following posterior tibilal nerve stimulation. Electroencephalogr., Clin. Neurophysiol., 1995; 95: 463-474..

[8] Hoshiyama, M., Kakigi, R., Koyama, S., Kitamura, Y., Shimojo, M. and Watanabe, S. Somatosensory evoked magnetic fields after mechanical stimulation of the scalp in humans. Neurosci. Lett., 1995, 195: 29-32.

[9] Hoshiyama, M., Kakigi, R., Koyama, S., Kitamura, Y., Shimojo, M. and Watanabe, S. Somatosensory evoked magnetic fields following stimulation of the lip in humans. Electroencephalogr. Clin. Neurophysiol., 1996, in press.

[10] Shimojo, M., Kakigi, R., Hoshiyama, M., Kitamura, Y., Koyama, S. and Watanabe, S. Intracerebral interactions caused by bilateral median nerve stimulation in man: a magnetoencephalographic study. Neurosci. Res., 1996; 24: 175-181.

[11] Shimojo, M., Kakigi, R., Hoshiyama, M., Kitamura, Y., Koyama, S. and Watanabe, S. Differentiation of receptive fields in the sensory cortex following stimulation of various nerves of the lower limb in man. J. Neurosurg., 1996, in press.

[12] Hari, R., Kaukoranta, E., Reinikainen., K, Huopaniemie, T. and Mauno, J. Neuromagnetic localization of cortical activity evoked by painful dental stimulation in man. Neurosci. Lett., 1983; 42: 77-82.

[13] Huttunen, J., Kobal, G,. Kaukoranta, E. and Hari, R. Cortical responses to painful CO_2 stimulation of nasal mucosa; A magnetoencephalographic study in man.   Electroencephalogr. Clin. Neurophysiol., 1986, 64: 347-349.

[14] Howland, E.W., Wakai, R.T.,   Mjaanes, B.A., Balog, J.P. and Cleeland, C.S. Whole head mapping of magnetic fields following painful electric finger shock. Cogn. Brain Res., 1995, 2: 165-172.

[15] Laudahn, R., Kohlhoff, H. and Bromm, B. Magnetoencephalography in the investigation of the cortical pain processing system. In: Bromm, B., Desmedt, J.E., Pain and the Brain. From nociception to Cognition. Advances in Pain Research and Therapy, Vol. 22, New York, Raven Press, 1995, 267-282.

[16] Burton, H. Second sensory cortex and related areas. In: Jones, E.G., Peters, A.,. The Cerebral Cortex, Vol. 5, New York: Plenum Press, 1986: 31-98.

[17] Tarbot, J.D., Marrett, S., Evans, A.C., Meyer, E., Bushnell, M.C., Duncan, G.H. Multiple representations of pain in human cerebral cortex. Science, 1991; 251: 1355-1358.

Normalized N100m Latency of the Auditory Evoked Fields After Surgical Removal of Temporal Lobe Gliomas

Kanno, A.[1], Nakasato, N.[1,2], Ohtomo, S.[2], Seki, K.[2], Kawamura, T.[2], Fujita, S.[4], Kumabe, T.[2], Kayama, T.[3], and Yoshimoto, T.[2]

MEG Laboratory, Kohnan Hospital[1], Sendai, Japan., Department of Neurosurgery, Tohoku University School of Medicine[2], Sendai, Japan; Department of Neurosurgery, Yamagata University School of Medicine[3], Yamagata, Japan; and R&D Center, Osaka Gas Co. Ltd.[4] Osaka, Japan

Introduction

Auditory evoked potentials have been measured in patients with several temporal lobe diseases. The N100 responses disappear in patients with lesions on bilateral superior temporal gyrus[1]. However, separation of a unilateral abnormality in evoked potentials, is difficult due to the spearing effect by tissue layers with inhomogeneous electric conductivities. Activity on the normal hemisphere may interfere with abnormal activity on the diseased hemisphere. Magnetoencephalography (MEG) is known to be less influenced by the inhomogeneous head conductivity. We have found that the whole head MEG is especially suitable to identify differences in bilateral cerebral function [2-4].

MEG detects N100m as the most prominent peak of auditory evoked fields (AEFs). Both left and right monaural stimuli cause a two-dipole pattern of N100m over the bilateral hemispheres with a latency of about 100 ms. In normal subjects, the contralateral N100m is known to appear earlier than the ipsilateral response to the stimulated ear. N 100m dipoles are localized on the upper surface of temporal lobes [2-5].

There have been only few reports on clinical application of the AEFs. Disappearance of the unilateral N100m response was reported in patients with temporal lobe infarctions[5]. In this study. we measured AEFs to evaluate auditory function in two patients with temporal lobe gliomas.

Materials and Methods

Two patients with left temporal gliomas participated in the present study.

Case 1 is a 27-year-old man with mild aphasia. MRI revealed a cystic tumor in the left temporal lobe. After the first surgery and radio-chemotherapy, the patient returned to work with no language or hearing disturbances. Histological examination indicated anaplastic astrocytoma. Six month after the surgery, recurrence of the tumor was discovered. Left anterior temporal lobectomy was performed for total removal of the tumor. The patient had no hearing or language disturbances afterwards.

Case 2 is a 45-year-old man with mild sensory aphasia. MRI revealed a cystic tumor in the left temporal lobe. The tumor was totally removed and histological examination indicated anaplastic astrocytoma. After radio-chemotherapy, the patient returned home with persistent mild aphasia.

AEFs were measured before and after each surgery, using a helmet-shaped 66-channel MEG system (CTF Systems-Osaka Gas) linked to MRI[2,3].The MEG signals were averaged based on 50 monaural stimuli of 2000 Hz tones. We measured peak latency of the N100m response in each hemisphere. All N100m latencies were compared to our normal range obtained from 20 male volunteers: i) for the left hemispheric N100m, the contralateral response was 86.9±9.4 ms (mean±s.d.) and the ipsilateral response was 100.1±7.3 ms; and ii) for the right hemispheric N100m, the contralateral response was 84.2±9.0 ms and the ipsilateral response was 98.2±9.3 ms [2]. Finally, source locations of the N100m was estimated using single- or double-dipole models. Dipole positions were superimposed on the MRI for each subject.

Results

All AEF measurements detected clear, bilateral N100m responses in both patients. Dipole estimation indicated the bilateral temporal planes on MRI (Figs.1 and 3). Before the first surgery, both patients had delayed N100m latencies in their left, lesion side, hemispheres. After the first surgery, left N100m latencies shortened into the normal range (Figs. 2 and 4). The contralateral N100m latency indicated more clear-cut recovery than the ipsilateral response.

Discussion

The present study demonstrated the unilateral abnormality of the auditory function in patients with temporal lobe gliomas. Although several pathways have been proposed for N100m responses, we believe the cortical lesion introduced delay in the N100m latencies. At least in these two patients, delayed N100m was due to mass effect rather than invasion of the tumors, because the N100m latencies normalized after removal of the tumors. Our results further suggest that AEF can be applied to evaluation of auditory function in patients with other types of brain lesions. Our helmet-shaped MEG system is especially useful for clinical application, because of the simultaneous coverage of two cerebral hemispheres and short measurement time.

References

[1] Woods, D.L., Clayworth, C.C., Knight, R.T., Simpson, G.V., and Naeser, M.A. Generators of middle and long-latency auditory evoked potentials: implications from studies of patients with bitemporal lesions. Electroencephalography and Clinical Neurophysiology, 1987, 68: 132-148.
[2] Nakasato, N., Fujita, S., Seki K., Kawamura T., Matani, A., Tamura, I., Fujiwara, S., and Yoshimoto, T. Functional localization of bilateral auditory cortices using an MRI-linked whole head magnetoencephalography (MEG) system. Electroencephalography and Clinical Neurophysiology, 1995, 94: 183-190.
[3] Nakasato, N., Fujita, S., Matani, A., Tamura, I., Fujiwara, S., and Yoshimoto, T. Clinical application of the whole head MEG: Auditory evoked response in patients with intracranial structural lesions. In: Baumgartner, C., Deecke, L., Stroink, G., and Williamson, S.J. Biomagnetism: Fundamental Research and Clinical Applications, Amsterdam, Elsevier/IOS Press, 1995, pp 186-190.
[4] Kanno, A., Nakasato, N., Fujita, S., Seki, K., Kawamura, T., Ohtomo, S., Fujiwara, S., and Yoshimoto, T. Right hemispheric dominance in the auditory evoked magnetic fields for pure-tone stimuli. Electroencephalography and Clinical Neurophysiology, in press.
[5] Mäkelä, J.P. Auditory evoked magnetic fields in stroke. Physiological Measurement, 1993, 14: 51-54.

Acknowledgements

This work has been supported in part by Grants-in-Aid for Scientific Research No.03404042 and No.06404050 from the Ministry of Education, Science and Culture of Japan; and Magnetic Health Science Foundation.

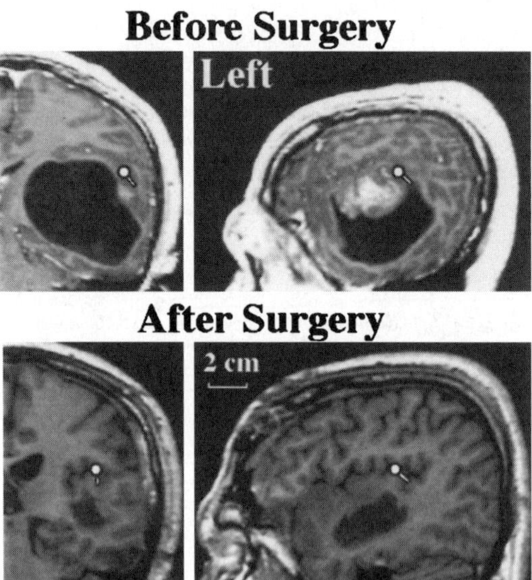

Fig. 1 N100m dipole positions before and after the first surgery in Case 1. Circles and bars indicate dipole positions and orientations of the left hemispheric N100m after contralateral ear stimuli. Note that stable dipole localizations on the upper surface of the temporal lobes, both with and without cystic tumor, and before and after craniectomy.

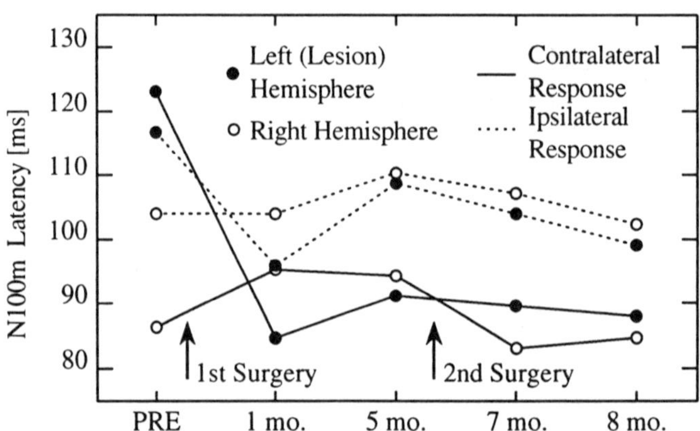

Fig. 2 N100m latency of AEFs in Case 1. Latency of the left hemispheric N100m due to contralateral ear stimuli (123.2 ms) was longer before the first surgery but normalized (88.0 ms) afterwards.

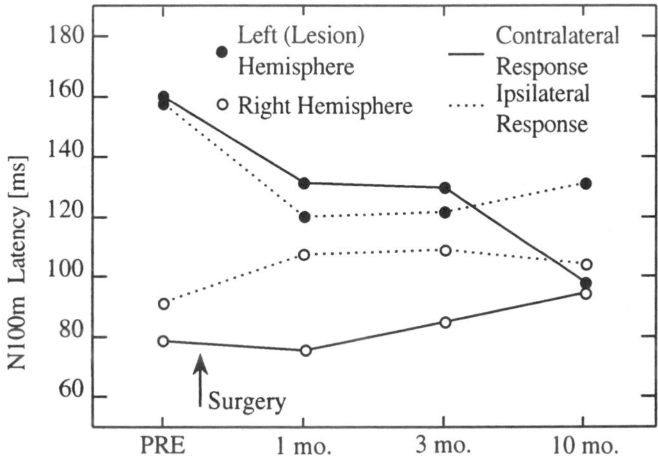

Fig. 3 N100m dipole positions before and after the surgery in Case 2. Circles and bars indicate dipole positions and orientations of the left hemispheric N100m due to contralateral ear stimuli. Note that stable dipole localizations on the upper surface of the temporal lobes, both with and without cystic tumor, and before and after craniectomy.

Fig. 4 N100m latency of AEFs in Case 2. Latency of the left hemispheric N100m due to contralateral ear stimuli (161.6 ms) was longer before the surgery but normalized 10 months after the surgery (99.2 ms).

Reversed Hemispheric Asymmetry of Auditory N100m in Schizophrenics

Karhu, J.[1], Tiihonen, J.[2], Pekkonen, E.[3,4], Katila, H.[5], Huotilainen, M.[3,4], Virtanen, J.[3,4], and Ilmoniemi, R.[3]

Department of Clinical Neurophysiology[1], Kuopio University Hospital; Department of Forensic Psychiatry[2], Niuvanniemi Hospital, Kuopio; BioMag Laboratory[3], Helsinki University Central Hospital; Cognitive Psychophysiology Research Unit[4], University of Helsinki; Department of Psychiatry[5], Helsinki University Central Hospital, Finland

Introduction

Human brain hemispheres are asymmetric. Individual anatomical differences involve widespread cortical and subcortical areas and are more prominent among males [1,2]. The interhemispheric asymmetry is particularly clear in temporal lobe structures and speech areas [3,4]).

Schizophrenia has been recently associated with a large number of anatomical abnormalities. In most structural imaging studies, the only well established abnormality is lateral ventricle enlargement with very limited diagnostic usefulness. Practically all imaging studies have related a single symptom or category of symptoms to a single brain region, with the exception of a magnetic resonance imaging (MRI) study showing thalamic disorganization in schizophrenia and attempting to explain multiple symptoms by dysfunction of this main relay station in brain [5]. So far, the most clearly defined *in vivo* findings show alterations in temporal lobe and limbic structures and have been associated with positive symptomatology of schizophrenia, which often modulates perception at early stages. For example, schizophrenics have decreased volumes of superior temporal gyri (STG) [6], and auditory N100m, which is generated mainly in STG, is altered by auditory hallucinations [7].

Post mortem [8], structural imaging [9], and magnetoencephalographic (MEG) studies [10] have shown a loss of temporal area asymmetry in some subgroups of schizophrenia. To explain this finding, it has been suggested that schizophrenics may fail to develop normal hemispheric asymmetry due to an early failure in neural development, possibly because of a gene defect influencing the hemispheric dominance [8]. Several structural studies have given contradictory results [11], and currently the implications and individual diagnostic value of possible abnormal lateralization remain elusive.

We evaluated the lateralization of primary auditory areas in temporal cortex by comparing the source locations of magnetic auditory N100m in schizophrenics and in normal subjects. MEG and electrocorticography recordings have shown that auditory evoked responses at 18–50 ms are generated in the primary auditory cortex in Heschl's gyrus and that magnetic 100 ms-responses have major source areas in or near primary auditory cortex [12]. Since N100m reflects the coherent activation of a functional neuronal network, it may in principle be more sensitive to subtle disorganization of functional anatomy than other structural and metabolic *in vivo* measures are. In addition to the hypothesis of abnormal hemispheric lateralization, we also wanted to test the diagnostic usefulness of simple and rapid MEG recording and analysis methods in scizophrenia.

Methods

We compared the source locations of auditory N100m in 10 schizophrenics (age range 19–34 years; 6 males; recordings were made within 10 days from hospitalization) and in 14 normal subjects (age range 19–50 years; 8 males). Psychiatric diagnoses were made by a senior psychiatrist (H. K.). The study protocol was accepted by the ethical committees in Kuopio and Helsinki University Hospitals and volunteers provided their written informed consent after a description of the study.

Each auditory stimulus block contained standard (80%) tone bursts (duration 50 ms; 0.8 kHz) and randomly embedded deviant tones (duration 25 ms; 0.8 kHz), which were delivered monaurally wih an interstimulus interval of 2.5 s. The subjective hearing threshold was measured before the recordings and the stimulus intensity was adjusted to 60 dB over the threshold for each ear. The subjects watched silent video presentations during the experiments and were instructed to ignore the stimuli.

Auditory evoked fields (AEFs) were measured with a 122-channel whole-scalp covering magnetometer (Neuromag Ltd.) in a magnetically shielded room (Euroshield Ltd.). The recording passband was 0.03–100 Hz with a sampling rate of 397 Hz. The accurate position of the subject's head inside the helmet-shaped magnetometer was detected by measuring the magnetic field produced by three marker coils attached to the scalp. Source locations were modelled by single equivalent current dipoles (ECDs) [13], which were determined with a least-squares fit at 1-ms-intervals from 80–130 ms for the responses elicited by contralateral standard tone stimulation. The calculation was done for each hemisphere separately utilizing a spherical head model and a subset of 40 channels over temporal brain areas (*cf.* Fig. 1).

Only ECDs which explained more than 85% of the recorded dipolar magnetic field and had a stable field pattern for over 10 ms around N100m deflection were accepted. The analysis was done blindly to the entire studied group and the subjects were divided to "patient" and "control" groups for further analysis only after completing the source localization routine.

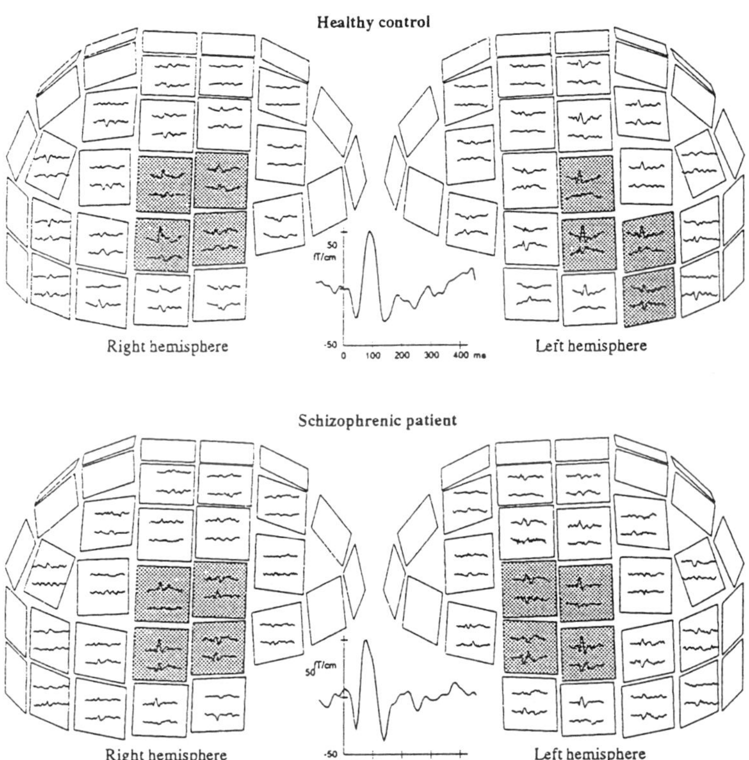

Figure 1. The averaged AEFs elicited by contralateral stimulation in a normal subject (above) and in a schizophrenic (below). The number of averages is about 100 and the responses have been digitally low-pass filtered at 30 Hz. The channel with the largest N100m is shown in the insert. The subsets of channels which show AEF waveforms in the figure were utilized to calculate ECDs for N100m deflections. Four channel pairs with the largest total signal in each hemisphere have been shaded.

Results

Fig. 1 shows the AEFs elicited by contralateral standard tone stimulation over both hemispheres in a normal subject (above) and in a schizophrenic (below). Since the maximum total signal ($\sqrt{(\partial B_r/\partial x)^2 + (\partial B_r/\partial y)^2}$) of planar gradiometers is observed directly above a cerebral source area, the location of the channels with largest N100m gives a rough estimate of its cortical source location. In the normal subject, these channel pairs are slightly posterior and inferior in the left hemiphere when compared with the right hemisphere, which suggests more posterior source(s) for N100m in the left hemisphere. In the schizophrenic, the same channel pairs show the largest N100m in both hemispheres.

N100m sources could be localized by a single ECD reliably in both hemispheres in 8 schizophrenics and in 12 normal subjects (Goodness-of fit values were in all subjects over 78% for N100m sources. Two normal subjects and two patients were rejected from further analysis because of the preset 85% limit for the acceptance of ECD.) All normal subjects showed more posterior N100m sources to contralateral ear stimulation in the left temporal lobe (mean antero-posterior difference 6.5 mm; range 3–15 mm; antero-posterior was defined as a line passing the nasion and perpendicular to the line connecting preauricular points), whereas 5 patients showed a reversed interhemispheric asymmetry between the temporal lobes (mean antero-posterior difference 6.0 mm; range 1–8 mm). Both schizophrenic females had more anterior location of N100m sources in the left temporal lobe. The individual interhemispheric antero-posterior differences in the N100m source locations are shown in Fig. 2.

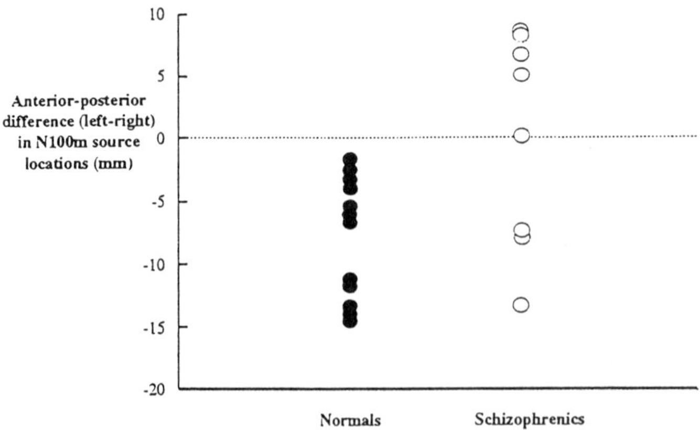

Figure 2. Interhemispheric differences (left minus right, mm) in the anterior-posterior location of the N100m sources. Normal subjects are depicted by filled circles and schizophrenics with open circles.

Discussion

These preliminary results suggest strongly that abnormalities in the individual functional anatomy of temporal structures can be detected by MEG in non-chronic schizophrenics. The interhemispheric asymmetry in the anterior-posterior location of auditory N100m equivalent sources was reversed in 50% of the studied schizophrenics. Since these sources are activated by any auditory input and probably reflect the activation of primary auditory cortices, the finding may reflect a fine-grained anatomical anomaly in the auditory areas of temporal lobes, caused for example by a neural migration disorder.

Studies of larger patient groups and an evaluation of schizophrenics with different spectra of clinical symptoms are obviously essential. We modelled the N100m deflection as a single point-like source, and thus we cannot rule out the possibility of diminished or increased activity in one of several underlying sources which could in principle lead to a source location shift in schizophrenia. Since this would require changes in one hemisphere only to cause the observed interhemispheric source location reversal, the explanation does not seem probable. If the specificity of our finding remains high in the future studies (at present 100 %), this simple auditory test may prove to be very helpful in the diagnostics and screening of high-risk populations.

References

[1] Galaburda, A.M., LeMay, M., Kemper, T., and Gershwind, N. Right-left asymmetries in the brain, Science, 1978, 199: 852–856.

[2] Weinberger, D.R., Luchins, D.J.,Morihisa, J., and Wyatt, R.J. Asymmetrical volumes of the right and left frontal and occipital regions of the human brain, Annals of Neurology, 1982: 97–100.

[3] Gerschwind, N. and Levitsky, W. Human brain: left-right asymmetries in the temporal speech region, Science, 1968, 161: 186–187.

[4] Witelson, S.F. and Kigar, D.L.Sylvian fissure morphology and asymmetry in men and women: bilateral differences in relation to handedness in men, Journal of Comparative Neurology, 1992, 323: 363–340.

[5] Andreasen, N.C., Arndt, S., Swayze II, V., Cizadio, T., Flaum, M., O'Leary, D., Ehrhardt, J.C., and Yuh, W.T.C. Thalamic abnormalities in schizophrenia visualized through magnetic resonance image averaging, Science, 1994, 266: 294–298.

[6] Shenton, M.E., Kikinis, R., Jolez, F.A., et al. Abnormalities of the left temporal lobe and thought disorder in schizophrenia. A quantitative magnetic resonance imaging study, New England Journal of Medicine, 1992, 327: 604–612.

[7] Tiihonen, J., Hari, R., Naukkarinen, H., Rimon, R., Jousmäki, V., and Kajola, M. Modified activity of the human auditory cortex during auditory hallucinations, 1992, American Journal of Psychiatry, 149: 255–257.

[8] Crow, T.J., Ball, J., Bloom, S.R., et al. Schizophrenia as an anomaly of development of cerebral asymmetry, Archives of General Psychiatry, 1989, 46: 1145–1150.

[9] Luchins, D.J., Weinberger, D.R., and Wyatt, R.J. Schizophrenia and cerebral asymmetry detected by computed tomography, American Journal of Psychiatry, 1982, 139: 753–757.

[10] Reite, M., Teale, P., Godstein, L., Whalen, J., and Linnville, S. Late auditory magnetic sources may differ in the left hemisphere of schizophrenic patients, Archives of General Psychiatry, 1989, 46: 565–572.

[11] Kuluych, J.J., Vladar, K., Fantie, B., Jones, D., and Weinberger, D.R. Normal asymmetry of planum temporale in patients with schizophrenia, British Journal of Psychiatry, 1995, 166: 742–749.

[12] Hari, R., Aittoniemi, K., Järvinen, M.-L., Katila, T. and Varpula, T. Auditory evoked transient and sustained magnetic fields of the human brain. Experimental Brain Research, 1980, 40:23724.

[13] Hämäläinen, M., Hari, R., Ilmoniemi, R.J., Knuutila, J., and Lounasmaa, O.V. Magnetoencephalography - theory, instrumentation, and applications to noninvasive studies of the working human brain. Reviews of Modern Physics, 1993, 65: 413–498.

Neuromagnetic Identification of the Somatosensory Cortex in Cases with Arteriovenous Malformation Adjacent to the Central Sulcus

Kawamura, T.[1], Nakasato, N.[1,2], Ohtomo, S.[1], Seki, K.[1], Kanno, A.[2], Fujita, S.[3], Fujiwara, S.[1,2] and Yoshimoto, T.[1]

[1]Department of Neurosurgery, Tohoku University School of Medicine, Sendai, Japan; [2]MEG Laboratory, Kohnan Hospital, Sendai, Japan; [3]R&D Center, Osaka Gas Co. Ltd., Osaka, Japan

Introduction

It is critical to preserve eloquent cortices during surgery of cerebral arteriovenous malformations (AVMs). Although MRI scans may identify "anatomical" central suclus, it is controversial whether the "functional" central sulcus can be shifted due to AVMs. Cortical recording of somatosensory evoked potentials (SEPs) can be used to recognize the central sulcus during open surgery. However, the cortical SEPs are not available during surgery for large AVMs, intravascular surgery, or stereotaxic radiosurgery. In the present study, somatosensory evoked fields (SEFs) were measured to localize the "functional" central sulcus non-invasively in cases with AVM adjacent to the central sulcus.

Method

Fourteen patients, 7 males and 7 females, aged 9 to 67 year-old (mean 39.5), participated in the study (Table.1). All cases had AVM adjacent to the central sulcus. All patients underwent three-dimensional MRI scans. Anatomical central sulcus was defined as the one-anterior-sulcus to the marginal branch of the cingulate sulcus on a paramedian sagittal MRI image. For median nerve stimuli, SEFs were measured using a helmet-shaped 66-channel MEG system (CTF Systems - Osaka Gas) linked to MRI [1]. At the peak latency of the first component (N20m), signal source was estimated using a single-dipole model. Location of the N20m dipole was superimposed on MRI. Finally, the dipole deviation, in the anterior-posterior direction, was measured from the posterior bank of the anatomical central sulcus.

Results

In 13 cases, the N20m dipoles in the lesion hemispheres were localized within one gyrus to the anatomical central sulcus (Fig.1). In only one case, case 3 with fronto-parietal subcortical hemorrhage, the N20m dipole disappeared in the lesion hemisphere. Even in two cases with a large AVM which occupied almost entire frontal lobe, N20m dipole was localized on the anatomical central sulcus (Fig.2).

Case	Sex	Age	Lesion	Initial Symptom
1	F	30	Rt. F	Headache and lt. facial palsy due to ICH
2	F	53	Lt. P	Rt heminumbness
3	F	47	Lt. F	Headache due to ICH
4	M	17	Lt. F	Attack of consciousness loss
5	M	50	Rt. P	Lt heminumbness due to ICH
6	F	50	Rt. F(Large)	Lt hand weakness and numbness
7	F	49	Lt. F(Large)	Rt hemiparesis and hemihypesthesia
8	F	9	Rt. F	Seizure
9	M	50	Lt. F	Headache due to ICH
10	M	23	Rt. P	Incidental
11	F	52	Lt. FP	Memory disturbance, Dysarthria
12	M	13	Rt. F	Partial seizure
13	M	67	Rt. P	Attack of consciousness loss
14	M	25	Lt. P	Incidental

Table.1 Summary of the cases with AVM adjacent to the central sulcus. F: frontal lobe, P: parietal lobe, ICH: intracerebral hemorrhage

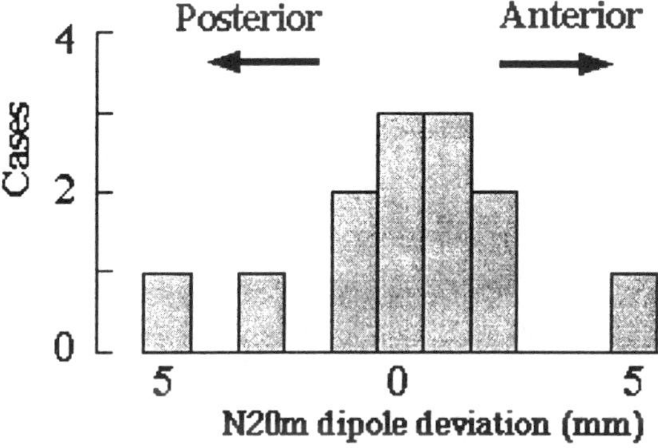

Fig.1 Anterior or posterior deviation of the N20m dipole from the surface of the posterior bank of the "anatomical " central sulcus.

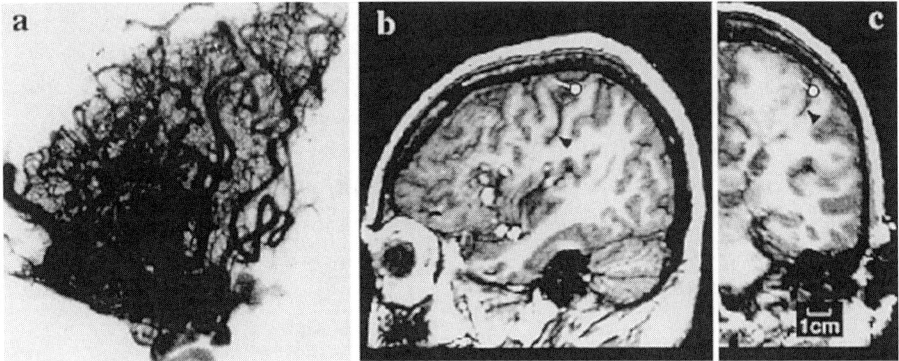

Fig.2 N20m source localization in a 49 year-old female patient (case 7) with a large AVM in the left frontal lobe. a: Lateral view of the left carotid angiogram revealing a large AVM in the left frontal lobe. b and c: N20m dipole position estimated on the "anatomical" central sulcus (arrow heads).

Discussion

Previous SEF studies indicated that the origin of the N20m is in the posterior wall of the central sulcus "area 3b" [2,3]. It is well known that mass lesions, such as brain tumors, may shift the central suclus. However, our results suggest that AVM does not shift the localization of somatosensory function. We believe cortical functions may be preserved in parenchyma adjacent to the nidus or among abnormal vessels of AVM. This non-invasive method can be applied for clinical decision making before not only conventional surgery but also intravascular surgery or stereotaxic radiosurgery.

References

[1] Kawamura, T., Nakasato, N., Seki, K., Shimizu H., Fujiwara, S., and Yoshimoto, T. Locating the central sulcus using three dimensional image fusion system of helmet shaped whole head MEG and MRI. No Shinkei Geka 22: 737-741, 1994

[2] Wood, C.C., Cohen, D., Cuffin, B.N., Yarita, M. and Allison, T. Electrical sources in human somatosensory cortex: Identification by combined magnetic and potential recordings. Science 227: 1051-1053, 1985

[3] Kawamura, T., Nakasato, N., Seki, K., Kanno, A., Fujita, S., Fujiwara, S., and Yoshimoto, T. Neuromagnetic evidence of pre - and post - central cortical sources of somatosensory evoked responses. Electroencephalography and Clinical Neurophysiology, in press.

Acknowledgements

This research was supported by Grants-in-Aid for Scientific Research No.03404042 and No.06404050 from the Ministry of Education, Science and Culture of Japan.

Time Course of Abnormal MEG Activity Associated with Transient Ischemic Attacks

Kober, H., Vieth, J. B., Stippich, C., Kassubek, J.R., Hopfengaertner, R.
Dept. of exp. Neuropsychiatry, University of Erlangen-Nürnberg, Germany

Introduction

Cerebrovascular emboli may cause irreversible deficits (brain infarcts) or reversible ones (transient ischemic attacks (TIA)). If the reason is a high grade stenosis (>70%) of the cervical internal carotid artery, this symptomatic stenosis should be treated surgically according to large North-American and European joint studies [3]. According to the location in the brain clinical symptoms may or may not occur. But not only clinically "silent"

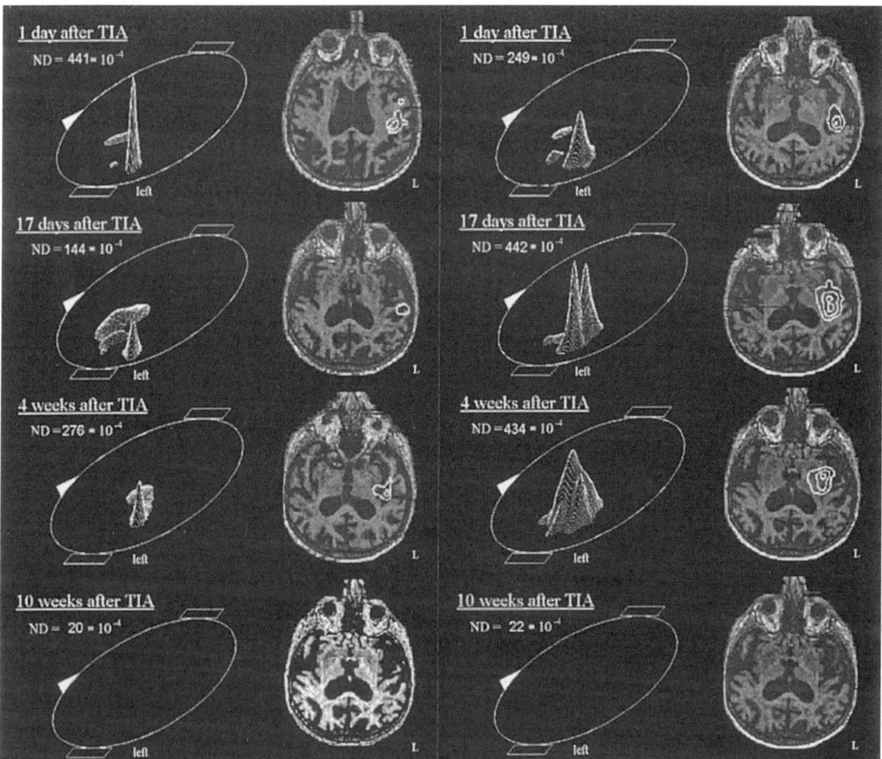

Fig. 1: Dipole-density-plot (DDP, c.f. methods) evaluation of the slow (2-6 Hz, left column) and beta wave activity (12.5-30 Hz, right column) of a 85 years old, right handed male (hkc) suffering from a TIA with an aphasia which lasted for half an hour. Upper row to lower row: time course of the normalized dipole density (ND, c.f. methods) in dipoles per ccm showing abnormal activity still four weeks after the TIA for both slow and beta wave activity. In the last measurement of the follow up study (lowest row, 10 weeks after the TIA) slow and beta wave activity have become normal.

infarcts [1] - shown by imaging techniques - but also reversible functional impairments ("silent TIA") may occur. So a "silent TIA" is another cause that a carotid stenosis is symptomatic.

It is well known that cerebrovascular lesions of the brain are the source of neuronal dysfunctions, which can produce focal slow wave activity (2-6 Hz). Recently, we could show that also focal beta wave activity (12.5-30 Hz) can be associated with these lesions [4,9]. Slow wave activity has not only been observed with *structural lesions* (infarcts), but also with TIAs [7,8]. Both, slow and beta wave activity, can be localized by magnetoencephalography (MEG) [2,7,8]. Therefore MEG is a suitable mean to investigate the time course of neuronal dysfunction due to cerebrovascular lesions.

In order to have a kind of screening test for "silent TIAs", it is of great clinical interest to know whether a "silent TIA" has occurred and where the corresponding abnormal activity is located. In order to use the MEG for this purpose we have to know how reliable the activity can be shown of a structural lesion (brain infarct) and of the non structural lesion, the TIA with clinical symptoms, in comparison to a base line group and how long the reversible effect of the TIA lasts.

Methods

Spontaneous measurements of 19 cases with TIAs (no structural lesions in the CT or MRI) and of 14 normal subjects (baseline group) were recorded with the biomagnetic 37-channel system KRENIKON[R] (Siemens). The recordings (10 min) were done sequentially over both hemispheres. In 2 cases with a TIA follow up studies were performed: 1 day, 1 week, 2.5 weeks, 4 weeks and 10 weeks after the cerebrovascular event.

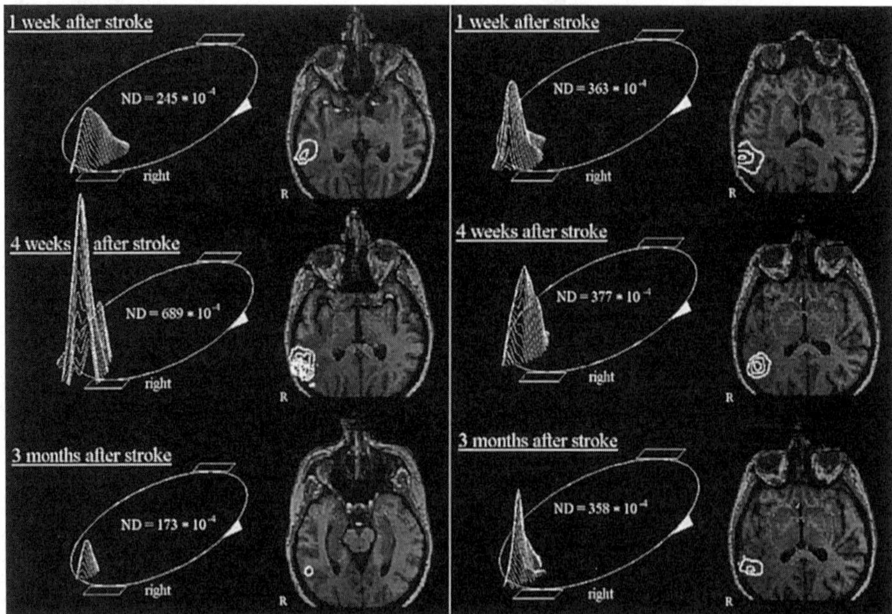

Fig. 2: Dipole-density-plot (DDP, c.f. methods) evaluation of the slow (2-6 Hz, left column) and beta wave activity (12.5-30 Hz, right column) of a 54 years old female (msf) suffering from a right temporo occipital brain infarct followed by a left side hemiparesis. Upper row to lower row: time course of the normalized dipole density (ND, c.f. methods) in dipoles per ccm. In contrast to the TIA patient (c.f. Fig. 1) abnormal activity is still found three month after the brain infarction.

After digitally filtering the whole data set from 2 to 6 Hz (delta/theta activity) and from 12.5 to 30 Hz (beta activity) the multisource problem of localizing spontaneous activity was handled by the Dipole-Density-Plot (DDP) [6], a temporal integration of localization results showing concentrations of dipoles over time in a standardized form. To use the single dipole model adequately, a Principal Component Analysis (PCA) first selects the time epochs with one dominant component (typically: 5 seconds selected out of 10 minutes).

The standardization is achieved by normalizing the dipole density to the number of dipoles with a minimal correlation of measured and calculated field distribution, used for the calculation of the 3-dimensional dipole-density distribution. Thus yielding a measure for the predominance of the focal abnormal activity over the background activity. The DDP results were inserted in form of isocontour lines of the activity´s distribution into MRI slices using a transformation obtained from a contour fit algorithm [5]. The accuracy of this very precise coregistration method is better than 2 mm.

Results

Figures 1 and 2 show location and time course of the abnormal activity in the case of a TIA and a brain infarction respectively. Fig. 3 shows that the stroke activity is at least 4 times higher and the TIA activity at least two times higher than the activity of the base line group. The DDP yielded abnormal activity still 4 weeks after the event for the TIA and 3 months for the infarction. The DDP localizations of the slow and the beta wave activity are well in accordance with the clinical symptoms in both cases presented in Fig. 1 and Fig. 2. For the infarct patient suffering from a left side hemiparesis the localizations were found close to the right temporo occipital infarction area in all measurements. For the right handed TIA patient suffering from an aphasia the localizations were found in the speech dominant (left) hemisphere of the brain.

In all TIA cases we found either the center of the slow and the fast wave activity in the correct hemisphere with a different time course and with a different extent and shape of the DDP. But in all cases the TIA activity reached normal values between 4 and 10 weeks.

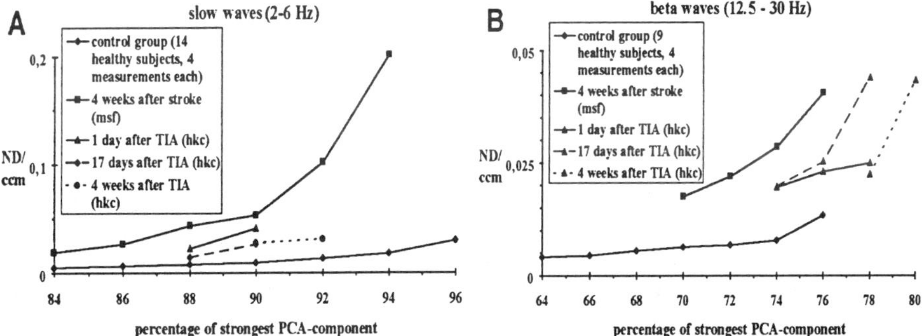

Fig. 3: Normalized dipole density (ND, c.f. methods) of a stroke patient four weeks after the brain infarction (c.f. Fig. 2) and a TIA patient one day, 17 days and four weeks after the incident (c.f. Fig. 1) compared to the average dipole density of a healthy baseline group for different percentages of predominace of the strongest principal component. A: Theta/delta activity. B: Beta activity. The pathologic brain activity induced by the brain infarct has the highest intensity. For the TIA patient the intensity of the dipole density is above the baseline group still four weeks after the occurence of the cerebrovascular lesion. For all cases an increase of ND with the PCA-percentage of the strongest signal component (dominance of a single source) can be observed. This is a consequence of the more adequately applied dipole model to the PCA-selected time slices. Thus reducing the scattering ("biological noise") of dipole locations.

Discussion

Our study shows abnormal activity for at least 4 weeks after a TIA. This would be sufficient for a screening of "silent TIAs" using the MEG in combination with the DDP. Thus this method could be used clinically as an additional indication for the surgical treatment of patients with a stenosis of the carotid artery. But further studies must show when the normal values are really reached by applying a more dense pattern of measurements between 4 and 10 weeks. In addition the different behavior of the slow and the fast wave band should be studied as well as the correlation to PET results.

References

[1] *Caplan L.R.* Silent brain infarcts, Cerebrovasc. Dis., 1994, 4 (Suppl 1):32-40.

[2] *Gallen C.C., Schwartz B.J., Pantev C., Hampson S., Sobel D., Hirschkoff E., Rieke K., Otis S., Bloom F.* Detection and localization of delta frequency activity in human strokes, In: *Hoke M et al.* Biomagnetism: Clinical Aspects, Amsterdam, Elsevier, 1992.

[3] Guidelines for Carotid Endarterectomy. A multidisciplinary consensus statement from the ad hoc committee, American Heart Association, Stroke, 1995, 26:188-201.

[4] *Grummich P., Vieth J.B., Kober H., Pongratz H. and Ulbricht D.* Focal spontaneous beta wave activity localized close to structural lesions, Biomed. Engineer., 1993, 38 (Suppl):197-198.

[5] *Kober H., Grummich P., and Vieth J.B.* Fit of the digitized head surface with the surface, reconstructed from MRI-tomography, Biomagnetism: Fundament. Res., Amsterdam, Elsevier, 1995.

[6] *Kober H., Vieth J.B., Grummich P., Daun A., Weise E. and Pongratz H.* The factor analysis used to improve the dipole-density-plot (DDP) to localize focal concentrations of spontaneous magnetic brain activity, Biomed. Eng. (Berlin), 1992, 37 (Suppl 2):164-165.

[7] *Vieth J.B.* Magnetoencephalography in the study of stroke (cerebrovascular accidents), In: *Sato S (Ed)* Magnetoencephalography, Adv. in Neurol. Vol. 54, New York, Raven Press, 1990.

[8] *Vieth J.B., Kober H., Weise E., Daun A., Moger A., Friedrich S. and Pongratz H.* Functional 3D localization of cerebrovascular accidents by magnetoencephalography (MEG), Neurol. Res., 1992, 14:132-134.

[9] *Vieth J.B., Kober H., Grummich P., Pongratz H. and Ulbricht D.* Slow wave and beta wave activity associated with white matter structural brain lesions, localized by the Dipole-Density-Plot (DDP), Biomagnetism: Fundament. Res., 1995, Amsterdam, Elsevier.

Preoperative Localization Of Sensorimotor Cortex: MRI versus MEG

Lewine, J.D., Baldwin N.G., Bucholz R.D., Sanders J.A., Halliday, A.L., Anson J.A., Shih, J., Stearley J., Davis, J.T., Astur, R.S., Paulson, K., and Orrison, W.W., Jr.

The New Mexico Institute of Neuroimaging The New Mexico Regional Federal Medical Center, Albuquerque, New Mexico, USA.

INTRODUCTION

Accurate knowledge of cortical functional organization is important in the surgical treatment of brain disorders, because the relevant pathology is often within, or adjacent to, cortical regions that are critical for essential sensory, motor, or cognitive functioning. Avoidance of sensorimotor cortex is especially important because damage to this area can cause a hemiparalysis that reduces the quality of a patient's post-operative life. It is well established that the central sulcus normally demarcates the boundary between primary motor and somatosensory areas, but presurgical identification of the central sulcus by magnetic resonance imaging (MRI) can be difficult, especially in neurosurgical patients where lesions can cause significant distortion of the local neuroanatomy. The present study compared central sulcus identifications using MRI versus MEG in patients that underwent subsequent intraoperative monitoring for validation of preoperative inferences.

METHODS

<u>Subjects</u>: Presurgical MSI mapping of sensorimotor cortex was performed with 52 neurosurgical candidates. Nine patients had been previously determined to have inoperable lesions because MR data were initially thought to indicate that the lesion could not be resected without induction of a significant hemiparalysis.

<u>MR</u>: One hundred twenty-eight, 1.5 mm thick, contiguous T1-weighted sagittal images were obtained using a 1.5 Tesla whole-body imager (Magnetom, Siemens, Islen, NJ). A neuroscientist (JDL) and a neuroradiologist (WWO) reviewed the full series of 2D images in an attempt to identify the central sulcus by anatomical methods. Three different 2D strategies were used, as described by Sobel and colleagues [1]. In addition, 3D surface rendered images were examined.

<u>MEG</u>: The MEG examination was performed using a 37-channel Magnes biomagnetometer system (Biomagnetic Technologies Incorporated, San Diego, CA) or a 122-channel whole head biomagnetometer (Neuromag Ltd., Helsinki, FI). Median nerve stimulation was accomplished using a constant current stimulator (model S88, Grass Electronics, Quincy MA). The stimulator delivered current pulses that were 200 usec in duration. The magnitude of each stimulus pulse was set to be just above the threshold needed to elicit a noticeable thumb twitch (typically 5-8 mA).

A minimum of 100 stimuli were delivered at a rate of 3/sec. MEG data epochs (50 msec prestimulus, 200 msec poststimulus) were recorded with a bandpass of 1-800 hz, at a digitization rate of more than 2Khz. After signal averaging of the data epochs, baseline correction, and additional digital filtering (10-150 hz), a dipole-in-a-sphere model was used to characterize the magnetic field pattern at the time of the 20 msec (M20) neuromagnetic response. Common fiducials were identifed for MEG and MRI data and magnetic source localization images showing the anatomical location of the M20 dipole were generated.

<u>ECoG</u>: Intraoperative electrocorticographic monitoring of the median nerve somatosensory evoked potential (SEP) was performed using an electrode strip or grid. Data were examined for the line of "phase-reversal" for the N20 component, and the point along the line at the intersection of a line connecting regions of maximal positive and negative potential was taken as the position of the hand representation of the somatosensory cortex. In 10 patients the spatial coordinates of this point were determined in MEG space, using a frameless stereotaxy system.

<u>Comparison of MEG and ECoG</u>: For all surgical patients, MEG and ECoG data were compared qualitatively to determine if the methods agreed on identification of the central sulcus. For the 10 patients where stereotaxic coordinates were obtained for the ECoG data, a quantitative comparison was made between the surface projected M20 dipole position and the surface position for the ECoG determined location of the somatosensory hand area.

RESULTS

Complete agreement in the central sulcus identifications of all 2D and 3D MR anatomical methods was found for 18 patients. For the other 34 patients, one or more of the anatomical methods failed to provide useful data for identification of the central sulcus, because the lesion obscured critical landmarks (e.g., the marginal ramus and central sulcus notch for the mid-sagittal method). Nevertheless, in 14 of these cases, there was sufficient congruence in the results from the non-obscured methods that there was confidence in the overall neuroanatomical identification of the central sulcus. The correctness of these identifications was validated by corticography in 13 cases, the 14th case not going to surgery because the lesion was located in primary motor cortex. For the remaining 20 patients (40% of the total population), the lesions caused so much distortion of the anatomy at the parietal-frontal border that identification of the central sulcus by anatomical methods was impossible, or uncertain, occasionally with disagreements between various methods (Fig. 1).

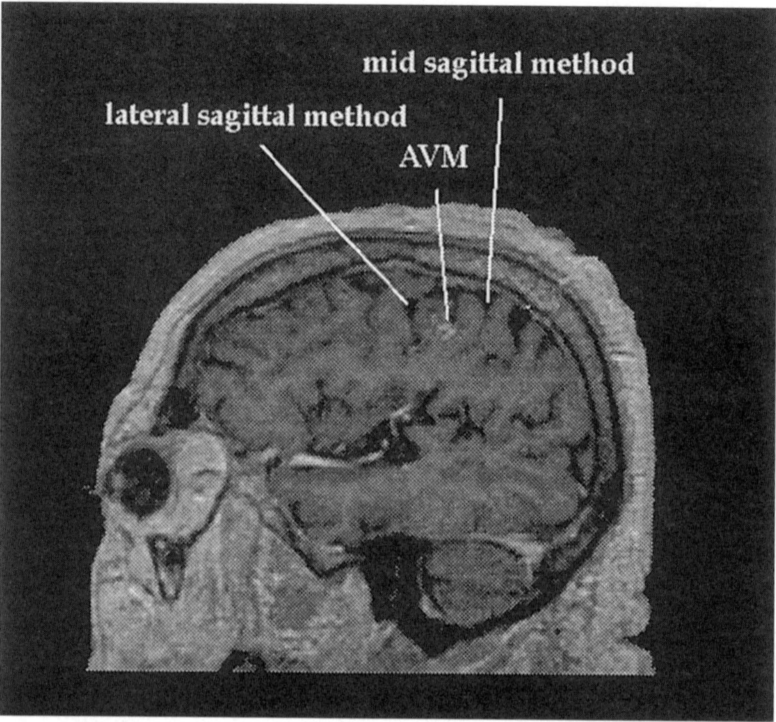

Fig 1: MRI from a patient with an arteriol-vascular malformation [AVM]. Lateral sagittal and mid-sagittal strategies for identifying the central sulcus yield conflicting results.

MEG signals with a good (>5) signal-to-noise ratio for the M20 component were obtained in 50 of the 52 cases. The magnetic field pattern at the time of the M20 was invariably dipolar, with a simple dipole-in-a-sphere model accounting for more than 95% of the variance in the field pattern. The MEG data from 2 patients were severely contaminated by high-amplitude magnetic artifacts that could not be filtered out. In one of these cases, CT and MRI revealed the patient to have a wire-mesh below a previous craniotomy site. The data in the other case were contaminated by magnetic artifacts believed to have originated from the patient's dental work which included a reconstructed tooth thought to have a metal post inside. Fortunately, in both cases, neuroanatomical methods provided for correct identification of the central sulcus as confirmed by corticography.

Forty-eight of the 50 patients with good MEG data underwent resective surgeries with electro-corticographic mapping of somatosensory cortex. In each case, ECoG validated MEG based identification of the post-central gyrus. Two patients did not have invasive surgery, because MEG indicated the patient's lesion to invade primary motor cortex. Ten patients were initially thought to have been inoperable because radiological data suggested that their lesion encroached upon primary motor cortex. MEG "confirmed" this suspicion in one case, but in the other nine, MEG indicated that the the pathological tissue could be resected without inducing significant motor deficits, because the lesion had displaced (rather than invaded) the motor cortex.

In 10 patients, precise stereotaxic coordinates were obtained for the ECoG determined location of the hand representation of somatosensory cortex. Surface projected MEG dipole locations showed good agreement with the corticography determined location of hand somatosensory cortex (mean 4.5 mm, range 2-8 mm, see Fig 2)

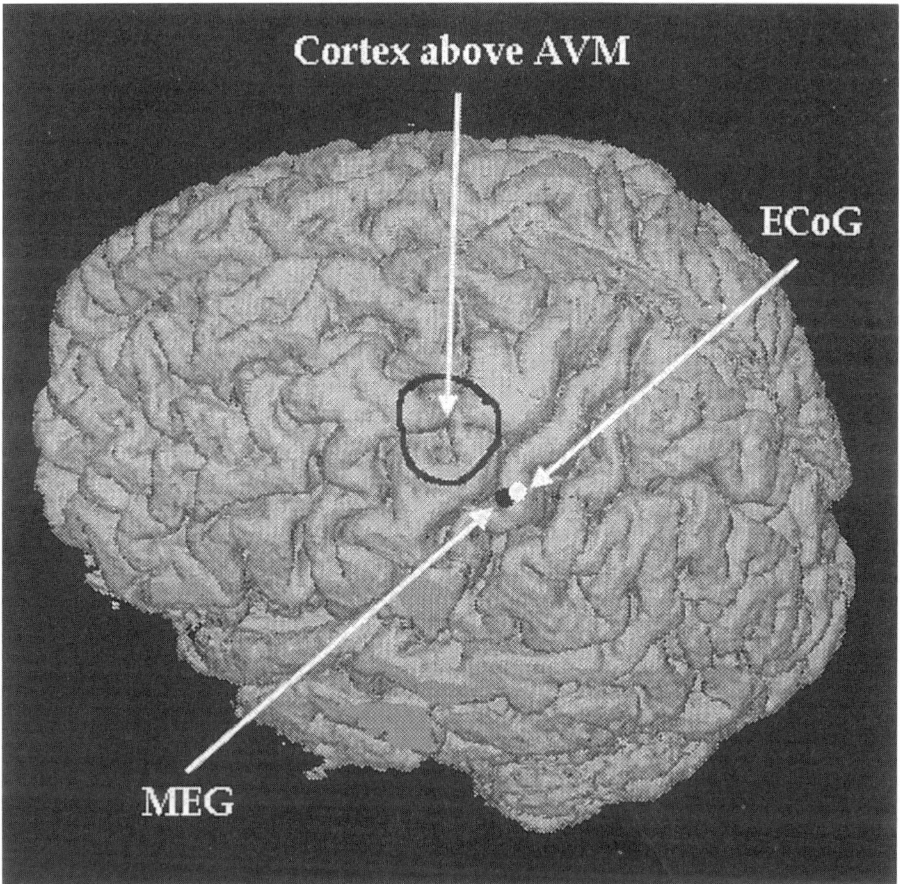

Fig 2: Preoperative MEG localization of the somatosensory cortex (black circle) shows excellent agreement with intraoperative ECoG findings (white circle). The data are from the same patient as shown in Fig 1, and they indicate that the AVM is anterior to the central sulcus and that the mid-sagittal anatomical method was correct.

DISCUSSION

MRI anatomical methods are often assumed to be useful in the noninvasive identification of the central sulcus, but the situation is complex in neurosurgical patients with lesions because of distortions of the local neuroanatomy. In this study, neuroanatomical methods were inadequate for 20 patients (38%). This rather poor ability of MRI to localize the central sulcus in patients with frontal and parietal mass lesions was markedly contrasted by the excellent MEG results in which the central sulcus was identified in 50 of 52 patients (97%) -- the two failures being caused by obvious artifacts in the data.

The clinical utility of MEG was clear in the 20 patients for whom anatomical methods had failed. Ten of these patients were originally judged to be inoperable, but in nine cases MEG indicated that the pathological tissue could be excised without induction of significant motor deficits. That the tenth patient should not have surgery was an equally important finding.

MEG is not the only functional imaging method that may be capable of providing for accurate presurgical localization of the sensorimotor areas of the brain. Individual case reports are now available for positron emission tomography (PET), functional magnetic resonance imaging (fMRI), and high-density scalp recordings of the somatosensory evoked potential. However, there have yet to be any large-scale validation studies of these methods, and each method has its limitations. For example, fMRI is very sensitive to movement artifacts, and success rate in patients with vascular malformations has been low. Dipole modeling of high density recordings of the scalp somatosensory evoked potential is often suggested as a cost-effective alternative to MEG evaluation of sensorimotor function, and in some cases this is probably true. However, neuroelectric evaluations are likely to be more severely compromised by electrical conductivity inhomogeneities at a lesion boundary, and also by skull defects from prior surgeries. Skull defects cause very severe distortions of scalp electrical potential patterns, but minimal distortions of the MEG field pattern. Of the 50 patients in the present study, 10 had significant skull defects.

CONCLUSIONS

MEG mapping of the median nerve evoked field is more effective than structural MRI for identification of the central sulcus in neurosurgical patients with lesions near the fronto-parietal junction. Preoperatively obtained MEG identification of the hand somatosensory area shows good agreement with intraoperative identification of this region and it serves as a valuable presurgical planning tool. The MEG portion of the examination takes less than 30 minutes to execute and magnetic source localization images can be made available to surgeons within minutes of completion of MEG and MRI examinations.

REFERENCES

[1] Sobel, D.F., Gallen C.C., Schwartz, B.J., et al., Central sulcus localization in humans: comparison of MRI-anatomic and magnetoencephalographic functional methods, AJNR, 1993, 14: 915-925.

Clinical MEG I: Towards a Standardized Examination

Lewine, J.D., Davis, J.T., Davis, L.E. Canive, J., Roberts, B., Graeber, D., Shih, J., Edgar, J.C., Provencal S.L., Paulson, K., Meyers, J., Christner, R., Silveri, J., Rawcliffe, N., Tessman, C., Espinosa, M., Depper, M., Sanders, J.A., and Orrison, W.W., Jr.

The New Mexico Institute of Neuroimaging, The New Mexico Regional Federal Medical Center, Albuquerque, New Mexico, USA

INTRODUCTION

Recent availability of large-array and whole-head biomagnetometer systems has finally allowed magneto-encephalography to begin to develop into the clinical tool foreshadowed by initial single-channel SQUID systems. Whereas clinical experiments used to take several hours or even days to complete, the newest generation of sensors allows for a basic neurodiagnostic examination to be completed in less than two hours.

The New Mexico Institute of Neuroimaging (NMIN), initially (1992-1994) equipped with a 37-channel biomagnetometer (BTi), and now equipped with a 122-channel whole-head biomagnetometer (Neuromag Ltd.), employs magnetoencephalography as part of the standard of diagnostic care for a wide range of patients with neurologic and psychiatric disorders. Dominant areas of direct clinical utility are in (1) presurgical mapping of sensorimotor functions, (2) characterization of epileptiform activity, and (3) characterization of the neurobiological effects of stroke and minor head trauma. Clinical research activities include investigations of (1) psychiatric conditions, (2) progressive dementias, (3) CNS effects of pharmacological agents, and (4) fetal cardiac and brain activities. Since 1992, the facility has performed examinations with more than 1500 patients.

Recently, we have developed a standardized neurodiagnostic examination that is used with most patients. This examination is divided into several sub-tests that provide for assessment of each of the three major sensory systems as well as spontaneous brain activity. In some cases, this standard examination is supplemented by additional specialized testing (e.g., addition of a P300 odd-ball paradigm in the testing of schizophrenic patients). To date, approximately 150 subjects have received the complete battery of sub-tests, with another 350 subjects having received most of the tests.

STRATEGY

The standardized examination has been developed for implementation of the 122-channel whole-head biomagnetometer system developed by Neuromag Ltd.

<u>Staff</u>: At NMIN, the dedicated staff is comprised of 4 technologists, 1 electronics technician, and 2 senior scientists. In addition, this staff is supported by the part-time efforts of 2 neuroradiologists, and 5% effort by each of two electroencephalographers.

<u>Subject Preparation (time 10-30 minutes)</u>: Following completion of an informed consent document (approved by the Human Research Review Committee of the University of New Mexico) and a brief clinical interview, the patient is fitted with a flexible cap that has holders for 3 small coils that can be energized during the experiment. The positions of the coils, relative to a head coordinate system defined by preauricular points and the naision, are defined using a 3D-digitizer (polhemus). For patients with a seizure history, a specially designed electrode cap is used. This cap has 62 electrodes and can be applied in about 20 minutes. When applicable, electrode positions are also digitized. For all patients, cardiac and EOG electrodes are also applied. EEG data are collected using SynAmplifiers (NeuroScan) interfaced directly with the Neuromag unit and also with the SCAN software and hardware package of NeuroScan Inc. Following completed preparation, the subject is led into a magnetically shielded chamber and positioned under the sensor unit. Because small children often have difficulty staying still in the sensor unit, we routinely sedate with chloral hydrate or elavil.

<u>Experimental Protocols (time 60-90 minutes)</u>: As specified below, each subject participates in 7 sub-tests. These are: (1) spontaneous - eyes open, (2) spontaneous - eyes closed, (3) spontaneous - reactivity, (4) visual, (5) contrast, (6) auditory, and (7) somatosensory. Depending upon a patient's clinical history, additional testing may be performed. All subjects also receive an MRI examination following the MEG examination. When appropriate, neuropsychological testing and SPECT examination may be performed.

<u>Stimuli</u>: Stimuli for evoked response studies are delivered using the STIM software package of NeuroScan Inc. For somatosensory experiments, electrodes are attached to right and left median and tibial nerves. Electrical

pulses are delivered using Grass Stimulators, controlled through the STIM GenTask program. For auditory stimulation, the GenTask program is used with signals relayed through etymotic speaker units within the shielded room. The STIM GenTask program controls visual stimulation which is effected by a monitor positioned outside of a 12x12 inch hole in the shielded room.

Data Analysis: Data analysis is a two step procedure. The initial step is performed by a group of technologists (which is also responsible for patient handling and data collection). The technologists perform pre-specified spectral analyses of the continuous data. Dipole analyses for pre-specified signal components in each sensory domain is performed also.

Reporting: The technologists relay a data packet containing spectral maps, contour maps, dipole fits, activity profiles, and magnetic source localization images to senior scientists who (1) evaluate the accuracy and clinical significance of the initial analyses, (2) visually inspect continuous data for epileptiform discharges and delta activity (with multiple dipole spatio-temporal modeling and MRI mapping as needed), and (3) generates special images and a clinical report for clinicians. For presurgical cases, a complete analysis of all data (including 3D surface rendered MRI images) is made available to physicians within 24 hours. For routine clinical cases (e.g., stroke, trauma, psychiatric), a complete analysis is available within 48 hours. Evaluation of data sets with epileptiform activity is especially complex because spikes must be visually identified and then characterized using multiple dipole spatio-temporal models in order to specify propagation pathways. These cases may take as much as 1 week for complete processing.

SUBTESTS 1 & 2:

FIVE MINUTES OF CONTINUOUS DATA IN EACH OF EYES OPEN AND EYES CLOSED CONDITIONS

Objective: Identification of abnormal resting physiology, with an emphasis on epileptiform transients and abnormal low frequency magnetic activity in the delta band.

Acquisition Parameters

Length of Record:	5 minutes	(additional recording is common for patients with epilepsy)
Digitization Rate:	300 hz	
On-line Filters:	0.1-100 hz	

Data Analysis Procedures

1. Spectral analysis of data: Average of one hundred, 3.4 sec-long artifact free epochs with 50% shift between successive epochs. Spectra are assessed for abnormal dominance by low frequency activity, and unusual topographies or asymmetries. Significant delta activity in the eyes open state is a definite sign of pathology.

2. Visual inspection of data: Data are visually inspected for epileptiform transients (spikes and sharpwaves) and bursts of delta activity. Visual inspection of data is imperative because spectral analyses involving averaging of epochs recorded over long time periods may fail to reveal intermittent abnormalities. Utilization of compressed spectral arrays rather than average spectra may help here.)

3. When abnormal activity is identified, spatio-temporal dipole modeling is applied.

SUBTEST 3:

Reactivity Protocol

Objective: To assess reactivity of spontaneous rhythms to changes in subject state.

Acquisition Parameters

Length of Record:	4 minutes -	30s eyes open; 30s bilateral fists; 30s eyes open;30s eyes closed; 30s eyes open; 30s eyes closed; 30s bilateral fists; 30s eyes closed.
Digitization Rate:	300 hz	
On-line Filters:	0.1-100 hz	

Data Analysis Procedures

1. Spectral analysis of data in 30s blocks: Check for reactivity of alpha rhythm to eye opening and closing and reactivity of mu rhythm to hand movements.

2. Visual inspection of data: Data are also visually inspected for epileptiform transients and bursts of delta activity. The reactivity paradigm distinguishes mu rhythm from fronto-parietal sharpwaves.

1042

SUBTEST 4:

Visual Mapping

Objective: To assess the integrity of the visual system with special reference to field cuts caused by lesions and delayed responses associated with optic neuritis or MS

Stimuli : 100 trials each, randomized, with random ISI of 200-300 msec. Stimulus duration is 50 msec. Subjects count the number of stimuli presented at fixation.

location	type	size	eccentricity	spatial frequency
fixation	blue/yellow	0.5 degree	0 degree	10 cycles/degree
URVF	red/green	1.0 degrees	1 degree	05 cycles/degree
ULVF	red/green	1.0 degrees	1 degree	05 cycles/degree
LRVF	red/green	1.0 degrees	1 degree	05 cycles/degree
LLVF	red/green	1.0 degrees	1 degree	05 cycles/degree
URVF	black/white	2.0 degrees	3 degrees	2.5 cycles/degree
ULVF	black/white	2.0 degrees	3 degrees	2.5 cycles/degree
LRVF	black/white	2.0 degrees	3 degrees	2.5 cycles/degree
LLVF	black/white	2.0 degrees	3 degrees	2.5 cycles/degree

Acquisition Parameters

Digitization Rate: 150 hz
On-line Filters: 1-50 hz
Epoch: -100 to 400 msec

Data Analysis Procedures

    1. Digital filter 3-30 hz, baseline correct -100 to 0

    2. Single dipole analysis of 90-110 msec component for periphery and 110-140 msec component for parafoveal stimuli. Rapid stimulus presentation biases system to dipolar responses from primary visual cortex with clear retinotopic organization.

SUBTEST 5:

Contrast Following

Objective: To assess the ability of the visual system to follow contrast reversing stimuli. Previous EEG data suggest that following ability may be a good predictor to response to medication in some populations.

Stimuli: 90 seconds of contrast reversal at each of 4, 8, 12, and 16 hz, for an 8x10 degree fixated black and white checkerboard pattern (1 cycle/degree).

Acquisition Parameters

Digitization Rate: 300 hz
On-line Filters: .1-100 hz

Data Analysis Procedures

    1. Spectral analysis at each frequency.

    2. Examination of amplitude and topography of spectra with assessment of harmonic responses and asymmetries.

SUBTEST 6:

Auditory Mapping

Objective: To assess the integrity of the auditory system with special reference to asymmetries in response topography, latency, amplitude or source location.

1043

Stimuli: 100 trials each, randomized, with random ISI of 700-800 msec. Stimulus duration is 50 msec. Subjects count the number of low intensity tones.

type	frequency	intensity
binaural	500 hz	85 dB
binaural	2000 hz	85 dB
binaural	4000 hz	85 dB
right ear only	2000 hz	85 db
left ear only	2000 hz	85 db
binaural	2000 hz	70 dB

<u>Acquisition Parameters</u>

Digitization Rate:	150 hz
On-line Filters:	1-50 hz
Epoch:	-100 to 400 msec

<u>Data Analysis Procedures</u>

1. Digital filter 1-30 hz, baseline correct -100 to 0
2. Dipole analysis of M100 response using bilateral dipole model. Modeling of other components is possible but these show high variability of response in normals. Data are assessed for abnormal latencies, amplitudes, asymmetries and topographies. A special interest is placed on abnormal topographies for particular tonal frequencies and monaural versus binaural conditions.

<u>SUBTEST 7:</u>

<u>Somatosensory Mapping</u>

Objective: To assess the integrity of the somatosensory system with special reference to asymmetries in response latency, amplitude or source location. Additivity in bilateral condition is of special interest in patients with simultaneous extinction. Also of interest is the spatial location of central sulcus relative to a lesion.

Stimuli: 100 trials each, randomized, with random ISI of 450-550 msec. Stimulus duration is 200 usec and adjusted to just above twitch threshold. In patients with poor cortical responses because of lesions, an increased ISI is recommended.

type	location
unilateral pulse	right median nerve
unilateral pulse	right tibial nerve
unilateral pulse	left median nerve
unilateral pulse	left tibial nerve
bilateral pulse	right and left median nerves

<u>Acquisition Parameters</u>

Digitization Rate:	1250 hz
On-line Filters:	1-400 hz
Epoch:	-100 to 400 msec

<u>Data Analysis Procedures</u>

1. Digital filter 10-150 hz (for SI), baseline correct -100 to 0. For SII, digital filter 1 - 50 hz.
2. Dipole analysis of M20 response for localization of primary somatosensory cortex.
3. Multiple dipole analysis required for assessment of later responses because of recurrent activation of SI.

1044

Two components of giant somatosensory evoked magnetic fields in patients with cortical myoclonus

Mima, T.[1], Mikuni, N[2], Nishitani, N[1], Nagamine, T.[1], Ikeda, A.[1], Fukuyama, H.[1], Takigawa, T.[3], Kimura, J.[4], and Shibasaki, H.[1]

Department of Brain Pathophysiology[1], Neurosurgery[2] and Neurology[4], Kyoto University, School of Medicine, Kyoto, Japan; National Utano Hospital, Kyoto, Japan[3]

Introduction

Pathologically enlarged cortical somatosensory evoked potentials (giant SEPs) are one of the important features of cortical myoclonus [1-4]. Magnetoencephalographic studies have revealed that the somatosensory evoked magnetic fields (SEFs) to median nerve stimulation are also enlarged, and that their generator sources can be localized at the somatosensory cortex [5; 6]. However, based on the SEP scalp topography using EEG, we have reported that the tangential (N30-P30) and "radial" (P25-N35) components can be distinguishable in giant SEPs [7]. Since giant somatosensory evoked responses (SERs) in cortical myoclonus are regarded as the indicator of the hyperexcitability of sensorimotor cortex in those patients, studies of the generator mechanisms of giant SERs will be of great help to understand the pathogenesis of cortical myoclonus. The purpose of this study is to identify MEG correlate of each component of giant SEPs, and to locate the generator sources on individual MRI.

Methods

We investigated 3 patients with cortical myoclonus (Patient 1; Lafora disease, Patients 2 and 3; familial cortical myoclonic tremor). Their clinical features are summarized in Table 1. All subjects gave informed consent before the experiments according to the approval by the Ethical Committee of Kyoto University School of Medicine.
 SEFs were recorded by a helmet-shaped magnetometer array with 122 first-order planar gradiometers. The planar gradiometer detects the largest signal above the local current source. MEG signals were recorded by the sampling rate of 0.9 kHz (bandpass; 0.03 to 300 Hz). SEPs were simultaneously recorded from at least 4 sites (C3, P3, C4 and P4). All electrodes were referenced to linked ear lobe electrodes. The electrooculogram (EOG) at the right inferior lateral canthus referred to the right earlobe was simultaneously recorded. The median nerve trunk at wrist was stimulated with electric pulse of 0.2 ms duration. MEG and EEG signals were averaged with respect to the stimulus onset. Any response with amplitude exceeding 150 μV in the EOG channel or 3000 fT/cm in any of MEG channels was rejected from the average. The analysis time window was 600 ms including the prestimulus 100 ms. In each session, 2 trials of 150 responses each were averaged separately to confirm the reproducibility of the responses. Amplitude of each recognizable peak of SEPs was measured from the baseline determined by averaging the prestimulus segment. N20 was identified as an initial local negative potential around 20 ms at the contralateral parietal area (P3/P4). P30 was identified as a positive peak at the same site. P25 and N35 were recognized at the contralateral central area (C3/C4). The terminology of the peaks followed the studies of normal subjects [8].
The MEG channel with the local extreme signal in each record was chosen for the measurement of the latency. The magnetic fields contour patterns were constructed by using a minimum-norm estimate. Parameters of an equivalent current dipole (ECD) that best explain the most dominant source at each peak were determined by a least-squares search, using subsets of channels over the response areas. ECDs explaining more than 80 % of the fields over the selected period of time were used for further analysis. MEG coordinates were matched with MRI by using the 3 head position indicators. ECDs were superimposed on individual brain MRI.

Results

EEG recordings demonstrated four sets of early components of SEPs (N20, P25, P30 and N35). Peak latency and amplitude of each component are summarized in Table 2.
In MEG recording, the initial cortical component of SEFs was N20m at the contralateral sensorimotor area, which was followed by a peak of opposite polarity at the same MEG channel (P30m). Another MEG peak between N20m and P30m could be recognized at the different channel at the same or nearby site (referred to as P25m). P25m was followed by the next peak of opposite polarity (N35m) (Fig. 1).
Contour maps of magnetic fields and the parameters of ECDs were calculated for each recognizable SEF component. Magnetic field pattern of N20m showed clear single dipolar pattern around the contralateral sensorimotor area in all 3 patients. Magnetic field contours of P30m were similar to those of N20m, though the flux direction was inverted. ECDs of N20m and P30m had the same location posterior to the central sulcus, and their directions were inverted each other (anterior-posterior orientation). Magnetic field contours of P25m were clearly different from those of N20m or P30m. Magnetic fields of N35m were similar to those of P25m in 5 nerves

TABLE 1 Clinical features of 3 patients

Patient	Diagnosis	Sex	Age	Clinical Symptoms	Medication (per day)
1	PME (Lafora disease)	M	20	seizure, mental deterioration	VPA: 1000 mg, CZP: 3 mg
2	Cortical Tremor	M	68	none	none
3	Cortical Tremor	F	58	none	VPA: 600 mg, CZP: 4 mg, zonisamide: 400 mg

F: female, M: male, PME: progressive myoclonus epilepsy, VPA: valproic acid, CZP: clonazepam

TABLE 2 Parameters of ECD for each SEF and SEP component

Patient (side)	SEF latency (ms)	source location x axis (mm)	y axis (mm)	z axis (mm)	dipole moment (nAm)	g-value (%)	confidence volume (mm^3)	SEP latency (ms)	amplitude (μV)
N20m								**N20**	
1(L)	22	39.3	19.5	100.8	38.2	95.3	0.9	20	−1.2
(R)	23	−47.1	17.6	98.6	19.3	93.1	8.2	20	−0.1
2(L)	22	43.8	−2.2	92.7	39.7	98.3	0.6	21	−4.8
(R)	22	−45.5	−3.3	97.0	23.1	96.0	3.8	21	−1.2
3(L)	20	41.0	17.9	94.9	22.6	95.9	10.5	20	−2.0
(R)	21	−38.4	6.0	91.1	35.5	98.5	2.7	20	−2.0
P25m								**P25**	
1(L)	27	33.4	17.0	115.2	18.0	95.1	2.5	25	11.3
(R)	28	−38.7	11.4	92.4	43.1	87.2	5.4	25	7.2
2(L)	25	39.4	−6.4	98.3	26.4	92.8	1.6	24	27.3
(R)	25	−34.8	−7.8	103.4	35.5	95.1	1.2	25	21.1
3(L)	24	34.0	22.7	97.0	55.4	96.1	1.2	23	20.2
(R)	23	−31.0	13.1	101.8	29.6	93.4	2.7	23	18.6
P30m								**P30**	
1(L)	31	37.4	17.8	106.1	46.0	97.7	0.3	28	9.1
(R)	40	−43.4	11.3	93.4	61.9	96.9	0.8	40	4.2
2(L)	27	43.0	−5.9	94.3	113.7	98.2	0.0	26	19.2
(R)	28	−44.0	−8.2	95.0	131.2	98.4	0.0	27	18.3
3(L)	26	37.1	19.1	91.5	72.9	96.0	0.8	25	11.1
(R)	26	−38.4	2.1	94.9	55.4	98.9	0.4	25	14.2
N35m								**N35**	
1(L)	50	30.7	17.1	99.4	37.9	94.3	3.7	49	−3.5
(R)	52	−46.3	12.8	98.4	17.7	85.0	16.5	52	−8.3
2(L)	40	32.1	−12.5	99.1	52.0	90.2	0.4	39	−60.8
(R)	39	−38.8	−7.1	103.0	57.7	91.8	0.2	39	−35.6
3(L)	30	33.5	25.4	99.1	36.1	96.7	3.1	29	−16.9
(R)	29	−30.6	15.3	92.5	53.6	94.4	1.8	29	−11.1

In this table, EEG amplitudes were measured from the baseline determined by the prestimulus segment. The 3-dimensional locations (x, y and z) of ECDs were measured from the midpoint of the two preauricular points. The x axis runs from the left to the right preauricular point. The y axis runs from back to the front through nasion. The z axis runs from the base to the top of the head through the center of the two preauricular point.

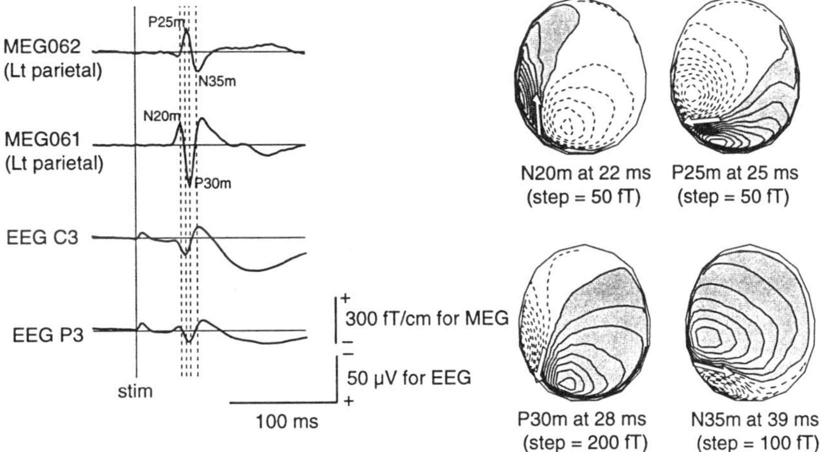

Fig. 1

    Left figure: Giant SEFs to the right median nerve stimulation in Patient 3. Time window is 100 ms including the prestimulus segment of 20 ms. The vertical line indicate the stimulus onset. Representative MEG channels and EEG electrodes are demonstrated. Negative deflexion of the EEG signals are shown upward. Each vertical dashed line indicates each SEF component (N20m, P25, P30m and N35m, respectively).

    Right figure: Magnetic field patterns in the same patient. The shadowed areas indicate magnetic flux out of, and the white area flux into the head. Each step of isocontour and the latency are shown below the contourmap. White arrow shows the locations and orientations of the ECDs.

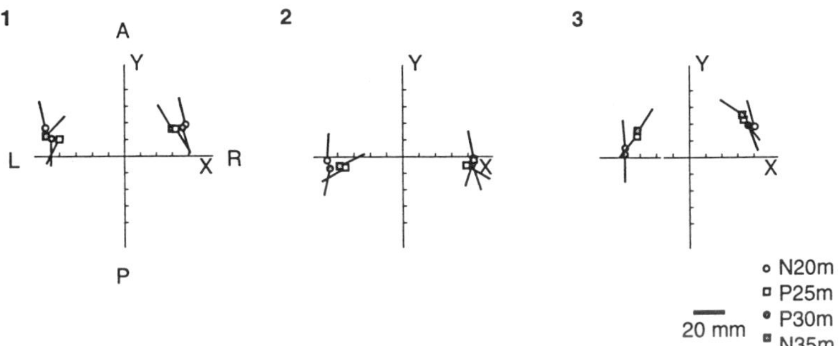

Fig. 2

    Source location and orientation of each SEF component projected on the topview. N20m (open circle), P25m (open square), P30m (filled circle) and N35m (filled square) are presented. The direction fo the ECDs are shown as a bar.

except for the left median nerve stimulation of Patient 2. In Patients 1 and 2, ECDs of P25m and N35m were located posteromedial to that of N20m (postcentral gyri), whereas they were located anteromedial to that of N20m (precentral gyri) in Patient 3. ECDs of P25m had the posterolateral direction, and those of N35m showed anteromedial direction in the other subjects except for that to left median nerve stimulation in Patient 2 which showed posteromedial direction. Parameters of ECDs are summarized in Table 2. The source locations and orientations of these SEF components were projected onto the topview of the head (Fig. 2).

Discussion

We could discern MEG counterparts of four EEG components of giant SEPs (N20m, P25m, P30m and N35m). Both EEG and MEG reflect the same electromagnetic activities in human brain, but MEG detects the radial components of the magnetic fields which are generated by tangentially oriented intracellular currents. The magnetic counterpart of P25 and N35 in SEPs were not always recognized clearly, possibly because these SEP components are generated by the electrical current radial to the brain surface, which were suggested by the analysis of scalp topography of SEPs [8; 9]. However, even in normal subjects, MEG correlates of P25 and N35 (P25m and N35m, respectively) could be identified, which are thought to reflect the tangential components of the so-called "radial" electrical dipoles described in EEG scalp topography [10; 11]. This is the first report of P25m and N35m components of giant SEFs in patients with cortical myoclonus.

In MEG recording, the strongest response was P30m in all subjects, whereas P25 was the largest in SEPs. Therefore, the largest positive potential in giant SEP does not always correspond to the largest MEG peak, which suggests that "radial" electrical dipole has some contribution to the generation of the exaggerated cortical responses to somatosensory stimuli. ECDs of N20m were localized at the posterior wall of the central sulcus (Brodmann area 3b). Those of enlarged P30m were located at the same site as those of N20m. ECDs of P25m and N35m were localized at the postcentral sensory cortices in Patients 1 and 2, whereas in Patient 3 they were at the precentral motor cortices. Recent monkey studies suggested that the monkey counterpart of P25 was caused by the simultaneous activation of both motor and sensory cortices [12]. It is most likely that P25 and N35 on EEG are also the summation of the potentials generated at both the precentral and postcentral gyri, and that the pathological enhancement of the cortical responses may occur separately at each site.

References

[1] Rothwell, J.C., Obeso, J.A. and Marsden, C.D. On the significance of giant somatosensory evoked potentials in cortical myoclonus, Journal of Neurology, Neurosurgery, and Psychiatry, 1984, 47: 33-42.

[2] Shibasaki, H., Kakigi, R. and Ikeda, A. Scalp topography of giant SEP and pre-myoclonus spike in cortical reflex myoclonus, Electroencephalography and Clinical Neurophysiology, 1991, 81: 31-37.

[3] Kakigi, R. and Shibasaki, H. Generator mechanisms of giant somatosensory evoked potentials in cortical reflex myoclonus, Brain, 1987, 110: 1359-1373.

[4] Hallett, M., Chadwick, D. and Marsden, C.D. Cortical reflex myoclonus, Neurology, 1979, 29: 1107-1125.

[5] Kahru, J., Hari, R., Pateau, R., Kajola, M. and Mervaala, E. Cortical reactivity in progressive myoclonus epilepsy, Electroencephalography and Clinical Neurophysiology, 1994, 90: 93-102.

[6] Uesaka, Y., Ugawa, Y., Yumoto, M., Sakuta, M. and Kanazawa, I. Giant somatosensory evoked magnetic field in patients with myoclonus epilepsy, Electroencephalography and Clinical Neurophysiology, 1993, 87: 300-305.

[7] Ikeda, A., Shibasaki, H., Nagamine, T., Xu, X., Terada, K., Mima, T., Kaji, R., Kawai, I., Tsatsuoka, Y. and Kimura, J. Peri-rolandic and fronto-parietal components of scalp-recorded giant SEPs in cortical myoclonus, Electroencephalography and Clinical Neurophysiology, 1995, 96: 300-309.

[8] Allison, T., McCarthy, G., Wood, C.C. and Jones, S.J. Potentials evoked in human and monkey cerebral cortex by stimulation of the median nerve, Brain, 1991, 114: 2465-2503.

[9] Desmedt, J.E., Nguyen, T.H. and Bourguet, M. Bit-mapped color imaging of human evoked potentials with reference to the N20, P22, P27 and N30 somatosensory responses, Electroencephalography and Clinical Neurophysiology, 1987, 68: 1-19.

[10] Tiihonen, J., Hari, R. and Hämäläinen, M. Early deflexions of cerebral magnetic responses to median nerve stimulation, Electroencephalography and Clinical Neurophysiology, 1989, 74: 290-296.

[11] Kakigi, R. Somatosensory evoked magnetic fields following median nerve stimulation, Neuroscience Research, 1994, 20: 165-174.

[12] Peterson, N.N., Schroeder, C.E. and Arezzo, J.C. Neural generators of somatosensory evoked potentials in the awake monkey, Electroencephalography and Clinical Neurophysiology, 1995, 96: 248-260.

Acknowledgements

This study was partly supported by Grant-in-Aid for International Scientific Research 07044258 from the Japan Ministry of Education, Science and Culture.

Deviations from the Tonotopic Map are Correlated with Tinnitus Strength

Mühlnickel, W.[1], Flor, H.[1], and Elbert, T.[2]

[1]Department of Psychology, Humboldt-University of Berlin, Germany; [2]Department of Psychology, University of Konstanz, Germany

INTRODUCTION

Tinnitus is characterized by phantom perception of acoustic signals without any external source of sound. Based on previous studies that reported reorganization of the somatosensory cortex in amputees with phantom sensation [1], we hypothesized that tinnitus might be based on plastic changes in the tonotopic map in primary auditory cortex. In accordance with the finding that phantom limb pain is positively related to reorganization in primary somatosensory cortex, it was assumed that tinnitus strength should be positively correlated with amount of cortical reorganization.

METHOD

Seventeen right-handed tinnitus patients (tonal tinnitus, maximal hearing loss 25 dB HL) and seventeen matched (age, tender) right-handed normal-hearing controls without tinnitus participated in the study. See table 1 for clinical aspects of the patients in the tinnitus group.

Table 1: Clinical aspects of the tinnitus group.

Pat-No.	Age	Sex	Strength	T.ear	T. freq.
2	31	m	1.0	left	4256
3	38	m	-	right	4000
4	35	w	4.3	left	8000
5	32	m	-	both	6000
6	35	m	4.0	both	6000
7	18	w	3.3	right	4000
8	16	m	-	right	200
9	19	w	2.6	both	noisiform
10	27	m	3.0	both	250
11	36	w	2.0	right	4000
12	44	w	3.0	left	1000
13	31	w	4.6	both	6000
14	33	m	2.6	right	8000
15	29	m	3.6	right	3000
16	31	w	-	both	1000
17	32	m	4.6	both	4000
18	20	m	4.3	left	3680
$\bar{x}$	30.3±2.4		3.4±1.05		4024±2485

Pure tone pulses (500 ms duration, 70 dB HL, four different carrier frequencies in random order) were sequentially presented to both ears, with an interstimulus interval (ISI) of 2 s. In the control group, carrier frequencies of 1000, 2000, 4000 or 8000 Hz were employed. In the tinnitus group three of these standard frequencies were employed. The tinnitus frequency was introduced as the fourth frequency and the standard frequency that was closest to the tinnitus frequency was deleted from the list of stimuli.

Magnetoencephalographic recordings were carried out in a magnetically shielded room using a 37 channel neuromagnetometer (Magnes^{TM}, Biomagnetic Technologies). 200 stimulus-related epochs of the magnetic field were recorded over the contralateral hemispheres. The magnetic signals were digitally filtered using a 20 Hz low pass filter and selectively averaged for each stimulus frequency. The location of wave N1m of the stimulus-related field was estimated using a equivalent current dipole model (ECD) in a homogeneous spherical volume conductor [2].

The ECD locations of the healthy controls that corresponded to the four different frequencies formed a trajectory in three-dimensional space representing the tonotopic map. The ECD locations of the standard frequencies are a logarithmic function of the stimulus frequencies with a main orientation in medial-lateral direction [3]. In the tinnitus group, the dipole locations of the three standard frequencies were used to reconstruct the map.

Next, the Euclidian distance of the location of the tinnitus frequency and the trajectory of the standard frequencies was computed. Reorganization of primary auditory cortex related to tinnitus might result in a shift of the tinnitus frequency into the upper or lower frequency range or in a shift into adjacent cortical aereas. Since tinnitus frequencies were not uniform it was impossible to test the shift into the higher or lower frequency range. It was, however, possible to compute the deviation of the tinnitus frequency from the trajectory of standard frequencies as a measure of invasion of the tinnitus frequency into neighbouring regions. This shift was modeled by computing the Euclidian distance of the location of the tinnitus frequency from the trajectory of standard frequencies. Figure 1 shows an enlarged twodimensional projection of the standard frequencies in the auditory cortex in one tinnitus patient. The filled black circles represent the ECD locations of the magnetic field distribution to the three standard stimuli. The open circle represents the ECD location of the tinnitus frequency. The deviation of the tinnitus frequency from the trajectory of the standard frequencies was determined by computing the Euclidian distance of the tinnitus frequency to the trajectory. The distancies were separately computed for both hemispheres. This procedure computes the minimum deviation of the tinnitus frequency from the normal tonotopic map. It is conceivable that the true pre-tinnitus frequency might has been further removed on this trajectory, the Euclidian distance would in this case underestimate the shift and is thus a conservative measure of reorganization. In a subject with chronic tinnitus the ECD location of the magnetic response of the tinnitus frequency is shifted outside of the area of the putative tonotopic map.

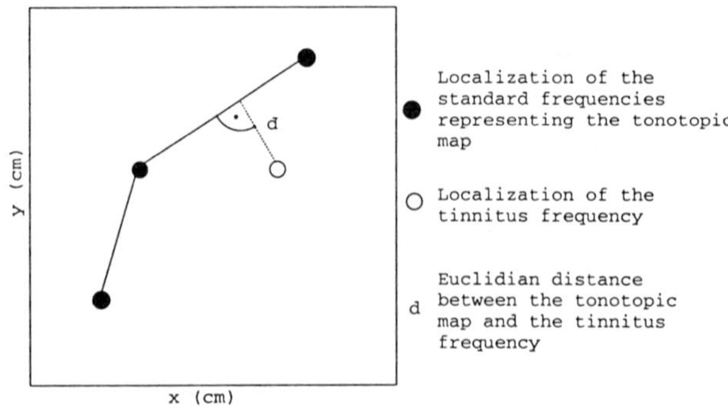

Figure 1: A model of computing the Euclidian distance d.

RESULTS

The deviation from the tonotopic map of the tinnitus patients was compared with the shift of the tonotopic map of the controls. Based on the typical tinnitus frequency range (see table 1) of the patients, tinnitus frequencies of 2,000 Hz and 4,000 Hz were "created" for the controls. The deviation of the 2 kHz or 4 kHz tone from the trajectory formed by the standard tones was calculated by using the Euclidian distance measure. It was expected that the Euclidian distance of the controls should be significantly smaller than those of the patients, thus indicating a lack of cortical reorganization. An ANOVA showed a significant difference between the Euclidian distance measures of the patients and the controls ($F(1,31) = 6.36$, p $<$ .05). When the deviations of 2 kHz and 4 kHz tones were summarized, there was no significant difference between them. Whereas the Euclidian distance of the tinnitus patients was on the average of 4.15 ± 2.2mm, the Euclidian distance of the controls amounted to 2.8 ± 1.0mm. This confirms the hypothesis of a shift of the cortical representation of the tinnitus frequency outside the area of the tonotopic map.

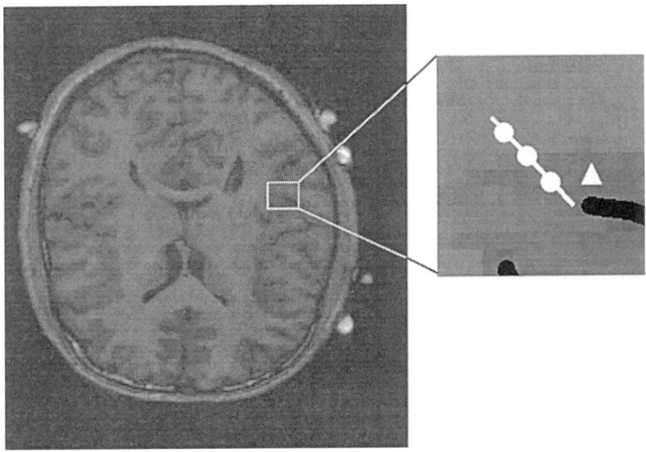

Figure 2: A horizontal MRI slice of Brodman area 41 of the right hemisphere. The results of one subject are shown. The line represents the trajectory of the dipole locations of three tones. The triangle represents the location of the corresponding tinnitus frequency.

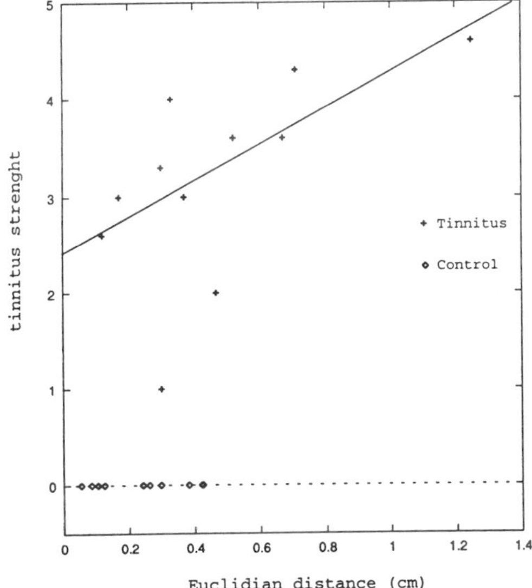

Figure 3: Scatterplot of the tinnitus strenght vs. the Euclidian distance. The distances of the controls were the distances representing the 4 kHz tone with respect to the average tinnitus frequency of the tinnitus group.

In order to determine the relationship of the cortical map and the subjective tinnitus strength a correlation coefficient was computed between the cortical distance measure and a tinnitus inventory. Tinnitus strength was assesed by a tinnitus-specific version of the West Haven-Yale Multidimensional Pain Inventory (WHYMPI) [4], where the expression *pain* was replaced by *earsound*. The correlation of the subjective tinnitus strength and the Euclidian distance was significant at $r(10) = 0.817$, $p < .05$, for the stimulation of the left ear and recordings of the right hemisphere. Right ear tinnitus was to infrequent ($n = 3$, see table 1) to compute correlation coefficients.

DISCUSSION

This study yielded two important results. First, tinnitus seems to be related to reorganization of primary auditory cortex as indicated in the shifts. The shifts have no preferred direction, e.g. the medial-lateral direction, suggesting an distortion rather than a mere shift of the tinnitus frequency into upper or lower frequency ranges of the auditory cortex. Second, in striking concordance with the results reported on phantom limb pain, tinnitus strenght and amount of cortical reorganization seem to be positively related.

These preliminary results need to be further substantiated. It would have been desirable to use the same frequencies in the patient and control group (e.g. by employing yoked control design). Moreover although the plastic changes were observed in auditory cortex, in the origin of the reorganization might be on the thalamic or lower levels.

The following hypothesis might be derived: The abnormal activity which is generated at the site of lesion (for instance, in the auditory nerve or in the cochlea) results in a loss of lateral inhibition in more centrally locted neural structures, which then in turn would generate the neural activity that underlies the sensation of tinnitus. After a period of extended abnormal input to these central structures, neural connections become strengthened, so that the abnormal activity continues, even after cessation of the abnormal input that originally caused the problem. This zone would be no longer be available for the representation of the tones at or near the vicinity of the tinnitus frequency. The representational zone for this frequency should consequently be displaced. Alternatively, one might argue that a distortion of the map has occured in the first place and tinnitus, as an auditory phantom perception, might be a consequence of such alterations.

This research was supported by the Deutsche Forschungsgemeinschaft (Fl 156/16; Ho 847/6) and is part of W.M.'s PhD thesis.

REFERENCES

[1] Flor, H., Elbert, T., Knecht, S., Wienbruch, C., Pantev, C., Birbaumer, N., Larbig, W. and Taub, E. Phantom-limb pain as a perceptual correlate of cortical reorganization following arm amputation, Nature, 1995, 375: 482-484.

[2] Hämäläinen, M.S. and Ilmoniemi, R.J. Interpreting measured magnetic fields of the brain. Estimates of current distributions, Report TKK-F-A559, 1984.

[3] Pantev. C, Eulitz, C., Elbert, T. and Hoke, M. The auditory evoked sustained field: origin and frequency dependence, Electroencephalography and clinical Neurophysiology, 1994, 90: 82-90.

[4] Flor, H., Rudy, T.E., Birbaumer, N., Streit, B. und Schugens, M.M. Zur Anwendbarkeit des West Haven-Yale Multidimensional Pain Inventory im deutschen Sprachraum, Der Schmerz, 1990, 4: 82-87.

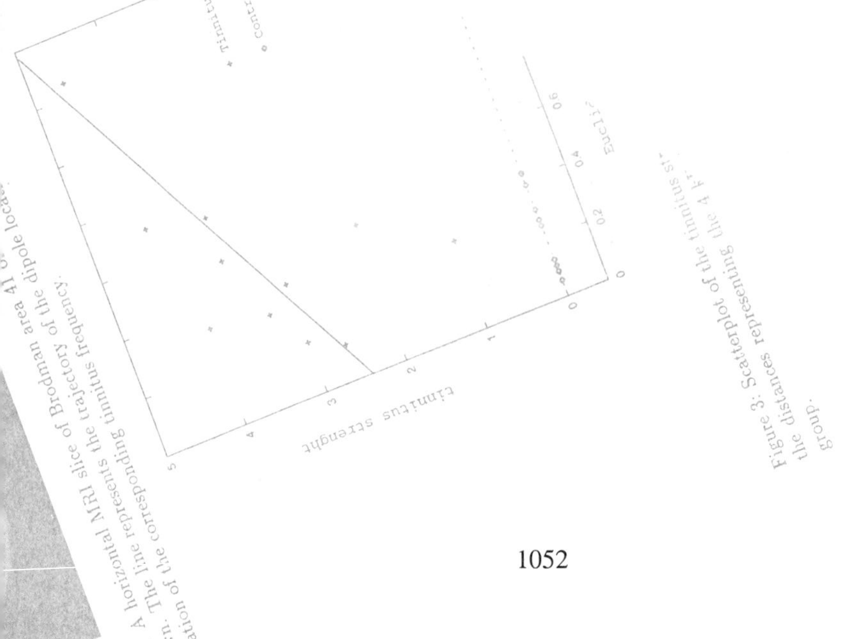

Figure 2: A horizontal MRI slice of Brodman area 41 of the dipole location. The line represents the trajectory of the tinnitus frequency. The corresponding tinnitus frequency location is shown.

Figure 3: Scatterplot of the tinnitus strength and the Euclidian distances representing the distances representing the 4 h group.

The Determinism of MEG in Tinnitus Patients

Mühlnickel, W.[1], Hoke, E.S.[2], Hoke, M.[2], and Flor, H.[1]

[1]*Department of Psychology, Humboldt-University of Berlin, Germany; [2]Institute of Experimental Audiology, University of Münster, Germany*

INTRODUCTION

Research during the last decade has shown that specific aspects of brain dynamics can be determined by using tools from nonlinear system's theory [1]. Based on a suggestion of Kaplan and Glass [2] it is possible to obtain information about the determinism of a time series even for short measurement intervals. The present study compares the determinism of the MEG in tinnitus patients with that of controls. Given that the auditory cortex is the final link in a chain producing tinnitus, then different patterns of brain dynamics may appear in the auditory cortex of patients. The MEG is ideally suited to measure the fluctuations in the tangentially oriented current flow of cortical areas. Therefore, we can expect to see differences in the determinism of the MEG between tinnitus sufferers and controls, when recorded over temporal regions.

THE DETERMINISTIC TEST OF A TIME SERIES

The goal of the presently introduced method for EEG analyses is to obtain information about the determinism of a time series, even for a short time interval. Kaplan & Glass [2] developed a direct test for determinism in a time series. This method is based on the fact that the tangent of the trajectory of a deterministic system is a function of the region in the phase space. The estimation is briefly described in the following steps:

- If the attractor is embedded in a d-dimensional phase space, then the phase space is divided into n^d boxes (where n is an integer)

- If a trajectory passes through a box j, then a vector v_{kj} (k pass) is constructed. This vector is defined by the coordinates of the starting-point and the end-point of that part of the trajectory inside the box j.

- The sum of all v_{kj} yields a vector $\mathbf{V}_j = \frac{1}{n_j}\sum_k \mathbf{v}_{k,j}$. The V_j characterizes the attractor.

- The average $\bar{L}_n^d = \langle \|\mathbf{V}_j\| \rangle_{n_j=n}$ over each box containing n passes, shows a statistic over all V_j with n passes (to gain deterministic pattern in a time series this last step was modified in so far, as $\bar{L}_n^d$ was calculated each time the trajectory passes a **given** box).

$\bar{L}_n^d = 1$ is valid only for deterministic systems with infinite time series and a box edge lenght $\to 0$. White noise falls off with $\bar{L}_n^d = n^{-\frac{1}{2}}$.

To gain deterministic patterns in a system the statistic term $L(d,n) = \bar{L}_n^d$ was replaced by a dynamic version $LD(d,n) = \bar{LD}_n^d$ [3]. The static version estimates $L(d,n) = \bar{L}_n^d$ after the complete reconstruction of the attractor. If, for example, the number of passes is $n = 5$, then in $L(d,5)$ there would be boxes with $n = 5$ passes. Another possibility would be to consider boxes that could contain more than $n = 5$ passes, meaning that the sum vector V_j was calculated subsequently for each pass. If one given box was passed $n = 20$ times, then this box was considered for each pass up to $n = 20$. The result for $LD(d,n)$ shows a gain of determistic patterns. Figure 1 shows the statistic term $\bar{LD}_n^d$ for a threedimensional embedding for different time series. The three time series have the same autocorrelation function and consited of 68,000 points. The delay time was $\tau = 80$ data points and the phase space was divided into $n = 16^3$ boxes.

As can be seen, the deterministic Glass-Mackey-System and the noise are identical. They differ if the embedding dimension increases, see Fig. 2.

Table 1 compares the statistic terms $\bar{L}_n^d$ and $\bar{LD}_n^d$ for an embedding dimension of $d = 7$ and a delay time of $\tau = 5$. The presented values were averages of the interval $u = [11,20]$ passes. Each time series contains 3,000 data points and their autocorrelation functions are identical. Table 1 includes results from

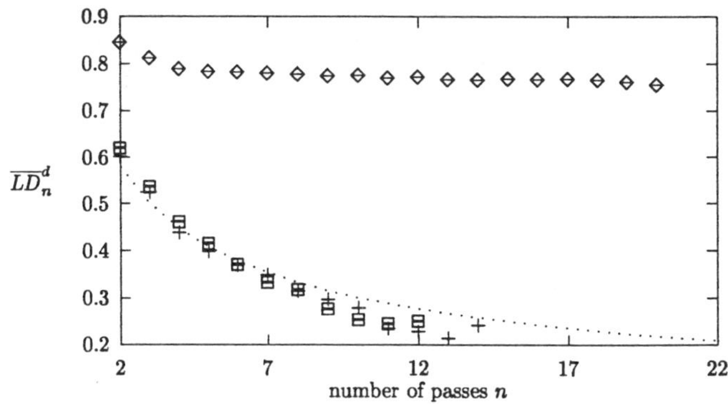

Figure 1:

Statistic term $\bar{LD}_n^d$ for a threedimensional embedding of: (a) $\diamond$ - the Lorenz-, (b) $\square$ - the Glass-Mackey- and (c) $+$ - a noisiform trajectory. The doted line represents the theoretically decrease of white noise - $1/\sqrt{n}$.

Table 1: Comparisons for the static and the dynamic version of the test for determinism.

Nonlinear system	Static $\bar{L}_n^d$	Dynamic $\bar{LD}_n^d$
Lorenz	.943 $\pm$.001	.977 $\pm$.001
van der Pol-type	.601 $\pm$.01	.779 $\pm$.007
Glass-Mackey	.542 $\pm$.002	.545 $\pm$.001
MEG (epileptic p.)	.338 $\pm$.021	.389 $\pm$.001
EEG (epileptic p.)	.402 $\pm$.030	.378 $\pm$.001
MEG (control)	.365 $\pm$.002	.381 $\pm$.001
filtered noise	.300 $\pm$.005	.303 $\pm$.002

time series of corresponding length generated by (a) the Lorenz attractor, (b) the van-der-Pol-like oscillator, (c) the Glass-Mackey equations with the result for (d) the noise and (e) the EEG and MEG-recordings measured over central locations (C_4). Two important outcomes are obvious: (1) a higher deterministic value for the MEG and for the EEG than for the white noise and (2) an increase in the dynamic patterns in a system in the dynamic version $\bar{LD}_n^d$. Furthermore, the results for the epileptic patient were even higher than those for a control subject, an observation that is in agreement with the literature [1]. MEG-analyses tend to produce higher values for determinism than EEG-based computations, a result that confirms the higher specificity of MEG.

METHOD

The data were taken from an investigation especially designed for the study of the contigent negative variation (CNV) in tinnitus patients [4]. In the present study dynamic aspects of these data were investigated. The estimation of the determinism of a given time series was chosen because this analysis requires only a short time series. 23 patients (tonal and noisiform tinnitus together) and 9 controls participated in the study. MEG signals were recorded from both temporal regions using a Magnes 37-channel system (sampling frequency 122 Hz). The test for determinism, adapted for the analysis of biosignals [3], was applied to separately examine the MEG for the three levels of the imperative stimulus (1,3 and 5 dB, resp.). The selected time interval of 10-13 s started 1 s before the onset of the warning stimulus (S1) and ended 1 s prior to the next S1. The embedding dimension was set to d = 7 and the delay time to t = 5 points. The values

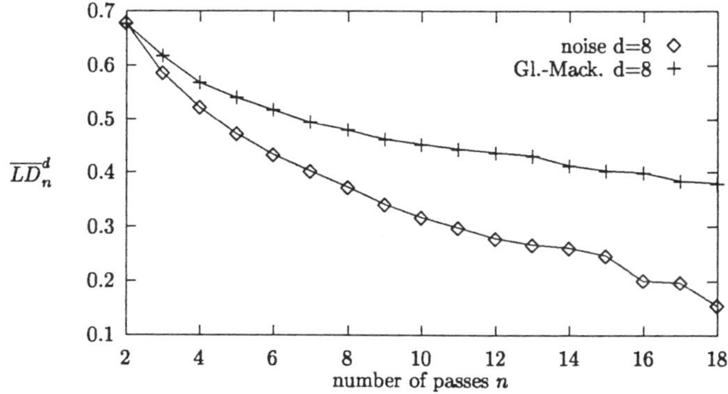

Figure 2:
Statistic term $\overline{LD}_n^d$ for a Glass-Mackey- and a noisiform trajectory. The embedding dimension was $d = 8$.

of $\overline{LD}_n^d$ were averaged within the interval of $u = [15, 25]$ passes.

RESULTS

Results averaged over five runs are presented in fig. 3. A greater value corresponds to a higher determinism of the system. The group difference of the averages is not significant (F(1,27)=2.4). As can be seen, however, the patient group displays a considerably larger variability than controls.

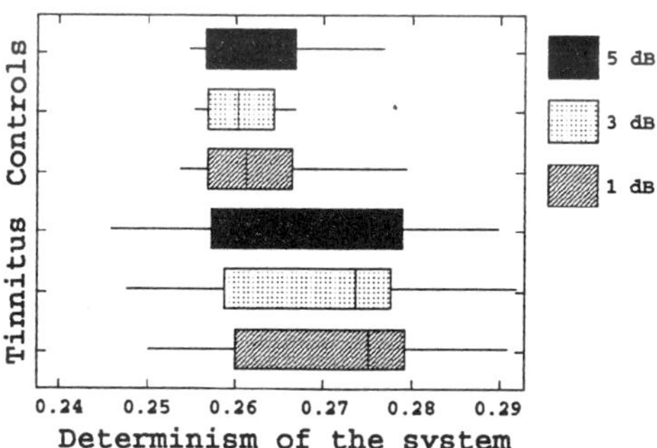

Figure 3: Boxplot of the averaged values of determinism.

But a trend can be seen in an ANOVA comparing the different levels of imperative stimulus (F(2,27)=3.16, p=.053). The lowest level (1 dB) of the imperative stimulus generates a more deterministic MEG than the higher level tones. This is consistent for both groups. Further analysis considering left or right ear tinnitus, the tinnitus frequency and the type of tinnitus were not significant.

DISCUSSION

One important result of the present study is the increased determinism of the MEG signal when the loudness of the imperative stimulus was decreased. This may indicate an increased attention of the subjects to the lower level tones. This differentiation between levels of loudness was not obtained when the spectral power was computed. This suggests that tools developed in the theory of nonlinear dynamics may yield important additional information on brain dynamics. Since the motor reaction to the imperative stimulus is an identical process to any level, the trend is an endogenous response reflecting the more difficult decision-making process in perceiving near-threshold tones.

In general, a larger variability of the deterministic measure of the MEG recorded over the auditory cortex in tinnitus patients than in the control group may indicate that different types of pathological processes are involved in the symptom of tinnitus. Therefore it could be of interest to see whether subjects with an atypical dynamical pattern can also be differentiated on the basis of traditional audiological and psychophysical methods. This investigation is currently under way.

This research was supported by the Deutsche Forschungsgemeinschaft (Ho 847/6) and is part of W.M.'s PhD thesis.

REFERENCES

[1] Elbert T., Ray W.J., Kowalik Z.J., Skinner J.E., Graf K.E., Birbaumer N. Chaos and physiology - Deterministic Chaos in excitable cell assemblies, Phys. Rev., 1994, 74: 1-47.

[2] Kaplan, D.T. and Glass, L. Direct Test for determinism in a time series, Phys. Rev. Lett., 1992, 68: 427-430.

[3] Mühlnickel, W, Rendtorff, N., Kowalik, Z.J., Rockstroh, B., Miltner, W. and Elbert, T. Testing the Determinism of EEG and MEG, Integrative Physiological and Behavioral Science, 1994, 29: 262-269.

[4] Proefrock, E.S. and Hoke, M. Contingent Magnetic Variation (CMV) studied with stimuli close to the hearing Threshold in normal subjects and tinnitus patients, In: Baumgartner, C., Deecke, L., Stroink, G. and Williamson, S.J. Biomagnetism: Fundamental Research and Clinical Applications, Amsterdam, Elsevier, 1995, pp. 234-239.

Auditory Signal Processing is Impaired in Alzheimer's Disease: A Whole-Head MEG Study

Pekkonen, E.[1,2], Huotilainen, M.[1], Virtanen, J.[1], Näätänen, R.[1], Ilmoniemi, R.J.[2] and Erkinjuntti, T.[3]

[1]*Cognitive Brain Research Unit, Department of Psychology, University of Helsinki, Finland;*
[2]*BioMag Laboratory, Medical Engineering Centre, Helsinki University Central Hospital, Finland;* [3]*Department of Neurology, Helsinki University Central Hospital, Finland*

Introduction

Auditory stimuli elicit in or near the primary auditory cortex a prominent magnetic response called N100m that is often preceded by the smaller P50m response [1, 2]. The N100m, which together the P50m reflects automatic stimulus processing, is larger and earlier over the contralateral than ipsilateral auditory cortex with respect to the ear stimulated [3, 4]. In addition, aging seems to delay ipsilateral, but not contralateral N100m in the auditory cortex [5].

The aim of this study was to investigate whether Alzheimer's disease (AD) impairs parallel auditory processing between the hemispheres. The 122-channel whole-head magnetometer offers a unique facility to measure activity over both hemispheres simultaneously without interference of vigilance changes.

Methods

Eleven patients with AD and 11 age-matched healthy subjects were studied with approval of the local Ethical Committee. The Mini-Mental State Examination (MMSE) [6] was performed to all AD patients and to control subjects over 60 years of age. The control subjects had no history of neurologic, psychiatric, or any other severe diseases. Subjects in each group had normal hearing and none of them used drugs affecting the central nervous system. In all patients, the head-MRI had been performed to exclude ischemic changes.

Recordings were performed in a magnetically shielded room (Euroshield Ltd.) and the subject sat under the helmet-shaped dewar with his/her head against the bottom of the instrument. The subject watched a silent video and was instructed not to attend to the 700-Hz tones presented monaurally. The stimulus intensity was adjusted to 60 dB over the subjective hearing threshold. Each stimulus block consisted of standard (80%) and randomly embedded deviant tones (20%) differing in duration, 50 and 25 ms, respectively. Stimulus blocks with 0.5- and 2.5-s interstimulus intervals (ISIs) were separately presented to the left ear. For 18 subjects, an additional stimulus block with 0.5-s ISI was presented to the right ear.

Auditory evoked fields (AEF) were recorded using a 122-channel whole-head magnetometer (Neuromag Ltd.). The analysis period was 750 ms, including a 150-ms prestimulus period. The recording passband was 0.03-100 Hz and the sampling rate 397 Hz. The electro-oculograms were recorded to exclude eye artefacts. Over 300 responses were averaged for the standard tone in each condition. Peak latency and amplitude values of the standard responses were measured from the channel showing the largest response. AEF sources were modelled as equivalent current dipoles (ECDs) found by a least-squares fit to the 2.5-s ISI data from a subset of 34 channels over each auditory cortex, separately. A spherical head model was used for the dipole fitting [7].

The MEG data were analyzed with the multivariate analysis of variance for repeated measures (MANOVA). Additional unpaired and paired *t*-tests were employed when appropriate.

Results

Figure 1 shows the evoked magnetic fields (AEFs) of one AD patient and one control subject to the standard tones in the 2.5-s condition. In each subject, the standard tone elicited a clear N100m response over both hemispheres, the responses being larger in the control subject. The ipsilateral, but not the contralateral, N100m is clearly delayed in the AD patient.

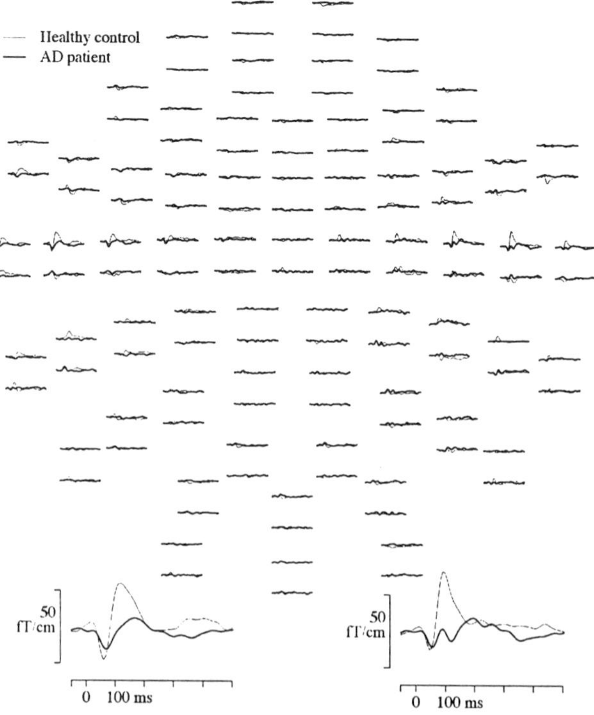

Fig. 1. Auditory evoked magnetic fields to the standard tones presented to the left ear in one healthy control subject and one AD patient with 2.5-s ISI. AEFs were recorded from 61 measurement sites, and in each pair the upper and lower AEFs show the latitudinal and longitudinal derivatives of the normal field component, respectively. Maximum responses over each auditory cortex of the hemisphere are shown enlarged. Note the delayed ipsilateral (left) N100m in the AD patient.

The ipsilateral P50m was significantly delayed in the AD group compared with the control group (MANOVA, F (1, 20) = 5.43, p < 0.05). Patients had significantly delayed ipsilateral P50m with 0.5-s ISI (unpaired t-test, p < 0.05), whereas the contralateral P50m latency was about the same in both groups.

AD patients had significantly delayed ipsilateral N100m to the standard tone (MANOVA; F (1, 20) = 14.56, p < 0.01). In contrast, the contralateral N100m latency was about the same in the two groups. In the 2.5-s condition, each response was significantly earlier contralaterally in both groups (P50m and N100m in each group, p < 0.001, paired t-test). No significant differences were observed in the amplitudes and the source locations of the P50m and N100m between the groups.

Discussion

This is the first MEG study demonstrating that ipsilateral, but not contralateral, auditory processing is impaired in patients with AD. It is proposed that impaired sound localization in AD reflects the dysfunction of the primary auditory cortex [8]. P50m and N100m, as previously stated, seem to be mainly generated in the primary auditory cortex [1]. Therefore, impaired parallel auditory processing between hemispheres might contribute to attenuated sound localization in AD.

In AD, there is both cortical and subcortical loss of neurons and accumulation of amyloid plaques. These degenerative changes also appear in the primary auditory cortices and in the subcortical auditory structures [9]. In the auditory system, a majority of the ascending fibers reach the contralateral auditory cortex, whereas fewer fibers reach the ipsilateral auditory cortex. If one assumes that degenerative changes in AD are rather symmetrical, then the smaller, ipsilateral auditory pathway will be first impaired, leaving the larger auditory pathway more intact.

In conclusion, MEG seems to be a useful tool to objectively study functional changes of the auditory system in AD.

Acknowledgements

This study was supported by the Academy of Finland and the Finnish Medical Foundation. We wish to thank Ms. Suvi Heikkilä for her important contribution in the measurements.

References

[1] Hari, R., Aittoniemi, K., Järvinen, M.-L., Katila, T. and Varpula, T. Auditory evoked transient and sustained magnetic fields of the human brain, Experimental Brain Research, 1980, 40: 237–240.

[2] Näätänen, R., Ilmoniemi, R.J. and Alho, K. Magnetoencephalography in studies of human cognitive brain function, Trends in Neuroscience, 1994, 17: 389–395.

[3] Reite, M., Zimmermann, T. and Zimmermann, E. Magnetic auditory evoked fields: interhemispheric asymmetry, Electroencephalography and Clinical Neurophysiology, 1981, 51: 388–392.

[4] Mäkelä, J.P., Ahonen, A., Hämäläinen, M., Hari, R., Ilmoniemi, R., Kajola, M., Knuutila, J., Lounasmaa, O.V., McEvou, L., Salmelin, R., Salonen, O., Sams, M., Simola, J., Tesche, C. and Vasama, J.-P. Functional differences between auditory cortices of the two hemispheres revealed by whole-head neuromagnetic recordings. Human Brain Mapping, 1993, 1: 8–56.

[5] Pekkonen, E., Huotilainen, M,. Virtanen, J., Sinkkonen, J., Rinne, T., Ilmoniemi, R.J. and Näätänen, R. Age-related functional differences between auditory cortices: A whole-head magnetic study, NeuroReport, 1995, 6: 1803–1806.

[6] Folstein, M., Folstein, S. and McHugh, P.R. "Mini-mental status." A practical method for grading the cognitive state of patients for the clinician, Journal Psychiatry Research, 1975, 12: 189–198.

[7] Hämäläinen, M., Hari, R., Ilmoniemi, R.J., Knuutila, J. and Lounasmaa, O.V. Magnetoencephalography - theory, instrumentation, and applications to noninvasive studies of the working human brain, Rev Modern Physics, 1993, 65: 413–497.

[8] Kurylo, D.D., Corkin, S., Allard, T., Zatorre, R.J. and Growdon, J.H. Auditory function in Alzheimer's disease, Neurology, 1993, 43: 1893–1899.

[9] Terry, R.D., Katzman, R. and Bick, K.L. Alzheimer disease, New York, Raven Press, 1994.

SEFs to Median Nerve and Finger Stimulation: Analysis of Interhemispheric Asymmetries in Healthy Subjects and Following Hemispheric Lesion

Pizzella, V.[1], Rossini, P.M.[2,5], Rossi, S.[2], Tecchio, F.[2], Lopez, L.[4] and Romani, G.L.[3,4]

[1]*Institute of Solid State Electronics, CNR, Rome, Italy;* [2]*Ospedale "Fatebenefratelli", Rome, Italy;* [3]*Institute of Medical Physics and* [4]*Institute of Advanced Biomedical Technologies, Gabriele D'Annunzio University, Chieti, Italy;* [5]*IRCCS "S. Lucia", Rome, Italy*

Introduction

A systematic investigation of the asymmetries between Equivalent Current Dipoles (ECDs) activated in the two hemispheres after left and right median nerve, thumb, and little finger electric stimulation has been performed on 20 healthy subjects and 10 patients with unilateral hemispheric lesion. We studied the location and strength of the ECDs activated in the primary somatosensory cortex contralateral to the stimulated side corresponding to the first cortical component of the Somatosensory Evoked magnetic Field (SEF), namely N20m. Aim of this work is twofold: i) define normal limits for the parameters of the ECD generating N20m component on normal population; ii) use these limits to look for sign of brain plasticity in patients following an hemispheric lesion. On this line a previous paper by ours was devoted to the analysis of interhemispheric asymmetries between ECDs elicited by median nerve stimulation [1]. In the present study we enlarged the sample of our previous control population and, in order to make the test more accurate, we added the stimualtion of single digits (thumb and little finger). In this way the "hand extension", that is the dimension of the parietal region receiving the sensory input from the contralateral hand, was also caluated. In fact, we calculated the "hand extension" as the distance - expressed in millimeters - of the generators responsible for the N20m cortical responses to separate stimulation of thumb and little finger in the contralateral hand.

Materials and Methods

The magnetic field distribution over the scalp following electrical stimulation of the hand was recorded in 20 healthy volunteers (9 males, 11 females, age range 27-60 years) and in 10 patients with unilateral hemispheric lesion. Among the 10 patients, 7 had a striatocapsular lesion, and 3 had a cortical lesion. All examined patients were in the chronic stage (between 2 and 7 months after the stroke), 7 had a complete recovery, while 2 evidenced residual hypoesthesia and the last one had a complete loss of hand sensation.

Measurements were performed using the 28-channel system [2], available at IESS-CNR in Rome. This system features 16 first-order axial gradiometers (1.8 cm coil diameter and 8 cm baseline) and 9 magnetometers (pick-up area 81 mm^2) plus 3 balancing magnetometers for noise cancellation coupled to low noise dc-SQUIDs, with an overall sensitivity of about 5-7 fT /Hz. The 25 measuring sites are regularly distributed on a spherical surface covering an area of about 180 cm^2. All the measurements were performed inside a magnetically shielded room (Vacuumschmelze GMBH).

Left and right little finger, thumb (via ring electrodes) and median nerve at wrist (via surface disks) were separately stimulated by 0.2 ms electric pulses with an interstimulus interval of 641 ms. Stimulus intensities were set at about twice the subjective threshold of perception for fingers and just above motor threshold for the median nerve. About 300 artefact-free trials were acquired (0.48 - 250 Hz bandwidth, 1 KHz of sampling rate) and averaged for each stimulation modality. The entire recording procedure lasted about 1 hour. The amplitude of SEFs was calculated for each channel with respect to a baseline level chosen as the mean value of the 5-15 ms post-stimulus epoch.

We analysed the initial deflection (N20m) of the SEFs, which is quite stable and repeatable, independent of subject attention and therefore more suitable for follow-up studies both in normals and in patients with hemispheric lesions. Since the brain activity at these post-stimulus latencies are exclusively generated by contralateral sources located close to the central sulcus, proper dewar positioning on the hemisphere contralateral to the stimulated side allowed to record the response with a single position of the recording system. The center of the dewar tail was placed over a scalp site approximately corresponding to the C3 and C4 positions of the 10-20 International EEG System. Exact dewar position with respect to the subject head was detected by using five coils positioned on nasion, the two preauricular points, vertex and inion, whose 3D-positions were digitized (Polhemus Isotrak) at the beginning and the end of the recording session.

ECD characteristics (spatial coordinates and strength) were calculated at 1 ms intervals in the 15-50 ms post-stimulus epoch using a moving dipole model inside an homogeneously conducting sphere. The localisation result was accepted only if the explained variance was above 90%. The ECD locations are expressed using a reference system built for each subject/patient according to the following steps (Fig. 1):

- XZ plane is defined by the vertex and the two pre-auricular points.
- Y axis passes through the nasion and hits normally the XZ plane defining in this way the origin. The nasion y value is assumed to be positive.
- Z axis goes from the origin to the vertex (positive toward the vertex).
- X axis is identified from the previous two using the right hand rule (positive toward right ear).

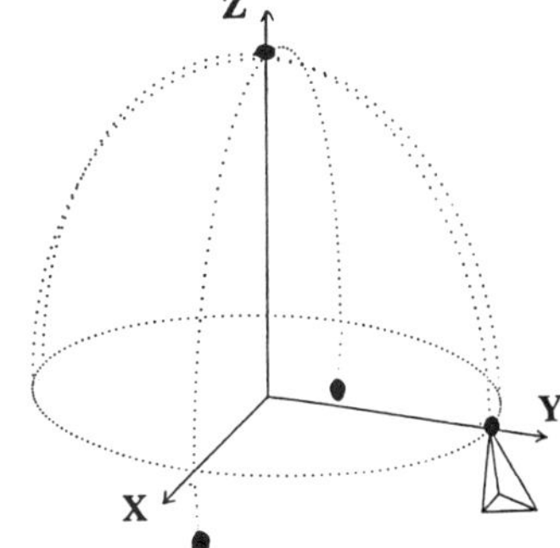

Figure 1 Used reference system. Anatomical landmarks are indicated with black a dot.

Results

For each normal subject the ECD locations relative to median nerve, thumb and little finger were found to be in agreement with the sensory homunculus mapping. The average ECD parameters, together with their standard deviation, relative to all stimulation modalities for the first cortical component (N20m) are reported in Table I.

Table I Average ECD location for N20m component in normal subjects

LEFT hemisphere					RIGHT hemisphere			
x (mm)	y (mm)	z (mm)	s (nA·m)		x (mm)	y (mm)	z (mm)	s (nA·m)
-44 ± 7	-6 ± 9	66 ± 7	6 ± 3	thumb	46 ± 7	-2 ± 6	70 ± 6	6 ± 4
-39 ± 6	-13 ± 6	69 ± 7	15 ± 5	median nerve	41 ± 6	-8 ± 6	72 ± 4	14 ± 7
-37 ± 7	-12 ± 9	77 ± 8	6 ± 5	little finger	36 ± 9	-9± 7	77 ± 8	5± 3

Four parameters were identified to define a normative data set:

loc: distance between the individual ECD and the average of the same ECD across all subjects for each stimulated district (Fig. 2a).

loc_asy: distance between homologous ECDs (i.e. thumb with thumb) in left and right hemispheres after mirroring left hemisphere into the right one, i.e. taking the same positive sign for the x coordinate (Fig. 2b).

hand: the "hand extension" calculated as the distance - expressed in millimeters - between the ECDs relative to thumb and little finger of the same hand (Fig. 2c).

hand_asy: the difference between left and right hand extension.

For each of the above parameters we identified the parameter interval including at least 98% of the normal population. In order to do so, the specific distribution of each parameter was considered. For instance the *loc* parameter is distributed as χ^2 with three degrees of freedom (one for each coordinate x, y, z). The *loc* values which can be considered normal are therefore included between 0 and 2.79×σ, where σ is the maximum value of the standard deviation among the three coordinates. Similarly for the *loc_asy* and *hand* parameters. The *hand_asy* parameter is normally distributed with zero mean (no difference has been found between left and right hand extension) and the normal range is simply $0.\pm1.96\times\sigma$.

The normal parameters calculated from the normative database are summarized in Table II. This table is to be used as follows: if, for instance, the *loc_asy* parameter for the thumb is greater than 27 mm, there is an "abnormal" asymmetry in the location of the left and right thumb ECDs. This is a sign of a possible brain plasticity. Only one set of normal intervals were identified for *loc* and *hand* parameters since left and right hemispheres were merged together.

Table II Normal parameter intervals

	normal parameter intervals				
	loc (mm)	loc_asy (mm)		hand (mm)	hand_asy (mm)
thumb	0 to 24	0 to 27		0 to 27	-13 to 13
median nerve	0 to 18	0 to 26			
little finger	0 to 26	0 to 32			

These parameter limits were used for the identification of possible brain plasticity in patients following an hemispheric lesion. In 8 out of the 10 studied patients we found that SEFs preserved their morphology, while in two patient no clear N20m response was found from the lesioned hemisphere. These patients were therefore excluded from the analysis. Moreover other four ECD location could not be succesfully identified (the explained variance was below 90%) leaving out a total of 44 successfully identified ECDs. In summary, in 5 cases out of the 10 studied, an abnormal value of one or more different parameter was found. In detail, the first two parameters evidenced the following results:

- subcortical lesions (5 cases): in 2 cases all the three districts (thumb, median nerve, little finger) showed "abnormal" values both for *loc* and *loc_asy* parameters, in one case only loc parametsr for median nerve was out of normality, while in the last two cases all values were within normal limits.
- cortical lesions (3 cases): 2 cases exhibit abnormal parameter values: one case had displacement of little finger ECD and the other of median nerve ECD, with a marked asymmetry respect to the healthy hemisphere.

The last two parameters evidenced the following results:

- subcortical lesions (5 cases): in 1 case the extension of the region reserved to the hand in the lesioned hemisphere was larger than normal. All other cases had parameter values within normal limits.
- cortical lesions (3 cases): in one case the hand extension was enlarged in both hemispheres.

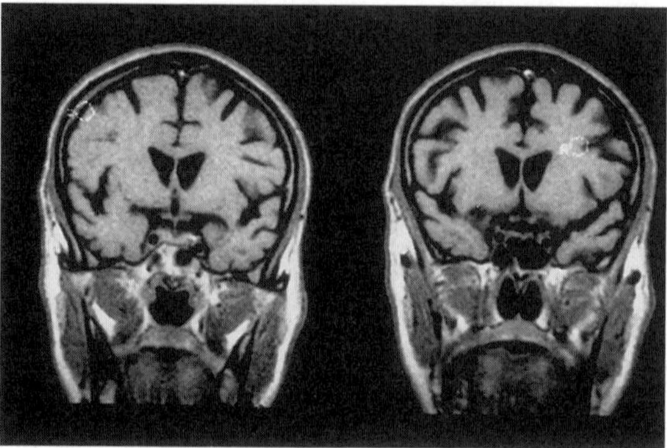

Figure 2 - Example of *loc_asy* "abnormal" value for thumb on a patient with subcortical left lesion. left) ECD location in the right hemisphere relative to left thumb stimulation; right) ECD location in the left hemisphere relative to right thumb stimulation.

Discussion

As it is known from pioneering Penfield's studies the extension of the somatotopical representation in SI of a given body district is proportional to the role it plays for sensory perception and is based upon the "density of cortical neurones directly connected with that body district" [3]. Additionally the short-latency generator sources strictly depict transient or stable reorganization of sensorimotor cortex following transient or stable sensory and motor deficits [4,5].

N20m ECD localisation for thumb and little finger showed the expected "homunculus" distribution [6]; this finding is consistent and repeatable at different times intra-subjectively, and it emphasises the reliability of the spatial ECDs coordinates as a parameter characterising the individual somatotopical organisation of the sensorimotor cortex.

Our findings show that the examined parameters are differently affected by subject's characteristics and instrumental errors. For what concerns the inter-subject variability of loc parameter, it may be ascribed not only to the variability of the sensory cortex somatotopy, but also to the differences in head dimensions as well as of the two central sulci position with respect to the anatomical landmarks. The instrumental error includes sensor noise, sensor position evaluation, limited spatial extension of the biomagnetic sensor and model inadequacy. Some of these elements are removed when absolute ECD coordinates are subtracted to each other. In fact, the loc_asy parameter is not affected by the left and right central sulci positions respect to the anatomical landmarks and the hand parameter is unaffected the sensor position. This justifies the use of several parameter when screening patients looking for brain plasticity.

At the present time, we only evaluated the cortical representation of the hand in the two hemispheres, while the analysis of possible relationships between the parameter "hand extension" and hemispheric "dominance" remained unexplored and beyond the scope of this investigation. Also work is in progress to include following components - such as P30m - and ECD strenght in the analysis to further improve the accuracy of the method.

In conclusion, we have defined the spatial properties of the neural generators embodied in the region of the primary sensory cortex devoted to hand in normals. Additionally, we showed that this non-invasive and safe technique permits to evaluate the dimension of the cortical region receiving sensory information from the hand. The properties of these generators have been used to screen patients following hemispheric lesion looking for sign of brain plasticity. This, in our opinion, will provide a unique opportunity also to follow-up reorganising processes in the sensory cortex somatotopy after unilateral hemispheric lesion and may help in the assessment of the subsequent rehabilitative therapy.

References

[1] Rossini, P.M., Narici, L., Martino, G., Pasquarelli, A., Peresson, M., Pizzella, V., Tecchio, F., and Romani G.L. Analysis of interhemispheric asymmetries of somatosensory evoked magnetic fields to right and left median nerve stimulation. Electroencephalogr. clin Neurophysiol., 1994, 91:476-482.

[2] Foglietti V, Pasquarelli A, Pizzella V, Torrioli A, Romani GL, Casciardi S, Gallagher WJ, Ketchen MB, Kleinsassers AW, Sandstrom RL. Operation of a hybrid 28-channels neuromagnetometer. IEEE Trans. Appl. Sup, 1993, 3: 1890-1893.

[4] Rossini, P.M., Martino, G., Narici, L., Pasquarelli, A., Peresson, M., Pizzella, V., Tecchio, F., Torrioli, G., Romani, G.L. Short-term brain plasticity in humans: transient finger representation changes in sensory cortex somatotopy following ischemic anesthesia, Brain Research, 1994, 642: 169-177.

[5] Flor, H., Elbert, T., Knecht, S., Wienbruch, C., Pantev, C., Birbaumer, N., Larbig, W., Taub, E. Phantom-limb pain as a perceptual corelate of cortical reorganization following arm amputation, Nature, 1995, 375:482-484.

[6] Narici, L., Modena, I., Opsomer, R.J., Pizzella, V., Romani, G.L., Torrioli, G., Traversa, R., Rossini, P.M. Neuromagnetic somatosensory homunculus: a non-invasive approach in humans, Neuroscience Letters, 1991, 121: 51-54.

Acknowledgments

We would like to thank Alberto Pasquarelli for helpful discussions and management, Iole Indovina for technical assistance during data acquisition, Patrizio Pasqualetti, Simone Rossi and Roberto Carluccio for statistical evaluations and preparation of the manuscript.

Effects of Acute, Oral Administration of Phenobarbital (1.6 mg/kg) to Healthy Subjects

Sannita, W.G.[2,3], Lewine, J.D.[1], Maclin, E.L., Orrison, W.W., Jr.[1] and Robinson, S.

[1]New Mexico Institute of Neuroimaging and Department of Radiology, UNM Medical School, Albuquerque, NM, USA; [2]Department of Psychiatry, State University of New York, Stony Brook, NY, USA; [3]Center of Neuropsychoactive Drugs, DISM-Neurophysiopathology, University of Genova, Genova, Italy

INTRODUCTION

The effects of barbiturates on the background EEG signal depend on the administered dose, with increments of fast frequency (peculiarly in the ~15.0-25.0 Hz range) or low frequency EEG activities following administration at small to moderate or at large doses respectively. The increment of fast frequency EEG activity is referred to as distinctive of barbiturates [1-3] and has been documented in healthy volunteers 30-120 min after single, oral doses Phenobarbital (PB) as low as of 50 to 100 mg [4,5], in spite of low drug plasma concentrations (below 6.9 µg/ml) and with pharmacokinetics and brain dynamics that were consistent with available evidence of early CNS bioavailability [6]. The ~15.0-25.0 Hz frequency interval appears to be a spectral parameter with peculiar sensitivity to PB, with substantial correlation between EEG effect and drug plasma concentration [4-5]. The power increase in this frequency segment occurs earlier and at lower doses on central electrodes, therefore allowing inference about possible generators

METHOD

Seven healthy volunteers (5 males and 2 females; age: 30-47 yr. [mean 38.7 ± 6.8 yr.]; body weight: 59-80 kg) were recruited among faculty and staff after excluding any history or clinical evidence of relevant pathology, allergy, or long-term drug use/abuse. Screening EEGs were normal, with the "alpha" rhythm being evident at the visual inspection for at least 60% of recording time in standard resting conditions. Subjects were aware of the study rationale and acquainted with equipment and data acquisition procedures; they agreed to avoid neuroactive compounds in the 72 h before experimental session. MEG signals (BTi System, 38 sensors; 3 antero-lateral locations of the probe) were obtained in baseline conditions and at regular intervals during the 2-h period following oral administration of PB, 1.6 mg/kg. Data acquisition was on the 1.0-100.0 Hz bandwidth, with sampling rate of 297.6 Hz. The raw signal was band-pass filtered at 13.0-23.0 Hz, and the best fitting dipole was determined on every one of 8-ms consecutive data segments by a sequential dipole fitting algorithm. For data analysis, the constrain was artificially set that the root mean square (RMS) signal strength recorded across all sensors at relevant time points be at least 100 fT to focus analysis on signals of relevant strength; in addition, 97% variance criteria were applied to isolate dipolar events under data conditions suggesting that a focal cortical area dominate the recorded signal. A temporal constrain allowed only one valid dipole fit each 40 ms of data, i.e. approximately one dipole per cycle of the beta activity waveforms.

RESULTS

An increase of current density and beta bursting was evident at anterior locations after PB (with drug plasma concentrations of 3.3 ± 0.8 µg/ml), though with limited pre- vs. postdrug differences in absolute power. Most MEG activity in the 13.0-23.0 Hz frequency interval could be modelled by dipoles with approximately 80-90% of variance being accounted for by the model. However, at several instances in time (especially in baseline conditions) the dipole model accounted for less than 80% of the variance, conceivably depending on erroneous

TABLE: number of dipole fits in baseline condition prior to and after administration of Phenobarbital (1.6 mg/kg)

subject	baseline (predrug)	postdrug (time: 30 min)
1	9	64
2	58	161
3	91	175
4	356	371
5	104	427
6	7	98
7	35	69
mean (sd)	94.3 (121)	195.0 (164.4) paired t-test: p < 0.01

source localization with a single dipole model or implying multiple generators. A reasonable dipole fit was obtained under the specific constraints involving temporal selection, RMS signal strength and variance criteria. The number of valid dipole fits increased in all subjects after PB (predrug: 94.3 ± 121; postdrug: 195.0 ± 164.4; paired t-test: $p < 0.01$). In the subjects whose 13.0-23.0 Hz activity in baseline was accounted for by a sufficient number of dipole fits, these were distributed in superior temporal and inferior frontal-parietal locations. After drug administration, the increase in the number of frontal-parietal dipole fits was preponderant compared to superior temporal locations.

DISCUSSION

These preliminary observations suggest that PB action results in increased number and stronger activation of multiple cortical generators of fast activity, particularly those that in baseline are at superior temporal and frontal-parietal locations. At therapeutic doses and with moderate plasma levels of barbiturate: i) the transmembrane resistance and threshold to direct stimulation are increased in <u>in vitro</u> models, while the resting membrane potential is virtually unchanged; these effects are conceivably due to inactivation of Na carrier mechanism, with reduction of Na conductance change, and are common to other compounds active on the CNS; ii) GABAergic inhibition is enhanced, with characteristics (i.e. enhancement of postsynaptic GABA-receptor current with binding at sites different from those of benzodiazepines and by means of regulation of different properties of the GABA-receptor channel) that may in part be peculiar of PB; and iii) barbiturates reduce presynaptic neurotransmitters release and affect Ca channels, including Ca-regulated T-currents involved in the generation of pacemaker currents in the thalamus. A dual role of $GABA_B$ receptors in excitation and inhibition of thalamocortical cells (conceivably mediated by Ca potentials) is also known and is suggested to play a role in the generation of 'burst' firing excitation [8-10]. The experimental conditions of the study and the preliminary results do not allow inference favouring any of these mechanisms of Phenobarbital action on structures/functions generating/modulating the spontaneous electrophysiological activities of the brain. Further studies with selective pharmacological manipulation of these mechanisms are needed for this purpose. Consistent with theory and experimental evidence, the peculiarities of MEG techniques potentially favour application in (human) neuropharmacology.

REFERENCES

1. Fink, M EEG and human psychopharmacology. Ann. Rev. Pharmacol. 1969, 9: 241-258,
2. Sato, H. Relationship between serum levels and fast EEG activities in rats by a single administration of Phenobarbital. Electroenceph. clin. Neurophysiol. 1980, 50: 509-514,
3. Schwartz, J., Feldstein, S., Fink, M., Shapiro, D.M. and Itil, T.M. Evidence for a characteristic EEG frequency response to thiopental. Electroenceph. clin. Neurophysiol. 1971, 31: 149-153

4. Sannita, W.G., Rapallino, M.V., Rodriguez, G. and Rosadini, G. EEG effects and plasma concentrations of Phenobarbital in volunteers. Neuropharmacology 1980, 19: 927-930

5. Sannita, W.G., Balbi, A., Giacchino, F. and Rosadini, G. Quantitative EEG effects and plasma concentration of Phenobarbital, 50 mg and 100 mg single-dose oral administration to healthy volunteers: evidence of early CNS bioavailability. Neuropsychobiol. 1990, 205-212

6. Viswanathan, C.T., Booker, H.E. and Welling, P.G. Pharmacokinetics of Phenobarbital following single or repeated doses. J. clin. Pharmacol. 1979, 3: 382-389

8. MacDonald, R. Antiepileptic drug actions. Epilepsia 1989, 30:19-28

9. Miller, J.W. and Ferrendelli, J.A. Characterization of GABAergic seizure regulation in the midline thalamus. Neuropharmacology 1990, 29: 649-655

10. Crunelli, V. and Leresche, N. A role for $GABA_B$ receptors in excitation and inhibition of thalamo-cortical cells. Trends Neurosci. 1991, 14: 16-21

Visual Evoked Fields for Pattern Reversal Stimuli in Patients with Occipital Lobe Lesions

Seki, K.[1], Nakasato, N.[1,2], Fujita, S.[3], Kanno, A.[2], Hatanaka, K.[4], Kawamura, T.[1], Ohtomo, S.[1], Fujiwara, S.[1,2], Takahashi, A.[2] and Yoshimoto, T.[1]

[1]Department of Neurosurgery, Tohoku University School of Medicine, Sendai, Japan; [2]MEG Laboratory, Kohnan Hospital, Sendai, Japan; [3]R&D Center, Osaka Gas Co. Ltd, Osaka, Japan; [4]KRI, Kyoto, Japan

Introduction

The "cruciform model" of the visual cortex suggests that the P100 generators of the visual evoked responses to pattern reversal (PR) stimuli are located throughout the entire striate cortex, including the interhemisphere surface and calcarine fissures. Our recent studies of visual evoked fields (VEFs) to PR stimuli [1-3] have localized the P100m sources near the lateral end of the calcarine fissures. We suggested that a smaller part of the striate cortex contributes to the P100m response than in the cruciform model. In the present study, we measured PR-VEFs in patients with homonymous hemianopsia due to unilateral occipital lobe lesions, to demonstrate the correlation between P100m patterns and occipital lesions.

Methods

Five patients, 4 males and a female, aged 17 to 68 (mean 42.6) years old, with occipital lesions (four arteriovenous malformations [AVMs] and one brain tumor) participated in this study. All patients had homonymous hemianopsia due to the lesions. Informed consent was obtained after the experimental procedures were explained.

Green-black checkerboard PR stimuli, reversed once per second, were sequentially delivered through a fiber optics system to the full-visual fields. Patients were instructed to watch the fixation point on the screen monocularly. Individual checks and full-field size subtended visual angles of 1.4 and 17 degrees at the patient's eye, respectively. Mean luminance level of the green checks was about 20 cd/m², with a contrast greater than 98% to the black checks.

VEFs were measured using a helmet-shaped MEG system (CTF - Osaka Gas) [1-3]. The MEG signals were filtered with 50 and 100 Hz notch filters and direct current to 100 Hz low pass filters and digitized at 625 Hz. The resulting data were averaged based on 200 PR presentations, including a 200 ms pre-trigger baseline and 300 ms after the PR. The P100m peak in the averaged signal was identified within a latency window of 90 to 140 ms after the PR. When a clear two-dipole pattern appeared, the source location was estimated using a two-dipole model. When a one-dipole pattern appeared, the source location was estimated using a one-dipole model. All subjects also underwent three-dimensional MR image. Three fiduciary points, nasion and preauricular positions were marked with small coils for MEG and small oil-containing capsules for MRI. These fiduciary points were used to coordinate the localized dipole position and the MR imaging anatomy.

Results

The P100m peaks of PR-VEFs were observed in all patients (Table 1). In four patients, a one-dipole pattern of P100m appeared over the normal occipital area, corresponding to their homonymous hemianopsia. The P100m dipole positions were estimated at the lateral end of the calcarine fissures in the normal hemisphere (Fig. 1).

In the other patient (*Case 4*), a two-dipole pattern of P100m appeared over the bilateral occipital areas, contradictory to his homonymous left hemianopsia. In this patient, the lateral part of the calcarine fissure was spared where the P100m dipole was localized; only the medial part of the calcarine fissure was occupied by the AVM (Fig. 2).

Visual evoked fields in patients with occipital lobe lesions.

Case No.	Age /Sex	Diagnosis /Side	Side of homonymous hemianopsia	P100m latency	Map pattern /Side
1	31/F	AVM /Rt	Lt	105.6 ms	One-dipole /Lt
2	48/M	AVM /Lt	Rt	91.2 ms	One-dipole /Rt
3	17/M	AVM /Lt	Rt	112.0 ms	One-dipole /Rt
4	49/M	AVM /Rt	Lt	116.8 ms	Two-dipole /Bilateral
5	68/M	Brain tumor /Rt	Lt	128.0 ms	One-dipole /Lt

AVM: brain arteriovenous malformation

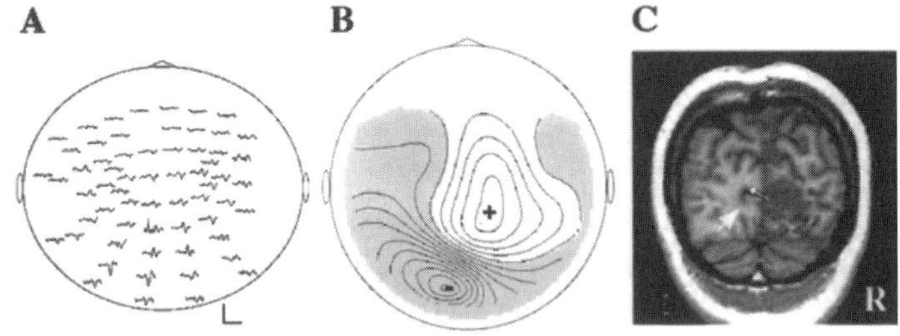

Fig. 1: Visual evoked fields for the pattern reversal full-field stimuli in *Case 1*.
A: All 64 channel data are shown as a view from above. Calibrations are 400 fT and 200 ms.
B: Isofield map of the P100m indicates a one-dipole pattern over the normal left occipital area (15 fT/step).
C: The P100m dipole is estimated near the lateral end of the left calcarine fissure (arrow) on the coronal slice of MRI. Note that the large AVM occupies the right calcarine fissure. Calibration is 1 cm.

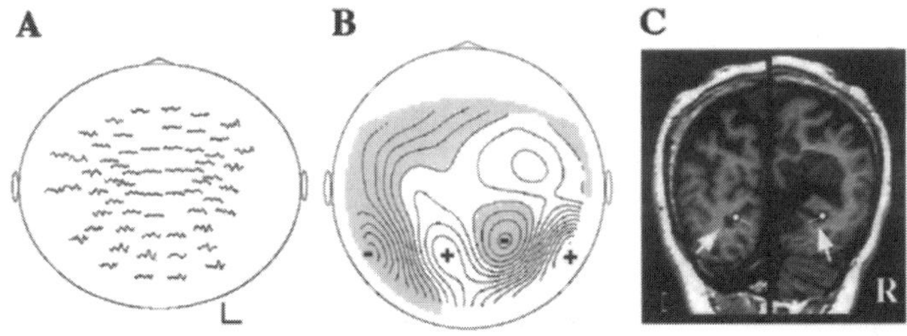

Fig. 2: Visual evoked fields for the pattern reversal full-field stimuli in *Case 4*.
A: All 64 channel data are shown as a view from above. Calibrations are 400 fT and 200 ms.
B: Isofield map of the P100m indicates a two-dipole pattern over the bilateral occipital areas (10 fT/step).
C: The P100m dipoles are estimated near the lateral ends of the bilateral calcarine fissures (arrows) on the coronal slice of MRI. Note that the AVM occupies medial part of the right calcarine fissure, and that the lateral part is spared where the P100m dipole is localized. Calibration is 1 cm.

Discussion

Our recent studies [2] have demonstrated that a two-dipole pattern of the P100m appears over the bilateral occipital areas due to PR stimuli of the full visual fields in normal subjects. The present study found a unilateral one-dipole pattern in four patients with homonymous hemianopsia. Although we showed the lesion-side P100m disappeared due to occipital lesions, it is difficult to confirm the specific cortical area generating P100 responses. However, the other patient had the normal two-dipole pattern, despite homonymous hemianopsia due to a large AVM occupying the medial part of the calcarine fissure and the interhemispheric occipital cortex. We believe this finding supports our hypothesis that the P100m is generated by the lateral part of the calcarine fissure only. PR-VEF may not completely represent the retinotopy of the visual cortex, because of the discrepancy between bilateral P100m responses and homonymous hemianopsia. Further studies are necessary to confirm the cortical origin of PR-VEFs.

References:

[1] Nakasato, N., Seki, K., Fujita, S., Kanno, A., Kawamura, T., Ohotomo, S., Fujiwara, S., and Yoshimoto, T. Functional localization of visual cortex using an MRI-linked whole head MEG system, Human Brain Mapping, 1995, Suppl 1: 63.

[2] Nakasato, N., Seki, K., Fujita, S., Hatanaka, K., Kawamura, T., Ohtomo, S., Kanno, A., Ikeda, H., and Yoshimoto, T. Clinical application of visual evoked fields using an MRI-linked whole head MEG system, Frontiers of Medical and Biological Engineering, in press.

[3] Seki, K., Nakasato, N., Kawamura, T., Fujita, S., and Yoshimoto. T. Source localization of P100m in pattern reversal visual evoked magnetic response. Japanese Journal of Electroencephalography and Electromyography, 1994, 22: 369-374.

Acknowledgements

This work has been supported in part by Grants-in-Aid for Scientific Research No. 03404042 and No. 06404050 form the Ministry of Education, Science and Culture of Japan; and Magnetic Science Foundation.

MEG and fMRI Localisation of Primary Sensory-Motor Cortex in Normals and a Patient with a Low-Grade Astrocytoma

Singh, K.D.[1], Furlong, P.L.[1], Weale, P.[2], Harding, G.F.A.[1] and Walsh, A.R.[2]

[1]Clinical Neurophysiology Group, Dept. of Vision Sciences, Aston University, Birmingham, UK; [2]Queen Elizabeth Medical Centre, Birmingham, UK

Introduction

The aim of this study was to compare MEG and fMRI localisations of primary sensory-motor cortex, in normal subjects and patients scheduled for neurosurgery. The study is a precursor to the development of a pre-surgical mapping program. The results indicate that in both normals and patients, fMRI and MEG localisations are confirmatory and match well with localisations obtained during surgery. However, in one subject, the MEG localisations were displaced significantly from the central sulcus. We discuss the implications and possible generating mechanisms of this displacement.

Methods

MEG Acquisition

The subjects were six healthy control volunteers and one surgical candidate. The patient was a 44 year old right handed male who presented with a history of tonic-clonic seizures and partial seizures involving myoclonic jerks of the right hand and arm. An MRI scan revealed what appeared to be a large cystic tumour in close proximity to the sensory-motor cortex. The seizures were partially controlled with anti-convulsants but the seizure frequency remained unacceptable. A surgical option was offered.

The proximity of the lesion to the sensory-motor cortex was a cause for concern and in order to investigate the precise juxtaposition of the functional anatomy, Magnetoencephalography (MEG) and Functional Magnetic Resonance Imaging (fMRI) were performed. The same procedures were also performed in the control volunteers for comparison.

MEG evoked signals were measured using a 19 channel magnetometer in response to median nerve and posterior tibial stimulation using surface mounted electrodes. The stimulus duration was 0.1ms with a 600ms inter-stimulus interval. In each measurement, 250 epochs were collected with the dewar optimally positioned over the relevant portion of the head. The data collection rate was 2kHz for an epoch length of 64ms - 128ms. Analogue filters were used with a lowpass range of 0-424Hz. In addition, noise rejection magnetometers were used for spatial filtering. After collection and averaging, digital filtering was used with a bandpass of 2-250Hz.

Precise head location with respect to the magnetometer was achieved using bitebar and head based fiducial markers, which contained oil filled reservoirs for MRI co-registration. MEG source localisations were performed using a single equivalent current dipole model. Once a valid localisation had been obtained at a particular latency, Monte-Carlo analyses [1] were used to assess the estimated cortical region of activation as a 3D ellipsoid [2]. These ellipsoids were then displayed on appropriate MRI views.

Functional MRI

fMRI data were collected using a standard Siemens 1.5T clinical scanner. A FLASH imaging protocol was used (TE = 60 msec, TR = 75 msec, Flip Angle = 40 degrees). The acquisitions were made in a similar manner to those of Jack et al. [3].

The activated state for the subject was finger-thumb opposition, which contains both sensory and motor stimulation. In each run, 10 rest images and 10 activated images were acquired, each image consisting of seven seconds of rest or stimulation. The runs were interleaved in blocks of 5.

Functional data was collected from 4 slices, each 5mm thick, and positioned to encompass the region of interest. The functional data was analysed by subtraction of the averaged rest image from the averaged activated image, and thresholding to leave only high intensity pixels in the functional image. This final image was then superimposed on a high resolution (1mm x 1mm x 1.5 mm) anatomical MRI. Co-registration of function and anatomy was automatic, as the anatomical image was acquired in the same run as the functional image.

Results

MEG Evoked Responses

In the case of the patient, stimulation of the right median nerve at the wrist yielded a recognisable evoked response (Fig. 1a) which was normal in latency and amplitude in comparison with the control group. Two clear peaks of activation were noted at 21ms and 40ms and single equivalent dipole source localisations were calculated at these latencies. Ellipsoid confidence volumes for the source localisations were superimposed onto the MRI (Fig. 1b). The white grid lines were derived from the co-ordinate system of Talairach and Tournoux [4] with the grid origins passing through the anterior and posterior commissures.

Data from stimulation of the right posterior tibial nerve was computed in the same manner at two latencies (43 and 78ms) and yielded similar findings (Fig. 2a and Fig. 2b). Both datasets implicated a location on the posterior bank of the lesion for the site of primary sensory-cortex. This is confirmed by the source localisations being between the two commissural axes of the Talairach and Tournoux co-ordinate system.

Functional Magnetic Resonance Imaging

Analysis of pixel image intensity in relation to the activation sequence revealed areas of significant activation in the sensory-motor cortex region in three control subjects and the patient.

In the control subjects, the fMRI locus was either adjacent to the MEG localisation or anterior to it, as seen in Fig. 3a. In the patient data (Fig. 3b) the MEG localisation is clearly anterior to the fMRI localisation.

Surgical validation

A large craniotomy was performed over the left fronto-tempero-parietal regions to expose both the lesion and the expected area encompassing the sensory-motor cortex. Once the cortex was exposed, the patient was aroused from general anaesthesia to respond to direct electrical stimulation of the cortex. A bipolar probe delivering an alternating current at 50Hz was used. Verbal feedback to sensory stimulation and observation of motor signs enabled mapping of the cortex which was labelled with numeric indicators. For validation purposes, the surgical area was photographed and the gyral patterns compared with 3 dimensional surface rendered images of the cortex derived from the MRI (Fig. 4b). Using this technique, the Rolandic fissure was determined as the first sulcus posterior to the lesion and the appropriate landmark is indicated by the bold black lines in Fig. 4a and Fig. 4b.

Histological section of the lesion determined it to be a low grade Astrocytoma. This was successfully excised and the patient made a complete recovery with no permanent neurological deficit.

Discussion

In this work we have found that in the control subjects, 95% confidence ellipsoids for the MEG dipole fits for median and posterior tibial stimulation were visually identified to be close to the expected location of the central sulcus. fMRI data yielded more widespread activation, but the primary activity was again close to the central sulcus and was therefore consistent with the MEG localisations.

In a single patient, the MEG localisation was displaced anteriorly from the fMRI localisation. This was confirmed by intra-operative cortical mapping which suggested that the MEG data was displaced anteriorly by 8mm from the sensory cortex, for both median and posterior tibial nerve data. In this individual case, therefore, fMRI appeared to be a more accurate predictor for the location of primary sensory-motor cortex.

A possible explanation for this displacement is that the Astrocytoma was a large conductivity perturbation and so the true conductivity profile may not be the homogeneous sphere model used in the source localisation algorithm. If this is so, this has important implications for the use of MEG/EEG pre-surgical mapping in subjects who have large tumours (for example) close to the region of interest. It may be that in these circumstances, realistic conductivity models of the head and any lesions may be required. Alternatively, the displacement might reflect cortical re-organisation of primary sensory-motor cortex which has occured during the development of the Astrocytoma.

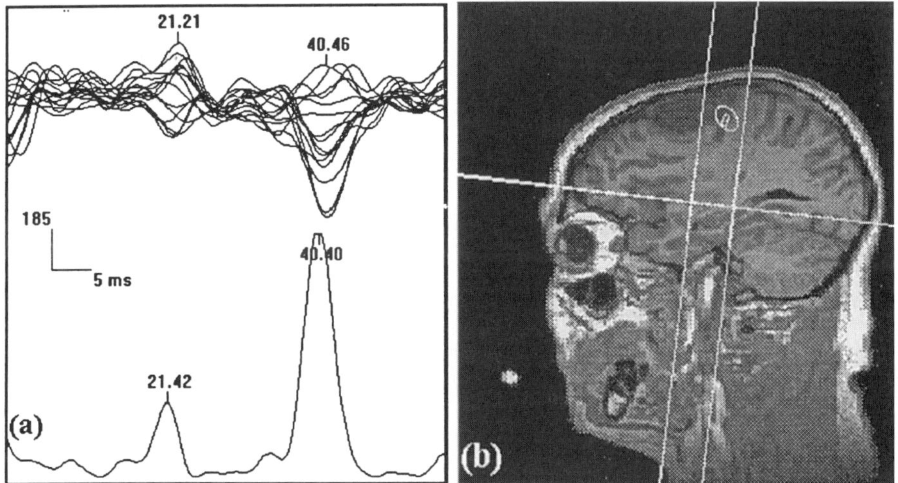

Fig. 1a *(left upper) shows the MEG evoked responses, from the patient, following Right Median nerve stimulation. The lower trace shows the integrated Global Field Power of the same dataset, indicating two latencies of maximum activation.* Fig. 1b *(right) shows the 95% confidence ellipses for source localisations at the two latencies indicated. The white grid lines show the axes of the co-ordinate system of Talairach and Tournoux.*

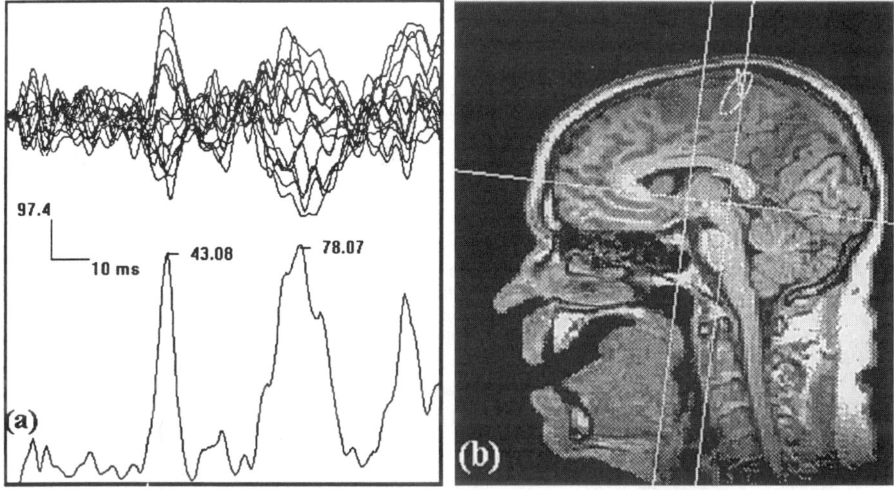

Fig. 2a *(left upper) shows the MEG evoked responses, from the patient, following Right Posterior Tibial nerve stimulation. The lower trace shows the integrated Global Field Power of the same dataset indicating two latencies of maximum activation.* Fig. 2b *(right) shows the 95% confidence ellipses for source localisations at the two latencies indicated. The white grid lines show the axes of the co-ordinate system of Talairach and Tournoux.*

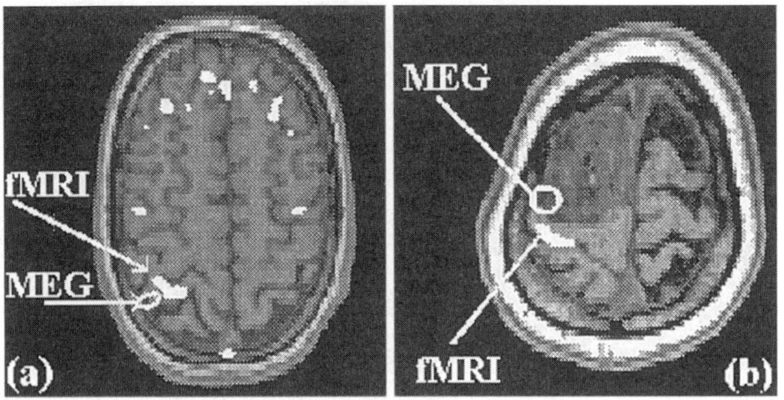

*Fig. 3a (left) indicates the location of maximal fMRI activation following a finger-thumb opposition sequence, together with the MEG localisation following Median nerve stimulation, in a control subject. **Fig. 3b** (right) shows the corresponding dataset from the patient.*

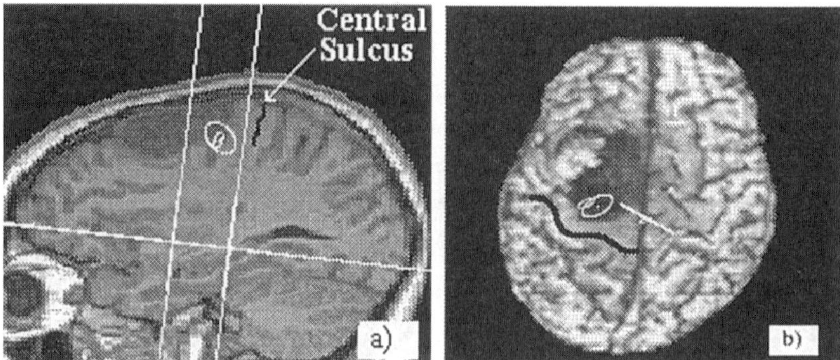

*Fig. 4a (left) shows the position of the central sulcus as determined by peroperative cortical stimulation (bold black line) together with the MEG localisation following Median nerve stimulation (white ellipses). **Fig. 4b** (right) shows both Median nerve and Posterior Tibial nerve MEG localisations superimposed onto a 3D surface rendered cortical image with the central sulcus indicated by the bold black line.*

References

[1] Medvick, P., Lewis, P.S., Aine, C.J., and Flynn, E.R. Monte Carlo analysis of localisation errors in magnetoencephalography. In: Williamson, S.J. et al (Eds), 1990, Advances in Biomagnetism. Plenum Press, New York, 543-546.

[2] Singh, K.D., Harding, G.F.A. This Volume.

[3] Jack, C.R., Thompson, R.M., Butts, R.K., Sharbrough, F.W., Kelly, P.J., Hanson, D.P., Riederer, S.J., Ehman, R.L., Hangiandreou, N.J., and Cascino G.D. Sensory motor cortex: correlation of presurgical mapping with Functional MR imaging and invasive cortical mapping. Radiology, 1994, 190:85-92.

[4] Talairach, J., and Tournoux, P., Co-planar stereotaxic atlas of the human brain. Stuttgart, Thieme, 1988.

Comparison Between the Processing of Static Images of Human Faces in High Functioning Autistic Subjects and Normal Controls

Swithenby, S.J.[1], Bräutigam, S.[1], Bailey, A.J.[2], Jousmaki, V.[3] and Tesche, C.[3]

[1]The Open University, Department of Physics, Milton Keynes, MK7 6AA, UK; [2]MRC Child Psychiatry Unit, 16 De Crespigny Park, London SE5 8AF, UK; [3]Low Temperature Laboratory, Helsinki University of Technology, 02150 Espoo, Finland

1 Introduction

We have studied neural processing of photographs of faces by normal and autistic individuals using the Neuromag 122^{TM} channel helmet system.

Autism is a severe neurodevelopmental disorder that involves a variety of fundamental cognitive, psychological and neurobiological abnormalities and is not associated with a characteristic brain lesion. The abnormalities which tend to occur in an autistic individual are;

- deficits in social interactions, specifically a lack of social reciprocity

- a delay in language development and continuing communication problems

- a tendency to stereotyped and repetitive patterns of behaviour.

Studies on autistic subjects using other techniques have revealed deficits in face processing abilities including impairments in recognition of emotional expression and identity (for recent reviews [1], [2]).

Previous studies on normal subjects with both MEG and EEG have revealed face specific activation in the extrastriate cortex in association with presentation of images showing faces and non-face objects[4,5]. Guided by these observations we developed four experimental tasks to probe face specific responses in normal subjects and possible abnormalities in autistic individuals, specifically in the early stages of neural processing of images. The subjects had to;

1. identify faces (or control images) in a sequence of randomly presented images (adult faces, motor bikes, abstract and dot patterns, familiar objects and animals).

2. identify identical faces (or control images) amongst a series of image pairs (boys' faces, wallpaper and dot patterns, motor bikes and mugs).

3. identify neutral (smiling) faces in a series of faces.

4. identify the same individual amongst pairs of faces expressing different emotions.

In each task, stimuli were followed by a visual cue requesting the subject to respond by pressing an appropriate key on a non-magnetic keypad. The combination of visual cue and keypad response was used to control attention conditions, to record behavioural data and to separate motor activity from visually evoked responses.

2 Method

Subjects
10 normal (6 males and 4 females, ages between 24 and 50) and 4 autistic (3 males and 1 female, ages between 24 and 35) subjects were tested. The autistic subjects were classified as high functioning and recruited from Great Britain. No significant attempt has been made to match these two groups with respect to age, gender and psychological measures, such as verbal or non-verbal IQ.

MEG

All experiments were carried out at the Low Temperature Laboratory in Helsinki using a 122 whole-head system. Data was sampled at a rate between 370 and 400 Hz, with an anti-aliasing filter set at 130 Hz. One-line averaging (after artefact rejection) was performed for the different stimulus categories within a given task. Prior to further analysis, the data sets were clipped in time range, baseline corrected and digitally filtered (0.3 - 45 Hz).

Task Protocols

Image presentation, trigger signals for the data acquisition system, and driving of a non-magnetic keypad to record subject responses were controlled by a laptop computer and custom software. All images were presented as grey-scale 8-bit bitmaps via a projection monitor. The images were controlled for luminance, contrast and size. The diagram below shows one cycle in the procedure for Task 1 (similar timings and sequences were used for the other tasks)

$$\text{Image} \longrightarrow \text{Pause} \longrightarrow \text{Prompt Cue} \longrightarrow \text{Response}$$

100ms   1s +/- 200ms   200ms   max 1.8s

Analysis

MEG data were analysed in several steps comprising 1) visual inspection of channel output $S_i(t)$, 2) tabulation of the statistics of latencies and amplitudes of easily recognisable peaks, 3) visual inspection of signal topography via rotated gradient plots, and 4) examination of the statistics of integrated power summed over groups of channels. Behavioural data (correctness of key-presses and reaction times) were analysed by standard statistical methods. Integrated power ratios for a given interval are defined as

$$IPR = \frac{\Sigma_{i=1}^{N} \int_{t_1}^{t_2} S_i^2(t)dt}{\Sigma_{i=1}^{N} \int_{-b}^{0} S_i^2(t)dt}$$

where N denotes the number of channels in a given channel group, and b determines the stretch of data used for baseline correction. All integrals were numerically estimated by a Gill-Miller algorithm. The channel groups were carefully selected to sharpen the power content over a given brain region without blurring the analysis due to overlapping areas of activation.

3   Results

Behavioural Data

Task performance, as measured by correct responses, was essentially perfect for normal subjects in all tasks. The four autistic subjects completed tasks 1 and 2 equally well. However, their performance on tasks 3 and 4 (tasks involving classification of emotion) was significantly impaired. There are no simple and clear trends in the reaction times

Qualitative MEG Observations

In normal subjects, there is a clear and characteristic face specific response in the right occipito-temporal region at 130 to 140 ms. This response appears to be independent of the specific nature of a task, as shown in Fig. 1a. This observation holds true in 8 out of 10 normal subjects. One subject showed an equivalent signal pattern but in the left occipito-temporal region. The signal topography, given by rotated gradient plots, between 130 and 140 ms is similar across this subject group but has varying degree of complexity. In most subjects, the dominant current contribution is consistent with activation in the fusiform gyrus. The averaged signals did not reveal any significant evidence for face specific activity at other latencies.

In our 4 autistic subjects, qualitative observations failed to show any clear evidence for face specific responses. However, evoked responses, clearly distinguished from noise, were observed in the right occipito-temporal region in all cases (see Fig. 1b).

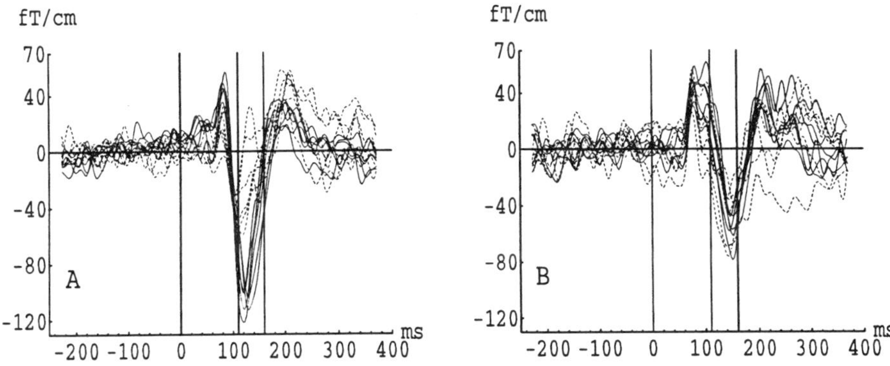

Figure 1: Face (solid) versus non-face (dashed) responses across all tasks for a normal, A, and an autistic, B, subject.

Quantitative MEG Observations

Summing signal power of a channel group sharpens specific features in the evoked responses (Fig 2). In the right channel group alone, there is a clear peak at approximately 130 ms, denoting the face specificity. Integrating over this peak (between 110 and 160 ms) provides a numerical measure which we have computed for all subjects. Fig. 3 shows relevant cohort means. There is a clear difference between face and non-face responses in normal but not autistic subjects.

An analysis of variance[1] for the data shown in Fig. 3 confirmed the clear distinction between face and non-face stimuli ($F_{3,36} = 4.3$, $p = 0.01$) in normal subjects. This pattern was further confirmed by the task 2 data, but with slightly less significance ($F_{2,18} = 3.3$, $p < 0.07$). This distinction could not be established for either task in the autistic subjects ($F_{3,12} = 0.77$, $p > 0.5$ and $F_{2,9} = 0.16$, $p > 0.8$). In both subject groups, in task 1, there is an hemispherical asymmetry in the response to faces (normals : $F_{1,14} = 4.5$, $p \approx 0.05$; autistics : $F_{1,6} = 3.8$, $p < 0.1$).

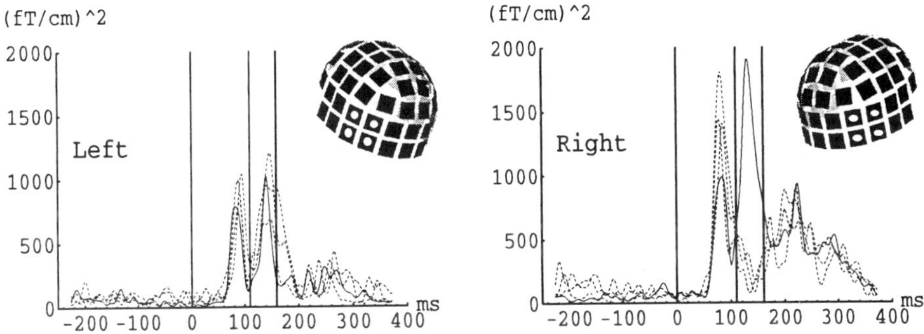

Figure 2: Channel-summed signal power in a normal subject (face:solid, non-face:dashed). The insets show the channel groups used for summation.

---

[1] Skewness was acceptable in most cases. In cases with higher skewness, logarithmic transformation did not alter the statistics significantly and analysis was performed on the raw ratios

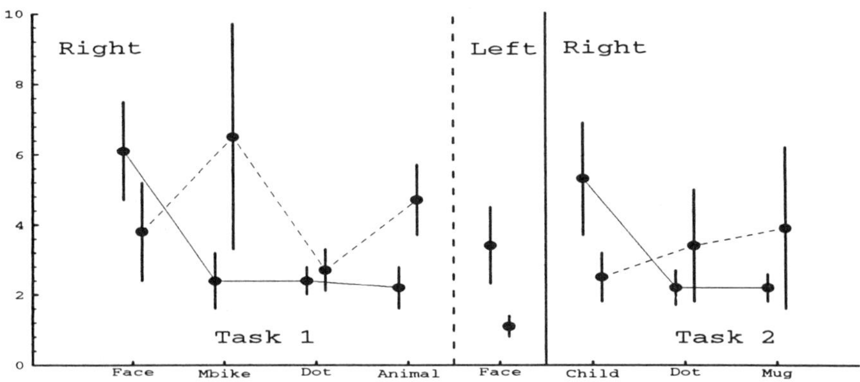

Figure 3: Cohort mean values of power ratios for normal (solid) and autistic (dashed) subjects.

4 Concluding Remarks

In this study, we have departed from common practice in not attempting any detailed source identification. The complexity of the signal topography and the results of preliminary modelling studies suggested that localised generator models would not be appropriate. Instead we have used a measure based on integrated regional power to compare responses of different subject groups and tasks. In normal subjects, our results are broadly consistent with face specific processing predominantly in the right hemisphere, as shown in previous studies. However, the associated latencies are generally earlier than reported. The underlying activity may reflect the earliest task independent stages in Bruce and Young's face processing model[3].

The autistic group was able to carry out those face processing task not involving classification of emotion, but there was no significant evidence for face specific activity. These observation suggest they use an abnormal strategy. It is unclear how this relates to current suggestions that autistic people analyse in terms of components parts, rather than using whole face representations.

Further work involving a larger group will emphasise possible differences between normal and autistic individuals at later latencies, specifically in pair discrimination tasks. It will also include source distribution analysis.

This study was supported by the EC HCM program BIRCH at the Low Temperature Laboratory, Helsinki, and by the MRC and EPSRC, UK.

[1] Bailey A, Phillips W and Rutter M   J. Child Psychol. and Psychiatry, 1996, Vol 36 No 1

[2] Boucher J and Lewis V   J. Child Psychol. and Psychiatry, 1992, Vol 33 No 5, pp 843-859

[3] Young A W   Phil. Trans. R. Soc. Lond. B, 1992, 335, pp 47-54

[4] Lu et al.   Neuroscience, 1991, Vol 43 No 2/3, pp 287-290

[5] K. Bötzel, S. Schulze and S. Stodieck   Experimental Brain Research, 1995, Vol 104, pp 135-143

A Comparison of MEG and EEG for the Assessment of Traumatic Brain Injury

Weinberg, H.[1], Gaetz, M.[1], Jantzen, K.[1] Cheyne, D.[1] and Deecke, L.[2]

[1]Brain Behaviour Laboratory, School of Kinesiology at Simon Fraser University;
[2]Neurological University Clinic, Vienna, Austria

Introduction

The current series of experiments are designed to establish methods for using MEG to assess information processing deficits after mild head injury and for the identification of which functional neurophysiological systems are impaired. The eventual goal of these experiments is to facilitate rehabilitation by accurately targeting the functional anatomy of the impaired systems through analysis of MEG and EEG. One important aspect of these procedures is to determine the plastic capacity of impaired systems. What are reported here are initial results with respect to the utilisation of MEG in EEG paradigms we have been using for the assessment of head injury.

Methods

EEG data was collected with 21 electrodes placed according to the international 10/20 electrode placement system. Third gradient MEG data was collected using a CTF 143 channel whole head MEG system in an unshielded environment.

<u>Stimulation parameters:</u> Stimulation parameters were kept as similar as possible in the EEG and MEG paradigms. However, due to the magnetic properties of the stimulus transducers, the auditory stimuli for MEG recordings were delivered through plastic tubes, resulting in a stimulus delay of 23 ms. Visual stimuli were presented at a distance of 1 meter for EEG recordings whereas in the MEG paradigm, a computer monitor had to be positioned at a distance of two meters from the MEG sensors. Adjustments in stimulus size were made to maintain a constant visual angle for both EEG and MEG recordings. Auditory mid latency responses were evoked with alternating clicks presented at 70 dB every 100 ms. Clicks were presented to the left and right ears separately; a masking noise (-20 dB) was presented in the non stimulated ear. Stimulation for the pattern visual evoked responses was monocular full visual field presented seperately to the left and right visual field of each eye. The stimulus consisted of a black and white checkerboard pattern which reversed at 3 Hz. Each check subtended a visual angle of approximately 35.93 minutes of arc.

Visual and auditory p300 event related responses were collected using the standard oddball paradigm. Three types of visual stimuli were used, shapes, numbers and words. The auditory stimuli were two sine wave tones. The stimulus duration was 100 ms with an ISI of 1000 +- 100 ms. The infrequent stimulus was presented according to a pseudo-random schedule in 20 percent of the trials.

For CNV's, as with the P300's, one auditory and three visual conditions were collected. The stimuli for the visual conditions consisted of shapes, numbers and words; the auditory stimuli were sine wave tones. The stimulus duration was 100 ms., The interval between S1 and S2 was 2000 ms and the ISI was 2000 +- 100 ms. The "go" stimulus, in a standard go-no-go CNV paradigm was presented on 50 percent of trials according to a pseudo-random schedule.

<u>Data Collection Parameters:</u> Mid-latency Auditory Evoked Potentials (MAEP) were digitized at a sample rate of 4000 Hz for a period of 64 ms and filtered online between 10 and 1000 Hz. The MAEPs were digitized at a sample rate of 1250 Hz for 100 ms. and were filtered online between 0 and 500 Hz. Averages consisted of 700 to 1000 artifact free trials. Pattern Visual Evoked Potentials (PVEP) were sampled at 1000 Hz and filtered online between 1 and 100 Hz. The PVEP were sampled at 1250 hz for 200 ms. and filtered online between 0 and 300 Hz. Each average consisted of 150 to 200 artifact free single trials. EEG P300 data was sampled at 250 Hz for 1024 ms and filtered online between 1 and 30 Hz. MEG P300's were sampled at 250 for 1000 ms and lowpass filtered below 35 Hz. Each average consisted of 15 to 30 artifact free single trials. MEG CNV's were sampled at 62.5 Hz for 4096 ms. and lowpassed online below 30 Hz. The Magnetic counterpart of the CNV was sampled at 125 Hz for 3504 ms and lowpassed online below 25 Hz. Each average consisted of 15 to 30 single trials.

Results

<u>Auditory and Visual P300 MEG and EEG</u>: Fig. 1 (auditory) and 2 (visual) are P3 responses that were significantly delayed in both EEG and MEG when compared to the normative EEG data base (greater than 2.5 SD). Although the EEG shows a vertex maximum distribution suggesting a single   source, the MEG shows two dipolar like distributions, one in each hemisphere, for both auditory and visual modalities. Further, the MEG data suggest a different distribution for visual and auditory P300's whereas the EEG distributions are the same for both modalities.

Figure 1. Auditory P300 Responses

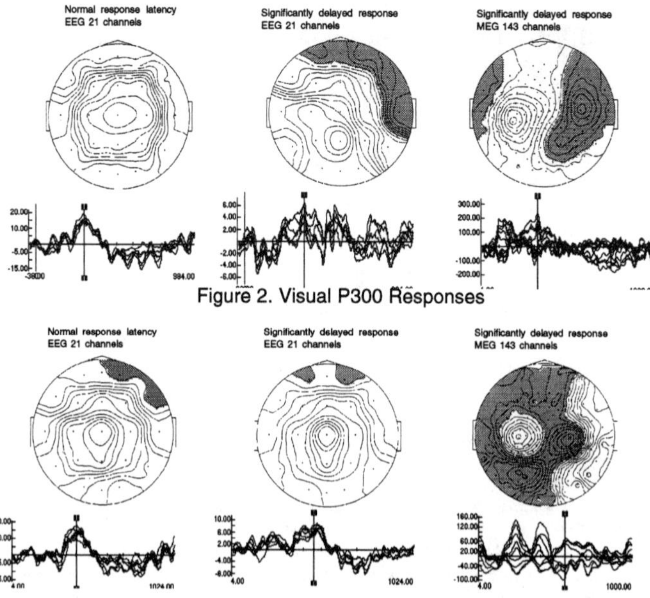

Figure 2. Visual P300 Responses

<u>Checkerboard Reversal MEG and EEG</u>: Fig. 3 shows a checkerboard response for right monocular full field stimulation. Both EEG and MEG showed a delayed response compared to the EEG normal data base. The EEG shows the usual paradoxical localisation [1] resulting from the orientation of sources in calcarine fissure. However, the MEG checkerboard response, shows the correct lateralisation. A comparison of   PVEPs using EEG and a whole head MEG in the same subject has not been previously reported as far as we know although PVEPs using a whole head system has been reported [2].

<u>Middle-Latency Responses</u>: Figure 4 shows one of the first reports of MEG mid-latency response using a whole head gradiometer system. Kakela et al., 1994 [3] reported a midlatency response using a planar gradiometer. In the four head inured patients used there were no abnormalities in latencies in either EEG or MEG. EEG mid-latency delays are seen in head injured patients when there is clear thalamic involvement in sensory deficits. However, in these patients there were no such delays. The EEG distributions shows no localized activity whereas the MEG reveals a clear bilateral dipolar response in posterior temporal lobe.

<u>Contingent Negative Variation (CNV)</u>: The EEG CNV maps (figure 5) shows a typical vertex maximal distribution which is not uncommon in both normal and head injured subjects, unless there is significant frontal

Right Monocular Full Field

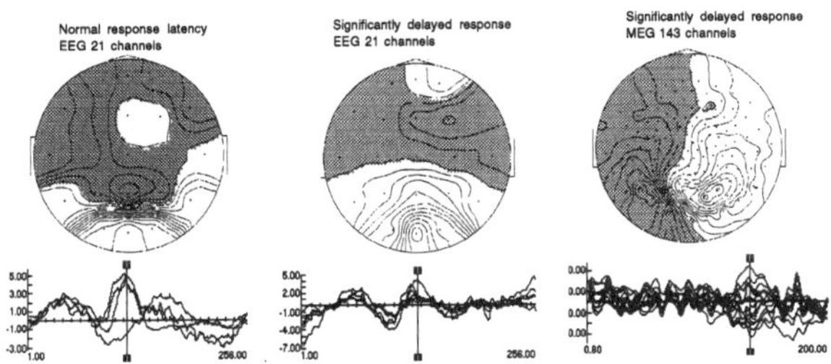

Right Monocular Right Visual Field

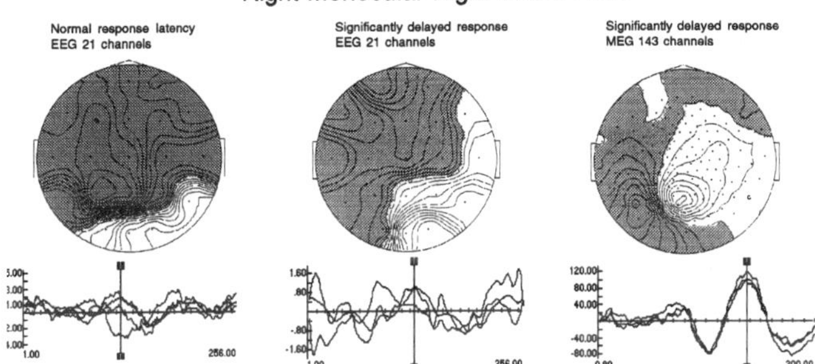

lobe damage. However, the MEG, unlike the EEG, shows clearly defined bilateral source fields for visual stimuli in S1 and S2. Compared to EEG, the whole head MEG recorded, appears to provide similar information regarding response latencies, but a much greater definition of localised activity. The MEG adds additional information about the distribution and strength of various components of evoked and event related potentials; although the EEG generally shows a similar distribution for

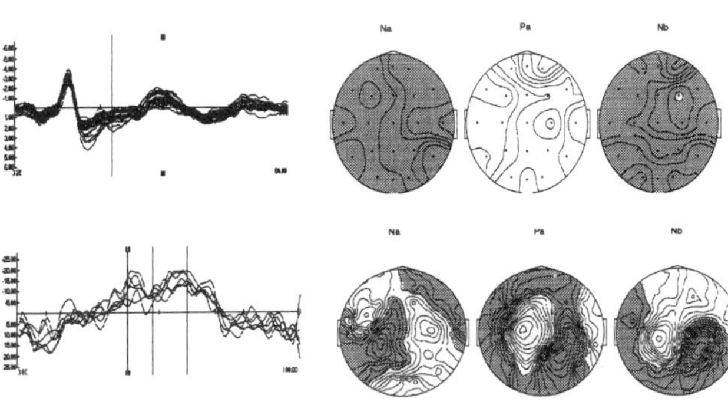

different components of an evoked potential.   This is particularly important in mid-latency responses where the MEG results in dipolar-like bilateral fields whereas the EEG does not. The checkerboard MEG response is similar to EEG for full field stimulation, however monocular half-field stimulation resulted in a clearly defined activity in the correct hemisphere and did not display paradoxical localisation - a result which will have important advantages for the analysis of abnormalities resulting from the thalamo-cortical systems responsible for this response. For example, the N2-P2 latencies observed in the EEG and MEG were similar whereas the EEG resulted in a distribution with a vertex-parietal maximum. The distribution of

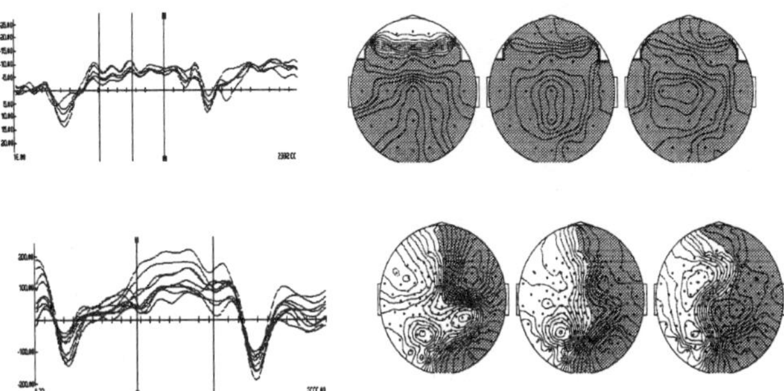

the MEG clearly indicated bilateral sources. This increased definition of the complexity of distributions is also demonstrated in the measurement of MEG and EEG, CNVs

Discussion

Compared to EEG, the whole head MEG recorded, appears to provide similar information regarding response latencies, but a much reater defintion of localised activity. The MEG adds additional information about the distribution and strength of various components of evoked and event related potentials; although the EEG generally shows a similar distribution for different components of an evoked potential. This is particularly important in mid-latency responses where the MEG results in dipolar-like bilateral fields whereas the EEG does not. The checkerboard MEG response is similar to EEG for full field stimulation, however monocular half-field stimulation resulted in a clearly defined activity in the correct hemisphere and did not display paradoxical localisation - a result which will have important advantages for the analysis of abnormalities resulting from the thalamo-cortical systems responsible for this response. For example, the N2-P2 latencies observed in the EEG and MEG were similar whereas the EEG resulted in a distribution with a vertex-parietal maximum. The distribution of the MEG clearly indicated bilateral sources. This increased definition of the complexity of distributions is also demonstrated in the measurement of MEG and EEG, CNVs.

These experiments are reported as initial results in an ongoing program of research intended to use the MEG for the assessment of processing deficits associated with mild head injury and the definition of which functional source systems are impaired.

References

1] Barret G., Blumhardt L., Halliday AM., Halliday E and Kriss A. (1979). A paradox in the lateralization of the visual evoked response. Natrue, 261: 253-255.

[2] Seki, K., et al. P100m of pattern reversal visual evoked magnetic field: source localization using an MRI - linked helmet-shaped magnetoencephalography system. EEG Clin. Neurophysiol. (Society Proceedings), 1995, 95: 86-94

[3] Makela J.P., Hamalainen M., Hari R., and McEvoy L. (1994). Whole-head mapping of middle-latency auditory evoked magnetic fields. Electroencephalography and Clinical Neurophysiology, 92, 414 - 421.

Somatosensory Evoked Magnetic Fields in Acute Cerebral Infarction

Wikström, H.[1,2], Huttunen, J.[2], Roine, R.O.[1], Salonen, O.[3], Aronen, H.J.[3] and Ilmoniemi, R.J.[2]

[1]*Department of Neurology; [2]BioMag Laboratory, Medical Engineering Centre; [3]Department of Radiology, Helsinki University Central Hospital, Helsinki, Finland*

Introduction

Cerebral infarction interfering with sensorimotor functions causes abnormalities in the somatosensory evoked potentials (SEPs) [1,2]. Somatosensory evoked fields (SEFs) [3] reflect partially the same phenomena as SEPs, but allow more accurate spatial discrimination of the underlying neural generators. We present preliminary results of SEFs from the primary sensorimotor cortex (SMI) in the acute stage of cerebral infarction with an emphasis on correlating SEF abnormalities with clinical impairments.

Methods

Sixteen patients with their first unilateral cerebral infarction were studied within 1 week from the onset of symptoms. The patients were categorized according to their clinical picture to those with purely motor paresis, those with selectively sensory symptoms and those with sensorimotor paresis. On the basis of clinical testing of touch, vibration, position, pain and temperature senses, sensory dysfunction was categorized as mild, moderate, or severe. Motor paresis was considered mild if almost normal function of the affected limb was possible, only skilled movements being impossible. A moderate paresis was characterized with preserved ability to lift the affected limb and in severe paresis the affected limb could not oppose the force of gravitation.

Left and right median nerves were stimulated alternately at the wrist with 0.2-ms constant-current pulses causing a thumb twitch. The interstimulus interval at each side was 3 s. The patient sat motionless in a magnetically shielded room watching a movie while SEFs were recorded with a 122-channel helmet-shaped first-order planar gradiometer array (Neuromag Ltd.) [4].

The signals were band-pass filtered at 0.03–310 Hz and digitized at 940 Hz. 100–200 responses were averaged for each run. Vertical and horizontal electro-oculograms (EOG) were monitored and sweeps with EOG signals exceeding 150 μV were automatically rejected.

An individual cartesian coordinate system was defined for each patient using a digitizer so that the x-axis passed through the preauricular points with the positive direction to the right; the positive y-axis passed through the nasion; thus, the positive z-axis pointed upwards. Three position indicator coils were attached to the head; their coordinates were also determined with the digitizer. Before each SEF trial, when the subject's head was positioned under the magnetometer dewar, current was passed through the coils; the resulting magnetic fields were recorded to locate the coils, allowing for a transformation between head and magnetometer coordinate systems.

The amplitudes of N20m, P35m, and P60m were expressed as dipole moments of equivalent current dipoles (ECDs). The strengths and locations of the ECDs were determined with least-squares searches using data from 18–22 channels covering both field extrema of the corresponding SEF deflection. The head was modelled with a sphere best fitting the inner surface of the skull in each subject's magnetic resonance (MR) images. The transformation between head and MR coordinate systems was done on the basis of the three anatomical landmarks easily identified from the MR images: the nasion and the two preauricular points. All ECDs obtained in this way explained over 80% of the field variance and were accepted as models of local cortical activity.

Results

There were no significant asymmetries in the ECD locations between the hemispheres. Neither did we observe changes in SEF laterality, i.e., no abnormally enhanced SEFs from the ipsilateral hemisphere after stimulation of the affected hand were seen. The data from individual patients are presented in the Table 1.

SEF amplitude asymmetries and clinical picture

Four patients had purely sensory symptoms. Three of them had symmetric SEFs, in the fourth patient they were missing on the affected side.

Three patients suffered from selectively motor paresis. Two of them had severe and one mild paresis. All shared mild SEF asymmetries.

Nine patients suffered from sensorimotor paresis. Four had severe paresis; three of them had also severe asymmetry of SEFs. However, one patient with severe paresis had no SEF asymmetry at all. Three patients had moderate symptoms. One of these had severe SEF asymmetry and the two remaining had moderate asymmetry. Two subjects with mild paresis had either mild or moderate SEF asymmetry.

Four patients, one with sensory, one with motor, and two with sensorimotor paresis had an enlarged P35m or P60m deflection in the non-affected hemisphere. Their pareses were classified as moderate or severe. Three of the patients had large cortical-subcortical infarcts and one had a small pontine lesion. The data from patient EW, featured by enlarged P60m deflection on the healthy side, is shown in the Fig. 1.

Table 1. Patient characteristics and SEFs in the infarcted hemisphere compared with the unaffected side. C = cortical, SC = subcortical, 0 = no asymmetry, – = moderately reduced, — = severely reduced, + = enhanced, a.b = absent bilaterally. In the symptoms column, the size of the letters indicate magnitude of involvement of the corresponding system, for example, Sm indicates dominance of sensory symptoms over motor symptoms.

Patient	Age	Sex	Side	Infarction	Symptoms	Severity	N20m	P35m	P60m	SEF abnormality
KL	65	M	L	C SC	S	moderate	0	0	0	none
JP	51	M	R	SC	S	moderate	—	—	—	severe
HG	48	M	L	C SC	S	severe	0	0	0	none
AP	59	F	L	C SC	S	mild	0	0	0	none
ML	56	M	L	C SC	M	mild	0	–	+	mild
SS	49	M	L	SC	M	severe	–	–	0	mild
PE	59	M	R	SC	M	severe	+	0	0	mild
VR	77	M	L	C SC	SM	moderate	—	—	a.b	severe
SH	26	F	L	SC	sM	severe	+	0	0	mild
UH	53	M	L	SC	SM	mild	0	+	0	mild
AW	62	M	R	SC	SM	mild	—	–	a.b	mild
JR	48	M	L	C SC	SM	severe	—	—	—	severe
HV	63	M	R	C SC	Sm	moderate	—	–	—	mild
RL	66	M	R	C SC	Sm	moderate	0	—	a.b	mild
AV	56	M	R	C SC	SM	severe	—	a.b	—	severe
EW	64	M	L	C SC	SM	severe	—	—	—	severe

Discussion

In the present study we had 16 patients with their first unilateral cerebral infarction. The data of four of them with only sensory dysfunction suggests that in purely sensory paresis the SEF amplitudes can be symmetric. Purely motor paresis, observed in three patients, was associated with mild amplitude asymmetries. Sensorimotor paresis had more profound effects on the patients' SEFs. Of our nine patients with sensorimotor dysfunction, all had asymmetric SEFs and in four the asymmetry was severe. The severity of SEF amplitude asymmetry correlated poorly, however, with the severity of sensory symptoms; even severe sensory paresis could be associated with symmetric SEFs. An interesting finding is the enhancement of P35m or P60m in the non-affected hemisphere of four patients with moderate to severe paresis. Three of them had large cortical infarctions and one small pontine lesion. This enhancement can be in association with impaired tonic interhemispheric interactions in patients with extensive infarctions (2).

The present results did not show any changes in the laterality of sensory functions in the acute stage of infarction; i.e. SEFs after stimulation of the affected limb, if present, had source locations at SMI on the affected side, and no major activity was seen in the unaffected hemisphere. A follow-up study will reveal, whether changes in the laterality occur in the long run, especially in those patients who recover from their clinical symptoms.

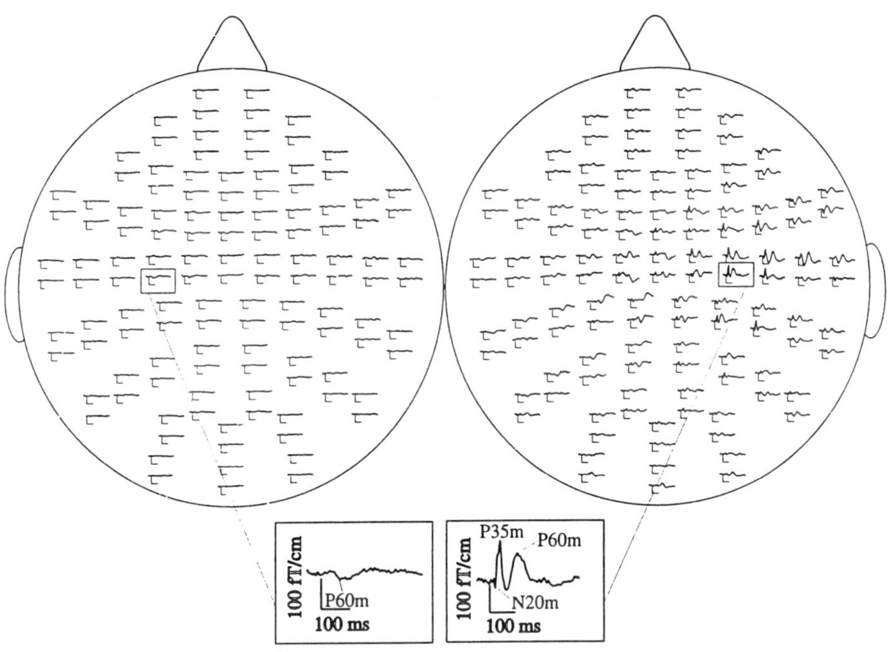

Figure 1. 122-channel SEFs in patient EW. Large left-sided cortical-subcortical infarction with severe sensorimotor paresis. Left: right median nerve stimulation. Right: left median nerve stimulation. On the infarcted side, only a small P60m could be detected. Note the enlarged P60m deflection on the healthy side.

References:

[1] Mauguière, F., Desmedt, J.E., and Courjon, J. Astereognosis and dissociated loss of frontal or parietal components of somatosensory evoked potentials in hemispheric lesions, Brain, 1983, 106: 271–311.

[2] Nakashima, K., Kanba, M., Fujimoto, K., Sato, T., and Takahashi, K. Somatosensory evoked potentials over the non-affected hemisphere in patients with unilateral cerebrovascular lesions, Journal of the Neurological Sciences, 1985, 70: 117–127.

[3] Hari, R., and Kaukoranta, E. Neuromagnetic studies of somatosensory system: principles and examples, Progress in Neurobiology, 1985, 24: 233–256.

[4] Ahonen, A.I., Hämäläinen, M.S., Kajola, M.J., Knuutila, J.E., Laine, P.P., Lounasmaa, O.V., Parkkonen, L.T., Simola, J.T., and Teche, C.D. 122-channel SQUID instrument for investigating the magnetic signals from the human brain, Physica Scripta, 1993, 49: 198–205.

Magnetoencephalographic vs electrocorticographic source localization in patients with epilepsy

Aung, M. [1], Sobel, D. [1], Schwartz, B. [2], Squires, K. [2], and Hirschkoff, E. [2]

1.The Scripps Clinic and Research Foundation, La Jolla, CA , USA
2. Biomagnetic Technologies, Inc, San Diego, CA , USA

Introduction

Surgical resection of cortical tissue responsible for seizure activity has proven to be an effective therapy for patients with intractable epilepsy. Its success, however, depends on the patient having one or a few focal seizure sources located in areas for which excision can be carried out without creating undue deficit. Conventional non-invasive methods for localizing sources of interictal or ictal epileptogenic activity generally provide only crude spatial information, and many surgery candidates without localized structural lesions undergo invasive electroencephalographic (iEEG) monitoring with implanted grid or depth electrodes to localize epileptogenic sources. In some cases, even after extended observation with iEEG, intraoperative monitoring with depth electrodes preparatory to surgery will indicate that resection would be too much of a risk to eloquent cortex. (One such case is included in our study). Magnetic source imaging (MSI) offers the potential to localize these sources non-invasively in a short out-patient procedure.

In this study, 100 epilepsy patients judged by their physicians to be candidates for surgery, and who showed frequent interictal epileptogenic activity on routine surface EEG studies, and who were likely to undergo invasive EEG monitoring were recruited to add a MSI to their evaluation program. Patients were referred from eleven epilepsy centers throughout the U.S. and Canada. Non-invasive tests at those centers prior to MSI typically included EEG, MRI, neurological examination, neuropsychological testing, and phase I video EEG monitoring. In many cases, PET and/or SPECT test results were also available.

Methods

Subsequent to the MEG measurement, each of the patients underwent invasive EEG "phase II" monitoring with implanted electrode grids and/or strips. For the MEG phase of the study, each patient underwent a concurrent MEG and EEG examination utilizing a 74 channel biomagnetometer (Biomagnetic Technologies Magnes II$^{\circledR}$) and a 21 channel EEG recording system (Nihon Khoden) to record spontaneous incidents of interictal epileptogenic activity. A radio frequency 3-D digitizer was used to determine the position of the individual gradiometers relative to a head-centered 3-D orthogonal coordinate system defined by the nasion and by markers in the right and left outer ear canal. Each patient underwent an MR scan with fiducial markers attached, providing a common coordinate system to overlay MEG source localizations [1].

Patient anti-seizure medication levels were not reduced for the MEG study. With the patient lying on one side and the head voluntarily stabilized, the two 37-channel probes were positioned over homotopic areas of the two hemispheres. A sequence of four to five dual probe placements and measurements assured coverage of the whole head, and afforded the ability to distinguish between synchronous and asynchronous bilateral sources. Recording sessions typically lasted from two to four hours, depending on the patient's level of interictal activity. EEG and MEG waveforms were monitored in real-time by observers trained to recognize and distinguish the morphology of epileptogenic events as against artifacts. (EEG data were later reviewed and interpreted separately by an experienced epileptologist.) Twenty 5-6 second epochs of data were obtained at each recording site. Raw data were digitized at 860 Hz and band-pass filtered at 0.1-200 Hz. Data epochs were then digitally filtered off line with a bandpass of 3-70 Hz. Off line data analysis proceeded with visual identification of MEG and EEG interictal spikes or sharp waves. Temporal windows including these events were then analyzed at each time point in the data sample with a single equivalent current dipole (ECD) model. Each event for which the dipole model correlated with the recorded data to at least 98% and for which the equivalent dipole strength was physiologically realistic was represented by a single focal source plotted on an MRI image of the patient's brain.   Images showing all ECD modeled epileptiform events (Fig 1) were presented along with waveform information and the EEG tracings for review by an experienced epileptologist. The epileptologist indicated the likely source of seizure activity and target for surgical resection, if such an inference could reasonably be drawn from the MSI data. The diagnostic conclusion drawn from the MSI assessment was then compared with that from the consensus of all other measures including invasive EEG monitoring undertaken during the patient's normal course of care [2].

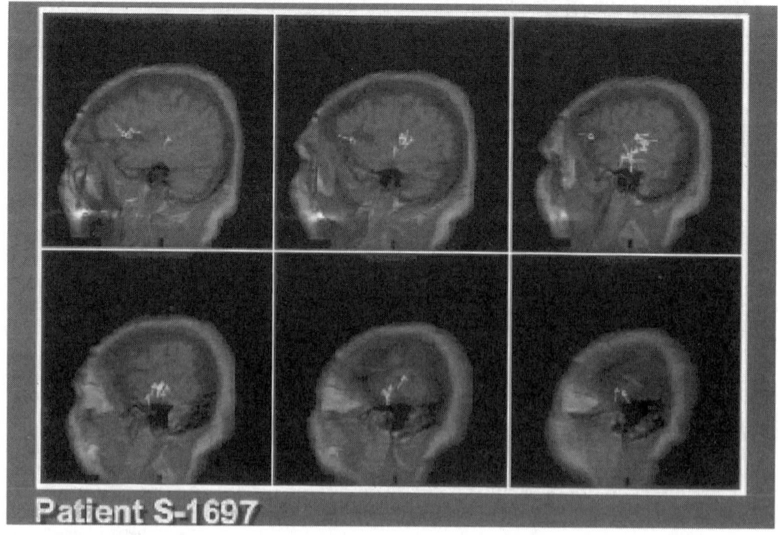

Fig. 1. MEG dipole overlays on sagittal images for patient #S1697.(Spike sources are indicated as yellow triangles, with lines indicating orientation.) The patient had suffered complex partial seizures with secondary generalization from age 12. Invasive EEG monitoring had shown left mesiotemporal sources. Subsequently, the patient underwent a left temporal lobectomy, and was seizure free in a follow up 11 months later.

Results

Of the total of twenty patients studied, MSI detected and localized interictal spikes in 16 and slow waves in another two cases. In eight of the cases MSI and invasive EEG ("Phase II") results indicated the same focal area. In five other cases, MSI and invasive EEG were in agreement in indicating multi-focal or bilateral epileptogenic zones. In two cases, MEG spikes were detected over a broader area which included that indicated by the interpretations based on invasive EEG. One case produced a discordant result, with MEG indicating a region inconsistent with that based on invasive EEG. Data recorded from the remaining four patients produced insufficient MEG spike signals for source localization.

Eighteen of the twenty cases went on to have surgery, of whom follow up information has been obtained for seventeen. Twelve patients have been seizure free for the 3-12 months following the surgery. One additional patient ha been seizure free but has suffered visual field loss . Three patients have shown a 75-90% reduction om seizure occurance, and one patient has suffered an increase in seizure frequency since surgery.

Discussion

MSI is often consistent with results from a consensus of current methods including invasive EEG. Given its concordance with invasive EEG in particular, MSI has the potential to aid in guiding invasive monitoring, especially in patients with no structural lesions, to ensure adequate cortical coverage. MSI may even be able in some cases to obviate the need for invasive EEG prior to surgery. Such a change would save not only considerable cost to the patient and the insurer, but would provide a much less intrusive procedure for the patients, many of whom are quite young.

References

[1] Gallen, C.C., Hirschkoff E.C., Buchanan, D.S. Magnetoencephalography and Magnetic Source Imaging. Neuroimaging Clinics of North America, 5(2), May, 1995.
[2] Smith, J.R., Schwartz, B.J., Gallen C.C., Orrison, W., Lewine, J., Murro A.M., King, D.W. Park, Y.D. (1995) Utilization of Multichannel Magnetoencephalography in the Guidance of Ablative Surgery. Journal of Epilepsy 8:119-130.

Acknowledgements

The authors would like to thank the following collaborators for their assistance: Dr. Izak Fried and Dr. Sachar Eliashiv (UCLA Medical Center), Dr. Vicente Iragui and Dr Evelyn Tecoma (UCSD Medical Center, La Jolla, CA), Dr. Brien Smith (Henry Ford Hospital, Detroit, MI), Dr. Joseph Smith and Dr. Yong Park(Medical College of Georgia, Augusta, GA), Dr. Thomas Waltz (Scripps Clinic Research and Foundation, La Jolla, CA, Dr. James Wheless and Dr. Valerie Curtis (Hermann Hospital, Houston, TX)

Propagation of epileptic activity during single (unaveraged) interictal spikes as studied with Magnetic Field Tomography (MFT)

Bamidis, P.D.[1], Hellstrand, E.[2], Ioannides, A.A.[1,3]

[1]*Dept. of Physics, The Open University, Milton Keynes, U.K.;* [2]*Dept. of Clin. Neurophysiology, Karolinska Hospital, Stockholm, Sweden;* [3]*IME, Research Centre Jülich, Germany*

Introduction

The potential in multichannel MEG systems to analyse interictal epileptic data directly, without signal averaging, has been underlined by many MEG investigators [1]. Since spikes of similar shapes may arise from different cortical areas, analysis of individual, non-averaged spikes would be preferable to that of averaged spike signals [2]. An averaging technique might give good localisation results at the maximum of the spike (in consensus with some unaveraged spikes), but one cannot guarantee that the propagation of epileptic activity as appearing in the averaged record would represent the actual epileptic propagation. The study of unaveraged epileptic events, however, necessitates a robust analysis method that: (i) does not include any preconceptions on the number and position of sources; (ii) allows for distributed sources, since extended and/or multiple brain areas may be involved in the generation of epileptiform activity [3,4]. Magnetic Field Tomography (MFT) meets these requirements [5]. Earlier studies exploited the feasibility of analysing unaveraged interictal multichannel MEG data (i.e. spikes, sharp waves) with MFT [6]. More recently, MFT was used to study the spatio-temporal evolution of interictal activity during single spike events [7]. In this work, emphasis is drawn on the interactions between neocortex and deep temporal structures, as well as the spatio-temporal characteristics of interictal activity in the depth of the temporal lobe of two epileptic patients.

Materials and Methods

Patient #1 was an 18-year old girl with a history of pharmaco-resistant complex partial epilepsy (CPE) since the age of 14, with daily seizures of dizziness and impaired consciousness. Patient #2 was a 42-year-old male, suffering from CPE since the age of 13 without known aetiology. His seizures started with visual symptoms, with more acoustic sensations, like "noise", dominating later on. Generalisations of seizures occurred frequently.

The MEG recordings were made with the 37-channel KRENIKON®, Siemens system placed over the temporal area. The data were band-pass filtered on-line through 0.5-70.0 Hz, digitised at 0.4 kHz (Patient #1) or 0.5 kHz (Patient #2). The present analysis has been limited to those datasets off-line filtered through 1-20 Hz.

Three dimensional MFT estimates of the primary current density, $\mathbf{J}^p$, were obtained within a partial hemispherical volume encompassing the left (Patient #1) or right (Patient #2) hemisphere. A succinct description of MFT has been recently given in [5]; more details can be found in other reports of this volume as well as in [8]. In short, MFT uses continuous probabilistic estimates of $\mathbf{J}^p$, in order to construct tomographic displays of brain activity. In our work, the instantaneous MFT estimates were studied, as well as the temporal integrals of the intensity, $|\mathbf{J}^p|^2$. We use integrals of intensity in space (activation curves) to study the temporal evolution of activity within a region of interest (ROI), as well as displays of temporal integrals of intensity to obtain summaries of activations within a given time window. Whenever activation curves within specific ROIs or levels of the source space (e.g. at the depth of the temporal lobe - hippocampal level) indicate strong activation, the spatial changes in the maximum activity are studied; such changes within a given image/level are displayed by connecting areas of strong activity with arrows. This provides an image of the various anatomical regions of the brain which are directly or indirectly functionally connected, allowing for propagations and/or sequential shifts of activity to be studied.

Results

In earlier publications [6,7] we have observed that certain cortical areas seemed to be activated both before and after deep temporal activation. The interplays between deep and superficial activation occurred with various time constants. For the two patients involved in this study, the time differences of the activations from cortical to deep or deep to cortical areas varied between 10 to 60 ms. An example from each patient is shown in Fig. 1. During the initial, rising phase of the spike, the activity appeared in superficial, cortical areas of the temporal lobes (a); around the peak of the spike (b) in the amygdala-hippocampal complex, and then back to the cortex (c).

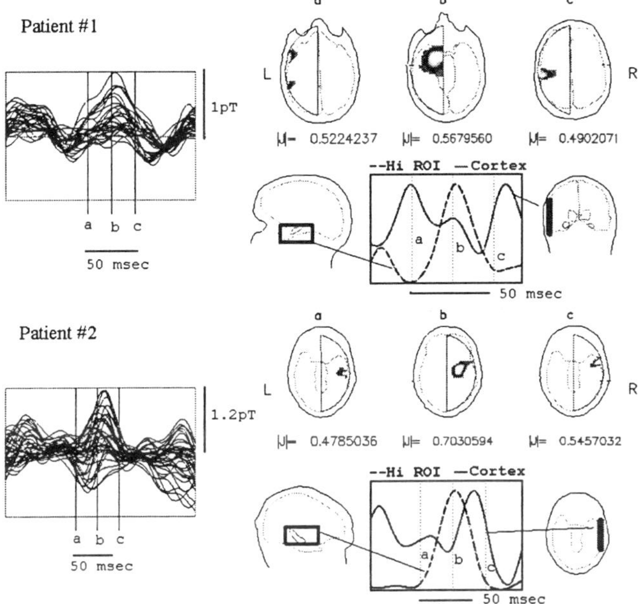

Fig 1: Top half; Patient #1; Left: left-hemisphere interictal spike with superimposed signals of all channels (note time bars a-c). Right; *top row:* instantaneous MFT estimates of intensity on axial MR images (note time instants a-c and |J^p| values in arbitrary units); *bottom row; middle:* graph of temporal evolution of activity in the left hippocampal area (dotted line) and temporal cortex (full line) (note time bars a-c); *bottom row; left and right:* MR images of ROIs at the level of hippocampus (sagittal view) and temporal cortex (coronal view). Bottom half: similar, but for the right hemisphere of patient #2. See text.

However, as different cortical areas are involved in the pre- and post-deep activation, one could easily assume that the observed cortical activations could just be random variations of noise. To exclude this possibility, we back-averaged a number of single epoch events, time-locked to the peaks of strong hippocampal activations. This procedure selectively amplifies activations from areas which maintain a consistent time relation with the activation in the hippocampus. The results are summarised in Fig. 2, where 18 single event solutions were back-averaged for patient #1, while 19 for patient #2.

The involved cortical ROIs remain "silent" during the interval of hippocampal activation, but, they do fire before and after this interval. This is clearly shown in the cortical activation curves, which reveal distinct peaks above the noise level before and after the hippocampal firing. Similar patterns occur for both patients. Note also that the cortical areas appear extended rather than "focal". Furthermore, as the schematics at the far right of Fig. 2 demonstrate, there seems to be an interplay of activation between the different regions of the cortex. In both patients, the hippocampal areas appear to be interacting mostly with the anterior cortical regions.

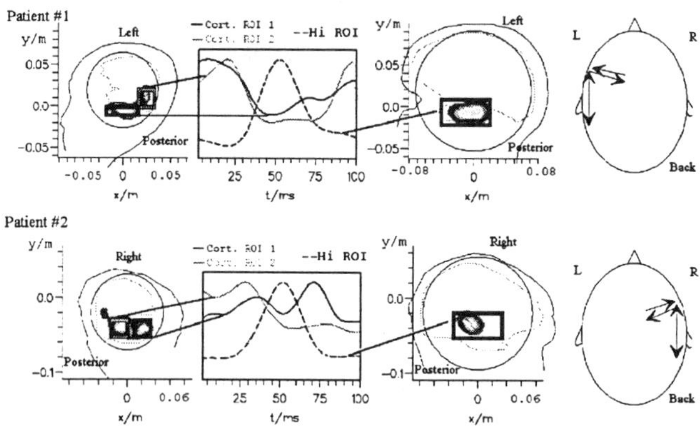

Fig 2: <u>Top row: (Patient #1)</u>; *Left*: superficial integrals of intensity over 100 ms, after back-averaging 18 single event solutions of strong hippocampal activation. Two main ROIs are revealed in the sagittal view: one fronto-infero-temporal and one more posterior in the temporo-parietal region. *Middle*: graph of activations of the cortical and hippocampal ROIs. Note the pre- and post-hippocampal activation of the cortex. *Right*: sagittal view of deep intensity integrals after back-averaging. Hippocampal activation is clearly shown. *Extreme Right*: schematic axial view of the neocortico-mesiobasal interactions emerging from the back-averaged solutions. <u>Bottom</u>: Similar, but after back-averaging 19 solutions for Patient #2.

For both patients, the MFT estimates at the peaks of the spikes indicated activation at the depth of the temporal lobe. The shapes of the solutions were consistent with activation of the hippocampal areas. Studying the spatial changes in the maximum activity in a sagittal section through the amygdala and hippocampus complex, we noticed that for each patient, activity propagated mostly in the anterior-posterior direction, suggesting spread from amygdala to hippocampus (12/26 events for patient #1, 7/15 for patient #2). The effect was more prominent for patient #1, for whom the shapes of the integrated intensity during such intervals appeared usually along an arch, indicating propagation from structures anterior to the hippocampus along its main axis to its posterior end (cf. Fig. 3a and [7]). However, shifts of activity (Fig 3b) from anterior to posterior and then back to anterior and vice versa were also observed (8/26 for patient #1, 4/15 for patient #2). Propagations from posterior to anterior were less frequent for both patients (4/26 for patient #1, 4/15 for patient #2).

Discussion and Conclusions

There has been an increasing interest in the study of unaveraged epileptic data [9]. The ECD model is mostly used to explain small segments of the unaveraged record, usually around the peaks of spikes or sharp waves. This leaves part of the event unexplained. We are using a more general method (MFT) to obtain estimates of activity throughout the whole time period of the various events, that is from the rising phase of a spike to its peak, and then the repolarisation. In doing so, we have observed the interactions between neocortical areas and deep temporal structures during a number of unaveraged interictal spikes in 2 epileptic patients. Superficial activity was seen to precede and follow deep activity. The time differences of the activations from cortical to deep or deep to cortical areas varied in the range between 10-60 ms. Back-averaging of the individual solutions was performed to enhance any orderly series of activations. Two rather extended cortical areas in each patient emerged after back-averaging. Interplays of activity were consistently found between these two cortical sites, and between the most anterior cortical area and the hippocampal complex. The image revealed from the back-averaged solution (cf. schematics in Fig. 2) may not necessarily represent the actual sequence of activations in each individual event, but it does mean that it occurs frequently (more than half of the events studied for each patient), and certainly above chance.

The persistent shifts in the maximum of the MFT estimates over time during an interictal spike occur consistently over different epochs. However, the evolution of activity in different areas of the brain such as superficial cortical structures and deep situated amygdala-hippocampus areas may not necessarily indicate an established anatomical pathway - intermediate steps may be missing (e.g. activity on the contralateral side [7]), partly due to the set-up of the sensors. MFT simulations and tests of the validity of this analysis could not be included here due to space limitations, but are provided elsewhere [8].

In conclusion, MEG measurements and MFT analysis offer the opportunity to non-invasively recover the functional connectivities of the various anatomical structures from single epoch data, avoiding the dangers associated with averaging.

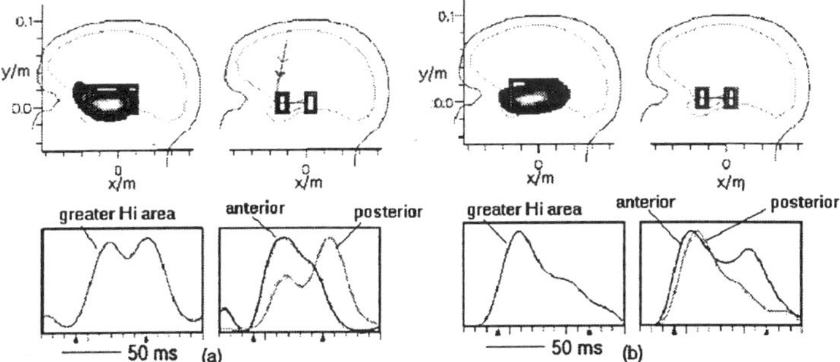

Fig 3: Patient #1. (a) Sagittal view of integrated intensity over 60 ms in a sagittal level through hippocampus (left). Instantaneous shifts of maximum activity computed every 2.5 ms (right). Propagation from anterior to posterior hippocampus is observed. Note the shape of the integrated intensity and the activation curves for the two indicated ROIs (bottom right). Activation of the greater hippocampal area is given in bottom left. (b) Similar but with anterior-posterior-anterior sequential shifts.

References:

[1]   Sato, S. Current status of biomagnetic research in epileptology. In: Hoke, M., Erne, S.N., et al. Biomagnetism: Clinical Aspects, Amsterdam: Elsevier, 1992.

[2]   Witte, O.W., Dorn, T., and Uhlig, S. Contribution of different areas of epileptic foci to the generation of interictal epileptic discharges. In: Hoke, M., Erne, S.N., et al. Biomagnetism: Clinical Aspects, Amsterdam: Elsevier, 1992.

[3]   Barth, D.S., Sutherling, W.W., Engel, J., et al. Neuromagnetic evidence of spatially distributed sources underlying epileptiform spikes in the human brain. Science 1984, 223:293-296.

[4]   Graf, M., Niedermeyer, J., Schiemann, J., et al. Electrocorticography: information derived from intra-operative recordings during seizure surgery. Clin. Electroenceph., 1984, 15(2):83-91.

[5]   Ioannides, A.A. Estimates of 3d brain activity ms by ms from biomagnetic signals: method (MFT), results and their significance, In: Eiselt, E., Zwiener, U., et al. Quantitative and Topological EEG and MEG analysis, Jena, Universitätsverlag Druckhaus-Maayer Gmbh, 1995.

[6]   Ioannides, A.A., Hellstrand, E., Bamidis, P.D., et al. Estimates of brain activity from unaveraged interictal multichannel MEG signals, In: Baumgartner, C., Deecke, L., et al. Biomagnetism: Fundamental research and Clinical Applications, Amsterdam, Elsevier, 1995.

[7]   Bamidis, P.D., Hellstrand, E., Lidholm, H., et al. MFT in complex partial epilepsy: spatio-temporal estimates of interictal activity, Neuroreport, 1995, 7(1):17-23.

[8]   Bamidis, P.D., PhD Thesis, The Open University, Milton Keynes, UK, 1996.

[9]   Lewine J.D., Orrison, W.W. Jr, Astur, R.S., et al. Explorations of pathophysiological spontaneous activity by magnetic source imaging. In: Deecke L, Baumgartner C, et al. Biomagnetism: Fundamental research and Clinical Applications, Amsterdam, Elsevier, 1995.

Presurgical Evaluation of Methohexital–Induced Epileptiform Activity: Comparison of EEG–, MEG–, and Simultaneous EEG/MEG Models

Diekmann, V.[1], Becker, W.[1], Jürgens R.[1], Grözinger, B.[1], Kleiser, B.[2], Wollinski, K.H.[3], Richter, H.P.[4]

[1] *Sektion Neurophysiologie and* [2] *Abteilung Neurologie, Universität Ulm, 89081 Ulm, Germany;*
[3] *Abteilung Anästhesiologie, Rehabilitationskrankenhaus Ulm, 89081 Ulm, Germany;*
[4] *Abteilung Neurochirurgie, Universität Ulm, 89312 Günzburg, Germany*

Introduction

We have developed a method of *simultaneous* recording of EEG and MEG and localizing the origin of epileptic spikes by calculating single equivalent current dipoles (ECDs) on the basis of electric, magnetic, and *simultaneous* electric–magnetic models. We here compare the results of these 3 methods in view of the question whether one of these models presents specific advantages in the search for foci of simple or complex partial epilepsy (SCPE).

Methods

Our report draws on a series of 32 SCPE patients (Ps) referred to the Dept. of Neurology for presurgical diagnosis and a total of 37 sessions (some Ps were repeatedly examined) in which EEG (25 channels) and MEG (Dornier; 22 signal/6 compensation channels software gradiometer [1, 2]) were simultaneously recorded inside a shielding room (Vacuumschmelze Type AK3a). In 2 of these Ps, ECoG (from intra–cranial strip electrodes), instead of EEG, was recorded in conjunction with MEG. Also, in some Ps, EEG was complemented by foramen ovale or sphenoidal recordings. EEG electrodes were arranged according to the 10–20–system, and 6 additional electrodes were placed over the suspected epileptogenic area where electrode distance was thus reduced to less than 2 cm. Similarly, the two cryostats of the MEG system were conjointly centered over the suspected area.

Patients' head was immobilized by vacuum pads. Since the frequency of spontaneous epileptiform activity can be low in SCPE patients, it is often not possible to collect a sufficient number of spikes within the limited time a patient can be immobilized. In 22 of the above 32 patients we therefore resorted to provocation by i.v. injection of the barbiturate methohexital. Methohexital induces a brief anesthesia accompanied by seizurelike brain activity with foci matching those of spontaneous seizures [3]. During anesthesia, patients were given oxygen while respiration, heart rate, blood–pressure , blood–O_2–saturation, and expiratory CO_2 were monitored. On average, during the 3–min epoch preceding methohexital injection, less than 1 spike (range 0–5) could be clearly recognized in surface recordings (EEG, MEG), whereas there were 15 (0–58) such events in the 5–min period following injection.

Spikes were identified off–line by visual inspection and isolated by subtracting a linear interpolation of pre– and post–spike background activity. Using a 4–layer sphere fitted to the patient's head on the basis of MR–images, single electric and magnetic ECDs were first calculated separately from electric (EM) and magnetic (MM) models; EEG and MEG then were weighted with the goodness of fit of EM and MM, respectively, by taking EEG/σ_e and MEG/σ_m (σ_e^2 and σ_m^2 , residual variance of EM and MM). Finally, the weighted data were fitted with a simultaneous electric–magnetic model (SM) [4]. ECD positions and moments were transferred into the patients' MRI for visualization (Fig. 1). Observed spike activity was summarized by calculating the spatial dipole density distribution [5]. To this end, each ECD was associated with a 3D–Gaussian distribution centered at its locus with variances reflecting the Cramer–Rao lower bounds [6] of the locus; iso–density lines then were derived from the sum of all Gauss–distributions and transferred into the patient's MRI (see Figs. 2 – 4).

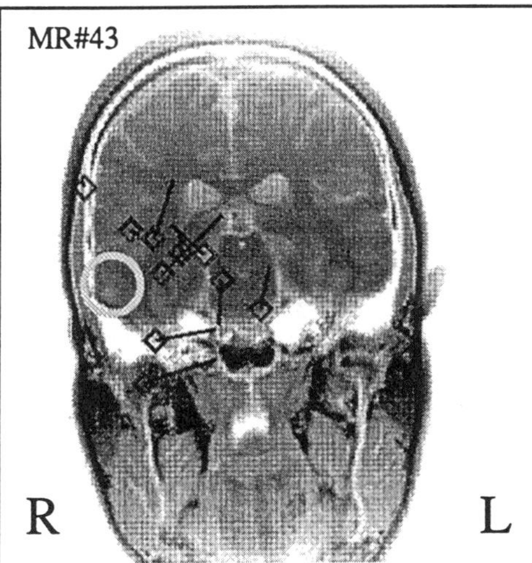

Fig. 1. ECDs obtained by SM in a patient with temporobasal tumor (circled). Locations (◊) and moments (bars) projected into coronal MRI–slice containing centre of tumor (R right, L left).

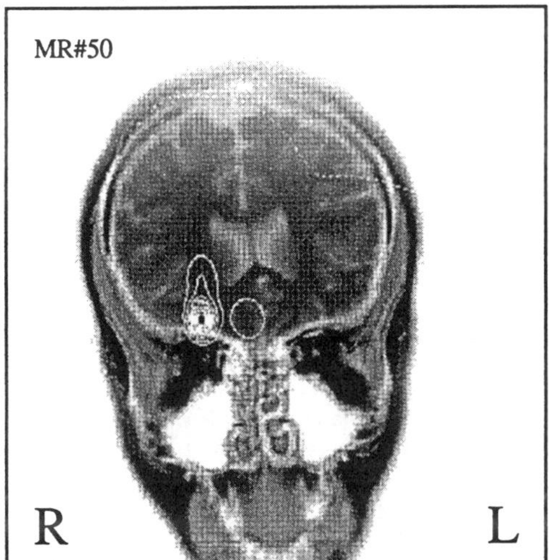

Fig. 2. Same P as in Fig. 1, dipole density obtained from EM. Coronal MRI–slice cutting through maximum of EM dipole density with isodensity curves (density Δ = 0.5 cm^{-3})

Results

Among our 32 patients we selected candidates for a verification of ECD localization by successively applying the following criteria:

(1) More than 5 localizable spikes. Spikes were considered localizable if SM explained more than 75% of the EEG+MEG variance (although this condition may appear weak by the standards for EM and MM fits, it is a strong one for the combined data). Criterion (1) reduced the number of cases to 16.

(2) "Classical" clinical examinations (video-monitoring, intra-cranial electrodes, semiology) suggest either a single epileptogenic focus or two clearly distinct foci (one in each hemisphere); criterion (2) left ten cases.

(3) Clinical reasoning and postoperative outcome associates foci with a clearly recognizable structural change in MRI (neoplasm, infarction, etc.); criterion (3) left over six cases, which on average yielded 27 localizations.

In these six Ps, the locus of peak dipole density was established for each of the three models (EM, MM, SM) and its 3D-distance to the nearest recognizable lesion border in MRI was measured (Fig. 2). The distances obtained in this way from EM and MM averaged 3.0 and 3.3 cm respectively, whereas the average distance indicated by SM was clearly smaller (2.1 cm). At the individual level, SM distance was closest to the lesion border in four of the six Ps. Interestingly, in each of these four patients MM was worse, in terms of distance from lesion border, than EM - an observation which underscores the possible pit-falls of relying solely on MEG. The patient shown in Figs. 1-4 constitutes a typical example for this type of behaviour. He suffered from a ganglioglioma in his right temporobasal cortex (Fig. 1). Peak dipole density derived from EEG occurred at a distance of 4.0 cm from the tumor border (Fig. 2; note that consecutively numbered MRI slices have 6 mm distance; hence, the coronal plane shown in Fig. 2 is 4.2 cm anterior to the centre of the tumor shown in Fig. 1). The P's MM dipole density peaked at 5.9 cm from the nearest tumor border (Fig. 3), whereas the maximum of SM density was at 2.4 cm (Fig.4).

In the remaining two Ps, SM-distance was intermediate to those of EM and MM with one case of EM and one of MM yielding the closest localization.

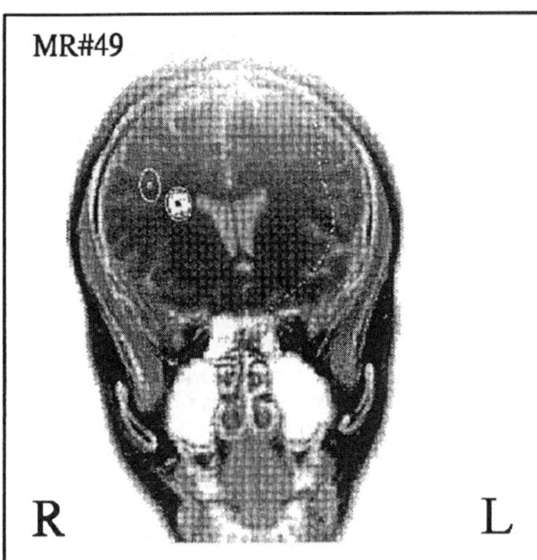

Fig. 3. Same P as in Fig. 1, dipole density obtained from MM. Coronal MRI–slice cutting through maximum of MM dipole density with isodensity curves (density $\Delta = 1\ cm^{-3}$)

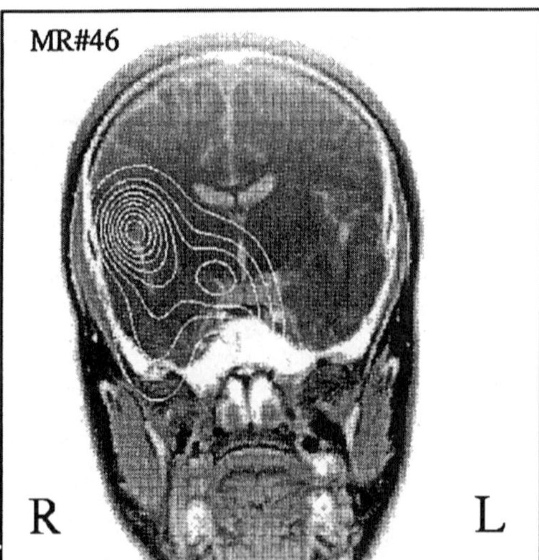

Fig. 4. Same P as in Fig. 1, dipole density obtained from SM. Coronal MRI–slice cutting through maximum of SM dipole density with isodensity curves (density $\Delta = 0.05\ cm^{-3}$)

In terms of their contribution to the over all evaluation of a patient's presurgical clinical data, our SM localizations have led to a significant reappraisal in two cases; in the remaining four Ps, including the P of Figs. 1-4, the outcome of SM localization confirmed the conclusions reached independently on the basis of standard clinical examinations.

Discussion

Although several recent studies report successful MM localization of interictal and ictal spike activity to regions congruent with those indicated by other diagnostic procedures [e.g. 7, 8], an appraisal of the clinical importance of MEG as compared to EEG is still difficult.

What is obvious is that MEG cannot simply replace EEG since it would miss activity emanating from predominantly radial sources. One of our patients provided an instructive example hereof: A focus in her extra-mesial temporal cortex radiated large EEG spikes accompanied by very small MEG activity which, when viewed in isolation, would have been difficult to interpret; EM indicated an almost radial source. Thus, EEG remains a mandatory method in the search for epileptic activity.

On the other hand, in view of its cost, few institutions will afford MEG if it does not provide more than only a confirmation of EEG. The basic question therefore is whether *adding* MEG will *improve* the quality of localization.

In order to try a first answer to this question, we have selected six SCPE patients (out of a population of 32) by applying strict criteria as to both the mathematical quality of spike localization (S/N-ratio >4) and the verification of the epileptogenic region so that we could quantitatively compare the results of EM, MM, and SM. The results of this comparison suggest that the combination of EEG and MEG in a *simultaneous electric magnetic model* may indeed improve localization in that it tends to point to a locus closer to the border of the lesion in cases of lesion-related SCPE than either EEG or MEG alone. However, this conclusion is expressed with several reservations:
(1) The number of well documented cases is still very small. (2) By combining EEG and MEG we have increased the spatial sampling density in comparison to the situation where either only EEG or MEG is considered. One could argue that

with double the number of EEG-channels, EM might yield results comparable to SM. Yet, we feel that the resulting $\sqrt{2}$-fold increase of sampling density cannot engender shifts in localization by up to 3.5 cm (maximum in our group of Ps), and this all the more as we had spaced electrodes overlying the presumed spiking area more closely (cf. Methods) than called for by the Nyquist frequency of superficial sources (0.5 cm^{-1}). (3) We have arbitrarily considered the border of the lesions (as discernible in MRI) as the most likely area of origin of the spike activity (unless there was clear clinical evidence to the contrary). Clearly, a better delineation on the basis of intra-cranial grid recordings is needed which, ideally, would be accompanied by simul-taneous surface EEG and MEG measurements. On the other hand, the MEG machine used in the present experiments had only a limited angle of view on the Ps' heads. It may well turn out that with a whole head system, SM will give even better results.

Finally, independent of the questions regarding the specific contribution of MEG, provocation by metho-hexital proved to be extremely useful for simultaneous MEG/EEG measurements in SCPE patients. Methohexital induces quasi-ictal activity [3] (which is more relevant than interictal activity for the identification of the epileptogenic area), and often reduces the time required to collect a sufficient number of spikes from hours to minutes.

References:

[1] Becker, W., Diekmann, V., Jürgens, R. First experiences with a multichannel software gradiometer recording normal and tangential components of MEG. *Physiol. Meas.,* 1993, 14: A45-A50

[2] Diekmann, V., Jürgens, R., Becker, W., Elias, H., Ludwig, W., and Vodel, W. RF- to DC-SQUID upgrade of a 28-channel magnetoencephalography (MEG) system. *Measurement Science and Technology,* in press

[3] Hufnagel, A., Burr, W., Elger, C.E., Nadstawek, J., Hefner, G. Localization of the epileptic focus during methohexital-induced anesthesia, Epilepsia, 1992,33: 271-284

[4] Diekmann, V., Becker, W., Grözinger, B., Jürgens, R., Westphal, K.-P., Kornhuber, H.H. Localization of focal epileptic activity with a simultaneous EEG and MEG model, In: Baumgartner, C., Deecke, L., Stroink, G., and Williamson, S.J., Biomagnetism: Fundamental Research and Clinical Applications, Amsterdam, IOS-Press, 1995, pp 23-27

[5] Vieth, J., Kober, H., Weise, E., Daun, A., Moegner, A., Friedrich, S. and Pongratz, H. Functional 3D localization of cerebrovascular accidents by magnetoencephalography, *Neurological Research,* 1992, 14, Suppl.: 132-134

[6] Mosher, J.C., Spencer, M.E., Leahy, R.M., Lewis, P.S. Error bounds for EEG and MEG dipole source localization, *Electroenceph. clin. Neurophysiol.,* 1993, 86: 303-321

[7] Smith, J.R., Gallen, C., Orrison, W., Lewine, J., Murro, M., King. D.W., Gallagher, B.B. Role of multichannel magnetoencephalography in the evaluation of ablative seizure, *Stereotact. Funct. Neurosurg.,* 1994,62: 238-244

[8] Stefan, H., Schneider, S., Feistel, H., Pawlik, G., Schüler, P., Abraham-Fuchs, K., Schlegel, T., Neubauer, U., Huk, W.J. Ictal and interictal activity in partial epilepsy recorded with multichannel magneto-encephalography: correlation of electroencephalography/electrocorticography, magnetic resonance imaging, single photon emission computed tomography, and positron emission tomography findings, *Epilepsia,* 1992, 33(5): 874-887

Intracortical Spread of Epileptic Spikes - Simultaneous Measurement of Magnetic Field and Electric Potential in Rabbits

[1]Eiselt, M., [1]Zwiener, U., [2]Gießler, F., [1]Wagner, H., [2]Haueisen, J., [2]Huonker, R., [2]Nowak, H., and [3]Schack, B.

[1]Inst. Pathophysiology; [2]Biomagnetic Centre; [3]Inst. of Medical Statistics, Informatics and Documentation, Clinic of the Friedrich Schiller University, 07740 Jena, Germany

Introduction

In many experiments we observed differences between simultaneously measured ECoG and MEG signals concerning the temporal relationship of characteristic features like maxima or minima. This was not surprising despite MEG and extracellular electric potential are functionally related. They are caused by different bioelectric processes. The MEG is mainly caused by intracellular currents of neuronal cell populations. The extracellular electric potential is caused by currents through the cellular membrane. These currents can be elucidated by current source density analysis (CSD).

The following questions should be answered: How do different cortical depths of spike foci influence the temporal relationship between ECoG and MEG ? Which part of spike generation processes is detected by the MEG and which part may influence the electric potential at the cortical surface ?

Methods

Initiation of epileptic spikes

Interictal spikes were investigated by CSD in 5 and the MEG in 12 adult rabbits. Additionally, in all animals the ECoG was recorded. Spikes were initiated by stereotactic, intracortical application (600 - 750 μm below the cortical surface) of small amounts of penicillin (200 I.E.). For CSD investigations penicillin was injected in superficial brain surface (Regio retrosplenialis). For MEG investigations two locations of penicillin application were used: superficial cortical area (Regio retrosplenialis; 3 animals; main current dipole with radial direction) and Gyrus cinguli of the left or right hemisphere 2000 μm below the superficial brain surface (9 animals; main current dipole tangential direction to the cortical surface; fig. 1). Both regions had similar architecture of histology.

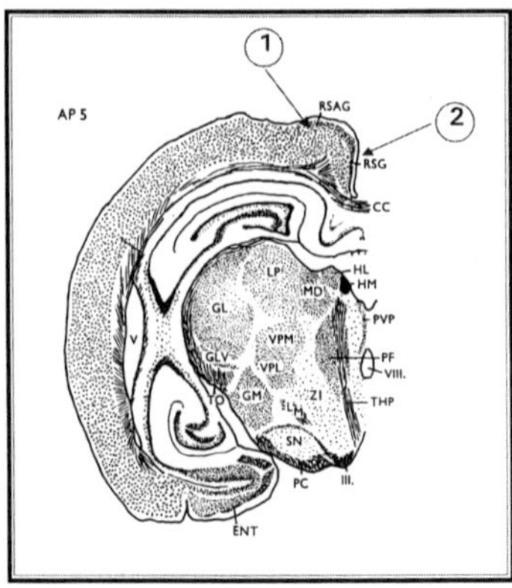

Fig. 1. Rabbit brain with cortical regions of penicillin application: 1 = Regio retrosplenialis - current dipole in radial direction; 2 = Gyrus cinguli - current dipole in tangential direction.

Measurements

Intracortical electric potential distribution perpendicular to the cortical surface was detected by 16 electrodes spaced 150 μm apart. ECoG was measured on the brain surface immediately above the region of penicillin application. The magnetic field was detected by a 7 channel 1st order gradiometer system (sensitivity 30 fT/ √Hz, diametre 9.5 mm). The measurement was repeated five times and thus, 35 channel MEG measurement could be performed.

MEG and ECoG data were sampled at a rate of 1000 Hz, intracortical potential at a rate of 512 Hz. They were filtered with band pass 3 - 30 Hz. This has changed the moment of maximum and minimum appearance in ECoG and MEG in the same way. In comparison to raw data the time shift caused by filtering was smaller than 2 msec. To improve signal to noise ratio (> 30 : 1) 200 spikes detected by MEG were averaged.

Testing the recording systems with an artificial signal the frequency dependent phase shift between magnetic signal and electric signal was at 20 Hz 1.5° which corresponds to 0.2 msec. The electric signal preceded the magnetic signal independent from the frequency band investigated.

Calculations

One dimensional CSD analysis was based on the 2nd spatial derivative of intracortical electric potential. Calculation of CSD and fitting of data sets were done by algorithms published by Rappelsberger et al. [5] and the Neville algorithm [7].

For CSD and the magnetic field hilliness (which corresponds to global field power [4]) and global similarity (a parameter similar to the parameter published by Desmedt and Chalkin [1]) were calculated millisecond by millisecond (fig. 2). With the help of these parameters the occurrence of greatest amplitude of the whole signal at each instant and the strongest changes over time could be detected.

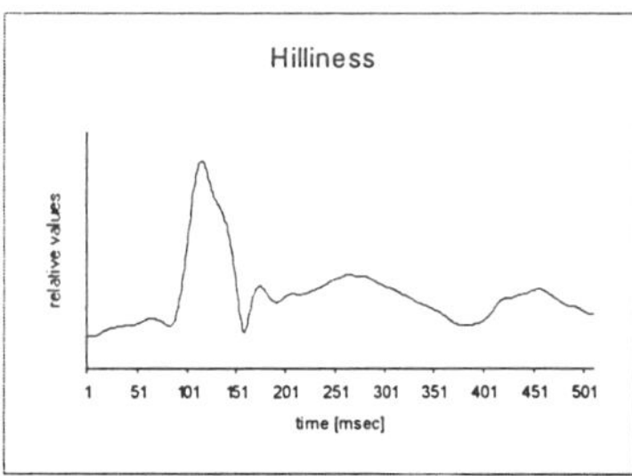
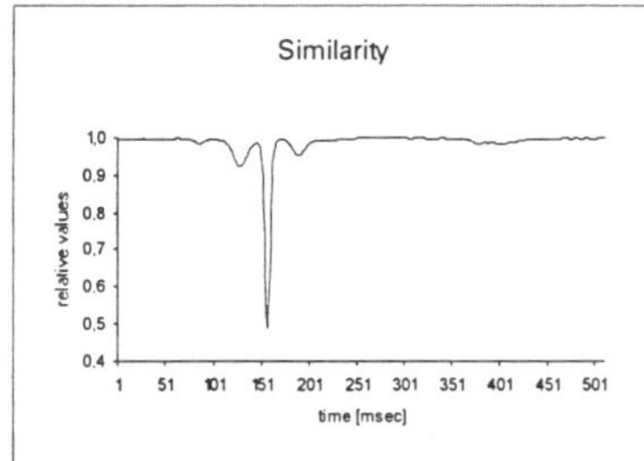

Fig. 2. Hilliness and similarity calculated from the magnetic field. In the left diagram a peak of hilliness can be observed at about 120 msec followed by three further peaks with lower amplitude. In the right diagram the reduction of similarity indicates strong change within the magnetic field. Both may indicate that there are multiple processes and the first is different from the following.

For CSD and MEG the temporal relationship between peaks of hilliness were compared with the occurrence of the first negative maximum and the following positive maximum of ECoG spikes. Events occurring earlier were marked by negative time.

Results

There were no differences between spikes detected by ECoG in the three groups (CSD, MEG with radial directed dipole and MEG with tangential directed dipole) concerning the temporal relationship of the negative and positive maximum occurrence.

The maxima of CSD occurred simultaneously with the negative (1 ± 2.6 msec; not significant - n.s.) or positive maximum (1 ± 2.7 msec; n.s.) of the ECoG spike (fig. 3).

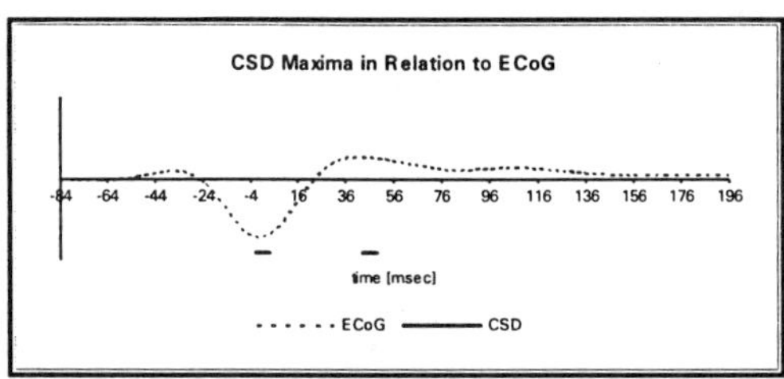

Fig. 3. The averaged moments of occurrence of CSD maxima in relation to ECoG spike (n = 5; lines = mean ± SD of the moments of occurrence of CSD maxima; dashed line = electric potential of ECoG).

Similar observations were made between spikes detected by MEG and ECoG when penicillin was injected in the superficial cortical area. There were no significant temporal differences between ECoG and MEG (negative ECoG maximum - 1st maximum MEG: -2.1 ± 5.3 msec, n.s.; positive ECoG maximum - 2nd MEG maximum: -2.9 ± 8.6 msec, n.s.; fig. 4).

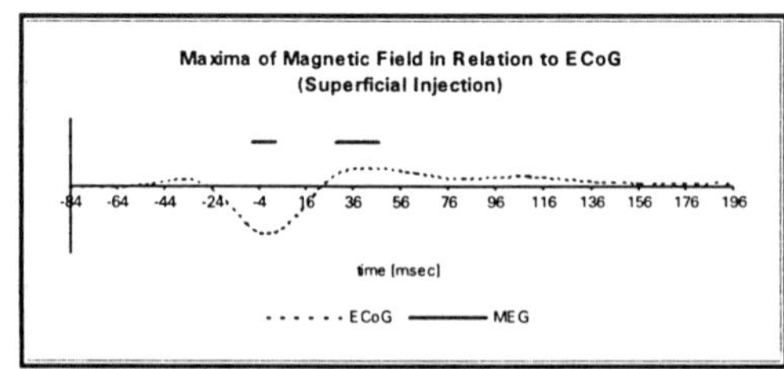

Fig. 4. The averaged moments of occurrence of magnetic field maxima in relation to ECoG spike when penicillin was injected in superficial brain surface (n = 3; lines = mean ± SD of the moments of occurrence of magnetic field maxima; dashed line = electric potential of ECoG).

Spikes detected by MEG appeared earlier than spikes detected by ECoG when penicillin was injected in the region of Gyrus cinguli (2000 μm below the brain surface). The first MEG maximum occurred -19.3 ± 14.1 msec (p < 0.001) before the negative maximum of the spike detected by ECoG and the second MEG maximum -10.0 ± 21.7 msec (n.s.) before the positive maximum (fig. 5).

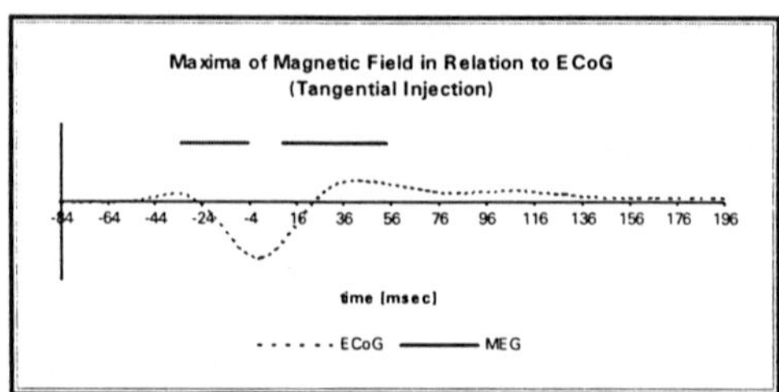

Fig. 5. The averaged moments of occurrence of magnetic field maxima in relation to ECoG spike when penicillin was injected in Gyrus cinguli 2000 μm below the brain surface (n = 9; lines = mean ± SD of the moments of occurrence of magnetic field maxima; dashed line = electric potential of ECoG).

The mean difference between areas of penicillin application and the location of the current dipoles by the method of spatial filtering [6, 8] was between 0.7 - 1 mm.

Discussion

In this study neuronal processes initiated by focal penicillin application were investigated by three different methods. The one dimensional CSD perpendicular to the cortical surface describes neuronal processes immediately influenced by penicillin application. These are synchronised excitation of great amounts of neurones followed by inhibitory processes responsible for avoidance of repetitive excitations observable during seizures. During these periods intracellular currents are directed towards cortical depth or cortical surface. Simultaneously to this process neuronal activity propagates towards the surrounding tissue involving other neuronal populations and initiate there inhibition which limits the further spread [2]. If the location of the spike generation is near the cortical surface the ECoG is mainly influenced by the first excitatory processes. If spike generation occurs in deeper cortical areas the ECoG is mainly influence by the events caused by the spread of neuronal activity within the surrounding tissue. The MEG will be influenced by the whole neuronal activity but mainly by the synchronised excitation involving many neurones in the same way. This will cause a signal strong enough to be detected outside the brain.

There are different time shifts between ECoG and MEG during the initial process describing the excitation and the following process. We suppose that neuronal activity during the negative spike occurs in restricted regions and propagates towards the cortical surface and thus causing a time delay between ECoG spike and MEG. Neuronal activity is more distributed during the positive wave [3]. Thus, it reaches earlier the cortical surface and the time delay between ECoG and MEG is smaller.

References

[1] Desmedt, J.E., and Chalkin, V. New method for titrating differences in scalp topographic patterns in brain evoked potential mapping, Electroencephalography and clinical Neurophysiology, 1989, 74: 359-366.

[2] Eiselt, M., Zwiener, U. , Wagner, H. Conditions for a precise analysis of penicillin-induced spikes by CSD-analysis, In: Eiselt, M., Zwiener, U., and Witte, H. Quantitative and topological EEG and MEG analysis, Jena, Universitätsverlag Druckhaus Mayer, (1995) 231 - 236.

[3] Goldensohn, E.S. and Salazar, A.M., Temporal and spatial distribution of intracellular during generation and spread of epileptogenic discharges, In: Delgado-Escueta, A.V., Ward, A.A., Woodbury D.M. and Porter, R.J., Advances in Neurology, New York, Raven Press, 1986, 44: 559-589.

[4] Lehmann, D., Ozaki, H. and Pal, I. EEG alpha map series: brain micro-states by space-oriented adaptive segmentation, Electroencephalography and clinical Neurophysiology, 1987, 67: 271 - 288.

[5] Rappelsberger, P., Pockberger, H. and Petsche, H. Current Source Density analysis: Methods and application to simultaneously recorded field potentials of the rabbit's visual cortex, Pflügers Archiv, 1981, 389: 159 - 170.

[6] Robinson, S.E. and Rose, D.F. Current source image estimation by spatially filtered MEG. In: Hoke, M., Erné, S.N., Okada Y.C. and Romani, G.L., Biomagnetism: Clinical aspects, Amsterdam: Excerpta medica, 1992, 337 - 338.

[7] Stoer, G. Einführung in die numerische Mathematik I., Berlin, Heidelberg, New York, Springer, 1976.

[8] Wagner, H., Eiselt, M. Zwiener, U. Investigation of the exactness of the source analysis from biomagnetic signals by the method of spatial filtering with particular consideration of the analysis of interictal spike activity - a computer simulation. Medical & Biologica Engineering & Computing 1995 (submitted)

supported by BMBF P 01 ZZ9104

Effects of Volume Current on Localization of Focal Epileptic Activity

Ikeda, H.[1], Takeuchi, R.[1], Osozawa, S.[1], Ishiyama, A.[1], Kasai, N.[2], and Minami, T.[3]

Department of Electrical Engineering, Waseda University[1], Tokyo, Japan
Electrotechnical Laboratory[2], Tsukuba, Japan
Faculty of Medicine, Kyushu University[3], Fukuoka, Japan

Introduction

In the localization of focal epileptic activity, there are some cases that the estimated positions scatter. In these cases, the improvement on the localization of focal epileptic activity is expected. Usually the foci of epilepsy are localized with the spherical model assuming a single current dipole for a focus from MEG (Magnetoencephalogram). However, skull and palate are in a head, then a head is not spherical and is inhomogenous. We previously investigated the effects of volume current on the source localization by experiments and computer simulation with a single-layer hemispherical phantom. From the results, it was found that the contribution of volume current to the MEG increases when the current dipole exists on the region near the bottom and in the deep place of the head [1][2][3]. In this paper, we investigate the possibility of improvement on the localization of focal epileptic activity with the MEGs of three subjects of focal epilepsy by considering the effects of the volume current.

Methods

The MEGs of subject A (1 year old, female) and subject B (20 years old, male), and the MEG of subject C (10 years old, male) were recorded at Kyushu University (at rate of 520.83 samples/sec) and Tokyo University (at rate of 1041.7 samples/sec) respectively. All MEGs were recorded with a Magnes™ 37-channel Biomagnetometer (Biomagnetic Technologies Inc.). After recording, the MEGs were filtered to retain only the frequency band 1-30 Hz for subject A and 1-60 Hz for subject B, and 8-48 Hz for subject C to select for focal epileptic waves. An example of the focal epileptic waves for subject B is shown in Fig.1. It is difficult to know the zero-level in MEG. We take the output of each magnetometer at the beginning of the wave, t1 as the zero-level. The time t1 is shown by an arrow in Fig.1. We localized the focus of epilepsy using the MEG at the time near the peak of the waves. The isofield map at t2 of the waves shown in Fig.1 is shown in Fig.2.

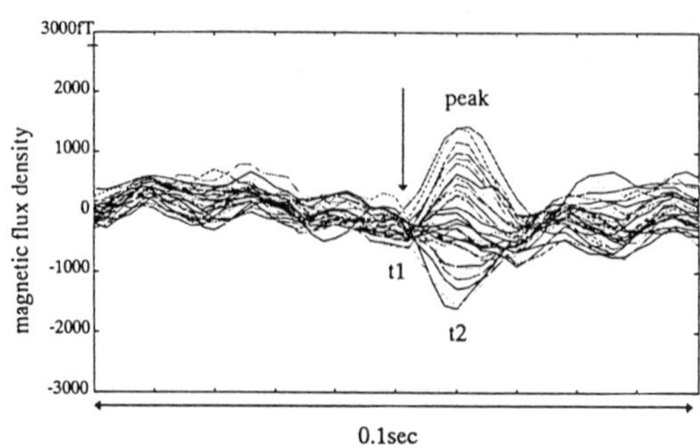

Fig.1 Focal epileptic waves for subject B
The all waves measured are shown together.

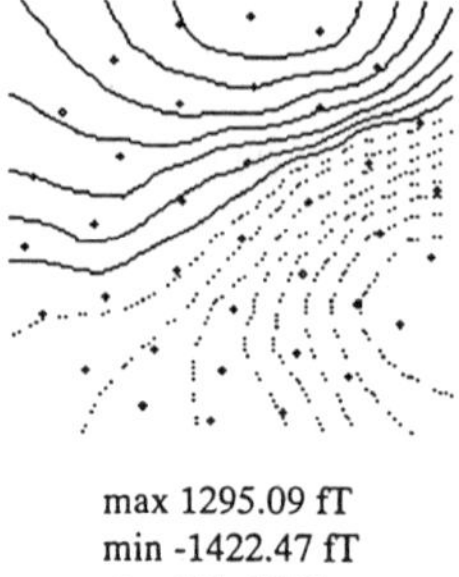

Fig.2 Isofield map at t2
The solid lines show plus values and the dotted lines show minus values, and the circles show the position of pick up coils.

We assumed the focus of epilepsy as a single current dipole. In the source localization, the volume current was considered by the equation (1) [4].

$$B(r_P) = \frac{\mu_0}{4\pi}\int_v J_i \times \nabla \frac{1}{|r - r_P|}dv - \frac{\mu_0}{4\pi}\sum_j \left(\sigma_j^{in} - \sigma_j^{out}\right)\int_{S_j} Vn_j \times \nabla \frac{1}{|r - r_P|}dS \tag{1}$$

where σ_j^{in} and σ_j^{out} are the conductivities of two different adjacent regions separated by the surfaces S_j, J_i is the current sources at r, n_j is the outer unit normal to the surface S_j, and V is the electrical potential. In above equation, the first term is the field due to the current sources and the second term denotes the field due to the volume current. The models for boundary elements were constructed by extracting the outline of the brain from the MRI or X-ray CT of each subjects to calculate the second term in equation (1). To compare our model with the spherical model, the localization of the focus is also carried out by the spherical model. The equation (2) calculated by Sarvas was used in the localization for the spherical model [5].

$$B(r_p) = \frac{\mu_0}{4\pi F^2}\left(FQ \times r - Q \times r \cdot r_p \nabla F\right) \tag{2}$$

where $F = a\left(r_p a + r_p^2 - r \cdot r_p\right), a = r_p - r, a = |a|, r_p = |r_p|, Q = \int_v J_i(r)dv$

The models for boundary elements are shown in Fig.3 with the position of pick up coils that are shown as black dots. Subject A has a chipped region in her head. The models which consist of two parts, brain and cerebrospinal fluid were used as shown in Figs.3 (a) and (b). The conductivities of cerebrospinal fluid and the brain ,and the outside of the brain were assumed to be 1.0 S/m, 0.3 S/m, and 0.0 S/m, respectively. (c) is the model for subject B, and (d) is the model for subject C. The CT of subject C is the only upper side of his brain, therefore the model for subject C looks like a hemisphere as shown in (d). All models are looked from the left in Fig.3.

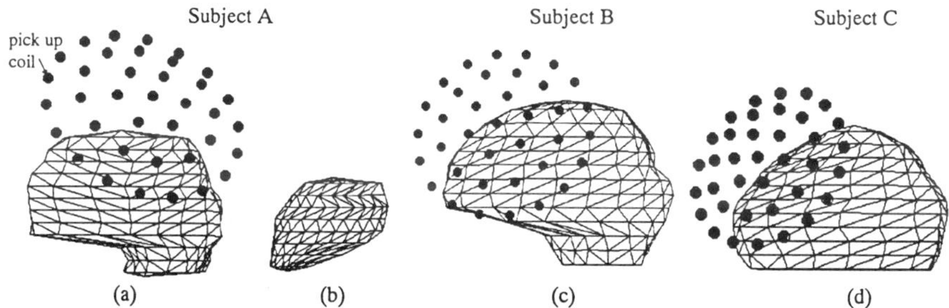

Fig.3 Models for boundary elements and the position of pick up coils
(a): the boundary elements of the brain with subject A, (b): the boundary elements
of the chipped region with subject A , (c): the boundary elements of the brain with
subject B , and (d): the boundary elements of brain with subject C

Results

The localizations of the foci were carried out with 35 waves for subject A and 79 waves for subject B, and 28 waves for subject C. The results of the localization of the focus are shown on the outline of the brain below. Only the results which have the dipolarity higher than 98% are shown. We applied the equation (3) as the dipolarity (DP).

$$DP = \sqrt{1 - \frac{\sum_i \left(B_i^c - B_i^m\right)^2}{\sum_i \left(B_i^m\right)^2}} \times 100 \quad [\%] \tag{3}$$

where B_i^c is the computed flux density value at the point of the pick up coil i, B_i^m is the measured flux density value at the point of the pick up coil i.

The results for subject A are shown in Fig.4. (a) is the result localized using the spherical model and (b) is the result localized using boundary element method (BEM). The results are looked from the left. The dashed line shows the outline of chipped region.

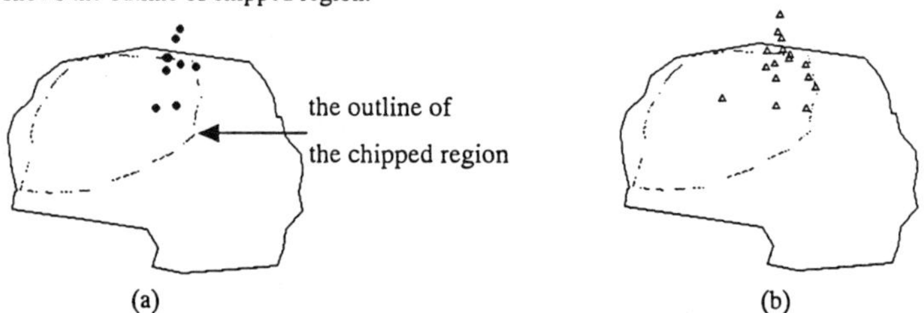

Fig.4 Localization of the focus with subject A
(a): results localized using the spherical model and (b): results using BEM

Both of results seems to be similar. It shows that the inhomogenity slightly influences the localization of the focus.

The results for subject B and subject C are shown in Fig.5 and in Fig.6 respectively. (a) is the result localized using the spherical model and (b) is the result localized using BEM in Fig.5 and in Fig.6. These results are looked from the left.

Fig.5 Localization of the focus with subject B
(a): results localized using the spherical model and (b): results using BEM

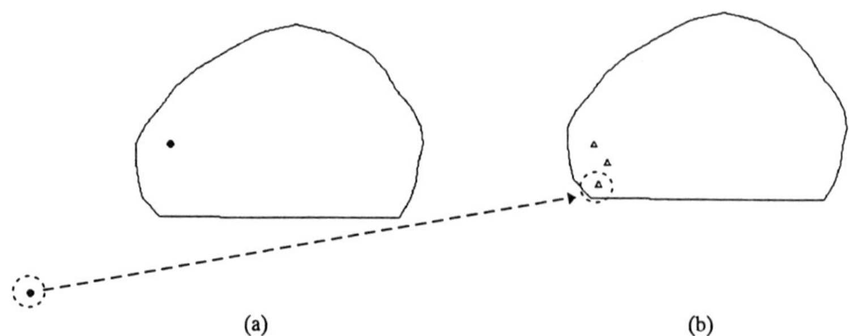

Fig.6 Localization of the focus with subject C
(a): results localized using the spherical model and (b): results using BEM

In the case of subject B, the estimated positions using BEM are almost same as the estimated positions using the spherical model.

Though, in the case of subject C, one of the estimated positions using the spherical model is out of the brain as shown by the dotted circle in Fig.6, however the estimated positions using BEM are in the brain. It suggests that there is a possibility of improvement on the localization of the focus using BEM.

Discussion

We investigated the contribution of volume current to the MEG depending the dipole position by a computer simulation with a single-layer hemispherical phantom model. The model we used is shown in Fig.7. We simulated the contribution of volume current to the MEG by changing positions of the current dipole. d is the depth from the surface of hemispherical model. The result of the simulation is shown in Fig.8 [2][3]. To estimate the contribution of volume current to the MEG, the equation (4) is used.

$$f_t = \frac{\sum_i (B_i^{vol} - B_i^{source})^2}{\sum_i (B_i^{sourcet})^2} \qquad (4)$$

where B_i^{vol} is the field due to the current sources and volume (return) current, B_i^{source} is the field due to the current sources. From the result, it is found that the contribution of volume current to the sum of magnetic fields at all measurement points increases when the current dipole exists on the region near the bottom and in a deep place of the model. It suggests that the localization of the focus considering the volume current using BEM will improve the accuracy when the current dipole exists in the frontal lobe or temporal lobe. The results of 3 subjects are in agreement with the result of the simulation.

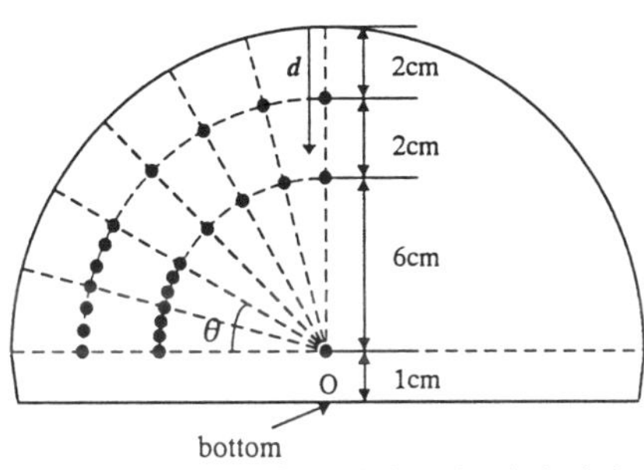

Fig.7 Cross section of a single-layer hemispherical phantom

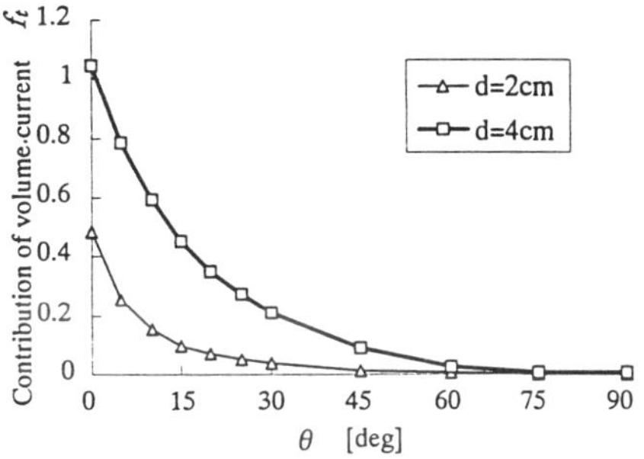

Fig.8 Contribution of volume current to the MEG depending the dipole position

Summary

We have investigated the possibility of improvement on the localization of focal epileptic activity with the MEGs of three subjects of focal epilepsy by considering the effects of volume current. From the result, it suggests that the method of the localization considering the volume current has a potential to improve the accuracy when the current dipole exists in the frontal lobe or temporal lobe.

Acknowledgements

The authors would like to thank Dr. Kaneko for his helpful suggestions and discussions.

References:

[1] I. Kanai, Y. Nagasaka, A. Ishiyama, Source Estimation by a Method Combines MEG and EEG, The Journal of Japan Biomagnetism and Bioelectromagnetics Society (in Japanese), 1992, 5: 204-207.

[2] M. Kobayashi, I. Tonegawa, A. Ishiyama, N. Kasai, Influence of Volume Current on MEG Using a Semi-spherical Phantom Model, The Journal of Japan Biomagnetism and Bioelectromagnetics Society (in Japanese),1993,6: 102-105.

[3] R. Takeuchi, S. Osozawa, T. Miyashita, A. Ishiyama, N. Kasai, Influence of Volume Current and Dipole Position on MEG Source Localization, The Journal of Japan Biomagnetism and Bioelectromagnetics Society (in Japanese), 1994, 7: 154-157.

[4] F. Grynspan, D.B.Geselowitz, Model studies of the magnetocardiogram, Biophys.J., 1973, 13: 911-925.

[5] J.Sarvas, Basic mathematical and electromagnetic concepts of the biomagnetic inverse problem, Phys. Mad. Biol, 1987, 32, 11-22.

Evaluation Of Abnormal Low Frequency Magnetic Activity (ALFMA) In Epilepsy

Lewine, J.D., Davis, J.T., and Orrison, W.W. Jr.

The New Mexico Institute of Neuroimaging, The New Mexico Regional Federal Medical Center, Albuquerque, New Mexico, USA.

INTRODUCTION

MEG attempts to define regions of epileptic pathology in patients with a seizure history are compromised by several factors. First, most studies are interictal in nature, an issue of concern to many epileptologists. Also, in some cases epileptic magnetic field patterns are highly complex and not amenable to simple single or even multiple dipole modeling. A final concern is the observation that most patients have seizures only very intermittently, with their interictal EEG often devoid of epileptic spikes and sharpwaves.

We have therefore sought to determine if aspects of MEG signals beyond spikes and sharpwaves might provide insights into epileptic pathology. Of particular interest were preliminary observations that many patients with a history of seizures, including those with interictal MEG and EEG devoid of spikes and sharpwaves, show abnormal low frequency magnetic activity (ALFMA) that can provide clinically useful information on the location of an epileptic focus [1,2]. Presented herein, are preliminary results of a large scale assessment of ALFMA in epileptic patients.

METHODS

Subjects: Sixty sequential patients with a history of seizures but without mass lesions (tumors of vascular malformations) or other gross neuroanatomical pathology (e.g., cortical dysplasia) were included in this study. The ages of the patient population ranged from 14 to 66 years.

MEG Examination: Neuromagnetic data were collected using a 37-channel, axial-gradiometer system (Magnes, Biomagnetic Technologies Inc., San Diego, CA). The study was conducted in a magnetically and radio-frequency shielded room. For the characterization of ALFMA, three minutes of continuous data (collected as ninety sequential two-second long epochs) were collected at each of eight placements of the array. The placements were over the following cortical regions: (1) midline frontal, (2) left parietal, (3) left temporal-occipital-parietal junction, (4) left temporal/inferior frontal, (5) right parietal, (6) right temporal-occipital-parietal junction, (7) right temporal/inferior frontal, and (8) midline occipital. Additional data were collected over regions of interest for evaluation of spike and sharpwave, but the data from these recordings were not included in the analyses reported below.

Data were collected at a digitization rate of 300 Hz, with on-line filtering from 1-100 Hz. The data set from each probe position was analyzed separately. The dipole analysis of ALFMA involved four steps: (1) filtering and peak selection; (2) dipole modeling; (3) cluster analysis; and (4) generation of magnetic source localization images. First, MEG data were digitally filtered with a bandpass of 1-4 Hz in order to isolate ALFMA in the delta band. Independent analyses were performed on the data collected at each dewar position. After identification of the latencies of local amplitude peaks where the magnetic field strength was greater than 200 femtoTesla, single dipole modeling was performed every 8 msec during a 40-msec long window spanning each amplitude peak. In most cases, more than 5000 dipole fits were calculated for the data stream collected at each dewar position. Only the most dipolar of these fits ($r > 0.97$) were considered in further analyses. As a consequence of this strict criteria for dipolarity, more than 99% of the total amount of data collected is rejected from further consideration. A spatial clustering algorithm was used to determine if a particular cortical region was responsible for generation of multiple dipolar low frequency events. This is a critical aspect of the analysis procedure because available data suggest that it is the repetitive generation of dipolar low frequency activity that provides the most reliable sign that a region of tissue is dysfunctional. A 1-cm radius volume element was considered pathological only if it was responsible for generation of fifteen distinct low frequency events. This analysis strategy has been used to assess several clinical populations and more than 40 neurologically normal subjects. It yields a low false positive rate of only 8%.

MRI Examination: Magnetic resonance imaging was performed with a 1.5 Tesla whole body imaging system (Magnetom, Siemens Medical Systems, Islen, NJ). Images were obtained using modified gradient and spin echo sequences.By identifying the MEG coordinate fiducials on corresponding MR images it was possible to reconcile MEG and MR spatial frames to generate magnetic source localization images where the locations of the dipole sources of ALFMA are marked on appropriate MR sections

RESULTS

The MEG data from ten patients were uninterpretable because of low frequency magnetic artifacts from dental work or excessive eye movements. For the remaining fifty patients, interictal spikes and sharpwaves were found for twenty (40%) whereas ALFMA was found for forty-one (82%). ALFMA activity tended to be aperiodic, as shown in the below neuromagnetic waveforms recorded over the left temporal cortex of a sixteen-year-old female patient [Fig 1]. The early portion of epoch 30 show a relatively normal pattern of neuromagnetic activity. In contrast, the later portion of epoch 30 and all of 31 are punctated by low frequency delta and theta bursts. Whereas MEG power spectra for this recording position are normally dominated by 8-13 hz activity with little delta, the average power spectrum (across epochs and sensors) for this patient showed abnormal dominance of delta activity. Most slow waves have complex magnetic field patterns, but some have highly dipolar patterns. This suggests relatively focal generation, and single dipole modeling strategies can provide a first order localization of the relevant neuronal sources. In many cases the generators of ALFMA are found to localize to a single brain lobe, a result strongly suggestive of focal dysfunction [Fig 1D].

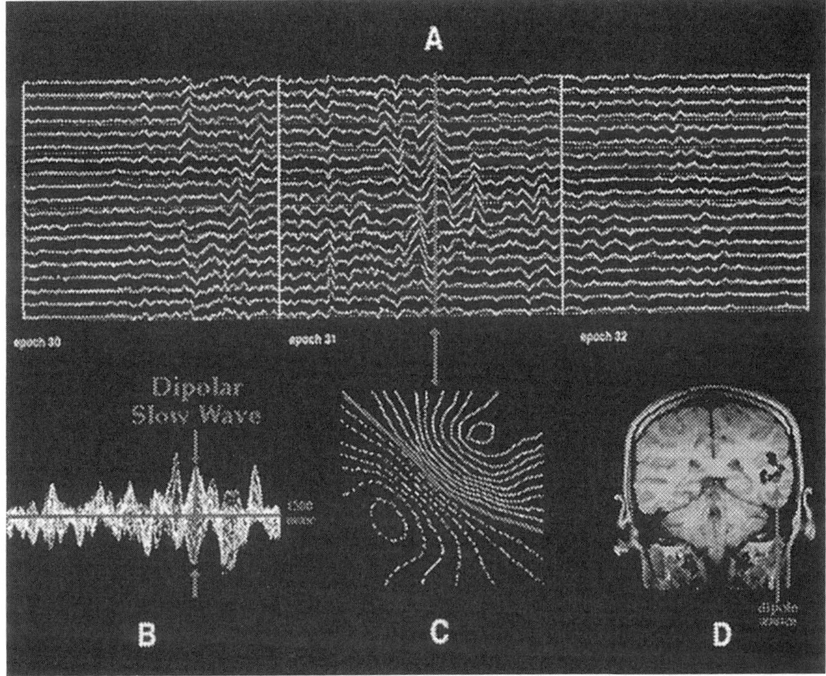

Fig 1: (A) Continuous data showing aperiodic slow waves. (B) Data overlay of 37-channels. Some slow waves have a dipolar field pattern, as shown in (C). (D) The sources of dipolar abnormal low frequency magnetic activity often cluster to a common area, a result that suggests the region to be dysfunctional. [reproduced from Lewine and Orrison, Neuroimaging Clinics of North America, 1995, 5(4), page 593]

Individual ALFMA source sometimes localize to white matter, a result suggestive of mis-modelling of activity generated by an extended cortical sheet. Nevertheless, even a mis-localized deep source provides information on the side and lobe of dysfunction. The distribution of ALFMA sources within a lobe may be scattered, and each source position is probably accurate to within only 1-2 centimeters. Nevertheless, in most cases ALFMA analyses clearly implicate a particular hemisphere and lobe in the epileptic pathology of the patient, and the result is consistent with clinical history and data from other neuroimaging procedures.

Several observations support the ALFMA approach to characterization of epilepsy. ALFMA is present for about 90% of those patients that do show interictal MEG spikes or sharpwaves, and when present, there has always been co-localization of spikes and some ALFMA. Admittedly, in many cases, several sites of ALFMA may be identified, although the most intense focus invariably correlates with spike locations [Fig 2].

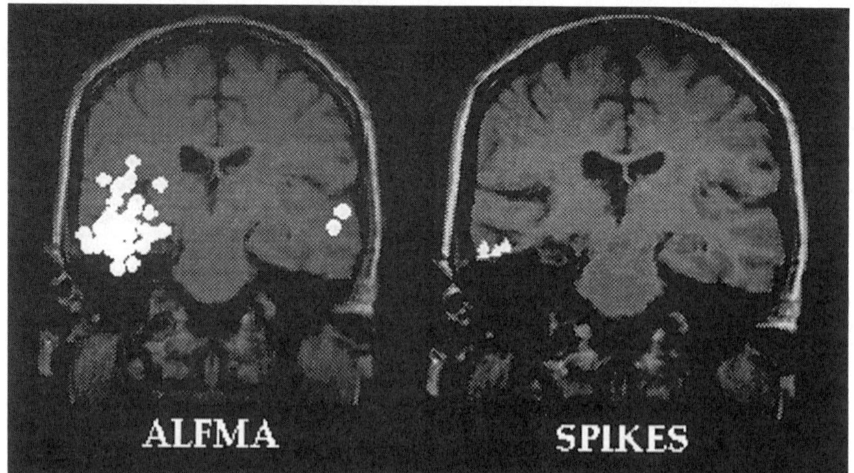

Fig 2: Data are from a 36 year-old female with seizures that proved resistance to partial right anterior temporal lobectomy. Dipolar sources of ALFMA are identified bitemporally. More activity is identified on the right. Several epileptic spikes were recorded also, with sources localizing to the right temporal lobe, just posterior to the original resection zone.

DISCUSSION

Biomagnetic characterization of patients with epilepsy is a challenging problem because examinations are mostly limited to the interictal period. At present, there continues to be tremendous debate over the utility of interictal data relative to ictal data. It is generally accepted that ictal data is preferred, but this is not always available. Within the interictal domain, MEG is faced with the problem that many patients fail to show sufficient interictal spiking to allow for confident identification of a focus. In fact, when non-selected patients are considered, only about 40% of patients demonstrate spikes and sharpwaves during a typical 2 hour MEG examination. Fortunately, as presented herein, greater than 80% of epileptic patients demonstrate abnormal low frequency magnetic activity during the interictal period. These patients also demonstrate slow waves in EEG recordings, but the activity tends to be relatively diffuse, even in cases suspected to have focal epilepsy. In contrast, slow waves in MEG are often focal and sources associated with separate slow waves cluster into circumscribed brain regions.

On the one hand, it is surprising that the ALFMA analysis technique, which is unable to adequately characterize more than 99% of the collected data, nevertheless provides clinically useful information. Single dipole modeling of spontaneous brain activity is considered by many to be inappropriate, but the present results indicate that this approach can be useful in the clinical identification of pathophysiology.

The fundamental rationale for this strategy is that the abnormal signals from focal dysfunctional regions are so large that they dominate the recorded magnetic field pattern, with the simple dipole model being only minimally perturbed by the lower amplitude activity from nearby normal tissue. In evoked response studies where MEG is used to identify the location of primary somatosensory cortex, source locations are thought accurate to within a few millimeters. This is almost certainly not the case in ALFMA studies where there is no signal averaging and the magnitude of the extraneous background brain activity is high. Dipole sources of ALFMA often localized in the white matter, rather than at the cortical mantle as expected on the basis of biophysical considerations. This most likely indicates that an extended region of the overlying cortex was actually responsible for ALFMA generation, with dipole modeling of this extended activity giving an erroneously deep source location (the magnetic field pattern of an extended cortical sheet and a deeper dipole source are often indistinguishable). Nevertheless, ALFMA source locations are probably an accurate reflection of the physiology to within 1-2 cm. Whereas this could be inadequate resolution for some neurosurgical planning, it is more than sufficient for identifying which gross brain regions are compromised by epilepsy.

At present the nature of the pathological processes that cause ALFMA are not fully understood. ALFMA is found in many conditions including mild head trauma, ischemic disease, and dementia. Multiple physiological mechanisms may all have similar biomagnetic manifestations.

In considering the clinical impact of the examination of ALFMA it is clear that the presence of ALFMA alone is inadequate for directing surgical intervention. However, when ALFMA data agree with data from other noninvasive methods, a strong case can be made for surgical intervention without the need for invasive monitoring. In those cases where invasive monitoring is still indicated, data on ALFMA can be used to guide the placement of subcortical grids and depth probes.

CONCLUSIONS

The ALFMA strategy outlined here has promise for augmenting the clinical utility of MEG evaluations of patients with epilepsy. ALFMA is present in a high percentage of patients, even in the absence of interictal spikes and sharp waves. Preliminary results suggest that the clinical effectiveness of the ALFMA investigation is superior to that for interictal SPECT or PET.

REFERENCES

[1] Lewine, J.D., and Orrison, W.W., Spike and slow wave localization by magnetoencephalography, Neuroimaging Clinics of North America - Epilepsy, 1995, 5(4): 575-596.

[2] Vieth, J., Grummich, P., Kober H., et al. Localization of slow and beta MEG waves associated with epileptogenic lesions. Epilepsia, 1993, 34 (supplement 6): 143.

Event-Related Magnetic Fields Associated with Auditory Oddball Paradigm - Postoperative Study in Patients with Temporal Lobe Epilepsy

Nishitani, N.[1], Nagamine, T.[1], Mikuni, N.[1,2], Ikeda, A.[1], Kimura, J.[3] and Shibasaki, H.[1]

[1]Departments of Brain Pathophysiology; [2]Neurosurgery; [3]Neurology, Kyoto University School of Medicine, Kyoto, Japan

Introduction

Regarding the late positive component of event-related potentials (ERPs) associated with auditory oddball paradigm which is now well known as P300, many studies have been done about its characteristics and neural generators. Since P300 is considered to reflect higher brain functions, some investigators have studied the topographical distribution of the scalp-recorded P300 as well as invasive recording using intracranial depth electrodes or chronically implanted subdural electrodes to identify its neural generators and functional significance [1-4]. In addition, there are several studies using magnetoencephalography (MEG), which has been developed as one of the non-invasive technology, to elucidate the brain function. Two regions have been suggested regarding the sources of the event-related magnetic fields corresponding to P300; the mesial region of the temporal lobe, possibly the hippocampus [5,6], and the superficial temporal lobe [7]. Rogers et al. [8] also suggested that the responses to the target stimuli arose in the deep regions of brain initially and were transferred to the auditory cortex. However, since these studies recorded the magnetic responses from the restricted area with one or several channels of magnetometer and analyzed them with one dipole model, it might not be enough to disclose the cognitive processing which activates many regions simultaneously and parallel. Therefore, in order to elucidate the generator sources of P300, it would be better to acquire the responses of all the areas in the brain simultaneously and to analyze them by adopting the multidipole model. We studied the event-related magnetic fields (ERFs) associated with the auditory oddball paradigm in normal volunteers and in patients after the operation for the temporal lobe epilepsy by using a whole head magnetometer with 122-channel planar first-order gradiometers to clarify the possible brain areas involved in the cognitive processing.

Methods

Six healthy volunteers (4 males, 2 females, 20-38 years, rt-handed) and seven patients after the operation for the intractable temporal lobe epilepsy (1 male, 6 females, 18-38 years, rt-handed) participated in this study. The lesion of the epileptic foci lay on the unilateral temporal lobe. Out of seven patients, five were operated with standard anterior temporal lobectomy, one with lateral temporal resection and one with selective amygdalohippocampectomy. An informed consent was obtained from the subjects after the full explanation of this study.

The subjects sat in a glassfiber chair in a quiet and magnetically shielded room. They were requested to keep their eyes open. A standard auditory oddball paradigm was used as a stimulus sequence (50 ms duration, 10 ms rise/fall and 30 ms plateau). The intensity of both tones was adjusted to 85 dBSPL at the exit of plastic tube. The order of tone presentation was randomized with a varying rate of once per 2.2 s on the average (2.0-2.4 s). The subject was requested to count the number of the presented target stimuli silently. All the measurement of magnetic field was done by using the whole head coverage magnetometer with 122-channel planar first-order gradiometers at 61 measurement sites. The simultaneous electroencephalographic (EEG) recording in all subjects was obtained from 4 electrodes according to the International 10-20 System (Fz, Cz, Pz and Oz) referenced to linked earlobe electrodes. To monitor eye movements and blinks, the electrooculogram (EOG) was recorded from an electrode placed below the right outer canthus and referenced to linked earlobe electrodes. The recording passband was 0.03-100 Hz for MEG, and 0.07-120 Hz for EEG, and sampling rate was 404 Hz. The responses were averaged on-line with respect to the onset of target and non-target stimuli independently after excluding epochs containing big artifacts such as eye blinks. Analysis window for the on-line averaging was from 200 ms before to 800 ms after the stimulus. One session consisted of 100 epochs. At least 3 sessions were performed to confirm the reproducibility. Amplitude measurement was done from the baseline, which was determined by averaging the 100 ms segment before the stimulus onset for each channel of EEG and MEG. The peak latency of N100 component was visually determined at Cz, and that of N1m component was also visually determined at the channel showing maximum magnetic deflection with the peak latency around 100 ms after the onset of the stimulus, on each hemisphere. P300 component was determined as the positive deflection of the response seen only following the target stimuli at Cz or Pz at the latency between 250 and 600 ms. The magnetic deflections corresponding to P300 were detected in the response to the target stimuli at the latency between 250 and 600 ms. Only the magnetic responses to the target stimuli were analyzed for the following dipole analysis of the late magnetic component corresponding to the P300 component. To identify the current source underlying the measured signals, a single spherical head model, which best fitted to each subject's brain surface determined on one's own MRI, was adopted for the calculation of the equivalent current dipoles (ECDs). ECDs were assumed from the sub-region consisting of 20-30 channels showing significant difference between deflections to the target and those to the non-target stimuli. ECDs indicated the most dominant source acquired by the least-squares search during several segmental time periods by using a subset of channels over the interested areas. The goodness of fit of the model was also calculated, to demonstrate the ratio of correspondence between the estimate magnetic field distribution by ECD and the actual magnetic field distribution. Only ECDs with the goodness of fit more than 60 % at selected periods of time were adopted for the further analysis. Then the analysis time and channel were extended to the whole recording time period and all channels. At first ECD for N1m was determined on each hemisphere independently. The estimated waveforms by introducing two dipole model for N1m were superimposed on the measured responses. Channels showing the significant discrepancy between the simulated responses and the actually measured ones were, if any, selected visually for the next subset channels to estimate ECDs. Every time after obtaining a new ECD, simulation was performed with various combinations of the estimated ECDs. These analysis processes were repeated until the discrepancy between the simulated and measured responses became small and the goodness of fit could not get significantly higher value. Estimated dipoles by these processes were superimposed on the subject's own MRI.

Results

All subjects performed well the task of counting the number of target stimuli silently. N1m was recognized clearly in response to both target and non-target stimuli at about 100 ms after the onset of stimuli, which was almost the same as the latency of N100. In all patients and all healthy subjects, P300 was recognized with the maximum amplitude at Pz at the mean latency of 331 ms and 343 ms, respectively, between which there was no statistically significant difference (t-test; $p>0.05$)(Fig. 1). In healthy subjects, responses to the target stimuli showed an additional magnetic deflection at around 400 ms on the anterior temporal, middle temporal and parietal regions on each hemisphere, but not in the responses to the non-target stimuli (M400). Their latency was longer than that of P300 by 20-100 ms. Dipoles of M400 lay in the bilateral mesial temporal, superior temporal and parietal regions. The simulated signals assumed by 6 dipole model for M400 could explain the measured signals. Further attempt to detect an additional dipole was failed. On the other hand, in 4 out of 5 patients after standard temporal lobectomy, M400 could not be detected over the anterior temporal area of the operated side, whereas M400 over other areas could be estimated (Fig. 2). In the patient after lateral temporal resection, M400 could be observed over the operated temporal area as well as other areas. The latency of M400 in patients was longer than that of P300 by 100-150 ms. In the remaining two patients, M400 was not detected due to large artifacts around the region of craniotomy.

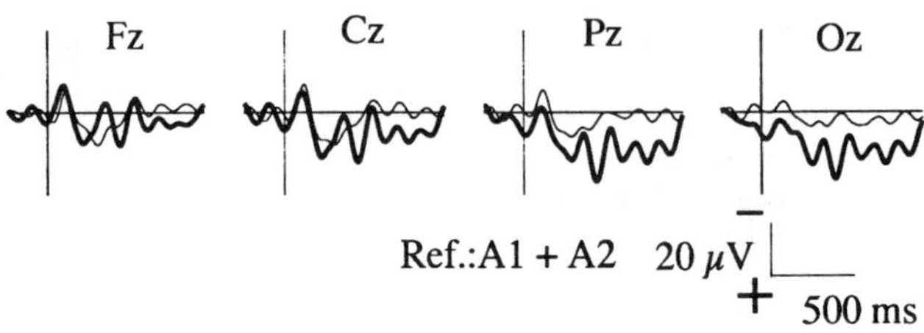

Fig. 1

    Group average responses of auditory ERP after the left standard temporal lobectomy in a patient with temporal lobe epilepsy. Thick and thin lines show wave forms in response to target and non-target stimuli, respectively. P300 is clearly seen 340 ms after the target stimuli.

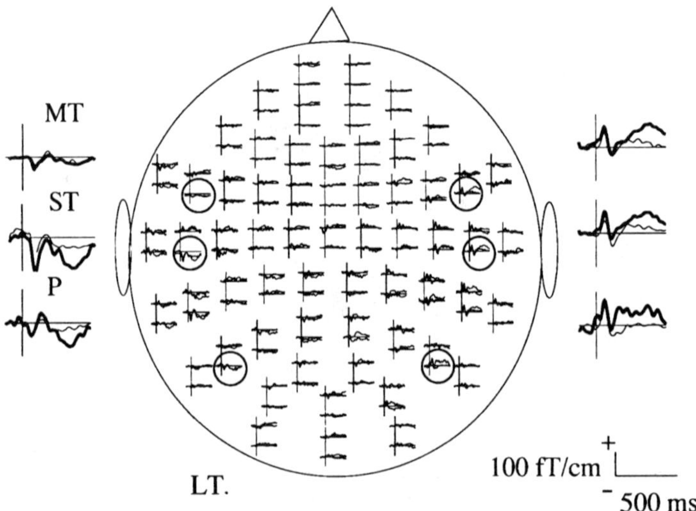

Fig. 2

    Group average responses of auditory ERF in the same patient as shown in Fig.1. Selected channels from the mesial temporal (MT), superior temporal (ST) and parietal (P) regions on each hemisphere are magnified. The early component was detected in the bilateral temporal areas. During the later period between 400 and 500 ms after the onset of the target stimuli, there are significant differences between deflections to the target and to the non-target stimuli in the channels except for the one on the left MT area.

Discussion

In the present study M400 recorded by the whole head type MEG is considered to reflect the cognitive processing like P300, because it was recognized only in the responses to the target stimuli. The estimated dipoles of M400 were located in the mesial temporal, superior temporal and parietal regions on each hemisphere. These areas are compatible with the previous EEG studies showing the possible sources of P300 in the mesial temporal, superior temporal and temporoparietal regions [1,3 and 4]. In the previous MEG studies, only the hippocampus and the superior temporal region were detected as the source areas of ERFs [5 - 8], possibly because they were estimated by one dipole model based on the recordings with   one or several channels magnetometer.

Although P300 could be detected in all patients, M400 was not observed over the anterior temporal region of the operated side in the patient after the standard temporal lobectomy. In the patient after the lateral temporal lobe resection, M400 was detected in all six areas including the ipsilateral temporal area. Polich and Squire [2], who studied P300 in amnesic patients with bilateral hippocampal lesions, concluded that the hippocampal lesion did not contribute significantly to the scalp-recorded P300. These results indicate the minimal contribution of the mesial temporal region to the scalp recorded P300, even if it might still be one of the generator sources of P300. Therefore, it can be concluded that the cognitive processing in the mesial temporal region would be detected more clearly by MEG.

Conclusions

We studied the ERFs in normal subjects and in patients after the operation for the intractable temporal lobe epilepsy, and clarified the possible generators of M400. These findings suggested the followings. 1) The possible sites involved in of the cognitive processing associated with the auditory oddball paradigm lie in the bilateral mesial temporal, superior temporal and parietal regions. 2) Cognitive function may be carried out not only by the mesial temporal region but by the combination of these multiple areas.

References

[1] Knight, R.T., Scabini, D., Woods, D.L. and Clayworth, C.C.   Contributions of temporal-parietal junction to the human auditory P3, Brain Res., 1989, 502: 109-116.

[2] Polich, J. and Squire, L.R.   P300 from amnesic patients with bilateral hippocampal lesions, Electroenceph. clin. Neurophysiol., 1993, 86: 408-417.

[3] McCarthy, G. and Wood, C.C.   Intracranial recordings of endogeneous ERPs in humans,   Electroenceph. clin. Neurophysiol., 1987, 39: 331-337.

[4] Halgren, E., Baudena, P., Clarke, J.M., Heit, G., Liegeois, C., Chauvel, P. and Musolino, A. Intracerebral potentials to rare target and distractor auditory and visual stimuli. I Superior temporal plane and parietal lobe, Electroenceph. clin. Neurophysiol., 1995, 94: 191-220.

[5] Okada, Y.C., Kaufman, L. and Williamson, S.J.   The hippocampal formation as a source of the slow endogenous potentials, Electroenceph. clin. Neurophysiol., 1983, 55: 417-426.

[6] Lewine, J.D., Roeder, S.B., Oakey, M.T., Arthur, D.L., Aine, C.J., George, J.S. and Flynn, E.R.,   Localization of the generators of the magnetic P3, In: Williamson, S., Hoke, M., Stroink, G. and Kotani, M.   Advances in Biomagnetism. Plenum Press, New York, 229.

[7] Gordon, E., Slogget, G., Harvay, I., Kraiuhin, C., Pennie, C., Yiannikas, C. and Meares, R.   Magnetoencephalography: locating the source of P300 via magnetic field recording,   Clin. Exp. Neurol., 1987, 23: 101-110.

[8] Rogers, R.L., Baumann, S.B., Papanicolaou, A.C., Bourbon, T.W., Alagaarsamy, S. and Eisenberg, H.M.   Localization of the P3 sources using magnetoencephalography and magnetic resonance imaging, Electroenceph. clin. Neurophysiol., 1991, 79: 308-321.

Use of Magnetic Source Imaging in Surgical Planning for Epilepsy Patients who have Undergone Prior Resection

Sobel, D.[1], Aung, M.[1], Schwartz, B.[1,2] Squires, K.[2,] and Hirschkoff, E.[1,2]

1. The Scripps Clinic and Research Foundation, La Jolla, CA , USA
2. Biomagnetic Technologies, Inc, San Diego, CA , USA

Introduction

While surgical intervention often substantially reduces seizure activity, a number of cases arise in which the seizure activity following resection is either not reduced as much as desirable or is initially reduced but increases in frequency and severity over time. In these cases, further surgery may be indicated. A quick, non-invasive means to assess the location of remaining epileptogenic activity would be a valuable tool for this purpose. A key issue is whether the continuing seizures originate from zones adjacent to prior surgery, or whether they represent multifocal regions or areas not adjacent to prior surgery which were previously undetected which had developed since the surgery.

The present study was undertaken to determine the utility of magnetic source imaging as a tool to evaluate candidates for surgery in cases where initial resection has not reduced seizure activity to an acceptable level or where seizure activity has resumed. The cases evaluated in this study were part of a larger study evaluating 100 candidates for epilepsy surgery referred from eleven epilepsy centers throughout the U.S. and Canada. Evaluations at those centers prior to MSI typically included EEG, MRI, neurological examination, neuropsychological testing, and video EEG (Phase I) monitoring, and in some cases invasive EEG (Phase II) monitoring.. In many cases, PET and/or SPECT test results were also available. Earlier work had demonstrated the capacity of Magnetic Source Imaging (MSI) to localize focal sources of interictal epileptogenic activity, in agreement with more invasive methods and with electrocorticographic monitoring during and after surgery.[1]

Methods

Twenty-four patients who had had resective surgery but who now were experiencing uncontrollable seizure activity underwent MEG recording during which interictal epileptiform events were recorded concurrently on a 74-channel biomagnetometer *(Biomagnetic Technologies Magnes II$^{®}$)* and a 21 channel EEG recording system (Nihon Khoden) to record spontaneous incidents of interictal epileptogenic activity. A radio frequency 3-D digitizer determined the position of the individual gradiometers relative to a head-centered 3-D orthogonal coordinate system defined by the nasion and by markers in the right and left outer ear canal. Each patient underwent an MR scan with fiducial markers attached, providing a common coordinate system to overlay MEG source localizations [2]. Normal seizure control medications were maintained.

With the patient lying on one side and the head voluntarily stabilized, the two 37-channel probes were positioned over homotopic areas of the two hemispheres. A sequence of four to five dual probe placements and measurements assured coverage of the whole head, and afforded the ability to distinguish between synchronous and asynchronous bilateral sources. Recording sessions typically lasted from two to four hours, depending on the patient's level of interictal activity. EEG and MEG waveforms were monitored in real-time by observers trained to recognize and distinguish the morphology of epileptogenic events as against artifacts. (EEG data were later reviewed and interpreted separately by an experienced epileptologist.) Twenty 5-6 second epochs of data were obtained at each recording site. Raw data were digitized at 860 Hz and band-pass filtered at 0.1-200 Hz. Data epochs were then digitally filtered off line with a bandpass of 3-70 Hz. Off line data analysis proceeded with visual identification of MEG and EEG interictal spikes or sharp waves. Temporal windows including these events were then analyzed at each time point in the data sample with a single equivalent current dipole (ECD) model. Each event for which the dipole model correlated with the recorded data to at least 98% and for which the equivalent dipole strength was physiologically realistic was represented by a single focal source plotted on an MRI image of the patient's brain.

Images showing all ECD modeled epileptiform events (Fig 1) were presented along with waveform information and the EEG tracings for review by an experienced epileptologist. The epileptologist indicated the likely source of seizure activity and target for surgical resection, if such an inference could reasonably be drawn from the MSI data. The diagnostic conclusion drawn from the MSI assessment was then compared with that from the consensus of all other measures including invasive EEG monitoring undertaken during the patient's normal course of care [2].

Five of these patients later underwent invasive EEG Phase II monitoring with implanted grid or depth electrodes. The source locations suggested by the MEG data were then compared with the results of conventional diagnostic testing (MRI, Phase I or Phase II monitoring).

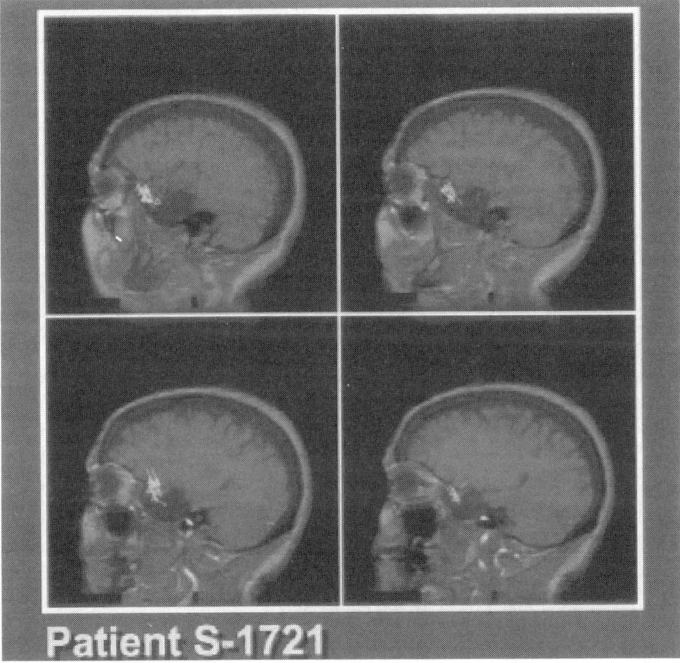

Fig. 1. MEG dipole overlays on sagittal images for patient #S1721.(Spike sources are indicated as yellow triangles, with lines indicating orientation.) The patient had suffered complex partial since age 12. Prior surgery was a right anterior temporal lobectomy, evident in sagittal slices. Spike sources are seen at the superior and anterior margins of prior surgery.

Results

MSI showed abnormal spike activity in 21 of the 24 patients. In three cases there was insufficient data for MSI to localize spike activity. MSI localizations agreed with the consensus of Phase I and Phase II EEG indications in 14 cases, in the sources indicated were adjacent to the region of prior surgery in 12 patients. In an additional 4 cases, MSI and conventional measures were concordant in indicating multi-focal or bilateral sites of activity. An additional two cases were only in partial concordance, with MSI indicating a wider and more diffuse area of abnormal activity in relation to Phase I and Phase II EEG measures. There was one case with an inconsistent localization between MSI and EEG measures.

Of the 24 patients measured, 10 have gone on to a second surgery. Of these cases, 6 are seizure free after 3-6 months, with 2 cases experiencing visual or motor deficits. One other patient has continuing auras 6 months post-surgery, and two more cases show only partial or minimal improvement over 3-6 months. One case had an increase in seizure frequency as of 3 months after surgery.

Discussion

MEG localization has the potential to indicate quickly and non-invasively whether or not epileptogenic source localizations are consistent with regions adjacent to prior surgery. Where bilateral or contralateral MEG signals are found, MEG may help indicate or even guide further invasive monitoring. Where MEG indicates abnormalities adjacent to prior resection, it may help increase confidence or reduce the time needed to determine whether further surgery is feasible or advisable.

References

[1] Gallen, C.C., Hirschkoff E.C., Buchanan, D.S. Magnetoencephalography and Magnetic Source Imaging. Neuroimaging Clinics of North America, 5(2), May, 1995.
[2] Smith, J.R., Schwartz, B.J., Gallen C.C., Orrison, W., Lewine, J., Murro A.M., King, D.W. Park, Y.D. (1995) Utilization of Multichannel Magnetoencephalography in the Guidance of Ablative Surgery. Journal of Epilepsy 8:119-130.

Acknowledgements

The authors would like to thank the following collaborators for their assistance: Dr. Izak Fried and Dr. Sachar Eliashiv (UCLA Medical Center), Dr. Vicente Iragui and Dr Evelyn Tecoma (UCSD Medical Center, La Jolla, CA), Dr. Brien Smith (Henry Ford Hospital, Detroit, MI), Dr. Joseph Smith and Dr. Yong Park(Medical College of Georgia, Augusta, GA), Dr. Thomas Waltz (Scripps Clinic Research and Foundation, La Jolla, CA, Dr. James Wheless and Dr. Valerie Curtis (Hermann Hospital, Houston, TX)

Methohexital-Induced Changes in Spectral Power of Neuromagnetic Signals: Reduced β-Band Enhancement Over the Hemisphere Ipsilateral to the Epileptogenic Focus

Wienbruch, C.[1], Eulitz, C.[3], Lehnertz, K.[2], Brockhaus, A.[2], Elger, C.E.[2], Elbert, T.[3] and Hoke, M.[1]

[1]*Institute of Experimental Audiology, University of Münster, Germany;* [2]*Clinic of Epileptology, University of Bonn, Germany;* [3]*Department of Psychology, University of Konstanz, Germany*

Introduction

Effects of anesthetics on electroencephalographic (EEG) activity are used to localize brain lesions as well as epileptogenic areas. Small doses of barbiturates (up to 200 mg) result in an increase of fast activity in the β-band of the EEG. This increase of β-band activity is known to be larger over normal brain areas as compared to areas with cerebral lesions (e.g. [1]). Furthermore, it has been reported that some barbiturates as amobarbital, thiopental, or methohexital or other narcotics as propofol may provoke epileptiform activity resulting in spike activity and spike-burst-suppression patterns. As an additional feature, various authors ([1] [2]) described qualitatively a loss of methohexital-induced β-band activity in the electrocorticogramm (ECoG) over the epileptogenic area. A quantitative analysis of the spectral contents of the ECoG is mandatory before the clinical use of methohexital induced changes in β-band activity can be rated.

Drug-induced activation of the epileptogenic focus provides a rapid and safe method to obtain a sufficient amount of information relevant for the lateralization and localization of the primary epileptogenic area. We investigated the value of methohexital induced activity for the presurgical lateralization and localization of the epileptogenic zone using MEG. We measured simultaneously the MEG and EEG during methohexital induced narcosis in 13 patients all suffering from temporal lobe epilepsy (TLE) of temporo-mesial origin. The aim of this study was to quantify anesthesia-induced changes in the spectral power characteristics of MEG and EEG signals and to analyze the topographical aspects of these changes with respect to the side of the primary epileptogenic area in patients suffering from temporal lobe epilepsy (cf. [3]).

Method

Thirteen patients suffering from pharmacoresistant unilateral temporal lobe epilepsy were investigated. Before performing this study, all patients underwent invasive presurgical evaluation (cf. Bonn protocol of presurgical evaluation, [4]) using electrocorticographic and stereo-EEG recordings for the localization and delineation of the primary epileptogenic area. As confirmed by the post-operative outcome a left temporal seizure origin was found in 5 patients and a right temporal seizure origin in 8 patients. Selective amygdalo-hippocampectomy (sAHE) was performed in 8 cases, AHE with partial temporal resection in 4 cases and a resection of a ganglioglioma in one case. After surgery all but one patient are completely free of seizures for at least 6 month.

Neuromagnetic data were recorded using a 37-channel neuromagnetometer (Magnes^{TM}; Biomagnetic Technologies, Inc.). Measurements were carried out in a magnetically shielded room. The sensor array was centered sequentially over the left and the right supra-temporal cortex about 1.5 cm superior to T3 or T4 electrode position of the international 10-20 system. The head position relative to sensor pickup coils was measured by a sensor position indicator. EEG data were simultaneously recorded (SYN-AMPS^{TM}; Neuro Scan, Inc.) from temporal, frontal, central and parietal electrode positions according to the international 10-20 system with the nose used as a reference. Subjects were lying on their side with their head and body fixated by a vacuum cushion. They were instructed to avoid eye blinks and head movements. Continuous data were recorded from left and right hemisphere in 10 minute blocks using a sampling rate of 297.6 Hz and a passband of 0.03 to 100 Hz.

After a 2 min baseline recording a dose of 100 mg methohexital was applied intravenously (i.v.) in a bolus injection according to the conditions used during the invasive presurgical evaluation [2]. After a break of approximately 20 min exactly the same procedure was repeated on the opposite hemisphere. This break is sufficient for the recovery of the patient with respect to consciousness and recovery of MEG and EEG back to the baseline due to the short half life time of approximately 2 min. Anesthetic monitoring and standby was provided, and precautions were taken to counteract anesthetic complications. During the measurement an anesthesist was within the magnetically shielded room. Additional video monitoring of the patient and the anesthesist was performed. All patients were informed about the procedure before the investigation and declared their informed consent.

Off-line the data were digitally bandpass filtered using a 4-48 Hz recursive Butterworth-filter (2nd order). The spectral power was estimated for each magnetic and electric channel. Data sets of 300 seconds length starting 20 seconds before i.v. injection of methohexital were analyzed using an overlapping segmentation technique. The length of the Parzen-windowed segments was 512 points which corresponds to 1.720 s. According to [5] the segments were overlapped by one half of their length to obtain the smallest spectral variance per data point. The power spectra of these segments were averaged within 19.785 seconds time windows and normalized by dividing each mean spectrum by the spectrum of the time before the injection of methohexital separately for each channel. The spectral band power in the θ-, α-, β- and γ-band represents the integrated power within the border frequencies (θ (4.07-7.55 Hz), α (8.13-12.79 Hz), β (13.37-23.83 Hz) and γ (24.41-47.66 Hz)). To further reduce the data, spectral band power was averaged across neighboring channels. MEG channels were collapsed over frontocentral, temporoparietal and frontotemporal brain regions, the EEG-channels were averaged across frontocentral and temporal electrodes separately for the left and the right hemisphere.

The resulting normalized spectral band power values of each frequency band were statistically tested for differences between hemispheres and brain regions as dependent variables. For MEG data we used two-way univariate analyses of variance (ANOVA) for a 3 *brain-regions* (frontocentral vs. frontotemporal vs. temporoparietal) x 2 *side-of-focus* (side of epileptogenic focus vs. contralateral to the side of epileptogenic focus) design. For the statistical analyses of EEG data the factor *brain-regions* had only two steps (frontocentral vs. temporal). Wherever appropriate, a Greenhouse-Geisser correction of the degrees of freedom was applied.

Results

No adverse effects during or after anesthesia were noticed by either the patients or the investigators. Patients recovered quick and had no clinical excitatory effects or signs like vomiting and nausea. No seizure was elicited.

In contrast to the baseline recording, a high amplitude β-band activity is obtained around 2 minutes after administration of methohexital. A well pronounced difference of the MEG β-band spectral power between the hemispheres ipsi- and contralateral to the epileptogenic focus can be observed approximately two minutes after the injection of methohexital for a period of 1-2 minutes, while the EEG β-band spectral power on the ipsilateral side is even more enhanced compared to the contralateral side. The normalized spectral power estimates derived from MEG signals of the third minute after methohexital activation measured over frontocentral, frontotemporal and temporoparietal brain regions of both hemispheres were used as dependent variables for a two-way ANOVA. Only main effects of the *side-of-focus* or *side-of-focus* by *brain-regions* interactions will indicate a usefulness of changes in spectral power for the lateralization of the side of the primary epileptogenic focus. In the β-band changes in normalized spectral power were influenced by the factor *side-of-focus* (ipsilateral vs. contralateral side) which revealed a statistically significant main effect ($F(1/12) = 10.56; p < 0.007$). The *side-of-focus* effect did not differ between the three separated brain regions.

As illustrated in Fig. 1, during the third minute after methohexital injection, normalized β-band spectral power was more pronounced over the hemisphere contralateral to the epileptogenic focus as compared to the ipsilateral side in ten of thirteen patients, which holds true for both the left and right TLE patients. Only one patient (R16) showed a clear reversal of that typical lateralization pattern. In this particular patient methohexital induced high-amplitude spike activity followed by short intervals of isoelectric line or very slow δ-activity with a rate of approximately one per second was seen in MEG and EEG recorded over the hemisphere ipsilateral to the epileptogenic focus. The *side-of-focus* effect was evident over all brain regions with a slight predominance over frontocentral regions, without preferring one of them.

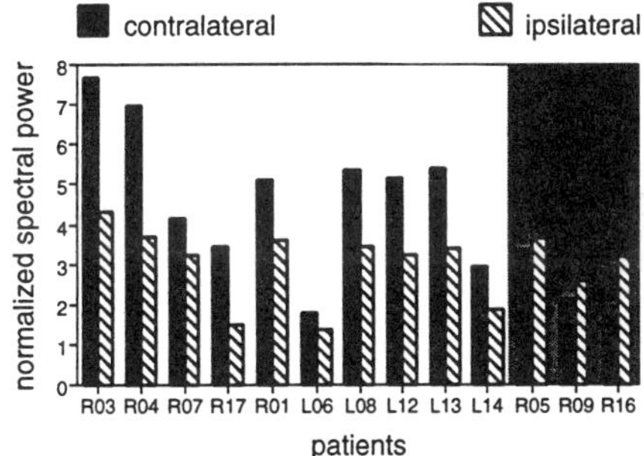

Figure 1: Methohexital induced changes of the normalized spectral power in the β-band of the MEG data for each individual patient. The figure shows the mean normalized β-band spectral power averaged across all brain regions for the third minute after the injection of methohexital. The first letter of the patient identification indicates whether the hemisphere with the primary epileptogenic focus was the right (R) or the left (L) one. The gray shaded area marks those patients showing ipsilateral to the epileptogenic focus more pronounced enhancement of the mean normalized β-band spectral power.

For the θ-, α- and γ-band exclusively effects of *brain-regions* were found while *side-of-focus* effects were not evident in these frequency bands. The fine structure of differences between brain regions was different for the various frequency bands as illustrated in Fig. 2.

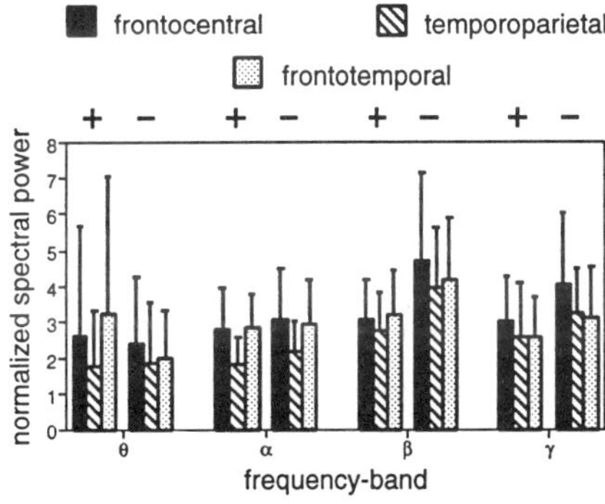

Figure 2: Normalized spectral power of the MEG data over three separated brain regions for the θ-, α-, β- and γ-band. Each column represents the mean ± standard deviation of the normalized spectral power across all patients. The hemisphere ipsilateral to the side of the epileptogenic focus is marked with a (+), the contralateral side with a (−).

For the θ-band a main effect of *brain-regions* was found ($F(2/24) = 4.55; \epsilon = 0.82; p < 0.03$). Post-hoc Scheffé comparisons revealed that the effect of *brain-regions* was mainly caused by the difference between frontotemporal areas where the normalized θ-band spectral power exceeded 2.6 times the baseline level, and the temporoparietal regions where the spectral power was enhanced only 1.8 times ($p < 0.04$). A similar pattern was found for the α-band (main effect of *brain-regions*: $F(2/24) = 14.60; \epsilon = 0.94; p < 0.0001$). Post-hoc Scheffé comparisons confirmed that in this frequency band the normalized spectral power over the temporoparietal brain region (2.0 times) was remarkable lower than over both the frontotemporal (2.9 times: $p < 0.0005$) and the frontocentral area (2.9 times: $p < 0.0005$). The *brain-regions* effect in the γ-band ($F(2/24) = 6.22; \epsilon = 0.86; p < 0.008$) was mainly originated by the marked enhancement of normalized spectral power over the frontocentral regions (3.5 times) as compared to the temporoparietal (2.9 times: $p < 0.03$) as well as the frontotemporal brain area (2.8 times: $p < 0.02$).

The EEG data did not show lateralized effects in the β-band. Brain region effects occurred only in the θ-band ($F(1/12) = 13.14; p < 0.004$) and the α-band ($F(1/12) = 19.17; p < 0.002$). The pattern of differences between brain areas was similar to that obtained for the MEG data.

Based on the differences in normalized β-band spectral powers of MEG recordings between the hemispheres ipsi- and contralateral to the epileptogenic area during the third minute after methohexital injection we were able to correctly lateralize the primary epileptogenic focus in 77% (10) of our patients.

Nevertheless, false positive decisions would have been made in 23% (3) of our patients.

Discussion

The aim of the presurgical evaluation of patients with pharmacoresistant epilepsies is the correct localization and delineation of the primary epileptogenic area. Our results, based on a spectral analysis of simultaneous MEG and EEG recordings, confirm a loss of methohexital-induced enhancement of β-band spectral power in the MEG over the side of the primary epileptogenic area. This could be established in 10 of the 13 investigated patients. The effect was clearly expressed during a 1-2 minute interval starting approximately two minutes after the administration of methohexital. No significant differences between fronto-central, fronto-temporal or temporo-parietal brain regions were found. Although a general induction of fast β-activity could also be observed in our EEG recordings, a reduction of this activity over the ipsilateral hemisphere is not statistically significant.

Our results indicate that the loss of methohexital-induced enhancement of β-band spectral power in the MEG over the side of the primary epileptogenic area might contribute to the non-invasive lateralization of the primary epileptogenic area in patients suffering from pharmacoresistant temporal lobe epilepsy. This holds true particularly in those cases not exhibiting an increase of drug-induced spike activity or spike-burst-suppression (SBS) patterns in the MEG or EEG recordings, the latter pattern being of high significance for the lateralization and localization of epileptogenicity to the ictogenic temporal lobe based on invasive ECoG recordings via subdural strip electrodes [2]. Although the number of patients investigated so far is still too small in order to determine the value of the presented method for non-invasive presurgical evaluation, we would be able to correctly lateralize the primary epileptogenic area in 77 % of the cases. Nevertheless, false positive decisions about the side of the ictogenic temporal lobe would have been made in 23 % of our patients.

In conclusion, the reduced enhancement of β-band spectral power of MEG recordings can be used as an additional non-invasive tool for the presurgical lateralization of the primary epileptogenic area. In order to determine the value of this method for non-invasive presurgical evaluation, additional investigations on a larger number of cases, especially with single or multiple epileptogenic foci in brain regions other than temporo-mesial structures are required.

References

[1] G. Pampiglione. Induced fast activity in the EEG as an aid in the location of cerebral lesions. *Electroenceph clin Neurophysiol*, 4:79–82, 1952.

[2] A. Hufnagel, W. Burr, C. E. Elger, J. Nadstawek, and G. Hefner. Localization of the epileptic focus during methohexital-induced anesthesia. *Epilepsia*, 33:271–284, 1992.

[3] C. Wienbruch, C. Eulitz, K. Lehnertz, A. Brockhaus, C. E. Elger, T. Elbert, and M. Hoke. Methohexital-induced changes in spectral power of neuromagnetic signals: β-augmentation is smaller over the hemisphere containing the epileptogenic focus. *submitted to: Electroenceph clin Neurophysiol*.

[4] J. Engel Jr., editor. *Surgical Treatment of the Epilepsies*, pages 740–742. Raven Press, 2 edition, 1993.

[5] W. H. Press, B. P. Flannery, S. A. Teukolsky, and W. T. Vetterling. *Numerical Recipes in C*. Cambridge University Press, Cambridge, 1988.

Acknowledgment

This research was supported by grants from the Deutsche Forschungsgemeinschaft (Klinische Forschergruppe Biomagnetismus & Biosignalanalyse, Ho 847/6) and is part of a PhD thesis.

The authors wish to thank A. Kowalik MD, A. Baborie MD, M. Möllmann MD, N. Mertes MD, L. Spinne, G. Westermeier, C. Goeters MD, M. Reuss MD, W. Frebel MD, U. Ruta MD, T. Jakob, and S. Lorenz for their technical assistance.

VII. Multimodality Comparisons And Integration

Evaluation of Source Localization Models for Somatosensory Evoked MEG With PET Verification

Akhtari, M.[1], McNay, D.[1], Levesque, M.[1], Rogers, R.L.[1], Brown, C.[2], Mandelkern, M.[2,3], Babai, A.[4] and Sutherling, W.W.[1]

[1] Neuromagnetism Laboratory, Epilepsy & Brain Mapping Center, Hospital of the Good Samaritan, Los Angeles, CA; [2] Department of Nuclear Medicine, West Los Angeles VA Medical Center, Los Angeles, CA; [3] Department of Physics, University of California at Irvine, Irvine, CA; [4] UCSD School of Medicine, San Diego, CA

Introduction

We used MRI to segment the gray/white parenchymal surface and applied these results to constrain the cortical surface to solve for the location of the current sources generating externally measured magnetic field patterns evoked by somatosensory stimuli [1]. This method was applied to four normal subjects . We obtained somatosensory evoked PET data from one subject under identical stimulus conditions that were used to obtain the MEG data. Using prominent anatomical landmarks, the results of the MRI overlaid on the PET study show good agreement with the cortically constrained MEG localization. Furthermore, we compared this MRI constrained model to random search sphere model analyses and obtained similar localization results. The results from cortically constrained boundary element method (BEM) showed agreement with all the above methods of localization.

Method

Four neurologically normal subjects (1 female and 3 males, ages 22 to 31 years old) were studied. Somatosensory evoked responses were measured during non-simultaneous stimulation of the right median and the posterior tibial nerves using a Grass S88 stimulator and SIU7 DC current stimulus unit. 7-15 mA stimuli were delivered at 3.1 Hz, with 0.3 ms duration. Data were recorded and digitized at a rate of 4096 Hz. The first 256 time points following each stimulation were saved and two independent sets of 512 recordings each were averaged and stored separately (Figure 1). Magnetic fields were measured at 63 locations using a 7 channel SQUID-coupled second order gradiometer (Biomagnetic Technologies, Inc., Model 607). Eighteen overlapping channels were deleted and data from 45 channels were used for analysis. The measurements were recorded over the left hemisphere, and over the calvarium, for median nerve, and posterior tibial nerve stimulations, respectively.

Sixty-four sequential 3-mm cranial images were acquired on a 1.5 T G. E. Signa 5 MRI machine using a double spin echo sequence (TR=4000-6000 μs; TE= 60-80 μs; FoV=220-280 mm). We used custom molded bite plates and/or vitamin E capsules to mark the common fiducials between MRI and MEG coordinate systems. The MRI segmentation was performed using the General Electric workstation software package. The gray/white interface was digitized and the information was used as coordinates and orientations of $2-3 \times 10^5$ possible current sources. We used the conventional sphere model [2] with Nelder Mead simplex algorithm to randomly search for the location of the best fitting dipole. One subject's segmentation data was used to triangulate the brain/skull, skull/scalp, and scalp/air interfaces.(Fig. 2) We input the segmentation and triangulation data along with the best fitting source element from the cortically constrained primary model to a BEM code developed by the General Electric company. Several cortically constrained surfaces of various extents, all centered on the best fitting element were also evaluated as potential active extended sources using BEM.

The PET data were obtained after bolus injections of four nonconsecutive doses of H2150 during electrical stimulation of the right median nerve. The PET data from left posterior tibial stimulation were obtained in the same manner. The PET data were overlaid on standard MRI available on PET analysis software. The overlays were performed using transformations utilizing the Talairach atlas.

Results

The location of the MRI constrained sources are displayed on the centroid of the highlighted area in Figure 1 for sagittal, axial and coronal views for all four subjects. Note that the locations of these sources are posterior and in close proximity to the central sulcus. Location of central sulcus was determined by visually tracking the cingulate sulcus on the sagittal images.

Table 1 compares the MRI constrained free space model with the spherical model using 1024 trials per averaged waveform. The goodness of fit values, (i.e. % variance accounted for), ranged from 87.1% to 90.7% for the MRI constrained model, and 86.4% to 95.4% for the spherical model. Intra-subject Euclidean difference in localization between the two models ranged from 2.0 to 15.5 mm. The coordinate system is right handed with the x-axis toward the left ear and the z-axis directed superiorly; the center of the best sphere was located 4-5 cm above the nasion-pa plane in all four subjects.

Figure 2 represents the results of MRI constrained MEG localization of right and left posterior tibial stimulation of subject #2. The active element is located at the cross-hair.
Figure 3 represents the locations of the cortical representation of right median, and left tibial nerves of subject #1 as obtained by PET scan.

The different boundaries of the head as obtained from reconstructed MRI of subject #1, and the corresponding triangulations are represented in Figure 4. The goodness of fit for MRI constrained best fitting element improved by approximately 4%. The goodness-of-fit values for the extended sources reached a peak value of 94% for an extended source that had an area of approximately 140 mm2. This area is displayed in Figure 1), subject #1. The extended sources of lesser or higher area had lower goodness of fit values.

Discussion

Results obtained using the MRI constrained model are in good agreement with the PET results. These response locations are also consistent with the known functional neuroanatomy, based on human [3] studies. In the present results, the goodness-of-fit values were similar for the three single element models. The discrepancy between the results of the sphere model and MRI constrained model (Table 1) for subjects 2 and 4, suggests that the results of the sphere model may be in error. The sphere model localized the sources in the two subjects by more than ten mm posterior to the central sulcus. Somatosensory localization of the median nerve has never been observed to be more than ten mm posterior to the central sulcus in nonpathological subjects. Moreover, with cortical constraints, extended active sources with realistic conformity to the cortical shape can be readily studied. For MEG localization of these rather superficial sources, the results of single element BEM analysis indicate only a slight improvement of goodness of fit compared to the cortically constrained primary model. However, the comparison of cortically constrained free space extended sources, and that of BEM remains to be carried out.

References

[1] Akhtari M, McNay D, Mandelkern M, Teeter B, Cline HE, Mallick J, Clark G, Tatar R, Lufkin R, Chan K, Law C, Sutherling WW. Somatosensory evoked response source localization using actual cortical surface as the spacial constraint. Brain Topography, 1994 Fall, 7 (1) : 63-9.
[2] Stok CJ. The influence of model parameters on EEG/MEG single dipole source estimation. IEEE Trans. Biomed. Eng. 1987;BME-34:289-296.
[3] Sutherling WW, Crandall PH, Darcey TM, Becker DP, Levesque MF, Barth DS. The magnetic and electric fields agree with intracranial localizations of somatosensory cortex. Neurology 1988;38:1705-1714.

Acknowledgements

Supported by grant NS20806 from the National Institute of Neurological, Communicative Disorders and Stroke, Epilepsy Branch, the Visions of the Brain foundation, and the Mathers charitable foundation. We are greatly indebted to Ms. Winifred L. Monaco B.A. for her invaluable contribution to this project.

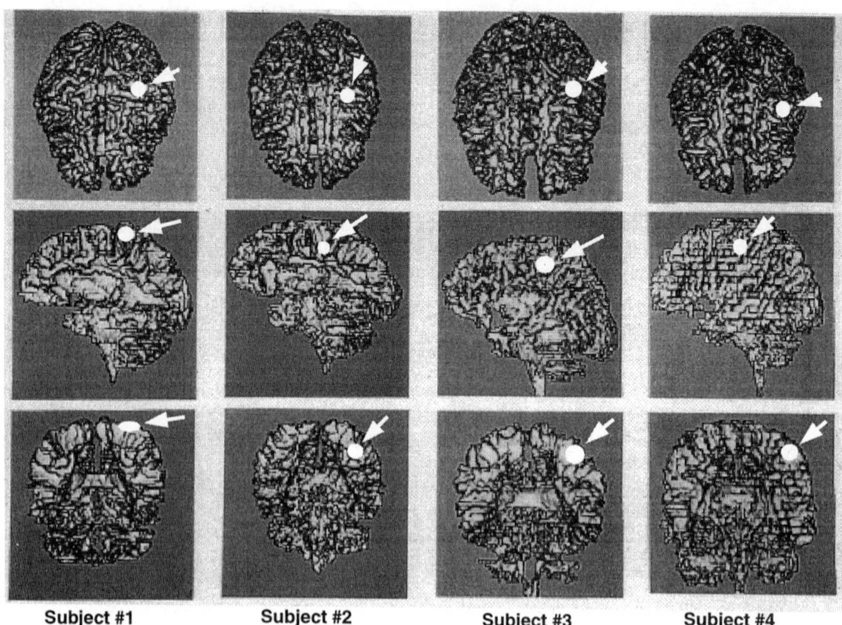

Subject #1   **Subject #2**   **Subject #3**   **Subject #4**

Fig. 1) Somatosensory evoked MEG localization of the right median nerve of four subjects (centroid of the highlighted area, arrow)

	MRI CONSTRAINED MODEL					SPHERICAL MODEL					MRI vs. SPHERE
SUBJECT	LAT	X	Y	Z	%V	LAT	X	Y	Z	%V	
1	19.6	54	-7	8	87.1	20.3	52	-7	8	93.4	2.0
2	26.4	74	5	-6	90.7	26.4	63	6	-3	86.4	11.4
3	23.4	53	-1	4	89.7	23.2	52	1	1	95.4	3.7
4	23.9	82	-2	0	89.8	23.9	68	4	-3	91.1	15.5

Latency in ms.
All distances are measured in mm.
%V: percent variance accounted for.

Table 1) The locations and the goodness-of-fit values of the best fitting MRI constrained current elements, and the best fitting single dipoles using the sphere model.

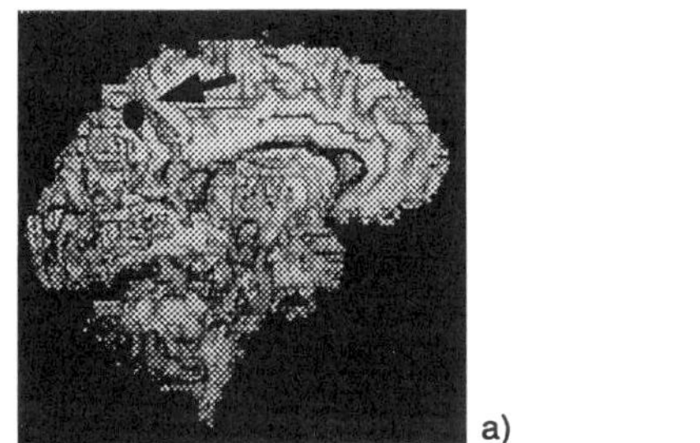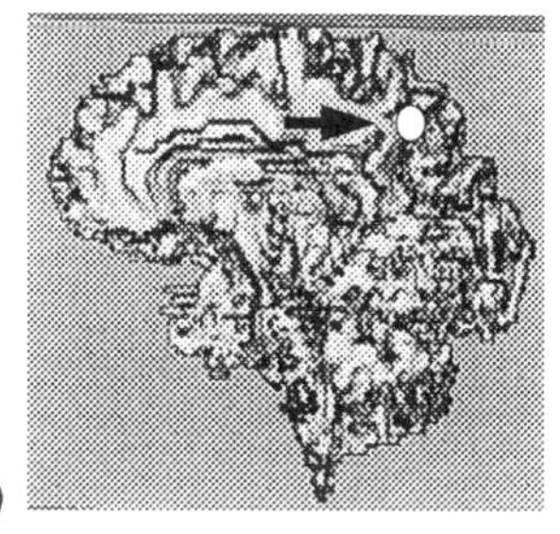

Fig. 2) a) Somatosensory evoked MEG localization of the right posterior tibial nerve (centroid of the highlighted area, arrow); and b) the corresponding localization of the left posterior tibial nerve (centroid of the highlighted area, arrow).

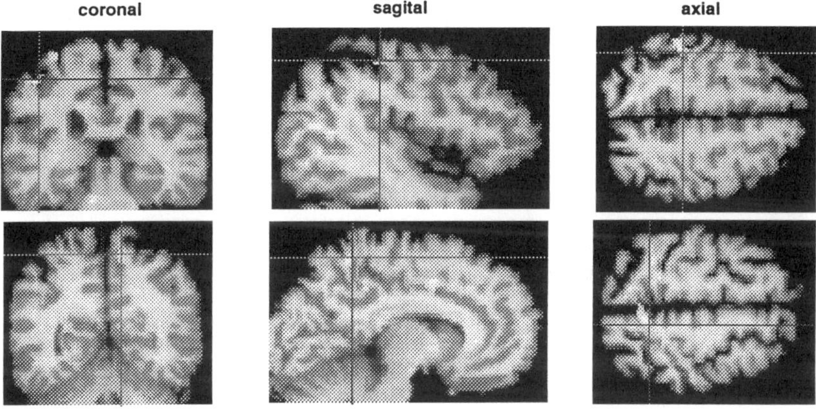

Fig. 3) Somatosensory evoked PET localization of the right median nerve (cross-hair, top pannel), and the left posterior tibial nerve (cross-hair, bottom pannel).

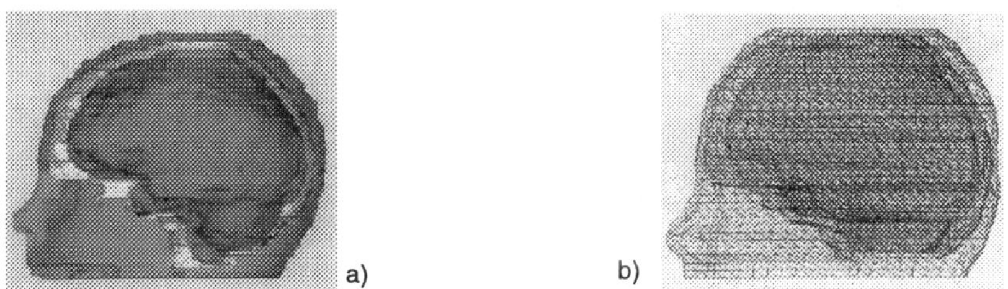

Fig. 4) a) MRI reconstruction of the three cranial surfaces; b) direct tessellation of the corresponding surfaces from a).

Combination of point and surface matching techniques for accurate registration of MEG and MRI

Bamidis, P.D.[1], Ioannides, A.A.[1,2]

[1]*Physics Department, The Open University, Milton Keynes, U.K.;*
[2]*Institute of Medicine, Research Centre Jülich, Germany*

Introduction

The easiest and most commonly used method to register the MRI and MEG datasets is point matching, where a few 'fiducial' points, i.e. nasion, 2 preauriculars, are marked during the acquisition of either technique. Another methodology involves the use of a stereotactic frame. In either case, the registration of the two modalities requires the definition of the correct transformation matrix with respect to scaling, translation and rotation parameters. We have exploited both the above strategies and, in general, root mean square (RMS) mismatches per point of less than 1 mm (simulations) or 6 mm (general practice with real data) were achieved.

However, we faced situations where a nominal accuracy of millimetres was obtained for the fiducial points, but the surface of the head as outlined in the MEG (with a 3D digitizer) did not fit with the MRI obtained head surface, as a result of inconsistent definitions of the fiducials. In this work, we show how we resolved such problems, and how accuracy can be improved by combining point matching with a more general surface matching approach, thereby practically demonstrating the need to move to a more robust registration methodology.

Motivation

Fig. 1 displays the situation where the fit of the pre-auricular and the nasion points (marked with arrows in (a)) is excellent (crosses in (b)) but the description of the digitised scalp in the MEG system (dots in (b)) does not fit with the outlines as extracted from the presumably aligned MRIs (light contours in (b)). This was probably due to the inconsistent definition of the fiduciary skull landmarks in each acquisition (MEG and MRI). Note that similar discrepancies occurred in two individual cases (subjects). In such cases, an inaccuracy on the order of 1-2 cm ruined the possibility to relate the MEG solutions to precise anatomical sites. The need to move to a more robust and less subjective alignment procedure is, therefore, obvious.

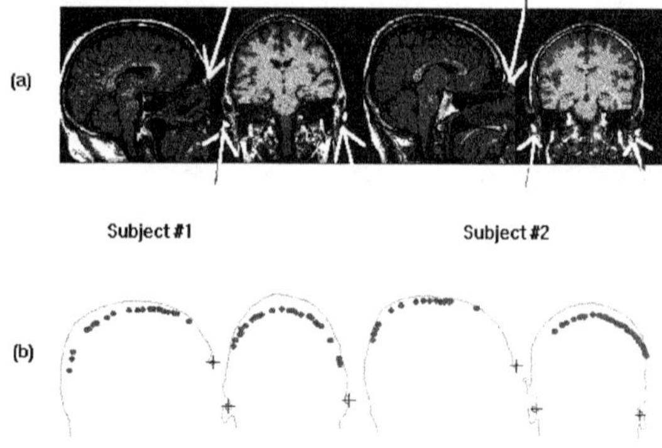

Fig. 1: (a): the vitamin pills used to perform the point-matching (arrows). (b): The fit of the MEG (crosses) and MRI fiducials (circles) is perfect. However, the head shape outlines in the MEG system (dots) do not match the MRI scalp outlines (light contours).

Methods

Motivated from such problems, we developed a system that combines the point matching with a surface matching approach. The head-hat analogy [1] is followed for a rigid body translation and rotation, in a way similar to that used by others [2,3]. A cost function, F, is formed by the sum of the squared distances of the MEG points from the MRI head surface, and Powell's method is employed to minimise F. We have adopted the following strategy in an attempt to optimise the registration results.

In the initial stage, a point matching is performed (3-5 points) and the accuracy is evaluated not in terms of the RMS mismatch per point, but by visualisation of the fitting of MEG head outlines on the MRI scalp. Sometimes satisfactory results are obtained and the process stops there. Usually, however, a stepwise surface matching is attempted: the program is given the parameters found in the first stage as initial approximations. Then the translation is roughly fixed by making the centroids of MEG (hat) and MRI (head) surfaces to coincide, and a 3D minimisation of F with respect to the three angles of rotation is tried (scaling is assumed to be known [3]). Finally, F is minimised with respect to all six parameters together, to fine-tune the fitting.

Results

Simulations have shown an overall accuracy in the range of 1-3 mm, that is, less than 1 mm translation and less than 2 degrees rotation errors (see Table 1). Given the intrinsic noise in real data, 3-5 mm RMS mismatches per point have been considered quite satisfactory. Whenever point registration was considered as unreliable, surface matching improved the accuracy by 1-2 cm, while in the general case, an improvement of 1-4 mm was achieved. The number of MEG points used, varied from a few hundred up to a few thousand. The fit was not significantly dependent on the number of points as such, but often points obtained in perpendicular directions (e.g. midsagittal, midcoronal lines) accelerated the minimisation and improved the overall performance.

Table 1: Results from three simulated MEG/MRI registrations. The starting positions (initial errors) are varied, and so is the outcome. In all cases, however, the final misfit per point is less than 3 mm.

	Translation (mm)			Rotation (degrees)			
Errors	b_x	b_y	b_z	θ_x	θ_y	θ_z	RMS misfit/point (mm)
initial	+10.0	-10.0	+10.0	-5.0	+2.0	+6.0	14.18
final	-0.14	-0.78	-0.10	-1.4	-0.1	-0.4	2.46
initial	-15.0	+10.0	+15.0	+8.0	+4.0	-10.0	12.7
final	-1.23	+0.08	+1.76	+1.02	+2.1	-0.18	3.0
initial	+5.0	+5.0	+5.0	-2.0	-2.0	-2.0	6.40
final	-0.1	+0.66	-0.1	-1.3	-0.05	+1.0	1.40

In Fig. 2 we show the final registration product from a real experimental set-up where the final mismatch per point was some 3 mm, having started from some 23 mm per point. Different visualisation angles reveal that the fit between MEG head shape points and MRI scalp outlines is excellent throughout. Finally, Fig. 3 provides a summary of the whole registration procedure in another real case. On the most left are the initial datasets (unregistered), approximation by point matching in the middle, and final result after surface matching on the right.

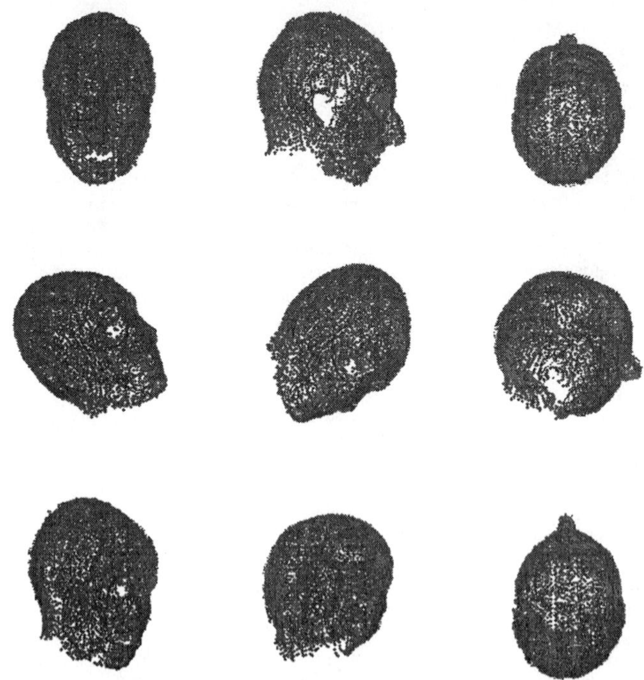

Fig. 2: Visualisation of registration results in different perspective angles. MEG headshape points (dark grey) are superimposed on MRI outlines (light grey). The fit is excellent throughout. The RMS mismatch/point is some 3 mm.

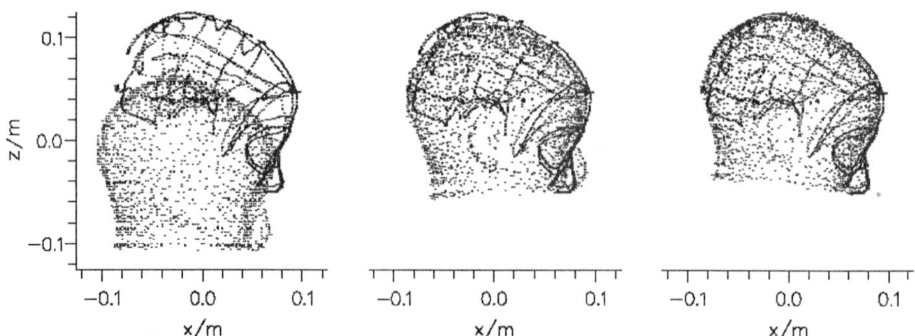

Fig. 3: Skull outlines in MRI (light grey) and MEG (dark grey) co-ordinate systems: before registration (left), after point matching (centre), and after surface matching (right).

Discussion and Outlook

This is not the first time that the surface matching registration technique is applied on MEG/MRI datasets. However, it is the first time the need to move from the most common and easy point matching strategy to the more robust and less subjective methodology of surface matching is shown through the use of practical examples. The improvement in accuracy obtained from the combination of the prior and the latter technique was not shown before through visual examples.

The accuracy we have achieved is comparable with that achieved by others [3]. Our simulations gave better results than the applications to actual data. However, the real data we have been using were always part of an actual MEG experiment (e.g. auditory odd-ball, CNV etc). In other words, we have not attempted any model experiments as such, in which all the individual steps would have been carefully followed to allow for a proper registration accuracy evaluation.

Furthermore, the software we have developed does not include a proper Graphics User Interface (GUI) yet. Efforts to combine it with a proper GUI as well as to optimise the code for faster minimisations are currently being undertaken.

High registration accuracy is the first necessary step for any MEG modelling. For Magnetic Field Tomography (MFT) [4], our method of choice, accuracy is mandatory. First, more complex source spaces [5] can be designed to follow the anatomy closer. Second, the source space itself can also be changed iteratively [6,7] to model deep generators. These are intermediate steps before constraining the source space to the cortical mantle or other important "deeper" structures so that physiologically more interesting questions can be tackled.

However, before the latter can occur, one should aim for optimisation of all the individual steps taken during a MEG recording. These are, avoidance of subject/patient movements, more accurate acquisition of MEG headshape points through the use of better defined MEG/Dewar systems, and automation of the procedure.

As the MEG research is indulging onto its new era of whole head cover systems and multimodality imaging, accuracy in registration becomes more important. We believe, that our experiences will provide a useful reference for the MEG researcher, and more importantly will help solving problems by combining the simple (point matching) with the more difficult and general (surface matching) techniques.

References:

[1] Pelizzari, C.A., Chen, G.T.Y., Spelbring, D.R., et al. Accurate 3d correlation of CT, PET, and/or MR images of the brain, J. Comp. Assist. Tomography, 1989, 13: 20-26

[2] Abraham-Fuchs K., Lindner T., Wegener P. Fusion of biomagnetism with MR or CT images by contour-fitting, Biomed. Eng., 1991, 36(Suppl): 88-89

[3] Kober, H., Grummich, P., Vieth, J. In: Baumgartner, C., Deecke, L., et al. Biomagnetism: Fundamental research and Clinical Applications, Amsterdam, Elsevier, 1995.

[4] Ioannides, A.A. In: Ho, M.W., Popp, F.A. and Warnke, U., Bioelectrodynamics and Biocommunication, Singapore, World Scientific, 1994.

[5] Liu, M.J., and Ioannides, A.A. In Deecke L, Baumgartner C et al. Biomagnetism: Fundamental Research and Clinical Applications, Elsevier, Amsterdam, 1995.

[6] Bamidis, P.D., Hellstrand, E., Lidholm, H., et al. MFT in complex partial epilepsy: spatio-temporal estimates of interictal activity, Neuroreport, 1995, 7(1): 17-23.

[7] Bamidis, P.D., PhD Thesis, The Open University, Milton Keynes, UK, 1996.

EEG and MEG Source Analysis of Somatosensory Evoked Responses to Mechanical Stimulation of the Fingers

Cheyne, D.[1], Roberts, L.E.[2], Gaetz, W.[2], Bosnyak, D.[2], Weinberg, H.[1], Johnson, B.[1], Nahmias, C.[3] and Deecke, L.[4]

[1]Brain Behaviour Laboratory, Simon Fraser University, Burnaby, Canada;
Departments of [2]Psychology and [3]Nuclear Medicine, McMaster University, Hamilton, Canada;
[4]Neurological Clinic, University of Vienna, Vienna, Austria.

Introduction

Previous MEG and EEG studies have successfully utilized electrical stimulation of the digits in order to study generators of early responses in the human primary somatosensory cortex [1-3]. Natural stimulation of the somatosensory system using vibratory pulses to the tips of the digits has been found to elicit qualitatively different responses in the EEG [4, 5] presumably due to the activation of different neuronal pathways. Similar responses have been observed in the MEG to transient mechanical stimuli producing large responses at latencies of 50 msec, corresponding to the electrical P50 response [6, 7]. A more recent study [8] also noted MEG responses at 70 msec latencies in some subjects resembling the N70 component described by Hämäläinen and co-workers [4]. The earlier, P50 response is presumed to reflect activation of primary sensory cortex (SI). A study of mechanically evoked epidural and single unit responses in waking monkeys [5] found that this component was associated with a period of inhibitory input to neurons in areas 3b and 1, whereas the later, slow component (N70) was not associated with activity in SI and appeared to arise from other cortical areas, such as SII. The magnetically recorded P50 (P50m) appears to be the largest and most consistent MEG event recorded during transient tactile stimulation in humans and its potential application for somatotopic mapping studies [6, 7] warrants further investigation of its neural generation. Moreover, the marked orthogonality of the EEG and MEG topographies of the P50 response makes it an ideal candidate for the comparison and/or combination of EEG and MEG localization methods. The current study compared separate high-density, 32 channel EEG and 143 channel MEG recordings in two subjects using identical stimulation paradigms in order to compare dipole source locations in somatosensory cortex obtained separately for each method. In addition, 3-dimensional MRI was obtained for both subjects in order to constrain source model information as well as aid in the integration of coordinate systems for the MEG and EEG source models.

Methods

EEG recordings were made during mechanical stimulation of the finger tips of the right hand in two subjects using a high density 32 channel EEG system (NeuroScan, Herndon, VA) consisting of a 19 channel grid overlying the left Rolandic fissure and 13 distributed 10-20 locations. MEG recordings were obtained separately using a whole cortex MEG system (CTF Systems, Inc) consisting of 143 1st-order gradiometers covering the whole head. Recordings were performed without magnetic shielding with the aid of software 3rd-order gradient formation with an overall white noise level of about 6 fT/√Hz [9]. Stimuli consisted of 5 msec duration pulses produced by an electromechanical stimulator with an inter-stimulus interval of 750 msec. Acoustic artifacts from the stimulator were masked with white noise and for MEG recordings, the electromechanical stimulator was driven from a distance of several meters to reduce magnetic artifact which was noticeable in some posterior channels. Trials of 256 msec duration were averaged using a 64 msec pre-stimulus baseline and off-line bandpass filtered between 1 and 100 Hz. For EEG data two blocks of 800 trials (1000 samples/sec) were collected and for MEG one block of 600 trials (1250 samples/sec). Comparison of responses for the first and second blocks of EEG data indicated good replicability and dipole fits were performed on the first block of data for comparison with the MEG. For both subjects, three-dimensional proton density magnetic resonance images were obtained with vitamin E markers placed at fiduciary landmarks used to create the head-based MEG coordinate system and selected 10-20 locations. An off-line procedure allowed the identification of these markers in the MRI for the co-registration the MRI and MEG coordinate systems.

Results

MEG responses showed a clear field reversal peaking at 50-60 msec and a slower component peaking at latencies of 70-90 msec, similar to the P50m and N70m responses observed previously [7]. The P50 component

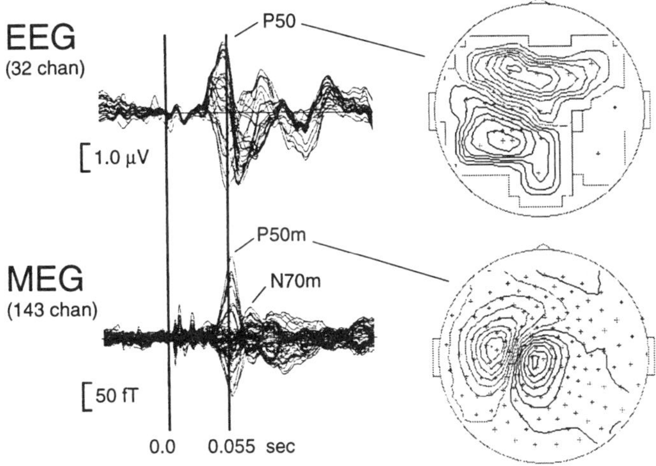

Fig. 1. Left: Superposition of 32 EEG channels (upper) and 143 MEG channels (lower) for averaged response to mechanical stimulation of the right index finger in one subject. Right: Isocontour maps of the electrical and magnetic field patterns at the peak of the P50 response. Small crosses indicate gradiometer/electrode location and darker contours represent outgoing flux and positive potential in steps of 23 fT and 0.37 µV, respectively. Small deflections visible in early phase of MEG response are due to electromechanical stimulator artifact.

was characterised by orthogonal dipolar patterns of magnetic flux and electrical potential at similar latencies suggestive of a tangential dipole source with opposite orientation of intracellular and extracellular current measured by MEG and EEG, respectively (Fig. 1). The EEG also showed monopolar distributions of peak responses at latencies of 60 - 70 msec in some conditions, indicative of radially oriented sources not observed in the MEG.

Localization of single dipole sources was obtained for the MEG using a least-squares fitting algorithm with corrections for non-radial field contributions [9]. A best fit sphere was chosen for each subject based on the structural MRI data, aided by the MRI markers of fiduciary and 10-20 locations. EEG source analysis was achieved using a 4-shell spherical model after Stok [10], based on digitized locations of the electrode montage projected onto a sphere with the same radius. For comparison with MEG results, EEG dipole locations in the spherical model were transformed to a common head-based coordinate system by appropriate translation by the MEG sphere origin (no rotation was assumed). The overall approach for the integration of these three modalities is summarized in Figure 2. Since MEG source localization which includes non-radial measurements (due to helmet geometry) is dependent on the sphere origin (i.e., definition of radial direction) confirmation of the choice of best-fit sphere was attempted by a trial and error procedure, comparing changes in residual error with slight deviations in the sphere location (indicated by "adjust X,Y,Z" step in Fig. 2). Most interestingly, we found that when the sphere origin was iteratively adjusted in this manner, the initial sphere location based on the MRI produced the lowest error in almost all cases and was therefore accepted as the default origin in subsequent analysis. The sphere origin also determines the location of the EEG sources relative to the MEG sources, since EEG source modelling was performed relative to a sphere with the same radius onto which digitized locations of the electrodes had been projected. Appropriate translation of the EEG source locations by this origin, therefore allows direct comparison of EEG and MEG source locations. As also indicated by Fig. 2, when comparing final source locations based on the two modalities, MEG sources are assumed to provide more accurate information regarding dipole moment (for cases where radial contributions are negligible) and might therefore be used to adjust the EEG spherical model conductivity values.

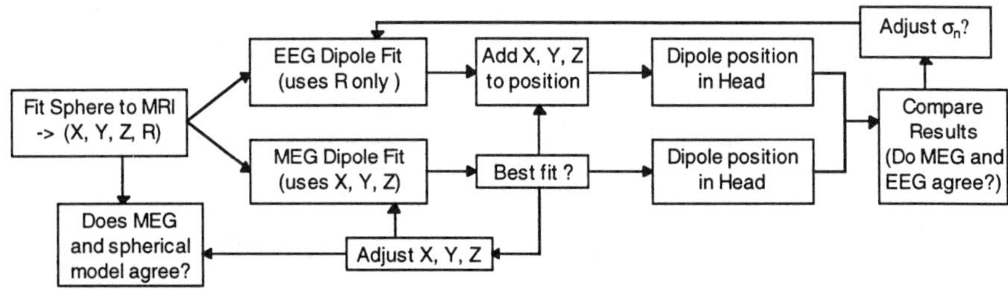

Fig. 2. Flow chart of iterative steps used in the integration of EEG and MEG source modelling procedures with 3-dimensional MRI data. X, Y, Z = sphere location in head coordinate system. R = sphere radius. σ_n = conductivity for spherical model (shells 1.. n).

Tangential dipole sources were localized from MEG data for the peak of the P50m component with residual errors ranging from 2.1% to 6.1%. EEG source localization yielded fits of sources for the P50 with errors ranging from 1.5% to 5.5%. Comparisons of EEG locations to those obtained with MEG were made in the common coordinate system after translation by the sphere origin as described above. Table 1 shows dipole locations obtained for three digits (1 = thumb, 3 = middle, 5 = pinky) for both subjects obtained from EEG and MEG data and the resulting differences in source location and strength indicating generally good agreement between source locations for the P50 component. One notable difference was the degree of lateral dispersion of the sources which tended to be greater for EEG fits. Additionally, there is some indication of systematic vertical and lateral shifts between EEG and MEG sources in subject 1. Other differences did not show any systematic trend and therefore do not likely reflect incorrect specification of the sphere origin. Constraining the EEG dipoles to a tangential orientation resulted in only minor changes in location and error indicating that this component is generated by a largely tangential source in SI, most likely area 3b. Dipole moment for EEG fits tended to be lower than that of the MEG by approximately 25 to 50%. Since higher moments would be expected for any contributions of a radial source component, this suggests a possible correction to the spherical model conductivities for the EEG based on the MEG dipole moments.

Fits attempted for the peak of the N70 component in both MEG and EEG data tended to be less consistent and higher in error, possibly due to the lower signal to noise ratio of these responses. These sources also tended to locate lateral and deeper to those for the P50, consistent with an SII origin but were not analyszed further in the current study. Dipole locations for the P50 projected onto the appropriate MRI slice indicated location in the region

Table 1. Comparison of dipole source parameters for EEG and MEG dipole solutions in 2 subjects.

Subject 1 Digit	Latency (ms)	Position X (cm)	Y (cm)	Z (cm)	Moment (nA-m)	Error (%)
(EEG)				(+5.5)		
1	59.0	0.52	3.21	9.07	6.3	1.7
3	58.0	0.59	3.27	9.39	7.1	2.8
5	59.0	0.23	3.16	9.63	6.0	1.5
(MEG)						
1	59.2	0.45	4.12	9.96	10.3	6.1
3	60.0	0.93	3.76	9.84	15.4	4.9
5	61.6	0.91	4.17	10.28	8.9	3.5
Difference (EEG - MEG)						
1		0.08	-0.91	-0.89	-4.0	
3		-0.35	-0.49	-0.45	-8.3	
5		-0.68	-1.01	-0.65	-2.9	
Mean		-0.32	-0.80	-0.66	-5.0	

Subject 2 Digit	Latency (ms)	Position X (cm)	Y (cm)	Z (cm)	Moment (nA-m)	Error (%)
(EEG)		(-0.5)		(+5.2)		
1	56.0	0.69	4.52	8.88	13.9	5.5
3	54.0	0.02	3.21	9.09	19.1	2.8
5	54.0	-0.03	3.13	8.27	17.5	3.0
(MEG)						
1	57.6	-0.32	4.06	8.83	17.5	4.5
3	57.6	-0.04	3.98	8.84	23.7	2.3
5	60.0	-0.09	4.16	9.01	17.5	2.1
Difference (EEG - MEG)						
1		1.01	0.46	0.05	-3.6	
3		0.06	-0.78	0.25	-4.6	
5		0.06	-1.03	-0.74	0.0	
Mean		0.38	-0.45	-0.15	-2.7	

of the postcentral gyrus with a tangential orientation (anterior directed for EEG, posterior directed for MEG) consistent with surface negativity at the posterior bank of the central sulcus, due to either deep inhibitory, or superficial excitatory input to neuronal populations in area 3b. Sources for different fingers tended to be distributed along a line in agreement with the known somatotopy of the hand although these differed in their extent for the MEG and EEG solutions. Also, sources tended to move downward and medial in one subject, and in the other subject upward and medial, suggesting differences in gyral folding which was also evident in the individual MR images.

Discussion

Comparisons of early components of electrically evoked MEG and EEG somatosensory responses have suggested that MEG sensitivity to only tangential sources may provide additional constraints for EEG source analysis [3]. In the current study, somatosensory responses to tactile stimulation of the digits indicated that EEG and MEG responses produce highly orthogonal field patterns at latencies of 50 msec, most likely due to tangential generators in SI, with little contribution from radial sources, allowing for direct comparison of dipole modelling results from EEG and MEG data independently. This result is similar to that found by Buchner et al., [3] for the 30 msec (N30/P30) response evoked by electrical stimulation. In addition, we tested strategies for combination of the MEG and EEG coordinate systems, which confirmed structural MRI as a useful tool for estimating the best fit sphere, both for determining contributions of non-radial magnetic fields for MEG dipole solutions, as well as for the translation of EEG dipole solutions to the MEG coordinate system. This approach yielded similar results for absolute EEG and MEG source locations with overall deviations of about half a centimeter, although some systematic discrepancies remained. Combined solutions for EEG and MEG sources using simultaneously acquired data may provide more accurate source location estimates, although the correct integration of the two coordinate systems will be an important factor for the successful combination of these two methods.

Acknowledgments

This project was supported in part, through NSERC operating grants (Roberts-OGP0000132), (Weinberg-OGP0008322), an NSERC CRD grant (Weinberg-661044492), the NATO Division of Scientific Affairs (Roberts/Elbert-940648), and a Deutsche Forschungsgemeinschaft Fellowship (DFG BE1538) to L. Roberts.

References

[1] Hari, R. and Kaukoranta E. Neuromagnetic studies of somatosensory system: principles and examples. Prog. Neurobiol., 1985, 24: 233-256.

[2] Buchner, H., Adams, L., Müller, A., Ludwig, I., Knepper, A., Thron, A., Niemann, K. and Scherg, M. Somatotopy of human hand somatosensory cortex revealed by dipole source analysis of early somatosensory evoked potentials and 3D-NMR tomography. Electroenceph. clin. Neurophysiol., 1995, 96: 121-134.

[3] Buchner, H., Fuchs, M., Wischmann, H-A., Dössel, O., Ludwig, I., Knepper, A. and Berg., P. Source analysis of median nerve and finger stimulated somatosensory evoked potentials: Multichannel simultaneous recording of electric and magnetic fields combined with 3D-MR tomography. Brain Topog., 1994: 299-310

[4] Hämäläinen, H. Kekoni, J., Sams, M., Reinikainen, K. and Näätänen R. Human somatosensory evoked potentials to mechanical pulses and vibration: contributions of SI and SII somatosensory cortices to P50 and P100 components. Electroenceph. clin. Neurophysiol., 1990, 75:13-21.

[5] Gardner, E.P., Hämäläinen, H.A., Warren, S., Davis, J. and Young W. Somatosensory evoked potentials (SEPs) and cortical single unit responses elicited by mechanical tactile stimuli in awake monkeys. Electroenceph. clin. Neurophysiol., 1984, 58: 537-552.

[6] Suk, J., Ribary U., Cappell, J., Yamamoto T. and Llinás R. Anatomical localization revealed by MEG recordings of the human somatosensory system. Electroenceph. clin. Neurophysiol., 1991, 78:185-196.

[7] Yang, T.T., Gallen C.C., Schwartz, B.J., and Bloom, F.E. Noninvasive somatosensory homunculus mapping in humans by using a large-array biomagnetometer. Proc. Natl. Acad. Sci., 1993, 90: 3098-3102.

[8] Cheyne, D., Weinberg H., Hattori, H., Gordon, R., Vrba, J. and Burbank M. In: N. Tepley and G. L. Barkley (Eds.) Characterisation of human sensorimotor cortex using whole-cortex MEG: implications for clinical use. Proceedings of the North American Biomagnetism Action Group Annual Meeting. Detroit, 1994.

[9] Vrba J., Angus, V., Betts K., Burbank M., Cheung T., Fife A., Haid G., Kubik P., Lee S., McCubbin J., McKay J., McKenzie, Robinson, S., D., Spear P., Taylor B., Tillotson M., Cheyne D. and Weinberg H. A 143 channel, whole cortex MEG system (these proceedings).

[10] Stok C. J. The inverse problem in EEG and MEG with application to visual evoked responses. Ph.D. Thesis, 1986, Krips Repro Meppel.

Dynamic Neuroimaging by MEG, Constrained by MRI and fMRI

George, J.S.[1], Schmidt, D.M.[1], Mosher, J.C.[1], Aine, C.J.[1], Ranken, D.M.[1], Wood, C.C.[1], Lewine, J.D.[2], Sanders, J.A.[2] and Belliveau, J.W.[3]

[1]Los Alamos National Laboratory, Los Alamos, New Mexico, USA; [2]New Mexico Regional Federal Medical Center, Albuquerque, New Mexico, USA; [3]Massachusetts General Hospital NMR Center, Charlestown, Massachusetts, USA

Introduction

MEG is a direct measure of the electrical activity of populations of neurons, with excellent temporal resolution. However, the spatial characterization of sources depends on the solution of an ill-posed inverse problem. MRI provides high resolution volumetric data defining the anatomy of the head and brain; alternative MRI data acquisition strategies allow mapping of hemodynamic correlates of neural function. However, functional MRI (fMRI) and related techniques are limited by the slow timecourse of the hemodynamic response and the ill-defined relationship between neural activation and associated hemodynamic changes. Given such complementary strengths and weaknesses, the integration of multiple imaging technologies should provide dynamic functional neuroimaging capabilities with optimal spatial and temporal resolution. We are exploring a range of strategies for the integrated analysis of data from MRI, fMRI and MEG.

Anatomical and functional MRI provide constraints on the location and configuration of putative neural sources. We assume that the neural sources measured with MEG are confined to the thin sheet of gray matter comprising neocortex and similar structures in the human brain. For some analyses we further assume that the net current moment in any patch of cortex is oriented normal to the local surface. The geometry of cortex is defined from MRI using software tools developed in our laboratory. Analysis of fMRI data depends on the identification of clusters of significant activation in statistical parametric images. Our general stratgey is to use MEG to define the detailed temporal dynamics of sources localized (at least in part) on the basis of MRI data.

Methods

MRI data described here were obtained on a 1.5 T Siemens Magnetom at the New Mexico Regional Federal Medical Center (NMRFMC). Anatomical data were acquired using a 3-D volume acquisition technique, MP-RAGE. In-plane resolution was typically 1mm (256^2 voxels) with slice resolution of 1.5 or 2 mm (128 slices). The volume was typically resampled to an isotropic resolution of 2 mm for segmentation and other analyses. Multislice functional MRI was acquired with the same instrument, using a T2* weighted FLASH aquisition. Typically, 6 slices were acquired with 2 mm in-plane resolution, and with 5 or 6 mm between contiguous slices. Interleaved blocks of 4 control or stimulated image sets (2 minutes per block) were acquired. Volume data segmentation, analysis and visualization was performed with MRIVIEW [1] and other codes developed in our laboratory. These systems are written in IDL, a high level programming and analysis system developed by Research Systems Inc. Figure 1 illustrates several features of our analysis of anatomical and functional MRI data.

MEG data were collected on a 37 channel neuromagnetometer (BTi, Magnes) at the NMRFMC employing 2 to 4 replications of the experiment at different sensor array placements. Because of differences in the physics and underlying physiology of fMRI and MEG measurements, we initially compared these alternative measures of neural activation. Although correspondence between identified sources was not perfect, and optimal experimental conditions for MEG and fMRI may vary considerably, the results encouraged us to pursue integrated analyses [2,3].

Statistically significant voxels in fMRI (95% confidence level), were identified based on t-statistic difference values, taking into account the (large) number of independent measurements. Analyses were repeated for 3-D clusters of contiguous voxels of increasing size, which lowered the threshold value for statistical significance. The average value of a voxel or region of interest in a stimulated trial was required to exceed the average value of adjacent control periods, across all stimulus/control cycles. This method is similar to the widely used temporal correlation techniques but is applicable even given the limited temporal sampling of our conventional MRI acquisition scheme. This method is particularly useful for eliminating large artifacts that may arise through transient movement of the subject.

Results

Our first approach for integrated analysis employed simplified equivalent sources to account for observed MEG responses. Given a cluster of active voxels derived from fMRI, current orientations defined by MRI geometry, and an estimate of the relative activation of voxels within the cluster based on functional MRI, we found an equivalent current dipole that best accounted for the fields associated with the composite source. Dipole parameters

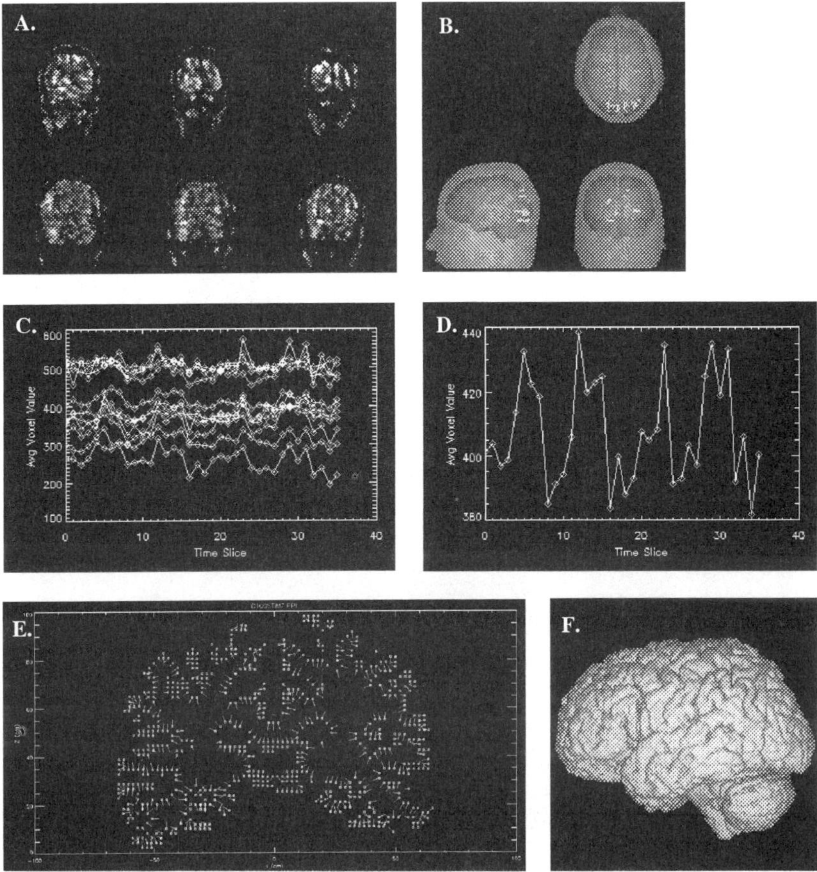

Figure 1.   Structural and functional MRI data
A. A visual fMRI experiment with six coronal slices through occipital and adjacent cortex.
B. Projection images illustrating the location and extent of each of the fMRI clusters
C. fMRI timecourses for each of 11 identified clusters.
D. The average timecourse across all 11 clusters showing ~7% signal over most stimulus periods.
E. Locations and normals of cortical surface voxels from segmentation of 3-D data with MRIVIEW.
F. Reconstruction of the cortical surface from data as in E.

were determined by fitting the predicted field distributions, or locations were set equal to the center of mass of the cluster. A set of such sources was fit to MEG data to establish activation timecourses. An example of this sort of analysis is illustrated in figure 2, A and B. Dipole orientations were either estimated and fixed based on MRI data, or were allowed to rotate during the fitting procedure. While rotating dipoles produced a better fit (with many more parameters), we did not see striking differences between optimized and center of mass equivalent dipoles. However, many studies suggest that the best equivalent source may not correspond to the actual region of activation when the underlying current distribution is complex or extended. The early peaks in the timecourse data (0-50 ms latency) reflect patterns in the field maps arising from "wrap-around" due to the fixed and inadequate interstimulus interval used in these MEG experiments for compatibility with fMRI protocols.

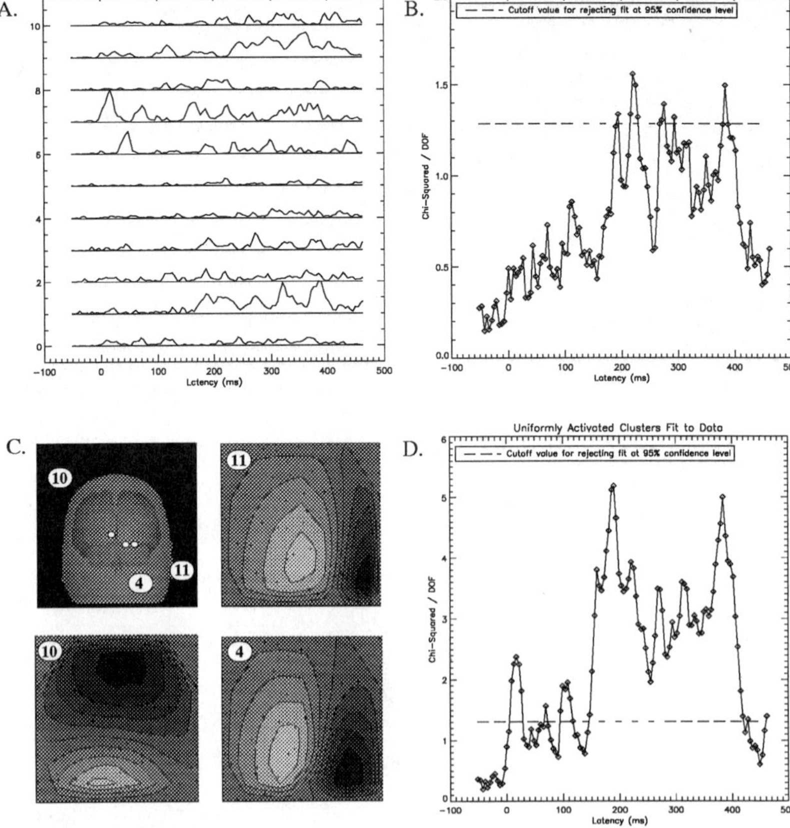

Figure 2. Parametric estimates of MEG source activation constrained by MRI
A. Significant currents for rotating dipoles at the centers of mass of the fMRI clusters.
B. Reduced Chi-square error across time for the rotating dipoles.
C. Basis fields associated with three of the 11 uniformly activated clusters identified from fMRI.
D. Reduced Chi-square error across time in composite model reconstruction.

The second approach similarly employed anatomical and functional MRI to define the location and geometry of the source, but did not employ a simplified equivalent source. Instead, a detailed forward calculation employing all active voxels was used to define a single basis vector for the entire cluster. A set of such sources was scaled to produce a best fit to the observed data. This approach explicitly accounts for distributed sources and is physically interpretable. In the analysis illustrated, our automatic cluster-finding procedure was used to identify 11 significant clusters in the fMRI data, and field distributions corresponding to composite sources were constructed assuming uniform activation within each cluster. Time courses were estimated by linear regression on the MEG spatial temporal data using the normalized forward calculations (such as illustrated in figure 2c) as basis vectors. However, basis vectors derived in this manner did not form an orthogonal set, so that the decomposition of the MEG data into a set of timecourses was ambiguous, consistent with the observation that similar temporal peaks appeared in several clusters. Further, in this study, uniformly activated clusters did not adequately fit the data. This might reflect physiological heterogeneity in activation levels within an area, or missing sources. The limited extent (in depth) of the fMRI data in this study insured that some visual areas were missed.

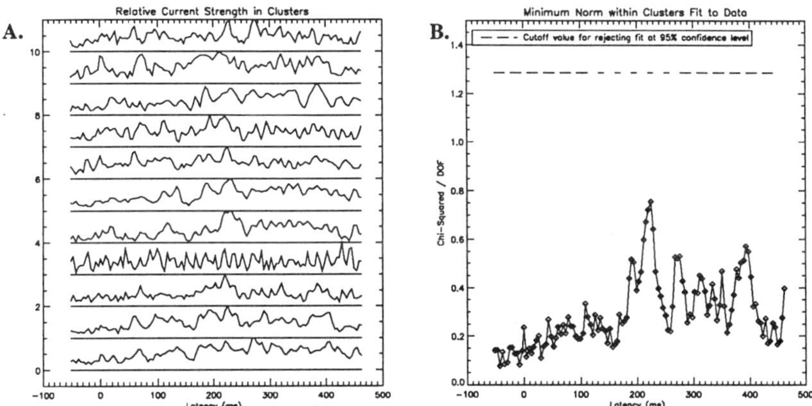

Figure 3. Constrained Minimum Norm Reconstruction
A. Reconstructed current timecourses from MEG data for each of the clusters constrained by MRI.
B. Reduced Chi-square error in the constrained MEG reconstruction as a function of time.

Our third approach was an extension of methods we have developed for distributed source reconstruction from MEG data incorporating anatomical constraints. Linear estimator procedures based on the Moore-Penrose pseudoinverse allow reconstructions of distributed current distributions of minimum Euclidean norm that account for the observed data. MRI derived constraints on source location and orientation may provide more accurate estimates of source location and extent. In the analysis illustrated in figure 3, "active" voxels identified by statistical analysis of fMRI data were used as a mask to select a small subset of the cortical voxels identified from anatomical segmentation. Currents were reconstructed using a weighted pseudoinverse procedure. The resulting inverse problem remained underdetermined (requiring more parameters than MEG sensors) yet might be overconstrained by MRI data. A more flexible paradigm would allow fMRI and MEG data to each contribute to the estimation of source spatial configuration. An alternative approach employs fMRI data used as a soft constraint, increasing the a priori estimate of the probability of current within a region. This prior probability together with experimental data can be used to compute a spatial probability metric based on Bayesian decision theory (Schmidt and George, this volume).

Discussion
The complementary strengths of multiple imaging techniques are required for mapping the structure and function of the human brain. Neural electromagnetic techniques provide excellent temporal resolution of neural population dynamics, but the requirement to solve an ill-posed inverse problem limits the resolution of source location and extent, particularly when multiple sources may be active simultaneously. MRI provides excellent spatial resolution. However, the physiology of the underlying hemodynamic changes limit the temporal resolution of activation measurements to a few seconds, and several problems can produce artifactual regions of apparent activation. A knowledge of individual anatomy can improve neuromagnetic current reconstructions by providing constraints on parameters. Sophisticated computational tools now exist to allow these prospects to be realized. We have provided an initial demonstration of several techniques for the integrated analysis of human MRI, fMRI, and MEG data. Future work will be necessary to establish the validity and scientific utility of these methods.

References

[1] Ranken, D., George, J. MRIVIEW: An interactive computational tool for investigation of brain structure and
       function, In: Nielson, G.M., Bergeron, D. IEEE Visualization '93, IEEE Computer Society Press, 1993
[2] George, J.S., Sanders, J.S., Lewine, J.D., Caprihan, A., and Aine, C.J. Comparative Studies of brain
       activation with MEG and Functional MRI. In: Deecke, L., Baumgartner, C., Stroink,G. and Williamson,
       S.J., Biomagnetism: fundamental research and clinical applications. Elsevier/IOS, Amsterdam, 1995
[3] George, J.S.,Aine, C.J., Mosher, J.C., Ranken, D.M. Schlitt, H.A., Wood, C.C., Lewine, J.D., Sanders, J.A.,
       Belliveau, J.W. Mapping function in the human brain with MEG, anatomical MRI and functional MRI, J.
       Clin.Neurophysiol., 1995, 12(5):406-431

This work was supported by NIH grants EY08610 and DA/MH09972 and DOE contract KP0601000.

Combining Modalities to Investigate Cortical Retinotopy

Hurdal, M.K., McElwain, D.L.S.

Queensland University of Technology, School of Mathematics, Brisbane, Australia

Introduction

The visual cortex is often used in investigations of the human brain. The primary visual cortex, or V1, is mainly buried in the calcarine fissure, a deep sulcus which is situated on the medial surface of the occipital cortex. Virtually all information in the visual system is recognized as being processed by V1 first, and then passed out to higher order systems [1].

It is well accepted that there is a precise map of how segments of the retina are related to areas of V1. V1 has a retinotopic map in that one spot in the visual field maps directly to a spot on V1 [2]. The overall details of this map are well known from animal studies and the study of lesions in humans [3, 4]. It is agreed that more area of V1 is devoted to representing central vision rather than peripheral vision. Thus, the scale of this mapping changes as a function of retinal location. The term 'cortical magnification' [5], denoted by M, is used to describe this mapping and refers to an area of cortex that is devoted to representing a small area of visual field.

In addition, upper visual cortex receives signals from the lower visual field and similarly, lower visual cortex processes information from the upper visual field. The right visual cortex process the left field of view and vice versa. Central vision is represented at the pole of occipital cortex [3]. This general representation is referred to as the cruciform model [6].

Beyond this general agreement of this mapping, there is disagreement as to the amount of cortex that is allocated to representing central vision or other portions of the visual field [7]. There is much controversy regarding the precise details of cortical magnification and this retinotopic map. Furthermore, individual anatomical variations in the calcarine fissure seem to indicate that the cruciform model in not adequate to describe cortical retinotopy [8, 9].

Combining modalities is an ideal approach to investigate cortical retinotopy. Dipole source localization [10, 11] is one method of non-invasively determining which region of cortex is active. A neural source of magnetic and electrical activity can be modeled by a single dipole. The location of a source can then be determined by estimating a dipole that explains observed MEG or EEG data. Cortical anatomy also needs to be considered [12] and magnetic resonance (MR) images are well suited to this purpose. In order to be able to use these different modalities together effectively, a model of the visual cortex is required.

Methods

The Theoretical Model

Let the visual field be represented in polar co-ordinates (r, θ) and let the surface of the cortex be described in Cartesian co-ordinates (x, y). Cortical magnification is a function of retinal location and this relates a small area of cortex to a small area of visual field. Mathematically, this can be represented by derivatives. A small area $dxdy$ in the cortical plane corresponds to a small area of visual field $rdrd\theta$ adjusted by the cortical magnification parameter $M(r, \theta)$ [13]:

$$dxdy = M(r, \theta)rdrd\theta. \tag{1}$$

In addition, cortical magnification is a point to point mapping which transforms the visual field into its neuronal representation [14]. This means that the number of points in the retina is equivalent to the number of points in the cortex. Let $C(x, y)$ represent the density of the points in the cortex and let $\rho(r, \theta)$ represent the density in the retina. Then

$$C(x, y)dxdy = \rho(r, \theta)rdrd\theta \tag{2}$$

Equating (1) and (2) yields

$$M(r, \theta) = \frac{\rho(r, \theta)}{C(x, y)}, \tag{3}$$

giving an expression for cortical magnification in terms of physiological properties.

The Practical Model

In order to incorporate cortical anatomy into the investigation of cortical retinotopy, a practical and realistic model of V1 is required. Cortical features were obtained from T1 weighted MR scans taken in a 1.5 Tesla field. Scans were taken 1.5 mm apart with contiguous spacings of 1.5 mm. Points located along the calcarine fissure were systematically extracted from the MR scans (see Fig. 1a). These points formed the basis of the model of V1.

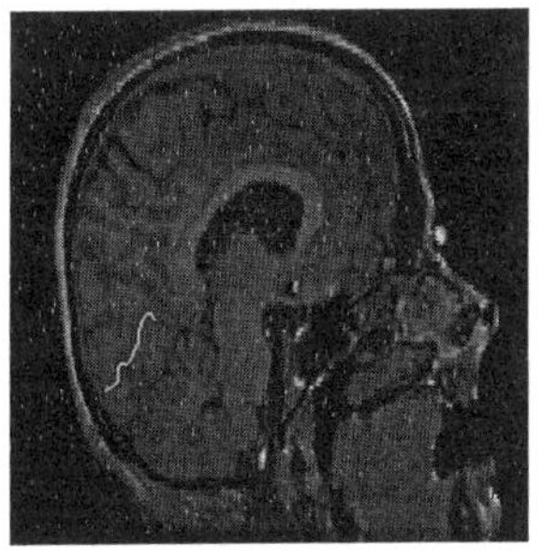

(a) MRI with Calcarine Fissure Highlighted

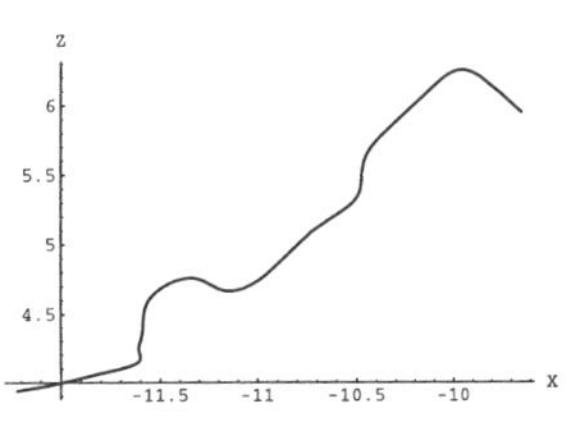

(b) Spline Curve which Models Calcarine Fissure

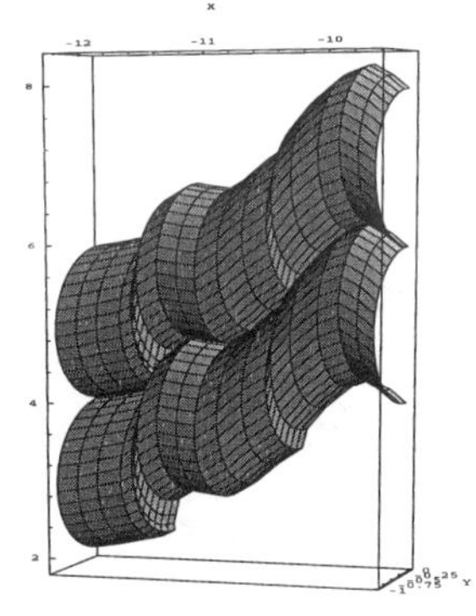

(c) Cylinder Model of V1

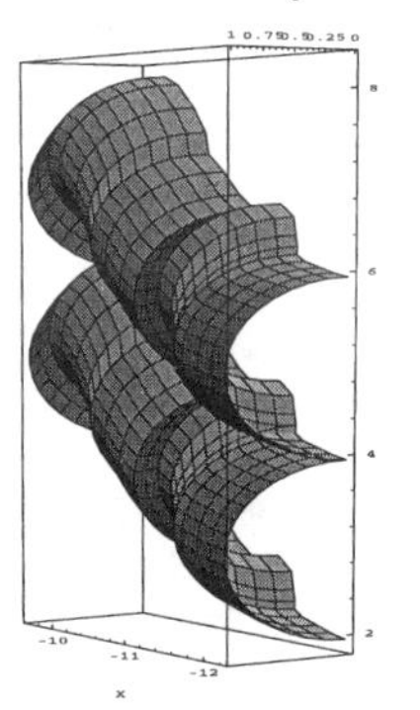

(d) Different Orientation of Cylinder Model

Fig. 1: Points representing the calcarine fissure were extracted from MRI data (Fig. 1a). A cubic spline was then fitted to the data (Fig. 1b). V1 is modeled by two cylinders representing the upper and lower visual cortex in one hemisphere of the brain. The cylinders lie beside each other and abut along the calcarine fissure. Since the calcarine fissure is modeled by the spline curve, the cylinders follow the realistic shape of the calcarine fissure (Fig. 1c and Fig. 1d).

Results

Cylindrical Model of Primary Visual Cortex

The points that were extracted from the MR scans were fitted to a cubic spline curve, resulting in the curve shown in Fig. 1b. As is apparent from comparing Fig. 1a and Fig. 1b, this curve approximates the shape of the calcarine fissure extremely well.

The visual cortex in one hemisphere of the brain was then modeled as two adjacent cylinders. The upper visual cortex is represented by one cylinder, as is the lower visual cortex. The two cylinders lie along side each other and abut along the length of the calcarine fissure. Since the fissure is modeled by the spline curve, the cylinders also follow the shape of the spline curve (see Fig. 1c and Fig. 1d). The result is a three dimensional cylindrical model of V1 which incorporates the realistic shape of the calcarine fissure. Fig. 1c and Fig. 1d show different orientations of this model.

Discussion

A mathematical definition for cortical magnification and a procedure for estimating human cortical magnification have been discussed in this paper. A theoretical model has been presented which is a formal and novel mathematical definition and has been obtained through new equations. By incorporating physiological properties of the point to point mapping between the visual field and the visual cortex, the concept of cortical magnification has been placed on a firmer mathematical footing.

A new and realistic model of the calcarine fissure has also been developed. The upper and lower visual cortex are modeled as two cylinders which follow the shape of the calcarine fissure. This cylinder model can now be used in conjunction with dipole source localization to estimate cortical magnification. Visual evoked potential data has been collected from a checkerboard pattern reversal experiment and the neural sources have been estimated using a dipole source localization algorithm [9]. The cylindrical model can now be used to relate the position of the estimated sources to the positions of the visual stimuli. In this manner, a truly human estimate for cortical magnification can be obtained.

The advantage of this approach is that this methodology can be applied to other areas of brain research. For example, the motor cortex and auditory cortex both have their own unique mapping [15]. By combining the modalities and techniques of dipole source localization, MEG / EEG data collection, MR images and mathematical modeling, similar approaches can be used to define and estimate a cortical mapping in these other regions.

References

[1] Zeki, S. The visual image in mind and brain, Scientific American, 1992, 267: 42–50.

[2] Schwartz, E.L. Local and global functional architecture in primate striate cortex: outline of a spatial mapping doctrine for perception, In: Rose, D., and Dobson, V.G. Models of the Visual Cortex, New York, John Wiley and Sons, Ltd., 1985.

[3] Holmes, G. The organization of the visual cortex in man, Proceedings of the Royal Society of London, Series B. Biological Sciences, 1945, 132: 348–361.

[4] Zeki, S. A Vision of the Brain, Oxford, Blackwell Scientific Publications, 1993.

[5] Daniel, P.M., and Whitteridge, D. The representation of the visual field on the cerebral cortex in monkeys, Journal of Physiology (London), 1961, 159: 203–221.

[6] Ahlfors, S.P., Ilmoniemi, R.J., and Hämäläinen, M.S. Estimates of visually evoked cortical currents, Electroencephalography and Clinical Neurophysiology, 1992, 82: 225–236.

[7] Horton, J.C., and Hoyt, W.F. The representation of the visual field in human striate cortex: a revision of the classic Holmes map, Arch. Opthalmol., 1991, 109: 816–824.

[8] Chorlton, M.C., Hurdal, M.K., Fulham, W.R., Finlay, D.C., and McElwain, D.L.S. Visual evoked potentials to small stimuli presented along a vertical meridian: Individual differences and dipole modelling, Australian Journal of Psychology, 1994, 46: 87–94.

[9] Hurdal, M.K. Dipole modelling for the localization of human visual evoked scalp potential sources, Master of Science thesis, University of Newcastle, Australia, 1994.

[10] Rush, S., and Driscoll, D.A. Current distribution in the brain from surface electrodes, Anesthesia and Analgesia, 1968, 47: 717–723.

[11] Ary, J.P., Klein, S.A., and Fender, D.H. Location of sources of evoked scalp potentials: correction for skull and scalp thickness, IEEE Transactions on Biomedical Engineering, 1981, 28: 447–452.

[12] Aine, C., Supek, S., George, J., Ranken, D., Best, E., Tiee, W., Vigil, V., Flynn, E., and Wood, C. MEG studies of human vision: retinotopic organization of V1. In: Baumgartner, C. et al. Biomagnetism: Fundamental Research and Clinical Applications, Amsterdam, Elsevier Science, 1995.

[14] Murray, J.D. Mathematical Biology, Heidelberg, Springer-Verlag, 1989.

[13] Fischer, B. Overlap of receptive field centers and representation of the visual field in the cat's optic tract, Vision Research, 1973, 13: 2113–2120.

[15] Schwartz, E.L. Spatial mapping in the primate sensory projection: analytic structure and relevance to perception, Biological Cybernetics, 1977, 25: 181–194.

Acknowledgements

The authors wish to acknowledge Dr. Krish Singh of the Department of Psychology, Royal Holloway, University of London for the use of his software in displaying and presenting MR data.

fMRI Constrained Linear Estimation of Cortical Activity from MEG Measurements: A Model Study

Liu, A.K.[1,2], **Sereno, M.I.**[3], **Rosen, B.R.**[1], **Belliveau, J.W.**[1] and **Dale, A.M.**[1]

[1]*Massachusetts General Hospital NMR Center, Charlestown, MA*
[2] *Harvard-MIT Division of Health Sciences and Technology, Cambridge, MA*
[3]*University of California, San Diego Department of Neurosciences, La Jolla CA*

Introduction

The localization of activity within the brain based on magnetoencephalography (MEG) measurements is inherently ill-posed. It has been suggested that information from functional magnetic resonance images (fMRI) can be used to constrain the ill-posedness [1,2,3,4]. Using the MEG temporal data and a linear estimation approach [1] or weighted minimum norm, we wish to compute the time courses of activated areas observed by fMRI.

One can foresee situations where the hemodynamic and magnetic measurements differ, although the actual region of cortical activation is the same. In these modeling studies, we have examined the effects of several "mis-specifications" on various types of weightings of the linear approach.

The three mis-specifications examined were:
1) mis-registration - the number of areas of activation on fMRI correspond to the number of generators of magnetic fields, but the two are offset by some distance
2) extra fMRI source - an area of fMRI activation does not generate magnetic fields
3) missing fMRI source - a generator of magnetic fields is not measured as an area of fMRI activation.

The spatial weightings were:
1) minimum norm - weights all locations equally
2) depth-weighted minimum norm - weights locations approximately inversely proportional to depth
3) fMRI constrained - spatial information from fMRI is used.

Methods

The derivation of the linear inverse operator [1] begins with a set of measurements
$$\mathbf{x} = \mathbf{As} + \mathbf{n}$$
where $\mathbf{x}$ are the measurements, $\mathbf{A}$ is the gain matrix, $\mathbf{s}$ is the strength of each dipole component, and $\mathbf{n}$ is the noise vector. We would like to calculate a linear inverse operator $\mathbf{W}$ that minimizes the expected difference between the estimated and the correct source solution. The expected error can be defined as

$$Err_W = \left\langle \left| \mathbf{Wx} - \mathbf{s} \right|^2 \right\rangle$$

Assuming both $\mathbf{n}$ and $\mathbf{s}$ are normally distributed with zero mean and covariance matrices $\mathbf{C}$ and $\mathbf{R}$, respectively, we can rewrite the expected error as

$$Err_W = \left\langle \left| \mathbf{W}(\mathbf{As} + \mathbf{n}) - \mathbf{s} \right|^2 \right\rangle$$

$$= \left\langle \left| (\mathbf{WA} - \mathbf{I})\mathbf{s} + \mathbf{Wn} \right|^2 \right\rangle$$

$$= \left\langle \left| \mathbf{Ms} + \mathbf{Wn} \right|^2 \right\rangle \text{ where } \mathbf{M} = \mathbf{WA} - \mathbf{I}$$

$$= \left\langle \left| \mathbf{Ms} \right|^2 \right\rangle + \left\langle \left| \mathbf{Wn} \right|^2 \right\rangle$$

$$= tr(\mathbf{MRM}^\mathsf{T}) + tr(\mathbf{WCW}^\mathsf{T})$$

This last expression can be explicitly minimized by taking the gradient, setting it to zero and solving for $\mathbf{W}$. This gives a linear inverse operator

$$\mathbf{W} = \mathbf{R}\mathbf{A}^{\mathrm{T}}(\mathbf{A}\mathbf{R}\mathbf{A}^{\mathrm{T}}+\mathbf{C})^{-1}$$

Information from fMRI, or any other spatial activation information, is incorporated into the $\mathbf{R}$ matrix. Noise characteristics of the sensors may be included in the $\mathbf{C}$ matrix.

For these model studies, we used a spherical head model with a radius of 10 cm and a 122 channel MEG sensor description (Neuromag, Inc.). Noise was assumed to be spatially uncorrelated and white with a SNR of 10.

The fMRI weighting was systematically varied from no weighting (i.e. minimum norm) to complete fMRI weighting (all locations not specified by fMRI were set to 0). A depth-weighted minimum norm was also computed for comparison [3]. The depth-weighted minimum norm weighted each location by the inverse of the power at the detectors given a unit activation at that location. Only the optimum fMRI weighting case, chosen to minimize the effect of the missing dipole, is presented below.

Either 5 or 10 sources were located at random within a single hemisphere or a single octant of the head model. Then the average percentage of incorrectly specified power per dipole pair was computed for each type of spatial weighting (minimum norm, depth-weighted minimum norm, and fMRI constrained) and mis-specification between fMRI and actual electromagnetic source. This random placement was repeated 10 times in the hemisphere and 100 time in the octant, and the average $\pm$ standard error of the mean of the percentage of incorrectly specified power was computed.

Results

Shown below are the percentages of incorrectly specified power for each of the spatial weightings (Minimum Norm, Depth-Weighted Minimum Norm, and fMRI Constrained) due to the following mis-specifications: no mis-specification between fMRI and the actual sources, an extra fMRI source that is not an electromagnetic generator, a missing fMRI source that is generating electromagnetic fields, and a spatial mis-registration between fMRI and the actual sources. The percentages of incorrectly specified power $\pm$ the standard error of the mean for 5 dipoles and 10 dipoles placed in a single hemisphere are shown separately in Tables 1 and 2. The percentages of incorrectly specified power $\pm$ the standard error of the mean for 5 and 10 dipoles placed in a single octant are shown in Tables 3 and 4.

Each column within the table describes slightly different values:
None: percentage of power that is mis-localized among locations that are correctly specified by fMRI.
Extra fMRI source: percentage of power from the correct locations that is placed on the extra fMRI source.
Missing fMRI source: percentage of power from the missing fMRI source that is placed on the locations correctly specified by fMRI.
Spatial: percentage of power that is mis-localized when there is a 1cm spatial discrepancy between fMRI and the actual electromagnetic generators

Spatial Weightings	Mis-specifications			
	None	Extra fMRI source	Missing fMRI source	Spatial (1cm)
Minimum Norm	25±2.5	28±5.0	28±5.3	26±2.5
Depth-Weighted	20±2.3	14±1.4	21±5.7	21±2.3
fMRI Constrained	6±1.6	6±1.3	28±4.8	15±1.6

Table 1: 5 Dipoles in Hemisphere. Percentage of incorrectly specified power (mean ± standard error of the mean). n=10

Spatial Weightings	Mis-specifications			
	None	Extra fMRI source	Missing fMRI source	Spatial (1cm)
Minimum Norm	28±1.5	28±2.9	31±1.8	28±1.5
Depth-Weighted	19±1.7	17±3.8	21±1.9	20±1.7
fMRI Constrained	9±1.4	9±1.4	27±2.1	16±1.2

Table 2: 10 Dipoles in Hemisphere. Percentage of incorrectly specified power (mean ± standard error of the mean). n=10

Spatial Weightings	Mis-specifications			
	None	Extra fMRI source	Missing fMRI source	Spatial (1cm)
Minimum Norm	30±0.5	30±1.3	29±1.3	30±0.5
Depth-Weighted	26±0.6	19±1.0	26±1.5	27±0.6
fMRI Constrained	9±0.5	8±0.5	29±1.4	19±0.4

Table 3: 5 Dipoles in an Octant. Percentage of incorrectly specified power (mean ± standard error of the mean). n=100

Spatial Weightings	Mis-specifications			
	None	Extra fMRI source	Missing fMRI source	Spatial (1cm)
Minimum Norm	31±0.3	30±1.1	30±1.1	31±0.3
Depth-Weighted	27±0.4	21±0.9	27±1.1	28±0.4
fMRI Constrained	11±0.3	11±0.5	27±1.1	20±0.2

Table 4: 10 Dipoles in an Octant. Percentage of incorrectly specified power (mean ± standard error of the mean). n=100

Conclusions

There is only a slight difference between 5 and 10 sources in the amount of incorrectly specified power, suggesting that the number of sources will not be problematic. Also, when the sources are placed in closer proximity (i.e. placement in one octant versus placement in a hemisphere), there is a small increase in the amount of incorrectly specified power. More sources and sources that are closer together will both increase the error in power specification. However, the increase in incorrectly specified power due to the number of sources and the proximity of sources would, on average, be small.

Regarding the various spatial constraints, minimum norm was the least successful at dealing with the various types of mis-specifications we examined. The addition of a depth-dependent weighting on the minimum norm produces very little improvement (< 5% for the various metrics). However, the addition of the fMRI weighting to the minimum norm correctly localizes more of the activation, produces less crosstalk, and also places less activation on an extra fMRI source. Unfortunately, none of these various weighted minimum norm schemes handles a source that is missing from fMRI. Although the inclusion of fMRI weighting does not worsen the situation, in even the best cases a hemodynamically undetected source would still perturb the other sources by approximately 28%, on average.

In conclusion, with only spatial mis-specifications or extra fMRI areas of activation, these simulations demonstrate that the use of a priori fMRI information allows us to correctly predict spatiotemporal brain activity. Additional refinements will be required to adequately describe the spatiotemporal record in the presence of hemodynamically inactive, but magnetically active, sources.

References

[1] Dale, A.M., Sereno, M.I. Improved Localization of Cortical Activity by Combining EEG and MEG with MRI Cortical Surface Reconstruction: A Linear Approach. Journal of Cognitive Neuroscience, 1993, 5: 162-176.
[2] Fox, P.T., Woldorff M.G. Current Opinion in Neurobiology, 1994, 4:151-6.
[3] Hamalainen, M., Hari, R., Ilmoniemi, R.J., Knuutila, J., Lounasmaa, O.V. Magnetoencephalography - theory, instrumentation, and applications to noninvasive studies of the working human brain. Review of Modern Physics, 1993, 65: 413-497.
[4] Belliveau J.W. In: Uemura K., Lassen N.A., Jones T., and Kanno I., eds. MRI techniques for functional mapping of the human brain: integration with PET, EEG/MEG and infrared spectroscopy. Quantification of Brain Function. Amsterdam: Elsevier Science Publishers, 1993: 639-667.

Acknowledgements

This study was supported by grants from the United States Public Health Service: MH50054, EY07980, by grants from the Human Frontier Science Program, and conducted during the tenure of an established investigatorship from the American Heart Association to JWB.

Multiple Modality Biomagnetic Analysis System

Martin, K.[1], Erne, S.[2], Law, C.[1], Conforto, S.[3], Mallick, J.[1] and Tatar, B.[1]

[1]GE Corporate R&D, Schenectady, New York, USA; [2]Institute for Biomedical Engineering, University Ulm, Ulm, Germany; [3]Institute of Advanced Biomedical Technologies, Chieti, Italy

Introduction

As an emerging new modality, biomagnetic analysis presents exciting opportunities while raising some difficult issues. During the past four years researchers at General Electric, The University of Ulm and The University of Chieti have created a state of the art biomagnetic analysis system, and have addressed many of these difficulties. One of the critical aspects of a biomagnetic analysis system is how it handles different types of data. Many of the clinical requirements for a successful system require fast, efficient and flexible data management. This paper describes what those requirements are and how we addressed them. We start by providing a high level overview of our system. We then list some of the pertinent clinical requirements, present our solutions and describe how well they have worked. Finally we discuss future directions.

Before going into detail on the clinical requirements, we will provide a high level overview of our system. The five main components of our system are; the sensor head, the BSS (biomagnetic sensor system), the DAPS (digital acquisition processing system), the BOC (biomagnetic operators console) and the BAC (biomagnetic analysis console). The sensor head measures the signals, while the BSS performs analog to digital conversion and a variety of control functions. The BSS and DAPS communicate through a high speed fiber optic connection. The DAPS performs additional data processing, and records the data to disk. The BOC communicates to the DAPS through ethernet and shared memory. The BOC coordinates the data acquisition and is the control center for the sensor head, BSS and DAPS. The end product of the BOC is a runfile, and the process is described in Corby [1]. A runfile contains all the configuration information and data from an acquisition. The BAC is responsible for analyzing the runfile once it has been acquired. The BAC can obtain a runfile from a variety of sources ranging from mounting disks off of the BOC to inserting a magnetic tape. These five components are designed to support generating a runfile with 256 channels of 16 bit data being acquired at 10,000 Hertz.

Clinical Requirements
1) Handle large amounts of data.
2) Handle high data rates.
3) Display raw and filtered data in real time.
4) Support multiple modalities.
5) Quick turnaround time.
6) Multiple platform support.
7) Handle different sensor heads.

The first requirement we will address is handling large amounts of data. From the above discussion, you can see that a runfile can consist of a gigabyte of data. At this point in time, handling a gigabyte of data is a significant issue for most PCs and workstations. Similarly, displaying 256 channels of 10KHz data in real time is a challenge. A clinician wants to see a high quality stripchart of the acquired data, displayed in real time and properly subsampled. As the third item in our list indicates, most clinicians want to perform band pass filtering or base line correction in real time as well.

The fourth requirement demands that we support MR, CT, and 3D surface data in addition to the temporal data from the sensor head. This is important for displaying dipole positions within image data, or relative to a 3D model. The fifth requirement stems from a desire to perform a quick analysis before the patient has left the facility. The sixth and seventh requirements are important due to the variety of hardware available. We do not want to lock the clinician into a specific computer platform or a specific sensor head.

Methods

To meet the needs of a biomagnetic analysis system we had to design a specialized data architecture. Traditional techniques that load in the data and then overwrite it when filtering would not work. Additionally, constraints that the system handle multiple runfiles being open at once, required us to use no global variables for data storage. Our solution consists of three data types, two data pipelines, a complete runfile and a hardware independent abstraction. We will first discuss the three data types: display, image and signal.

Display data can be thought of as 3D geometric data, typically consisting of polygons, lines, triangle strips and points. Display data is used to store and represent torso and skull models, sensor arrays, dipole and min-norm results, fiducial points and iso-surfaces. Display data can have attributes such as normals, vectors, scalars, texture coordinates and time. Image data is used to store a series of 2D scalar slices such as MR or CT. Image data is assumed to be static. Signal data is used to store temporal signals such as those from an EEG or MEG. There may be any number of channels of data and any sample rate. The patient position, electric and magnetic data are stored as signal data.

There are two data pipelines: the display data pipeline and the signal data pipeline. There are a number of qualities that they have in common. Both pipelines are object oriented and represent filtering operations as objects (or nodes) in the pipeline. These pipelines have multiple inputs and outputs, are demand driven and rely on implicit control and conditional execution. Schroeder [2] provides a good introduction to data pipelines. Fig. 1 shows a simple

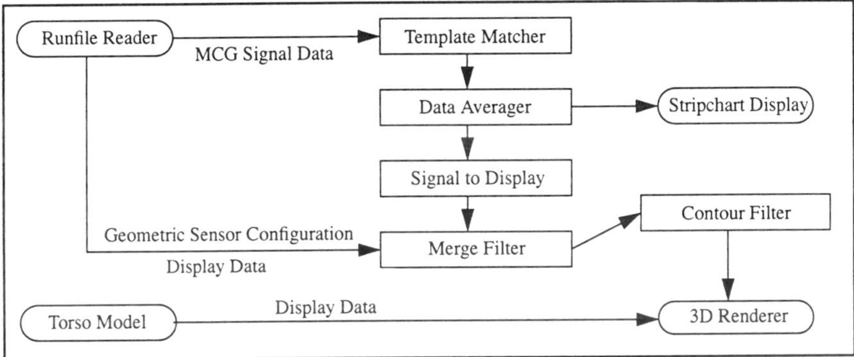

Fig. 1, Sample example data pipeline.

example of a data pipeline. The runfile reader and torso model are data sources for the pipeline. The runfile reader generates both display data and signal data. The signal data the goes through a template matcher followed by a data averager. This will average a number of heartbeats into a composite heartbeat with an improved signal to noise ratio. This result is then displayed on a stripchart and also sent to a signal to display filter. This converts signal data into scalar values for display data. These scalar values are then merged with the geometric configuration of the sensor array. This then gets passed through a contour filter to generate contour lines which are then displayed by a 3D renderer. The torso model is simple read in and then displayed in the 3D renderer along with the contour lines.

The display data pipeline provides little beyond what was discussed above. In contrast the signal data pipeline supports multiple ring buffer output caches. The caching allows us to handle very large amounts of data since only a small portion resides in memory. Given the demand driven nature of the pipeline, only the portions of data required for a given operation are read in. Likewise, filtering is also a demand driven process. The following example will serve to illustrate the caching operation. The clinician is looking at a stripchart showing two seconds of raw MEG data. As the stripchart scrolls forward in real time, it requests the additional data it needs to display from its input. The input checks to see if all or part of the data is already cached. It then requests whatever data it needs, if any, from its input. Eventually the request may get back to the runfile reader where the required data will be read from the disk.

Now the clinician turns on band pass filtering, this results in a filter being inserted between the stripchart and the runfile reader. The stripchart will refresh its entire display, requesting two seconds of data. The band pass filter then checks its cache to determine if any of the data is already present. Nothing will be there, so it will request the full two seconds from the runfile reader which should already have that data cached. The band pass filter operates on the data and stores its result in its cache. The stripchart then displays the result and continues to scroll forward. To perform its scrolling the stipchart may need about a tenth of a seconds worth of data. The actual amount will depend on the load on the machine and the desired rate of scrolling. The band pass filter then needs to generate this additional tenth of a seconds worth of data. Due to the limitations of band pass filtering, the filter may require a minimum of a half second worth of input data to properly work. To handle this the filter requests the minimum amount of data it needs to work, performs the filtering, and then writes the entire result into the cache. This means that the band pass filter will already have cached the next few updates worth of data for the stripchart. If the clinician were to change the frequencies of the band pass filter, its entire output cache would be considered invalid and cleared. But the band pass filter's input would still have the required data cached for quick recomputation.

The multiple output caching is used to support comparative analysis. A clinician may want to perform two different operations of the same data and then compare the results. By supporting multiple output caches, a data pipeline can be split off into two paths at any point without a loss in performance. The multiple output caches effectively prevent thrashing of the data. The same approach can be used within a filter that requires frequent access to different parts of the raw data, such as analyzing evoked response data for a particular stimulus.

Now that we have discussed the three types of data and the two data pipelines, we will consider the complete runfile and hardware abstraction. The runfile is designed to contain all the information necessary to analyze its data. Any MSI system could generate the runfile and it would be sufficient to be processed by the BAC. A fringe benefit of this is that changes in the sensor head do not require changes in the analysis software. In addition to the information contained in this file, its organization is very important.

Many of our analysis routines start by preprocessing the raw data over time to generate an averaged result. This preprocessing step frequently relies on a few reference channels of data for selecting templates and performing temporal correlation. These reference channels are stored in a contiguous space in the runfile. This enables the initial scanning of the reference channels, selection of templates, template matching and correlation tree generation to be done without having to scan through the entire runfile. Then the averaging step can only read from disk the portions that are required. This can create a huge savings in time. We go from having to scan the entire runfile twice, to only having to read select parts of it. Given that scanning through a gigabyte file takes about five minutes this is an important issue.

The hardware abstraction is what we use to maintain platform independence. Since we have restricted ourselves to three primary data types, we have created a hardware independent abstraction for displaying them. This allows us to run under WindowsNT displaying our 3D graphics (display data) using OpenGL, or run on a HP UNIX workstation and display it using Starbase graphics.

Results

We have built a biomagnetic analysis console (BAC) on top of the data architecture described above. The resulting system supports a cornucopia of features and is flexible enough so that trying out new algorithms does not require significant changes to the software. As the field continues to develop this is an important feature. The BAC has been run on HP, Sun, SGI and WindowsNT platforms. With the current hardware we can properly display and filter a few stripcharts with a sample rate of about 10,000 Hz. For a larger number of channels the display must be slowed down or the display can be approximated by subsampling the data to approximately 0.1 pixel resolution.

The BAC can work with multiple CT/MR, runfile, and 3D data sets at once and different data sets can be registered so that dipoles appear as markers on the MR/CT slices. Initial 3D surface data can be obtained through Marching Cubes iso-surface extraction [3] or laser range scanners [4]. To generate closed surfaces required for some forms of analysis we use the surface construction algorithm of Yamrom [5]. The system supports an electronic notebook for recording pictures and results. Most anything that the clinician sees on the screen can be stored into this electronic notebook. This is made possible by the restriction of our data to the three primary data types. There is support for both neuro and cardiac analysis, continuous and evoked recording. Pipeline filters have been created for: baseline correction, band pass filtering, template matching, temporal averaging, correlation tree calculation, artifact rejection, geometric channel selection, power spectrum, cross correlation, minimum norm, multiple dipole fit, rigid body fiducial registration, contour line extraction and a selection of traditional MR/CT operations.

We have found that the resulting system can perform a quick analysis of a runfile within a few minutes after the acquisition has been completed. Further analysis of the runfile can occur while a new data set is being acquired by the BOC, avoiding a potential bottleneck.

Discussion

Throughout the medical community, the software demands of data acquisition systems have become considerable. The BAC consists of over 200,000 lines of code with tough performance demands. Without a good software foundation such a system quickly degenerates into an unmanageable mess. Within the MSI community this is compounded by a lack of clearly defined and accepted clinical procedures. While much progress has been made on this front, a successful MSI system must still provide a very high degree of flexibility.

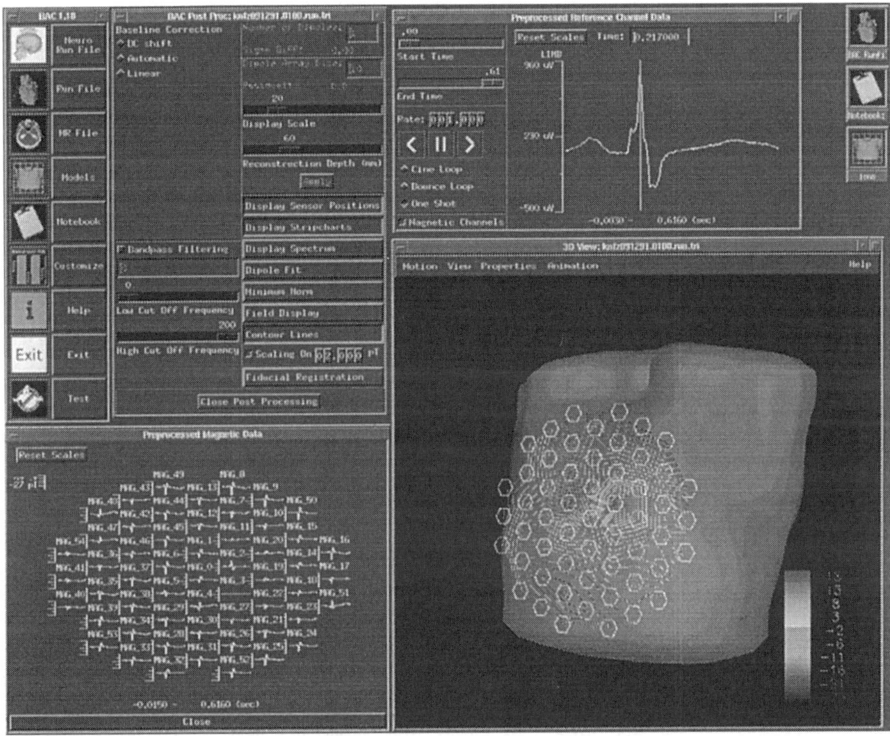

Fig. 2, Sample screen from the BAC.

References

[1] Corby, N.R., Jr., Miller, P.D., Hogle, R.A., "A realtime, high-bandwidth, programmable digital signal-processing system for monitoring, acquisition, filtering, and archival storage of biomagnetic signals" Biomag '96, Santa Fe NM, February 16-21, 1996.

[2] Schroeder, W., Martin, K., Lorensen, W., "The Visualization Toolkit: An Object Oriented Approach to 3D Graphics" Prentice Hall 1996.

[3] Lorensen, W.E., Cline, H.E., "Marching cubes: a high resolution 3D surface construciton algorithm," Computer Graphics, 1987, 21 (3): 163-169.

[4] Cyberware, 2110 Del Monte Avenue, Monterey, CA 93940.

[5] Yamrom, B. "Modelling closed surface geometry for skull and torso using snapping" Biomag '96, Santa Fe NM, February 16-21, 1996.

Measurement of Visually Evoked EEG and MEG Associated with Verbal Cognitive Processes

Nakagawa, S.[1], Ueno, S.[1], Iramina, K.[1] and Weinberg, H.[2]

[1]*Institute of Medical Electronics, Faculty of Medicine, University of Tokyo, Tokyo 113, Japan;*
[2]*Brain Behaviour Laboratory, Simon Fraser University, B.C., Burnaby, Canada*

Introduction

Magnetoencephalography (MEG) is a useful method for the diagnosis of functional brain diseases as well as for the study of higher brain function, such as cognition, memory and blending of brain functions. In recent years, substantial advances in understanding the functional organization of the human brain have been made through the use of MEG. Most of these investigations were concerned with evoked magnetic fields. Only a few neuromagnetic studies on higher brain function have been reported [1],[2].

In this study, we focused on the measurement of verbal activities in the human brain. Kutas et al. reported a late negative slow wave elicited by words, which is attenuated if the word is repeated or primed by a related word or by sentence context [3],[4]. We observed verbal cognitive activities by measuring electroencephalograms (EEGs) and MEGs evoked by visually cognitive tasks. We analyzed the EEG and MEG data based on topographic maps [5], and discussed the source estimation using MEG data.

Methods

Seven male volunteers with no history of neurological disease took part in this experiment. Their ages ranged between 21 - 30 years. All the subjects were right handed. Visual evoked potentials (VEPs) and magnetic fields were recorded simultaneously. Visual stimulation was projected on the screen. (Fig.1) Four letter English words, meaningless sets of four letters (nonsense words) and random dots were presented for subjects as visual stimuli. (Fig. 2) The visual angle of each letter was 1.3 degrees. The intensity of the stimulation of random dots was adjusted to be identical to the stimuli of letters. Each stimulus was randomly presented for 1.4 sec at an interval of 2.0 ~ 3.0 sec. Subjects were asked to maintain fixation of their eyes at center of screen.

The EEG data were also recorded from 14 locations over the surface of the head, corresponding to the points defined by the international 10-20 system. The magnetic field components perpendicular to the surface of the head were measured using a 7-channel DC-SQUID first order gradiometer and a 64-channel third order gradiometer system. After being filtered with a bandpass filter (0.5 - 30 Hz), the EEG and MEG data were averaged. 110 - 300 responses were averaged at each position. The analysis time was 1.0 sec including 0.2 sec prestimulus baseline.

Results and Discussion

Figure 3 shows simultaneous measurements of EEG and MEG. The bold solid lines denote responses evoked by English words, the bold dotted lines denote responses evoked by nonsense words, and the fine lines denote responses evoked by random dots. All waves have a large peak at about 150 msec . The differences among the three stimulus conditions are observed between latencies of 300 and 600 msec.

Figure 4 shows EEG topographies of each stimulation. Topography at 152 msec shows that the activity

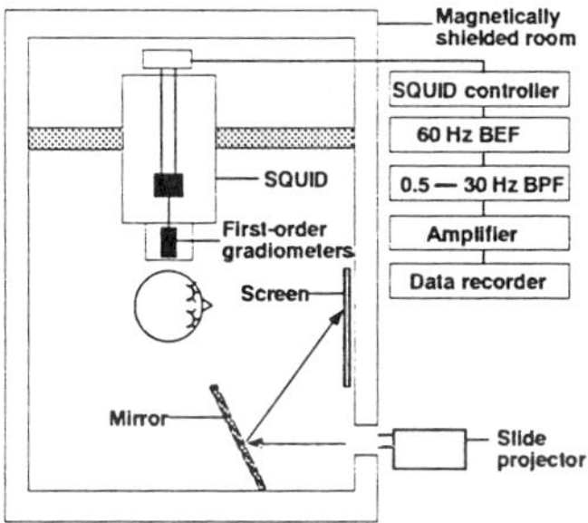
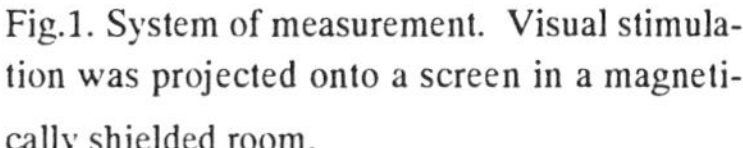

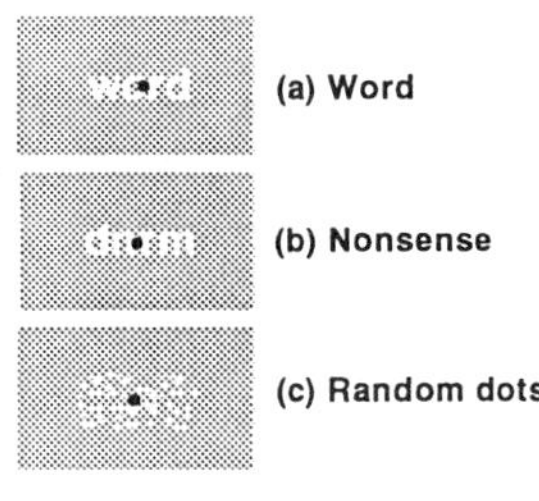

Fig.1. System of measurement. Visual stimulation was projected onto a screen in a magnetically shielded room.

Fig.2. Patterns of visual stimulation: (a) English
word, (b) Nonsense word, (c) Random dots. Subjects were asked to maintain fixation of their eyes
at center of screen.

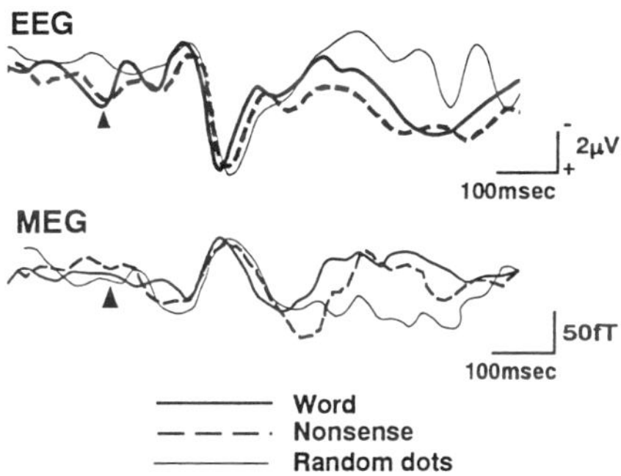

Fig.3. Averaged wave forms of visual evoked potentials and magnetic fields recorded at C3
(subject A). The bold solid lines denote responses evoked by English words, the bold dotted
lines denote responses evoked by nonsense words, and the thin solid lines denote responses
evoked by random dots.

is dominant in the occipital region. The sources at a latency of about 160 msec were estimated with a 2-dipole
model using MEG data. The results are shown in Figure 5. Two equivalent current dipoles are localized in or
near the primary visual cortex.

We estimated the sources for recognition of English words from the MEG topographic data. The MEG
topographic patterns obtained here were complicated, and it was difficult to estimate the source with a single-
dipole model. A 2-dipole model was used to estimate sources as shown in Figure 6. In subject A, an asym-

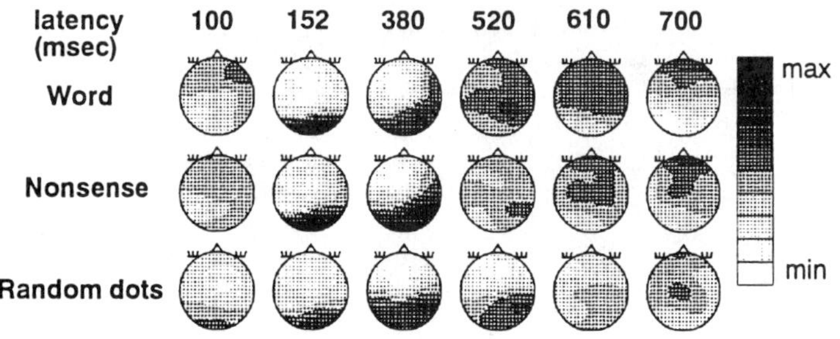

Fig. 4. The EEG topographies of VEPs evoked by English words, nonsense words, and random dots (subject A). The potentials are quantified into 10 levels. The topographies of English words and random dots show dominant activity in the frontal and central regions between latencies of 520 and 700 msec.

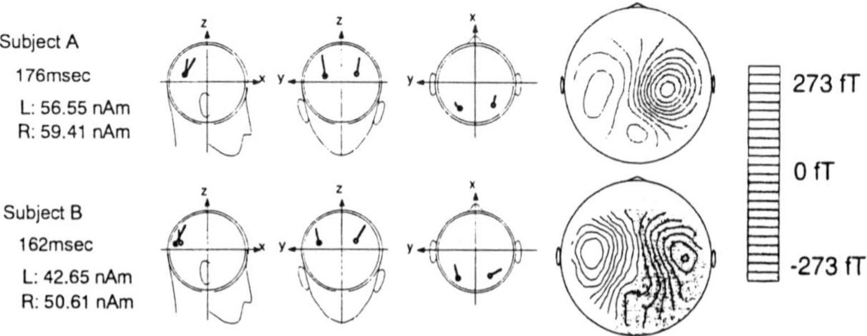

Fig. 5. The estimated sources of visual evoked MEG at a latency about 160 msec. Two equivalent current dipoles are localized in or near the primary visual cortex.

metrical distribution of dipoles was estimated. In subject B, the estimated source was located near the center in the head. In three of the four subjects for which we estimated sources, the dipole in the left-hemisphere was larger than the dipole in the right hemisphere.

In estimation of source with a latency of 160-170 msec, the sources with 50 nAm seems rather intense. These intense sources can be caused by undermodeling the data with too few sources. Aine et al. estimated the sources of visually evoked magnetic fields with 5-dipole model [6]. It is essential to estimate sources by taking into account anatomical and functional organization of the brain.

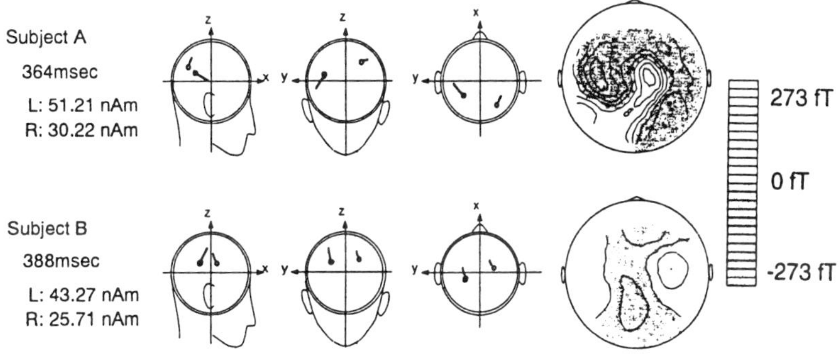

Fig.6. The estimated sources of two subjects for recognition of English words.

References :

[1] Lu, S.T., Hämäläinen, M., Hari, R., Ilmoniemi, R., Lounasmaa, O., Sams, M., and Vilkin, V., Seeing faces activates three separate areas outside the occipital visual cortex in man, Neuroscience, 1991, 43: 287-290.

[2] Gordon, E., Rennie, C., and Collins, L., Magnetoencephalography and late component ERPs, Clinical & Experimental Neurology, 1990, 27 : 113-20.

[3] Kutas, M., Hillyard, S. A., Reading senseless sentences : Brain potentials reflect semantic incongruity, Science, 1980, 207: 203-204.

[4] Kutas, M., Hillyard, S. A., Brain potentials during reading reflect word expectancy and semantic association, Nature, 1984, 11: 161-163.

[5] Ueno, S., Matsuoka, S., Mizogochi, T., Nagashima, M., and Chen, C., Topographic computer display of abnormal EEG activities in patients with CNS disease, Memoirs of the Faculty of Engineering, Kyushu Univ., 1975, 34 : 165-209.

[6] Aine, C., Supek, S., George, J., Ranken, D., Best, E., Tiee, W., Vigil, V., Flynn, E., and Wood, C., MEG studies of human vision: Retinotopic organization of V1, In: Baumgartner, C., Deecke, L., Stroink, G., Williamson, S.J., Biomagnetism; Fundamental Research and Clinical Applications, Amsterdam, IOS Press, 1995.

Superadditive Information from Simultaneous MEG/EEG Data

Pflieger, M.E.[1], Simpson, G.V.[2], Ahlfors, S.P[2,3] and Ilmoniemi, R.J.[4]

[1]*Neuro Scan, Inc., El Paso, USA;* [2]*Albert Einstein College of Medicine, New York City, USA;*
[3]*Low Temperature Laboratory, Helsinki University of Technology, Espoo, Finland;* [4]*BioMag
Laboratory, Helsinki University Central Hospital, Helsinki, Finland*

Introduction

Although MEG and EEG measurement modalities are empirically distinct, there is a theoretically unified statistical-biophysical framework for posing and (in a limited sense) solving the electromagnetic inverse problem. In addition, biomagnetic and bioelectric measurements are differentially sensitive to the same type of intracranial signals, i.e., macroscopic ionic source currents. Since MEG and EEG share a common foundation and probably provide complementary information about the same kind of source signals, a truly integrated electromagnetic source imaging (EMSI) modality appears theoretically immanent. In practice, EMSI requires combined analysis of MEG and EEG acquired (in the ideal case) simultaneously.

Combined MEG/EEG analysis must first surmount the *measurement scaling problem* which arises because the physical units for MEG (femtotesla) and EEG (microvolts) are not directly comparable. Consequently, the direct sum of an N-channel MEG measurement space with an M-channel EEG measurement space does not yield an $(N+M)$-channel MEG/EEG measurement space with a usable metric. Greenblatt [1] showed how to resolve the scaling problem by using background noise variance-covariance statistics to normalize the combined data. Rescaling all data channels with respect to the inverse symmetric square root of the background noise matrix (see below) produces a common frame of reference in which magnetic field and electric potential units are converted to *unitless* signal-to-noise ratios (SNRs). This multichannel normalization operation is called here an *SNR transform* of the data.

Simultaneous acquisition of MEG and EEG is clearly advantageous from the standpoint of experimental design since between-session variables (e.g., subject state and response variability) are eliminated by constructing average event-related field (ERF) and potential (ERP) waveforms from exactly the same set of trials. Moreover, from the standpoint of the SNR transform, note that *between-modality* background noise covariance statistics cannot be computed unless acquisition is simultaneous. In this paper we show that there are substantial benefits to be gained by incorporating these inter-modal background noise covariances in the SNR transform.

To quantify such benefits, we compute a measure of *empirical information available for source estimation* (based on the SNR transform) that meets several criteria. First, for the sake of generality, the information measure does not use specific modeling assumptions that are otherwise required to actually estimate sources. In particular, the computation does not use a head model (compare [2]), nor does it make specific assumptions about the nature or configuration of the current sources. Second, the information measure is logarithmic following Shannon's rationale [3]. Third, the computed quantity of information is invariant with respect to any nonsingular linear reformatting of the multichannel data; i.e., all such reversible transformations are "information preserving" (in a linear sense). As a corollary, it follows that information is independent of the scale utilized for each channel; i.e., the units of physical measurement need not be uniform across channels. Fourth, information computed for a noise-only data sample is close to zero. Fifth, information is additive for independent sources of signal variance when the background noise is uncorrelated across sensors. Sixth, redundant sampling of a signal at the sensors increases information just to the extent that uncertainty due to noise is reduced. And seventh, from the perspective of ERFs and ERPs, information ultimately measures our ability to *spatially distinguish* event-related signals from the residual background noise. A procedure for computing an information measure that satisfies these criteria is described below.

The phenomenon of *superadditive* information is theoretically possible for a channel partition $A+B$ (i.e., the information of $A+B$ can exceed the information of A plus the information of B) if the event-related A signal is sufficiently uncorrelated with the event-related B signal while at the same time the background noise is sufficiently correlated between A and B. Superadditivity cannot occur if the noise covariances between A and B are neglected. Thus, superadditivity for an MEG+EEG partition is *a priori* possible only if acquisition is simultaneous.

Methods

Data acquisition. The analyzed data were derived from 122 Neuromag MEG channels (retaining 120 good channels) with 29 simultaneous EEG channels (using silver electrodes and copper leads) sampled at 319 Hz (0.03 to 80 Hz bandpass) from one subject (GVS) in 1993 at the Low Temperature Laboratory of the Helsinki University of Technology. Event-related averages were replicated 6 times (~70 epochs each) for 4 visual stimuli consisting of wedge-shaped checkerboard pattern onsets (50 ms duration) in four quadrants (upper right, upper left, lower left, and lower right). The event-related averages were baseline adjusted and 40 Hz lowpass filtered offline.

Computation of event-related information: 8 special cases. The composite ERF/ERP signal for each of the 4 stimulus conditions was estimated by the grand average across the 6 replications. The grand average $\bar{y}_i(t)$ for each condition i consisted of 149 channels by 118 time points in a window of 30 ms to 400 ms post-stimulus. The associated residual background MEG/EEG noise covariance $\mathbf{C}_i$ for each condition i was estimated by computing spatial covariance (149×149) statistics for each time sample across the 6 replications, dividing by 6 to rescale single replication covariances to the level of residual grand average covariances, and averaging over the 118 points in the time interval. Time averaging is necessary in this case since each point sample covariance matrix is singular (5:149) whereas the lumped interval covariance matrix is not (590:149). The partitioned MEG/EEG grand average signal $\bar{y}_i(t)$ and noise covariance $\mathbf{C}_i$ statistics were restricted and constrained to derive 8 special cases:

M0: MEG only with off-diagonal noise covariances forced to zero;
M1: MEG only utilizing all estimated between-channel noise covariances;
E0: EEG only with off-diagonal noise covariances forced to zero;
E1: EEG only utilizing all estimated between-channel noise covariances;
ME0: MEG+EEG with off-diagonal covariances forced to zero;
ME1: MEG+EEG utilizing all estimated between-channel covariances;
ME01: MEG+EEG with zero inter-modal covariances, utilizing intra-modal covariances; and
ME10: MEG+EEG utilizing inter-modal covariances, with zero intra-modal covariances.

For each special case and stimulus condition, the inverse symmetric square root $\mathbf{C}_i^{-0.5}$ of the noise spatial covariance matrix was obtained via singular value decomposition (SVD) of $\mathbf{C}_i$ by taking the reciprocal square root of the singular values (which are all positive since $\mathbf{C}_i$ is positive definite). There are other senses of taking the "square root" of a positive definite matrix, such as Cholesky decomposition, but only the symmetric square root is invariant with respect to channel number reorderings. The grand average ERF/ERP spatial vectors $\bar{y}_i(t)$ were then normalized by the SNR transform $\bar{z}_i(t) = \mathbf{C}_i^{-0.5}\bar{y}_i(t)$ which yielded unitless signal-to-noise ratio vectors. For each special case, spatial SVD was performed on 472 $\bar{z}$ vectors (4 stimulus conditions $\times$ 118 time points) in order to span the signal space for the experimental conditions and time window of interest. The resulting singular values w_j are signal-to-noise ratios for their respective spatial components. Components with SNRs less than 1.0 were discarded. Finally, total information was computed in units of "bits above noise" [4] for all R retained components as $H = \sum_{j=1}^{R} \log_2 w_j$.

Random channel permutations: How typical is the MEG/EEG partition? From the standpoint of computing information, is there anything special about the distinction between MEG and EEG channels? What information quantities would be obtained for arbitrary partitions of 120+29 channels made at random without regard for modality? Note that of the 4 special cases that include all 149 channels, the computations for cases ME0 and ME1 are unaffected by the distinction between MEG and EEG. However, cases ME01 and ME10 depend on an explicit modal partition of 120 MEG and 29 EEG channels: case ME01 forces 6960 inter-modal covariances to zero, and case ME10 forces 15,092 intra-modal off-diagonal covariances to zero. We guessed that MEG and EEG channels "specially complement" each other from an information perspective. Specifically, we hypothesized that (a) the information computed for ME01 (obtained by *deleting* the MEG/EEG noise interactions from case ME1) should be *less* than would be observed for a typical partition, and (b) the information computed for ME10 (obtained by *adding* the MEG/EEG noise interactions to case ME0) should be *greater* than would be observed for a typical partition. To test this hypothesized relation, 750 arbitrary partitions were generated by randomly permuting the channels but retaining (incorrectly) the original MEG/EEG channel designations. ME01 and ME10 information computations were performed for each random partition. Each 120+29 partition X (either the actual MEG/EEG partition or a random partition) was assigned a score equal to the number of partitions for which the hypothesized relation was true for X. The typicality of the MEG/EEG partition was estimated by the fraction of random partitions with scores greater than or equal to its score.

Results

Computation of empirical information in units of bits above noise for the 8 special cases yielded: 46.3 for M0, 132.7 for M1, 14.7 for E0, 37.0 for E1, 53.3 for ME0, 194.3 for ME1, 142.2 for ME01, and 69.5 for ME10. These results are graphed in Fig. 1.

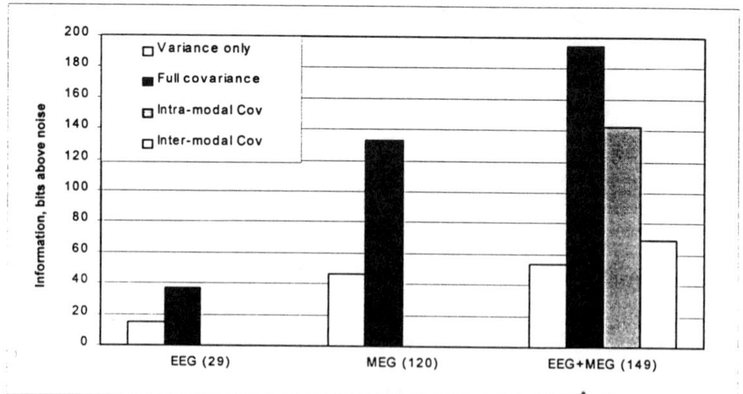

Fig. 1. Empirical event-related information computed for the 8 special cases.

From these numbers, the benefit of using the full noise covariance matrix in the SNR transform versus variances only can be expressed as information ratios. Thus, the full covariance versus variance only benefit ratios are: 2.52 for EEG only, 2.87 for MEG only, and 3.65 for MEG+EEG.

The *additivity ratio* α for a channel partition $A+B$ is defined as $\alpha \equiv H(A+B)/(H(A)+H(B))$, where $H(A+B)$ is the information computed for all channels, $H(A)$ is the information computed for the A channels only, and $H(B)$ is the information computed for the B channels only. Information is said to be *additive* if $\alpha=1.0$ ("the whole equals the sum of the parts"), *subadditive* if $\alpha<1.0$ ("the whole is less than the sum of the parts"), and *superadditive* if $\alpha>1.0$ ("the whole is greater than the sum of the parts"). The results above give rise to the following additivity ratios: for the variance only cases, ME0/(M0+E0) = 0.87 (subadditive); for the full covariance cases, ME1/(M1+E1) = 1.14 (superadditive); for the mixed case where inter-modal noise covariances are deleted from the full covariance matrix, ME01/(M1+E1) = 0.84 (subadditive); and for the mixed case where inter-modal noise covariances are added to the variance only matrix, ME10/(M0+E0) = 1.14 (superadditive).

The estimated typicality of the MEG/EEG partition was 0.052 -- 39 out of 750 random partitions had scores that equaled or exceeded the MEG/EEG partition score (=527).

Further computations. A few exploratory computations were made to further investigate the conditions for superadditivity. The 120 MEG channels were 60 pairs of tangential gradiometers at each location. The odd-numbered channels covered all 60 locations with one tangential orientation, and the even-numbered channels had the other orientation. Full covariance information for a 60+60 odd-even partition of the 120 MEG channels was superadditive ($\alpha=1.15=132.7/(57.3+58.5)$). A 60+60 (odd-even)-(even-odd) partition with equal representation of the two tangential orientations in both parts was also superadditive ($\alpha=1.12=132.7/(58.9+59.3)$). Further subdivisions were made. A 30+30 partition of the 60 odd-numbered MEG channels was subadditive ($\alpha=0.95=57.4/(29.7+30.6)$); likewise, a 30+30 partition of the 60 even-numbered channels was subadditive ($\alpha=0.93=58.5/(31.1+32.1)$). Four 30MEG+29EEG partitions were all closer to being additive: $\alpha=1.00$ ($66.5/(29.7+37.0)$), $\alpha=1.00$ ($67.3/(30.6+37.0)$), $\alpha=0.98$ ($67.0/(31.1+37.0)$), and $\alpha=0.97$ ($67.0/(32.1+37.0)$).

Discussion

Without using specific modeling assumptions that are required to actually estimate sources, the foregoing analysis has been made to study the potential benefits of combining MEG and EEG for the purpose of source estimation. The method has been to compare the raw information empirically available at the sensors for 8 special cases derived from an actual simultaneous 120-channel MEG plus 29-channel EEG visual event-related recording.

An SNR transform of the bimodal event-related data converts the MEG and EEG data to a common frame of reference derived from the covariance matrix of the residual background noise. Although the noise matrix can be simplified by considering within-channel variances only, this assumes that the background MEG/EEG is statistically independent at the sensors. As a consequence, the total noise in the multichannel system is overestimated since the background activities picked up at two channels are actually correlated (to some extent) due to common neural origins. Overestimation of the total noise results in underestimation of SNRs which directly reduces the information estimates. For a single modality, incorporation of between-channel covariances resulted in information yields 2.5-2.9 times the variance only yields. The improvement in yield was even greater (over 3.6 fold) for the combined MEG+EEG data. Therefore, there is considerable *common* background noise picked up at the sensors.

Superadditive information was observed for the MEG/EEG partition, and there was a 27% information loss when inter-modal covariances were neglected. Superadditivity was also observed for 60+60 within-MEG partitions, but not for 30+30 partitions. In a 256-channel ERP study that used the same stimuli, we did not observe superadditivity for 128+128 EEG partitions [5]. Thus, superadditive event-related information has been shown to be *empirically possible* for MEG/EEG partitions and for within-MEG partitions. It of course remains to be seen how typical these results are.

As for the question of whether MEG and EEG channels complement each other in an atypical way, there is a hint (but no conclusive evidence) that they do. Only about 5% of random partitions were more complementary than the actual MEG/EEG partition; however, further discernment is needed to tease apart the effects of modality from location (e.g., occipital versus frontal sensors). In this regard, it is certainly possible that location could take precedence over modality. Another bit of evidence in favor of a special relationship between MEG and EEG is the observation that the 30+29 MEG+EEG partitions had higher additivity ratios than the 30+30 within-MEG partitions.

Finally, there are grounds for expecting that the empirical information gain obtained by considering between-channel noise covariances will translate to improved source estimation. For example, a quasi-realistic simulation study using 64-channel EEG data demonstrated improved accuracy (2×) and stability (20×) of estimated ERP source waveforms for the full covariance versus the variance only cases [6]. However, the actual benefits of combined MEG/EEG source estimation remain to be determined in the context of specific modeling assumptions. Moreover, the issue of different accuracies for magnetic versus electric head models needs to be considered.

References

[1] Greenblatt, R.E. Combined EEG/MEG source estimation methods. In: Baumgartner, C., Deecke, L., Stroink, G., and Williamson, S.J. *Biomagnetism: Fundamental Research and Clinical Applications*, Amsterdam, Elsevier/IOS-Press, 1995, pp. 402-405.

[2] Kemppainen, P.K., and Ilmoniemi, R.J. Channel capacity of multichannel magnetometers. In: Williamson, S.J., Hoke, M., Stroink, G., and Kotani, M. *Advances in Biomagnetism*, New York, Plenum Press, 1989, pp. 635-638.

[3] Shannon, C.E., and Weaver, W. *The Mathematical Theory of Communication*. Urbana, University of Illinois Press, 1949.

[4] Pflieger, M.E., and Sands, S.F. Information growth in multichannel EEG. *Human Brain Mapping*, 1995, Supp. 1: 99.

[5] Pflieger, M.E., and Sands, S.F. 256-channel ERP information growth. *NeuroImage* 3(3): S10, 1996.

[6] Pflieger, M.E., Simpson, G.V., and Vaughan, H.G., Jr. Improved estimation of ERP source activities in the presence of realistic background EEG. *Human Brain Mapping*, 1995, Supp. 1: 101.

Multimodal Functional Imaging of the Sensorimotor Cortex with fMRI and Magnetic Source Imaging (MSI)

Roberts, T.P.L. and Rowley, H.A.

University of California San Francisco, San Francisco, USA

Introduction

The clinical advent of echo planar imaging (EPI) combined with the appreciation of the blood oxygen level dependent (BOLD) contrast mechanism has given rise to rapid development in 'functional magnetic resonance imaging' (fMRI) of the cortical response to peripheral stimulation and cognitive task performance [1,2]. Studies tend to involve stimulation paradigms lasting from seconds to minutes, with high speed MR images being acquired at a rate of the order of 1/sec. The BOLD contrast mechanism relies on the fact that a regional cortical blood flow increase occurs in response to stimulation but that this is not accompanied by a concomitant increase in local tissue oxygen extraction [3]. This leads to a relative increase in the proportion of oxygenated to deoxygenated blood in the capillary bed. Since deoxyhemoglobin is paramagnetic (and thus accelerates transverse magnetization dephasing) such a change will lead to an attenuation of this dephasing effect and somewhat increased signal intensity on imaging sequences such as gradient-recalled-echo (GRE) and echo-planar imaging (EPI); small signal increases of up to 4% have been reported at clinical field strengths [1,2,4]. This hemodynamic response to stimulation is not instantaneous but exhibits a gradual approach towards a plateau level, with a time constant of the order of a few seconds. So hemodynamic rise-time essentially limits the meaningful temporal resolution achievable in such studies.

However, electrophysiological studies demonstrate that significant neural events occur within milliseconds of stimulus presentation. Particularly if the goal of the study is to track the propagation of signal from one brain area to others, significantly higher temporal resolution is required. Electroencephalography (EEG) studies with scalp-placed electrodes, while providing such temporal resolution, may not be adequate for the localization of the neuronal current source. Magnetoencephalography (MEG) shares the temporal resolution of EEG, limited only by analog-to-digital conversion rates (typically 1-4 kHz). However, since MEG detects the magnetic rather than the electric component of the extracranial field, it can form the basis of a more robust neuronal source localization method, provided an array of magnetic field detectors can cover sufficient spatial extent to allow adequate modeling of the current source. When MEG source models are combined with high-resolution MRI, magnetic source images (MSI) display functional information in anatomic context [5,6].

An important clinical role for functional brain imaging is presurgical mapping to allow definition of eloquent cortex in relation to mass lesions which may be treated by resection or alternative non-surgical approaches, such as focused irradiation (gamma knife) or chemotherapy/radiotherapy, according to the functional nature of nearby brain tissue [7].

This study evaluated two functional imaging techniques, fMRI and MSI, for mapping of the sensorimotor cortex. The methods were compared in normal volunteers for rate of achieving localization with comparable stimuli. Assessing the comparability and relative merits of each technique in the spatial and temporal domains leads to the tantalizing opportunity to combine information derived from each examination to obtain a more accurate or more complete description of the underlying brain functional activity.

Methods

Studies were performed with the approval of the UCSF human studies committee. Eight normal volunteers were studied using both MRI and MEG (3F, 5M, mean age 30 yrs, range 25-36 yrs). Painless tactile stimulation was performed in both MRI and MSI environments using a balloon diaphragm, clipped to the fingertip. The diaphragm was driven with bursts of compressed air (15-30 p.s.i.) lasting approximately 30ms and repeated at an inter-stimulus-interval of 0.5-1s. Stimulation was applied sequentially to the left and right index fingers. Additionally, fMRI was used to observe the cortical activity associated with the performance of a simple motor task, involving the flexion of the index finger.

fMRI was performed using a GE SIGNA 1.5T system (GE Medical Systems, Milwaukee, WI), equipped with gradient coils that can produce ±20 mT/m with a rise time of 230 mT/m/ms. For fMRI studies, periods of activation were interleaved with periods of rest. Each period was of 20s duration. Throughout the protocol, multislice echo planar images (128x128 matrix, 40cmx20cm field-of-view) were acquired with the interval between successive images of the same slice being 2.5s. Thus 8 multislice image sets were acquired during each 20 second period. In each multislice set, 5 axial slices of 7mm thickness and 3mm interslice interval were collected, covering the extent from the corpus callosum to the vertex. In addition to gradient-recalled echo EPI (TE=60ms), spin-echo (TE=100ms) EPI sequences were used in three subjects.

The fMRI time-series of images were analyzed as follows: the first four images were discarded to eliminate signal intensity variations arising from progressive saturation. Subsequently, on a pixel-by-pixel basis, signal intensity variations over the time-course of the image series were correlated with the stimulus function and both r, the correlation coefficient and the corresponding t-statistic were determined. A t-test was then performed to determine the correlation significance (threshold: $p<0.01$). Further a spatial constraint was imposed: to eliminate spurious activation-noise, it was required that at least 5 contiguous pixels (with no geometric constraints) be similarly correlated.

MEG was performed using a 37-channel biomagnetometer (MAGNES, BTi., San Diego, CA). Data were collected in 300ms epochs, centered on the stimulation trigger, with a sampling rate of 297.8 Hz per channel. 256-512 epochs, collected with identical stimuli and pseudo-random inter-stimulus intervals in the range 500ms-1s were averaged to improve signal-to-noise ratio. Latencies within the range 30ms to 70ms post stimulus were examined. Extracranial magnetic fields were modeled using the single equivalent dipole method to obtain the spatial coordinates of the neuronal current source. Co-registration of MEG localizations with high-resolution 3D SPGR MRI was performed (using skin landmarks) to depict the anatomic origin of the activity.

Results

In eight subjects with a total of sixteen stimulation sites, fMRI achieved true positive localization in eleven cases; of the stimulations which failed to elicit a detectable fMRI response, one case was a bilateral failure and the remaining three were unilateral. MSI achieved satisfactory localization in all cases (Fig. 1).

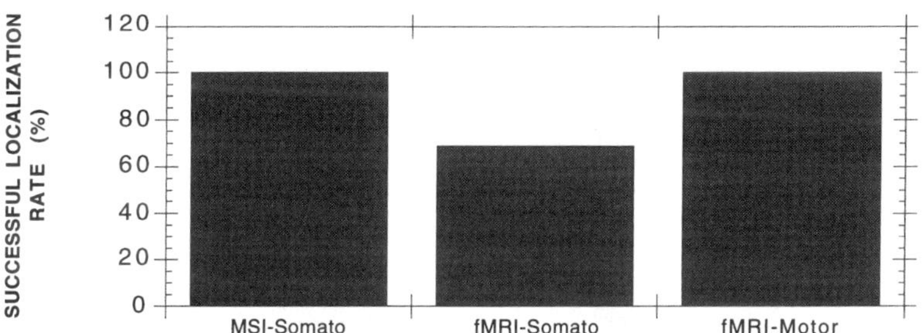

Figure 1   Localization Rate : fMRI vs. MSI with Somatosensory Stimulation

Ipsilateral and frontal/pre-motor activation was detected with fMRI in three cases; this was inherently avoided using MSI by the spatial position of the detector probe and its restricted field-of-view. In the majority of cases, fMRI and MSI localizations appear in accordant gyri and at similar axial levels (Fig.2).

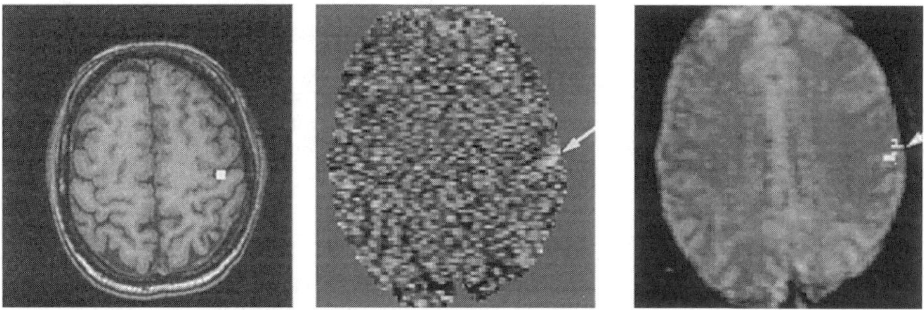

Figure 2   fMRI and MSI localization in good anatomic agreement.
Right index finger stimulation: (left) MSI; (middle) fMRI correlation coefficient map; (right) fMRI overlay

There was an observed tendency for the fMRI localization to lie more superficially than the MSI localization. A typical time course of signal intensity changes from pixels, identified on fMRI overlay maps is illustrated in figure 3. fMRI also has the potential for false negatives; some subjects demonstrate no clearly identifiable activation (5/16); however, MSI provides a reliable post-central localization with a similar stimulus. In the three subjects in whom spin-echo echo planar imaging was performed with a similar stimulus protocol, no significant activation was detected using the above statistical approach.

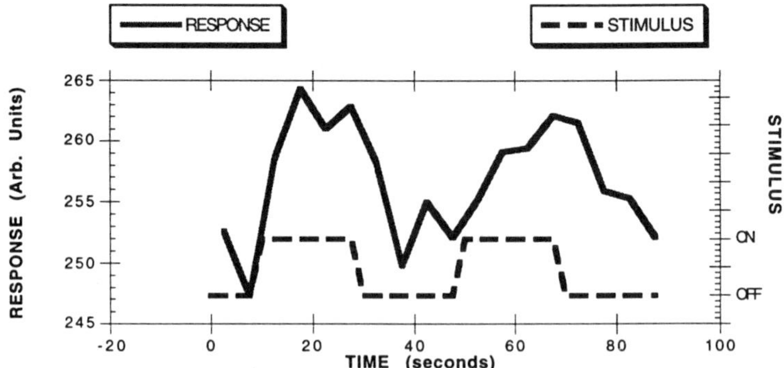

Figure 3 fMRI signal intensity change with stimulus

Discussion

Mapping the sensorimotor cortex provides the neurosurgeon with vital information regarding the functional organization of cortical tissue, in the planning of surgical approach route for the resection of mass lesions, epileptogenic tissues and other intracranial anomalies [4,7]. Currently intraoperative mapping by direct cortical stimulation is required; this is expensive, time-consuming and requires a craniotomy. A non-invasive, pre-operative mapping alternative could help indicate the inherent risk of surgical intervention and suggest alternative therapeutic strategies. Both fMRI and MSI offer this potential; the purpose of this study was to compare the two modalities for the reliability with which they depicted the sensorimotor cortical organization.

fMRI offers attractive benefits for clinical cortical mapping. Most neurosurgery candidates undergo a pre-operative MRI scan and therefore a functional mapping protocol could simply be incorporated into the standard, avoiding the need for a separate scan and associated errors of image co-registration (as well as penalties associated with time and cost). Furthermore the sensitivity of fMRI does not vary with the depth or geometric extent or orientation of the activated source and so it is amenable for use with a wide variety of stimuli to map a range of primary and associated functional activities. Some regions (e.g. brainstem and frontal pole) may be less successfully imaged with fMRI since the technique is inherently sensitive to magnetic susceptibility artifacts commonly present in these anatomic regions. To achieve clinical utility, it is necessary to have fMRI coverage of more than a single anatomic plane or slice. This is particularly important in cases where functional areas might be displaced from their expected anatomical site, either by mass lesions or adaptation. To this end, multislice or 3D approaches should be advocated. In order to achieve this and maintain adequate temporal resolution (of the order of 1-2s), EPI sequences are required. Conventional gradient-recalled-echo imaging, while providing appropriate contrast, cannot satisfy such multislice capabilities without incurring a loss of temporal resolution.

In this study, we employed a gradient-recalled-echo EPI sequence with a TE of 60ms to give strong magnetic susceptibility sensitivity. As discussed by Weisskoff and others, this sequence tends to preferentially highlight larger vessels, such as veins and venules rather than simply capillaries. In 3 of the subjects we additionally therefore employed a spin-echo echo-planar imaging sequence, which might be predicted to be rather insensitive to larger vessels and therefore reflect only changes in oxygenation present at the capillary level. However, no statistically significant correlation with the stimulus function was found in these subjects with this sequence. This was true even when regions of interest were targeted using the gradient-recalled-echo activation response.

A number of statistical approaches for the analysis of fMRI data have been proposed. All seek to identify pixels within the image that respond to stimulation or task performance. Since the observed response is only of the order of a few percent and image signal to noise itself may be poor, these methods must attempt to provide a rigorous basis for the unequivocal identification of cortical activation. Simple subtraction of "resting-state" from "activated-state" images (similar to the method used in PET studies of stimulation and task performance) generally suffers from inadequate signal intensity difference compared to random image noise. To counter this, several acquisitions in each state may be averaged. However, this necessarily reduces temporal resolution. We performed a correlation analysis to allow identification of pixels whose signal intensity variations correlated with the stimulus presentation. However, a simple temporal correlation is insufficient (with a threshold of p<0.01, an image of matrix 128x128 may contain many "random" or false positive correlations). We invoked a requirement for spatial connectivity, reducing the incidence of such correlation "noise".

Neuronal localizations based on MEG data were derived using the single equivalent dipole (SED) approach. This simplification of the neuronal environment, although not a faithful description of human brain activation, has been widely used to provide an estimate of a single current source that might give rise to the measured extracranial fields. Several studies have indicated that in the case of simple cortical processes, the model provides adequate accuracy of localization, compared to alternative standards such as invasive electrocorticography. The modeling process involves the fitting of dipole strength, localization and orientation, to minimize the difference between observed and predicted extracranial fields. The fitting process provides a measure of data-model correlation. Following previous investigators, we accepted "dipole localizations" that satisfied the observed data with a confidence volume of less than $1cm^3$ and with r>0.98. Changing such selection criteria influences the rate of localization "success".

Somatosensory stimulation with a painless pneumatic tapping to the fingertips proved inadequate for routine use with fMRI (69% statistically acceptable localizations), although in the 11/16 cases where localizations were achieved, good correspondence with MSI and anatomical expectation was found. Thus it seems that failed localizations result from poor signal-to-noise and statistical power rather than an inherent inability of the technique. Simple motor task performance seems more reliable, providing 16/16 successful localizations with fMRI. MSI was not performed with motor task performance, due to the complexity of the motor evoked neuromagnetic field. Also, although attempts were made to provide identical somatosensory stimuli for fMRI and MSI studies, the pressure and duration of the stimulus pulse, as well as the inter-stimulus-interval may have varied in the different environments (the MSI stimulus was computer generated and regulated; to operate in the fringe fields of the magnet during fMRI, the stimulus was manually-driven).

The general agreement between the two distinctly different modalities is encouraging, particularly when considering the rather different physiological aspects they probe (hemodynamic versus electrical). The lack of precise colocalization might be anticipated especially since to achieve detectable fMRI localizations we adopted the GRE-EPI sequence with its dominant venous contribution. Furthermore the simplistic MSI modeling adopted limits the confidence of MSI source localization. Improved MEG modeling strategies with more complex current source descriptions might become feasible if limited in range by some spatial prior knowledge, provided for example by fMRI. However, clearly strict spatial constraints imposed by fMRI should be avoided due to underlying physiological differences. Although not specifically required in the application of presurgical mapping of the sensorimotor strip, it appears that the temporal resolution of MSI might allow the elucidation of temporal sequences of multiple sites of activation identified by fMRI (and inherently time-averaged and thus unresolvable in time). Thus the combination of techniques, even without the requirement for strict spatial colocalization, might provide valuable information about neural signal propagation, unobtainable from either method alone: the combination might provide information which is "more than the sum of the parts".

References

1. Kwong KK. Functional MRI with echo planar imaging. Magnetic Resonance Quarterly 1995; 11(1):1-20.
2. Kwong KK, Belliveau JW, Chesler DA, et al. Dynamic magnetic resonance imaging of human brain activity during primary sensory stimulation. Proceedings of the National Academy of Sciences, USA 1992; 89:5675-5679.
3. Fox P, Raichle M. Focal physiological uncoupling of cerebral blood flow and oxidative metabolism during somatosensory stimulation in human subjects. Proceedings of the National Academy of Sciences, USA 1986; 83:1140-1144.
4. Hammeke TA, Yetkin FZ, Mueller WM, et al. Functional magnetic resonance imaging of somatosensory stimulation. Neurosurgery 1994; 35(4):677-681.
5. Baumgartner C, Doppelbauer A, Deecke L, et al. Neuromagnetic investigation of somatotopy of human hand somatosensory cortex. Exp Brain Res 1991; 87(3):641-8.
6. Gallen CC, Sobel DF, Schwartz B, Copeland B, Waltz T, Aung M. Magnetic source imaging. Present and future. Invest Radiol 1993; 28 Suppl 3:S153-7.
7. Roberts TPL, Rowley HA, Kucharczyk J. Applications of magnetic source imaging to presurgical brain mapping. In: Kucharczyk J, Roberts TPL, Moseley ME, Orrison W, ed. Neuroimaging Clinics of North America "Functional Neuroimaging" Saunders, 1995; 5(2): 251-266.

Acknowledgments

We are grateful to BTi for equipment support and to Susanne Honma, R.T. for technical help.

Quantitative Comparison of Motor Control Studied with MEG and fMRI

Russell, D.P.[1], Schwartz, B.J.[2,3] and Hirschkoff, E.C.[2,3]

[1]The Neurosciences Institute, La Jolla, CA, USA; [2]Biomagnetic Technologies, La Jolla, CA, USA; [3]The Scripps Research Institute, La Jolla, CA, USA

Introduction

We have compared the data resulting from magnetoencephalography (MEG) and functional MRI (fMRI) mapping of the same subjects under the same stimulus conditions in an evoked motor response paradigm. The goals of this study were threefold: 1) to compare the activation foci revealed by each modality as a cross-validation check; 2) to compare the quantitative aspects of the signal yielded by each modality, in this case parametric changes in signal strength as a function of stimulus intensity; and 3) to lay a foundation for future work in which results from both modalities will be combined in such a way as to enhance the information content of both.

Methods

The motor task studied was movement of the dominant-hand index finger. The subject's hand was placed in a holder which helped maintain a constant position of the other fingers, and which provided a trigger bar against which the index finger was placed. During MEG acquisition movement initiation/conclusion was indicated by the leading/trailing edges of the trigger signal, and movement onset provided the reference point for around which MEG epochs were averaged. The same arrangement sans trigger signal was used in the MRI suite to provide a uniform mechanical environment across imaging modalities. The finger was flexed and relaxed against an elastic load. Loads of four different magnitudes (three in MEG) were employed, the lowest (force level 1) being approximately 1.4 Newtons at 1.5 cm displacement, and ranging to level 4, each unit adding another 1.4 N at 1.5 cm displacement. Level 1 generally felt quite modest to subjects, while level 4 (5.6 N) was moderately strenuous. The amplitude of finger movement was mechanically limited. Subjects were instructed to concentrate on applying only as much force as was needed to reach the stop point, and to keep the movement time as constant as possible.

Seven subjects (two female, five male; six right-handed, one left) were each submitted to both MEG and fMRI within a two-month period. Additional subjects were studied using only one modality or the other.

MEG was performed on a Biomagnetic Technologies Magnes system with 37 SQUID detector channels. Finger movement was initiated by a visual cue (LED) with a single flexion/relaxation repeated at an average interval of two seconds which was randomly varied within a half-second window. Each force-level trial consisted of 256 repetitions, which were averaged, bandpass filtered at 1-20 Hz, and submitted to equivalent current dipole (ECD) source analysis [1] and other quantitative and statistical measures as described in the results. Whenever possible, all three force levels (1, 2, and 4) were studied, and selected levels repeated for increased within-subject statistics. A separate trial in which the index finger was pneumatically stimulated was also performed in order to unambiguously localize somatosensory response.

Functional MRI was performed on a General Electric Signa 1.5-tesla scanner retrofitted for echo-planar imaging (EPI) with the Advanced NMR Instascan system. A set of eight spatially consecutive gradient-echo EPI axial images through the relevant region enclosing areas M1/S1 was collected every two seconds (the TR, or repetition time) for 4 minutes, alternating between the stimulated and resting conditions every 32 seconds. A blood-oxygenation-level-dependent (BOLD) contrast was achieved by using an echo time (TE) of 50 msec. Each of the four force levels was applied for a 4-minute scan session. Finger movement was gradual and at a rate of one flexion cycle per second as timed by the EPI scan sounds. These images were submitted to a time-course analysis to detect pixels with statistically significant activation and their activation amplitude. MR images for anatomical localization of MEG-derived sources were also collected in the same session so that the coordinate transformation placing the MEG results in the MRI reference frame would apply to the fMRI maps as well.

Results

Good accordance was found between ECD locations as determined by MEG and the activation regions found in fMRI, with both motor (M1) and sensory (S1) components identified in each. An example from one subject is shown in Fig. 1. In Fig. 1(a) three separate dipoles are shown localized on MRI anatomical views of the same subject, with the orientation indicated by the line segment. The black open square indicates the dipole at a typical latency of 30 msec prior to movement onset, consistent with a location within the precentral gyrus. The white open square is at a latency of 48 msec post-movement-onset. Note the more posterior location and opposite orientation, consistent with the postcentral gyrus, as is typically the case. The white filled square is taken from the somatosensory-stimulation run, again indicating a postcentral-gyrus location quite similar to the so-called somatosensory component of the motor task. All dipole fits showed high correlation ($r > 0.95$) with the point-dipole model.

Fig. 1(b) shows the two motor-task dipoles, now localized on the fMRI activation maps from the same subject and force level. The significantly activated pixels (dark tones) are shown on a sample lower-resolution echo-planar image from the time series (brighter gray tones). Significance is defined on an individual-pixel basis as Student's $t > 4$. Note the close correspondence between the MEG dipole locations and the fMRI activity. Activity is seen primarily in the contralateral S1/M1 region of this right-handed subject.

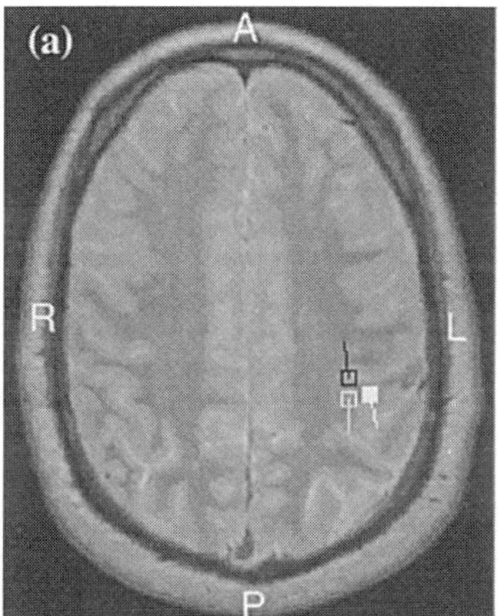
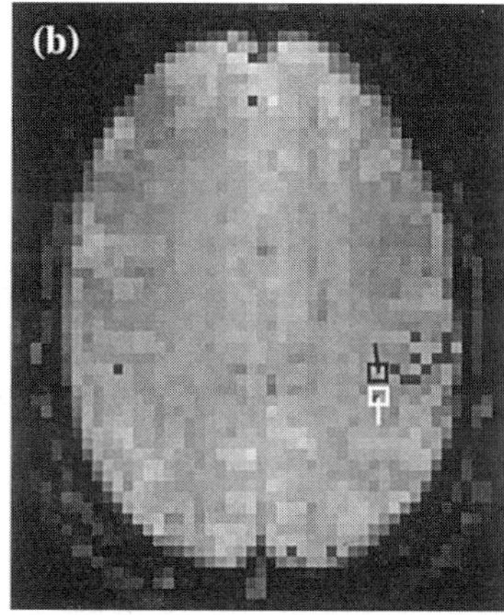

Figure 1: Sample MEG and fMRI localizations of motor activity in one subject

The amplitude of both MEG and the fMRI activation amplitude varied consistently with applied force level and with movement rate. The data for all nine subjects studied with MEG are shown in Fig. 2. Shown is the magnitude $|Q|$ of the dipole identified as the somatosensory component, found in a latency range of 20-100 msec post-trigger in all subjects, the particular latency identified as that characterized by the highest correlation coefficient to the dipole model. This typically coincided closely with the peak RMS power across all detectors of the averaged waveform. The data have been normalized by subtracting from each subject's data the mean across all force levels for that subject in order to remove intersubject variance and emphasize force-dependent behavior.

The mean across subjects of the (unnormalized) MEG data at each force level is plotted in Fig. 3(a). The corresponding fMRI data is shown in Fig. 3(b). Error bars are the standard error on the mean, which now includes the variance due to intersubject differences in average activation level. The fMRI index of activity is taken here as the sum of activation amplitude from all pixels within a region of interest (ROI) surrounding the S1/M1 region and which includes all pixels significantly activated at *any* force level. Note the striking similarity in the quantitative behavior of these two activity indices, taken from two imaging modalities derived from distinct physiological substrates.

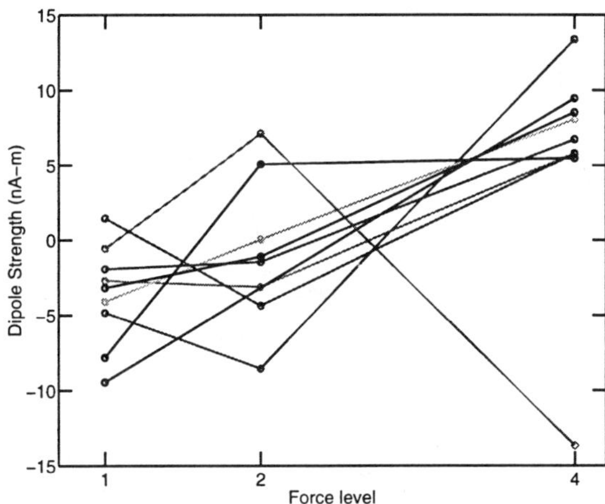

Figure 2: MEG dipole magnitude as a function of force load in all subjects

This similarity is further emphasized in Fig. 4(a). Here the average MEG dipole magnitude at each force is plotted against the corresponding average fMRI amplitude for the three forces common to both. The plotted line is simply drawn from the origin to the highest data point (force level 4) to make the point that the data are consistent with a zero intercept, implying that the two indices are functionally analogous as a measure of differential activation specific to the task (force control). If, however the data from each subject are averaged over all force loads and similarly analyzed for correlations on a subject-to-subject basis, no such result emerges.

Another index of MEG activation strength is proposed as a more likely candidate of subject-specific activation. In Fig. 4(b) the data are pooled across force loads within each subject, and each subject's MEG activation amplitude is plotted against the fMRI amplitude. Here, however, the MEG activation is measured as the RMS power (across all detectors) of the averaged (across epochs) MEG waveforms, integrated over the interval of 0-300 msec post-movement onset. The correlation of these data is 0.52, not statistically significant. Nevertheless, the correspondence, which is quite good with the exception of two subjects, is suggestive and worth further investigation.

Discussion

The MEG and fMRI signals obviously originate from rather distinct substrates of neural physiology. Hence it is important to characterize the degree to which these and other modalities offer common as well as complementary information about the underlying neural processes. Beyond simply cross-validating the results of each modality and utilizing the obvious spatiotemporal synergy that exists between them, it is hoped that combining data from such distinct techniques will provide information about an expanded subset of the rich ensemble of neural activity subserving such mental processes as motor control.

The most obvious validation consists of comparing the spatial information obtained from each, as is done here and in other studies. In addition, it is interesting to compare the quantitative (i.e., activation amplitude) aspects of the data for several reasons, one of which is the usefulness such knowledge will be in planning and carrying out any further studies in which the quantitative aspects of the signal are relevant. Such might be the case in attempts to observe more subtle mental state changes than the simple "on/off" pattern employed in blood-flow-mapping techniques such as fMRI, or in observing transient brain states that cannot be robustly averaged over many trials as is the case in evoked electromagnetic-field methods such as MEG.

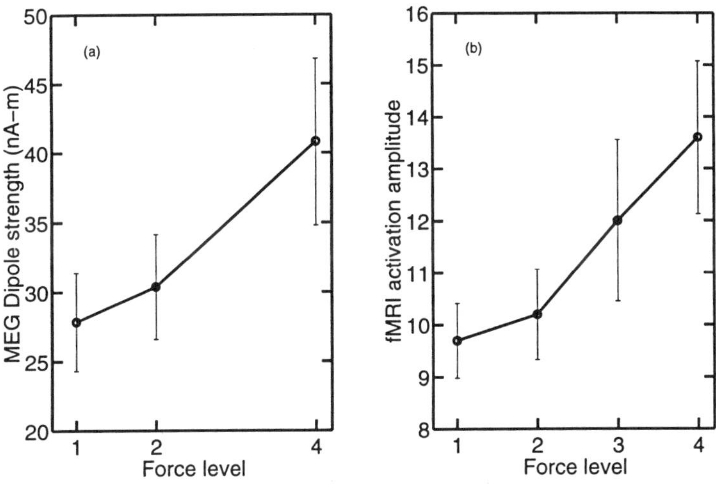

Figure 3: Average MEG and fMRI activation amplitudes as a function of force load, averaged over all subjects

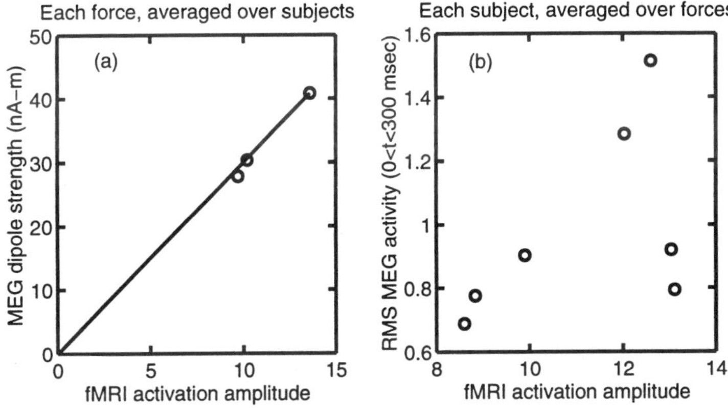

Figure 4: Quantitative correlations between MEG and fMRI activation amplitudes

References

[1] Sarvas, J. Phys. Med. Biol., 1987, 32: 11–22.

Generating Cortical Constraints for MEG Inverse Procedures Using MR Volume Data

Sandor, S.R.[1], Leahy, R.M.[2] and Timsari, B.[2]

[1]TRW Inc., Redondo Beach, California, USA; [2]University of Southern California, Los Angeles, California, USA

Introduction

Accurate knowledge of the location of the cortical surface is important in solving forward and inverse problems in EEG and MEG. Since the sources of event related surface potentials and magnetic fields are generally assumed to be confined to the cortex, knowledge of the shape of the cortex can be used to constrain the location of sources when solving the inverse problem [3]. Accurate knowledge of the cortical surface can also be used in solving the forward problem. By assuming that the head consists of a set of homogeneous regions of constant conductivity (e.g. brain, skull, scalp) the forward problem relating the measured EEG or MEG signal and the primary current sources can be solved using a boundary element method (BEM). The brain region for the BEM can be found using the method described here.

Our primary objective is to develop a technique that is automatic and accurate in extracting the cortical surface and which can track the major sulci. The MR images we use are T_1-weighted where cerebrospinal fluid (CSF) appears dark compared with white and gray matter. In these images the cortex is a ribbon with an outer boundary defined by the surface of the brain and an inner boundary by gray-white matter transitions. In general the outer cortical surface is clearly defined by a gray-CSF border, but due to noise, finite resolution and partial volume effects, surface extraction based on thresholding or simple edge detection typically results in links between the brain and other structures. Also, cortical structure may not be apparent from any one 2-D slice through the brain. Therefore, our approach processes these images in 3-D and separates the cortex from surrounding structures. We use the following pre-processing steps: 1) anisotropic filtering, an optional step which is used to sharpen image boundaries and 2) 3-D Marr-Hildreth filtering which locates surfaces within the head volume. The filtered image is then processed using a sequence of 3-D morphological operations which extract a binary brain volume from the skull and connective tissues.

Edge Detection

In our work, anatomical boundaries in MR brain images are located by edge detection. In order to select an edge detector capable of extracting accurate detailed information of brain anatomy from the gray-scale MR images we performed a subjective comparison between detectors which have been reported in the literature as giving the best results when applied to medical images - the Canny[2], Deriche[6], and Marr-Hildreth[5] methods. An ideal operator for our purpose is one which has a low computational cost and accurately detects closed boundaries of important anatomical structures. Our morphological algorithm requires as input a binary image of the closed boundaries of tissues inside the head. The algorithm can deal with some degree of noise in the form of disconnected edge elements and scattered false contours. However, if too many artifacts of this type occur, then preprocessing as described in the following section is required.

The Canny and Deriche edge finding methods can both produce disconnected edge segments and require costly edge linking procedures to produce a set of closed boundaries. In contrast, the Marr-Hildreth operator always produces a set of closed contours. While the Canny and Deriche methods gave better localization of edge segments in some regions of the image, these segments are often too dense to produce accurate closed boundaries using a simple edge linking procedure, consequently we decided to use the Marr-Hildreth operator for edge finding. While it is easy to show that this operator can produce spurious boundaries, we have found that these are removed when used in conjunction with the morphological processing step. Our experimental experience indicates that the spatial resolution of a Marr-Hildreth detector can be chosen so that it is effective in finding the outlines of different tissues. We have found that at a resolution appropriate for extracting the brain surface, this processor detects sulci boundaries corresponding to white-gray matter transitions rather than gray-CSF transitions [1]. Thus the final cortical surface that we extract has a tendency to move between the outer cortical boundary on the gyri and the inner cortical surface in the sulcal folds.

Anisotropic Filtering

The accuracy of anatomical boundaries obtained by edge detection is a direct consequence of the quality of the input image. An edge enhancement preprocessor appears to be essential when dealing with low contrast MR images with a low signal to noise ratio. Conventional linear spatial filtering methods for image enhancement reduce noise at the cost of reduced resolution. This results in edge blurring and is not a good preprocessor for our purpose. Instead we have found nonlinear anisotropic diffusion filtering to be especially appropriate for enhancing MR image data because of its edge preserving properties. In this method a filtered image is viewed as the solution to the anisotropic diffusion equation [7]. In this equation, the diffusion coefficient adaptively controls the diffusion strength to smooth the image within a region and stop smoothing across a boundary. This coefficient is a monotonically decreasing function of some measure of edge strength, usually the magnitude of the gradient of the image intensity. By selectively smoothing the image while preserving edges, we generate a result which gives improved edge definition with fewer spurious edges when processed with a conventional edge detector.

To demonstrate the efficiency of anisotropic filtering in reducing noise and enhancing region boundaries in MR images, in figure (1-a) and (1-c) we show a transaxial slice of a volume MR image before and after applying 3D anisotropic filtering with an exponential diffusion coefficient. Homogenous structures in the filtered image appear smoother and their boundaries are more clearly defined. The improved differentiation between adjacent regions of similar intensity characteristics facilitates the performance of the edge detector [figure (1-b,d)].

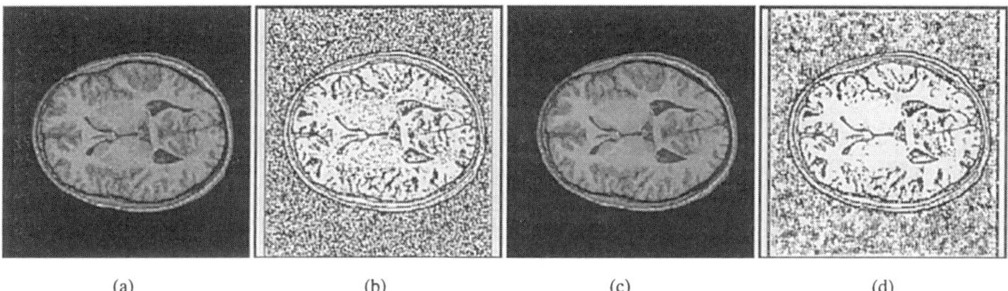

(a)   (b)   (c)   (d)

Fig. 1. (a) Transaxial slice of an MR volume (b) Edges detected for that slice by the 3-D Marr-Hildreth operator (c) The same slice after applying anisotropic filtering (d) Edges detected for the filtered slice by the 3-D Marr-Hildreth operator

A Morphological Algorithm for Cortex Extraction

Our morphological algorithm requires as input a binary image volume, which we generate by applying the Marr-Hildreth edge detector to the MR brain data. In the resulting edge image voxels are either black or white depending on whether or not they represent region boundaries. We represent a binary head image by a 3-D set, denoted X, which is made up of two types of voxels: $X = X_{Int} \cup X_{Bound}$ where X_{Int} represents the object interior voxels which are connected only to other object interior voxels, and X_{Bound} represents object voxels that are connected to the background. In our work, "connected" is defined as 6-connectedness, i.e. voxels are connected only if they share a face, but not an edge or a corner. Set X represents other anatomical regions, not just the brain. The brain is a connected 3-D subset of X, where any two points in this subset can be joined by a 3-D path entirely contained within the brain volume. Because of noise, partial volume effects, or true anatomical connections, X may contain voxels that link brain regions to extraneous surrounding structures, such as dura matter or skin.

To sever unwanted connections and extract only the brain, we first perform an erosion, which shrinks the brain volume and eliminates all regions smaller than the structuring element. We have chosen a 3-D rhombus structuring element of discrete size one, which is a 3-D cross three voxels wide in the x, y, and z directions. This element deletes narrow connections without globally damaging or distorting an image. In particular, erosion by a rhombus will eliminate all boundary voxels X_{Bound} from X and leave X_{Int} unchanged. For definitions of morphological operations and their sizes see [4]. For an isotropically sampled MR volume with $1mm^3$ voxels, transforming by R1 eliminates regions which have a size of 2 mm or less in any direction and shrink a majority of the brain surface by 1 mm. After erosion we carry out a 3-D flood filling operation. This step is necessary because although erosion eliminates undesirable connections, non-brain regions, such as eyeballs or skull, still remain. We want to select only brain voxels. A 3-D flood filling routine finds all voxels connected to a seed point in the brain (figure (2-a)). We will denote this eroded and flood filled set of brain voxels as X_{EBrain}, where $X_{EBrain} \subseteq X_{Int}$. Since

1167

erosion shrinks the brain surface and widens holes in the volume, X_{EBrain} is not a true representation of an entire brain. To expand the brain from the eroded image we first dilate it by R1 to restore some voxels eliminated by erosion and denote the result X_{DBrain} (figure (2-b)). Then we fill holes in X_{DBrain} by closing the set with an octagon structuring element of size two, denoted O2. This isotropic element, which is a digital approximation to a sphere in Euclidean space, has a width of 9 voxels. We choose size two because an octagon of size one is only 5 voxels wide and may not be large enough to close all holes that were widened by erosion. Besides filling image holes, a closing operation replaces portions of an object's boundary with the structuring element's boundary. Therefore, this operation will result in a brain volume in which boundaries are smooth and all holes less than 9 voxels wide are closed (figure (2-c)). We will denote the result of this operation X_{CBrain}. Set X_{CBrain} is a smooth image that serves as a template on which we can reintroduce fine details from the original binary volume, X. When we take the set difference $X_{CBrain} \backslash X$ we obtain a set which represents all gaps and holes in the brain which include those corresponding to the internal cavities of the brain as well those resulting from the foldings of the brain cortex. We denote this set X_{Holes}. A difference operation between the closed brain and its set of holes yields the brain volume X_{VBrain} (figure (2-d)). The current X_{VBrain} contains some holes, not connected to the outer brain surface, which clearly do not contribute to the cortex. To remove these hole, we apply the 3-D flood filling process to $X^c{}_{VBrain}$, which corresponds to the background of X_{VBrain}. The output, X_{Back} can be considered as a mold for brain, and its complement gives us the solid brain volume, X_{Brain}. The outer contour of X_{Brain} is the cortical surface. Figure (2-e) shows a slice from a brain extracted by this morphological processing.

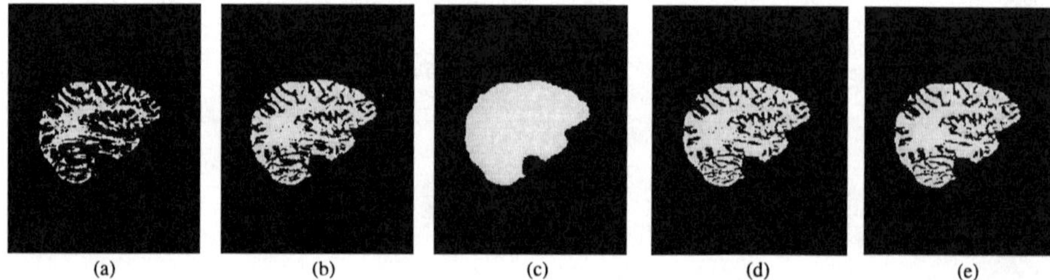

(a)   (b)   (c)   (d)   (e)

Fig. 2. Example: (a) a slice through the volume after erosion by a 3-D rhombus structuring element and applying flood-filling (X_{EBrain}), (b) the slice after the set of eroded brain voxels has been dilated by a 3-D rhombus structuring element (X_{DBrain}), (c) the slice after 3-D closing of the dilated image by an octagon of size 2 (X_{CBrain}), (d) the result of taking a difference operation between the set of brain holes and the closed image (X_{VBrain}), (e) the final result of the morphologic extraction algorithm(X_{Brain}).

Below, we summarize the steps of our algorithm for extracting a brain from an MR head image. The symbols $\ominus, \oplus, \bullet$ represent morphological erosion, dilation, and closing respectively.

1. $X_{Int} = X \ominus R1$
2. Flood-fill X_{Int} to arrive at X_{EBrain}
3. $X_{DBrain} = X_{EBrain} \oplus R1$
4. $X_{CBrain} = X_{DBrain} \bullet O2$
5. $X_{Holes} = X_{CBrain} \backslash X$
6. $X_{VBrain} = X_{CBrain} \backslash X_{Holes}$
7. Flood-fill $X^c{}_{VBrain}$ to arrive at X_{Back}
8. $X_{Brain} = X^c{}_{Back}$
9. Find the outer boundary of X_{Brain} to get X_{Cortex}

Experimental Results

We have implemented the brain extraction method described above, including edge detection and anisotropic filtering. The entire procedure is written in C and takes approximately 10 minutes on a 60MHz SPARC20 workstation to extract the brain from a typical stack of 120 images. An additional 10 minutes are required for 3 iterations of the anisotropic pre-filtering. Our implementation gives the user the option to: 1) select whether or not anisotropic filtering is desired before edge detection and 2) chose the size of the Gaussian operator used in Marr-Hildreth edge detection, where a larger operator smooths the image and detects fewer fine details. With this software we have thus far processed 30 different MR data sets, and in each instance, a visual inspection of the extracted brain surface indicated that our algorithm was successful in automatically extracting the cortex from these head images. These results show that the major sulci are found using this method. We have noted in a few cases that the cerebellum is removed from the brain volume.

Shown in figure (3) is an example of a cortical surface that was automatically extracted from a T_1-weighted volume acquisition (0.5T Picker Outlook) with a 2.5mm slice thickness. In figure (3-b) a typical slice through the brain volume is shown with its corresponding cortical boundaries superimposed on the slice. A rendering of the cortical surface is displayed in figure (4-a). Figure (4-b) shows the smoothed brain surface obtained by applying a morphological closing operation to the high resolution cortical surface.

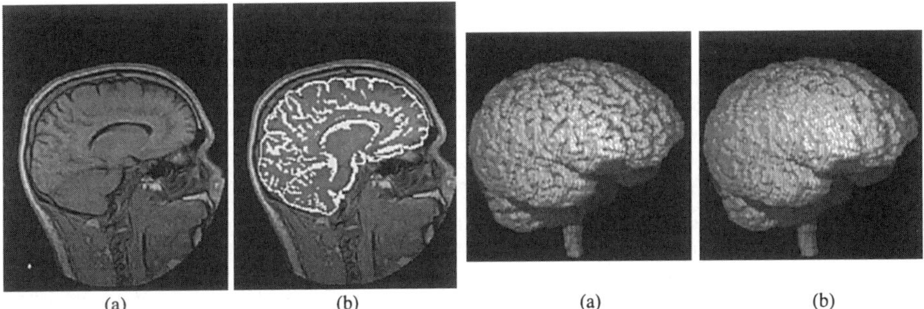

<table>
<tr><td align="center">(a)</td><td align="center">(b)</td><td align="center">(a)</td><td align="center">(b)</td></tr>
</table>

Fig. 3. (a) a slice through the original binary MR volume, (b) outer boundaries of the same slice superimposed on it

Fig. 4. (a) a rendering of the cortical surface extracted with morphological processing, (b) a rendering of smoothed cortex

Conclusion

We have described an image processing technique which will automatically find the cortical surface from T_1-weighted volume magnetic resonance (MR) data. This method uses a combination of non-linear smoothing, edge finding and binary morphology procedures to find the surface. It requires no user interaction other than specification of a few input parameters. We have tested the method on approximately 30 data sets with a range of slice thicknesses (~1mm-2.5mm) and scan parameters. The method successfully extracts the cortical surface and tracks the major sulci.

REFERENCES

[1] M. Bomans, K-H Hohne, U. Tiede and M. Riemer, "3-D segmentation of MR images of the head for 3-D display," IEEE Trans. Medical Imaging, vol. MI-9, pp 177-183, 1990.

[2] J. Canny, "A Computational Approach to Edge Detection," IEEE Transactions on Pattern Analysis and Machine Intelligence, vol. PAMI-8, pp. 679-698, November, 1986.

[3] A. M. Dale, M. I. Sereno, "Improved Localization of Cortical Activity by Combining EEG and MEG with MRI Cortical Surface Reconstruction: A Linear Approach", Journal of Cognitive Neuroscience, vol. 5, pp. 162-176, 1993.

[4] P. Maragos, "Pattern Spectrum and Multiscale Shape Representation," IEEE Transactions on Pattern Analysis and Machine Intelligence, vol. 11, pp. 701-716, July, 1989.

[5] D. Marr, E. Hildreth, "Theory of Edge Detection,"Proceedings of Royal Society of London, vol. 207, pp. 187--217, 1980.

[6] O. Monga, R. Deriche, J. M. Rocchisani, "3-D Edge Detection Using Recursive Filtering: Application to Scanner Images," CVGIP: Image Understanding, vol. 53, no. 1, pp. 76-87, Jan 1991.

[7] P. Perona and J. Malik, "Scale-Space and Edge Detection Using Anisotropic Diffusion," IEEE Transactions on Pattern Analysis and Machine Intelligence, vol. 12, pp. 629-639, July, 1990.

Acknowledgements

This work supported by NIMH, Grant No. R01-MH53213, and by the TRW Inc. Doctoral Fellowship Program.

Correlation Between Depth and MEG/EEG Surface Recordings in Auditory Cortex. Contribution of Numerical Simulations

Schwartz, D.[1], Badier, J.M.[2], Scarabin, J.M.[1] and Liégeois-Chauvel, C.[2]

[1]Laboratoire Signaux et Images en Médecine, Rennes, France; [2]Clinique Neurologique, CHR Pontchaillou, Rennes, France

Introduction

The main purpose of our study is to combine three different methods of brain activity recordings: depth recordings (stereotactic EEG: SEEG) and surface recordings (EEG/MEG) in order to improve localizations of the sources of the brain activity. We present here the first step of this study. We built a model of a region of interest (ROI) within the brain based on MRI information. Our goal was to obtain an accurate description of the surface of the ROI. According to this model, we defined realistic dipole layers in respect to the anatomy (orientation, shape, extension). By activating those dipole layers we produced SEEG, EEG and MEG theoretical signals. In order to validate this concept on a physiological point of view, we chose the auditory cortex as the ROI. We compared the simulations with in vivo signals (Auditory evoked potential (AEP)) for which Liégeois-Chauvel et al [1] studied the localization of the middle latency components using SEEG. We performed the validation by taking into account the auditory cortex anatomy and built the simulations for each patient studied. We then compared the theoretical results with the real SEEG data. Furthermore we used those simulations to test the reliability of several MEG sources modeling algorithms.

Materials

We studied patients with intractable temporal lobe epilepsy undergoing SEEG for localization of seizure foci [2]. Informed consents were obtained for all patients. Those patients underwent the following examinations:

□ A MRI examination after the withdrawal of SEEG electrodes. The original examination was reconstructed in 3D with a trilinear interpolation (isotropic volume, voxel size=$0.9375mm^3$).

□ A simultaneous MEG/EEG examination with the 37 channels BTi Magnes system and a standard 10-20 EEG electrical arrangement. The AEP and auditory evoked field (AEF) were recorded with pure tones: tone burst, Frequency=(500HZ, 1kHz, 2kHz), duration=40ms, rise and decay=5ms, 500 stimulations (1/sec), stimulus level=70 dB HL (hearing level). The averaged signals were filtered with a band-pass filter (1-120 Hz).

□ A SEEG examination (monopolar recording). Same stimulation as the MEG examination except duration=24ms and 100 stimulations. The averaged signals were filtered with a band-pass filter (1-350Hz).

Methods

ROI Segmentation

The auditory cortex and the surrounding areas were manually segmented on MRI by an anatomist. The segmentation was performed in a 3D environment allowing a constant checking of the accuracy in sagittal, coronal and axial directions. The description contained the Heschl gyrus, the superior-temporal gyrus, the planum temporale and the the Sylvius fissure.

Model description and simulations

The list of voxels obtained above was not suitable in our context. Therefore we chose a B-spline model to match the data. The local characteristics of the ROI such as normal to the actual surface were easier to compute with this mathematical description.

We defined an area of activity (AoA) as a dipole layer constrained on the segmented cortical surface, i.e. defined by a given part of the cortical surface (location and spatial extent) and a uniform density of single dipoles located on the surface with a direction normal to this surface. We added to this geometrical description three temporal parameters: the initial time of activation, a window of activation and the shape of the activation (mono or bipolar). Each dipole contribution was computed (accordingly to the SEEG leads position, and to the MEG/EEG sensors positions) by the following methods:

□ EEG/SEEG: The potentials were computed assuming the surface lying in a three shells sphere. An integration of potential was performed along each lead in order to take their size into account.

□ MEG: The magnetic field was computed assuming a surface lying in an homogeneous sphere.

The MEG, EEG and SEEG signals generated by a AoA were the sum of the contributions of each single dipole. In the same way we combined AoAs in order to generate the complete MEG, EEG and SEEG simulated signals. i) We simulated standing sources (standing AoA) with three of four AoA overlapping in time (parameters of those AoA: 100mm^2, 600 to 900 dipoles, time of activation: n°1: 30ms, n°2: 50ms, n°3: 60ms, n°4: 80ms, window of activation: ±30ms) (fig. 1). ii) We simulated activity propagating along the gyrus (moving AoA) with a AoA (50mm^2, 150 dipoles) moved along a trajectory (from the medial part to the lateral part of Heschl gyrus, Fig. 2) with a given speed (0.1mm/ms). In the comparisons with real signals, the SEEG leads and the MEG/EEG sensors were localized in their actual positions for a given patient. In this case an accurate MEG, EEG, SEEG to MRI registration were used.

Fig. 1: 3D view of the model from the patient DR with two actual SEEG electrodes. The location on medial to lateral axe of AoAs n°1, n°2 and n°4 were chosen according to the Liégeois-Chauvel's results [1]. The AoA n°3 was chosen to test the effect of a source interfering with the others.

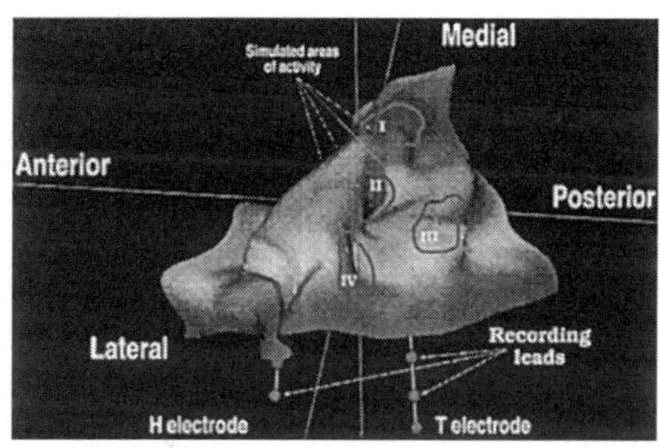

Sources modeling algorithms

We computed localizations on MEG simulated signals with the following algorithms: a monodipolar moving dipole fit (MDF) and a monodipolar spatio-temporal algorithm (SPT). Both algorithms searched for an equivalent dipole to match the MEG signals. The MDF performs a localization for a given latency. The SPT performs a localization by analyzing the signals on a time window around a given latency [3].

Results

Validation of simulations

We verified the capabilities of the model to reproduce SEEG like signals from simulations by analyzing simulated signals obtained on theoretical SEEG electrodes. Virtual electrodes were defined to explore the entire surface of the ROI. From moving AoA or standing AoA we calculated the shapes of simulated signals and we compared them with typical shapes of real signals recorded in AEP. We found typical shapes, curves presenting polarity inversions between two recordings leads of the same electrode and polarity inversions between the recording leads of two electrodes as well as time shift between peaks of activity. Those characteristics are used to localize the activities with depth recordings. Those inversions show the effects of an AoA lying between two leads of the same electrode or lying between two electrodes (Fig. 2). The polarity inversions between two leads of the same electrodes showed a gradient of large amplitude. On the contrary for a polarity inversion between two electrodes the gradient of amplitude was smaller (2 to 10 times inferior to the gradient between two leads of an electrode). Each inversion was studied in respect to the localization of the recording site (Fig. 2). The simulation showed that a lead is sensitive to only the small volume that

1171

surrounds it. The relative amplitude decreased very fast when the distance increases between the recording site and the sources. In real signals, it implied that with a small number of averages only the closest sources to a lead were recorded the others disappearing in physiological noise. All those properties can be used to establish the precise location of the AEP activities when recorded from the depth.

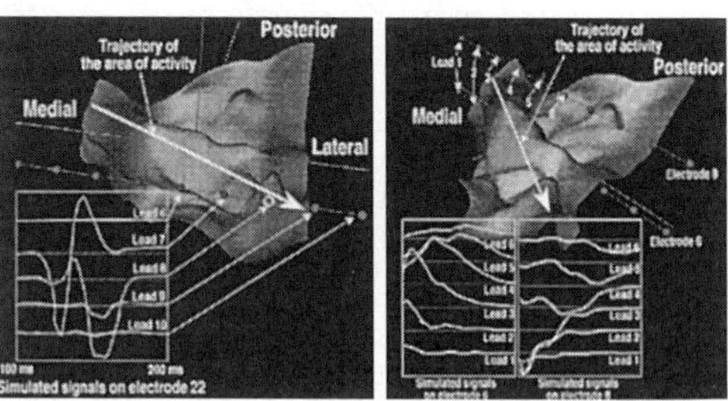

Fig. 2: Polarity inversions on simulated signals according to the location of the recordings sites (between two leads of an electrode (left) and between two electrodes (right)).

At this stage, we performed the simulations with the real position of two depth electrodes surrounding the Heschl gyrus of patient DR. The data recorded from the patient showed a inversion between the electrode T and the electrode H (Fig. 3). According to the latency of the inversion we defined three standing AoAs (the first in the medial part of Heschl (time of activation=55ms), the second 9 mm more laterally (time of activation 80ms) and the third on the lateral part of Heschl (time of activation=95ms). With those AoAs we retrieved the shape of the real signal especially the inversion (explained by the orientation of the second AoA on posterior to anterior direction). This simulation shows that we can reproduce useful part of the real signals. It implies that the assumptions for the model were correct in our context of studies on auditory cortex.

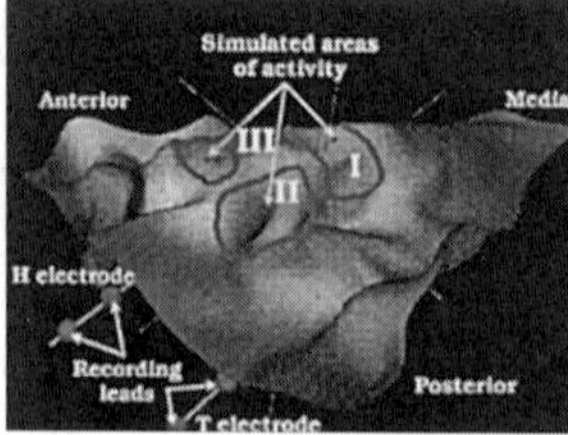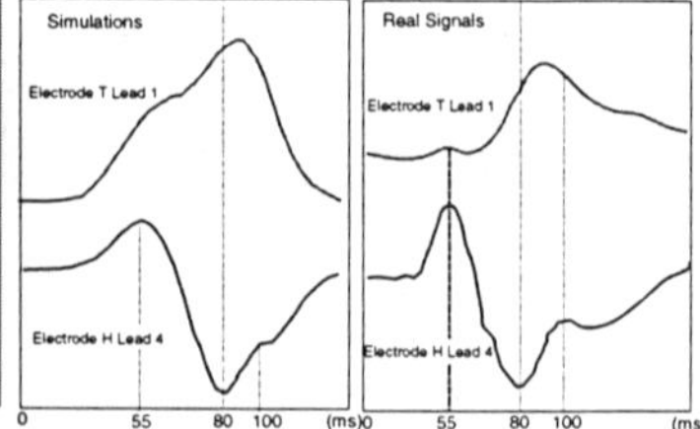

Fig. 3: Comparisons between simulation and real signals on SEEG electrodes H&T of patient DR. The locations of the AoAs used in the simulations are shown on left figure and right we present the resulting simulated signals and the real signals.

Evaluation of source modeling algorithms

Preliminary results of the evaluation of a moving dipole fit algorithm and a spatio-temporal algorithm in MEG context are presented here.

Tests with a moving AOA: We tested the effects of the shape of the ROI with a moving AoA. On both algorithms, we noticed an increase of the localization errors and a decrease of the stability of the localization when the shape of the structure supporting the dipole layer became more complex. When the AoAs had a part of their single dipoles almost radially oriented (on the medial to lateral direction) or when the contributions of the single dipoles of the AoAs cancelled each others the errors of localization increased.

Tests with standing AoAs: We tested the effects of AoAs overlapping in time. We combined three or four standing AoAs. With three AoAs (n°1, n°2 and n°4) (Fig. 1) and an optimal position of the sensor (i.e. magnetic field map showing a symmetrical inversion) the SPT algorithm showed a better stability in time and in space of the solution than the MDF. The separation between the activities was better and the time of stability for each AoA was greater i.e. each AoA was localized by the SPT algorithm during a larger amount of time and during this time the error of localization was stable (Fig. 4). Moreover the MDF algorithm produced equivalent dipoles localized between the AoAs. These localizations were consistent with the anatomy (i.e. they were located at 1 to 3 mm under or above the surface) but they were not relevant of the actual AoAs. We tested the effect of a shift of the sensor in the Cz direction. It increased the errors especially for the localization of the AoA n°2 which overlapped in time with the AoA n°1 and n°3.

With four AoAs (Fig. 1) the MDF algorithm didn't find a reliable localization of the AoA n°2 and n°3 because of the overlap in time between those AoAs. On the contrary the SPT algorithm succeeded in localize the four AoAs during a large amount of time with a good accuracy (Fig. 4).

Latency errors: We noticed that the time overlapping produced an important error in term of latency of the localization of the AoA (for both algorithms and especially the MDF, Fig. 4). A AoA was rarely localized at its peak of activity but before or after this peak according to the overlapping of the others AoAs. In conclusion, overlapping sources may produce spatial and temporal localizations errors. This result is of main interest in the study of the processes involved in AEP and others evoked potentials.

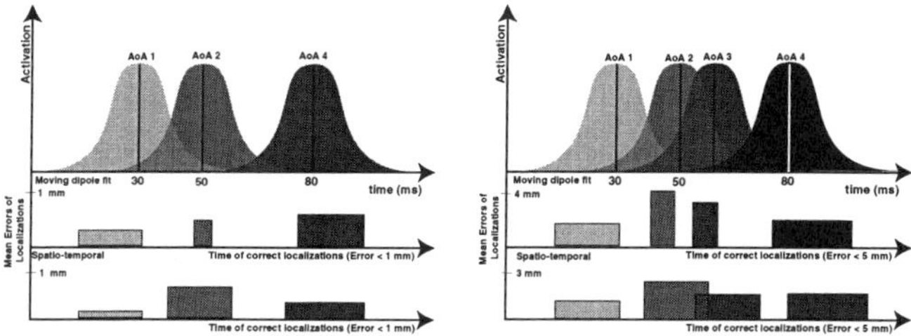

Fig. 4: Results of the localization for three standing AoAs (right) and four standing AoAs (left). The curves give the time of activation and the overlap of the AoAs. The graphics represent the results of the localization: The horizontal axis represents the time where the algorithms found a localization for the AoA with an error inferior to the following thresholds: 1 mm (3 AoAs) and 5 mm (4 AoAs). The vertical axis represents the mean errors localization for each AoA. The errors of localization calculated were the distance between the localization and the center of gravity of a given AoA.

Discussion and conclusion

Those simulations provide useful information for the analysis of AEP and AEF. The use of the anatomical model and the actual locations of depth electrodes for a given patient allowed the study of the different combinations of AoAs which reproduced the main characteristics of the real signal (especially inversions which may be explained accordingly to the ROI). In a following stage, it will be of main interest to integrate this information in localizations process from MEG or EEG signals. With the results of the evaluation of the source modeling algorithms we have the possibility to analyze our MEG localizations from real AEF in term of time and space stability to identify the sources of the main components of the AEF. In this evaluation we used only monodipolar algorithms. The SPT gave better results than the MDF in term of accuracy of localization on the spatial and temporal point of view. The SPT succeeded in localizing AoAs despite the overlap of the sources and the use of a single dipole model. The use of multi-dipolar algorithms may improve the localization results when the overlap in time between sources becomes more complicated. In conclusion, this method give the possibility to refine source spatial and temporal localization by the use of combined EEG/MEG as well as SEEG based on the individual anatomy.

References

1. Liégeois-Chauvel, C., Mussolino, A., Badier, J.M., Marquis, P. and Chauvel, P. Evoked potentials recorded from the auditory cortex in man: evaluation and topography of the middle latency components, Electroenceph. and clin. Neurophysiol., 1994, 7(4):618-625.
2. Bancaud, J., Talairach, J., Bonis, A., Schaub, C, Szikla, G., Morel, P. and Bordas-Ferrer, M. La Stéréo-encéphalographie dans l'épilepsie: Informations neurophysiopathologiques apportées par l'investigation fonctionnelle stéréotaxique. Masson, Paris, 1965.
3. Bouliou, A., Bihoué, P., Toulouse, P., Scarabin, J.M. Modelisation of the cerebral activity by a probability distribution upon the cortical surface from spatio-temporal MEG data. Human Brain Mapping, 1995, (Sup.1):105.

Brain Activities Evoked by Missing Auditory Stimuli: A Combined Study Using MEG and fMRI Data

Takino, R.[1], Miyauchi, S.[2], Sasaki, Y.[2], Pütz, B.[2] and Sekihara, K.[3]

[1]Shiraume Gakuen College, Kodaira-shi, Tokyo 187, Japan; [2]Communications Research Laboratory, Koganei-shi, Tokyo 184, Japan; [3]Central Research Laboratory, Hitachi Ltd., Kokubunji-shi, Tokyo 185, Japan

Introduction

In MEG studies, cortical activities around the human auditory cortex were found to be evoked by randomly occurring infrequent stimulus changes, e.g., in tone frequency or in tone intensity, among a monotonous sequence of tones [1]. Some studies using random stimulus omissions also found evoked 'endogenous' responses by the 'missing' stimulus itself [2, 3]. However, the location of the activities differed between the two types of studies. In this study, we employed the same 'missing' tone sequence in MEG and fMRI measurements and investigated the activities around the superior temporal auditory cortex.

Methods

Subjects and Stimulus

Seven (six males and one female, age 28–47y) and five (four males and one female, age 30–39y) healthy right-handed subjects participated in the fMRI and MEG experiments, respectively. They were instructed to just listen to the tones with their eyes open without exerting any mental activities, like counting tones. In both experiments, a simple repetition of 1kHz tones (ISI 300ms, duration 20ms) were presented dichotically with either a headset through a plastic tube (fMRI experiment) or a speaker at the outside of the magnetically shielded room (MEG experiment). Stimulus intensity was adjusted at a level that the subjects could clearly hear the tones, usually the audio-amplifier output level being set at 90dB. We employed three types of tone sequences: (1) *standard,* consisting of only 1kHz tones, (2) *missing,* consisting of 1kHz tones (85%) and randomly occurring stimulus omissions (15%) following more than three consecutive tones, and (3) *no tone,* in which no tone stimuli were presented.

In the MEG experiment, only the *missing* tone sequence was presented, and selective averaging for the missing tones and for the standard tones being presented just before the missing ones were performed. 150 epochs were averaged respectively over 350ms beginning 100ms before stimulus onset.

In the fMRI experiment, the subjects underwent two kinds of experiment. In order to define the location of the primary auditory cortex, the *standard* tone sequence was presented in the test condition and the *no tone* sequence was presented in the control condition. In the second experiment, we studied missing tone effects. The subjects listened to the *missing* tone sequence during the test condition while the *standard* tone sequence was presented during the control condition. The computed t-values that exceeded the statistical significance level ($p < 0.01$) were superimposed onto the corresponding anatomical images, and areas of activation were thus determined.

Data acquisition and analysis

MEG experiments

Magnetic fields were measured over both the left and right auditory cortices using a dual 37-channel SQUID system (second order gradiometer, BTi). The MEG signals were bandpass filtered 1–60Hz and digitized at a 500Hz sampling rate. Dipole fits with a correlation with the measured field of 0.95 or better were accepted. For one subject, we employed a Neuromag-122 whole-scalp magnetometer system and applied the spatio-temporal analysis of the Multiple-Signal Classification (MUSIC) algorithm [4] to the data with latencies between 0 and 250ms.

fMRI experiments

fMRI studies were carried out on a 1.5T Siemens scanner equipped with an EPI booster. We collected 6 to 8 EPI functional images (transverse, 5mm thick, FoV=200–250mm) capturing the sylvian fissure. The test and the control conditions were alternately repeated 20–30 times. EPI scans were triggered after 15 seconds of sequence presentation in either condition. Data were analyzed by pixel to pixel Student's paired t-test. For one subject, we employed one additional tone sequence in which, instead of omitting tones as in the *missing* tone sequence, deviant tones that only differ in frequency (2 kHz) from the other tones that were presented (*deviant* tone sequence). Each tone sequence was presented 24s and EPI images were collected every 8s. A set of four tone sequences was repeated 10 times. Data were analyzed via cross correlation calculation for each pixel [5].

Results
MEG experiments

All five subjects showed a distinctive peak following the standard stimuli with a peak latency of 60–83ms. The electric current dipoles (ECDs) were located in the superior temporal gyrus, pointing upwards, almost orthogonal to the sylvian fissure in both hemispheres. Four of the five subjects showed one stable peak after the presentation of missing stimuli with a peak latency of 105–148ms. The ECDs were found near the standard stimuli's ECDs locations, but pointing downwards in both hemispheres.

Figure 1 shows the results of calculating the MUSIC-Mosher localizer in a three-dimensional volume. The results are projected onto the transverse, coronal, and sagittal planes. The contours in Fig. 1 show the relative value of the Mosher MUSIC localizer; and the area where the localizer forms a peak is considered to be the location of the dipole source. These results revealed that a single source dominates in the right hemisphere for both the standard stimuli and the missing stimuli. However, the source for the missing stimuli is located nearly 1cm posterior and 0.5cm lateral to the source for the standard one.

Fig. 1   MUSIC source localization results

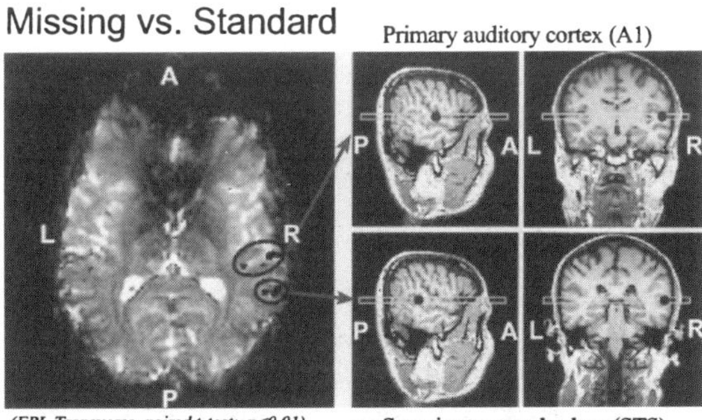

Fig. 2 Areas activated by missing auditory stimuli

fMRI experiments

During the first experiment defining the location of the primary auditory cortex, all subjects showed statistically significant activation in the superior temporal regions of both hemispheres, especially in the Heschl's gyri. In the second experiment, although the *missing* tone sequence has 15% fewer stimuli in number, it activated the area around the Heschl's gyri more than the *standard* tone sequence in all subjects. We found another activation in the auditory association cortex in the middle to posterior part of the superior temporal sulcus (STS; Fig. 2). The activation patterns were consistent when the same experiments were repeated using the same subjects. In five subjects, the activated area in the STS was found in the right hemisphere. The other two subjects showed the activation in the both hemispheres.

Cross correlation analysis shows that the small areas near the primary auditory cortex were also activated by the missing and the deviant tones compared to the standard tones, but did not clearly overlap with the activation areas determined by the comparison between the *standard* tone sequence and the *no tone* sequence. The most prominent activation was found in the right STS in the presentation of the *missing* tone sequence. The location of activation in the STS was almost the same as the one determined by the MEG MUSIC analysis (Fig. 3). The time course analysis for the ROI in the STS shows the activation by the missing tones. The deviant tones activated the same areas, but less effectively than the missing tones. The standard tones did not activate these areas. Rather, MR signals obtained during the presentation of the standard tones were the same level or less than the ones for the no tone sequence (Fig. 4). The cross correlation coefficient for the averaged data and the reference function that is shown as a shadow pattern in Fig. 3 is 0.893.

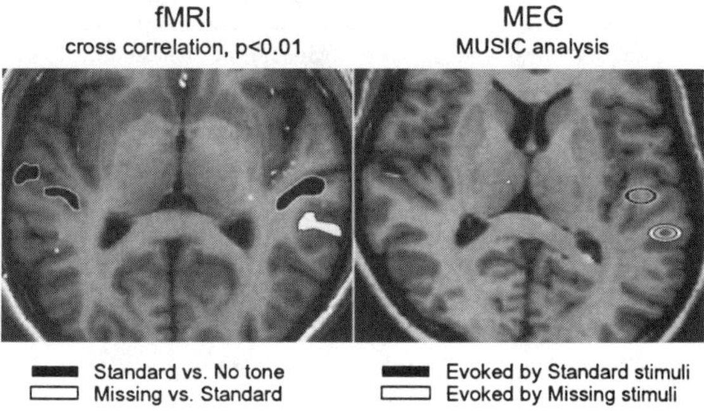

Fig.3   Comparison of fMRI and MEG results

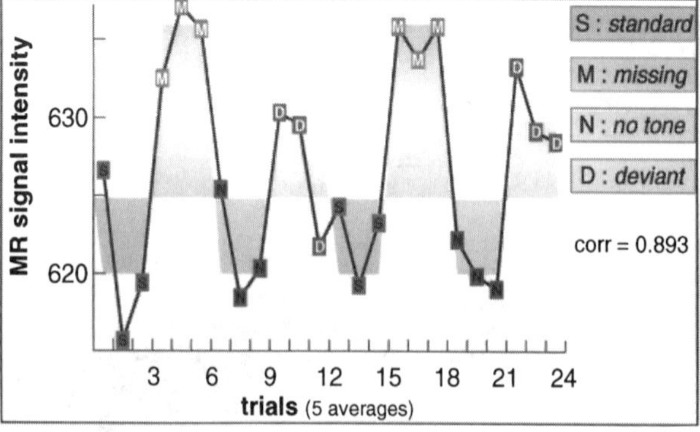

Fig. 4 Averaged Activation for ROI in STS

Conclusion

Both MEG and fMRI experiments results showed that the areas around the human auditory cortex were more activated by the missing (or deviant) stimuli than by monotonous tone stimuli. Our fMRI experiments showed that additional areas in the superior temporal sulcus are also involved in the detection of auditory stimulus changes. Although we had examined only one subject with the MEG MUSIC analysis, this method showed exactly the same areas activated during the presentation of the *missing* tone sequence.

The fMRI time course data showed that those areas were activated when there were some unexpected stimulus changes in the tone sequences (*missing* and *deviant*). The *standard* tone sequence had no effect on these areas. Activation levels were equal to or less than the ones by the *no tone* sequence. This would imply that these areas are not activated by a real stimulus, e.g., a tone presentation. They rather seem to be involved in the detection processes of global changes based on some endogenous neuronal activities.

References:

[1] Hari, R., Joutsiniemi, S.L., Hämäläinen, M., and Vilkman, V. (1989) Neuroscience. Letter, 99:164–168

[2] Joutsiniemi, S.L. and Hari, R. Omissions of auditory stimuli may activate frontal cortex, European J. of Neuroscience, 1989, 1:524–528

[3] Miyauchi, S., Takino, R., Sasaki, Y., Pütz, B., and Okamura, H. Missing auditory stimuli activate the primary and associate auditory cortices: a combined study MEG with fMRI, The 10th Tokyo Institute of Psychiatry International Symposium Abstract., 1995, pp.72–73

[4] Mosher, J.C., Lewis, P.S., Leahy, R. Multiple Dipole Modeling and Localization from Spatio-Temporal MEG Data, IEEE Trans. Biomed. Eng., 1992, 39:541–557

[5] Bandettini, P.A., Jesmanowicz, A., Wong, E.C., and Hyde, J.S. Processing strategies for time-course data sets in functional MRI of the human brain, Magn. Reson. Med., 1993, 30:161–173

Integration of Functional MRI and MEG of Median Nerve Stimulation

Verbiesen, M.J.R.[1], Peters, A.R.[2], den Boer, J.A.[2], Knösche, T.R.[1] and Peters, M.J.[1]

[1]*Biomagnetic Centre Twente, University of Twente, Enschede, The Netherlands;* [2]*MR Research Medisch Spectrum Twente, Enschede, The Netherlands*

Introduction

In this study we combined two non-invasive methods of brain imaging, functional magnetic resonance imaging (fMRI) and magnetoencephalography (MEG), in order to obtain a more detailed and integrated view of stimulus processing by the brain. MRI has conventionally been used to obtain an anatomical frame for interpreting EEG and MEG localisation results. In recent years, fMRI has been developed. It is based on the blood oxygenation level dependency (BOLD) effect of active brain regions, which is linked to neuronal activity. The effect of BOLD mostly reflects the microvasculature for venous blood [1]. It has been proven that there is a linear relation between the level of oxygenation and the measured T2 values [2]. fMRI has a good spatial resolution (1 mm) but poor temporal resolution (5 s). MEG reflects the currents within excited pyramidal neurons and has a high temporal (around 1 ms) but a rather poor spatial (several cm) resolution.

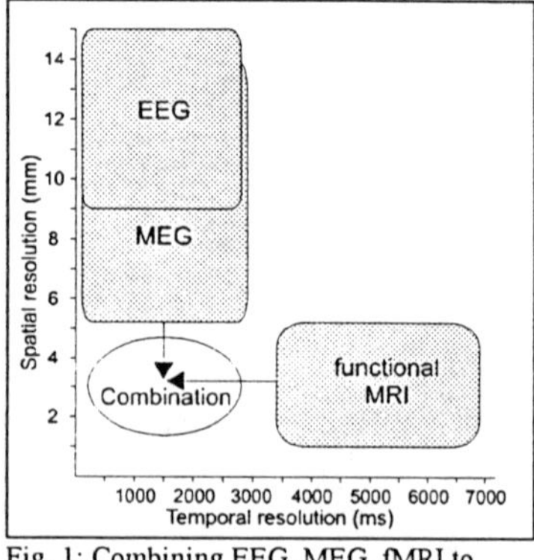

Fig. 1: Combining EEG, MEG, fMRI to improve both temporal and spatial resolution

In order to explore the combination of these techniques, we choose to use a well-known paradigm. Such a paradigm is the stimulation of one median nerve using an electric wrist stimulator. We used this stimulus for several reasons. First, it was feasible for both modalities. Second, much is known about the cortical responses to this stimulus from EEG, ECoG, and MEG studies.

Methods

In our study we performed repeated measurements of three right-handed healthy young volunteers (2M, 1F, 24-25 years). For fMRI we held several measurement sessions with a long time interval in between, in order to test the measurements' reproducibility. For MEG we did several measurements within one session.

We stimulated the median nerve (MN) of the right hand.. The stimulation sequences for fMRI and MEG had to differ, because of safety precaution during fMRI scanning. For this modality we used a special MR compatible stimulator (Digilog). PC-based software controls both the stimulation pattern and the timing of the MR scanner, to make sure that no stimulation is given when gradient fields are present in the MR scanner. The stimulation sequence consisted of 5 electric pulses with a duration of 500 µs and an interval of 50 ms during every scan cycle of 470 ms. The stimulation sequences for MEG consisted of a single electric pulse with a duration of 500 µs every 500 or 1000 ms. The intensity of the pulse was in all cases the maximum that could be tolerated by the volunteer.

For the fMRI scans we used a 1.5T ACS-II system (Philips M.S.) without any special hardware or software. The fMRI sequence is a Turbo Field Echo, this sequence is sensitive to the BOLD effect, but also sensitive to changes in blood flow, especially in the draining veins. With this sequence we covered a large volume of the brain (192x144x35 mm^3). The scan could be acquired in 30 s and consisted of 7 transversal slices (thickness=5 mm, scan matrix: 64^2, reconstruction matrix: 256^2). Consequently, we had a voxel size of 3x3x5 mm. With this scan, we obtained 30 dynamical frames in a regime of 3 without and 3 with stimulation (total duration 15 min). The dynamical scans during stimulation are averaged and their variance computed. So are the ones during rest periods. We identified the significant differences between the two resulting images with a two-sided t-test (pooled variance) for each pixel. A mask is made by threshholding the t-image at 95% significance. This way we defined the activated areas. We determined the actual effect of the stimulus from the intensity of the ratio between stimulation and rest periods. The masked ratio image was projected onto an anatomical T1-weighted (IR) MR image of the corresponding area, resulting in a colour image containing all the information. We reproduced position and orientation of the scans over several sessions with high accuracy (better than 2 mm) by computer matching of survey images. For each of our volunteers, we masked the two data sets with the highest response at P<0.05. These two masks were multiplied, resulting in the activated areas that could be reproduced in both sessions (P<0.0025).

The MEG measurements were performed with a 19-channel MEG system (University of Twente) at two locations: superior and at the left lateral side of the head, so at a total of 2x19=38 positions. Several measurements during one session were carried out applying two different interstimulus intervals: 500 ms (300 trials) and 1000 ms (200 trials). We made use of a 3D space tracker device and a special coil-cap on the head of the subject to determine the position of the head with respect to the gradiometer pickup coils. The positions of 3 marker points (left/right ear and nasion) with respect to the MRI co-ordinate system (visible in the MRI) were used to couple the co-ordinate systems of the two modalities. After averaging and low pass filtering (<100 Hz) the MEG data, we performed non-linear dipole fits using the Marquardt algorithm. We fitted rotating dipoles in several time intervals between 12 and 120 ms after the stimulus. The head model we used, was a realistic brain model extracted from the MRI of a different person. It was adjusted to match with the individual shape and position of the volunteer brains. We performed both single and double dipole fits. The program ASA (University of Twente) was used to perform these fits.

In order to combine both sorts of measurements we applied different strategies. First, we superimposed the positions of the current dipoles fitted to the MEG data, to the fMRI results. Second, we fixed dipoles at the locations found by fMRI and subsequently fitted their directions and strength to the MEG data. The residual variances gave an indication, how well and at which latencies these dipole positions match in the MEG data. Third, we used the fMRI results to estimate the number of active dipoles and use this information to obtain better dipole fits from MEG.

Results

The results from the fMRI data show that even with a stimulus as simple as the one employed in this study, differences in position and extent of the reproducibly activated areas for

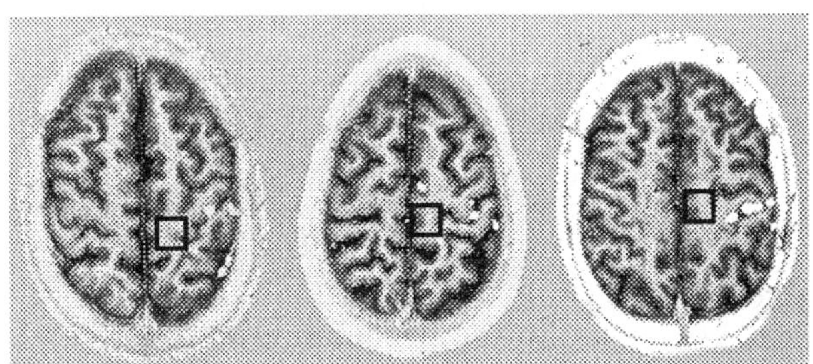

Fig. 2: One slice of reproducibly activated fMRI areas (white) for the three volunteers and projected position of one fitted rotating dipole (box) in time window 12-25 ms after the stimulus; 3D distances between closest fMRI area and dipole position ranges between 17-32 mm.

the different volunteers do occur (see Fig. 2). Also the intensities of the activated areas differ between two session with the same volunteer. The significant areas exhibit a change in signal amplitude between 1% and 8%. It seems that the larger changes are mostly due to the vein draining the cerebral tissue adjacent to the central sulcus. The cortex shows intensity changes up to 6%. With the MN nerve stimulation, there appears activity in both walls of the contralateral central sulcus. We observed also areas with significant signal decrease (1-2%) in both the ipsilateral and contralateral hemisphere of the brain. These locations are scattered and not reproducible among different session with the same volunteer. One of the volunteers consistently showed a large area of signal increase in the contralateral medial grey matter, presumably the supplementary motor area (SMA), which is remarkable for a sensory stimulus. Also the other volunteers had several small reproducible areas of signal increase outside the contralateral central sulcus.

The results of the analysis of the MEG data also show differences between the three volunteers. The time courses of the MEG data are consistent between the different measurements and volunteers. The location for the fitted dipoles vary among the volunteers (see Fig. 2). The qualities of the fits vary as well. For example, a fit of a rotating dipole in the time interval 12-25 ms yielded residual variances (RV) of 7-40 %. In the case of the high RV, the data could be explained much better with 2 dipoles (RV 11%). In all cases, the RV was the lowest in the time interval 45-70 ms for a single dipole fit. For other time intervals, a better fit could be achieved with more than one dipole.

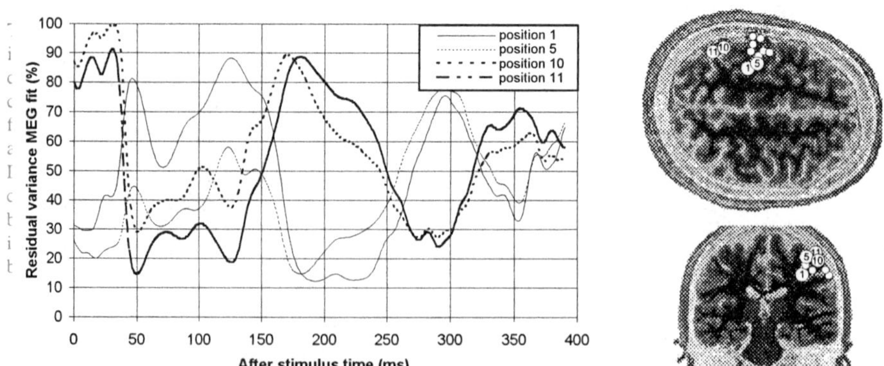

Fig. 3: Residual variance for a single rotating dipole fit in MEG data with a fixed position in four reproducible fMRI areas for different latencies. Left: anatomical positions of the fMRI areas

Results independently obtained from fMRI and MEG (dipole fit) analysis are superimposed in Fig. 2. The distance between the position of the fitted rotating dipole in the time interval 12-25 s and the nearest fMRI area is ranging between 17 and 32 mm.

Our next step was to use the non venous fMRI active areas as *a priori* known positions for a single rotating dipole, which is fitted to the MEG data. In Fig. 3 can be seen that in this volunteer two groups of positions (1+5 and 10+11) give alternately a low RV, so these two groups can tentatively be assigned to different areas which have a different sequence of activation. The first region is clearly in the primary somatosensory cortex, which is activated at short latencies. About the second region we do not wish to speculate.

Finally, we performed different multiple dipole fits with the following conditions.
1. Three rotating dipoles with their position fixed at the three fMRI positions with the largest distance between each other.
2. Three rotating dipole with their position fitted to the MEG data, while the initial guess (which is always necessary for a dipole fit) was extracted from the fMRI positions.
3. Three rotating dipole with their position fitted to the MEG data and random initial guess.

The results can be seen in Fig. 4-6. The fit with the overall lowest RV is found with the second condition. The change in position of the three dipoles when fitted (condition 2) can be seen in Fig. 5. This fit appears not to change very much the mutual position of the dipoles. This shift is ranging from 15-18 mm. In Fig. 6 one can see that the magnitude of the three dipoles have a different course in time. It seems that there are at least two major areas (represented by dipoles 1 and 2). Dipole 3 exhibits a much smaller contribution to the MEG data, which could be due to a radial orientation of the dipole.

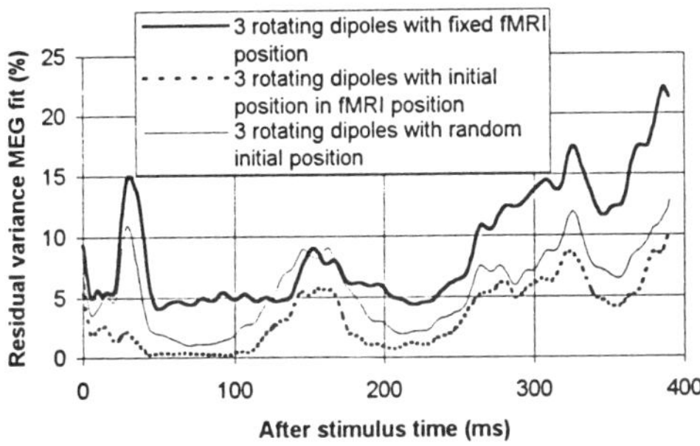

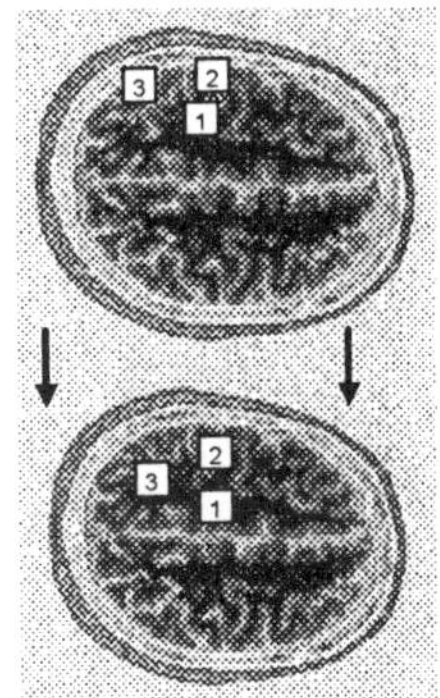

Fig. 4: RV for MEG fit with three rotating dipoles with fixed and fitted position and with random initial guess

Fig. 5: Change of projected position of dipole before (fMRI position; top) and after (bottom) position fit

Discussion

First of all we have to consider that the location of the active areas in fMRI, which is linked with the oxygen consumption by the nerve cells, and the location of activity from MEG data, which is linked to electric activity of the nerve cells, can differ. Keeping this in mind we tried to combine the two modalities. Our study showed that the differences in location between the fMRI areas and the dipole positions of the MEG range between 17 and 32 mm , which is rather large. In a similar study, John A. Sanders et al [3] compared the location for motor tasks. They found differences of 7-14 mm.

There are three possible reasons for the differences in location we found. First, the underlying brain processes detected by the two modalities may indeed have different locations. Second, matching the two co-ordinate systems may introduce an error up to 4 mm. Third, there may be a localisation error in one of the modalities. Looking at both techniques, MEG has presumably the largest spatial error (volume conductor model, positioning gradiometers, etc.). In our multiple dipole fits, we fitted three dipoles to the MEG data with an initial guess at the fMRI locations. In this fit we saw that the mutual position of the locations as extracted from the fMRI results was largely preserved after fitting to the MEG data (Fig. 5). Apparently we are seeing the same sources in both modalities. Considering that with MEG the depths of the sources is difficult to estimate, we might attribute the observed position differences to inaccuracy of the MEG based dipole fit. This leads us to the conclusion that, when using the spatial information of the fMRI for the position and number of active areas and the temporal information of the MEG for the temporal behaviour of these areas, we may obtain better results than with each of the modalities alone. Further research is of course necessary to substantiate these preliminary results.

Another interesting phenomenon appeared in this study. Even with a simple stimulus such as MN, there are a lot of differences in responses among the volunteers in spite of similar conditions. These differences were present in both fMRI and MEG

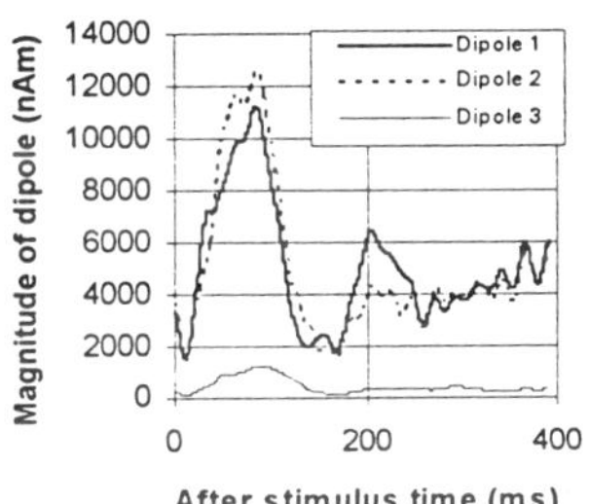

Fig. 6: Magnitude of the three position fitted rotating dipoles of Fig. 5

References

[1] Orrison, W.W., Lewine J.D., Sanders, J.A., Hartshorne, M.F.: Functional Magnetic Resonance Imaging, Functional Brain Imaging, 1995: 145-181

[2] Brooks, R.A., DiChiro, G.: Magnetic resonance imaging of stationary blood: a review, Med Phys 14, 1987: 903-913

[3] Sanders, J.A., Lewine, J.D., George, J.S. et al: Correlation of fMRI with MEG, Proceedings of 12th SMRM Annual Meeting, New York, 1993: 1418.

Comparison of Functional Localization in Human Visual Cortices Using MEG and fMRI: A Preliminary Report

Yoshikawa, K.[1], Takanashi, Y.[1], Iwamoto, K.[1], Yoshida, Y.[3], Ueda, M.[3,4], Tanaka, C.[5], Umeda, M.[5], Ebisu, T.[5], Fukunaga, M.[5], Naruse, S.[2], Sato, H.[6] and Nakajima, K.[1]

[1]Department of Neurology, Research Institute for Neurological Diseases and Geriatrics, Kyoto, Japan; [2]Department of Neurosurgery, Kyoto Prefectural University of Medicine, Kyoto, Japan; [3]Keihanna Research Laboratory, Shimadzu Corporation, Kyoto, Japan; [4]Superconducting Sensor Laboratory, Chiba, Japan; [5]Department of Neurosurgery, Meiji College of Oriental Medicine, Kyoto, Japan; [6]GE-Yokogawa Medical System, Tokyo, Japan

Introduction

The application of computer technology to functional magnetic resonance imaging (fMRI) and magnetoencephalography (MEG) has enabled visualization of human brain function in vivo and has revolutionized clinical neurology [1, 2, 3, 4, 5].

Although MEG can identify neuronal activity directly with higher temporal resolution in the millisecond range, this technique is limited because it requires an estimate of current distribution and an assumption of the number of electrical current dipoles associated with neuronal activities. Moreover, we have to superimpose the estimated data appropriately on computer tomography or magnetic resonance imaging of the human brain to decide the anatomical localization.

On the other hand, fMRI has high spatial resolution and can estimate multiple activated areas with ease. But it's temporal resolution is not better than 1second.

We had reported the comparison of information from MEG and fMRI in response to the same visual stimuli.[10]. In the present study, we attempted to make use of the comparison in order to overcome the weaknesses of the electromagnetic approach.

Methods

Three normal right-handed subjects aged between 23 to 24 years with normal visual acuity participated in this study following informed consent. Each subject was presented with a checkerboard pattern at 1 Hz during MEG testing and at 2 Hz during fMRI. The visual stimulation sets were composed of left hemifield, right hemifield, and whole field patterns. The subject's task was to gaze at a fixation point in the center of the screen on which visual stimuli were presented by a rear projecting computer that generated images using a multimodal stimulator (ST10, Medlec, England). During MEG, the screen was positioned about 1 m from the subject who was sitting on a non-magnetic chair. During fMRI, the screen was reflected, and the images were presented by a mirror.

Magnetic visual evoked fields (VEF) were recorded in a magnetic-shielded room using a novel 129 channel vector neuromagnetic imaging system, which has been developed recently by Shimadzu Corporation, Japan. In the present study, the signals were band pass filtered (0.7-100 Hz); digitized at 1024 Hz; and averaged off-line. Signals were also recorded 50 msec pre-stimulus and 300 msec following the stimulus to determine the baseline. To identify the source of evoked responses, the equivalent current dipole (ECD) with the highest correlation value was found by a least-squares algorithm based on a one- or two-dipole model. Generator sites and vectors of ECDs were superimposed on appropriate axial, sagittal and coronal MRIs.

FMRI was obtained by a conventional 1.5 T MRI scanner (Sigma, GE) in the same subjects. FMR images were obtained by using a one-shot gradient echo planar imaging sequence (TR/TE/FA=4000 ms/50 ms/90 degrees) at rest and with stimulation alternately. Axial plane images of 5mm thickness were acquired. One examination consisted of 10 trials (off-on-off-on-off-on-off-on-off-on) with 400 total slices. The statistical significance of differences was evaluated by a cross correlation thresholding method. The activated areas were indicated by colored dots according to their correlation values.

Results

Typical VEFs were obtained at a latency of 100 to 170 ms from all subjects following three stimulation sets. The best correlation of ECD estimation was derived between 150 and 165 ms. Generator sites and vectors of the ECDs from a subject (O.M., 23-year-old, man) were superimposed on appropriate axial, sagittal, and coronal

MRI scans (right half of Fig. 1, 2 and 3). The activated areas in the functional MRI study of the same subject were visualized in the left half of Figs. 1, 2 and 3.

On hemifield stimulation (Figs. 1 and 2), the ECD generator site was localized in the calcarine sulcus contralateral to the stimulated field, which corresponded anatomically to the activated area in the fMRI study, respectively.

On whole field stimulation, a single generator site was localized in the right occipital lobe with a correlation of 83.7% based on a one-dipole model, while two generator sites were localized in subcortical areas of the occipital lobe bilaterally with a correlation of 94.4% based on a two-dipole model (Fig. 3). In fMRI, activated areas with a high cross correlation coefficient were distributed along the medial sides of the occipital lobes bilaterally as seen in the left half of Fig. 3. However, these areas did not correspond to the ECD sites determined by MEG.

Discussion

Our results revealed that, in each hemifield visual stimulation, the generator site of the ECD based on a one-dipole model corresponded well to the activated visual cortical areas in our fMRI study [6, 7, 8]. However, with whole field stimulation, ECD generator site using a one- or two- dipole model did not correspond to the activated areas determined by fMRI. The fMRI data appeared to represent a simple summation of those obtained with right- and left- hemifield stimulation.

In analyzing the neuromagnetic data, we did not have any criterion to objectively evaluate the validity of the predicted ECD generator site except for the correlation value. Our data, especially that obtained with hemifield stimulation, indicate that the combination of MEG and fMRI may confirm the accuracy of ECD localization.

Furthermore, the MEG data are limited by the need to predict the number of ECDs associated with neuronal events. However, in the present study, we assumed that two dipoles, rather than one, should be estimated prior to ECD analysis in whole field stimulation because at least two areas were activated during fMRI with the same stimulation. FMRI may predict the number of ECDs to be estimated in MEG data analysis. This is the second advantage of combination of MEG and fMRI.

Although a relatively high correlation was derived in a two-dipole model, the estimated generator site of the ECD was found to be different from that activated in fMRI. This discrepancy may be explained by the mutual interaction between the two dipoles which were evoked simultaneously and located very closely to each other [9, 11]. We do not have an alternative explanation for this discrepancy between estimated generator sites in MEG and activated areas in fMRI at present. Future studies are required to resolve these localization problems.

Based on this preliminary report, we hypothesize that localization of neuronal events in MEG, properly confirmed by fMRI, will contribute to the research of higher human brain function.

References:

[1] Kwong, K.K., Belliveau, J.W., et al. Dynamic magnetic resonance imaging of human brain activity during primary sensory stimulation, Proc. Natl. Acad. Sci. USA, 1992, 89: 5675-5679.

[2] Ogawa, S., Tank, D.W., et al. Intrinsic signal changes accompanying sensory stimulation : Functional brain mapping with magnetic resonance imaging, Proc. Natl. Acad. Sci. USA, 1992, 89: 5951-5955.

[3] Cohen, D. Magnetoencephalography: evidence of magnetic fields produced by alpha rhythm currents, Science, 1968, 161: 784-786.

[4] Tayler, T.J., Cuffin, B.N., et al. The visual evoked magnetoencephalogram, Life Sci., 1975, 17: 683-692.

[5] Aine, C.J., Bodis-Wollner, I., George, J.S. Generators of visual evoked neuromagnetic responses, Advances in Neurology Vol.54: Magnetoencephalography, New York, Raven Press, 1990: 141-155.

[6] Holms, G. Disturbances of vision from cerebral lesions, with special references to the cortical representation of the macula, Brain, 1916, 39: 34-73.

[7] Phelps, M.E., Kuhl, D.E., Mazziotta, J.C. Metabolic mapping of the brain's response to visual stimulations: Studies in humans, Science, 1981, 211: 1445-1448.

[8] Sereno, M.I., Dale, A.M., et al. Borders of multiple visual areas in humans revealed by functional magnetic resonance imaging, Science, 1995, 268: 889-893.

[9] Tan, S., Roth, J., et al. The magnetic field of cortical current sources: the application of a spatial filtering model to the forward and inverse problems, Electroencephalogr. Clin. Neurophysiol., 1990, 76: 73-85.

[10] Takanashi, Y., Yoshikawa, K., et al. Comparisons of Two Different Functional Localizations in Human Visual cortices using MEG and fMRI, In: Book of Abstracts, Visualization of information Processing in the Human Brain: Recent Advances in MEG and Functional MRI, Tokyo, Tokyo Institute of Psychiatry, 1995.

[11] Lutkenhoener, B. A simulation study of the resolving power of the biomagnetic inverse procedure, Clin. Phys. Physiol. Meas., 1991, 12 (Suppl.A): 73-78.

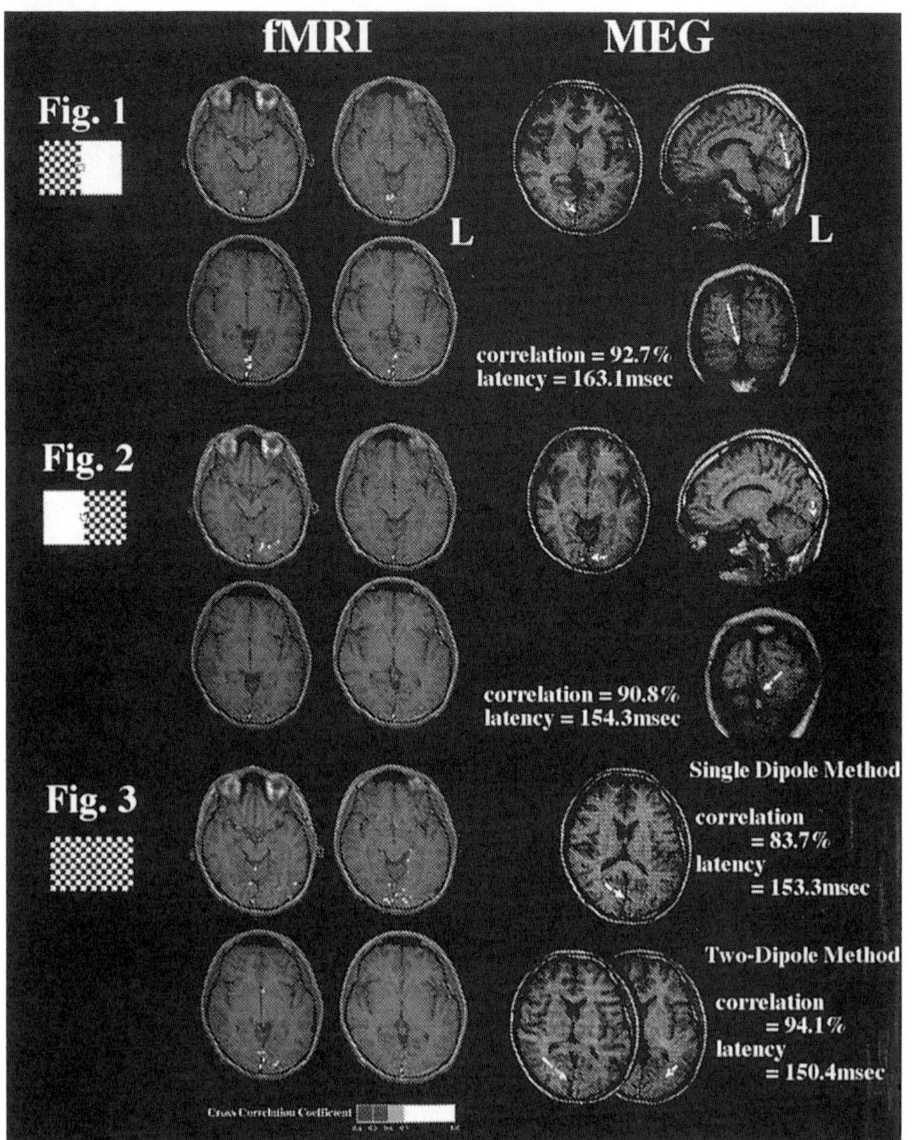

Figure Legends

Comparison of MEG and fMRI localization following visual stimulation. See details in the text.
Fig. 1: In left hemifield visual stimulation.
Fig. 2: In right hemifield visual stimulation.
Fig. 3: In whole filed visual stimulation.